The Definitive Handbook of *Azospirillum*

Editors
Fabricio Dario Cassán
Instituto de Investigaciones Agrobiotecnológicas (INIAB-CONICET)
Universidad Nacional de Rio Cuarto (UNRC)
Rio Cuarto, Córdoba, Argentina

Florence Wisniewski-Dyé
Laboratoire d'Ecologie Microbienne
Université Lyon 1
Villeurbanne, France

Luz de Bashan
Bashan Institute of Science
Auburn, AL, USA

Guillermo A. Maroniche
Facultad de Ciencias Agrarias, CONICET
Universidad Nacional de Mar del Plata
Mar del Plata, Argentina

ISBN 978-3-032-09221-2 ISBN 978-3-032-09222-9 (eBook)
https://doi.org/10.1007/978-3-032-09222-9

This Springer imprint is published by the registered company Springer Nature Switzerland AG
The registered company address is: Gewerbestrasse 11, 6330 Cham, Switzerland

Preface

The genus *Azospirillum* represents one of the most representative examples of scientific perseverance and interdisciplinary collaboration in the study of plant-microbe interactions. *The Definitive Handbook of Azospirillum: 100 Years of Research and Application Around the World* celebrates a full century since the first publication related to *Azospirillum*-like soil beneficial bacterium in 1925 by Martinus Beijerinck in the Netherlands. After that initial observation of a nitrogen-fixing spirillum-shaped bacterium in grassland soils, a remarkable scientific journey has unfolded, one that has redefined our understanding of microbial ecology, plant physiology, and sustainable agriculture. Over the last century, *Azospirillum* has evolved from a taxonomic interest into a model organism for studying plant growth-promoting rhizobacteria. Its path to global recognition was accelerated by the pioneering work of Johanna Döbereiner in Brazil, who isolated these bacteria from tropical grasses and demonstrated their potential to enhance crop productivity. Since then, decades of research have expanded the genus taxonomically, genetically, and ecologically, transforming *Azospirillum* into a keystone of agricultural microbiology across the globe, with deep roots in Latin America. The current volume brings together an extensive and rigorous update of what 100 years of science have revealed about this bacterium. Far from a simple retrospective, this book presents a forward-looking integration of foundational knowledge and new findings. The contributors, leading researchers from Latin America, Europe, Asia, and North America, have compiled and updated key information across several research areas: taxonomy, physiology, genetics, environmental resilience, plant signaling, and biotechnology. Their collective effort provides both a historical context and a methodological toolkit for researchers, students, and professionals interested in the multifaceted biology of *Azospirillum* and its practical applications. One of the defining features of this updated edition is its interdisciplinary design. The chapters are organized to follow the bacterium's progression from isolation and biochemical characterization to its interactions with plants and other microorganisms, and ultimately to its formulation and commercialization as a biostimulant, biofertilizer, or bioinput. The review begins with a panoramic view of the taxonomy, genomics, and phenotypic diversity of the genus. Tracking and quantification methods are explored in detail, with an emphasis on molecular, immunological, and imaging tools that support both basic research and regulatory assessment. The middle sections focus on the bacterium's lifestyle and communication strategies. Mechanisms such as chemotaxis, cell aggregation, biofilm formation, and cell–cell signaling are examined in light of recent advances in genetic manipulation and transcriptomic profiling. These chapters emphasize how *Azospirillum* coordinates its behavior within the rhizosphere and interacts dynamically with the host plant. It then delves into the physiology of nitrogen metabolism and stress tolerance, highlighting the metabolic versatility that allows *Azospirillum* these bacteria to thrive in diverse environments. Another major contribution of this volume is the comprehensive treatment of *Azospirillum* as a model for plant growth promotion beyond nitrogen fixation and ammonium excretion. Detailed chapters are dedicated to phytohormone biosynthesis, bacterial effectors, plant molecular responses, and interactions with other beneficial microorganisms. These chapters examine in detail how *Azospirillum* influence root architecture, nutrient uptake, and stress resilience in host plants, contextualized in both extensive and intensive cropping systems. In addition to its agricultural relevance, inoculant formulation and regulatory frameworks are assessed, describing advances in liquid and solid carriers, encapsulation techniques, quality control standards, and the emergence of international guidelines for commercialization. Importantly, the book highlights the transformative role of emerging technologies in reshaping how we study and apply *Azospirillum*. From high-throughput genomics and artificial intelligence to synthetic microbial consortia and precision agriculture, the contributors illustrate how scientific tools are rapidly evolving to address complex agroecological challenges. The closing chapter explores *Azospirillum*'s contributions to soil health, revegetation of degraded areas, bioremediation of polluted soils, and symbiotic interactions with wild plant species illustrate its potential in environmental biotechnology. These innovations open new avenues for integrating *Azospirillum* into sustainable development strategies, climate adaptation efforts, and global food security agendas.

This volume would not have been possible without the enduring legacy of researchers who paved the way for current and future generations. We pay tribute to Martinus Beijerinck for his foundational discovery, and to Johanna Döbereiner for recognizing and proving the agricultural value of these bacteria. We also wish to acknowledge the extraordinary contributions of Yoav Bashan, Yaacov Okon, Jesús Caballero Mellado, Fabio Pedrosa, Anton Hartmann, Ivo and Vera Baldani and Jos Vanderleyden, scientists whose vision, dedication, and leadership shaped the field of plant growth promotion research and helped position *Azospirillum* as a reference organism for microbial biotechnology. To all the authors who contributed their knowledge and time to this work, we offer our deepest gratitude. Your commitment to scientific excellence and collaboration has resulted in a comprehensive and valuable resource. After 100 years, the story of *Azospirillum* is far from being finished. On the contrary, it continues to evolve, rooted in science, driven by innovation, and enriched by the collective memory of those who believed in its potential.

Rio Cuarto, Córdoba, Argentina — Fabricio Dario Cassán
Villeurbanne, France — Florence Wisnieski-Dyé
Auburn, AL, USA — Luz de-Bashan
Mar del Plata, Argentina — Guillermo A. Maroniche

About the Editors

Fabricio Dario Cassán is a microbiologist and holds a PhD in biological sciences from the National University of Río Cuarto (UNRC). He is an associate professor in the area of plant physiology and a principal researcher at the Institute of Agrobiotechnology (INIAB), CONICET. Dr. Cassán has nearly 30 years of experience studying plant growth-promoting rhizobacteria (PGPR) and their impact on agriculture. He has authored over 100 scientific contributions in the field and has led numerous research projects in Argentina and in collaboration with researchers from Brazil, Spain, Switzerland, Germany, the UK, Canada, the USA, and Belgium—where he also completed part of his postdoctoral training. He has served as coordinator of the Inoculant Quality Control Network (REDCAI) in Argentina, member of the Advisory Committee on Agricultural Bio-inputs (CABUA) of the Argentine Ministry of Agriculture, and of the Ibero-American Biofertilizer Network (RedBIOFAG). He has also participated as an advisor for the IAEA to the Ministries of Agriculture of Chile and Costa Rica, as well as to the European Union. Dr. Cassán is an honorary member of the Argentine Chamber of Bio-inputs (CABIO). He currently leads a research group at UNRC focused on developing biotechnological alternatives for agriculture.

Florence Wisniewski-Dyé is a professor of microbiology and microbial ecology at the University of Lyon 1 (France). She earned a PhD in biology-biochemistry in 1995 from the University of Orleans, where she studied the induction of virulence genes in the phytopathogen *Agrobacterium tumefaciens*. She continued her career with a postdoctoral fellowship in the United Kingdom at the John Innes Centre, where she worked on quorum sensing signaling in Rhizobium in Prof. J. A. Downie's team. She was then recruited as a lecturer at the University of Lyon 1 at the end of 1999 and joined the Microbial Ecology Laboratory, where she began working on plant growth-promoting bacteria. Her work focused in particular on quorum sensing, comparative genomics, and functional genomics in Azospirillum, leading her to collaborate with researchers in the USA, Argentina, Israel, Belgium, and the UK. In addition to her teaching activities and responsibility of the Master's program in Microbiology at the University of Lyon 1, she is the author of some fifty publications, has coordinated and partnered in several research projects, and has supervised a dozen PhD theses. She is currently Deputy Director of the Microbial Ecology Laboratory.

Luz E. de-Bashan was born in Bogotá, Colombia. She is a biologist from Javeriana University in Bogota; she obtained her master's degree from the National University of Colombia; she has a doctorate from the University of Laval, Quebec, Canada, and a postdoctoral fellowship from the University of Arizona. From 2000 to 2022, she was a senior researcher and leader of the Environmental Microbiology Group at the Center for Biological Research of the Northwest (CIBNOR) in Mexico. Currently, she is the director of the Bashan Institute of Science in Alabama, USA, and an affiliate professor with the Department of Entomology and Plant Pathology at Auburn University, Alabama, USA.

Her research fields are the interaction between *Azospirillum* and microalgae for biotechnological applications, the restoration of degraded soils using plant growth-promoting bacteria and native plants, and the development of bacterial inoculants.

Dr. Bashan has published 122 peer-reviewed articles, a book, chapters in books, and popular scientific publications. In addition, she published four comprehensive websites. She serves as part of the Editorial Board of Biology and Fertility of Soils (since 2019), Section Editor of *Plant and Soil* (since 2019), Associate Editor of *Stresses* (since 2020), and has been Guest Editor of *Special Issues in Agronomy* (2021), *FEMS Microbiology Ecology* (2023–2024), *Frontiers in Plant Science* (2023–2024), and *BMC Microbiology* (2024–2025). Additionally, she is a reviewer for 49 scientific journals from 12 countries and five funding agencies.

Guillermo A. Maroniche holds a PhD in Biological Sciences from the University of Buenos Aires, Argentina. He is currently an independent researcher of Consejo Nacional de Investigaciones Científicas y Técnicas (CONICET), working at the Laboratory of Plant and Microbial Biochemistry, Faculty of Agronomic Sciences, National University of Mar del Plata, Balcarce, Argentina. His research focuses on understanding the processes that link the microorganism–plant–environment system, by studying the physiological and molecular mechanisms that modulate the interspecific relationships that establish beneficial bacteria, such as *Azospirillum* and *Pseudomonas*, between themselves and with plants, as well as their modes of action. In the long term, these investigations aim at improving the efficacy of biological inoculants based on synthetic microbial communities and developing new technologies for their formulation and application.

Part I
Isolation, Identification, Characterization and Genetic Manipulation of *Azospirillum* spp.

Chapter 1
Isolation, Identification, and Biochemical and Genetic Characterization of *Azospirillum* sp.

Veronica Massena Reis, José Ivo Baldani, Alberto F. de Oliveira Junior, and Jerri Edson Zilli

Abstract *Azospirillum* is a genus that possesses 37 species, and only one does not fix nitrogen. *A. brasilense* and *A. lipoferum* were the two species described and the former has been used in agriculture in many countries since its identification in 1978. *A. argentinense* is an inoculant recommended in Argentina for more than 30 years. These species are well-known bacteria that produce high amounts of auxins, affecting the root architecture by its inoculation and the visual symptoms associated with their application on plants. The main strategy used to isolate and count this bacterial genus is based on the utilization of semisolid medium without nitrogen addition. This recipe has been used by many authors since 1976. This chapter presents procedures used for diazotrophic counting and identification either from rhizosphere soil or on the surface of or within plant tissues. It also includes methods to measure nitrogenase activity in vials with the semisolid medium, carbon source characterization, and storage procedure. We also included analyses and discussion of techniques based on phylogeny, phylogenomics, average nucleotide identity (ANI), and mass spectrometry using type strains of *Azospirillum* species to guide the correct characterization of species in this genus.

Keywords Nitrogen fixation · Phylogeny · Culture medium · Auxins · Root development

1.1 Introduction

The *Azospirillum* genus currently comprises 37 species but not all are validated with the name. They belong to the α-subclass of proteobacteria and started with the reclassification of *Spirillum lipoferum (*Beijerinck 1925). The first report showed bacteria isolated from the root samples of several types of grass such as Digitaria,

V. M. Reis (✉) · J. I. Baldani · A. F. de Oliveira Junior · J. E. Zilli
Embrapa Agrobiologia, Laboratório de Gramíneas, Laboratório de Genética and Biological Research Centre "CRB-Johanna Döbereiner", Rio de Janeiro, Brazil
e-mail: veronica.massena@embrapa.br

F. D. Cassán et al. (eds.), *The Definitive Handbook of Azospirillum*,
https://doi.org/10.1007/978-3-032-09222-9_1

Panicum maximum, *Cynodon dactylon*, and grain crops (wheat, maize, sorghum, millet, and rice). Also, soil samples were used by Döbereiner and collaborators to isolate this organism. Bacteria were identified in several countries such as Brazil, Peru, Nigeria, Colombia, Senegal, Ecuador, Venezuela, USA and were compared by Tarrand et al. (1978). This group of 61 isolates from several plants and places was utilized in a detailed taxonomic study. Based on DNA homology, Tarrand et al. (1978) classified the bacteria into two groups, leading to the establishment of the genus *Azospirillum*, which was described with two species: *A. lipoferum* (group II) and *A. brasilense* (group I). Other species such as *A. rugosum* (Young et al. 2008) and *A. picis* (Lin et al. 2009) were isolated from oil-contaminated soil and discarded road tar, respectively. *A. thiophilum* was isolated from sulfide spring (Lavrinenko et al. 2010), *A. cavenae* from a water sample obtained from a cave in China (Zhu et al. 2021), *A. humicireducens* from fuel cell (Zhou et al. 2013), and *A. ramasamyi* from fermented bovine products (Anandham et al. 2019). This cosmopolitan bacterium can survive in various locations and plants, including grass and tobacco plants (Duan et al. 2022), such as *A. tabaci*, a species waiting for validation. Although most species are aerobics, *A. humicireducens* is a facultative anaerobic species (Zhou et al. 2013). Interestingly, one species, *A. thermophilum,* was isolated from a hot-spring soil sample, which can grow at 50 °C, its optimum growth temperature (Zhao et al. 2020). Some species were reclassified into a new genus, such as *A. irakense,* which is now *Niveispirillum irakense* (Lin et al. 2014), and *A. amazonense* which is *Nitrospirillum amazonense* (Lin et al. 2014).

Recently, two new species were described, leading to the reclassification of strains previously assigned to *A. brasilense*. The well-known strain Sp245 is now classified as *A. baldaniorum* (Dos Santos Ferreira et al. 2020), while Az39, which has been used as an inoculant in Argentina since 1996, was reclassified as *A. argentinense* (Dos Santos Ferreira et al. 2022). Interestingly, both species were isolated from wheat roots, in different countries. Both produce high amounts of indole-3-acetic acid in the presence of L-tryptophan and are recognized as plant growth promoters.

Several recipes of the media were used to isolate this genus. Most of the media were based on malate such as the NFb (Tarrand et al. 1978) (Table 1.1). Malate medium called SM with pH 8.5 and NaCl 1.5% was used to isolate *A. halopraeferens* (Reinhold et al. 1987). *A. melinis* was isolated using a recipe based on sucrose as a carbon source (Peng et al. 2006). Xie and Yokota (2005) changed malate to glucose (NFG medium) and isolated *A. oryzae*. Lavrinenko et al. (2010) used sodium succinate as a carbon source to isolate *A. thiophilum*. A modification of the culture medium to a high pH contributes to the discovery of two new ones: *A. canadense* (Mehnaz et al. 2007a) and *A. zeae* (Mehnaz et al. 2007b). The N-free NFb medium, which uses neutralized malic acid as a carbon source, has a pH near 7.0, and prepared with a low concentration of agar (1.75 g L^{-1}) to produce a semisolid condition in tubes or vials, was key to the discovery that plants from different soils, regions, and plant parts possess diazotrophs (Tarrand et al. 1978). This semisolid condition was important to produce the microaerophilic environment that allowed the organism to synthesize the nitrogenase complex and initiate nitrogen fixation.

Table 1.1 List of *Azospirillum* species isolated from plant samples using different media

Media	*Azospirillum*	Carbon source	Conditions	References
Semisolid medium				
NFb	*A. lipoferum*	Malate[a]	pH 6.5–6.8 + biotin	Tarrand et al. (1978)
	A. brasilense	Malate	pH 6.5–6.8 + biotin	Tarrand et al. (1978)
	A. doebereinerae	Malate	pH 6.5–6.8 + biotin	Eckert et al. (2001)
	A. melinis	Malate	pH 6.5–6.8 + biotin	Peng et al. (2006)
	A. formosense	Malate	pH 6.5–6.8 + biotin	Lin et al. (2012)
SM	*A. halopraeferens*	Malate	pH 8.5 + 1.5% NaCl. Incubation at 41 °C	Reinhold et al. (1987)
LGI	*A. melinis*	Sucrose	pH 6.0–6.2	Peng et al. (2006)
M	*A. oryzae*	Malate	pH 6.8 + yeast extract and NaCl 100 mg each	Xie and Yokota (2005)
NFg	*A. oryzae*	Glucose	pH 7.3 + $CaCO_3$ 5 g	
M	*A. zeae*	Malate	pH 7.2–7.4 without biotin	Xie and Yokota (2005)
M	*A. canadense*	Malate	pH 7.2–7.4 without biotin	Xie and Yokota (2005)
NFb or LGI	*A. melinis*	Malate	Depending on the recipe	Peng et al. (2006)
MPSS	*A. thiophilum*	Sodium succinate	pH 7.5 + FeS suspension	Lavrinenko et al. (2010)
Plates				
TYB	*A. palatum*	Rich medium	Yeast-beef-tryptone extract	Zhou et al. (2009)
TSA	*A. picis* *A. thermophilum* *A. soli* *A. ramasamyi* *A. palatum*	Rich medium	Tryptone soya agar	Lin et al. (2009), Lin et al. (2015), Anandham et al. (2019), and Zhou et al. (2009)
	A. rugosum	TSA	Brain-heat infusion	Young et al. (2008)
NB	*A. humicireducens*	Rich medium	pH 7.2	Zhou et al. (2013)
NB	*A. fermentarium*	Rich medium	Tryptone soya agar	Lin et al. (2013)
R2A	*A. tabaci*	Rich medium	Yeast, peptone, CA[b], glucose, SS[a] soluble starch	Reasoner and Geldreich (1985)

[a]Malic acid neutralized with KOH. [b]*CA* casamino acid, *SS* sodium pyruvate

1.2 Methods

In 2014, Baldani et al. described several methods of isolation and characterization of many diazotrophs. Media and procedure were described in detail.

The main base of the isolation is the use of a semisolid medium without N addition. Day and Döbereiner (1976) used this simple method, and several recipes have been described since that time. The semisolid media were not new and were used to measure chemotaxis (Adler 1966) or the reduction of nitrate to nitrite (ZoBell 1932). This semisolid condition must allow the bacteria that fix nitrogen to adapt to bacterial growth and the oxygen diffusion to synthesize the nitrogenase. Therefore, there are options for creating new semisolid media to survey new bacterial species and exploit the diversity of diazotrophic bacteria. Changing the carbon source of the NFb medium was a strategy to isolate the former *A. amazonense* (Magalhães et al. 1983), now *Nitrospirillum amazonense* (Lin et al. 2014). Also, the combination of more than one carbon source can be used to isolate new species, as done by Oliveira (1992), who replaced malic acid with a mixture of citric acid, oxalic acid, and glucose (ratio 1;1:3 g L^{-1}) at pH 5.5, calling that medium NFb COG.

1.2.1 Steps for the Isolation

1.2.1.1 Roots, Stems, Leaves, and Soil

Roots of the harvested field plants should be washed in tap water to remove rhizosphere soil. Then, they should be cut into 10 cm pieces, dried on paper towels, and weighed as 10 g samples. Afterward, it is ready for the dilution procedure.

Usually, rhizosphere soil (10 g) is also homogenized for 30–60 min in 90 mL of saline solution for dilution or maceration (ex: salts of NFb medium) or sucrose solution (4%) using a rotary shaker at 150 rpm. In this case, it is used directly in the dilution procedure.

1.2.1.2 Surface Disinfestation

In the case of surface-sterilized tissue (roots), they were immersed in a 1% Chloramine T solution ($CH_3.C_6H_4.SO_2.N(Na)Cl.3H_2O$). Time in this solution is determined by age and type of plant: usually, roots from maize and sorghum at the flowering stage are immersed for 30 min to 1 h, while rice and wheat only resist 5–15 min in the same solution. After this surface disinfestation, the roots are placed in sterile distilled water for one-third of the time they were immersed in the sterilization solution. This step is repeated for the same time using phosphate buffer (50 mM, pH 7.0) and again with distilled water, so that the total time amounting to that of the disinfestation in Chloramine T. Roots that are not to be disinfected should be soaked in distilled water for the same time as that used in the disinfestation procedure (Baldani et al. 2014).

In the case of stems, e.g., those on sugarcane and elephant grass (*Pennisetum purpureum*), it is first necessary to remove dust and waxes from the surfaces with tap water before proceeding with the disinfestation procedure. The stems are sprayed with 70% ethanol, flamed, and then peeled with a knife to remove the external tissue. To ensure that no bacteria from the plant surface are present in the stems, they are dipped into 70% ethanol and flamed again. Leaf samples should first be washed in tap water and then surface disinfested with 70% ethanol (Baldani et al. 2014). In cases where the diazotrophic bacteria colonizing the apoplast fluid are the target, the procedure described by Dong et al. (1994) is suggested.

1.2.2 Counts of Diazotrophic Microorganisms

Sterilized or washed roots, stems, and leaves (10 g) are blended in one of the above solutions (see Sect. 1.2.4) for 1–2 min. Then, after 15 min, aliquots of 1 ml are used to prepare serial dilutions (10^{-2} to 10^{-7}) in 9 mL saline or sucrose solution. Also, Tween 80 is added to this saline solution to improve the dilution efficiency (0.02% final concentration)

1.2.3 Counting Using Semisolid Medium

In most cases, the numbers of nonsymbiotic nitrogen-fixing bacteria (free-living, associative, and endophytic) are estimated by the Most Probable Number (MPN) method (Cochran 1950) using McCrady's probability tables (Okon et al. 1977). This classic method is based on the extinction of the population using a serial dilution procedure. Counts are made on semisolid medium by inoculation of 0.1 mL of each dilution in three semisolid vials (replicates) containing the medium specific for the organism in question, placing the inoculum from the bottom to the center of the tube, as these bacteria are sensitive to O_2. These vials are then incubated at 30–34 °C, without shaking, for 5–7 days, allowing pellicles to be formed. The MPN method is based on the probability of positive and negative growth of the target bacteria and can also be calculated using five vials instead of three. Using N-free semisolid media relies upon the appearance of a typical diazotrophic bacterial pellicle in the subsurface of the medium after incubation for 7–10 days at 30 °C. The presence or absence of the pellicle in the semisolid medium generates a number in McCrad's table (Table 1.2).

When dealing with pure cultures, it is possible to determine the number of cells by homogenizing 1 mL of the culture in 9 mL of saline solution, sterile water, or sucrose solution (4%) and proceeding serial dilutions up to 10^{-9} as previously described. It is also possible to use Tween 20 or Tween 80 detergents to disperse the organisms. By using Drigalsky's loop, aliquots of 0.1 ml are spread onto Petri dishes containing the specific medium, amended with N source, usually 50 mg yeast extract L^{-1}. If using a

Table 1.2 McCrady's table for Most Probable Number (MPN) bacterial counting using three vials per dilution

Dilution with growth	Diazotrophic number	Dilution with growth	Diazotrophic number	Dilution with growth	Diazotrophic number
000	0.0	201	1.4	302	6.5
001	0.3	202	2.0	310	4.5
010	0.3	210	1.5	311	7.5
011	0.6	211	2.0	312	11.5
020	0.6	212	3.0	313	16.0
100	0.4	220	2.0	320	9.5
101	0.7	221	3.0	321	15.0
102	1.1	222	3.5	322	20.0
110	0.7	223	4.0	323	30.0
111	1.1	230	3.0	330	25.0
120	1.1	231	3.5	331	45.0
121	1.4	232	4.0	332	110.0
130	1.6	300	2.5	333	140.0
200	0.9	301	4.0		

Example of counting: take the three highest dilutions with growth (10^{-5}, 10^{-6}, and 10^{-7})
10^{-5}—three positive vials
10^{-6}—two positive tubes
10^{-7}—one positive tube
Look at the Table to number 321, corresponding to the positive dilutions. In this case, the diazotrophic number is 15.0. Multiply it by the first dilution on counting and then by 10 (for the 0.1 mL for inoculation).
MPN (cells mL^{-1}) $= 15.0 \times 10^5 \times 10 = 1.5 \times 10^7$

rich medium (potato agar, etc.), the N source is not required. Calculation of the number of cells is just a matter of multiplying the number of colonies on the plate by the corresponding dilution factor and the sample volume (in case of 0.1 mL, this equals 10). Another alternative is to use the Drop Plate method as suggested for rhizobia (Hoben and Somasegaran 1982). For this method, each plate is divided into eight sectors, and one drop (10–30 µl per drop) is delivered to each sector with a calibrated pipette. The plates can also be divided into four sectors, and in this case, three drops are delivered to each sector. The Drop Plate method allows more counts to be made on one plate providing economy in materials and labor.

1.2.4 *How to Isolate* Azospirillum

Different media are used to isolate different bacterial species. For the isolation of *A. brasilense, A. lipoferum, and A. baldaniorum*, the semisolid NFb medium is inoculated with 0.1 mL of soil or root suspension and incubated at 30–35 °C. Pellicles will be formed 4–7 days after, and the cultures should be transferred to a new semisolid NFb medium and incubated again for 4–7 days. After this period, the culture is streaked on solid NFb medium supplemented with 50 mg of yeast extract and incubated 3–5 days at 30 °C when colonies from these species are white, dry, and small. Individual colonies should be transferred to a new semisolid medium and incubated for more 4–7 days. After pellicle formation, the culture should be streaked on a solid Potato Agar medium. Colonies are initially yellowish-white and eventually become pink.

These species may also be differentiated using a medium containing glucose as *A. lipoferum*, but not *A. brasilense,* and *A. baldaniorum* is capable of metabolizing glucose as the sole C source (Table 1.3). *A. halopraeferans* is identified by its growth at high temperature (41 °C) and is salt tolerant. Cells may grow to 1.2 μm in length and 0.7–1.4 μm in thickness if the pH turns alkaline. *A. largimobile* colonies grown on peptone yeast extract agar are 2 mm in diameter after 72 h of incubation at 28 °C, cream to buff-colored, opaque, low, convex, and round with an entire edge and smooth surface. Colonies grown on agar medium which includes Congo red have a scarlet color. Slower growth occurs in the presence of 2% NaCl as conglomerates form and have a pale pink color. Under 3% NaCl, growth is inhibited. *Azospirillum* species may be distinguished phenotypically as they show several physiological and morphological differences (Table 1.3).

1.2.5 Media

The isolation of *Azospirillum* spp. is based on the use of N-free semisolid media, containing a low concentration of agar. NFb medium for *A. lipoferum*, *A. brasilense,* and *A. baldaniorum* and later with some minor modifications was also useful to isolate *A. halopraeferans.* The recipe is described in Table 1.3. *A. doebereinerae* (Eckert et al. 2001) is isolated using the NFb semisolid medium after incubation for 3–5 days at 30 °C. Further purification is done on NFb (supplemented with 50 mg yeast extract L^{-1}). *A. halopraeferans* (Reinhold et al. 1987) cells grown in SM medium can reach 1.2 μm in length and 0.7–1.4 μm in thickness if the pH turns alkaline. It was also used as a medium containing malic acid and 1.2 g of NaCl and biotin as a sole source of vitamins.

A. oryzae can be isolated using the M medium (Table 1.3). The other medium used was NFg medium, composed of (g L^{-1}): glucose, 10.0; $CaCl_2.2H_2O$, 0.020; $MgSO_4.7H_2O$, 0.2; K_2HPO_4, 1.0; $CaCO_3$, 5.0; $FeSO_4.7H_2O$, 0.050; and $Na_2MoO_4.2H_2O$, 0.001. Complete the volume to 1000 mL with distilled water and adjust the pH to 7.3. *A. zeae* and *A. canadense* can be isolated on M medium by omitting the addition of biotin and adjusting the pH of the medium to 7.2–7.4 (Xie and Yokota 2005. *A. melinis* (Peng et al. 2006) is isolated in semisolid LGI medium or semisolid NFb medium after incubation at 28 °C for 3–5 days. Purification can be done by repeatedly streaking the isolates on plates of solid LGI or NFb medium.

A. palatum (Zhou et al. 2009) did not utilize the minimal medium to obtain the new species that was isolated using TYB medium containing 0.3% yeast extract, 0.2% beef extract, 0.6% tryptone, 0.3% NaCl, and 0.001% FeC13 at pH 7.0.

Table 1.3 Five semisolid media used for the isolation of some *Azospirillum* species

Ingredient (per litre)	NFb	LGI	FAM	SM	M
DL-malic acid	5 g	–	–	5 g	5 g
Cristal sugar	–	5 g	5 g	–	
K_2HPO_4	0.5 g	0.2 g	0.12 g	0.13 g	0.1 g
KH_2PO_4	–	0.6 g	0.03 g	0.17 g	0.4 g
$MgSO_4.7H_2O$	0.2 g	0.2 g	0.2 g	0.25 g	0.2
NaCl	0.1 g	–	0.1 g	1.2 g	0.1 g
$CaCl_2.2H_2O$	0.02 g	0.02 g	0.02 g	0.22 g	0.02
$Na_2MoO_4.2H_2O$		0.002 g	0.002 g	–	
Na_2SO_4	–	–		2.4 g	
$NaHCO_3$	–	–		0.5 g	
Na_2CO_3	–	–		0.09 g	
K_2SO_4	–	–		0.17 g	
FeEDTA, solution 1.64%	4 mL	4 mL	4 mL	4 mL	
$FeCL_3$	–	–	–	–	10 mg
KOH–control pH:	4.5 g	v		4.8 g	4.8 g
Bromothymol Blue (0.5% in 0.2N KOH)—pH indicator	2 ml	5 mL		–	
Microelements solution	2 mL	–	AD	AC	AE
$CuCL_2.2H_2O$	–		–	0.002 mg	–
Vitamin solution	[a]1 mL	[a]1 mL	[a]1 mL	[b]10 mL	
Biotin[a]	–	–	–	0.1 mg	2 µg
Yeast extract	–	–	–	–	0.1 g
Final pH	6.8	6.0–6.2	6.0	8.5	6.8
Agar-semi-soli	1.75 g	1.75 g	1.75 g	1.75 g	N.D.
Agar solid media + yeast extract	17 g	15 g	15 g	15 g	15 g
Solution prepared separated from the medium recipe					
Microelements (sol.–1000 mL)			AD	AC	AE
$CuSO_4.5H_2O$	0.04 g		0.08 mg	–	–
$ZnSO_4.7H_2O$	0.12 g		0.24 mg		–
H_3BO_3	1.4 g		2.8 mg	0.15 mg	–
$Na_2MoO_4.2H_2O$	1.0 g		2 mg	2 mg	2 mg
$MnSO_4.H_2O$	1.175 g		2.35 mg	0.2 mg	–
Vitamin (solution 100 mL)[a]					
Biotin[a]	10 mg				
Pyridoxal-HCl[a]	20 mg			–	
Vitamin solution (sol. 1000 mL)[b]					
D-biotin[b]	–	–	–	40 mg	–
Calcium pantothenate[b]	–	–	–	40 mg	–
Myo-inositol[b]	–	–	–	200 mg	–
Niacinamide[b]	–	–	–	40 mg	–
p-aminobenzoic acid[b]	–	–	–	20 mg	–
Pyridoxine hydrochloride[b]	–	–	–	40 mg	–

(continued)

Table 1.3 (continued)

Ingredient (per litre)	NFb	LGI	FAM	SM	M
Riboflavin[b]	–	–	–	20 mg	–
Thiamine hydrochloride[b]	–	–	–	4 mg	–
Incubation at	30 °C	30 °C	30 °C	41 °C	30 °C

[a]Vitamin solution described by Tarrand et al. (1978) for NFb medium. [b]Vitamin solution described by Reinhold et al. (1987). v—adjust the pH with a KOH solution 10%. ND, Not described (Xie and Yokota 2005). LGI and FAM medium (Magalhães et al. 1983)

After characteristic pellicles have formed, N_2 fixation can be checked by acetylene reduction activity, and active cultures are transferred to new vials containing the same medium. As soon as a new pellicle is visible, the cultures are streaked out on agar plates containing the same medium supplemented with yeast extract (20 mg L^{-1}). The small amount of yeast extract permits the growth of small colonies on the surface of plates. Characteristic individual colonies are then transferred again to N-free semisolid media, and those that grow well are streaked out on potato agar for final purification.

JNFb medium (*Herbaspirillum* spp.) was developed by J. Döbereiner from the NFb medium. Composition (in g L^{-1}): malic acid, 5.0; K_2HPO_4, 0.6; KH_2PO_4, 1.8; $MgSO_4.7H_2O$, 0.2; NaCl, 0.1; $CaCl_2.2H_2O$, 0.02; micronutrient solution (above), 2 mL; bromothymol blue (0.5% in 0.2N KOH), 2 mL; FeEDTA (1.64%), 4 mL; vitamin solution (above), 1 mL; and KOH, 4.5. Complete the volume to 1000 mL with distilled water. Adjust pH to 5.8 with KOH. To semisolid and solid medium, add, respectively, 1.8 and 17 g agar L^{-1}.

Media Used in the Characterization of Diazotrophs from Plant Samples

Potato medium (purification of *Azospirillum* spp. and *Herbaspirillum* spp.) composition: 200 g of potato, 2.5 g of malic acid, 2.5 g of sucrose, 2 mL of micronutrient solution (above), and 1 mL of vitamin solution (above). Weigh 200 g of potato, peal, wash, and boil them for 30 min. Use a large funnel and cotton wool to filter. Mix malic acid and white sugar in 25 ml of distilled water, adjusting pH to 6.5–7.0 using KOH. To this solution add the vitamin solution (NFb medium) and micronutrient solution. Complete the volume to 1000 ml by adding distilled water.

For rapid multiplication, the *Azospirillum* spp. can be grown in liquid media to which a combined nitrogen source has been added (NH_4Cl [1 g L^{-1}] or yeast extract [0.4 g L^{-1}]). Alternatively, complex media such as nutrient broth (NB) or DYGS medium (D, L-malate [1 g L^{-1}], yeast extract [2 g L^{-1}], glucose [1 g L^{-1}], glutamate [1.5 g L^{-1}], peptone [1.5 g L^{-1}], $MgSO_4$. $7H_2O$ [0.5 g L^{-1}]) can be used (Baldani et al. 2014). In such media, with rapid stirring or shaking, cell concentrations of 10^8 cells per ml are reached after 24–48 h. To stabilize the pH at the desired value upon prolonged growth, the addition of 50 mM MOPS (3-[N-morpholino] propanesulfonic acid) buffer (pH 6.8) or MES (2-[N-morpholino] ethane-sulfonic acid) buffer (pH 6.0) is recommended.

Alternatively, the *Azospirillum* minimal medium of Okon et al. (1977), which also contains high phosphate levels, can be used. The medium contains the following (per liter of distilled water): K_2HPO_4, 6.0 g; KH_2PO_4, 4.0 g (mixed in 0.1 the final volume and autoclaved separately from the other medium constituents; the phosphate solution is later mixed with the cold medium); $MgSO_4.7H_20$, 0.2 g; NaCl, 0.1 g; $CaCl_2$, 0.02 g; DL-malic acid, 5.0 g; NaOH, 3.0 g; $FeCl_3$, 10.0 mg; $NaMoO_4.2H_2O$, 2.0 mg; $MnSO_4$, 2.1 mg; H_3BO_3, 2.8 mg; $CuSO_4.5H_2O$, 0.04 mg; and $ZnSO_4.7H_2O$, 0.24 mg. The final pH is adjusted to 6.8. For growth on N_2 under microaerophilic conditions, no NH_4Cl or yeast extract is added to the medium, but 0.5 g of agar is added per liter.

1.3 Nitrogen Fixing Capacity

The nitrogenase activity can be estimated using the acetylene reduction assay (ARA) (Boddey 1987) on vials (10 mL) containing 5 mL of modified N-free semisolid media (without yeast extract, vitamins, and bromothymol blue—condition necessary to use the Lowry protein measurement) inoculated with 50 μL of adjusted cell suspension. After growth at 30 °C for 48 h, the vials were sealed with rubber stopper (SubaSeal®), and 1 mL of acetylene gas was injected to a final concentration of 10% (v/v) of the flask volume. Samples were removed after a 1-h incubation period and assayed for ethylene using a gas chromatograph, equipped with a flame ionization detector (Perkin Elmer—Autosystem II). Ethylene concentration was determined using a standard calibration curve and a digital integrator (PE Nelson model 1022). For the quantification of protein in cultures grown on semisolid medium, a 1 M NaOH solution was used for cell lyses. Total cell extracts were diluted twice and were used for Lowry reaction. The cell sediments of these cultures were used for protein quantification by the Lowry method (Lowry et al. 1951). The cultures (10 mL) were centrifuged (5000 g, 15 min), resuspended an equal volume of sterilized water, and diluted (100 μL cell suspension: 400 μL water). This cell suspension (500 μL) was mixed with NaOH 1 M (500 μL) and boiled (100 °C, 5 min) to disrupt the membranes and release the total cellular contents. To the cellular extracts (500 μL) were added 2.5 mL of fresh solution (100:1:1) of Na_2CO_3 (5%):$CuSO_4$ (1%):$KNaC_4H_4O_6.4H_2O$ (2%). After 10 min of reaction, 500 μL Folin-Ciocalteu reagent (1 M) was added. After 30 min of incubation in the dark, the protein content of each sample was measured at 750 nm using a spectrophotometer (PE Lambda Bio, Model 11). A standard curve of serum albumin was used for calibration of the protein method and protein quantification.

1.4 Other Physiological Tests

Bacteria belonging to the genus *Azospirillum* are Gram-negative, motile, generally vibroid in shape, and contain poly-β-hydroxybutyrate (PHB) granules. Three species were classified as facultative anaerobic: *A. melinis*, *A. thiophilum*, and

A. humicireducens. Almost all strains were classified with one polar flagellum; a single species was not described as a nitrogen-fixing bacterium, *A. palatum*. Based on several new isolates obtained from soil and other environments, it is confirmed that this genus is widespread in nature and different regions of the planet. *Azospirilla* have a typically aerobic metabolism, with O_2 as the terminal electron acceptor. In *A. brasilense* and *A. lipoferum*, NO_3^- can replace O_2 in respiration, and under alkaline conditions, N_2O and N_2 are formed (Stephan et al. 1984). Under these conditions, anaerobic growth and N_2 fixation have been shown to occur (Nelson and Knowles 1978; Scott et al. 1979). The weak fermentative ability has been observed with *A. lipoferum* grown on glucose or fructose (Tarrand et al. 1978). Bacteria of this genus also utilize several carbon sources such as sugar, amino acids, and sugar alcohols (Table 1.4). The pattern of carbon utilization has been used for discriminatory purposes between species of the genus. The versatile C and N metabolism in this genus makes it well adapted to the rhizosphere and soil samples, also being important as a plant growth promoter bacterium—specially *A. brasilense* and *A. lipoferum* strains (Table 1.4).

1.5 Storage

Storage of the cultures for many years at −80 °C or in liquid N_2 is also possible after adding 50% glycerol or dimethylsulfoxide (DMSO) to an exponentially growing culture. The cells can also be preserved by lyophilization according to the following protocol (Döbereiner 1995). The cultures are grown to late log-phase in the following medium: K_2HPO_4, 0.5 g; $MgSO_4 \cdot 7H_2O$, 0.2 g; NaCl, 0.1 g; DL-malate (for *A. brasilense* and *A. baldaniorum*) or glucose (for *A. lipoferum*), 5 g; yeast extract, 0.4 g; and 1 L of distilled water. The cells must then be collected by centrifugation and resuspended to a dense cell suspension with 10% sucrose solution containing 5% peptone. Then, 0.1 mL portions are transferred into lyophilization ampoules, which are frozen and lyophilized according to the procedures recommended for *Rhizobium* spp. (Vincent 1970).

The simplest and cheapest way to preserve bacteria is in their isolation medium or potato agar. Tubes may contain slopes of medium, which facilitate inoculation with a loop and also enhance the area for bacterial growth, although a flat medium can be used. After inoculation with the organism and incubation for 2–3 days, it is possible to either seal the flask with a rubber cap, or add 2–3 ml of sterile mineral oil, glycerol, or liquid glycerin before sealing with a cotton wool or rubber cap. Several tubes should be prepared for each bacterium, as once a tube is opened, the bacteria will usually die. It happens more frequently when cultures are stored without any protection over the medium surface and the tube is not completely sealed. *N. amazonense* or *N. viridazoti* should be stored in medium with added sucrose; otherwise, the rise in pH to over 6.5—when the substrate is an organic acid—induces death of the bacteria within a few days. *A. lipoferum* is relatively sensitive to alkaline pH, and also, *A. diazotrophicus* does not survive long in Potato agar. After growing the microorganism 48–72 h at 30 °C in tubes containing an inclined specific

Table 1.4 Physiological differences among species of the genus *Azospirillum*

Azospirillum	Cell size (μm)	Fix N	Biot.	Op T	pH range	3% NACl	Denitr	Nitr. reduc.	Ind. produc.
A. lipoferum	1.0–1.7	+	+	37	5.0–8.0	–	+	+	+
A. brasilense	1.0–1.2	+	–	37	6.0–8.0	v	+	+	+
A. aestuarii[a]	Nd	+	nd	28	5.0–9.0	+	+	+	–
A. agricola	3.2–3.5	+	nd	25–37	7.0–8.0	+	nd	+	nd
A. argentinense	1.8–2.2	+	–	38	6.0–8.0	+	+	+	+
A. baldaniorum	1.6–2.1	+	–	30	4.0–8.0	+	+	+	+
A. canadense	0.9–1.8	+	–	25–30	5.0–7.0	–	nd	+	–
A. cavernae	4.0–6.0	+	nd	nd	nd	nd	nd	nd	nd
A. doebereinerae	1.0–2.0	+	–	30	6.0–7.0	–	+	+	+
A. endophyticum[a]	nd	+	nd	35	6.0–9.5	+	nd	+	nd
A. fermentarium	2.4–1.0	+	–	30	6.0–8.0	+	nd	+	nd
A. formosense	2.5–0.8	+	–	30	5.0–9.0	+	nd	+	+
A. griseum	3.2–3.6	+	nd	37	5.0–8.0	–	nd	+	–
A. halopraeferens	0.7–2.4	+	+	41	6.8–8.0	+	+	+	+
A. humicireducens	2.5–5.5	+	–	30	5.5–8.5	–	–	–	–
A. isscasi	nd	+		37	5.0–9.0	+	nd	nd	nd
A. largimobile	0.7–3.0	+	–	28	nd	–	–	+	+
A. melinis	1.0–1.5	nd	–	20–33	4.0–8.0	+	nd	+	
A. oryzae	1.0–1.5	nd	+	30	6.0–7.0	–	–	+	–
A. palatum[a]	2.0–2.6	nd	nd	30–47	6.0–8.0	nd	nd	–	–
A. palustre	1.2–2.5	+	–	5–40	5.5–8.5	–	nd	nd	nd
A. picis	nd	+	–	37	6.5–9.0	–	nd	+	–
A. ramasamyi	0.9–1.1	+	nd	28–30	6.0–10.0	+	nd	+	nd
A. rugosum	nd	+	nd	22–37	nd	–	nd	+	–
A. soli	2.3–2.6	+	nd	25–40	6.0–8.0	–	nd	+	nd
A. tabaci[a]	2.0–3.2	+	nd	30	6.0–8.0	+	nd	+	nd
A. thermophilum	1.0–3.0	nd	nd	50	5.0–7.0	+	nd	+	nd
A. thiophilum	nd	+	–	37	6.5–8.5	+	–	+	nd
A. zeae	1.9–6.8	+	–	30	5.0–7.0	–	nd	+	+

Legend: [a]Species described but not validated. 1. Cell size—length (μm); 2. Fix nitrogen; 3. Biotin requirement; 4. Op T, optimal temperature; 5. pH range; 6. Growth with 3% NaCl; 7. Denitrification; 8. Nitrate reduction; 9. Indole production. + positive, – negative, *nd* not determined

medium, or a variation of the C source related to the species to be lyophilized (*A. brasilense* grows using malic acid; *A. lipoferum* and *Herbaspirillum* spp. using glucose; *N. amazonense and N. viridazoti* using sucrose; *G. diazotrophicus* using medium SYP), SYP medium, as described by Caballero-Mellado et al. (1994), with the following composition per liter: sucrose (10 g), yeast extract (3 g), K_2HPO_4 (1 g), and KH_2PO_4 (3 g), can be used with bromothimol blue as an indicator at pH 6-6.25. Cultures are resuspended in 3 ml of lyophilization solution. Aliquots of 0.2 ml are distributed into lyophilization ampoules which are dehydrated as described for

Rhizobium species (Vincent 1970). Lyophilization solution is prepared with 10.0 g of sucrose plus 5 g of peptone. Complete to 100 mL adding distilled water.

1.6 Definition of *Azospirillum* Species Using Phylogeny, Phylogenomic, and Average Nucleotide Identity (ANI)

The extensive exploration of information on the genus *Azospirillum* has been unveiling the potential of these bacteria as true invisible allies of plants, performing essential biological functions that fascinatingly boost plant growth (Cassán et al. 2009; Pereg et al. 2011). With an impressive diversity of species, the genus *Azospirillum* inhabits various ecosystems around the world, adapting from lush tropical soils to more arid and challenging environments (Fibach-Paldi et al. 2012; Reinhold-Hurek and Hurek 2011).

Understanding this biodiversity has been further advanced through computational methodologies, like bioinformatics. With the progression of sequencing methods—encompassing both proteomic and genomic data—molecular data have become increasingly important in analyses of evolutionary ancestry and species diversity. The scientific literature highlights that each species brings its characteristics and benefits, influencing different types of plants (Fibach-Paldi et al. 2012) and adapting to various environmental conditions. This genetic and functional variety makes *Azospirillum* a significant focus of research in the quest for sustainable solutions in agriculture.

Figure 1.1 highlights how different species of the genus *Azospirillum* can be identified based on the 16S sequence of their ribosomes. Once sequences have been amplified and sequenced, data can be used by multiple alignment techniques, evolutionary model searches, and phylogenetic inferences to assess how the nucleotide configuration of sequences determines the evolutionary proximity of species.

In general, as shown in Fig. 1.1, the 16S rRNA sequence among the type strains of *Azospirillum* exhibits a high degree of conservation, except for *A. palatum*, which has been described based on a single strain (Zhou et al. 2009), and *A. massilliense*, which remains a candidate species, both of which are positioned distant from the others in the phylogram. More detailed studies should be conducted for a better understanding of the biodiversity and taxonomy of these species in light of current tools.

In summary, the 16S rRNA sequences within species are widely used for organism differentiation, particularly at the taxonomic level of genus, i.e., the phylogeny of this molecule is used to define the genus to which a given bacterium belongs. It is important to highlight that there are several databases currently available with millions of 16S rRNA sequences, making this molecule unparalleled for rapid taxonomic searches. However, a more comprehensive exploratory analysis, capable of reaching a deeper level of information among different species, will be

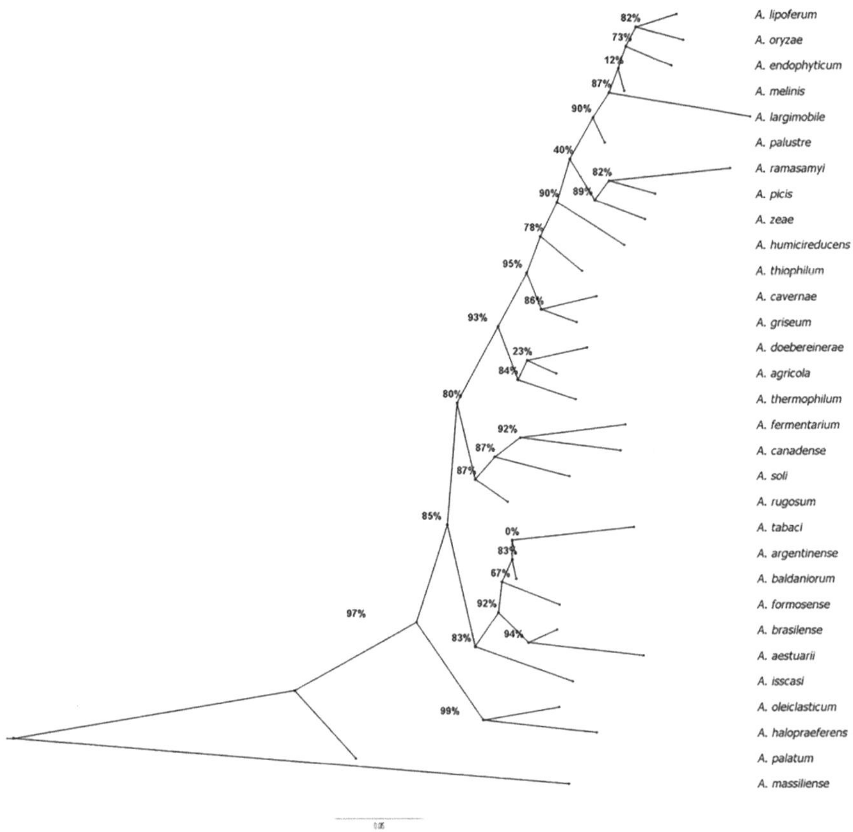

Fig. 1.1 Molecular phylogeny of the 16S gene for different species of the genus *Azospirillum*. The phylogenetic tree shows the formation of clades within the bacterial genus, highlighting the relationships between species that configure evolutionary divergences and convergences

better achieved with methods that consider the entire genome of the species, and not just some sequences like the 16S.

In contrast to phylogenetic approaches based on one or a few genes, phylogenomics utilizes data from complete genomes or large sets of genes, providing significantly higher resolution in phylogenetic analyses and allowing for a more detailed understanding of the evolutionary relationships among organisms. By employing a substantial amount of genomic data, phylogenomics minimizes sampling errors that can occur when only a few genes are analyzed, enabling more comprehensive analyses that consider genetic variation across the entire genome rather than being limited to a restricted set of molecular markers.

Figure 1.2 highlights the existing relationships among *Azospirillum* genomes of type strains and, together with data from the literature, sheds light on how different species' functions explain evolutionary convergences and divergences. With the genomes already available, we generally observe four main clades and some species

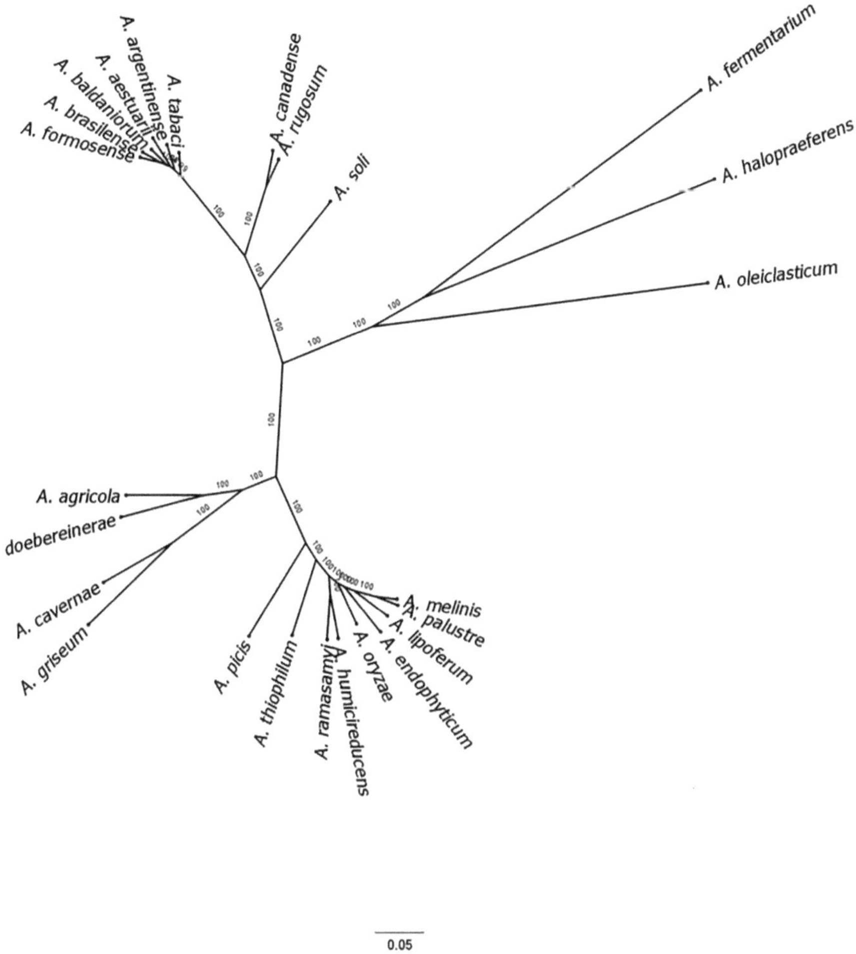

Fig. 1.2 Phylogenomic tree representation of the available type strains genome from genus *Azospirillum*. The tree illustrates the evolutionary relationships among the different species of the genus *Azospirillum*, highlighting those that are evolutionarily closer and more distant

that are more distantly distributed (*A. soli, A. fermentarium, A. halopraeferens,* and *A. oleiclasticum*). In the specific case of *A. brasilense*, the widely studied species (strain Sp7^T) show similarity to other species such as *A. formosense* (strain CC-NFb-7), *A. aestuarii* (strain YIM DDC1), *A. tabaci* (strain W712^T), *A. baldaniorum* (Sp245^T), and *A. argentinese* (Z39^T), the last two considered for many years as *A. brasilense*.

These observations underscore the importance of integrating genomic analyses with ecological data to understand the evolutionary dynamics of the *Azospirillum* genus. By examining the distinct phylogenomic clades formed by these species, researchers can delve deeper into how specific functional adaptations have driven their diversification. This insight is crucial not only for the taxonomy and evolutionary biology of these microorganisms but also for their strategic application in agriculture and environmental management.

Using the ANI (average nucleotide identity) technique, we can determine the emergence of new species based on the degree of identity observed among the genomes of different organisms. Generally, once the genomes are obtained through sequencing and assembly techniques, they undergo a fragmentation step into smaller pieces. These fragments are generally of fixed size, for example, 1000 base pairs (1 kb). Each fragment of one genome is aligned with the corresponding fragments of the other genome, performed by sequence alignment algorithms, such as BLAST (Basic Local Alignment Search Tool). For each pair of aligned fragments, the percentage of identical nucleotides is calculated. For instance, if 950 out of 1000 nucleotides in a pair of fragments are identical, the nucleotide identity for that pair is 95%. The average nucleotide identity (ANI) is then calculated as the nucleotide identity percentages of all aligned fragment pairs. ANI values above 95–96% generally indicate that the genomes belong to the same species, while lower values suggest that the organisms are of different species. The ANI technique is a powerful tool for microbial taxonomy and the identification of new species, providing a precise and quantitative way to compare genomes. Here, we display the set of reference species (genomes available) from the *Azospirillum* genus to perform the ANI evaluation.

Additional computational tools in bioinformatics include gANI (Genomic Annotation and Identification), Ortho ANI (Orthologous ANI), and digital DNA-DNA hybridization (dDDH) (Konstantinidis and Tiedje 2005; Meier-Kolthoff et al. 2013; Varghese et al. 2015; Lee et al. 2016), which have been used for the identification of bacterial species. gANI facilitates detailed genome annotation by highlighting specific genes and metabolic characteristics, thereby aiding in the identification and classification of bacteria. Conversely, Ortho ANI calculates genomic similarity between species by focusing on orthologous genes, making it extensively useful in phylogenetic studies (Fig. 1.3). Collectively, these tools offer a comprehensive understanding of the evolutionary relationships and genomic particularities of bacteria, supporting their differentiation and taxonomic classification.

1.7 Matrix-Assisted Laser Desorption/Ionization—Time-of-Flight (MALDI-TOF) Mass Spectrometry Analysis for Rapid Species Identification

In recent years, the use of mass spectrometry, particularly with the MALDI-TOF system (matrix-assisted laser desorption/ionization—time of flight), has been revolutionizing bacterial characterization. This method offers rapid, accurate, and cost-effective microorganism identification based on bacterial protein profiles (Singhal et al. 2015). By generating unique spectra that can be compared to specialized

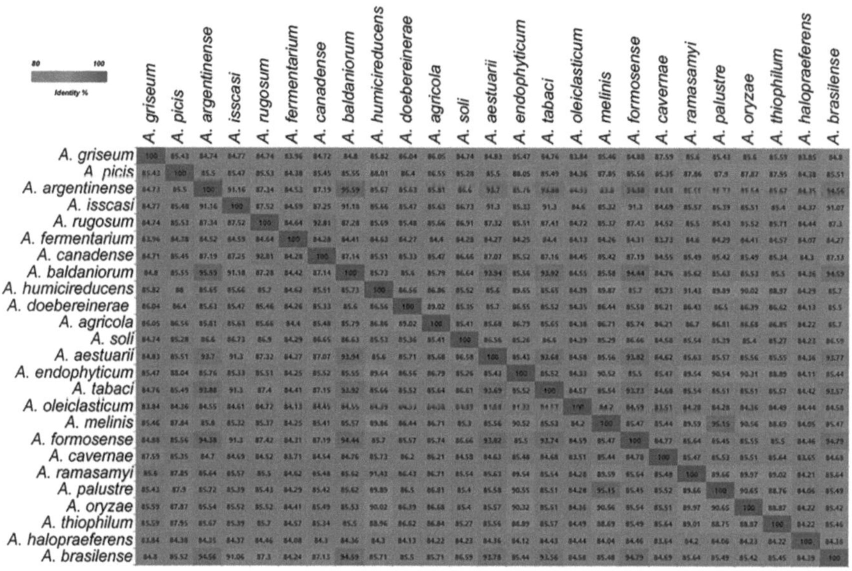

Fig. 1.3 Heatmap for ANI analysis of different species of the genus *Azospirillum* highlighting the reference strains. The graph demonstrates an all-versus-all comparative analysis, highlighting which genomes have a higher degree of identity among themselves and those with a lower degree of identity. None of the values exceeded the cutoff measure of 95–96%, indicating that all are distinct species

databases, MALDI-TOF enables bacterial identification within minutes, with high reliability. Its potential lies in its ability to analyze a wide range of organisms, reducing the time and costs associated with traditional methods, making it a promising tool in clinical diagnostics, environmental research, and agriculture (Schulthess et al. 2014).

To illustrate the application of this technique, the type strains of 12 species were considered (Fig. 1.4). The strains were cultivated on DYGS and LGI solid media in Petri dishes at 30 °C for 24, 48, and 72 h (Baldani et al. 2024). An isolated colony from each strain was directly applied to a stainless-steel MALDI plate, creating a thin layer. Subsequently, it was promptly covered with 1 μL of a saturated α-cyano-4-hydroxy-cinnamic acid matrix (CHCA, 5 mg/mL in a solution of acetonitrile/water 50:50 with 0.1% trifluoroacetate) and air-dried before analysis. Mass spectra were acquired using an Axima Performance iD Plus (Shimadzu Corp., integrated with the SARAMIS database) with a 60 Hz pulsed nitrogen laser. The analysis was conducted in the positive, linear SARAMIS extraction mode, with a laser repetition rate of 50 Hz, five shots accumulated per profile, and a mass-to-charge ratio (m/z) of 2000–20,000.

Protein mass fingerprints for each colony were generated by accumulating 500 laser pulse cycles and processed using Shimadzu Biotech Launchpad software before exportation for analysis with the SARAMIS package. SuperSpectras were constructed by scrutinizing mass spectra from multiple conditions of the same strains. This involved identifying consistent mass signals frequently appearing

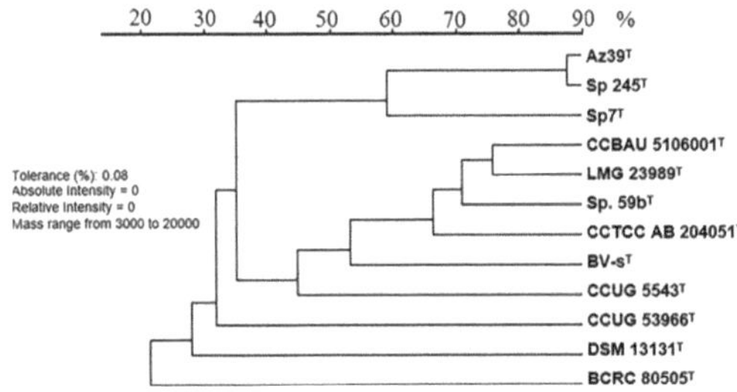

	Az39T	SP 245T	Sp7T	CCBAU 5106001T	LMG 23989T	Sp. 59bT	CCTCC AB 204051T	BV-sT	CCUG 5543T	CCUG 53966T	DSM 13131T	BCRC 80505T
A. argentinense	**41**	35	24	9	8	8	8	11	7	12	2	7
A. baladaniorum	36	**41**	24	10	8	7	9	10	7	10	3	8
A. brasilense	24	24	**41**	13	13	12	13	13	8	9	6	8
A. melinis	9	10	14	**41**	31	29	27	20	18	7	9	7
A. zeae	9	9	14	30	**41**	23	26	17	15	10	9	8
A. lipoferum	8	7	12	27	23	**41**	24	21	12	7	11	5
A. oryzae	8	9	13	25	25	24	**41**	20	14	7	10	6
A. thiophilum	11	10	13	19	17	21	20	**40**	10	9	8	8
A. picis	7	7	8	18	14	12	14	10	**41**	7	8	7
A. rugosum	12	10	9	7	9	7	7	9	7	**39**	3	7
A. dobereinerae	2	3	6	9	9	11	10	8	8	3	**41**	6
A. fermentarium	7	7	8	7	7	5	6	8	7	7	6	**41**

Fig. 1.4 Dendrogram of *Azospirillum* spp. constructed from the mass spectrum using the SARAMIS single-linkage cluster analysis. The samples were obtained from bacteria cultivated on LGI and DYGS culture media at 30 °C during 24, 48, and 72 h incubation and analyzed in triplicate. The scale bar indicates similarity (%). The accumulated number of peaks shown as data count is listed

across the mass spectra and assigning weights to the commonly agreed-upon peaks. Single-linkage cluster analysis was performed using SARAMIS parameters to produce a dendrogram of the strains (0.08% tolerance, range from 3000 to 20,000 m/z).

The MALDI-TOF MS analysis generated distinct mass spectra of whole-cell proteins for all studied strains, revealing 39–41 well-defined peaks of high quality. These peaks vary between 3097 and 11,352 in total.

Using protein profiles, a phylogram was constructed, revealing a high diversity of profiles among the type strains, with similarity ranging from 18% to 97% (Fig. 1.4). Overall, the groups formed resemble those observed in phylogenomics: *A. artinense*, *A. baldaniorum*, and *A. brasilense* clustered together; *A. melinis*, *A. lipoferum*, and *A. oryzae* grouped with *A. thiophilum* and *A. picis*, which were slightly more distant but still within the same group. *A. zeae* (LMG 23989^T) does not have an available genome and was therefore not analyzed by phylogenomics, but here it clustered near *A. melinis* (CCBAU 5106001^T) and *A. lipoferum* (Sp. 59b^T). *A. dobereinerae*, *A. rugosum*, and *A. fermentarium* were positioned far from other groups and each other, which was consistent with phylogenomic analysis, although we did not have access to additional strains for further analysis.

In summary, the mass spectrometry data showed a high correlation with the genomic analysis results, confirming that it is a robust technique for positioning strains within *Azospirillum* species. It is important to highlight that for closely related species, such as *A. argentinense* and *A. baldaniorum*, the discriminatory capacity of MALDI-TOF may be limited. This suggests that cultivating the bacteria under a wider range of conditions—cultural medium, incubation time, and temperature—may be necessary to detect differences between strains.

Acknowledgments The authors thank Embrapa Agrobiologia, the CNPq/INCT-FBN, FAPERJ, and CNPq for the fellowship to the researchers of Embrapa Agrobiologia.

Funding Embrapa (Brazilian Agricultural Research Corporation); INCT Plant Growth–Promoting Microorganisms for Agricultural Sustainability and Environmental Responsibility (CNPq 465133/2014-2, Fundação Araucária-STI-043/2019, CAPES); and FAPERJ (Fundação Carlos Chagas Filho de Amparo à Pesquisa do Estado do Rio de Janeiro), projects E-26/201.074/2022 and E-26/210.303/2021.

References

Adler J (1966) Chemotaxis in bacteria. Science 153:708–716

Anandham R, Heo J, Krishnamoorthy R et al (2019) *Azospirillum ramasamyi* sp. nov., a novel diazotrophic bacterium isolated from fermented bovine products. Int J Syst Evol Microbiol 69: 1369–1375

Baldani JI, Reis VM, Videira SS et al (2014) The art of isolating nitrogen-fixing bacteria from non-leguminous plants using N-free semi-solid media: a practical guide for microbiologists. Plant Soil 384:413–431

Baldani JI, Dos Santos Ferreira N, Shwab S et al (2024) *Nitrospirillum viridazoti* sp. nov., an efficient nitrogen-fixing species isolated from grasses. Curr Microbiol 81:144

Beijerinck MW (1925) Über ein *Spirillum*, welches freien Stickstoff binden kann. Zentralbl Bakteriol Parasitenkd Infektionskr Abt 63:353–359

Boddey RM (1987) Methods for quantification of nitrogen fixation associated with Gramineae. CRC Crit Rev Plant Sci 6:209–266

Caballero-Mellado J, Martínez-Romero E (1994) Limited genetic diversity in the endophytic sugarcane bacterium Acetobacter diazotrophicus. Appl Environ Microbiol. 60:1532–1537

Cassán F, Perrig D, Sgroy V et al (2009) *Azospirillum brasilense* Az39 and *Bradyrhizobium japonicum* E109, inoculated singly or in combination, promote seed germination and early seedling growth in corn (*Zea mays* L.) and soybean (*Glycine max* L.). Eur J Soil Biol 45:28–35

Cochran WG (1950) Estimation of bacterial densities by means of the "Most Probable Number". Biometrics 6:105–116

Day JM, Döbereiner J (1976) Physiological aspects of N2-fixation by a *Spirillum* from *Digitaria* roots. Soil Biol Biochem 8:45–50

Döbereiner J (1995) Isolation and identification of aerobic nitrogen-fixing bacteria from soil and plants. In: Alef K, Nannipieri P (eds) Methods in applied soil microbiology and biochemistry. Academic, San Diego, pp 134–141

Dong Z, Canny MJ, McCully ME et al (1994) A nitrogen-fixing endophyte of sugarcane stems. Plant Physiol 105:1139–1147

Dos Santos Ferreira N, Hayashi Sant' Anna F, Reis VM et al (2020) Genome-based reclassification of *Azospirillum brasilense* Sp245 as the type strain of *Azospirillum baldaniorum* sp. nov. Int J Syst Evol Microbiol 70:6203–6212

Dos Santos Ferreira N, Coniglio A, Puente M et al (2022) Genome-based reclassification of *Azospirillum brasilense* Az39 as the type strain of *Azospirillum argentinense* sp. nov. Int J Syst Evol Microbiol 72(8). https://doi.org/10.1099/ijsem.0.005475

Duan YQ, Zhou XK, Habib N et al (2022) *Azospirillum tabaci* sp. nov., a bacterium isolated from rhizosphere soil of *Nicotiana tabacum* L. Arch Microbiol 204:80. https://doi.org/10.1007/s00203-021-02688-7

Eckert B, Weber OB, Kirchhof G et al (2001) *Azospirillum dobereinerae* sp. nov., a nitrogen-fixing bacterium associated with the C_4-grass Miscanthus. Int J Syst Evol Bacteriol 51:17–26

Fibach-Paldi S, Burdman S, Okon Y (2012) Key physiological properties contributing to rhizosphere adaptation and plant growth promotion abilities of *Azospirillum brasilense*. FEMS Microbiol Lett 326:99–108

Hoben HJ, Somasegaran P (1982) Comparison of the pour, spread, and drop plate methods for enumeration of *Rhizobium* spp. in inoculants made from presterilized peat. Appl Environ Microbiol 44:1246–1247
Konstantinidis KT, Tiedje JM (2005) Genomic insights that advance the species definition for prokaryotes. Proc Natl Acad Sci 102:2567–2572
Lavrinenko K, Chernousova E, Gridneva ED et al (2010) *Azospirillum thiophilum* sp. nov., a diazotrophic bacterium isolated from a sulfide spring. Int J Syst Evol Microbiol 60:2832–2837
Lee I, Ouk Kim Y, Park SC, Chun J (2016) OrthoANI: An improved algorithm and software for calculating average nucleotide identity. Int J Syst Evol Microbiol 66:1100–1103
Lin S-Y, Young C-C, Hupfer H et al (2009) *Azospirillum picis* sp. nov., isolated from discarded tar. Int J Syst Evol Microbiol 59:761–765
Lin S-Y, Shen F-T, Young L-S, Zhu Z-L, Chen W-M, Young C-C (2012) Azospirillum formosense sp. nov., adiazotroph from agricultural soil. Int J Syst Evol Microbiol 62:1185–1190
Lin S-Y, Liu Y-C, Hameed A, Hsu Y-H, Lai W-A, Shen F-T, Young C-C (2013) Azospirillum fermentarium sp. nov., a nitrogen-fixing species isolated from a fermenter. Int J Syst Evol Microbiol 63:3762–3768
Lin S-Y, Hameed A, Liu Y-C, Hsu Y-H, Lai W-A, Shen F-T, Young C-C (2015) Azospirillum soli sp. nov., a nitrogenfixing species isolated from agriculture soil. Int J Syst Evol Microbiol 65: 4601–4607
Lin S-Y, Hameed A, Shen F-T et al (2014) Description of *Niveispirillum fermenti* gen. nov., sp. nov., isolated from a fermentor in Taiwan, transfer of *Azospirillum irakense* (1989) as *Niveispirillum irakense* comb. nov., and reclassification of *Azospirillum amazonense* (1983) as *Nitrospirillum amazonense* gen. nov. Antonie Van Leeuwenhoek 105:1149–1162
Lowry OH, Rosebrough N, Farr AL et al (1951) Protein measurement with the folin phenol reagent. J Biol Chem 193:265–275
Magalhães FM, Baldani JI, Souto SM et al (1983) A new acid-tolerant *Azospirillum* species. Ann Acad Bras Ci 55:417–430
Mehnaz S, Weselowski B, Lazarovits G (2007a) *Azospirillum canadense* sp. nov., a nitrogen-fixing bacterium isolated from corn rhizosphere. Int J Syst Evol Microbiol 57:620–624
Mehnaz S, Weselowski B, Lazarovits G (2007b) *Azospirillum zeae* sp. nov. a diazotrophic bacterium isolated from rhizosphere soil of *Zea mays*. Int J Syst Evol Microbiol 57:2805–2809
Meier-Kolthoff JP, Auch AF, Klenk HP (2013) Genome sequence-based species delimitation with confidence intervals and improved distance functions. BMC Bioinf 14:60
Nelson LM, Knowles R (1978) Effect of oxygen and nitrate on nitrogen fixation and denitrification by *Azospirillum brasilense* grown in continous culture. Can J Microbiol 24:1395–1403
Okon Y, Albrecht SL, Burris RH (1977) Methods for growing *Spirillum lipoferum* and for counting it in pure culture and in association with plants. Appl Environ Microbiol 33:85–88
Oliveira E (1992) Estudo da associação entre bactérias diazotróficas e arroz. MSc thesis, Federal Rural University of Rio de Janeiro, Seropédica, RJ, Brazil
Peng G, Wang H, Zhang G et al (2006) *Azospirillum melinis* sp. nov., a group of diazotrophs isolated from tropical molasses grass. Int J Syst Evol Microbiol 56:1263–1271
Pereg L, de-Bashan LE, Bashan Y (2011) Assessment of affinity and specificity of *Azospirillum* for plants. Plant Soil 356:45–60
Reinhold B, Hurek T, Fendrik I et al (1987) *Azospirillum halopraeferens* sp. nov., a nitrogen-fixing organism associated with roots of Kallar Grass (*Leptochloa fusca* (L.) Kunth). Int J Syst Bacteriol 37:43–51
Reinhold-Hurek B, Hurek T (2011) Living inside plants: bacterial endophytes. Curr Opin Plant Biol 14:435
Schulthess B, Bloemberg GV, Zbinden R et al (2014) Evaluation of the Bruker MALDI Biotyper for identification of Gram-positive rods: development of a diagnostic algorithm for the clinical laboratory. J Clin Microbiol 52(4):1089–1097. https://doi.org/10.1128/JCM.02399-13
Scott DB, Scott CA, Döbereiner J (1979) Nitrogenase activity and nitrate respiration in Azospirillum spp. Arch Microbiol 121:141–145

Singhal N, Kumar M, Kanaujia PK et al (2015) MALDI-TOF mass spectrometry: an emerging technology for microbial identification and diagnosis. Front Microbiol 6:791

Stephan MP, Zimmer W, Bothe H (1984) Denitrification by *Azospirillum brasilense* Sp7. II: growth with nitrous oxide as respiratory electron acceptor. Arch Microbiol 138:212–216

Tarrand JJ, Krieg NR, Döbereiner J (1978) A taxonomic study of the *Spirillum lipoferum* group, with descriptions of a new genus, *Azospirillum* gen. nov. and two species, *Azospirillum lipoferum* (Beijerinck) comb. nov. and *Azospirillum brasilense* sp. nov. Can J Microbiol 24: 967–980

Varghese NJ, Mukherjee S, Ivanova N et al (2015) Microbial species delineation using whole genome sequences. Nucleic Acids Res 43:6761–6771

Vincent J (1970) A manual for the practical study of root-nodule bacteria. Blackwell Scientific, Oxford, p 164. (IBP Handbook, 15)

Xie C, Yokota A (2005) *Azospirillum oryzae* sp. nov., a nitrogen-fixing bacterium isolated from the roots of the rice plant *Oryza sativa*. Int J Syst Evol Microbiol 55:13435–11438

Young C-C, Hupfer H, Siering C et al (2008) *Azospirillum rugosum* sp. nov., isolated from oil-contaminated soil. Int J Syst Evol Microbiol 58:959–963

Zhao Z-L, Ming H, Ding C-L et al (2020) *Azospirillum thermophilum* sp. nov., isolated from a hot spring. Int J Syst Evol Microbiol 70:550–554

Zhou Y, Wei W, Wang X (2009) *Azospirillum palatum* sp. nov. isolated from forest soil in Zhejiang province, China. J Gen Appl Microbiol 55:1–7

Zhou S, Han L, Wang Y et al (2013) *Azospirillum humicireducens* sp. nov., a nitrogen-fixing bacterium isolated from a microbial fuel cell. Int J Syst Evol Microbiol 63:2618–2624

Zhu HZ, Zhang ZF, Zhou N et al (2021) Bacteria and metabolic potential in karst caves revealed by intensive bacterial cultivation and genome assembly. Appl Environ Microbiol 87(6): e02440–20. https://doi.org/10.1128/AEM.02440-20. Erratum in: Appl Environ Microbiol 87(12):e0057721. https://doi.org/10.1128/AEM.00577-21

ZoBell CE (1932) Factors influencing the reduction of nitrates and nitrites by bacteria in semisolid media. J Bacteriol 24:273–281

Chapter 2
Tracking and Quantification of *Azospirillum* spp.

Willian Y. Takahashi, Carolina W. Galvão, Anton Hartmann, Anahi Coniglio, Fabricio Dario Cassán, and Rafael M. Etto

Abstract This chapter provides a comprehensive overview of cutting-edge methodologies for the precise localization and quantification of *Azospirillum* in plant-soil ecosystems. Accurate tracking of *Azospirillum* populations is pivotal for evaluating the efficacy of these plant growth-promoting rhizobacteria (PGPR) as biological inoculants and assessing their ecological impact on indigenous microbial communities. The chapter systematically examines three major methodological frameworks: (i) reporter gene-based systems (*lacZ, gusA*, and *gfp*); (ii) immunological techniques (ELISA and immunofluorescence); and (iii) nucleic acid-based methods, emphasizing quantitative PCR (qPCR) and fluorescence in situ hybridization (FISH). Additionally, emerging technologies are explored, including DNA biosensors, flow cytometry, and synthetic microbial communities (SynComs), offering new perspectives for monitoring *Azospirillum* dynamics in complex soil-plant systems. Future directions emphasize the integration of next-generation sequencing with modular bacterial tags (MoBacTag) for enhanced strain-level tracking capabilities. A key highlight is the presentation of a validated gold standard protocol for *Azospirillum* tracking through qPCR, providing researchers with standardized procedures for reliable bacterial enumeration across environmental samples, from rhizosphere to plant endosphere compartments.

Keywords Microorganism tracking · Bacteria quantification · qPCR · Reporter genes · *Azospirillum* monitoring

W. Y. Takahashi · C. W. Galvão · R. M. Etto (✉)
Microbial Molecular Biology Laboratory (LABMOM), State University of Ponta Grossa (UEPG), Ponta Grossa, Paraná, Brazil
e-mail: mazeretto@uepg.br

A. Hartmann
Faculty of Biology, Department of Microbiology, Microbe-Plant Interactions, Ludwig-Maximilian-University Munich, Planegg/Martinsried, Germany

A. Coniglio · F. D. Cassán
Instituto de Investigaciones Agrobiotecnológicas (INIAB-CONICET), National University of Rio Cuarto (UNRC), Río Cuarto, Córdoba, Argentina

F. D. Cassán et al. (eds.), *The Definitive Handbook of Azospirillum*,
https://doi.org/10.1007/978-3-032-09222-9_2

2.1 Introduction

One of the most remarkable characteristics of *Azospirillum* spp. is its ubiquity. Although typically isolated from plant fractions and considered root colonizers (Bashan et al. 2004), *Azospirillum* spp. can adapt to a wide range of environments (Nievas et al. 2023). Due to this extraordinary adaptability, along with its biochemical and physiological traits, this bacteria can positively influence the growth of hundreds of plant species. This has led to their use in biological products designed to promote plant growth and increase yields under field conditions in an environmentally friendly and cost-effective manner. These products are collectively known as inoculants or biofertilizers. The increasing use of *Azospirillum* inoculants has also raised several questions regarding their efficacy and their effects on the microbiomes of both soil and plants. Two major issues regarding the application of *Azospirillum* inoculants are: (i) their localization in the plant and the quantification of their populations over time and (ii) monitoring the impact of introduced *Azospirillum* populations on native microbial communities. For over 30 years, researchers have sought answers to these questions through various available approaches. This chapter summarizes the main methods used for the localization and quantification of *Azospirillum*, as well as new strategies and future perspectives. Finally, it presents a detailed protocol for the method considered the most accurate for tracking and quantifying *Azospirillum* in plants to date.

2.2 Reporter Gene-Based Methods

A reporter gene is an exogenous coding region attached to a promoter sequence or element within an expression vector, which is introduced into cells to enable the measurement of promoter activity and facilitate cell tracking. Ideally, the insertion of the reporter gene should not interfere with the plant growth-promoting traits of the organism (Rilling et al. 2019). To minimize biosafety risks, genetically modified reporter gene-tagged bacteria are typically used under controlled in vitro conditions, such as laboratory or greenhouse experiments (Webster et al. 1998; Kim et al. 2005). Reporter genes such as *lacZ* and *gusA* are detected through color changes, while *gfp* is identified via fluorescence. A summary of studies using reporter genes for tracking *Azospirillum* is provided in Table 2.1, with further discussion of selected studies below.

2.2.1 *β-Galactosidase (***lacZ***)*

The gene *lacZ*, derived *from Escherichia coli*, encodes β-D-galactosidase that cleaves X-gal (5-bromo-4-chloro-3-indolyl-β-D galactopyronaside), and generates

Table 2.1 Summary of *Azospirillum* articles using reporter gene-based methods

	1	2	3	4	5	6	7	8
	Target Azospirillum strain	Type of assay (in vitro or field)	Reporter gene-based methods used	Culture host (rice, maize, wheat, etc.)	Localization (where the bacteria was localized or quantified; e.g., root, rhizosphere, seed, soil, etc.)	Evaluation time (when the bacteria was localized or quantified; how many days or hours post-plant inoculation)	Quantification (number of CFU g^{-1} of soil, root, etc., if analyzed, or the note "Not counted" if only localization or quantification is provided)	References
1	*A. brasilense* Sp7 *A. brasilense*	In vitro	*nifA-lacZ*, *nifB-lacZ* and *nifH-lacZ* + MPN method	Wheat	Root	10 days	2.7×10^7 CFU g^{-1} 1.9×10^7 CFU g^{-1}	Arsene et al. (1994)
2	*A. brasilense* Sp7	In vitro	*nifA-lacZ* + MPN method	Wheat	Root	15 days	1.0×10^6 to 1.0×10^7 CFU $plant^{-1}$	Katupitiya et al. (1995)
3	*A. lipoferum* SpBr17	In vitro	*nifA-lacZ* + MPN method	Wheat	Root	14 days	1.0×10^7 to $1.0 \times 10^{7.5}$ CFU g^{-1}	Katupitiya et al. (1995)
4	*A. brasilense* Sp7	In vitro	*nifA-lacZ*	Cotton	Root (inside and surface)	14 days	Not counted	Feng and Kennedy (1997)
5	*A. baldaniorum* Sp245	In vitro	*kan-lacZ*	Wheat	Root (inside and surface)	14 days	Not counted	Webster et al. (1998)
6	*A. brasilense* Sp7 *A. baldaniorum* Sp245	In vitro	*kan-lacZ* and *nifA-lacZ*	Wheat	Root	10 days	Not counted	Gerk et al. (2000)
7	*A. brasilense* Cd	In vitro	*nifA-lacZ* + MPN method	Wheat	Root (endophytes) Root (total)	14 days	1.0×10^5 to 1.0×10^7 CFU g^{-1} 1.0×10^8 to 1.0×10^9 CFU g^{-1}	Fischer et al.

(continued)

Table 2.1 (continued)

	1	2	3	4	5	6	7	8
	Target Azospirillum strain	Type of assay (in vitro or field)	Reporter gene-based methods used	Culture host (rice, maize, wheat, etc.)	Localization (where the bacteria was localized or quantified; e.g., root, rhizosphere, seed, soil, etc.)	Evaluation time (when the bacteria was localized or quantified; how many days or hours post-plant inoculation)	Quantification (number of CFU g^{-1} of soil, root, etc., if analyzed, or the note "Not counted" if only localization or quantification is provided)	References
8	*A. brasilense* CDJA and A40	In vitro	*Tn5-lacZ* + CFU count	Wheat	Root (endorhizosphere)	30 days	1.0×10^6 CFU $plant^{-1}$	Kaushik et al. (2000)
9	*A. brasilense* Sp7	In vitro	kan-*lacZ* and *nifH-lacZ*	Wheat	Root	5 days	Not counted	Deaker and Kennedy (2001)
10	*A. baldaniorum* Sp245	In vitro	kan-*lacZ*	Rice	Root	14 days	Not counted	Jain and Gupta (2003)
11	*A. baldaniorum* Sp245	Field	*exoB1-lacZ* + CFU count	Sorghum	Root Soil	120 days 45 days	1.0×10^5 CFU g^{-1} $1.0 \times 10^{4.5}$ CFU g^{-1}	Basaglia et al. (2003)
12	*A. brasilense* Sp7	In vitro	kan-*lacZ* *flcA-lacZ*	Wheat	Root	10 d	Not counted	Pereg-Gerk (2004)
13	*A. brasilense* Sp7, CW301, CW903 and *A. lipoferum* CW1503	In vitro	*kan-lacZ*	Wheat	Root	10 days	Not counted	Kim et al. (2005)
14	*A. brasilense* Yu62	In vitro	*gfp* plasmid	Maize	Root (surface and inside)	8 and 12 days	Not counted	Liu et al. (2003)

15	*A. brasilense* Sp7 *A. baldaniorum* Sp245	In vitro	*gfp* plasmid + CFU count	Wheat	Root (total) Root (surface sterilized) Root (total) Root (surface sterilized)	4 weeks	1.0×10^7 to 1.0×10^8 CFU g^{-1} 1.0×10^5 to 1.0×10^6 CFU g^{-1} 1.0×10^7 to 1.0×10^8 CFU g^{-1} 1.0×10^4 to 1.0×10^5 CFU g^{-1}	Rothballer et al. (2003)
16	*A. brasilense* J-11	In vitro	*gfp* plasmid + CFU count	Pearl millet	Rizosphere	30 and 90 d	1.0×10^3 to 1.0×10^4 CFU mL^{-1}	Gera et al. (2003)
17	*A. brasilense* Yu62	In vitro	*gfp* plasmid + CFU count	Rice Tobacco	Root (total) Root (surface sterilized) Leaves Root (total) Root (surface sterilized) Leaves	18 days 40 days	1.0×10^6 CFU g^{-1} 1.0×10^5 CFU g^{-1} 1.0×10^2 CFU g^{-1} 1.0×10^7 CFU g^{-1} 1.0×10^4 CFU g^{-1} 1.0×10^5 to 1.0×10^7 CFU g^{-1}	Feng et al. (2004)
18	*A. lipoferum* JA4	In vitro	*gfp* chromosomal insertion + CFU count	Wheat	Root hair zone	11 and 21 days	1.0×10^5 to 1.0×10^7 CFU g^{-1}	Bacilio et al. (2004)
19	*A. brasilense* Ab-V5 and Ab-V6	In vitro	*gfp* plasmid	Rice	Root (surface and intercellular spaces)	10 and 20 days	Not counted	Hahn et al. (2016)
20	*A. brasilense* Ab-V5 *A. brasilense* Ab-V6	In vitro	*egfp/eyfp* + CFU count	Maize	Leaf	1 days/ 2 days	$1.0 \times 10^5/5.0 \times 10^2$ CFU cm^{-2} $5.0 \times 10^5/5.0 \times 10^2$ CFU cm^{-2}	Fukami et al. (2017)

(continued)

Table 2.1 (continued)

	1	2	3	4	5	6	7	8
	Target Azospirillum strain	Type of assay (in vitro or field)	Reporter gene-based methods used	Culture host (rice, maize, wheat, etc.)	Localization (where the bacteria was localized or quantified; e.g., root, rhizosphere, seed, soil, etc.)	Evaluation time (when the bacteria was localized or quantified; how many days or hours post-plant inoculation)	Quantification (number of CFU g^{-1} of soil, root, etc., if analyzed, or the note "Not counted" if only localization or quantification is provided)	References
21	*A. brasilense* Sp7	In vitro	Gfp-inner membrane protein + CFU count	Common bean (*Phaseolus vulgaris*)	Root Stem Leaf Seed	Early reproductive (6–7 weeks growth)	1.0×10^3 CFU g^{-1} 1.0×10 CFU g^{-1} 1.0×10^2 CFU g^{-1} 1.0×10^2 CFU g^{-1}	Malinich and Bauer (2018)
22	*A. brasilense* Sp7	In vitro	*gfp* (constitutive)	Wheat Alfalfa	Root	48 h	2.5–10 CFU area Zero CFU.area	O'Neal et al. (2020)
23	*A. brasilense* Sp7 *A. baldaniorum* Sp245	In vitro	Gfp or mCherry constitutive (plasmid and genome insertion) + CFU count	Wheat	Root	7 days	6.0×10^9 CFU g^{-1} 7.0×10^{11} CFU g^{-1}	Ramirez-Mata et al. (2018)
24	*Azospirillum* FAJ0011 (*A. baldaniorum* Sp245 derivative)	In vitro	*mTn5gusA–pgfp21*	Wheat	Root surface	7 days	Not counted	Xi et al. (1999)
25	*A. brasilense* FP2	In vitro	*gus/gfp* plasmid	Wheat	Root	15 and 30 days	Not counted	Ramos et al. (2002)
26	*A. brasilense* 1224	In vitro	Gus/Gfp; mCherry plasmid	Onion (*Allium cepa* L.)	Root Bulb (stem)	0, 3, 7, and 28 days	Not counted	Hong et al. (2019)
27	*A. lipoferum* T1371 and *A. lipoferum* 137		*mTn5SSgusA20*	White clover (*Trifolium repens*)	Root	5. 10 and 20 days	Not counted	Tchebotar et al. (1998)

28	*A. lipoferum* T1371 and *A. lipoferum* 137		*mTn5SSgusA20*	White clover	Root (surface and tips)	20 days	Not counted	Akao et al. (1999)
29	*A. baldaniorum* Sp245	In vitro	*gusA* (constitutive)	Wheat Rice	Root	5 or 7 days	Not counted	Steenhoudt et al. (2001)
30	*Azospirillum lipoferum* 137	In vitro	*gusA* + CFU count	Soybean	Root	5, 10, and 20 days	1.0×10^8 CFU g^{-1}	Chebotar et al. (2001)
31	*A. brasilense* Sp6	Field	*exoB1–gusA* + CFU count	Sorghum	Root soil	120 days 45 days	1.0×10^3 to 1.0×10^4 CFU g^{-1} $\tilde{}1.0 \times 10^4$ CFU g^{-1}	Basaglia et al. (2003)
32	*A. brasilense* Cd	In vitro	*gusA* (constitutive) + CFU count	Tomato	Root (endophytes) Stem Leaf	30 days	3.9×10^7 CFU g^{-1} 1.6×10^5 CFU g^{-1} 1.6×10^5 CFU g^{-1}	Botta et al. (2013)
33	*A. brasilense* FP2 *A. brasilense* HM053	In vitro	*gusA-nifH* + CFU count	Wheat	Root	96 h	1.0×10^7 CFU g^{-1} 1.5×10^7 CFU g^{-1}	Santos et al. (2017b)
34	*A. brasilense* FP2 and HM053	In vitro	*gusA-nifH*	Barley	Root	3, 7, and 12 days	Not counted	Santos et al. (2017a)
35	*A. baldaniorum* Sp245	In vitro	mCherry plasmid + MPN in a field-associated experiment	Elephant grass (*Pennisetum purpureum*)	Root	1 year	2.0×10^3 CFU g^{-1}	Camelo et al. (2021)
36	*A. baldaniorum* Sp245	In vitro	mCherry plasmid + CFU count	Wheat	Root	7 days	$\sim 1.2 \times 10^7$ CFU g^{-1}	Cruz-Pérez et al. (2021)
37	*A. argentinense* Az39	In vitro	DsRed (constitutive) + CFU count	Soybean	Leaves	1 day	2.9×10^3 CFU g^{-1}	Puente et al. (2021)
38	*Azospirillum* sp. B510	In vitro	DsRed (constitutive) + CFU count	Rice	Root	10 days	1.0×10^5 to 1.0×10^6 CFU g^{-1}	Naher et al. (2018)

5,5′-dibromo-4,4′-dichloro-indigo, which is an insoluble blue compound. The activity of β-galactosidase can be measured colorimetrically or visualized in situ. Thirteen studies have used *lacZ* as a reporter gene for tracking *Azospirillum* sp. in the rhizosphere, six of which also included bacterial quantification, primarily determined using the most probable number (MPN) method. The most frequently analyzed strains using this method were *A. brasilense* Sp7 and *A. baldaniorum* Sp245, employing either constitutive fusions (*kan-lacZ* and *Tn5-lacZ*) or regulated fusions (*nifA-lacZ* and *nifH-lacZ*) (Table 1). All studies focused on root colonization 5 to 30 days after *Azospirillum* inoculation, mostly on wheat.

Studies using *lacZ* have shown that *Azospirillum* primarily colonizes the root surface, particularly in areas where lateral roots emerge (Webster et al. 1998; Kim et al. 2005). Mutations affecting flocculation and exopolysaccharide (EPS) production reduced the ability of *A. brasilense* Sp7 to colonize the root surface (Pereg-Gerk 2004). On the other hand, the application of synthetic auxins, such as 2,4-dichlorophenoxyacetic acid (2,4-D), and certain flavonoids, like naringenin, stimulated root colonization by *Azospirillum lipoferum* SpBr17 (Katupitiya et al. 1995) and *A. baldaniorum* Sp245 (Webster et al. 1998), respectively. Despite these findings, since 2005, no studies have employed this method to track *Azospirillum* sp. during plant interaction, likely due to the occurrence of false positives caused by indigenous and endogenous β-gal activity from soil, rhizosphere, or endosphere bacteria (Akao et al. 1999; Ramos et al. 2002).

2.2.2 *β-Glucuronidase* (gusA)

Similar to *lacZ* and β-D-galactosidase, the *gusA* gene from *E. coli* encodes β-D-glucuronidase (β-glu), which forms a blue, insoluble compound in the presence of 5-bromo-4-chloro-3-indolyl-β-D-glucopyranoside (X-glu). A key advantage of *gusA* over *lacZ* is its low background activity in both plant tissues and indigenous soil bacteria (Akao et al. 1999; Ramos et al. 2002). Between 1998 and 2017, eleven studies have used *gusA* as a reporter gene to track *Azospirillum* strains inoculated into plants. Half of these studies also included CFU quantification, based on direct counting. The species *A. lipoferum, A. brasilense*, and *A. baldaniorum* were tracked using either constitutive fusions (*exoB1-gusA* and *mTn5SSgusA20*) or regulated fusions (*gusA-nifH*). These studies, conducted on white clover, wheat, rice, and soybean, focused on root colonization between 5 and 20 days after inoculation. The study on tomatoes also evaluated colonization of the stem and leaves.

Using the constitutive *gusA* gene expression, *A. brasilense* Cd colonized tomato roots, with populations reaching 1.0×10^7 CFU g^{-1} dry weight in the roots and 1.0×10^5 CFU g^{-1} dry weight in the stem and leaves after 30 days of inoculation (Botta et al. 2013). Similarly, Santos et al. (2017a, b) showed that *A. brasilense* FP2 and HM053 *pnifH-gusA* strains were capable of colonizing and expressing *nif* genes on the root surface of barley and wheat, respectively. Despite both strains reaching 1.0×10^7 CFU g^{-1} of fresh root weight after 96 h of inoculation, the ammonium-excreting strain HM053 consistently exhibited stronger GUS expression signals on the roots compared to the wild-type strain FP2 (Santos et al. 2017b). As with studies using *lacZ* as a reporter, most studies employing *gusA* did not account for the potential false positives resulting from β-glu activity in indigenous bacteria and plant roots (Akao et al. 1999).

2.2.3 Green Fluorescent Protein (GFP)

Green fluorescent protein (GFP) was isolated from the jellyfish *Aequorea victoria* and first introduced as a reporter gene by Chalfie et al. (1994). This protein emits fluorescent light when exposed to light ranging from near-blue to UV wavelengths, enabling detection and quantification of GFP-tagged strains through epifluorescence microscopy (Xi et al. 1999; Bacilio et al. 2004; O'Neal et al. 2020), confocal laser scanning microscopy (CLSM) (Liu et al. 2003; Malinich and Bauer 2018; O'Neal et al. 2020; Camelo et al. 2021), or UV light exposure on agar plates (Xi et al. 1999; Gera et al. 2003; Malinich and Bauer 2018). GFP offers several advantages over traditional *lacZ* and *gusA* reporter methods: (i) no substrate is required, (ii) expression occurs within the cell, (iii) there is no background interference from indigenous bacteria, and (iv) in situ detection enables studies at the single-cell level (Rilling et al. 2019). Additionally, it can be combined with other fluorescent methods, such as Discomona red fluorescent protein (DsRed) and its derived mCherry (Ramirez-Mata et al. 2018). In cases where dual fluorescence imaging systems are unavailable, reporter systems based on different techniques have proven valuable. For example, constructs carrying both *gfp* and *gusA* genes, inserted into mini-Tn5 transposons (Xi et al. 1999) or plasmids (Ramos et al. 2002), have been used to visualize *Azospirillum brasilense* on wheat roots.

Thirteen studies have utilized *gfp* as a reporter gene for tracking *Azospirillum* species, and eight of these studies also included bacterial quantification using CFU counting methods. The most frequently analyzed species was *A. brasilense*, including strains Sp7, Yu62, Ab-V5, Ab-V6, and FP2, with constitutive GFP plasmids. The studies were conducted in a variety of plant species, such as maize, wheat, pearl millet, rice, tobacco, common bean, alfalfa, and onion, with a predominant focus on root colonization (Table 2.1).

O'Neal et al. (2020) demonstrated that by using *A. brasilense* Sp7 constitutively expressing GFP, which is chemotactically attracted to the root hair and elongation zones of wheat, facilitating the movement of *A. brasilense* toward these root surfaces over time, promoting root surface colonization. In contrast, motile *A. brasilense* bacteria are repelled by conditions in the root tip zone, preventing colonization of this region (O'Neal et al. 2020). Additionally, during the development of *Phaseolus vulgaris* (common bean), *A. brasilense* Sp7 was shown to actively migrate to developing seeds, where it forms a significant intercellular population. GFP-expressing *A. brasilense* was demonstrated to be vertically transmitted to successive plant generations, indicating that the seed constitutes an effective inoculum (Malinich and Bauer 2018).

2.3 Immunological-Based Methods

Several immunological approaches have been applied to enumerate and localize *Azospirillum* spp. in soils and when colonizing the roots of crop plants. These methods are based on the use of antibodies, which are conjugated with enzymes, isotopes, nanoparticles, or fluorochromes to detect and/or quantify antigen-antibody interactions. Table 2.2 summarizes articles using ELISA for the quantification and immunofluorescence labeling methods for the identification and localization of *Azospirillum*.

2.3.1 Enzyme-Linked Immunosorbent Assay (ELISA)

The first antibodies against *Azospirillum brasilense* were polyclonal antibodies raised against whole cells of *A. brasilense* Cd by Levanony et al. (1987). In indirect and competitive ELISA tests, no cross-reactivity was found with the closely related *A. brasilense* Sp7 strain or other *Azospirillum* spp. strains. Quantification of root colonization of wheat seedlings on wet filter paper or in sandy soil revealed 2.0×10^4 to 5.0×10^4 cells g^{-1} dry root using indirect ELISA (Levanony et al. 1987). In another set of experiments, *A. brasilense* Cd was quantified to 2.1×10^3 and 4.5×10^4 cells g^{-1} dry root by ELISA (Levanony and Bashan 1989). A time-limited liquid enrichment procedure in semi-solid medium was applied to support the quantification of very few cells of *A. brasilense* by ELISA or MPN; comparable counts of 5.0×10^4 cells or CFU g^{-1} dry root were detected (Bashan et al. 1991). In a field experiment with pearl millet, Wani et al. (1988) inoculated with *A. lipoferum* ICM1001, and the colonization of rhizosphere soil and roots was quantified using direct (double-antibody) ELISA. At harvest, 40–45 days after inoculation, the 2.5×10^4 cells g^{-1} dry rhizosphere soil and 9.9×10^4 cells g^{-1} dry root were estimated.

While the standard ELISA colorimetric test uses 2,2′azino-di-(3-ethylbenzthiazoline) sulphonic acid to quantify anti-mouse IgG conjugated with horseradish peroxidase in the wells, the chemiluminescence ELISA (C-ELISA) uses Luminol reagent and enhancer (Amersham); finally, the plates are scanned by a luminometer (ML 1000 Dynatech). The standard ELISA counted cells down to 1.0×10^4 cells mL^{-1}, while the sensitivity of the C-ELISA is enhanced 100 times to 1.0×10^2 cells mL^{-1} (Schloter et al. 1992b).

Several monoclonal antibodies (mAbs) against *Azospirillum* were developed by Schloter et al. at the Helmholtz Center, Munich, Germany. The binding sites of the mAbs were characterized by competition chemiluminescence ELISA and cell

Table 2.2 Summary of *Azospirillum* articles using immunological methods

	1	2	3	4	5	6	7	8
	Target Azospirillum strain	Type of assay (in vitro or Field)	Immunological method used	Culture host	Localization (where the bacteria was localized or quantified; e.g., root, rhizosphere, seed, soil, etc.)	Evaluation time (when the bacteria was localized or quantified; how many days or hours post-plant inoculation)	Quantification (number of CFU g^{-1} of soil, root, etc. if analyzed, or the note "Not counted" if only localization or quantification is provided)	References
1	*A. brasilense* Cd	In vitro	ELISA	Cereals (wheat, corn, barley, triticale, sorghum, setaria, etc.)	Root	4–14 days	2.0×10^4 to 5.0×10^4 CFU g^{-1}	Levanony et al. (1987)
2	*A. lipoferum* (ICM 1001)	Field	ELISA + MPN	Pearl millet cv. BJ 104	Rhizosphere soil Roots	At harvest	2.5×10^4 CFU g^{-1} 9.9×10^4 CFU g^{-1}	Wani et al. (1988)
3	*A. brasilense* Cd	In vitro	ELISA	Wheat	Root	24 and 48 h	4.5×10^4 CFU g^{-1}	Levanony and Bashan (1989)
4	*A. brasilense* Cd	Pots (quartz sand)	ELISA + MPN time-limited liquid enrichment method	Wheat	Root	5 days	5.0×10^4 CFU g^{-1}	Bashan et al. (1991)
5	*A. brasilense* Sp7 and Wa3	In vitro	Immunofluorescence combined with FISH and LSM	Wheat	Root	9–14 days post-inoculation	Not counted	Aßmus et al. (1997)

(continued)

Table 2.2 (continued)

	1	2	3	4	5	6	7	8
	Target Azospirillum strain	Type of assay (in vitro or Field)	Immunological method used	Culture host	Localization (where the bacteria was localized or quantified; e.g., root, rhizosphere, seed, soil, etc.)	Evaluation time (when the bacteria was localized or quantified; how many days or hours post-plant inoculation)	Quantification (number of CFU g^{-1} of soil, root, etc. if analyzed, or the note "Not counted" if only localization or quantification is provided)	References
6	*A. brasilense* Sp7 *A. brasilense* Wa3 *A.baldaniorum Sp245*	In vitro (tubes with unsterile soil and sterile quartz sand)	Chemiluminescence ELISA, immunogold staining (strain-specific Mabs)	Wheat (2 cultivars)	Root Root Root Root endophytes	4, 6, 8, 10, and 14 weeks	1.0×10^5 to 1.0×10^6 CFU g^{-1} 1.0×10^4 to 1.0×10^5 CFU g^{-1} 1.0×10^5 to 1.0×10^7 CFU g^{-1} 5.0×10^5 CFU g^{-1}	Schloter and Hartmann (1998)
7	*A. brasilense* Sp7	In vitro	Chemiluminescence ELISA	In vitro	Lab culture	–	Sensitivity: 1.0×10^2 cells mL^{-1}	Schloter et al. (1992b)
8	*A. brasilense* Sp7	In vitro	Fluorochrome-labeled Monoclonal antibody	Wheat	Rhizosphere	1 days	Not counted	Schloter et al. (1993)
9	*A. brasilense* Sp7, SR55, SR14 (Saratov) *A. baldaniorum* Sp245	In vitro	Immunodiffusion Immunoelectrophoresis ELISA	In vitro	Soil	1 days	Not counted	Shirokov et al. (2015)

10	*A. brasilense* Cd	In vitro	Immunogold, TEM or immunofluorescence + CFU counts	Tomato	Root (surface and endophytic)	30 d	1.0×10^8 CFU g^{-1}	Caiola et al. (2004)
11	*A. lipoferum* Fr	In vitro	Immunofluorescence	*Poa pratensis* Wheat	Root hairs	4-5 weeks	4.0×10^5 cells g^{-1} 3.1×10^6 cells g^{-1}	Haahtela et al. (1988)
12	*A. brasilense* SpB14	In vitro	Immunofluorescence	Wheat	Rhizosphere soil Root	up to 6 months	1.1×10^5 to 5.0×10^2 cells g^{-1} 2.9×10^3 to 1.8×10^3 cells g^{-1}	De Coninck et al. (1988)
13	*A. brasilense* Cd	In vitro	ELISA Immunoblotting Immunogold, TEM	In vitro	Lab culture	1 days	Not counted	Canini et al. (2006)
14	*A. baldaniorum* Sp245	In vitro	Indirect ELISA Immunofluorescence microscopy	Wheat	Root	1 days	Not counted	Yegorenkova et al. (2015)
15	*A. brasilense* JM82	In vitro	Immunofluorescence microscopy	Kallar grass Rice	Roo	2 days/3 weeks	Not counted	Bilal et al. (1993)
16	*A. brasilense* Sp7 *A. baldaniorum* Sp245	In vitro	ELISA Immunofluorescence	Wheat	Roots	7 days	Not counted	Shelud'ko et al. (2010)
17	*A. baldaniorum* Sp245	In vitro	Antigen characterization	In vitro	Lab culture	–	Not counted	Schloter and Hartmann (1996)
18	*A. brasilense* Sp7	In vitro	Antigen characterization	In vitro	Lab culture	–	Not counted	Schloter et al. (1994)
19	*A. baldaniorum* Sp245 *A. brasilense* Sp7	In vitro	ELISA, Immunofluorescence	Wheat	Roots	7 days	not counted	Schloter et al. (1997)
20	*A. brasilense* Wa5	In vitro	Antigen characterization	In vitro	Lab culture	–	1.5×10^3 antigens $cell^{-1}$	Schloter et al. (1992a)

surface mutants of *A. brasilense* Sp7. The Mic2-class mAbs react specifically to the 100 kDa flagellin protein of the polar flagellum of strain Sp7 and the closely related strain Cd; the Mic3-mAbs react to an 85 kDa outer membrane protein, and the Mic4-mAbs to a lipopolysaccharide (LPS) component of the outer membrane of Sp7 (Schloter et al. 1992a, 1994). The mAbs Mic3 and Mic4 are strain specific for strain Sp7. A hybridoma cell line producing a rat monoclonal antibody (mAb Bo-33) was directed against the lipopolysaccharide of *A. brasilense* Wa5. The number of antigens of this mAb per cell is about $1.5 \times 10^{3,}$ and the sensitivity of detection is about 1.0×10^2 bacteria per mL (Schloter et al. 1997). For *A. baldaniorum* Sp245, the developed mAbs were classified into four groups: Class 1 corresponds to lipopolysaccharides as antigens, and Class 4 to an outer membrane protein, which is common to strains Sp245 and Sp246. Classes 2 and 3 correspond to two outer membrane proteins, which are strain specific for strain Sp245. The number of antigens per cell varies from 1.0×10^3 (class 1) to 1.0×10^2 (class 4). Immunoelectrophoresis and immunodiffusion analysis of polyclonal antibodies against whole intact cells of *A. brasilense* Sp7 revealed at least three conservative surface immunogenic proteins and LPS (Shirokov et al. 2015). These antibodies were used for the specific quantification of *Azospirillum* as a genus-specific probe conjugated with horseradish peroxidase in root colonization and soil studies.

Immuno-trapping is another immunological approach to get insights into the (micro)diversity of a bacterial species colonizing roots. Monoclonal or polyclonal antibodies targeting, e.g., highly expressed LPS- or protein-antigenic epitopes at the bacterial surface, having high specificity at the strain or species level, are needed. After binding a specific antibody/antiserum to protein A-coated microtiter plates, aliquots of soil or root suspensions are added to the wells. After incubation and intensive washing, a treatment with glycine/HCl disrupted the antibody-bacteria binding, and the specifically enriched cells were eluted. Purified bacterial isolates were characterized by ELISA, molecular fingerprinting, and polyphasic taxonomy (Lebuhn et al. 2000). The advantage of this approach is that even very minor numbers of specific bacteria are enriched from environmental samples and studied for their (micro)diversity.

2.3.2 Immunological Staining for Light and Transmission Electron Microscopy

A. brasilense Sp7 was localized using the mAb-Mic3 (Schloter et al. 1994) on the root surface of a wheat plantlet by immunofluorescence. The mAb was marked directly with FITC, and CLSM enabled an in situ localization and identification (Schloter et al. 1993). The colonization of wheat seedlings by a mixed inoculation with *A. brasilense* Sp7 and Wa3 in an axenic *in vitro* system was monitored with a combined application of fluorescence-labeled antibodies (mAbs), rRNA-oligonucleotides with fluorescence in situ hybridization (FISH) analysis, and confocal laser scanning microscopy (CLSM) (Aßmus et al. 1997). CLSM and epifluorescence microscopy with a charge-coupled device (CCD) camera were used for the detection of individual cells. This strategy gave insights into the relative abundance of an inoculated strain in the rhizosphere, besides the presence of other rhizobacteria. It enabled the exact localization at the single-cell level and allowed the estimation of its general metabolic activity. Using a strain-specific monoclonal antibody for *A. brasilense* Wa3, it was possible to identify this particular strain in root samples of inoculated wheat plantlets. Strain Wa3, as well as other bacteria colonizing the rhizoplane, was stained simultaneously with rRNA-targeted, fluorescence-labeled oligonucleotide probes. In a co-inoculation experiment with *A. brasilense* Sp7 and Wa3, it was demonstrated by in situ identification and quantitative chemiluminescence ELISA that strain Sp7 outcompeted strain Wa3. In the combined application of specific antibody staining and FISH analysis, it was necessary to begin with the immunological staining and perform the FISH analysis afterwards. Finally, a counterstaining with DAPI was performed (Aßmus et al. 1997). Methodological details for the FISH approach and immunostaining can be found in Hartmann et al. (2015).

Endophytic and surface colonization of wheat roots by different *A. brasilense* strains were investigated using strain-specific mAbs, chemiluminescence ELISA, and immunogold labeling (Schloter and Hartmann 1998). The *A. brasilense* strains Sp7 and Wa5 and *A. baldaniorum* strains Sp245 were inoculated into two different wheat cultivars under axenic conditions and in soil microcosms. The wheat cultivars were colonized with different efficiency by the inoculated strains. The wheat rhizoplane was colonized in high numbers by all strains four weeks after inoculation, as determined with strain-specific mAbs and chemiluminescence ELISA. Strain Sp245 showed the highest colonization activity and could also be detected in the inner root tissue. According to his immunogold labeling, *A. baldaniorum* Sp245 formed microcolonies in the intracellular spaces of the root endodermis. The other strains formed microcolonies only at the root surface, and the colonization density decreased with time until 14 weeks after inoculation (flowering phase). All strains colonized mostly the root tips in the highest numbers, but strain Sp245 was also detected in the middle part of the root. While the colonization by strains Sp7 and Wa5 of the total root continuously decreased, the numbers of strains of Sp245 inside the roots were constant at around 5.5×10^5 cells g^{-1} dry weight of roots over the duration of the experiment. Co-inoculation with all three strains together resulted in a comparable colonization pattern; strain Wa5 was even further decreased on the roots because of its weak competition abilities (Schloter and Hartmann 1998).

The colonization of *A. brasilense* SpBR14 of hydroponically grown wheat plants was studied by immunofluorescence. Peat-based inoculant *Azospirillum* spp. established well in the rhizosphere of field-grown winter barley and winter wheat (De Coninck et al. 1988). The colonization of *Poa pratensis* and spring wheat by *A. lipoferum* strain Fr was studied in pot experiments by immunofluorescence staining, and a major colonization was found on root hairs (Haahtela et al. 1988). Kallar grass (*Leptochloa fusca*) and rice (*Oryza sativa*) colonization by *A. brasilense* JM82 and *Enterobacter* spp. was studied with immunofluorescence, and the area around the point of emergence of lateral roots showed maximum colonization (Bilal et al. 1993). Colonization of roots of tomato plantlets by *A. brasilense* Cd was investigated using immunofluorescence- and immunogold labeling (Caiola et al. 2004); within roots, strain Cd formed cyst-like round forms, which did not further divide (Caiola et al. 2004). Immunofluorescence studies were also performed to study the colonization of spring wheat by *A. baldaniorum* Sp245 (Shelud'ko et al. 2010). The cells formed microcolonies on the root surface and exhibited changes in the composition of genus-specific protein antigens on the bacterial cell surface upon colonization of roots. During co-inoculation of spring wheat with *A. baldaniorum* Sp245 and *Paenibacillus polymyxa* 1465, immunofluorescence with polyclonal antisera against these strains revealed that both strains colonize similar sites; ELISA data revealed that the colonization by strain Sp245 was higher in the presence of strain 1465. However, the plant development was reduced in the presence of the two strains in combination (Yegorenkova et al. 2015).

2.4 Nucleic Acid-Based Methods

Several methodologies have been developed to detect and identify organisms based on differences in nucleic acid sequences. Some of these are PCR-independent methods; however, the most used technologies for the detection of *Azospirillum* are PCR-dependent. Table 2.3 summarizes the articles using the nucleic acid method for the identification and localization of *Azospirillum.* The sequences of the primers, target genes, and the main species and strains are presented in Table 2.4.

2.4.1 PCR-Independent Methods

Fluorescence in situ hybridization (FISH) uses labeled DNA probes designed to bind to specific nucleic acid sequences, allowing the detection of microorganisms of interest through fluorescence emission. For *Azospirillum*, a successful combination of FISH methodology with CLSM has enabled the detection of strains of *A. baldaniorum* Sp245 and *A. brasilense* Sp7 and NH in wheat under optimal and salt-stressed conditions 30 days after inoculation (Aßmus et al. 1995; Rothballer et al. 2003; Nabti et al. 2010). In maize roots at the field level, *A. lipoferum* CRT1 was detected after 140 days using a strain-specific probe targeting 16S rDNA (El Zemrany et al. 2006). Additionally, FISH has demonstrated that *A. brasilense* Cd can survive and colonize sorghum plants in desert soil when co-inoculated in alginate beads with *Chlorella sorokiniana* (Trejo et al. 2012; Lopez et al. 2013). Furthermore, FISH analysis has been used to document the colonization of the rhizoplane and root colonization by *A. brasilense* Sp6 in quailbush grown in compost-amended soil (de-Bashan et al. 2010).

Table 2.3 Summary of *Azospirillum* articles using nucleic acid-based methods

	1	2	3	4	5	6	7	8
	Target Azospirillum strain	Type of assay (in vitro or field)	Nucleic acid-based methods used	Culture host	Localization (where the bacteria was localized or quantified; e.g., root, rhizosphere, seed, soil, etc.)	Evaluation time (when the bacteria was localized or quantified; how many days or hours post-plant inoculation)	Quantification (number of CFU/g of soil, root, etc. if analyzed, or the note "Not counted" if only localization or quantification is provided)	References
1	*A. brasilense* Sp7 *A. baldaniorum* Sp245	In vitro	FISH + CFU count	Wheat	Root	4 and 7 weeks	1.0×10^4 CFU g^{-1}	Rothballer et al. (2003)
2	*A. lipoferum* CRT1	Field	Hybridization with strain-specific probe (targeting 16s rDNA)	Maize	Root	40 days 140 days	1.0×10^8 CFU g^{-1} 1.0×10^4 CFU g^{-1}	El Zemrany et al. (2006)
3	*A. lipoferum* CRT1	In vitro Field	qPCR strain specific (SCAR)	Maize	Root system (root and rhizosphere and root)	10 days 18 days	Between 1.0×10^4 and 1.0×10^6 eqCFU g^{-1} 1.0×10^4 eqCFU g^{-1}	Couillerot et al. (2010a)
4	*A. brasilense* UAP-154 *A. brasilense* CFN-535	In vitro	qPCR strain specific (SCAR)	Maize	Root system (rhizosphere and root)	10 days	1.0×10^6 eqCFU g^{-1} 1.0×10^5 eqCFU g^{-1}	Couillerot et al. (2010b)
5	*A. brasilense* FP2	In vitro Pots (Greenhouse)	qPCR genus specific (targeting *nifA*)	Maize	Root	1 day 4 days 7 and 10 days	1.0×10^6 DNA copy number g^{-1} 1.0×10^8 DNA copy number g^{-1}	Faleiro et al. (2013)

(continued)

Table 2.3 (continued)

	1	2	3	4	5	6	7	8
	Target Azospirillum strain	Type of assay (in vitro or field)	Nucleic acid-based methods used	Culture host	Localization (where the bacteria was localized or quantified; e.g., root, rhizosphere, seed, soil, etc.)	Evaluation time (when the bacteria was localized or quantified; how many days or hours post-plant inoculation)	Quantification (number of CFU/g of soil, root, etc. if analyzed, or the note "Not counted" if only localization or quantification is provided)	References
						1, 4 and 7 days 10 days	1.0×10^9 DNA copy number g^{-1} 1.0×10^8 DNA copy number g^{-1} 1.0×10^7 DNA copy number g^{-1}	
6	*A. brasilense* CFN-535 *A. brasilense* UAP-154 *A. lipoferum* CRT1	In vitro	qPCR strain specific (SCAR)	Maize	Rhizosphere	10 days/35 days	$1.0 \times 10^7/1.0 \times 10^5$ eqCFU g^{-1} $1.0 \times 10^6/1.0 \times 10^4$ eqCFU g^{-1} $1.0 \times 10^6/<$lim det eqCFU g^{-1}	Couillerot et al. (2013)
7	*A. brasilense* FP2	In vitro	qPCR strain specific (SCAR) qPCR species specific (SCAR)	Wheat	Root	1–13 days	1.0×10^7 to 1.0×10^8 eqCFU g^{-1} 1.0×10^7 to 1.0×10^8 eqCFU g^{-1}	Stets et al. (2015)
8	*A. brasilense* Sp7	In vitro	qPCR strain specific (SCAR)	Maize	Rhizosphere	15 days/60 days	1.0×10^6 eqCFU g^{-1} /1.0×10^5 eqCFU g^{-1}	Reddypriya et al. (2019)

9	*A. brasilense* Sp7	In vitro	qPCR genus specific	Common bean	Root, Stem, Leaf and seed	Vegetative, early reproductive and mature	Bacteria:plant cell ratio	Malinich and Bauer (2018)
10	*A. brasilense* Ab-V5 and Ab-V6	Field	qPCR species specific (SCAR)	Maize	Rhizosphere Root	1–7 days 11 days 13–75 days	1.0×10^5 eqCFU g^{-1} 1.0×10^5 eqCFU g^{-1} 1.0×10^3 eqCFU g^{-1} 1.0×10^4 eqCFU g^{-1}	Urrea-Valencia et al. (2021)
11	*A. brasilense* FP2	In vitro	qPCR strain specific (SCAR) with PMA	Maize	Root (endophytes)	7 days	1.0×10^6 eqCFU g^{-1}	da Cunha et al. (2020)
12	*Azospirillum* sp. B510	In vitro	CRISPR loci-PCR	Wheat	Root	14 days	Qualitative	Rilling et al. (2021)
13	*A. baldaniorum* Sp245	Field	qPCR strain specific (SCAR)	Brachiaria	Shoot Root	60 days	1.0×10^2 eqCFU g^{-1} 1.0×10^3 eqCFU g^{-1}	Soares et al. (2021)
14	*A. brasilense* FP2	In vitro	Colony count qPCR species specific (SCAR)	Wheat	Seed Root	48 h 21 days	1.0×10^6 CFU g^{-1} 2.0×10^6 eqCFU g^{-1}	Takahashi et al. (2022)
15	*A. lipoferum* CRT1	Field	qPCR strain specific (SCAR)	Maize	Root system (root and rhizosphere and root)	Six-leaf and flowering stages	Absent	Renoud et al. (2022b)
16	*A. lipoferum* CRT1	Field	qPCR strain specific (SCAR)	Maize	Root system (root and rhizosphere and root)	Six-leaf and flowering stages	Absent	Renoud et al. (2022a)
17	*A. brasilense* Ab-V5 *A. brasilense* HM053 *A. argentinense* Az39	Pots (greenhouse)	qPCR genus specific (SCAR)	Maize	Root Shoot	7, 14, 21, 28, 35, and 42 days	1.0×10^3 to 1.0×10^8 CFU plant $root^{-1}$ below LOD to 1.0×10^5 CFU plant $shoot^{-1}$	Takahashi et al. (2024)

Table 2.4 Summary of primers designed for *Azospirillum* identification and localization based on nucleic acid methodologies

	1	2	3	4	5	6	6
	Target Azospirillum Strain	Method (ELISA, Immunoblot, etc.)	Oligo name	Oligo Sequence (5′–3′)	Gen target	PCR-product size	References
1	*A. brasilense*	FISH	Abras 1420	CCACCTTCGGGTAAAGCCA	16S rRNA	–	Rothballer et al. (2003)
2	*Azospirillum* sp.	ERIC-PCR (Fingerprinting)	ERICR1 ERIC2	ATGTAAGCTCCTGGGGATTCAC AAGTAAGTGACTGGGGTGAGCG	Genome repetitive sequences	DNA banding pattern	Versalovic et al. (1994)
3	*Azospirillum* sp.	BOX-PCR (Fingerprinting)	BOXAR1	CTACGGCAAGGCGACGCTGACG	Genome repetitive sequences	DNA banding pattern	Versalovic et al. (1994)
4	*Azospirillum* sp.	ARDRA (Fingerprinting)	FD1 RP2	CCGAATTCGTCGACAACAGAGTTTGATCCTGGCTCAG CCCGGGATCCAAGCTTACGGCTACCTTGTTACGACTT	16S rRNA	DNA banding pattern	Weisburg et al. (1991)
5	*Azospirillum* genus	PCR or qPCR	Azo494-F Azo756-R	GGCCYGWTYAGTCAGRAGTG AAGTGCATGCACCCCRRCGTCTAGC	16S rRNA (hypervariable region)	263 bp	Lin et al. (2011)
6	*Azospirillum* genus	PCR or qPCR	Az16S–A-F Az16S–A-R	GCGGTAATACGAAGGGGGCK CTTGTCACCGGCAGTTCCACCAG	16S rRNA	646 bp	Shime-Hattori et al. (2011)
7	*Azospirillum* genus	PCR or qPCR	fAZO rAZO	GGCGCATCCCTTCTCACGG GCTTGCGCCACGCGCAGG	16S–23S rDNA-ISR	500 bp	Baudoin et al. (2010a)
8	*Azospirillum* genus	PCR	Azo.rpoD F Azo.rpoD R2	GAGATGGGCATCAACATCGT ACCTTCTCCAGCGGCATCATC	rpoD	1100 bp	Maroniche et al. (2017)

9	*A. brasilense*	PCR or qPCR	Azo2-F Azo2-R	GCGCGGGAAGTCCTGAAT CCCTTCACCATCCAGTCGAT	Genus-specific region (SCAR marker)	90 bp	Stets et al. (2015)
10	*A. brasilense*	PCR or qPCR	Azo-WM-F Azo-WM-R	GTCCACCAAGCCAACAATG TAACCATAAAAGCCCACCCTG	Genus-specific region (SCAR marker)	90 bp	Takahashi et al. (2024)
11	*A. lipoferum* CRT1	PCR or qPCR	CRT1-Q1 CRT1-Q2	ATCCCGGTGGACAAAGTGGA GGTGCTGAAGGTGGAGAACTG	Strain-specific region (SCAR marker)	294 bp	Couillerot et al. (2010a)
12	*A. brasilense* UAP-154	PCR or qPCR	U2 U7	TGACGGCCAACACCAACGACTC TGCCGTCGATGAACGACGCCATCTG	Strain-specific region (SCAR marker)	261 bp	Couillerot et al. (2010b)
13	*A. brasilense* CFN-535	PCR or qPCR	F12 F24	AAGCGATCCCGACCTTGAGGCA TGTCGATGCCGACAGGCTTGACCA	Strain-specific region (SCAR marker)	200 bp	Couillerot et al. (2010b)
14	*A. lipoferum* Az204	qPCR strain specific (SCAR)	SCAR-Az204-1-F SCAR-Az204-1-R	CTCCTGGGGATTCACTCCAACA ACTTCGCACCAGCGTATTCC	Strain-specific region (SCAR marker)	641 bp	Reddy Priya et al. (2016)
15	*A. lipoferum* Az204	PCR or qPCR	SCAR-Az204-2-F SCAR-Az204-2-R	GATTCACTCCAACATGAGCCGC GCGCAACAAGTCAGGTGAGA	Strain-specific region (SCAR marker)	464 bp	Reddy Priya et al. (2016)
16	*A. brasilense* Sp7	PCR or qPCR	SCAR-Sp7-1-F SCAR-Sp7-1-R	TTCGTACCGCCTGACACTTC GGACAGCCCGTGAACATACA	Strain-specific region (SCAR marker)	299 bp	Reddy Priya et al. (2016)
17	*A. brasilense* Sp7	PCR or qPCR	SCAR-Sp7-2-F SCAR-Sp7-2-R	TTGCGATGCTGTCGGAGTAA GGTTCTCAAGGCACTCCTGT	Strain-specific region (SCAR marker)	437 bp	Reddy Priya et al. (2016)

(continued)

Table 2.4 (continued)

	1	2	3	4	5	6	6
	Target Azospirillum Strain	Method (ELISA, Immunoblot, etc.)	Oligo name	Oligo Sequence (5′–3′)	Gen target	PCR-product size	References
18	*A. brasilense* FP2	PCR or qPCR	AzoR2.1-F AzoR2.1-R	CGCCACCATGCGATCAA GCATGCCCAGTACTGCAAGTC	Strain-specific region (SCAR marker)	90 bp	Stets et al. (2015)
19	*A. brasilense* FP2	PCR or qPCR	AzoR2.2-F AzoR2.2-R	CCTTCACCTGGACGGTTCAG CGCGGCCAGCAGACTT	Strain-specific region (SCAR marker)	90 bp	Stets et al. (2015)
20	*A. brasilense* FP2	PCR or qPCR	AzoR5.3-F AzoR5.3-R	AATTCTTTCCGTTGGCTTTCAAGCTTGCCGACCGGAGTATC	Strain-specific region (SCAR marker)	90 bp	Stets et al. (2015)
21	*A. argentinense* Az39	PCR	TP5-F TP5-R	GGCCGTTGTTCTCGGTCTCTAT GCGATCTCCATTTATTTCGCCCT	Strain-specific region (SCAR marker)	307 bp	Coniglio et al. (2020)
22	*A. baldaniorum* Sp245	PCR or qPCR	Sp245p10F Sp245p10R	CCTCTATCGGAACCGCAACG GCACTTGTTTGATGCCGAGA	Strain-specific region (SCAR marker)	203 bp	Soares et al. (2021)

In general, these studies used probes for the bacteria domain detection (EUB-338 I, II, and III) and specific probes for *A. brasilense* (Abras-1420 or KO 205) (Amann et al. 1990; Daims et al. 1999; Stoffels et al. 2001). These probes were labeled with different fluorochromes such as Cy3, Cy5, and FITC, allowing differentiation between general bacteria and *Azospirillum* (Stoffels et al. 2001). The samples are collected and fixed with paraformaldehyde (commonly used for Gram-negative bacteria) and stored in a PBS/ethanol mixture before analysis. Hybridization is performed at 46–48 °C in a specific buffer to ensure optimal binding to nucleic acid sequences. Once the probes bind to the target nucleic acids, the samples are visualized using CLSM or epifluorescence microscopy, with channels adjusted according to each label. The images obtained with different fluorescence channels are processed with specialized software to distinguish *Azospirillum* bacterial populations. This technique enables image overlay and differentiation of the observed elements: root surfaces appear in one color, general bacteria in another, and *Azospirillum* in a different one.

Since in environmental and plant samples the background fluorescence is much disturbing, the so-called "Gold-FISH" technique was developed (Schmidt et al. 2012). FISH-hybridization is performed with horseradish peroxidase-labeled oligonucleotide probes, followed by catalyzed reporter deposition (CARD) of biotinylated tyramides, which facilitates the amplification of binding sites for streptavidin conjugates labeled with nanogold particles. To finally detect labeled cells with SEM, an autometallographic enhancement of the nanogold particles is performed.

2.4.2 PCR-Dependent Methods

PCR-based amplification methodologies are essential tools for identifying and characterizing *Azospirillum* (Table 2.3). Fingerprinting techniques such as terminal restriction fragment length polymorphism (T-RFLP), Automated rRNA Intergenic Spacer Analysis (ARISA), amplified ribosomal DNA restriction analysis (ARDRA), and denaturing gradient gel electrophoresis (DGGE) have been widely used to assess DNA polymorphisms and detect *Azospirillum* in complex samples (Nievas et al. 2023). The main application of these tools is to evaluate the impact of *Azospirillum* inoculation on microbial diversity in different habitats, including maize (Herschkovitz et al. 2005) and the wheat rhizosphere (Baudoin et al. 2010b), as well as the tomato rhizoplane and phyllosphere (Correa et al. 2007), the quailbush rhizoplane (de-Bashan et al. 2010), and other plant species communities (Russo et al. 2008). They have also been used to assess the effect of inoculation on soil populations (Yuan et al. 2012) and to evaluate the impact of specific compounds on *Azospirillum* (Lo 2010). Additionally, these techniques facilitate the strain identification of *Azospirillum* members by comparing band patterns with a control and are typically used in combination with one or more methods (Fani et al. 1993; Gündisch et al. 1993; Eid and Sherwood 1995). However, they have some limitations, including labor-intensive procedures and high time consumption (Rilling et al.

2019). Notably, these methods allow the detection of unique DNA regions within genomes, which can be characterized and amplified by PCR. This process is known as the sequence characterized amplified region (SCAR) marker technique. SCAR markers represent unique, genetically defined loci identified through PCR amplification using specific primers derived from genomic DNA fragments.

For *Azospirillum*, SCAR marker development has been achieved through fingerprinting methodologies. Couillerot et al. (2010a, b) identified specific sequences for the identification and quantification of the strain *A. lipoferum* CRT1 via qPCR using the RFLP technique (Couillerot et al. 2010a), and for *A. brasilense* UAP-154 and CFN-535 using BOX-PCR (Couillerot et al. 2010b). Meanwhile, the identification of *Azospirillum* sp. Az204 and *A. brasilense* Sp7 through SCAR markers was achieved via sequence analysis obtained using BOX-PCR and ERIC-PCR (Reddy Priya et al. 2016). The laboratory procedure involves identifying a specific sequence, selecting, cloning, and sequencing it, and then comparing it against databases to design primers for unique sequences (Grover and Sharma 2016). However, with the current availability of genomic data and bioinformatics tools, unique sequences can also be identified through comparative genomic analysis using *in silico* approaches. Fingerprinting methodologies remain useful for strains whose genomes have not yet been sequenced. In the case of *Azospirillum*, bioinformatics strategies have also been employed to detect unique sequences. Stets et al. (2015) identified strain-specific DNA sequences in *A. brasilense* FP2 based on a comparison of the genomic sequences in public databases. Soares et al. (2021) developed a methodology for strain-specific identification and quantification of *A. baldaniorum* Sp245 by fragmenting its genome into 500-bp sequences and comparing them with closely related strains to identify distinct regions for primer design. Additionally, other computational tools have been used to simulate fingerprinting methodologies *in silico*, including AFLP simulation, singleton analysis, and SNV searches (Coniglio et al. 2020). These unique and characterized sequences, which can be strain specific, are detected using PCR, qPCR, or alternative variants such as most probable number-PCR (MPN-PCR), competitive PCR, quantitative PCR, and droplet digital PCR (dd-PCR) (Rilling et al. 2019). Quantitative PCR (qPCR) is the most widely used method for microorganism quantification in the rhizosphere, plant tissues, and soil due to its sensitivity, specificity, and speed (Sørensen et al. 2009). It involves the continuous detection of a fluorescence signal generated during amplification, which is converted into a concentration value for the target DNA in the original sample. For microorganism quantification in complex samples, absolute quantification is performed using a known concentration standard to construct a calibration curve (Klein 2002). One of the advantages of absolute quantification is that it is culture-independent and allows for the detection of unlabeled cells (Rilling et al. 2019). While it does not distinguish between viable and non-viable cells, it can be combined with compounds such as propidium monoazide (PMA) to detect only viable ones (da Cunha et al. 2020). Under laboratory conditions, *Azospirillum* could be detected by qPCR between 10 and 60 days after inoculation in the rhizosphere samples (Couillerot et al. 2010a, b; Stets et al. 2015; Reddypriya et al. 2019). However, in field conditions, they could not be detected for extended periods (Urrea-Valencia et al. 2021; Renoud et al. 2022a, b). In contrast, strains remained detectable in plant tissues like root samples for longer periods (Soares et al. 2021; Urrea-Valencia et al. 2021). The viability was also demonstrated in maize roots up to seven days after inoculation when samples were pre-treated with PMA (da Cunha et al. 2020). Additionally, qPCR analysis follows *Azospirillum* in plants and differentiates between strains that colonize the plant in a specific tissue from others that can colonize the plant in a systemic way (Takahashi et al. 2024). Finally, qPCR could be used as quality control of commercial products (Reddy Priya et al. 2016; da Cunha et al. 2020).

2.4.3 Specific Gene Markers for Azospirillum *Identification*

The identification of *Azospirillum* spp. using specific genes has been widely explored (Table 2.4). Lin et al. (2011) developed PCR primers targeting a hypervariable 16S rRNA region, successfully differentiating *Azospirillum* from other genera. The primer pair Azo494-F (5′-GGC CYG WTY AGT CAG RAG TG-3′) and Azo756-R (5′-AAG TGC ATG CAC CCC RRC GTC TAG C-3′) amplifies a 263-bp fragment and has been successfully used in several works (Nievas et al. 2023). Baudoin et al. (2010a) targeted the 16–23S ribosomal intergenic spacer region (ISR) for genus-specific detection. Shime-Hattori et al. (2011) designed primers based on 16S rRNA, *ipdC*, and nitrogen fixation genes (*nifA, nifH*), while Sahoo et al. (2014) targeted the *nif* gene cluster. Auxin-related genes have also been identified in several *Azospirillum* strains. More recently, Maroniche et al. (2017) proposed the *rpoD* gene as a reliable marker for species-level identification, developing a PCR assay that enables partial amplification of this single-copy gene.

The most accurate method for detecting microorganisms in a sample is through whole-genome sequencing or sequencing specific conserved genes that allow for precise identification. The most used gene for this purpose is the 16S rRNA subunit gene, which provides reliable taxonomic resolution for bacterial identification.

2.5 Novel Strategies for Tracking and Quantifying *Azospirillum* spp. and Future Perspectives

In addition to the methods previously described, other detection approaches for *Azospirillum* have also gained prominence, offering new strategies for determining its location and quantification in both plants and soil.

2.5.1 DNA Sensor

Zahrebelnei et al. (2025) describe an effective electrochemical DNA biosensor for detecting *Azospirillum brasilense*, based on specific primers designed by Stets et al. (2015). The biosensor consists of a glassy carbon electrode modified with a nanocomposite made from carbon nanotubes and gold nanoparticles capped with 3-n-propylpyridinium chloride silsesquioxane. This is followed by the immobilization of a thiolated probe oligonucleotide that binds specifically to the *A. brasilense* genome. The nanocomposite was characterized using spectroscopic and morphological techniques. Its presence on the biosensor's surface enhanced electrochemical responses due to its excellent electrocatalytic properties, as observed in electrochemical impedance spectroscopy and cyclic voltammetry experiments. The biosensor successfully detected the *A. brasilense* genome after hybridization, which altered the electrochemical response of the electrode, and this was rapidly measured by square wave voltammetry. The detection range of the bacterial genome was from 1.17 to 146.8 pmol L^{-1}, with a limit of detection (LOD) of 0.261 pmol L^{-1} and a limit of quantification (LOQ) of 0.322 pmol L^{-1}, as well as a sensitivity of $-$15.560 μA $\log^{-1}$ (*A. brasilense* genome in pmol L^{-1}). The biosensor demonstrated good selectivity and reproducibility, with a coefficient of variation of 5.69%, in addition to satisfactory sensitivity and stability for up to seven weeks. These promising analytical characteristics enabled the quantification of *A. brasilense* at low concentrations in metagenomic DNA samples.

2.5.2 Flow Cytometry

Valdameri et al. (2015) developed a novel approach based on flow cytometry to enumerate bacteria, applying it to rice plants 7 and 20 days after colonization with *Azospirillum brasilense* FP2. The results were compared with those obtained using the traditional plate count method. Flow cytometry revealed a number of viable *A. brasilense* cells associated with rice roots that were approximately ten times greater than those obtained with the plate count method. These results suggested that the plate count method may underestimate epiphytic populations. The flow cytometry assay overcomes two major limitations of the plate count method: the long incubation times, which can extend up to 48 h, and its ability to only count culturable cells. This flow cytometry assay offers an efficient, precise, and rapid tool for enumerating epiphytic cells.

2.5.3 Future Perspective

The rapid advancement of next-generation sequencing technologies has significantly advanced research on soil and plant microbiomes. Key questions in this field focus on how *Azospirillum* sp. interacts with other microorganisms and the host plant, how these interactions influence community formation and dynamics, and how they ultimately affect soil and plant health. However, due to the vast microbial diversity and the limitations of culturing techniques, studying these interactions comprehensively with all or most native microorganisms is nearly impossible. Despite its remarkable versatility and presence in diverse environments, *Azospirillum* is rarely emphasized or referenced in microbiome analyses of soil and plant samples following inoculation. One possible explanation for this paradox is the relatively low abundance of *Azospirillum* species in the soil microbiome compared to other bacterial genera, which makes their identification through next-generation sequencing methods more challenging (Nievas et al. 2023). In this context, the use of synthetic microbial communities (SynComs) containing *Azospirillum* sp. and other key genetically tagged bacteria presents a promising strategy for tracking and deepening our understanding of community formation and dynamics.

Most SynComs are composed of various bacterial species, as bacteria dominate plant microbiomes, and many can be easily cultured and genetically manipulated. Although SynComs are simplified and do not fully capture the complexity of natural interactions, their simplicity helps uncover causal relationships that likely occur in natural environments. In this regard, new genetic manipulation techniques have enabled the labeling and tracking of target bacteria in metagenomic analyses.

Recently, Ordon et al. (2024) developed and validated DNA barcodes, which are co-amplified with natural endogenous bacterial V5–V7 rRNA, fungal ITS, or plant ITS sequences by incorporating the respective primer binding sites into the DNA barcode. The modular bacterial tags (MoBacTag) encoding these DNA barcodes enable direct tracking of multiple near-isogenic strains during conventional bacterial 16S rRNA or fungal ITS community profiling. Since MoBacTag plasmids are based on modular cloning principles, each position within the multi-gene construct can be customized (Weber et al. 2011). For example, Tn7 elements can be replaced with genomic regions for homologous recombination, or additional community profiling primer binding sites can be inserted into the DNA barcode. Furthermore, it is possible to incorporate (1) an expression cassette for complementation approaches or (2) available modular cloning-compatible elements, such as intensity-optimized fluorescent markers (Iverson et al. 2016; Geddes et al. 2019). Using DNA barcodes, Ordon et al. (2024) examined the contributions of individual bacterial genes to the establishment of *Arabidopsis thaliana* root microbiota with synthetic communities that include MoBacTag-labeled strains of *Pseudomonas capeferrum*. Their work demonstrates how MoBacTags can be applied to assess the role of specific bacterial genetic determinants in shaping the plant microbiota.

The strategy using MoBacTags represents a valuable tool for monitoring the impact of introduced *Azospirillum* on native microbial communities in both plants and soil, providing a deeper understanding of bacterial interactions and offering new insights into the effects of *Azospirillum* populations on microbiome formation and dynamics.

2.6 Standard Methods

2.6.1 Azospirillum *Tracking and Quantification by qPCR*

2.6.1.1 Sample Acquisition, Handling, and Preparation

To quantify *Azospirillum* populations, different samples can be targeted, including soil, roots, stems, leaves, fruits, and seeds. These can also be further compartmentalized into more specific fractions (e.g., rhizospheric soil, root elongation zone, etc.). Samples for *Azospirillum* quantification via qPCR must be carefully collected to avoid cross-contamination.

2.6.1.2 Discrimination Between Live and Dead Cells and DNA Extraction

Sample treatment with propidium monoazide (PMA) is a critical technique for discriminating between live and dead cells. PMA should be dissolved in ultrapure water (20 mmol L^{-1}), stored frozen, and protected from light. Samples should be incubated in 1 mL of PMA at a final concentration of 50 μmol L^{-1} and exposed to a halogen light source for 15 min, with intermittent agitation to enhance effectiveness. After this, centrifugation is performed, and the pellet is freeze-stored until DNA extraction.[1]

An alternative to PMA treatment for separating extracellular DNA is filtration using a 0.22 μm pore size membrane. This process requires centrifugation at 10,000 × g for 10 min at 4 °C, then resuspending the material retained on the filter membrane. Resuspension is accomplished by gently pipetting the sample up and down ten times with Dulbecco's PBS (Jin et al. 2024).[1]

DNA extraction can be conducted using commercial kits. Ensuring the quality of DNA is essential, particularly regarding its quantity and purity, as these factors can greatly impact accurate quantification.

2.6.1.3 Standard Curve Preparation

To establish a reliable standard curve and minimize interference from environmental DNA, a controlled inoculation approach using *Azospirillum* will be implemented in representative matrices of the study environment, which may include macerated

[1] It is crucial to validate these additional steps using the target sample. Evidence indicates that DNA-intercalating dyes have both practical and theoretical limitations, particularly when applied to environmental samples. The successful application of these dyes depends on several factors, including dye concentration, incubation conditions, light source, presence of dead cells, high levels of suspended solids or biomass, salt concentration, pH of the reaction mix, target gene length, and gene sequence, among others (Fittipaldi et al. 2012).

plant tissues or soil. If the goal is to quantify the bacteria in plant tissues, non-inoculated grown plants will be used as the matrix. Non-inoculated soil will serve as the matrix for quantification in soil or the rhizosphere.

100 mg aliquots of the selected matrix will be separated for subsequent inoculation with the chosen *Azospirillum* strain. Additionally, 250 mg aliquots of soil will be prepared, ensuring the material is representative of the environment where the bacteria will be quantified.

For *Azospirillum* inoculation, a bacterial culture should be prepared, followed by serial dilutions covering a concentration range of 1.0×10^2 to 1.0×10^9 colony-forming units per milliliter (CFU mL^{-1}). It is vital to use cultures in the logarithmic phase of growth. Next, 100 μL of the bacterial suspension corresponding to each concentration level will be added to the samples. After inoculation, the samples should be carefully homogenized and incubated for 1 h at 4 °C to promote the interaction of bacterial cells with the sample and minimize bacterial DNA degradation. Negative control samples without inoculation should be prepared to confirm that there is no contamination in the extracted DNA.

Following DNA extraction, the quality and concentration of the DNA should be verified to ensure no amplification inhibition.

Amplification efficiency will be calculated using the linear regression equation obtained by plotting the Cq values against the log10 of *Azospirillum* concentration in each sample. The ideal efficiency is expected to range from 90% to 110%, ensuring the reliability of the results and minimizing interference from non-target DNA in the assay. For additional information, please verify the MIQE guidelines (Bustin et al. 2009).

2.6.1.4 qPCR Reaction Conditions

Depending on the required resolution and experimental constraints, the qPCR reaction for quantifying *Azospirillum* can be conducted using different detection strategies and primer specificity levels. Key considerations include the choice of DNA detection method and primer design.

2.6.1.4.1 DNA Detection: Fluorescent Probes or DNA Intercalating Dyes

There are two primary methods for detecting the amplified DNA:

- Fluorescent probes (TaqMan, Molecular Beacons, etc.)

  Fluorescent probes used in qPCR offer higher specificity, reducing the risk of false-positive signals and enabling multiplexing. Still, their use comes at a higher cost and requires specific probes for each target.
- DNA intercalating dyes

  DNA intercalating dyes, such as SYBR Green and EvaGreen, are cost-effective, easy to use, and suitable for quantification assays. However, they can exhibit non-specific fluorescence detection, which can be affected by primer-dimers and non-specific amplification. This limitation requires an additional step, specifically melt curve analysis, for verification.

2.6.1.4.2 Primer Specificity: Strain, Species, or Genus-Level Detection

The choice of primers significantly impacts the specificity of *Azospirillum* detection in qPCR. Primers can be designed to target:

- Strain-specific sequences are useful for tracking colonization and monitoring inoculant performance. For example, primers can target unique strain-specific sequences.
- Species-specific sequences: Designed to detect all strains within a particular *Azospirillum* species while differentiating them from other species.
- Genus-specific sequences: Amplify all *Azospirillum* species to achieve an overall quantification of the genus.

The choice of specificity level depends on the study's objectives. Genus-specific primers are recommended for total *Azospirillum* quantification in environmental samples, whereas strain-specific primers may be required for inoculation studies.

2.6.1.4.3 Standard qPCR Reaction Conditions

The qPCR reaction is carried out in a final volume of 20 μL, consisting of:

- qPCR Master Mix: 10 μL (containing DNA polymerase, buffer, dNTPs, and intercalating dye).
- Primers and probes (if applicable): Final concentration of 0.2–0.4 μmol L^{-1} for primers and 0.1–0.2 μmol L^{-1} for probes.
- Sample DNA: 1–5 ng for environmental samples or 10–100 pg for purified bacterial DNA.
- Nuclease-free water: To adjust the final volume.
- BSA can be used to challenging samples: 0.1–0.5 mg mL^{-1}.

A standard thermal cycling program consists of:

1. *Initial denaturation*: 95 °C for 3–5 min
2. *Amplification cycles* (typically 40 cycles):[2]
    - 95 °C for 10–15 s (denaturation)
    - 58–62 °C for 20–30 s (annealing)
    - 72 °C for 20–30 s (extension)

[2]The temperature and time of each qPCR step depend on factors such as primer and amplicon melting temperature and the chosen DNA polymerase.

- Two-step amplification may also be utilized (denaturation and annealing + extension)

3. *Melt curve analysis (for DNA intercalating dye assays)*: Gradual heating from 65 to 95 °C to assess amplicon specificity.

Further optimization may be required depending on primer efficiency, sample type, and the presence of inhibitors.

References

Akao S, Minakawa Y, Taki H et al (1999) Use of lacZ and gusA reporter genes to trace the infection process of nitrogen-fixing bacteria. Jpn Agric Res Q 33:77–84

Amann RI, Binder BJ, Olson RJ et al (1990) Combination of 16S rRNA-targeted oligonucleotide probes with flow cytometry for analyzing mixed microbial populations. Appl Environ Microbiol 56:1919–1925. https://doi.org/10.1128/AEM.56.6.1919-1925.1990

Arsene F, Katupitiya S, Kennedy IR, Elmerich C (1994) Use of lacZ fusions to study the expression of nif genes of Azospirillum brasilense in association with plants. Molecular Plant-Microbe Interactions 7:748–757. https://doi.org/10.1094/MPMI-7-0748

Aßmus B, Hutzler P, Kirchhof G et al (1995) In situ localization of *Azospirillum brasilense* in the rhizosphere of wheat with fluorescently labeled, rRNA-targeted oligonucleotide probes and scanning confocal laser microscopy. Appl Environ Microbiol 61:1013–1019. https://doi.org/10.1128/AEM.61.3.1013-1019.1995

Aßmus B, Schloter M, Kirchhof G et al (1997) Improved in situ tracking of rhizosphere bacteria using dual staining with fluorescence-labeled antibodies and rRNA-targeted oligonucleotides. Microb Ecol 33:32–40. https://doi.org/10.1007/s002489900005

Bacilio M, Rodriguez H, Moreno M et al (2004) Mitigation of salt stress in wheat seedlings by a gfp-tagged *Azospirillum lipoferum*. Biol Fertil Soils 40:188–193. https://doi.org/10.1007/s00374-004-0757-z

Basaglia M, Casella S, Peruch U, et al (2003) Field release of genetically marked Azospirillum brasilense in association with Sorghum bicolor L. Plant Soil 256:281–290. https://doi.org/10.1023/A:1026198522123

Bashan Y, Mitiku G, Ziv-Vecht O et al (1991) Estimation of minimal numbers of Azospirillum brasilense using time-limited liquid enrichment combined with enzyme-linked immunosorbent assay. Soil Biol Biochem 23:135–138. https://doi.org/10.1016/0038-0717(91)90125-4

Bashan Y, Holguin G, De-Bashan LE (2004) Azospirillum-plant relationships: physiological, molecular, agricultural, and environmental advances (1997–2003). Can J Microbiol 50:521–577. https://doi.org/10.1139/W04-035

Baudoin E, Couillerot O, Spaepen S et al (2010a) Applicability of the 16S-23S rDNA internal spacer for PCR detection of the phytostimulatory PGPR inoculant *Azospirillum lipoferum* CRT1 in field soil. J Appl Microbiol 108:25–38. https://doi.org/10.1111/J.1365-2672.2009.04393.X

Baudoin E, Lerner A, Mirza MS et al (2010b) Effects of *Azospirillum brasilense* with genetically modified auxin biosynthesis gene ipdC upon the diversity of the indigenous microbiota of the wheat rhizosphere. Res Microbiol 161:219–226. https://doi.org/10.1016/J.RESMIC.2010.01.005

Bilal R, Rasul G, Arshad M et al (1993) Attachment, colonization and proliferation of Azospirillum brasilense and Enterobacter spp. on root surface of grasses. World J Microbiol Biotechnol 9:63–69. https://doi.org/10.1007/BF00656519

Botta AL, Santacecilia A, Ercole C et al (2013) In vitro and in vivo inoculation of four endophytic bacteria on *Lycopersicon esculentum*. New Biotechnol 30:666–674. https://doi.org/10.1016/j.nbt.2013.01.001
Bustin SA, Benes V, Garson JA et al (2009) The MIQE guidelines: minimum information for publication of quantitative real-time PCR experiments. Clin Chem 55:611–622. https://doi.org/10.1373/CLINCHEM.2008.112797
Caiola MG, Canini A, Botta AL et al (2004) Localization of *Azospirillum brasilense* Cd in inoculated tomato (*Lycopersicon esculentum* Mill.) roots. Ann Microbiol 54:365–380
Canini A, Canuti L, Grilli Caiola M, Del Gallo M (2006) Immunocytochemical characterisation of endophytic bacteria Azospirillum brasilense, Herbaspirillum seropedicae, Burkholderia ambifaria and Gluconacetobacter diazotrophicus. Ann Microbiol 56:393–398. https://doi.org/10.1007/BF03175039
Camelo A, Barreto CP, Vidal MS et al (2021) Field response of two seed propagated elephant grass genotypes to diazotrophic bacterial inoculation and in situ confocal microscopy colonization analyses. Symbiosis 83:41–53. https://doi.org/10.1007/s13199-020-00730-8
Chalfie M, Tu Y, Euskirchen G et al (1994) Green fluorescent protein as a marker for gene expression. Science (1979) 52:1766–1767. https://doi.org/10.1126/SCIENCE.8303295
Chebotar VK, Asis Jr. CA, Akao S (2001) Production of growth-promoting substances and high colonization ability of rhizobacteria enhance the nitrogen fixation of soybean when coinoculated with Bradyrhizobium japonicum. Biol Fertil Soils 34:427–432. https://doi.org/10.1007/s00374-001-0426-4
Coniglio A, López G, Gualpa J et al (2020) Development of sequence characterized amplified region markers for identification of *Azospirillum brasilense* Az39 | Desarrollo de marcadores moleculares del tipo SCAR para la identificación de Azospirillum brasilense Az39. Rev Argent Microbiol 52:50–60. https://doi.org/10.1016/j.ram.2019.02.004
Correa OS, Romero AM, Montecchia MS et al (2007) Tomato genotype and *Azospirillum* inoculation modulate the changes in bacterial communities associated with roots and leaves. J Appl Microbiol 102:781–786. https://doi.org/10.1111/j.1365-2672.2006.03122.x
Couillerot O, Bouffaud M-L, Baudoin E et al (2010a) Development of a real-time PCR method to quantify the PGPR strain *Azospirillum lipoferum* CRT1 on maize seedlings. Soil Biol Biochem 42:2298–2305. https://doi.org/10.1016/j.soilbio.2010.09.003
Couillerot O, Poirier M-A, Prigent-Combaret C et al (2010b) Assessment of SCAR markers to design real-time PCR primers for rhizosphere quantification of *Azospirillum brasilense* phytostimulatory inoculants of maize. J Appl Microbiol 109:528–538. https://doi.org/10.1111/j.1365-2672.2010.04673.x
Couillerot O, Ramírez-Trujillo A, Walker V, et al (2013) Comparison of prominent Azospirillum strains in Azospirillum-Pseudomonas- Glomus consortia for promotion of maize growth. Appl Microbiol Biotechnol 97:4639–4649. https://doi.org/10.1007/S00253-012-4249-Z/FIGURES/3
Cruz-Pérez JF, Lara-Oueilhe R, Marcos-Jiménez C, et al (2021) Expression and function of the cdgD gene, encoding a CHASE–PAS-DGC-EAL domain protein, in Azospirillum brasilense. Sci Rep 11:. https://doi.org/10.1038/s41598-020-80125-3
da Cunha ET, Pedrolo AM, Paludo F et al (2020) *Azospirillum brasilense* viable cells enumeration using propidium monoazide-quantitative PCR. Arch Microbiol 202:1653–1662. https://doi.org/10.1007/s00203-020-01877-0
Daims H, Brühl A, Amann R et al (1999) The domain-specific probe EUB338 is insufficient for the detection of all bacteria: development and evaluation of a more comprehensive probe set. Syst Appl Microbiol 22:434–444. https://doi.org/10.1016/S0723-2020(99)80053-8
Deaker R, Kennedy IR (2001) Improved potential for nitrogen fixation in Azospirillum brasilense Sp7-S associated with wheat nifH expression as a function of oxygen pressure. Acta Biotechnol 21:3–17. https://doi.org/10.1002/1521-3846(200102)21:13::AID-ABIO33.0.CO;2-V
De Coninck K, Horemans S, Randombage S et al (1988) Occurrence and survival of *Azospirillum* spp. in temperate regions. Plant Soil 110:213–218. https://doi.org/10.1007/BF02226801

de-Bashan LE, Hernandez JP, Nelson KN et al (2010) Growth of quailbush in acidic, metalliferous desert mine tailings: effect of *Azospirillum brasilense* Sp6 on biomass production and rhizosphere community structure. Microb Ecol 60:915–927. https://doi.org/10.1007/S00248-010-9713-7

Eid MM, Sherwood JE (1995) Pulsed-field gel electrophoresis fingerprinting for identification of *Azospirillum* species. Curr Microbiol 30:127–131. https://doi.org/10.1007/BF00296196

El Zemrany H, Cortet J, Peter Lutz M et al (2006) Field survival of the phytostimulator Azospirillum lipoferum CRT1 and functional impact on maize crop, biodegradation of crop residues, and soil faunal indicators in a context of decreasing nitrogen fertilisation. Soil Biol Biochem 38:1712–1726. https://doi.org/10.1016/J.SOILBIO.2005.11.025

Faleiro AC, Pereira TP, Espindula E, et al (2013) Real time PCR detection targeting nifA gene of plant growth promoting bacteria Azospirillum brasilense strain FP2 in maize roots. Symbiosis 61:125–133. https://doi.org/10.1007/s13199-013-0262-y

Fani R, Bandi C, Bardin MG et al (1993) RAPD fingerprinting is useful for identification of Azospirillum strains. Microb Releases 1:217–221

Feng C, Shi-Hua S, San-Feng C, Yu-Xiang J (2004) Migration of Azospirillum brasilense Yu62 from Root to Stem and Leaves Inside Rice and Tobacco Plants. Acta Bot Sin 46:1065–1070

Feng L, Kennedy IR (1997) Biodegradation and plant protection from the herbicide 2,4-D by plant-microbial associations in cotton production systems. Biotechnol Bioeng 54:513–519. https://doi.org/10.1002/(SICI)1097-0290(19970620)54:6513::AID-BIT23.0.CO;2-M

Fittipaldi M, Nocker A, Codony F (2012) Progress in understanding preferential detection of live cells using viability dyes in combination with DNA amplification. J Microbiol Methods 91:276–289. https://doi.org/10.1016/J.MIMET.2012.08.007

Fukami J, Ollero FJ, Megías M, Hungria M (2017) Phytohormones and induction of plant-stress tolerance and defense genes by seed and foliar inoculation with Azospirillum brasilense cells and metabolites promote maize growth. AMB Express 7:1–13. https://doi.org/10.1186/S13568-017-0453-7/FIGURES/4

Geddes BA, Mendoza-Suárez MA, Poole PS (2019) A Bacterial Expression Vector Archive (BEVA) for flexible modular assembly of golden gate-compatible vectors. Front Microbiol 10:432669. https://doi.org/10.3389/FMICB.2018.03345/BIBTEX

Gera R, Pathak DV, Sharma PK (2003) Monitoring rhizosphere colonization by Azospirillum using constitutively expressed gfp gene. Indian J Microbiol 43:219–222

Gerk LP, Gilchrist K, Kennedy IR (2000) Mutants with enhanced nitrogenase activity in hydroponic Azospirillum brasilense-wheat associations. Appl Environ Microbiol 66:2175–2184. https://doi.org/10.1128/AEM.66.5.2175-2184.2000

Grover A, Sharma PC (2016) Development and use of molecular markers: past and present. Crit Rev Biotechnol 36:290–302. https://doi.org/10.3109/07388551.2014.959891

Gündisch C, Kirchhof G, Baur M et al (1993) Identification of *Azospirillum* species by RFLP and pulsed-field gel electrophoresis. Microb Releases 2:41–45

Haahtela K, Laakso T, Nurmiaho-Lassila E-L et al (1988) Effects of inoculation of *Poa pratensis* and *Triticum aestivum* with root-associated, N_2-fixing *Klebsiella*, *Enterobacter* and *Azospirillum*. Plant Soil 106:239–248. https://doi.org/10.1007/BF02371219

Hahn L, de Sá ELS, Osório Filho BD, et al (2016) Rhizobial inoculation, alone or coinoculated with Azospirillum brasilense, promotes growth of wetland rice. Rev Bras Cienc Solo 40:. https://doi.org/10.1590/18069657rbcs20160006

Hartmann A, James EK, DeBruijn FJ et al (2015) In situ localization and strain-specific quantification of *Azospirillum* and other diazotrophic plant growth-promoting rhizobacteria using antibodies and molecular probes, Chapter 3. In: Cassán FD et al (eds) Handbook for *Azospirillum*. Springer, Cham, pp 45–64. https://doi.org/10.1007/978-3-319-06542-7_3

Herschkovitz Y, Lerner A, Davidov Y et al (2005) *Azospirillum brasilense* does not affect population structure of specific rhizobacterial communities of inoculated maize (*Zea mays*). Environ Microbiol 7:1847–1852. https://doi.org/10.1111/j.1462-2920.2005.00926.x

Hong L, Orikasa Y, Sakamoto H, Ohwada T (2019) Plant tissue localization and morphological conversion of azospirillum brasilense upon initial interaction with allium cepa L. Microorganisms 7:. https://doi.org/10.3390/microorganisms7090275
Iverson SV, Haddock TL, Beal J et al (2016) CIDAR MoClo: improved MoClo assembly standard and new E. coli part library enable rapid combinatorial design for synthetic and traditional biology. ACS Synth Biol 5:99–103. https://doi.org/10.1021/ACSSYNBIO.5B00124/SUPPL_FILE/SB5B00124_SI_007.TXT
Jain V, Gupta K (2003) The flavonoid naringenin enhances intercellular colonization of rice roots by Azorhizobium caulinodans. Biol Fertil Soils 38:119–123. https://doi.org/10.1007/s00374-003-0599-0
Jin J, Yamamoto R, Shiroguchi K (2024) High-throughput identification and quantification of bacterial cells in the microbiota based on 16S rRNA sequencing with single-base accuracy using BarBIQ. Nat Protoc 19:207–239. https://doi.org/10.1038/S41596-023-00906-8
Katupitiya S, New PB, Elmerich C et al (1995) Improved N_2 fixation in 2,4-D treated wheat roots associated with *Azospirillum lipoferum*: studies of colonization using reporter genes. Soil Biol Biochem 27:447–452. https://doi.org/10.1016/0038-0717(95)98617-W
Kaushik R, Saxena AK, Tilak KVBR (2000) Selection of Tn5::lacZ mutants isogenic to wild type Azospirillum brasilense strains capable of growing at sub-optimal temperature. World J Microbiol Biotechnol 16:567–570. https://doi.org/10.1023/A:1008901331991
Kim C, Kecskés ML, Deaker RJ et al (2005) Wheat root colonization and nitrogenase activity by *Azospirillum* isolates from crop plants in Korea. Can J Microbiol 51:948–956. https://doi.org/10.1139/w05-052
Klein D (2002) Quantification using real-time PCR technology: applications and limitations. Trends Mol Med 8:257–260. https://doi.org/10.1016/S1471-4914(02)02355-9
Lebuhn M, Achouak W, Schloter M et al (2000) Taxonomic characterization of *Ochrobactrum* sp. isolates from soil samples and wheat roots, and description of *Ochrobactrum tritici* sp. nov. and *Ochrobactrum grignonense* sp. nov. Int J Syst Evol Microbiol 50(Pt 6):2207–2223. https://doi.org/10.1099/00207713-50-6-2207
Levanony H, Bashan Y (1989) Enhancement of cell-division in wheat root-tips and growth of root elongation zone induced by *Azospirillum brasilense* cd. Can J Bot-Revue Canadienne De Bot 67:2213–2216
Levanony H, Bashan Y, Kahana ZE (1987) Enzyme-linked immunosorbent assay for specific identification and enumeration of *Azospirillum brasilense* Cd. in cereal roots. Appl Environ Microbiol 53:358–364. https://doi.org/10.1128/AEM.53.2.358-364.1987
Lin SY, Shen FT, Young CC (2011) Rapid detection and identification of the free-living nitrogen fixing genus *Azospirillum* by 16S rRNA-gene-targeted genus-specific primers. Antonie Van Leeuwenhoek 99:837–844. https://doi.org/10.1007/S10482-011-9558-1
Liu Y, Chen S-F, Li J-L (2003) Colonization pattern of *Azospirillum brasilense* Yu62 on maize roots. J Integr Plant Biol 45:748–752
Lo CC (2010) Effect of pesticides on soil microbial community. J Environ Sci Health B 45:348–359. https://doi.org/10.1080/03601231003799804
Lopez BR, Bashan Y, Trejo A et al (2013) Amendment of degraded desert soil with wastewater debris containing immobilized *Chlorella sorokiniana* and *Azospirillum brasilense* significantly modifies soil bacterial community structure, diversity, and richness. Biol Fertil Soils 49:1053–1063. https://doi.org/10.1007/s00374-013-0799-1
Malinich EA, Bauer CE (2018) The plant growth promoting bacterium *Azospirillum brasilense* is vertically transmitted in *Phaseolus vulgaris* (common bean). Symbiosis 76:97–108. https://doi.org/10.1007/s13199-018-0539-2
Maroniche GA, García JE, Salcedo F et al (2017) Molecular identification of *Azospirillum* spp.: limitations of 16S rRNA and qualities of rpoD as genetic markers. Microbiol Res 195:1–10. https://doi.org/10.1016/J.MICRES.2016.11.009
Nabti E, Sahnoune M, Ghoul M et al (2010) Restoration of growth of durum wheat (*Triticum durum* var. waha) under saline conditions due to inoculation with the rhizosphere bacterium

Azospirillum brasilense NH and extracts of the marine alga Ulva lactuca. J Plant Growth Regul 29:6–22. https://doi.org/10.1007/S00344-009-9107-6

Naher K, Miwa H, Okazaki S, Yasuda M (2018) Effects of Different Sources of Nitrogen on Endophytic Colonization of Rice Plants by Azospirillum sp. B510. Microbes Environ 33:301–308. https://doi.org/10.1264/JSME2.ME17186

Nievas S, Coniglio A, Takahashi WY et al (2023) Unraveling Azospirillum's colonization ability through microbiological and molecular evidence. J Appl Microbiol 134. https://doi.org/10.1093/JAMBIO/LXAD071

O'Neal L, Vo L, Alexandre G (2020) Specific root exudate compounds sensed by dedicated chemoreceptors shape *Azospirillum brasilense* chemotaxis in the rhizosphere. Appl Environ Microbiol 86. https://doi.org/10.1128/AEM.01026-20

Ordon J, Thouin J, Nakano RT et al (2024) Chromosomal barcodes for simultaneous tracking of near-isogenic bacterial strains in plant microbiota. Nat Microbiol 9:1117–1129. https://doi.org/10.1038/s41564-024-01619-8

Pereg-Gerk L (2004) Expression of flcA, a gene regulating differentiation and plant interaction in *Azospirillum*. Soil Biol Biochem 36:1245–1252. https://doi.org/10.1016/j.soilbio.2004.04.005

Puente ML, Maroniche GA, Panepucci M, et al (2021) Localization and survival of Azospirillum brasilense Az39 in soybean leaves. Lett Appl Microbiol 72:626–633. https://doi.org/10.1111/LAM.13444

Ramirez-Mata A, Pacheco MR, Moreno SJ et al (2018) Versatile use of *Azospirillum brasilense* strains tagged with egfp and mCherry genes for the visualization of biofilms associated with wheat roots. Microbiol Res 215:155–163. https://doi.org/10.1016/j.micres.2018.07.007

Ramos HJO, Roncato-Maccari LDB, Souza EM et al (2002) Monitoring Azospirillum-wheat interactions using the gfp and gusA genes constitutively expressed from a new broad-host range vector. J Biotechnol 97:243–252. https://doi.org/10.1016/S0168-1656(02)00108-6

Reddy Priya P, Selastin Antony R, Gopalaswamy G et al (2016) Development of sequence-characterized amplified region (SCAR) markers as a quality standard of inoculants based on *Azospirillum*. Arch Microbiol 198:257–267. https://doi.org/10.1007/s00203-016-1187-7

Reddypriya P, Soumare A, Balachandar D (2019) Multiplex and quantitative PCR targeting SCAR markers for strain-level detection and quantification of biofertilizers. J Basic Microbiol 59:111–119. https://doi.org/10.1002/jobm.201800318

Renoud S, Abrouk D, Prigent-Combaret C et al (2022a) Effect of inoculation level on the impact of the PGPR *Azospirillum lipoferum* CRT1 on selected microbial functional groups in the rhizosphere of field maize. Microorganisms 10. https://doi.org/10.3390/microorganisms10020325

Renoud S, Vacheron J, Abrouk D et al (2022b) Field site-specific effects of an *Azospirillum* seed inoculant on key microbial functional groups in the rhizosphere. Front Microbiol 12:760512. https://doi.org/10.3389/FMICB.2021.760512/BIBTEX

Rilling JI, Acuña JJ, Nannipieri P et al (2019) Current opinion and perspectives on the methods for tracking and monitoring plant growth–promoting bacteria. Soil Biol Biochem 130:205–219. https://doi.org/10.1016/J.SOILBIO.2018.12.012

Rilling JI, Maruyama F, Sadowsky MJ, et al (2021) Crispr loci-pcr as tool for tracking azospirillum sp. Strain b510. Microorganisms 9:1351. https://doi.org/10.3390/MICROORGANISMS9071351/S1

Rothballer M, Schmid M, Hartmann A (2003) In situ localization and PGPR-effect of *Azospirillum brasilense* strains colonizing roots of different wheat varieties. Symbiosis 34:261–279

Russo A, Vettori L, Felici C et al (2008) Enhanced micropropagation response and biocontrol effect of *Azospirillum brasilense* Sp245 on Prunus cerasifera L. clone Mr.S 2/5 plants. J Biotechnol 134:312–319. https://doi.org/10.1016/j.jbiotec.2008.01.020

Sahoo RK, Ansari MW, Pradhan M et al (2014) Phenotypic and molecular characterization of native *Azospirillum* strains from rice fields to improve crop productivity. Protoplasma 251:943–953. https://doi.org/10.1007/S00709-013-0607-7

Santos ARS, Etto RM, Furmam RW et al (2017a) Labeled *Azospirillum brasilense* wild type and excretion-ammonium strains in association with barley roots. Plant Physiol Biochem 118:422–426. https://doi.org/10.1016/j.plaphy.2017.07.003

Santos KFDN, Moure VR, Hauer V et al (2017b) Wheat colonization by an *Azospirillum brasilense* ammonium-excreting strain reveals upregulation of nitrogenase and superior plant growth promotion. Plant Soil 415:245–255. https://doi.org/10.1007/s11104-016-3140-6

Schloter M, Hartmann A (1996) Production and characterization of strain-specific monoclonal antibodies against outer membrane components of Azospirillum brasilense Sp245. Hybridoma 15:225–232. https://doi.org/10.1089/HYB.1996.15.225

Schloter M, Hartmann A (1998) Endophytic and surface colonization of wheat roots (*Triticum aestivum*) by different *Azospirillum brasilense* strains studied with strain-specific monoclonal antibodies. Symbiosis 25:159–179. https://api.semanticscholar.org/CorpusID:88937995

Schloter M, Bode W, Hartmann A (1992a) Characterization of monoclonal antibodies against cell surface structures of *Azospirillum brasilense* Sp7 using ELISA techniques. Symbiosis 13:37–45

Schloter M, Bode W, Hartmann A et al (1992b) Sensitive chemoluminescence-based immunological quantification of bacteria in soil extracts with monoclonal antibodies. Soil Biol Biochem 24: 399–403. https://doi.org/10.1016/0038-0717(92)90201-8

Schloter M, Borlinghaus R, Bode W et al (1993) Direct identification, and localization of Azospirillum in the rhizosphere of wheat using fluorescence-labelled monoclonal antibodies and confocal scanning laser microscopy. J Microsc 171:173–177. https://doi.org/10.1111/J.1365-2818.1993.TB03371.X

Schloter M, Moens S, Croes C et al (1994) Characterization of cell surface components of *Azospirillum brasilense* Sp7 as antigenic determinants for strain-specific monoclonal antibodies. Microbiology (N Y) 140:823–828. https://doi.org/10.1099/00221287-140-4-823

Schloter M, Bode W, Hartmann A (1997) Characterization and application of a strain-specific monoclonal antibody against the rhizosphere bacterium *Azospirillum brasilense* Wa5. Hybridoma 16:183–187. https://doi.org/10.1089/hyb.1997.16.183

Schmidt H, Eickhorst T, Mußmann M (2012) Gold-FISH: a new approach for the in situ detection of single microbial cells combining fluorescence and scanning electron microscopy. Syst Appl Microbiol 35:518–525. https://doi.org/10.1016/J.SYAPM.2012.04.006

Shelud'ko AV, Shirokov AA, Sokolova MK et al (2010) Wheat root colonization by *Azospirillum brasilense* strains with different motility. Microbiology (N Y) 79:688–695. https://doi.org/10.1134/S0026261710050140

Shime-Hattori A, Kobayashi S, Ikeda S et al (2011) A rapid and simple PCR method for identifying isolates of the genus *Azospirillum* within populations of rhizosphere bacteria. J Appl Microbiol 111:915–924. https://doi.org/10.1111/J.1365-2672.2011.05115.X

Shirokov AA, Krasov AI, Selivanov NY et al (2015) Immunochemical detection of Azospirilla in soil with genus-specific antibodies. Microbiology (Russian Federation) 84:263–267. https://doi.org/10.1134/S0026261715020137

Soares IC, Pacheco RS, da Silva CGN et al (2021) Real-time PCR method to quantify Sp245 strain of *Azospirillum baldaniorum* on Brachiaria grasses under field conditions. Plant Soil 468:525–538. https://doi.org/10.1007/s11104-021-05137-y

Sørensen J, Nicolaisen MH, Ron E et al (2009) Molecular tools in rhizosphere microbiology—from single-cell to whole-community analysis. Plant Soil 321:483–512. https://doi.org/10.1007/S11104-009-9946-8

Stets MI, Campbell Alqueres SM, Souza EM et al (2015) Quantification of *Azospirillum brasilense* FP2 bacteria in wheat roots by strain-specific quantitative PCR. Appl Environ Microbiol 81: 6700–6709. https://doi.org/10.1128/AEM.01351-15/SUPPL_FILE/ZAM999116597SO1.PDF

Steenhoudt O, Ping Z, Vande Broek A, Vanderleyden J (2001) A spontaneous chlorate-resistant mutant of Azospirillum brasilense Sp245 displays defects in nitrate reduction and plant root colonization. Biol Fertil Soils 33:317–322. https://doi.org/10.1007/s003740000329

Stoffels M, Castellanos T, Hartmann A (2001) Design and application of new 16S rRNA-targeted oligonucleotide probes for the Azospirillum-Skermanella-Rhodocista-cluster. Syst Appl Microbiol 24:83–97. https://doi.org/10.1078/0723-2020-00011

Takahashi WY, Galvão CW, Urrea-Valencia S, et al (2022) Impact of seed-applied fungicide and insecticide on Azospirillum brasilense survival and wheat growth-promoting ability. Lett Appl Microbiol 74:604–612. https://doi.org/10.1111/LAM.13645

Takahashi WY, Galvão CW, Cassán FD et al (2024) Tracking maize colonization and growth promotion by *Azospirillum* reveals strain-specific behavior and the influence of inoculation method. Plant Physiol Biochem 215. https://doi.org/10.1016/j.plaphy.2024.108979

Tchebotar VK, Kang UG, Asis Jr. CA, Akao S (1998) The use of GUS-reporter gene to study the effect of Azospirillum-Rhizobium coinoculation on nodulation of white clover. Biol Fertil Soils 27:349–352. https://doi.org/10.1007/s003740050442

Trejo A, de-Bashan LE, Hartmann A et al (2012) Recycling waste debris of immobilized microalgae and plant growth-promoting bacteria from wastewater treatment as a resource to improve fertility of eroded desert soil. Environ Exp Bot 75:65–73. https://doi.org/10.1016/J.ENVEXPBOT.2011.08.007

Urrea-Valencia S, Etto RM, Takahashi WY et al (2021) Detection of *Azospirillum brasilense* by qPCR throughout a maize field trial. Appl Soil Ecol 160. https://doi.org/10.1016/j.apsoil.2020.103849

Valdameri G, Kokot TB, Pedrosa FDO et al (2015) Rapid quantification of rice root-associated bacteria by flow cytometry. Lett Appl Microbiol 60:237–241. https://doi.org/10.1111/LAM.12351

Versalovic J, Schneider M, Bruijn FJ De, Lupski JR (1994) Genomic fingerprinting of bacteria using repetitive sequence-based polymerase chain reaction. Methods Mol Cell Biol 5:25–40

Wani SP, Chandrapalaih S, Zambre MA et al (1988) Association between N_2-fixing bacteria and pearl millet plants: responses, mechanisms and persistence. Plant Soil 110:289–302. https://doi.org/10.1007/BF02226810

Weber E, Engler C, Gruetzner R et al (2011) A modular cloning system for standardized assembly of multigene constructs. PLoS One 6:e16765. https://doi.org/10.1371/JOURNAL.PONE.0016765

Webster G, Jain V, Davey MR et al (1998) The flavonoid naringenin stimulates the intercellular colonization of wheat roots by Azorhizobium caulinodans. Plant Cell Environ 21:373–383. https://doi.org/10.1046/j.1365-3040.1998.00278.x

Weisburg WG, Barns SM, Pelletier DA, Lane DJ (1991) 16S ribosomal DNA amplification for phylogenetic study. J Bacteriol 173:697–703. https://doi.org/10.1128/JB.173.2.697-703.1991

Xi C, Lambrecht M, Vanderleyden J et al (1999) Bi-functional gfp-and gusA-containing mini-Tn5 transposon derivatives for combined gene expression and bacterial localization studies. J Microbiol Methods 35:85–92. https://doi.org/10.1016/S0167-7012(98)00103-1

Yegorenkova IV, Tregubova KV, Burygin GL et al (2015) Assessing the efficacy of co-inoculation of wheat seedlings with the associative bacteria *Paenibacillus polymyxa* 1465 and *Azospirillum brasilense* Sp245. Can J Microbiol 62:279–285. https://doi.org/10.1139/cjm-2015-0647

Yuan H, Ge T, Chen C et al (2012) Significant role for microbial autotrophy in the sequestration of soil carbon. Appl Environ Microbiol 78:2328–2336. https://doi.org/10.1128/AEM.06881-11

Zahrebelnei F, Lima D, de Lara LS et al (2025) A sensitive electrochemical DNA biosensor for detecting the genome of a plant growth-promoting bacteria. Talanta 286:127484. https://doi.org/10.1016/J.TALANTA.2024.127484

Chapter 3
One Hundred Years of Progress Dedicated to Understanding the Evolution of Tools and Knowledge in *Azospirillum* Taxonomy

Gabriele Bellotti, Florence Wisniewski-Dyé, Trevor C. Charles, Fabricio Dario Cassán, and Edoardo Puglisi

Abstract *Azospirillum* is one of the most extensively studied genera of plant growth-promoting rhizobacteria (PGPR), widely recognized for its biological nitrogen fixation (BNF) ability and phytohormone production. The genus currently includes 36 proposed species, 25 of which have been validly published. Despite its agronomic relevance, the taxonomy of *Azospirillum* remains incomplete and, in some cases, inconsistent. This chapter traces the historical development of *Azospirillum* taxonomy and highlights how genome-based tools are reshaping our understanding of species boundaries within the genus. Using a curated dataset of 69 unique genomes, including two newly sequenced strains (*A. largimobile* Sp A3a, *A. zeae* N7), we applied whole-genome comparisons, average nucleotide identity (ANI), and core genome phylogeny to evaluate interspecies relationships. Results indicate high genomic plasticity, an open pangenome ($\gamma = 0.85$), and support multiple taxonomic reassignments. In addition, we examined key functional traits linked to plant growth promotion, such as nitrogen fixation and indole-3-acetic acid (IAA) biosynthesis, and analyzed their distribution across phylogenetic groups. These findings underscore the limitations of classical taxonomical approaches and emphasize the importance of using high-resolution genomic tools for taxonomic studies. As genome sequencing becomes faster, more accurate, and economically accessible, its routine application should become the standard for defining new

G. Bellotti (✉) · E. Puglisi
Department for Sustainable Food Production (DiSTAS), Università Cattolica del Sacro Cuore, Piacenza, Italy
e-mail: gabriele.bellotti@unicatt.it

F. Wisniewski-Dyé
Laboratoire d'Ecologie Microbienne, UMR CNRS 5557, UMR INRAE 1418, VetAgro Sup, Université Claude Bernard Lyon 1, Villeurbanne, France

T. C. Charles
Biology Department, University of Waterloo, Waterloo, ON, Canada

F. D. Cassán
Universidad Nacional de Río Cuarto, Río Cuarto, Argentina

F. D. Cassán et al. (eds.), *The Definitive Handbook of Azospirillum*,
https://doi.org/10.1007/978-3-032-09222-9_3

Azospirillum species. To facilitate similar studies integrating bacterial genomics and taxonomy, we report a reproducible protocol covering the full workflow from *Azospirillum* cell culture to the generation of high quality, annotated, and complete genomes.

Keywords Taxonomy · Genomics · ANI · Pangenome · Classification

Abbreviations

ANI	Average nucleotide identity
dDDH	Digital DNA-DNA hybridization
LPSN	List of prokaryotic names with standing in nomenclature
MAGs	Metagenome-assembled genomes
WGS	Whole-genome sequencing

3.1 Introduction

Over the past 100 years, the taxonomy of the *Azospirillum* genus has undergone considerable modifications and refinements. The genus was originally described by Beijerink in 1925, when he isolated a spirillum-shaped bacterium from Dutch grass soil. The bacterium was initially classified as *Azotobacter spirillum* but was soon renamed *Spirillum lipoferum* (Beijerinck 1925), reflecting early challenges in defining soil bacterial taxonomy. Despite its early characterization, the genus gained significant attention only 40 years later, when its ability to fix atmospheric nitrogen was confirmed using labeled dinitrogen gas (N_2^{15}) (Becking 1963). Around that time, many other researchers had isolated bacterial strains with similar characteristics from diverse geographic areas, particularly from the rhizosphere of grasses and other crops. Recognizing shared traits among these isolates, Tarrand and his team first established the genus *Azospirillum* in 1978, bringing together under the same taxonomic group these free-living, plant-associating diazotrophic bacteria. The following years saw a rapid expansion of the genus, with new species being described worldwide (Tyler et al. 1979). A few years later, in 1985, *Azospirillum* was officially recognized for its potential as a plant inoculant for sustainable agriculture (Okon 1985).

Today, *Azospirillum* is considered one of the most extensively studied genera among the plant growth-promoting rhizobacteria (PGPR). Its reputation comes not only from its nitrogen-fixing ability but also from its production of phytohormones, including auxins, cytokinins, and gibberellins. Particularly, the synthesis of the auxin hormone indole-3-acetic acid (IAA) enables *Azospirillum* to influence root system architecture, stimulating the formation of root hairs and lateral roots. Because of these benefits, *Azospirillum*-based products are widely used in agriculture, especially in South America, where many countries have adopted these plant inoculants for commercial use (Cassán et al. 2020).

According to the list of prokaryotic names with standing in nomenclature (LPSN) (Parte et al. 2020), the genus currently comprises 36 proposed species, of which 25 are validly published (see Table 3.1). The current classification is based on multiple factors, including phenotypic characterization, chemotaxonomic markers, genotypic traits, and whole-genome data. Among these, whole-genome-based approaches have become the preferred method for bacterial taxonomy. Compared to traditional single-gene or multilocus gene analyses, whole-genome sequencing (WGS) enables more robust and higher resolution phylogenetic comparisons between strains. This is especially relevant for *Azospirillum*, since these bacteria often carry multiple copies of the 16S rRNA gene, sometimes up to nine homologous copies which differ in nucleotide sequence, making it difficult to accurately determine their phylogenetic relationship (Maroniche et al. 2017). As more high-quality genomic data become accessible, it will be crucial to integrate them with curated phenotypic information. Such a multi-disciplinary approach will be essential for refining the taxonomy of *Azospirillum*, ultimately enhancing both our fundamental understanding and the practical application of these bacteria in sustainable agriculture.

3.2 History of *Azospirillum* Taxonomy

3.2.1 Pre-genomic Approaches to Azospirillum *Taxonomy*

The taxonomic classification of *Azospirillum* has evolved considerably over the last century. Early efforts relied on classical microbiological techniques centered around morpho-physiological characterization. These included assessments of colony morphology, cell shape, nutrient utilization profiles, estimation of nitrogen fixation, and the production of phytohormones. Such methods formed the foundation for the initial differentiation of *Azospirillum* species. The range of methodologies that have been historically adopted for the classification of *Azospirillum* can be grouped into four main categories: (i) traditional microbiology, (ii) chemotaxonomic, (iii) genotypic, and (iv) genomic categories (summarized in Table 3.2). However, over the past decade, the advent of genomic tools based on the complete genetic content of microorganisms, particularly WGS, has transformed the classification of bacterial species. This shift reflects the general trend in bacterial taxonomy in favor of tools that offer higher resolution, reproducibility, and phylogenetic accuracy.

Among the microbiological techniques, morpho-physiological characterizations of strains have historically played a key role in the differentiation of *Azospirillum* among other soil-dwelling bacteria. Members of this genus are typically described as vibrioid/spirilli-shaped cells, gram-negative, and motile with a single polar flagellum. Another typical characteristic of the genus is the ability to accumulate cytoplasmic inclusions of poly-hydroxy butyrate (PHBs) granules, which serve as an

Table 3.1 LPSN-recognized *Azospirillum* species, along with the number of corresponding publicly available genomes

Recognized species (36)[a]	Validity (LPSN)[b]	First publication that described the species	NCBI genomes[c]
Azospirillum aestuarii	Not validly published	Xu et al. (2023)	1
Azospirillum agricola	Validly published	Lin et al. (2016)	2
Azospirillum argentinense	Validly published	dos Santos Ferreira et al. (2022)	9
Azospirillum baldaniorum	Validly published	dos Santos Ferreira et al. (2020)	3
Azospirillum brasilense	Validly published	Tarrand et al. (1978)	23
Azospirillum brasiliense	Valid but inaccurate spelling	–	–
Azospirillum canadense	Validly published	Mehnaz et al. (2007a)	1
Azospirillum cavernae	Validly published	Zhu et al. (2021)	1
Azospirillum doebereinerae	Validly published	Eckert et al. (2001)	3
Azospirillum endophyticum	Not validly published	Zhan et al. (2022)	1
Azospirillum fermentarium	Validly published	Lin et al. (2013)	1
Azospirillum formosense	Validly published	Lin et al. (2012)	2
Azospirillum griseum	Validly published	Yang et al. (2019)	1
Azospirillum halopraeferens	Validly published	Reinhold et al. (1987)	1
Azospirillum humicireducens	Validly published	Zhou et al. (2013)	1
Azospirillum isscasi	Validly published	Wang et al. (2024)	1
Azospirillum largimobile	Validly published	Skerman et al. (1983)	0
Azospirillum largomobile	Valid but inaccurate spelling	–	-
Azospirillum lipoferum	Validly published	Beijerinck (1925)	2
Candidatus *Azospirillum massiliense*	Not validly published	Oren and Garrity (2022) and Pagnier et al. (2008)	0
Azospirillum massiliensis	Inaccurate spelling	–	0
Azospirillum melinis	Validly published	Peng et al. (2006)	2
Azospirillum oleiclasticum	Not validly published	Wu et al. (2021)	2
Azospirillum oryzae	Validly published	Xie and Yokota (2005)	2
Azospirillum palatum	Not validly published	Zhou et al. (2009)	0

(continued)

Table 3.1 (continued)

Recognized species (36)[a]	Validity (LPSN)[b]	First publication that described the species	NCBI genomes[c]
Azospirillum palustre	Validly published	Tikhonova et al. (2019)	1
Azospirillum picis	Validly published	Lin et al. (2009)	1
Azospirillum ramasamyi	Validly published	Anandham et al. (2019)	1
Azospirillum rugosum	Validly published	Young et al. (2008)	1
Azospirillum soli	Validly published	Lin et al. (2015)	1
Azospirillum tabaci	Not validly published	Duan et al. (2022)	1
Azospirillum thermophilum	Validly published	Zhao et al. (2020)	1
Azospirillum thiophilum	Validly published	Lavrinenko et al. (2010)	1
Azospirillum zeae	Validly published	Mehnaz et al. (2007b)	0
<u>*Azospirillum amazonense*</u>	Now *Nitrospirillum amazonense*	Falk et al. (1985)	<u>6</u>
<u>*Azospirillum irakense*</u>	Now *Niveispirillum irakense*	Khammas et al. (1989)	<u>1</u>

[a]Validly published species are indicated in bold, while species no longer classified as *Azospirillum* are underlined

[b]*LPSN* list of prokaryotic names with standing in nomenclature

[c]Complete and draft genomes have been integrated, whereas duplicate entries and metagenome-assembled genomes (MAGs) have been excluded. The table includes a total of 67 unique genomes currently assigned to the *Azospirillum* genus

intracellular stock of carbon. Another interesting feature is the ability to flocculate and form cellular aggregates in high C:N media (Burdman et al. 1998). Additionally, under certain conditions, *Azospirillum* can show pleomorphism (Becking 1985), with cells transitioning from the typical vibrioid shape to a coccoid-ovoid form, reflecting morphological plasticity in response to environmental or nutritional conditions.

Moreover, *Azospirillum* species are non-fermentative and facultatively microaerophilic, able to activate the BNF pathway in low-oxygen environments. These physiological traits, along with the use of semi-selective media, such as the semi-solid NFb medium designed for diazotroph isolation (Day and Döbereiner 1976), resemble early taxonomic efforts. While these methods were useful in establishing preliminary species assignments, their low resolution does not allow for accurate determination of evolutionary relationships or clear species boundaries. A significant step forward occurred in the late 1970s with the introduction of DNA-DNA hybridization (DDH), a method that quantifies genetic similarity between closely related bacterial taxa. Using this method, for the first time, a distinction among members of the *Azospirillum* genus was observed, with strains grouped into two major clusters based on DDH values (>70%): Group I (*A. brasilense*) and Group II (*A. lipoferum*) (Tarrand et al. 1978).

Table 3.2 Overview of the techniques employed over the past 100 years for the taxonomic classification of *Azospirillum* species

Category	Methodology	Application	References
Traditional microbiology techniques			
Isolation with nitrogen-free media	Isolation and characterization of *Azospirillum* and other diazotrophs	Used to selectively isolate nitrogen-fixing bacteria	Day and Döbereiner (1976)
Morpho-physiological characterization	Determination of colony morphology (color), cellular shape, motility, and pleomorphism	Basic characterization under various growth conditions	Eskew et al. (1977)
Phenotypic and biochemical tests	Use of different C/N sources, EPS, PHB, IAA production, or biofilm synthesis	Species delineation based on metabolic and biochemical traits	Burdman et al. (2000), Fallik et al. (1989), and Okon and Itzigsohn (1992)
Chemotaxonomic techniques			
Polyamine analysis	Intracellular polyamines as chemotaxonomic markers	Helps distinguish between closely related taxa	Hamana et al. (1994)
Phospholipids and fatty acids analysis	Evaluation of lipid profiles as biomarkers	Differentiates taxa based on membrane composition	Di Salvo et al. (2014)
MALDI-TOF	Protein profiling through mass spectrometry	Generates proteomic fingerprints for taxonomic identification	Stets et al. (2013)
Genotypic techniques			
DNA-DNA hybridization and homology	Wet-lab measurement of DNA relatedness and % of hybridization	Pre-genomic gold standard for species definition	Tarrand et al. (1978)
16S rRNA gene sequencing	Amplification and sequencing of the 16S rRNA gene	Widely used bacterial marker for phylogenetic and taxonomic profiles	Xia et al. (1994)
Multilocus sequence analysis (MLSA)	Combined phylogeny from multiple housekeeping genes	Improves resolution over 16S rRNA-based phylogeny and enables better resolution	Maroniche et al. (2017)
Genomic techniques			
Average nucleotide identity (ANI)	Whole-genome similarity index	Current standard for species-level genome comparisons	Richter and Rosselló-Móra (2009)
Digital DNA-DNA hybridization (dDDH)	In silico replacement of the traditional DNA-DNA hybridization technique	Used for accurate species delineation based on full genome sequences	Auch et al. (2010)
Genome-based phylogeny	Whole-genome alignments	Produces robust phylogenetic frameworks for Prokaryotes	Meier-Kolthoff and Göker (2019)

(continued)

Table 3.2 (continued)

Category	Methodology	Application	References
Core genome analysis	Study of core, accessory, and unique genes across strains	Captures full genomic diversity within and across species for phylogenetic trees based on the core genome	Chung et al. (2018)

3.2.2 Refinements in Azospirillum *Taxonomy Over Time*

The taxonomy of *Azospirillum* has undergone continuous revision throughout time, following the emergence of increasingly sophisticated classification tools. As methodologies advanced from phenotypic and biochemical analyses to molecular sequencing and whole-genome comparisons, so did our understanding of species relationships within the *Azospirillum* genus.

One of the earliest taxonomy reassignments involved *Conglomeromonas largimobile*, initially described by Skerman et al. (1983) as a pleomorphic, rod-shaped water-dwelling bacterium that formed cell agglomerates in sodium-rich media. The bacterium was originally assigned to the *Conglomeromonas* genus based on its morphological resemblance to other strains obtained from the same water samples. However, later studies, employing 16S rRNA gene sequencing, demonstrated a closer phylogenetic affinity between *C. largimobile* and known *Azospirillum* species such as *A. brasilense* and *A. lipoferum*, rather than other members of the *Conglomeromonas* genus. This phylogenetic evidence, based on 16S rDNA gene homology, led to the reclassification of the species as *Azospirillum largimobile* (Ben Dekhil et al. 1997). Although initially misspelled as *A. largomobile*, the correct and currently accepted name is *A. largimobile* (Table 3.1).

A different example led to the removal of two *Azospirillum* species in 2014. Specifically, *A. amazonense* and *A. irakense* were excluded from the *Azospirillum* genus list. This decision followed a polyphasic approach that included morphological and biochemical characterization (e.g., flagellar structure, PHB accumulation, carbon source utilization, and metabolic activities), BNF assays, and molecular analyses such as 16S rRNA gene sequencing, fatty acid methyl ester (FAME) profiling, polyamine content, and DNA-DNA hybridization. Additional support came from the use of the primer set Azo494-F/Azo756-R developed by Lin et al. (2011), which amplifies a 236 bp DNA fragment of the 16S rRNA gene specific to *Azospirillum*; this fragment was absent in representative strains of *A. amazonense* and *A. irakense*. Based on this comprehensive comparison, which encompassed many of the techniques listed in Table 3.2, Lin et al. (2014) proposed the reclassification of these species into two novel genera within the Rhodospirillaceae family: *Nievispirillum* irakense and *Nitrospirillum amazonense*.

Following this, further refinements emerged from two more recent polyphasic studies combining multilocus sequence analysis (MLSA), average nucleotide identity (ANI), digital DDH (dDDH), and core proteome alignments. In these studies, strains initially thought to belong to *A. brasilense* were shown to form distinct separate clades, leading to the recognition of *A. baldaniorum* and *A. argentinense* as separate species within Group I (dos Santos Ferreira et al. 2020, 2022).

Finally, a recent comparative study analyzing all publicly available *Azospirillum* genomes suggested potential misassignments at the species level, raising concerns about the actual taxonomic accuracy (Bellotti et al. 2025). The study applied core genome phylogenetics, ANI% and dDDH%, to evaluate the phylogenetic relationships among all the *Azospirillum* genomes available on NCBI. Based exclusively on in silico analyses, the authors reported that several genomes appear to be misclassified at the species level and proposed the following taxonomic revisions, summarized in Table 3.3. These limitations can be attributed to several factors, including the absence of whole-genome data for some proposed species such as *A. largimobile*, *A. massiliense*, *A. palatum*, and *A. zeae*. Furthermore, many described species are represented by a single genome sequence, making it difficult to identify species-specific traits and species boundaries. In addition, only 22% of the available genomes are fully assembled with all the 5–8 circularized replicons typical of *Azospirillum*, limiting the resolution of genome-based taxonomy.

These findings underscore how genome-based analyses can progressively elucidate *Azospirillum* taxonomy, revealing both over-simplifications in early species assignments and the limitations of relying on phenotypic traits alone. The progressive refinement of *Azospirillum* taxonomy not only reflects improvements in classification tools but also reinforces the need for continued curation and integration of genomic and phenotypic datasets. As the number of available high-quality genomes increases, such efforts will be essential for maintaining a robust and coherent taxonomic framework for this ecologically and agriculturally important bacterial genus.

Table 3.3 Proposed taxonomic reassignments of publicly available *Azospirillum* genomes

Current taxonomic assignment	Proposed taxonomic assignment[a]	GenBank acc. number
A. brasilense 2A1	*A. tabaci* 2A1	GCA_002245955.1
A. brasilense BR 11026	*A. aestuarii* BR 11026	GCA_007827795.1
A. brasilense BR 11669	*A. aestuarii* BR 11669	GCA_016622085.1
A. brasilense HAMBI3172	*A. argentinense* HAMBI3172	GCA_003349955.1
A. brasilense D7	*A. argentinense* D7	GCA_016652835.1
A. brasilense 2020WEIHUA K	*A. baldaniorum* 2020WEIHUA K	GCA_022023855.1
A. brasilense MTCC4039	*A. baldaniorum* MTCC4039	GCA_005222205.1
A. brasilense BR 11005	*A. baldaniorum* BR 11005	GCA_007827915.1
A. brasilense BR 11796	*Azospirillum* sp. BR 11796	GCA_007828645.1
A. brasilense BR 11650	*Azospirillum* sp. BR 11650	GCA_007827815.1
A. brasilense BR 11138	*Azospirillum* sp. BR 11138	GCA_016622105.1
A. formosense UWAZO 1	*Azospirillum* sp. UWAZO 1	GCA_040500525.1
A. lipoferum 4B	*Azospirillum* sp. 4B	GCA_000283655.1

[a]The reassignments are based on ANI and dDDH analyses reported in Bellotti et al. (2025). Strains for which no species is assigned show values below the commonly accepted thresholds of 95% ANI and 70% dDDH, relative to every recognized type strain

3.3 The Use of Whole-Genome Data for *Azospirillum* Taxonomy

As previously discussed, the full-length 16S rRNA gene (~1500 bp) has historically been the primary marker for genotypic bacterial identification. However, due to the frequent presence of multiple rRNA gene copies in *Azospirillum*, this marker lacks the resolution necessary to distinguish closely related species within the genus. To address this limitation, subsequent studies introduced MLSA of housekeeping genes such as *atpD*, *glnII*, *recA*, and *rpoD*, which offer improved taxonomic resolution and led to the adoption of genome-scale approaches (Maroniche et al. 2017). Today, WGS and core genome phylogenetics are considered the most robust tools for high-resolution taxonomic assessment, allowing the exploration of genetic diversity, evolutionary relationships (dos Santos Ferreira et al. 2020, 2022), and the detection of potential misclassifications (Bellotti et al. 2025).

To further refine the current taxonomy of *Azospirillum* using genome-wide data, we performed a comprehensive comparative analysis of all available non-redundant genomes. A total of 67 high-quality assemblies, covering a broad representation of the known species, were retrieved from the NCBI public database. In addition, two novel genome assemblies were generated: one for *Azospirillum zeae* N7, the type strain of the *A. zeae* species, and one for *Azospirillum largimobile* Sp A3a, representing the first genomes ever sequenced for these two species. Their inclusion expanded the dataset to 69 unique genomes, encompassing 32 of the 34 currently described *Azospirillum* species (excluding *Nitrospirillum* and *Niveispirillum*), leaving only the species *A. massiliense* and *A. palatum* without genomic representation.

3.3.1 *A Reproducible Workflow for Genome Sequencing of* Azospirillum *Strains*

The genome of *A. largimobile* Sp A3a is used in this chapter as a model to demonstrate a reproducible workflow for obtaining a high-quality hybrid genome assembly of *Azospirillum* strains. The genomic structure of *Azospirillum* typically comprises 5–8 replicons, including a chromosome, two chromids, megaplasmids, and plasmids. The full experimental and computational workflow, from culturing to genome assembly, is graphically summarized in Fig. 3.1. The phases and steps of the adopted protocol are described in more detail in the following sections.

3.3.1.1 Bacterial Culture Conditions and Genomic DNA Extraction

Bacterial cultures were grown on DSMZ "Medium 221" agar plates and incubated at 30 °C. A single colony was transferred to 5 mL of Medium 221 broth in 10 mL culture tubes and incubated overnight for approximately 12 h at 30 °C with shaking

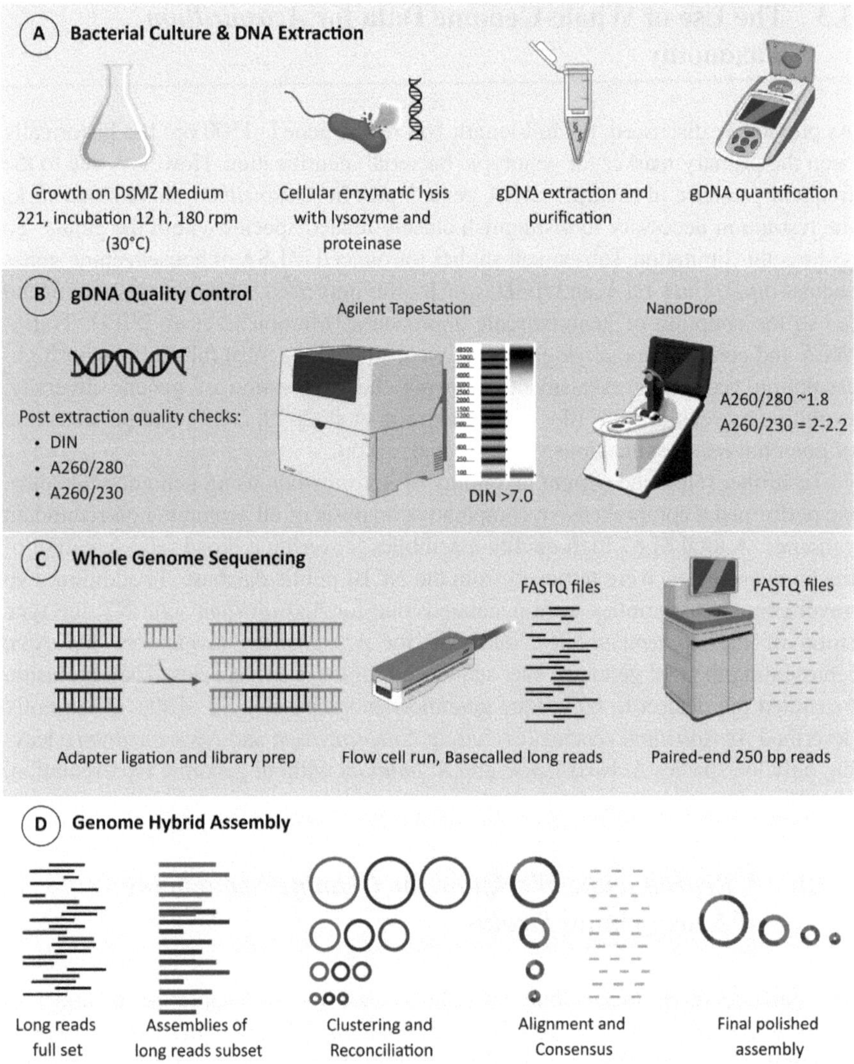

Fig. 3.1 Overview of the standardized experimental and computational workflow used for high-quality genome sequencing of *Azospirillum* strains. (**a**) culturing conditions and genomic DNA extraction; (**b**) quality control of gDNA using TapeStation and NanoDrop; (**c**) hybrid sequencing strategy combining long-read (Oxford Nanopore Technologies) and short-read (Illumina) platforms; and (**d**) hybrid genome assembly pipeline based on Trycycler and Polypolish, integrating long-read assemblies and polishing with short reads to obtain accurate and complete genomes suitable for accurate phylogenomic analysis

at 180 rpm. After incubation, cells were pelleted by centrifugation, and genomic DNA (gDNA) was extracted using the E.Z.N.A. Bacterial DNA Kit (Omega Bio-tek). The standard protocol was followed, with lysozyme and proteinase K incubation times extended to 1 hour each to improve gDNA yield (Fig. 3.1a).

3.3.1.2 gDNA Quality Control

To evaluate gDNA integrity, 1 μL of approximately 50 ng.μL of gDNA was loaded on a genomic DNA ScreenTape and analyzed using the TapeStation system (Agilent Technologies). This provides a DNA Integrity Number (DIN) from 1 (degraded) to 10 (highly intact), and values >7.0 were considered suitable for long-read sequencing. DNA purity was assessed with NanoDrop spectrophotometry (Thermofisher Scientific), evaluating $A_{260/280}$ and $A_{260/230}$ absorbance ratios. Values near 1.8 and 2.0, respectively, indicated high-quality DNA with minimal protein or solvent contamination (Fig. 3.1b).

3.3.1.3 Whole-Genome Sequencing

Starting from quality-checked gDNA, 200–400 ng of input DNA per sample were used for long-read sequencing, depending on the number of strains processed in a single run. Since Oxford Nanopore Technology (ONT) sequencing is typically performed using multiplexed samples to maximize throughput and reduce costs, libraries were prepared for multiple *Azospirillum*/bacterial strains simultaneously. Library preparation was carried out using the Rapid Barcoding Kit 24 V14 (Oxford Nanopore Technologies), which enables barcoding of up to 24 single strains using a PCR-free protocol. This approach minimizes DNA fragmentation and preserves the integrity of high-molecular-weight DNA, which is essential for long-read sequencing, without the introduction of PCR biases. Magnetic bead-based purification steps were used to remove residual enzymes, reaction buffers, and short DNA fragments. After quantification and pooling of barcoded libraries, the final library was loaded onto a MinION flow cell and sequenced on an ONT device. Sequencing was performed following the manufacturer's instructions and allowed to run for up to 72 h to ensure sufficient read depth. To complement ONT long-read data, Illumina short-read sequencing was performed externally using a NovaSeq 6000 platform (Illumina Inc.) with paired-end 250 bp reads (PE250). Genomic DNA was sent to a certified service provider, who handled library preparation, sequencing, and initial quality control (Fig. 3.1c).

3.3.1.4 Genomic Hybrid Assembly

Long reads were processed using the hybrid assembly pipeline proposed by Wick et al. (2021), based on Trycycler (https://github.com/rrwick/Trycycler). This approach began by dividing the long reads into multiple subsets. Each subset was independently assembled using three long-read assemblers based on three different algorithms: Flye, Raven, and Miniasm, generating a total of three draft assemblies per subset. These assemblies are then reconciled and circularized by Trycycler, which constructs a consensus genome by integrating the different drafts, effectively resolving structural inconsistencies, repeat regions, and closing potential gaps. To further improve sequence accuracy, the consensus genome obtained from long-read data was polished using Illumina short reads with Polypolish (https://github.com/rrwick/Polypolish), a tool designed to correct small-scale errors based on high-fidelity short-read alignment (Fig. 3.1d). This hybrid assembly strategy yielded a high quality, error-corrected genome with consistent coverage across all seven expected replicons, suitable for downstream phylogenomic and taxonomic analyses.

3.3.2 Phylogenomic and Genomic Similarity Analysis of 69 **Azospirillum** *Genomes*

To investigate species-level relationships and genome-wide relatedness across the 69 *Azospirillum* genomes, we performed a pairwise average nucleotide identity (ANI) analysis using FastANI (https://github.com/ParBLiSS/FastANI). The resulting ANI matrix was visualized with Morpheus (https://software.broadinstitute.org/morpheus/) to assess clustering patterns and delineate interspecies boundaries. The matrix revealed two distinct genomic clusters corresponding to the major taxonomic groups originally proposed by Tarrand et al. (1978), described as Group I and Group II (Fig. 3.2). Group I displayed higher intra-species ANI values, with many values close to the 95% threshold typically used for species demarcation, indicating substantial genomic conservation within the clade. This group includes the well-characterized *A. brasilense* and its closest relatives, *A. baldaniorum*, *A. argentinense*, *A. formosense*, *A. aestuarii*, *A. tabaci*, and *A. isscasi*. Within the same clade are also found *A. canadense* and *A. rugosum,* which exhibited comparable ANI values, while *A. soli* formed a more distinct sub-branch of Group I. Group II, by comparison, presented a broader range of ANI values and a more diffused clustering pattern. Species in this group generally exhibited lower genome-wide similarity to one another, highlighting a greater degree of evolutionary divergence. Within Group II, certain pairs, such as *A. cavernae* and *A. griseum*, or *A. agricola* and *A. doebereinerae*, clustered more closely together, suggesting potential shared evolutionary or ecological traits. The largest subcluster within Group II included a diverse array of species: *A. lipoferum*, *A. picis*, *A. thiophilum*, *A. endophyticum*, *A. melinis*, *A. palustre*, *A. oryzae*, *A. ramasamyi*, *A. humicireducens*, and the newly sequenced *A. zeae* and *A. largimobile*. In the same clade, *A. fermentarium*, *A. halopraeferens*, and *A. oleiclasticum* diverged significantly from the rest of Group II, forming distinct lineages characterized by consistently low ANI values when compared to all other *Azospirillum* genomes. This pattern based on ANI values suggests a more distant evolutionary trajectory for these species (Fig. 3.2).

To deepen the taxonomic resolution beyond ANI-based clustering and gain insights into the evolutionary relationships among *Azospirillum* species, we next focused on the genus-wide core genome using a pangenome-based approach. The pangenome analysis was conducted using the Roary pipeline (Page et al. 2015),

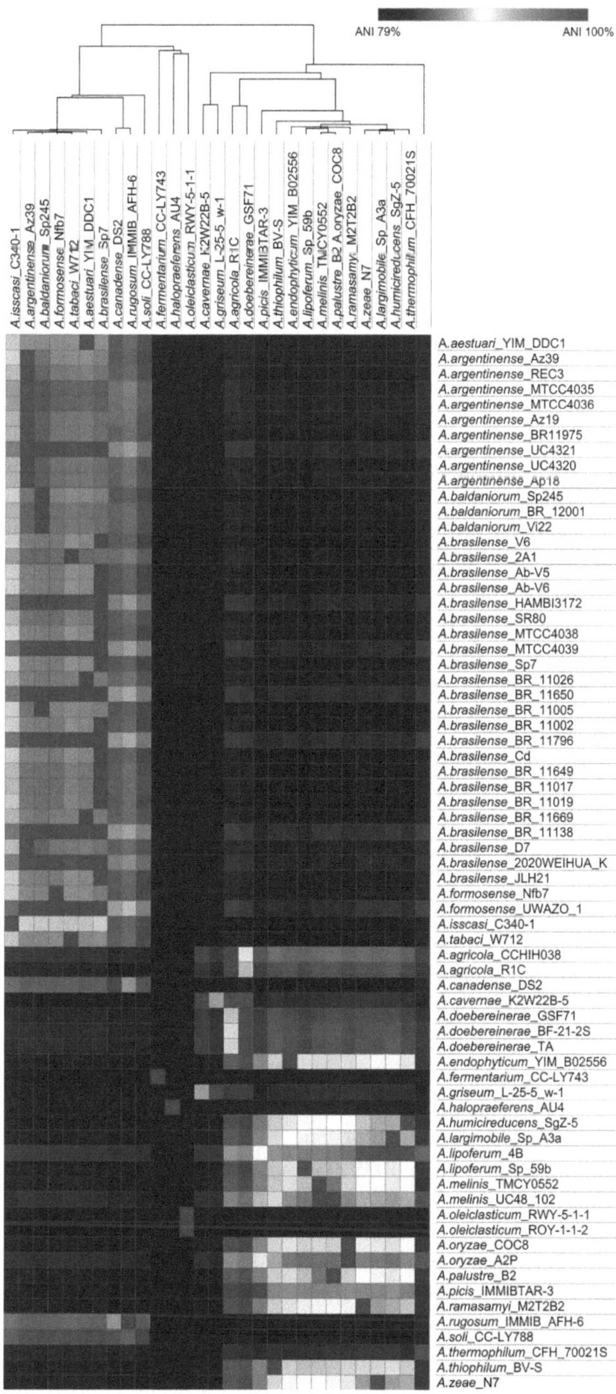

Fig. 3.2 Heatmap of average nucleotide identity (ANI) values among the 69 *Azospirillum* genomes. The matrix, visualized with Morpheus, shows pairwise genomic similarities (FastANI) between every genome vs. all the type strains, with color intensity ranging from blue (lower identity) to red (higher identity). The taxonomy shown reflects the current classification listed in the NCBI database

including all 69 unique genomes. Genes were grouped in clusters using a 70% sequence identity threshold, while core genes were defined as those present in at least 95% of the genomes. These thresholds represent a practical compromise due to the high genomic diversity and occasional incomplete assemblies that may result in the underrepresentation of otherwise conserved genes (Agarwal et al. 2023). Genes present in 94–95% of the genomes were considered part of the soft core, those found in 15–94% as shell genes, and genes present in fewer than 15% were classified as cloud genes.

As a result, the Roary pipeline identified 56,691 gene clusters across the datasct. Among these, only 1534 genes were present in ≥95% of the genomes and were classified as core genes, while 177 were categorized as soft-core. The majority of genes belonged to the accessory genome, represented by 7445 shell genes and 47,535 cloud genes. The high proportion of cloud genes reflects substantial genomic variability within the genus, likely associated with environmental adaptation and strain-specific functions (Maroniche et al. 2017; Wisniewski-Dyé et al. 2011). To resolve evolutionary relationships at a higher resolution, a phylogenetic tree was constructed based on the alignment of the 1534 core genes using RAxML-NG (https://github.com/amkozlov/raxml-ng), with 100 bootstrap replicates. The resulting tree, shown in Fig. 3.3, can be used to examine species boundaries and assess taxonomic consistency within the genus. The reconstructed phylogenetic relationships clearly reflect the division between the two major groups as previously identified through ANI analysis, with Group I and Group II forming well-supported clades. This structure aligns with the early genomic relatedness patterns originally proposed by Tarrand et al. (1978) and supports recent reassignments suggested by dos Santos Ferreira et al. (2020, 2022) and Bellotti et al. (2025), with most reclassified strains (Table 3.3) clustering alongside their newly designated type strains. In addition, the present analysis improved previous works by employing a more inclusive 70% identity threshold for core-gene clustering, which enabled higher resolution of species boundaries. For example, *A. argentinense* HAMBI 3172 and UC4320, previously unclustered in the core-gene tree, now clearly cluster within the *A. argentinense* clade, consistent with their high ANI similarity to the type strain. Similarly, *A. baldaniorum* strain 2020WEHUA, which had appeared phylogenetically distant in earlier analyses, now clusters with its conspecific group, aligning with ANI-based classification. The only exception that remains unresolved is *A. baldaniorum* strain MTCC4039 (GCA_005222205.1), originally reported as *A. brasilense* in public databases. While its ANI value supports reassignment to *A. baldaniorum* (ANI = 95.019 with the *A. baldaniorum* type strain vs. 94.491 with *A. brasilense*), its position in the core genome phylogeny (Fig. 3.3) does not reflect this relationship, suggesting possible limitations of current thresholds or the influence of horizontal gene transfer and genomic plasticity in soil-dwelling *Azospirillum*. As such, this case highlights the importance of interpreting ANI and core genome phylogenies in parallel, particularly for bacterial taxa with high genomic plasticity. This approach becomes especially critical for resolving taxonomy when ANI values fall close to the species delineation threshold.

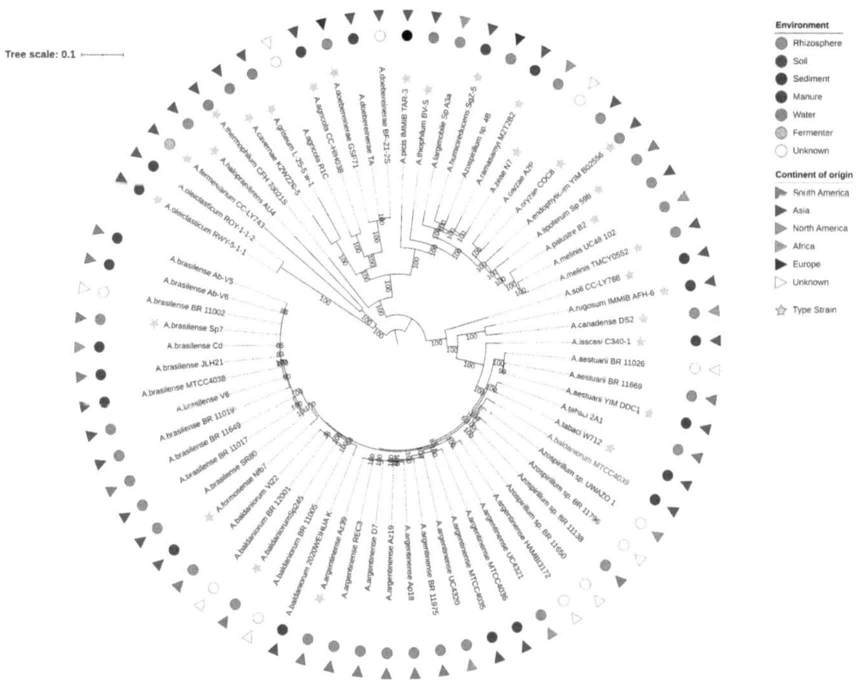

Fig. 3.3 Maximum likelihood phylogenetic tree of the 69 *Azospirillum* genomes analyzed in this chapter. The tree was constructed following the alignment of 1534 core genes identified using the Roary pipeline (≥95% presence across genomes). The tree was constructed with RAxML-NG using 100 bootstrap replicates. Tip labels are color coded to indicate the continent of isolation and the environmental origin of each strain. Type strains are also highlighted to show their position within each clade

Beyond species-level resolution, there is no evidence of clustering either by environmental origin (e.g., rhizosphere, soil, or aquatic environments) or geography. Nevertheless, an observable trend emerges: Group I strains (*A. brasilense* group) are predominantly associated with soil and rhizosphere environments. In contrast, Group II comprises isolates from a broader range of ecological niches, including soil, water, and other non-rhizosphere sources, suggesting a wider environmental adaptability within this clade. These ecological and geographical patterns may help explain the genomic differences observed between the two groups; specifically, the higher genomic conservation within Group I likely reflects its more specialized and stable niche bound to soil environments, while the broader habitat range of Group II may contribute to its greater genomic variability and lower inter-strain and intra-species similarity.

3.4 Pangenome Composition and Functional Divergence Across the *Azospirillum* Genus

To investigate the overall structure of the *Azospirillum* pangenome, we applied the power-law regression model commonly used in comparative genomic studies (Tettelin and Medini 2020):

$$G = cN^{\gamma}$$

where G is the total number of gene clusters, c is the number of core genes, N is the number of genomes included, and γ represents the openness coefficient.

Based on our dataset of 69 genomes and 1534 core genes over the total of 56,691 gene clusters, we estimated a γ value of 0.85, confirming that the Azospirillum pangenome is open. This high γ value likely reflects the limited representation of several species, particularly those in Group II, by only a single sequenced genome. As a consequence, many accessory genes appear to be strain-specific and are categorized as cloud genes in the absence of broader intra-species comparisons.

The gene presence/absence matrix generated by Roary, aligned with the core genome phylogeny tree, reveals some clustering patterns (Fig. 3.4). Species within Group I show higher conservation and share more gene clusters, consistent with their high ANI values, whereas Group II exhibits greater variability in terms of genetic content.

Within Group I, gene content is generally less diverse, with most strains sharing a higher proportion of gene clusters. While a degree of genetic variability is still observed, as expected from soil-dwelling bacteria, the pattern of gene presence/absence is more conserved across this group. This trend is particularly evident among strains classified as *A. brasilense*, *A. argentinense*, *A. baldaniorum*, *A. formosense*, *A. tabaci*, and *A. aestuarii*, along with several unclassified strains that cluster within the same phylogenetic clade. These taxa form a relatively cohesive genomic group, both in terms of accessory gene content and core genome phylogeny, suggesting a shared evolutionary origin and potentially similar ecological strategies, likely linked to their common association with soil and rhizosphere habitats. This contrasts with Group II, whose broader ecological range, including contaminated soil, manure, and tar, may contribute to the greater observed genetic diversity. Notably, *A. oleiclasticum*, *A. halopraeferens*, and *A. fermentarium* display distinct accessory gene repertoires and branch independently in the phylogenetic tree, suggesting ecological specialization. It would be valuable to further investigate the accessory genes found in these species to better understand the genomic features that may drive their adaptation to specific environmental niches. Conversely, other Group II species (e.g., *A. lipoferum*, *A. zeae*, *A. largimobile*, *A. doebereinerae*) show more overlapping gene clusters with fewer differences, possibly indicating shared evolutionary pressures and similar ecological constraints.

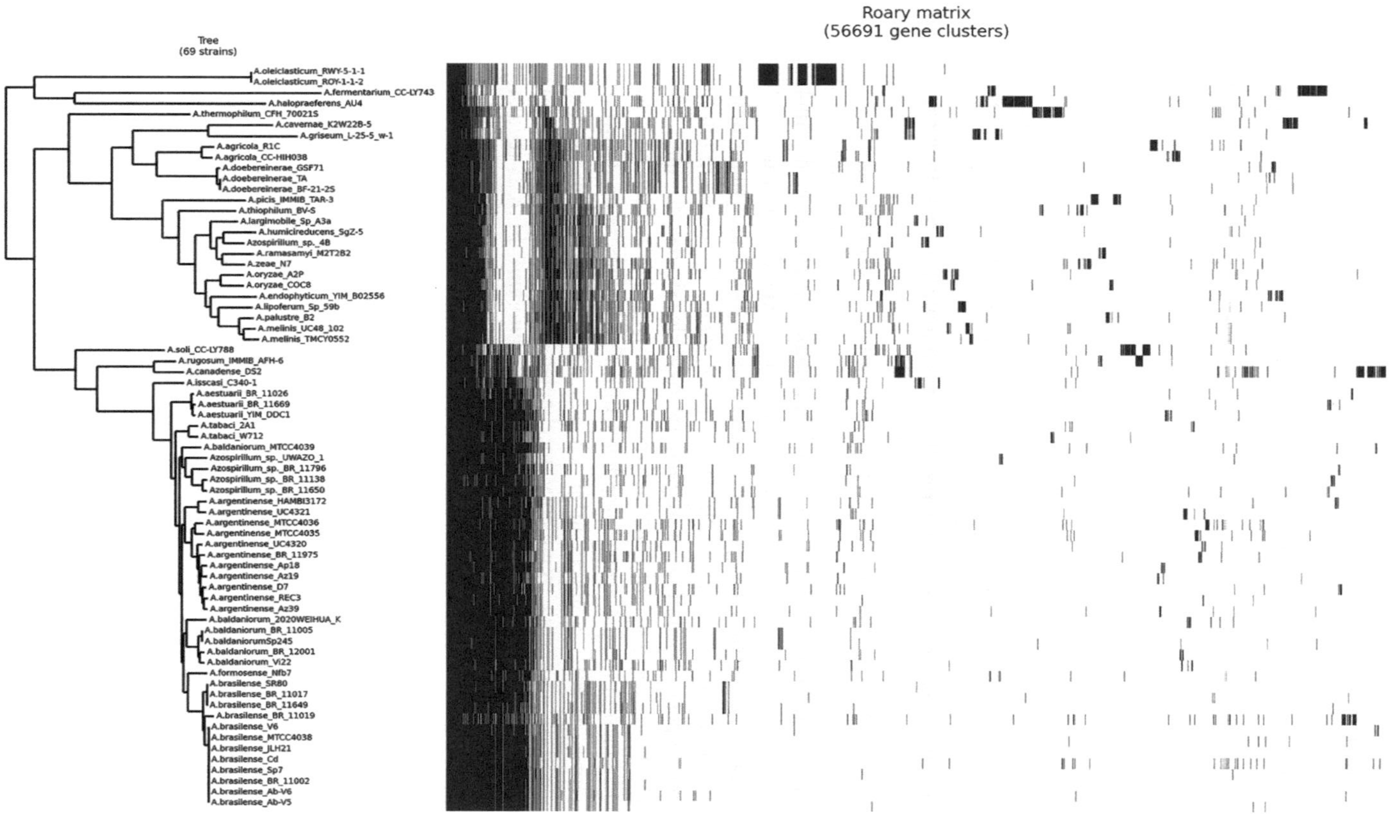

Fig. 3.4 Gene presence/absence matrix of the 69 *Azospirillum* genomes analyzed in this chapter. The figure combines a core-gene phylogenetic tree (left), constructed from 1534 core genes (≥95% presence across genomes), with a binary heatmap (right) showing the presence (dark blue) or absence (light blue) of the 56,691 gene clusters identified by Roary in each genome analyzed in this chapter

3.4.1 Conservation of Nitrogen Fixation Machinery Across the Genus

One of the most functionally significant traits conserved across the *Azospirillum* genus is the genetic capacity for BNF, including its complex regulatory systems (see Chap. 9). In our dataset, 68 out of 69 genomes, including both newly sequenced and public strains, encode the full complement of genes required to assemble and regulate the nitrogenase enzyme complex. The only exception is *A. griseum* strain L-25-5_w-1 (GCA_003966125.1), in which basic components are entirely absent. The canonical *nif* operon, comprising *nifH*, *nifD*, and *nifK*, is conserved across strains. These genes encode the two structural subunits of the nitrogenase enzyme complex, which catalyzes the ATP-dependent reduction of atmospheric dinitrogen (N_2) to ammonium (NH_4^+), a plant-available nitrogen form. The enzyme consists of two major components: (i) component I (MoFe-protein): encoded by the genes *nifD* and *nifK*, this catalytic core contains the FeMo cofactor and P-clusters and (ii) component II (Fe protein): encoded by *nifH*, serves as an ATP-dependent reductase, transferring electrons to component I during nitrogen reduction.

The nitrogenase complex is highly sensitive to oxygen, and even a brief exposure can result in irreversible inactivation. Diazotrophic microorganisms, including *Azospirillum*, have evolved various strategies to protect the enzyme, including microaerophilic respiration, oxygen-scavenging metabolism, and tight posttranscriptional regulation (Dixon and Kahn 2004). In *Azospirillum*, the control of nitrogenase expression appears to follow a regulatory cascade similar to that of other α-proteobacteria. The transcriptional activator NifA is widely conserved across the genus and functions as the primary regulator of *nif* gene expression. However, the typical repressor NifL is absent in all 68 genomes analyzed. This suggests the involvement of alternative regulatory mechanisms, for example, homologs of FixK, a known NifA activator in *Azorhizobium caulinodans*. FixK is regulated by the oxygen-sensing two-component system FixL-FixJ. In both genera, NifA but also NtrC function as σ^{54}-dependent enhancer-binding proteins (EBPs), interacting with σ^{54}-RNA polymerase to initiate *nif* operon transcription (Dixon and Kahn 2004). All key regulatory elements, *nifA*, *fixL*, *fixJ*, *fixK*, and *ntrC*, were identified in the 68 *Azospirillum* genomes examined.

These regulatory cascades are further modulated by metabolic feedback loops, particularly the nitrogen assimilation enzyme GlnA (glutamine synthetase, GS). GlnA participates in the GS-GOGAT cycle and indirectly regulates *nifA* expression in response to cellular nitrogen content (Steenhoudt and Vanderleyden 2000). The full nitrogenase activation in *Azospirillum* requires (i) proper transcription and assembly of structural and accessory components, (ii) nitrogen limitation, (iii) microaerobic conditions, and (iv) adequate ATP and reducing power for the catalysis.

Posttranslational regulation also plays a critical role. The *draTG* operon, conserved in all strains, encodes dinitrogenase reductase ADP-ribosyl transferase (*draT*) and glycohydrolase (*draG*). These enzymes mediate reversible ADP-ribosylation of

NifH in response to nitrogen availability. Under ammonium scarcity, DraG removes the ADP-ribose moiety from a conserved arginine residue on the NifH, restoring its electron transfer capacity. Overall, this tightly coordinated regulatory system, encompassing transcriptional, metabolic, and posttranslational regulators, reflects *Azospirillum*'s ecological specialization for survival in nitrogen-limited soils. Its consistent conservation across species underlines the genus's potential for supporting plant growth under low-nitrogen conditions (Guo et al. 2022).

3.4.2 *Conserved Phytohormonal Modulation Strategies in* **Azospirillum**

Next to nitrogen fixation, another specific plant-associated trait that was considered to be a hallmark of *Azospirillum* is the ability to synthesize the auxin indole-3-acetic acid (IAA), a phytohormone known to stimulate root elongation, lateral root formation, and influence overall plant development. Earlier genomic studies indicated that some *Azospirillum* strains were devoid of the indole-3-pyruvate (IPA) pathway, which is catalyzed by indole-3-pyruvate decarboxylase, encoded by *ipdC* (Wisniewski-Dyé et al. 2012).

Indeed, in the 69 genomes analyzed, *ipdC* was detected in all *Azospirillum* strains belonging to Group I, represented by the *A. brasilense* clade, and was entirely absent in Group II, with the sole exception of *A. oleiclasticum*, for which both available strains in our dataset carry the gene. The consistent presence of *ipdC* in these genomes obtained from soil-dwelling or rhizosphere isolates suggests that IAA production via the IPA pathway may represent a specialized adaptation associated with strains that colonize the rhizosphere. This trait likely enhances root colonization and promotes plant-microbe interaction by modulating host-root system architecture. As a result, IAA biosynthesis contributes meaningfully to the plant growth-promoting potential of *Azospirillum.*

However, among Group II, some strains are also rhizosphere isolates that have been reported to promote plant growth (Chamam et al. 2013; Veresoglou and Menexes 2010). This ability may be due to the presence of *acdS*, encoding the 1-aminocyclopropane-1-carboxylate (ACC) deaminase. This enzyme converts ACC into ammonia and α-ketobutyrate, thereby diminishing the amount of ACC available for the production of the phytohormone ethylene. As ethylene inhibits root growth, AcdS activity leads to an increase in root system development (Penrose and Glick 2003). Moreover, this activity prevents excessive increase in ethylene synthesis under various stresses, such as saline stress (Orozsco-Mosqueda et al. 2020). The *acdS* gene was found in only 9 of the 69 genomes analyzed, all of which belonged to Group II and were isolated from rhizosphere/soil (*A. endophyticum, A. largimobile, A. humicireducens, A. oleciclasticum, Azospirillum* sp. 4B, and *A. zeae*) or from other habitats (*A. picis, A. ramasamyi,* and *A. thiophilum*). Its presence suggests that *Azospirillum* species may rely on alternative phytohormonal modulation strategies, with different clades favoring either IAA biosynthesis or ethylene suppression as plant growth-promoting activity. This functional diversity supports the Multiple Mechanisms Hypothesis described earlier (Cassán et al. 2020), which posits that plant growth promotion by *Azospirillum* arises from a combination of independent yet complementary molecular pathways. Finally, *Azospirillum* strains may differ in their ability to synthesize other phytohormones (such as cytokinins and gibberellins) for which the genetic determinants are unknown or may rely on alternative mechanisms yet to be described.

3.5 Conclusion and Perspectives

This chapter presents a comprehensive genome-wide analysis of the *Azospirillum* genus, combining newly sequenced strains with all publicly available assemblies. Through comparative genomic approaches, we identified multiple taxonomic inconsistencies and proposed reassignments, while also documenting the genus's high heterogeneity in genetic content, which is characterized by a vast set of accessory genes that significantly outnumber the conserved core. These findings reinforce the importance of adopting whole-genome sequencing as a standard tool for the taxonomic classification of *Azospirillum* strains. Only through systematic and consistent genome analyses can we improve taxonomic resolution, gain functional insights, and ensure the safe and effective application of this genus in the development of plant inoculants.

In terms of conserved functions, nitrogen fixation genes were found to be universally present across the genus, while IAA biosynthesis via the IPA pathway was largely restricted to Group I strains, predominantly isolated from soil and rhizosphere environments that carried the *ipdC* gene. Conversely, a subset of Group II strains harbored the *acdS* gene, which enables the modulation of plant ethylene levels through ACC deaminase activity.

Altogether, these genomic insights not only clarify evolutionary relationships within the genus but also help distinguish functionally relevant traits that contribute to plant growth promotion. The combined action of biological nitrogen fixation (BNF) with either IAA production or ACC deaminase activity highlights the multifaceted role that *Azospirillum* can play in supporting plant health. When operating in direct association with plants, these mechanisms promote a model of sustainable agriculture by fostering beneficial plant-microbe interactions and offering potential for reducing external inputs, particularly under conditions of nutrient or abiotic stress.

References

Agarwal V, Stubits R, Nassrullah Z et al (2023) Pangenome insights into the diversification and disease specificity of worldwide *Xanthomonas* outbreaks. Front Microbiol 14. https://doi.org/10.3389/fmicb.2023.1213261

Anandham R, Heo J, Krishnamoorthy R, SenthilKumar M, Gopal NO, Kim SJ, Kwon SW (2019) Azospirillum ramasamyi sp. Nov., a novel diazotrophic bacterium isolated from fermented bovine products. Int J Syst Evol Microbiol 69(5):1369–1375. https://doi.org/10.1099/ijsem.0.003320

Auch AF, von Jan M, Klenk HP, Göker M (2010) Digital DNA-DNA hybridization for microbial species delineation by means of genome-to-genome sequence comparison. Stand Genomic Sci 2(1):117–134. https://doi.org/10.4056/sigs.531120

Becking JH (1963) Fixation of molecular nitrogen by an aerobic *Vibrio* or *Spirillum*. Antonie Van Leeuwenhoek 29(1):326–326. https://doi.org/10.1007/BF02046082

Becking JH (1985) Pleomorphism in *Azospirillum*. In: Klingmüller W (ed) Azospirillum III. Springer, Berlin/Heidelberg, pp 243–262. https://doi.org/10.1007/978-3-642-70791-9_23

Beijerinck M (1925) Uber ein *Spirillum*, welches frei en Stick-stoff binden kann? Zentralbl Bakteriol 63:353–359

Bellotti G, Cortimiglia C, Antinori ME et al (2025) Comprehensive genome-wide analysis for the safety assessment of microbial biostimulants in agricultural applications. Microb Genomics 11(4). https://doi.org/10.1099/mgen.0.001391

Ben Dekhil S, Cahill M, Stackebrandt E et al (1997) Transfer of *Conglomeromonas largomobilis* subsp. *Largomobilis* to the genus *Azospirillum* as *Azospirillum largomobile* comb. Nov., and elevation of *Conglomeromonas largomobilis* subsp. *Parooensis* to the new type species of *Conglomeromonas*, *Conglomeromonas parooensis* sp. Nov. Syst Appl Microbiol 20(1):72–77. https://doi.org/10.1016/S0723-2020(97)80050-1

Burdman S, Jurkevitch E, Schwartsburd B et al (1998) Aggregation in *Azospirillum brasilense*: effects of chemical and physical factors and involvement of extracellular components. Microbiology (Reading, England) 144(Pt 7):1989–1999. https://doi.org/10.1099/00221287-144-7-1989

Burdman S, Jurkevitch E, Soria-Díaz ME, Serrano AMG, Okon Y (2000) Extracellular polysaccharide composition of Azospirillum brasilense and its relation with cell aggregation. FEMS Microbiol Lett 189(2):259–264. https://doi.org/10.1111/j.1574-6968.2000.tb09240.x

Cassán F, Coniglio A, López G et al (2020) Everything you must know about *Azospirillum* and its impact on agriculture and beyond. Biol Fertil Soils 56(4):461–479. https://doi.org/10.1007/s00374-020-01463-y

Chamam A, Sanguin H, Bellvert F et al (2013) Plant secondary metabolite profiling evidences strain-dependent effect in the *Azospirillum–Oryza sativa* association. Phytochemistry 87:65–77. https://doi.org/10.1016/j.phytochem.2012.11.009

Chung M, Munro JB, Tettelin H, Dunning Hotopp JC (2018) Using Core genome alignments to assign bacterial species. mSystems 3(6). https://doi.org/10.1128/msystems.00236-18

Day JM, Döbereiner J (1976) Physiological aspects of N_2-fixation by a *Spirillum* from *Digitaria* roots. Soil Biol Biochem 8(1):45–50. https://doi.org/10.1016/0038-0717(76)90020-1

Di Salvo LP, Silva E, Teixeira KRS, Cote RE, Pereyra MA, García de Salamone IE (2014) Physiological and biochemical characterization of Azospirillum brasilense strains commonly used as plant growth-promoting rhizobacteria. J Basic Microbiol 54(12):1310–1321. https://doi.org/10.1002/jobm.201400135

Dixon R, Kahn D (2004) Genetic regulation of biological nitrogen fixation. Nat Rev Microbiol 2(8): 621–631. https://doi.org/10.1038/nrmicro954

dos Santos Ferreira N, Hayashi Sant' Anna F, Massena Reis V et al (2020) Genome-based reclassification of *Azospirillum brasilense* Sp245 as the type strain of *Azospirillum baldaniorum* sp. Nov. Int J Syst Evol Microbiol 70(12):6203–6212. https://doi.org/10.1099/ijsem.0.004517

dos Santos Ferreira N, Coniglio A, Puente M et al (2022) Genome-based reclassification of *Azospirillum brasilense* Az39 as the type strain of *Azospirillum argentinense* sp. Nov. Int J Syst Evol Microbiol 72(8):005475. https://doi.org/10.1099/ijsem.0.005475

Duan YQ, Zhou XK, Habib N, Gao SQ, Dong LM, Liu XF, Shi YM, Li WJ, Du G (2022) Azospirillum tabaci sp. Nov., a bacterium isolated from rhizosphere soil of Nicotiana tabacum L. Arch Microbiol 204(1):80. https://doi.org/10.1007/s00203-021-02688-7

Eckert B, Weber OB, Kirchhof G, Halbritter A, Stoffels M, Hartmann A (2001) Azospirillum doebereinerae sp. Nov., a nitrogen-fixing bacterium associated with the C4-grass Miscanthus. Int J Syst Evol Microbiol 51(1):17–26. https://doi.org/10.1099/00207713-51-1-17

Eskew DL, Focht DD, Ting IP (1977) Nitrogen fixation, denitrification, and pleomorphic growth in a highly pigmented spirillum lipoferum. Appl Environ Microbiol 34(5):582–585. https://doi.org/10.1128/aem.34.5.582-585.1977

Falk EC, Döbereiner J, Johnson JL, Krieg NR (1985) Deoxyribonucleic Acid Homology of Azospirillum amazonense Magalhães et al. 1984 and emendation of the description of the genus Azospirillum. Int J Syst Evol Microbiol 35(1):117–118. https://doi.org/10.1099/00207713-35-1-117

Fallik E, Okon Y, Epstein E, Goldman A, Fischer M (1989) Identification and quantification of IAA and IBA in Azospirillum brasilense-inoculated maize roots. Soil Biol Biochem 21(1):147–153. https://doi.org/10.1016/0038-0717(89)90024-2

Guo K, Yang J, Yu N et al (2022) Biological nitrogen fixation in cereal crops: progress, strategies, and perspectives. Plant Commun 4(2):100499. https://doi.org/10.1016/j.xplc.2022.100499

Hamana K, Sakane T, Yokota A (1994) Polyamine analysis of the genera Aquaspirillum, Magnetospirillum, Oceanospirillum and spirillum. J Gen Appl Microbiol 40(1):75–82. https://doi.org/10.2323/jgam.40.75

Khammas KM, Ageron E, Grimont PAD, Kaiser P (1989) Azospirillum irakense sp. Nov., a nitrogen-fixing bacterium associated with rice roots and rhizosphere soil. Res Microbiol 140(9):679–693. https://doi.org/10.1016/0923-2508(89)90199-X

Lavrinenko K, Chernousova E, Gridneva E, Dubinina G, Akimov V, Kuever J, Lysenko A, Grabovich M (2010) Azospirillum thiophilum sp. Nov., a diazotrophic bacterium isolated from a sulfide spring. Int J Syst Evol Microbiol 60(12):2832–2837. https://doi.org/10.1099/ijs.0.018853-0

Lin SY, Young CC, Hupfer H, Siering C, Arun AB, Chen WM, Lai WA, Shen FT, Rekha PD, Yassin AF (2009) Azospirillum picis sp. Nov., isolated from discarded tar. Int J Syst Evol Microbiol 59(4):761–765. https://doi.org/10.1099/ijs.0.65837-0

Lin SY, Shen FT, Young CC (2011) Rapid detection and identification of the free-living nitrogen fixing genus *Azospirillum* by 16S rRNA-gene-targeted genus-specific primers. Antonie Van Leeuwenhoek *99*(4):837–844. https://doi.org/10.1007/s10482-011-9558-1

Lin SY, Shen FT, Young LS, Zhu ZL, Chen WM, Young CC (2012) Azospirillum formosense sp. Nov., a diazotroph from agricultural soil. Int J Syst Evol Microbiol 62(Pt_5):1185–1190. https://doi.org/10.1099/ijs.0.030585-0

Lin SY, Liu YC, Hameed A, Hsu YH, Lai WA, Shen FT, Young CC (2013) Azospirillum fermentarium sp. Nov., a nitrogen-fixing species isolated from a fermenter. Int J Syst Evol Microbiol 63(Pt_10):3762–3768. https://doi.org/10.1099/ijs.0.050872-0

Lin SY, Hameed A, Shen FT et al (2014) Description of *Niveispirillum fermenti* gen. Nov., sp. Nov., isolated from a fermentor in Taiwan, transfer of *Azospirillum irakense* (1989) as *Niveispirillum irakense* comb. Nov., and reclassification of *Azospirillum amazonense* (1983) as *Nitrospirillum amazonense* gen. Nov. Antonie Van Leeuwenhoek 105(6):1149–1162. https://doi.org/10.1007/s10482-014-0176-6

Lin SY, Hameed A, Liu YC, Hsu YH, Lai WA, Shen FT, Young CC (2015) Azospirillum soli sp. Nov., a nitrogenfixing species isolated from agricultural soil. Int J Syst Evol Microbiol 65 (Pt_12):4601–4607. https://doi.org/10.1099/ijsem.0.000618

Lin SY, Liu YC, Hameed A, Hsu YH, Huang HI, Lai WA, Young CC (2016) Azospirillum agricola sp. Nov., a nitrogen-fixing species isolated from cultivated soil. Int J Syst Evol Microbiol 66(3): 1453–1458. https://doi.org/10.1099/ijsem.0.000904

Maroniche GA, García JE, Salcedo F et al (2017) Molecular identification of *Azospirillum* spp.: limitations of 16S rRNA and qualities of *rpo*D as genetic markers. Microbiol Res 195:1–10. https://doi.org/10.1016/j.micres.2016.11.009

Mehnaz S, Weselowski B, Lazarovits G (2007a) Azospirillum canadense sp. Nov., a nitrogen-fixing bacterium isolated from corn rhizosphere. Int J Syst Evol Microbiol 57(3):620–624. https://doi.org/10.1099/ijs.0.64804-0

Mehnaz S, Weselowski B, Lazarovits G (2007b) Azospirillum zeae sp. Nov., a diazotrophic bacterium isolated from rhizosphere soil of Zea mays. Int J Syst Evol Microbiol 57(12): 2805–2809. https://doi.org/10.1099/ijs.0.65128-0

Meier-Kolthoff JP, Göker M (2019) TYGS is an automated high-throughput platform for state-of-the-art genomebased taxonomy. Nat Commun 10(1):2182. https://doi.org/10.1038/s41467-019-10210-3

Okon Y, Itzigsohn R (1992) Poly-Î2-hydroxybutyrate metabolism in Azospirillum brasilense and the ecological role of PHB in the rhizosphere. FEMS Microbiol Lett 103(2–4):131–139. https://doi.org/10.1111/j.1574-6968.1992.tb05830.x

Okon Y (1985) *Azospirillum* as a potential inoculant for agriculture. Trends Biotechnol 3(9): 223–228

Oren A, Garrity GM (2022) CANDIDATUS LIST no. 3. Lists of names of prokaryotic Candidatus taxa. Int J Syst Evol Microbiol 72(1). https://doi.org/10.1099/ijsem.0.005186

Orozsco-Mosqueda M, Glick BR, Santoyo G (2020) ACC deaminase in plant growth-promoting bacteria (PGPB): an efficient mechanism to counter salt stress in crops. Microbiol Res 235: 126439. https://doi.org/10.1016/j.micres.2020.126439

Page AJ, Cummins CA, Hunt M et al (2015) Roary: rapid large-scale prokaryote pan genome analysis. Bioinformatics 31(22):3691–3693. https://doi.org/10.1093/bioinformatics/btv421

Parte AC, Sardà Carbasse J, Meier-Kolthoff JP et al (2020) List of Prokaryotic names with Standing in Nomenclature (LPSN) moves to the DSMZ. Int J Syst Evol Microbiol 70(11):5607–5612. https://doi.org/10.1099/ijsem.0.004332

Pagnier I, Raoult D, La Scola B (2008) Isolation and identification of amoeba-resisting bacteria from water in human environment by using an Acanthamoeba polyphaga co-culture procedure. Environ Microbiol 10(5):1135–1144. https://doi.org/10.1111/j.1462-2920.2007.01530.x

Peng G, Wang H, Zhang G, Hou W, Liu Y, Wang ET, Tan Z (2006) Azospirillum melinis sp. Nov., a group of diazotrophs isolated from tropical molasses grass. Int J Syst Evol Microbiol 56(6): 1263–1271. https://doi.org/10.1099/ijs.0.64025-0

Penrose DM, Glick BR (2003) Methods for isolating and characterizing ACC deaminase-containing plant growth-promoting rhizobacteria. Physiol Plant 118(1):10–15. https://doi.org/10.1034/j.1399-3054.2003.00086.x

Reinhold B, Hurek T, Fendrik I, Pot B, Gillis M, Kersters K, Thielemans S, De Ley J (1987) Azospirillum halopraeferens sp. Nov., a nitrogen-fixing organism associated with roots of Kallar grass (Leptochloa fusca (L.) Kunth). Int J Syst Evol Microbiol 37(1):43–51. https://doi.org/10.1099/00207713-37-1-43

Richter M, Rosselló-Móra R (2009) Shifting the genomic gold standard for the prokaryotic species definition. Proc Natl Acad Sci 106(45):19126–19131. https://doi.org/10.1073/pnas.0906412106

Skerman VBD, Sly LI, Williamson ML (1983) *Conglomeromonas largomobilis* gen. Nov., sp. Nov., a sodium-sensitive, mixed-flagellated organism from fresh waters. Int J Syst Evol Microbiol 33(2):300–308. https://doi.org/10.1099/00207713-33-2-300

Steenhoudt O, Vanderleyden J (2000) *Azospirillum*, a free-living nitrogen-fixing bacterium closely associated with grasses: Genetic, biochemical and ecological aspects. FEMS Microbiol Rev 24(4):487–506. https://doi.org/10.1111/j.1574-6976.2000.tb00552.x

Stets MI, Pinto AS, Huergo LF, De Souza EM, Guimarães VF, Alves AC, Steffens MBR, Monteiro RA, Pedrosa FDO, Cruz LM (2013) Rapid identification of bacterial isolates from wheat roots by high resolution whole cell MALDI-TOF MS analysis. J Biotechnol 165(3–4):167–174. https://doi.org/10.1016/j.jbiotec.2013.04.001

Tarrand JJ, Krieg NR, Döbereiner J (1978) A taxonomic study of the *Spirillum lipoferum* group, with descriptions of a new genus, *Azospirillum* gen. Nov. and two species, *Azospirillum lipoferum* (Beijerinck) comb. Nov. And *Azospirillum brasilense* sp. Nov. Can J Microbiol 24(8):967–980. https://doi.org/10.1139/m78-160

Tettelin H, Medini D (2020) The pangenome: diversity, dynamics and evolution of genomes. Springer, Cham. https://doi.org/10.1007/978-3-030-38281-0

Tikhonova EN, Grouzdev DS, Kravchenko IK (2019) Azospirillum palustre sp. Nov., a methylotrophic nitrogenfixing species isolated from raised bog. Int J Syst Evol Microbiol 69(9):2787–2793. https://doi.org/10.1099/ijsem.0.003560010-0543-7

Tyler ME, Milam JR, Smith RL et al (1979) Isolation of *Azospirillum* from diverse geographic regions. Can J Microbiol 25(6):693–697. https://doi.org/10.1139/m79-100

Veresoglou SD, Menexes G (2010) Impact of inoculation with *Azospirillum* spp. on growth properties and seed yield of wheat: a meta-analysis of studies in the ISI Web of Science from 1981 to 2008. Plant Soil 337(1):469–480. https://doi.org/10.1007/s11104-010-0543-7

Wang H, Jin H, Chen Z, Li W, Ma J, Hu T, Liu Q, Zhang Y, Lin X, Xie Z (2024) Azospirillum isscasi sp. Nov., a bacterium isolated from rhizosphere soil of rice. Int J Syst Evol Microbiol 74(1):006218. https://doi.org/10.1099/ijsem.0.006218

Wick RR, Judd LM, Cerdeira LT et al (2021) Trycycler: consensus long-read assemblies for bacterial genomes. Genome Biol 22(1):266. https://doi.org/10.1186/s13059-021-02483-z

Wisniewski-Dyé F, Borziak K, Khalsa-Moyers G et al (2011) *Azospirillum* genomes reveal transition of bacteria from aquatic to terrestrial environments. PLoS Genet 7(12):e1002430. https://doi.org/10.1371/journal.pgen.1002430

Wisniewski-Dyé F, Lozano L, Acosta-Cruz E et al (2012) Genome sequence of *Azospirillum brasilense* CBG497 and comparative analyses of *Azospirillum* core and accessory genomes provide insight into niche adaptation. Gene 3(4):Article 4. https://doi.org/10.3390/genes3040576

Wu D, Zhang XJ, Liu HC, Zhou YG, Wu XL, Nie Y, Kang YQ, Cai M (2021) Azospirillum oleiclasticum sp. Nov, a nitrogen-fixing and heavy oil degrading bacterium isolated from an oil production mixture of Yumen oilfield. Syst Appl Microbiol 44(1):126171. https://doi.org/10.1016/j.syapm.2020.126171

Xia Y, Embley TM, O'Donnell AG (1994) Phylogenetic analysis of Azospirillum by direct sequencing of PCR amplified 16S rDNA. Syst Appl Microbiol 17(2):197–201. https://doi.org/10.1016/S0723-2020(11)80007-X

Xie CH, Yokota A (2005) Azospirillum oryzae sp. Nov., a nitrogen-fixing bacterium isolated from the roots of the rice plant Oryza sativa. Int J Syst Evol Microbiol 55(4):1435–1438. https://doi.org/10.1099/ijs.0.63503-0

Xu CH, Chen G, Liu Y, Xiao W, Cui XL, Ding ZG, Liu WH, Wang YX (2023) Azospirillum aestuarii sp. Nov., a novel nitrogen-Fixting and aerobic denitrifying bacteria isolated from an estuary of a Freshwater River. Curr Microbiol 80(4):113. https://doi.org/10.1007/s00284-023-03213-3

Yang Y, Zhang R, Feng J, Wang C, Chen J (2019) Azospirillum griseum sp. Nov., isolated from lakewater. Int J Syst Evol Microbiol 69(12):3676–3681. https://doi.org/10.1099/ijsem.0.003460

Young CC, Hupfer H, Siering C, Ho MJ, Arun AB, Lai WA, Rekha PD, Shen FT, Hung MH, Chen WM, Yassin AF (2008) Azospirillum rugosum sp. Nov., isolated from oil-contaminated soil. Int J Syst Evol Microbiol 58(4):959–963. https://doi.org/10.1099/ijs.0.65065-0

Zhan PC, Mao RF, Li CJ, Zhang Z, Liu JR, Tang M, Zhi XY, Yang LL (2022) Azospirillum endophyticum sp. Nov., an endophyte of Paris Polyphylla smith var. Yunnanensis. Curr Microbiol 80(1):34. https://doi.org/10.1007/s00284-022-03142-7

Zhao Z, Ming H, Ding CL, Ji WL, Cheng LJ, Niu M, Zhang Y, Zhang LY, Meng XL, Nie GX (2020) Azospirillum thermophilum sp. Nov., isolated from a hot spring. Int J Syst Evol Microbiol 70(1):550–554. https://doi.org/10.1099/ijsem.0.003788

Zhou S, Han L, Wang Y, Yang G, Zhuang L, Hu P (2013) Azospirillum humicireducens sp. Nov., a nitrogen-fixing bacterium isolated from a microbial fuel cell. Int J Syst Evol Microbiol 63(Pt_7): 2618–2624. https://doi.org/10.1099/ijs.0.046813-0

Zhou Y, Wei W, Wang X, Xu L, Lai R (2009) Azospirillum palatum sp. Nov., isolated from forest soil in Zhejiang province, China. J Gen Appl Microbiol 55(1):1–7. https://doi.org/10.2323/jgam.55.1

Zhu HZ, Zhang ZF, Zhou N, Jiang CY, Wang BJ, Cai L, Wang HM, Liu SJ (2021) Bacteria and metabolic potential in karst caves revealed by intensive bacterial cultivation and genome assembly. Appl Environ Microbiol 87(6):e02440-20. https://doi.org/10.1128/AEM.02440-20

Chapter 4
Genetic Manipulation of *Azospirillum*

Hannah Hughes, Hahley Wiltse, Ishita Banerjee, Tasneem Ehsan, and Gladys Alexandre

Abstract Many, but not all, *Azospirillum* species are genetically tractable bacteria. Replicative and suicide vectors are available for allelic exchange, *in trans* gene expression, functional complementation, and reporter gene analyses. Because *Azospirillum* species are closely related to Rhizobiaceae, tools used in Rhizobia have been typically employed for the genetic manipulation of *Azospirillum* species. Insertional and deletion allelic exchanges have been optimized for *Azospirillum*. *Azospirillum* is not competent, and there is no phage transduction available for genome manipulation yet. The vast majority of genetic manipulation over the last several decades has occurred using conjugation. Generally, plasmids are constructed in donor *E. coli* strains and then conjugated into the acceptor *Azospirillum* strain. While this molecular kit can be further expanded, it is currently sufficient to address the molecular mechanisms underlying *Azospirillum* physiology and their beneficial association with the roots of many plants.

Keywords Genetic manipulation · Plasmids · Conjugation · Vectors · Gene expression

4.1 Overview of Genetic Manipulation Methods and Relevant Genome Characteristics

4.1.1 Genome Characteristics

Ever since the initial isolation of *Azospirillum*, genome characterization and manipulation have been a high priority. There are currently about 24 described species of *Azospirillum* with many more isolates awaiting classification as seen on this website: (http://www.bacterio.net/azospirillum.html) (Cassán et al. 2020). Initial studies

H. Hughes · H. Wiltse · I. Banerjee · T. Ehsan · G. Alexandre (✉)
Biochemistry & Cellular and Molecular Biology Department, The University of Tennessee, Knoxville, TN, USA
e-mail: galexan2@utk.edu

F. D. Cassán et al. (eds.), *The Definitive Handbook of Azospirillum*,
https://doi.org/10.1007/978-3-032-09222-9_4

found several distinguishing factors compared to more common model organisms. For example, *Azospirillum* genomes possess a high guanine (G) and cytosine (C) base pairs (GC) content (~70%) compared to *E. coli* K12 (~51%). The high GC content is a challenge when applying PCR methods or more generally, to engineer genetic changes. Additionally, *Azospirillum* spp. encode their genomes on a single chromosome, multiple large plasmids, and several chromids, which have features of both chromosomes and plasmids. While the precise interactions between these genome components have not yet been characterized, there is precedent for large-scale genetic rearrangement. For example, in *Azospirillum brasilense* and *Azospirillum lipoferum*, stable, nonmotile phase variants were isolated and shown to result from *recA*-dependent genetic rearrangements (Vial et al. 2006). Evidence of historical rearrangements can also be found on *A. brasilense*'s plasmid p5 (also known as pRhico). This plasmid encodes a series of phage genes that align to a prophage originally characterized in *A. brasilense* Cd (ΦAb-Cd) (Boyer et al. 2008; Vanbleu et al. 2004). The phage genes themselves are insufficient to produce phage particles and are thought to result from a genetic duplication event. *Azospirillum* genomes sequenced to date also evidence many mobile genetic elements in the form of insertion elements, transposases, and instances of significant lateral gene transfer (e.g., Wisniewski-Dyé et al. 2011). As a close relative to Rhizobiaceae, tools used in Rhizobia have been typically employed for the genetic manipulation of *Azospirillum* species. Conjugation, *in trans* stable expression of diverse genes, reporters from broad-host-range plasmids, and methods for allelic exchange are also available in *Azospirillum* species. This molecular kit is sufficient to address the molecular mechanisms underlying *Azospirillum* physiology and its beneficial association with the roots of many plants.

4.1.2 Techniques for Introducing Genetic Material

Historically, genetic manipulation of *Azospirillum* relied on techniques such as chemical mutagenesis and UV irradiation, followed by isolation and identification of suppressor mutations. More recently, whole-genome sequencing has allowed for publication of assembled sequences of many *Azospirillum* species, strains, and environmental isolates (Wisniewski-Dyé et al. 2011; Whitman et al. 2015; Rivera et al. 2014; Yu et al. 2018; Rayko et al. 2024; Kwak and Shin 2016). These tools have allowed for increasingly targeted genetic manipulation and direct investigation of molecular pathways.

The process of generating these targeted genetic manipulations requires the introduction of engineered DNA into *Azospirillum*. Since *Azospirillum* is not naturally competent, techniques such as transformation, electroporation, and conjugation to introduce plasmids have been used instead.

Transformation using heat-shock competent cells has previously been demonstrated for *A. lipoferum*, using a modified $CaCl_2$ method (Uozumi et al. 1986). *A. brasilense* has also been successfully transformed, although modification of the standard protocol was required, and transformants were only obtained in some of the strains tested (Fani et al. 1986).

Electroporation has also been used to introduce genetic material in *Azospirillum* spp., but it generally has low efficiency and is variably effective among different species and strains (Vande Broek et al. 1989; Vande Broek and Vanderleyden 2000). Another concern with the usage of electroporation is that it has been demonstrated to induce widespread, untargeted genome rearrangements (Nunes et al. 2000). Specifically, this occurred upon attempts to introduce and select for an antibiotic resistance gene-cartridge in place of a native gene. As there is precedence for a propensity toward genetic rearrangements, this method introduces concerns for the genetic integrity of the target strain. As such, electroporation is not typically used to generate mutations in *Azospirillum* species.

Instead, most of the genetic manipulation over the last several decades has occurred using conjugation. Generally, plasmids are constructed in donor *E. coli* strains and then conjugated into the acceptor *Azospirillum* strain. If the donor plasmid is self-transmissible, or if the donor strain carries transfer elements on its genome, a biparental conjugation technique can be used. Otherwise, triparental conjugation is carried out, where a third strain is included to mobilize the plasmid. A strain of *E. coli* (DH5-α, Top10, or HB101) containing the helper plasmid pRK2013 is commonly used for this purpose (Figurski and Helinski 1979). A typical protocol as used for *Azosprillum brasilense* Sp7 is outlined below.

4.1.3 Triparental Conjugation Protocol

- Grow up dense, overnight cultures of the donor *E. coli* strain, the helper *E. coli* strain, and the acceptor *Azospirillum* strain.
- Centrifuge 1 mL each *E. coli* strain and 4 mL of the *Azospirillum* strain to form dense pellets.
- Wash each strain three times with 1 mL 0.8% sterile KCl.
- Resuspend each strain in 100 μL 0.8% KCl.
- Mix 50 μL of each strain together and make five to six 25 μL spots on D-media agar plates (8 g Nutrient Broth, 0.25 g $MgSO_4 \cdot 7H_2O$, 1 g KCl, 0.01 g $MnCl_2$, 15 g agar, and dH_2O to 1000 mL) (Gullett et al. 2017). This will ensure drying of each spot before incubation. Avoid drying spots too long (indicated by cracking).
- Include spots of each strain individually (*E. coli* donor, *E. coli* helper, and *Azospirillum* recipient) and use these as negative controls for conjugation.
  - If growth appears on the *E. coli* negative control plates, it means that the antibiotic concentration used in the plate is insufficient to select against the acceptor *Azospirillum*. If growth appears on the *Azospirillum* negative control plate, it indicates that the antibiotic concentration used is not sufficient, as long as care was taken to verify that the acceptor *Azospirillum* strain is sensitive to the selective antibiotic prior to the experiment.

- Incubate the D-media plates in a closed container lined with wet paper towels to create a humid atmosphere to support mating at 28 °C for approximately 16–20 h. This incubation time is optimum for most vectors, but shorter times may be tested and appropriate for some vectors.
- Resuspend cells in 1 mL minimal medium with appropriate antibiotics. Use any antibiotics required for the selection of genetic material, along with 200 μg/mL ampicillin to allow the *Azospirillum* recipient to grow while selecting against the *E. coli*.
- Incubate at 28 °C with shaking for 4 h.
- Plate on minimal medium selection media with 200 μg/mL ampicillin and let grow at 28 °C for 3–5 days.
  - Additional selective pressures can be used to eliminate the *E. coli* strains, such as using nitrogen-free media. It is important to mention that with nitrogen-free media, noble agar should be used because most typically used commercial agar contains trace amounts of organic nitrogen.

4.2 Selection Markers for *Azospirillum*

Positive and negative selection markers are commonly used for genetic manipulation in most bacterial genetic studies. Most *Azospirillum* species are naturally highly resistant to β-lactam antibiotics (up to 500 μg/mL), as well as trimethoprim (with an Minimum Inhibitory Concentration (MIC) higher than 1024 μg/mL) (Lopez-Reyes et al. 1989; Pandey et al. 2022). Natural antibiotic resistance is common in bacterial isolates from the soil. The β-lactam resistance is an advantage in the lab, and ampicillin is commonly added to medium plates to keep pure cultures of *Azospirillum*. However, any vectors with ampicillin as the selection marker are not typically used for genetic manipulation with *Azospirillum*, given the natural resistance to this antibiotic.

4.2.1 Positive Selection Markers

The most common selection markers in vectors used with *Azospirillum* are tetracycline, chloramphenicol, gentamycin, and kanamycin due to *Azospirillum*'s general susceptibility to these antibiotics. *Azospirillum* is the most susceptible to tetracycline, chloramphenicol, and kanamycin (Lopez-Reyes et al. 1989), making these antibiotics generally good choices as selection markers in genetic manipulation. However, if the susceptibility of a particular strain to an antibiotic is unknown, it is best practice to perform an MIC to determine resistance of an antibiotic one wishes to use for genetic manipulation.

In Table 4.1, the gene and encoded protein that confer antibiotic resistance in *Azospirillum* naturally are listed, followed by commonly used vector selection

Table 4.1 Positive selection markers used in *Azospirillum*

Gene	Protein	Resistance	Conc. (μg/mL)	Plasmids	References
bla1	β-Lactamase 1 (Bla1)	Amp^R/ Cb^R	200	Natural resistance	Lopez-Reyes et al. (1989) and Pandey et al. (2022)
ltg1	Lytic transglygosylase (Ltg1)	Amp^R/ Cb^R	200	Natural resistance	Lopez-Reyes et al. (1989) and Pandey et al. (2022)
tetA	Tet efflux protein (TetA)	Tet^R	10	pFUS1, pRK415, pBBR1MCS-3	Kovach et al. (1995)
cat	Cm acetyltransferase (Cat)	Cm^R	20	pBBR1MCS-1	Dennis and Zylstra (1998) and Kovach et al. (1995)
aacC1	Gm-3-acetyltransferase (AacC1)	Gm^R	20	pBBR1MCS-5	Kovach et al. (1995)
aph	Aminoglycoside phosphotransferase (Aph)	Km^R	30	pBBR1MCS-2, pTS7, pTS9, pK18mobsacB	Dennis and Zylstra (1998) and Kovach et al. (1995)

Abbreviations: Amp^R ampicillin resistance, Cb^R carbenicillin resistance, Cm^R chloramphenicol resistance, Gm^R gentamycin resistance, Km^R kanamycin resistance, Tet^R tetracycline resistance

markers used for genetic manipulation and the typical concentrations used in different *Azospirillum* laboratories.

4.2.2 Negative Selection Markers

Negative selection markers are typically used in mutagenesis experiments for counter-selecting single-homologous recombinant strains that retain an integrated suicide vector and double homologues recombinant strains that have lost the integrated suicide vector. In *Azospirillum*, the vector pK18*mobsacB* has been successfully used in nonpolar mutagenesis by allelic exchange. The pK18*mobsacB* vector uses the *sacB* gene as a negative selection marker which selects against the presence of the vector backbone in the presence of sucrose. The *sacB* gene encodes levansucrase, which converts sucrose to levan. Levan builds up in the periplasm and is toxic to the cell, this causes cell death. Another example of a negative selection is the *pheS** gene in the pTS9 vector. The phenylalanyl tRNA synthetase gene, *pheS*, is required for specifically charging phenylalanine on its corresponding tRNA. PheS* contains an A294G substitution with the *pheS* gene, which results in the charging of toxic p-chloro-phenylalanine (p-Cl-Phe) into tRNA and then into proteins, ultimately causing cell death. The *pheS** gene was used successfully as a plasmid counter-selection in non-*Azospirillum* species first (Ishikawa et al. 2018), then was used successfully in pTS9 in *A. brasilense* (Schnabel and Sattely 2021).

4.3 Plasmids Used for Genetic Manipulation of *Azospirillum*

Historically, a wide range of vectors and their derivatives have been used for genetic manipulation across different species of *Azospirillum*. These vectors have been used to create genomic libraries, functional complementation, transposon mutagenesis, and expression of reporter genes including fluorescent tagging. Unstable suicide vectors have been used for allelic exchange and gene disruption.

4.3.1 Plasmid Vectors

The following Table 4.2 summarizes the most common vectors used and their roles in genetic manipulation of *Azospirillum*.

4.3.2 Stable Replicative Plasmids

Broad-host range replicative plasmids are used for expressing genes of interest for functional complementation, overexpression, generation of genomic libraries, or expression of reporter genes. The historically most widely used stable plasmids in *Azospirillum* have been pRK290 (Ditta et al. 1980; Liang et al. 1991), used for studying regulation of the *nif* gene (Liang et al. 1991), pLA2917 (Allen and Hanson 1985), used for studying wheat root colonization (Arsene et al. 1994), and gene expression of *flcA-lacZ* (Pereg-Gerk 2004).

pBBR1-MCS1-5 series of vectors with different antibiotic resistances and pRK415 have also been widely used as stable replicated broad-host-range plasmids in *Azospirillum*. The *lac* promoter drives the expression of cloned promoter-less genes on the plasmid. Genes and their own promoters can be cloned to drive native expression from the plasmid. We note that Plac appears to work mostly constitutively in *Azospirillum*, and Isopropyl β-D-Thiogalactopyranoside (IPTG) has limiting (if any) inducible effect at least when used in *A. brasilense* Sp7 strain. More recently, an inducible expression vector for *A. brasilense* was developed based on the tetracycline repression (TetR) system. This involves a tetracycline repressor encoding gene that represses a TetR promoter (PTetR) present on the plasmid. The PTetR is derepressed in the presence of anhydrotetracycline (analog of tetracycline) (Schnabel and Sattely 2021).

Table 4.2 Vectors used for genetic transformation in *Azospirillum*

Plasmids	Genotype	Characteristics and examples of uses in *Azospirillum*	References
Replicative stable vectors			
pRK415	pRK404 derivative carrying oriT, P*lac* promoter, Tet^R	Broad host range low copy plasmid to express gene of interest *in trans*	Keen et al. (1988)
pTS7	*oriV,* Neo^R, Km^R, P*TetR*	Broad host range inducible plasmid	Schnabel and Sattely (2021)
pBBR1-MCS1-5	P*lac* promoter, Km^R/Tet^R/Amp^R/Gm^R	Broad host range inducible plasmid	Kovach et al. (1995)
pLAFR3	Tet^R	Broad host range cosmid cloning vector	Milcamps et al. (1996)
pCZ750	pFAJ1700 containing the KpnI-AscI lacZ gene from pCZ367 plasmid; Tet^R, Amp^R	β-galactosidase encoding reporter gene	Dubey et al. (2020)
Suicide/mutagenesis vectors			
pKNOCK	pBSL63 derivative, RP4 oriT and oriR6Kγ Tet^R/Gm^R/Cm^R/Km^R/Amp^R	Mobilizable suicide vector by insertional mutagenesis by single homologous recombination	Alexeyev (1999)
pK18*mobsacB*	*oriV,* Km^R, Suc^S	Mobilizable suicide vector for allelic exchange by double homologous recombination	Katzen et al. (1999)
pSUP202	ColE1 replicon, mobilizable plasmid, Amp^R,Tet^R, Cm^R	Tn5 transposon mutagenesis, gene disruption	Simon et al. (1983)
pSUP2021	ColEl replicon Tra mob' Tn5 Donor, Amp^R $Cm^R$$Tet^R$ Km^R	Tn5 transposon mutagenesis	Simon et al. (1983)
pTS9	Neo^R, Km^R, PheS*	Mobilizable suicide vector for allelic exchange by double homologous recombination	Schnabel and Sattely (2021)
pRL27	Tn5-RL27 vector, Km^R, oriR6Kγ	Tn5 transposon mutagenesis	Larsen et al. (2002)
Reporter/fluorescent vectors			
pJBA21TC	pJB321; RK2oriV, constitutive *gusA* under P*nptII,* Tet^R	Stable reporter plasmid for constitutive expression of *gusA*	Wielbo and Skorupska (2001)

(continued)

Table 4.2 (continued)

Plasmids	Genotype	Characteristics and examples of uses in *Azospirillum*	References
pFUS1	pMP220 derivative containing a promoterless *gusA*, Tet^R	Transcriptional activity measurement by *gusA* reporter gene	Reeve et al. (1999)
pHRGFPGUS	pBBR replicon, *gfpmut3* and *gusA*, Km^R	Plant root colonization study	Ramos et al. (2002)
pRH004	Rk2oriV, Km^R Cm^R	Gateway-based plasmid with c-terminal CFP (cyan fluorescent protein) tagging	Hallez et al. (2007)
pRH005	Rk2oriV, Km^R Cm^R	Gateway-based plasmid with c-terminal YFP (yellow fluorescent protein) tagging	Hallez et al. (2007)
pBSK-mCherry	Cloning vector, containing mCherry gene, Amp^R	Visualization of biofilm associated with wheat root	Ramirez-Mata et al. (2018)
pMP2444	pBBR-1 origin, *egfp* under P*lac*, Gm^R	Visualization of biofilm associated with wheat root in *Azospirillum*	Bloemberg et al. (2000)
Vector for antibiotic resistance cassette			
p34S	Ori, Pc, Cm^R/Gm^R/Km^R/Tet^R	A template for antibiotic resistance cassettes for chloramphenicol, gentamicin, kanamycin, or tetracycline resistance	Dennis and Zylstra (1998)
Mobilization plasmid			
pRK2013	ColE1 replicon, *tra*, Km^R	Helper plasmid for mobilization during triparental conjugation	Figurski and Helinski (1979)

Abbreviations: Amp^R ampicillin resistance, Cm^R chloramphenicol resistance, Gm^R gentamycin resistance, Km^R kanamycin resistance, Neo^R neomycin resistance, *Pc* promoter constitutive, *Plac* lactose inducible promoter, *PTetR* promoter for tetracycline repressor, Suc^S sucrose sensitive, Tet^R tetracycline resistance

4.3.3 Suicide/Mutagenesis Vectors

Suicide vectors are unstable plasmids used for gene replacement by allelic exchange, transposon, and insertional mutations. pSUP2021 and its derivative pSUP202, which are mobilizable vectors, have been used as suicide vectors in *A. brasilense* and *A. lipoferum* because these vectors do not replicate and are not stable in these bacterial hosts (Vanstockem et al. 1987; Galimand et al. 1989; de Zamaroczy et al. 1993; Steenhoudt et al. 2001; Cruz-Pérez et al. 2021; Rai et al. 2025).

A hyperactive transposase-containing vector pRL27 has also been used for transposon mutagenesis. This vector encodes for the R6K-γ origin of replication which requires the π protein, encoded by the *pir* gene to replicate. The *pir* gene is cloned into a recombinant *E. coli* λpir strain that is used for plasmid transfer but is absent for the acceptor *Azospirillum* strain in which the vector acts as a suicide plasmid.

The pKNOCK suicide plasmids are a series of vectors offering different antibiotic resistance and are used for insertional mutagenesis via single or double homologous recombination in *Azospirillum*. For example, in *A. brasilense* Sp7, pKNOCK-Tc was used to generate a chemoreceptor mutant strain, *aerC::Km*^{R} mutant, by allelic exchange (Xie et al. 2010), and pKNOCK-Gm^R was used to generate an insertional mutant of the *tlp4b* chemoreceptor, *tlp4b*::pKNOCK-Gm^R (Ganusova et al. 2023).

4.3.4 Promoter Probe Vectors

Combining promoters of interest with a reporter gene has significantly improved the ability to investigate gene expression in laboratory settings and natural environments. Several transcriptional reporter systems have been widely used to study gene expression in *Azospirillum*. In these systems, the promoter region of the gene of interest is cloned upstream of the promoter-less reporter gene, and the promoter activity is measured by assessing the activity encoded by the reporter gene. Some reporter systems used in *Azospirillum* spp. include the chloramphenicol acetyltransferase enzyme encoded by the *cat* gene (Fani et al. 1988), the β-galactosidase enzyme encoded by the *lacZ* gene (Labes et al. 1990), the β-glucuronidase enzyme encoded by the *gusA* gene (Prell et al. 2002; Reeve et al. 1999), and the green fluorescent protein encoded by the *gfp* gene.

Examples of widely used promoter probe vectors to study gene expression in *Azospirillum* include the broad-host-range vector pAF300 that carries a promoter-less *cat* gene as a reporter. An active promoter of interest fused with the cat gene confers chloramphenicol resistance (Cm^R) to the *Azospirillum* recipient cells harboring the plasmid. The promoter activity can also be quantified using chloramphenicol acetyltransferase enzyme activity (Fani et al. 1988). Another plasmid vector, the pFUS1 vector (Table 4.2), carries a promoter-less *gusA* reporter gene. The promoter region of the gene of interest is cloned into the upstream region of the promoter-less *gusA* gene, and a colorimetric assay is then used to evaluate promoter activity. The pCZ750 plasmid vector (Table 4.2), which contains a promoter-less *lacZ* gene, is another widely used reporter gene used to study transcriptional activity in *Azospirillum* (Dubey et al. 2020).

4.3.5 Expressing Fluorescent Proteins from Plasmids

Creating a translational fusion, where a protein of interest is linked to a fluorescent protein at either its N-terminus or C-terminus, is a common method for examining subcellular localization and protein colocalization in bacterial cells. Various plasmid vectors are available that encode different fluorescent proteins such as cyan fluorescent protein (CFP), yellow fluorescent protein (YFP), and enhanced green fluorescent protein (EGFP), enabling the construction of translational fusions for studying specific proteins. One such vector, pRH005, is a Gateway-based destination vector designed to express proteins fused with YFP at their C-terminus (Hallez et al. 2007). For example, this plasmid was used to fluorescently label several chemotaxis proteins in *A. brasilense* to investigate their subcellular localization (O'Neal et al. 2019).

The broad-host-range stable plasmid pHRGFPGUS has been extensively utilized to track the colonization patterns of *Azospirillum*, whether on plant root surfaces or other abiotic surfaces both in a qualitative and quantitative manner (Ramos et al. 2002). Strong, constitutive expressions of GFP and GusA (β-glucuronidase) encoded on this plasmid are under the control of the gentamycin resistance gene promoter. Additionally, the plasmid pMP2444 is a stable, high-copy-number, broad-host-range vector that expresses enhanced green fluorescent protein (EGFP) under the control of the *lac* promoter (Bloemberg et al. 2000). Similarly, pMP2449-5, which expresses mCherry (encodes for a red fluorescent protein) under the *lac* promoter, was employed to investigate the root attachment and colonization pattern of *Azospirillum* (Ramirez-Mata et al. 2018). The use of genes encoding fluorescent proteins, such as EGFP or mCherry, proved beneficial since their fluorescence does not require additional cofactors for expression, high copy number ensures enhanced fluorescent protein production under certain conditions, and they are highly stable in the rhizosphere environment.

4.4 Mutagenesis in *Azospirillum*: Insertional and Allelic Exchange Mutagenesis

There are several different ways to knock out gene function in *Azospirillum*. These procedures range from random mutagenesis to site-specific allelic exchange. UV light exposure or chemical treatments were used commonly in the 1980s and the 1990s (Del Gallo et al. 1987; Holguin et al. 1999). Targeted mutagenesis and transposon-based libraries have become popularized with the advent of cheaper gene and genome sequencing. Each type of mutagenesis has advantages and shortcomings, but which one to use depends on the application. The various ways to mutagenize *Azospirillum* are detailed in protocols later in this chapter.

4.4.1 Insertional Mutagenesis

Disrupting a gene of interest by inserting a suicide vector containing a particular antibiotic cassette through single-homologous recombination is one of the simplest methods to disrupt a gene's function in the *A. brasilense* genome, and presumably, other *Azospirillum* genomes. These vectors can replicate in *E. coli* but not in *Azospirillum*.

A classic example of single-homologous insertion by a suicide vector is the pKNOCK series of vectors (Alexeyev 1999). The replicative properties of this vector are determined by the γ-replication origin of the R6K plasmid, which is only active in *E. coli* strains containing the trans π protein encoded by the R6K plasmid (Kolter and Helinski 1978). The detailed protocol for using pKNOCK is detailed below in Sect. 4.5.2. Insertional mutagenesis can also result in allelic exchange after double homologous recombination. This has been shown to be successful in pSUP202 (Mishra et al. 2021). The limitation on vectors like pSUP202 is the lack of a counter-selection marker, resulting in not being able to force a double homologous recombination.

Insertional mutations result in an interrupted gene and can result in polar effects downstream of the gene of interest, such as when the gene is in an operon. Insertional mutations can also lose the mutation over time and revert to wild type if not kept on selective pressure such as an antibiotic. This is due to the presence of the two halves of the original gene around the insertion.

4.4.2 Deletion-Insertion Mutagenesis

Deletion-insertion mutagenesis results in the allelic exchange of the wild-type copy of a gene usually with an antibiotic resistance marker gene. This type of mutagenesis can be done using pK18*mobsacB* and pTS9 vectors. Due to the deletion of the wild-type gene, these mutations are stable and do not need to be kept on a selective medium constantly. However, there is still the possibility of this type of mutation being polar, but it has less chance to be polar than an insertional mutant. A major advantage this method has over the marker-less method is the ability to select for mutants rather than screen for mutants. Given the relatively low efficiency of mutations in *Azospirillum*, screening colonies for mutations requires hundreds of picks depending on the gene and method used. The protocol for this type of mutagenesis using the pK18*mobsacB* vector is detailed in Sect. 4.5.3.

4.4.3 Marker-Less Deletion Mutagenesis

Marker-less deletion mutagenesis is done by an allelic exchange of a gene resulting in a gene deletion without a selectable marker in place of the gene. This type of mutagenesis can be done with vectors that contain counter-selection markers such as the *sacB* gene pK18*mobsacB* which was used in *A. brasilense* Sp7 (Mukherjee et al. 2019) and *pheS** in pTS9 (Schnabel and Sattely 2021). Marker-less deletions have been shown to be successful in *Nitrospirillum amazonense*, a closely related species belonging to the *Azospirillaceae* family (Sant'Anna et al. 2011). This type of mutagenesis leaves minimal changes to the DNA after allelic exchange is completed and has a minimal chance of having polar effects on downstream genes such as in operons compared to other types of mutants. This type of deletion is most useful when multiple, stable, gene knockouts are needed in a strain, or if the presence of an antibiotic resistance gene is an issue. This type of mutagenesis typically takes longer times to be completed because many more potential colonies need to be screened due to the lack of positive selection markers.

4.5 Golden Protocols for Genetic Manipulation in *Azospirillum*

4.5.1 Functional Complementation

Functional complementation is a genetic technique used to interpret a gene's function. Normally, it is done by introducing a functional copy of the gene of interest into a mutant strain to assess whether it can restore the mutant's phenotype to the wild type's, indicating there are no other effects of the mutation. This section describes a general protocol designed for functional complementation of *A. brasilense* Sp7 using stable replicative plasmids. Any gene of interest can be engineered into one of the broad-host-range replicative vectors to express genes *in trans* from a constitutive promoter encoded on the vector or from an inducible promoter.

- Design a pair of primers for your gene of interest, targeting cloning into a broad-host-range replicative plasmid relevant to your study. Clone the entire gene into the poly-linker part of the plasmid under the desired promoter followed by transformation into *E. coli* Top10 or equivalent cloning *E. coli* host.
- Select colonies of *E. coli* carrying the construct using selection with specific antibiotic resistance present on the selected plasmid. Verify the construct by isolating plasmids and performing PCR using a pair of primers designed to bind on one location inside in your gene of interest and one on the plasmid. You can also send the plasmid construct for sequencing to verify the proper insertion under the desired promoter.

- Once the correct construct in *E. coli* has been validated, carry out triparental conjugation (see protocol in Sect. 4.1.3), using *E. coli* with the construct as donor, the acceptor *Azospirillum* strain, and an *E. coli* helper strain carrying the pRK2013 vector (Table 4.2) for mobilization. Select exconjugant colonies on media suitable for *Azospirillum* such as the minimal medium for *Azospirillum* (MMAB) (per liter of dH_2O: 5 g sodium malate, 3 g K_2HPO_4, 1 g NaH_2PO_4, 1 g NH_4Cl, 0.3 g $MgSO_4$ $7H_2O$, 0.15 g KCl, 0.01 g $CaCl_2$ $2H_2O$, 0.0025 g $FeSO_4$ $7H_2O$, and 0.0001 g Na_2MoO_4 [adjusted to pH 6.85–7.0]) and antibiotic resistance provided by the plasmid backbone.
- Use appropriate controls by transferring the replicative plasmid lacking your gene of interest (empty vector) into your target *Azospirillum* strain as a negative control. This empty vector control will ensure that any phenotypic changes observed in the *Azospirillum* strain carrying the gene of interest expressed *in trans* is independent of the replicative vector.
- Make 50% (v/v) glycerol stocks of all strains and freeze at −80 °C.
- Interpret gene function by examining the phenotypic changes in the strains carrying the plasmid with the gene of interest versus the strain harboring an empty vector. Note that expressing a gene of interest from a plasmid often only partially restores the parent phenotype in a mutant strain because of differences in expression levels of the gene of interest, or because of copy number of the replicative plasmid. Lack of functional complementation or effect of expressing the gene of interest *in trans* could be due to additional uncontrolled for polar effects or lack of functionality of the resulting protein.

4.5.2 *Campbell Mutagenesis Using the pKNOCK Suicide Vectors in* **Azospirillum**

- Clone an internal part (300–500 bp) of the gene to be mutagenized into the polylinker region of the pKNOCK vector of your choice (Table 4.2). In this type of mutagenesis, a single-homologous recombination event in the recipient *Azospirillum* strain results in the formation of two mutant copies of the target genes, which are separated by the inserted pKNOCK plasmid.
- Carry out colony PCR to verify correct insertion of the construct on the plasmid.
- Introduce the resulting pKNOCK derivative into the recipient parent *Azospirillum* strain by using triparental conjugation (Sect. 4.1.3 protocol above).
- Select the exconjugant colonies on the plates supplemented with ampicillin (200 μg/mL for *Azospirillum*) and the antibiotic for which resistance is provided by the inserted cassette present in pKNOCK vector. Incubate the plates at 28 °C for 3–5 days.

  Note that it may take more than 3–5 days for the colonies to appear at this stage.
- Once colonies appear on the plate, individually transfer a few colonies on minimal media MMAB plates supplemented with ampicillin (200 μg/mL) and

the antibiotic selecting for resistance present on the pKNOCK vector backbone, and incubate at 28 °C for 3–5 days.

- Repeat individual colony streaks three times on the same media until there is no evidence of remaining *E.coli* contamination.
- Confirm insertion of the specific antibiotic resistance cassette present in pKNOCK vector by PCR (using primer pairs which specifically bind to the antibiotic cassette present on pKNOCK vector) and DNA sequencing.
- Grow confirmed mutant strains in MMAB liquid and make glycerol stocks.

4.5.3 Marker Deletion-Insertion Mutagenesis by Allelic Exchange Using Sucrose Counter-Selection Protocol

This protocol is used to generate a double homologous recombination targeting a specific site on the genome of *A. brasilense* Sp7, using an engineered version of the pK18*mobsacB* suicide plasmid and a chloramphenicol resistance (Cm^R) cassette that replaces most of the coding sequence of the target gene before being exchanged with the parental gene copy on the genome. Chloramphenicol is used at 20 μg/ml (Cm20). A 50 μg/mL of kanamycin (Km50) is used to maintain the suicide plasmid in *E. coli*, while only 30 μg/mL kanamycin (Km30) is used to maintain the plasmid in *A. brasilense*.

- Design a pK18mobsacB plasmid targeting your gene/site of interest. Include at least 1 kb upstream and downstream of your target mutation site. Using splicing-by-overlap-extension PCR, connect each homologous region to the Cm^R cassette (Bryksin and Matsumura 2010) and clone it into the pK18mobsacB plasmid. Transform the construct into *E. coli* Top10.
- Carry out colony PCR to verify correct insertion of the construct on the plasmid.
- For 1–3 positive hits, patch *E. coli* cells onto LB (yeast extract 5 g/L, tryptone 10 g/L, and NaCl 10 g/L) plates with Km50 Cm20 and onto LB plates with Cm20 and 10% sucrose. Choose a hit that did not grow on sucrose and stock the strain (the *sacB* gene is prone to suppressor mutations, so it is crucial to confirm sensitivity of the strains harboring the plasmid prior to transfer into *Azospirillum*).
- Carry out a triparental conjugation (see protocol in Sect. 4.1.3), using the *E. coli* carrying the pK18mobsacB as the donor, your target *Azospirillum* as the acceptor, and the *E. coli* helper strain.
- Pick at least 10 individual colonies that result from the conjugation and patch onto TY (yeast extract 5 g/L, tryptone 10 g/L) Amp200 Km30 Cm20 and TY + 10% sucrose. These should be single-homologous recombinants, and they should grow on Amp200 Km30 Cm20 but not on sucrose. Once grown, keep the nonsucrose plate in the dark at room temperature in case you need to pick from additional patches later.
- Choose three colonies from the Amp200 Km30 Cm20 plates that are still sucrose sensitive and inoculate each into 5 ml liquid TY Cm20. Incubate at 28 °C with

shaking overnight. This is the first passage without selection for the plasmid backbone.

- Use 15 μL from each culture tube and streak onto TY Cm20 +/− 10% sucrose. Use 50 μL from these same tubes to inoculate 5 mL fresh TY Cm20. This is the second passage without selection for the plasmid backbone.
- Continue passing and plating each passage until colonies appear on the 10% sucrose plate (~3–5 days). Generally, this should only take one to three passages.
- Patch colonies from the 10% sucrose plate onto TY Cm20 +/− Km30. If the growth on sucrose is due to a suppressor mutation, the colony will still grow on Km30. If, however, the sucrose resistance (Suc^R) is due to a second homologous recombination event, the colony will be sensitive to Km30 (Km^S) and only grow on Cm20.
- For Km^S and Suc^R colonies, carry out colony PCR to verify the correct replacement of the parental copy with the mutated copy.
- Stock a verified colony as your new construct.

4.5.4 *β-Glucuronidase (GusA) Transcriptional Reporter Assay to Measure Transcriptional (Promoter) Activity in Liquid Culture*

This β-glucuronidase enzyme assay measures promoter activity by using p-nitrophenyl-β-D-glucuronide (pNPG) as a substrate, and the intensity of the yellow-colored product is quantified by using a spectrophotometer.

- Grow *Azospirillum* harboring recombinant plasmid pFUS1 (Table 4.2) in which the promoter region (600 BP-1Kb) of the gene of interest is cloned upstream of the promoter-less *gusA* gene in 5 mL TY (yeast extract 5 g/L, tryptone 10 g/L) supplemented with tetracycline at 10 μg/mL (Tet10), overnight at 28 °C on a shaker (200 rpm).

  Note: Use cloning of the promoter region up to two to five nucleotides before the start codon for transcriptional fusions and cloning of the promoter region up to two to three codons after the start codon for translational fusions.
- Next day, pellet down the cell culture by centrifuging at 5000 RPM for 3 min and wash the culture three times with chemotaxis buffer (10 mM phosphate buffer [pH 7.0], 1 mM EDTA) (Stephens et al. 2006). Resuspend in 5 mL liquid MMAB with Tetracycline 10 μg/mL and grow at 28 °C in a shaker incubator overnight.
- Next day, record the optical density at 600 nm (OD 600) and aliquote 125 μL of the cultures into a 96-well flat-bottom plate (Falcon 96-well, nontreated, flat-bottom microplate).
- Incubate the plate in a −80 °C freezer 20 min followed by incubation on a heat block at 57 °C for 20 min (repeat this freeze-thaw cycle for three times) to lyse the cells and finally cool down on ice for 5 min.

- Dispense 4 µl of p-nitrophenyl-β-D-glucuronide (17 mg/mL stock dissolved in sterile water) (ACROS Organics) to each well at room temperature and mix five times by pipetting.
- Immediately place the assay microplate in the microplate reader. Take the reaction measurements at 37 °C every 2 min for 1 h using an 800 TS absorbance reader with the Gen5 software and a 405-nm filter (BioTek™ ELx808™ Ultra Microplate Reader).
- These settings are optimal for β-D-glucuronide produced as a result of the β-glucuronidase activity on the p-nitrophenyl-β-D-glucuronide (pNPG) substrate.

  Note for trouble shooting: If the reaction mixture produces yellow color too fast then either the cell density could be reduced by diluting the cell culture or the concentration of p-nitrophenyl-β-D-glucuronide (pNPG) could be decreased.
- To estimate promoter activity, calculate the rate of fluorescence increase over time using the fluorescence data reads. This can be done by plotting fluorescence (in arbitrary units) against time. Identify a linear section of the graph to determine the rate of fluorescence increase (the slope of the curve), which provides the rate of fluorescence per second (RFU/sec).
- Next, subtract the RFU/sec value of the empty vector control from all sample values. Finally, normalize the RFU/sec by the cell density (OD 600) recorded to obtain RFU/sec/OD600.

4.6 Conclusions and Outlook

The tool kit for manipulating *Azospirillum* species is comprehensive, and the tools have been largely validated. However, additional tools for facile manipulation of the genome would further benefit the application of *Azospirillum* species in agriculture or remediation. Despite *Azospirillum* genomes harboring insertion sequences and prophages, the availability of phages for genetic manipulation of *Azospirillum* genome by generalized transduction is lacking, and a few phages have been characterized to date for this genus (Boyer et al. 2008). Similarly, a neutral site within the chromosome to allow for the stable integration and expression of parental or foreign genes of *Azospirillum* species is lacking as is a set of versatile inducible promoters. The beneficial effects that inoculation by diverse *Azospirillum* species offer for plant growth promotion, bioremediation, aquaculture, and biofuel production should stimulate efforts to expand and develop these tools to catapult the next generation of applications of *Azospirillum* species in the bioeconomy.

Acknowledgments This research was supported by a National Science Foundation grant NSF-MCB 2130556 (to GA). Any opinions, findings, conclusions, or recommendations expressed in this material are those of the authors and do not necessarily reflect the views of the National Science Foundation.

References

Alexeyev MF (1999) The pKNOCK series of broad-host-range mobilizable suicide vectors for gene knockout and targeted DNA insertion into the chromosome of gram-negative bacteria. Biotechniques 26(5):824–828. https://doi.org/10.2144/99265bm05

Allen LN, Hanson RS (1985) Construction of broad-host-range cosmid cloning vectors: identification of genes necessary for growth of *Methylobacterium organophilum* on methanol. J Bacteriol 161(3):955–962. https://doi.org/10.1128/jb.161.3.955-962.1985

Arsené F, Katupitiya S, Kennedy IR et al (1994) Use of *lacZ* fusions to study the expression of *nif* genes of *Azospirillum brasilense* in associations with plants. Mol Plant-Microbe Interact 7:748–757. https://doi.org/10.1094/MPMI-7-0748

Bloemberg GV, Wijfles AHM, Lamers et al (2000) Simultaneous imaging of *Pseudomonas fluorescens* WCS365 populations expressing three different autofluorescent proteins in the rhizosphere: new perspectives for studying microbial communities. Mol Plant-Microbe Interact 13:1170 1176. https://doi.org/10.1094/MPMI.2000.13.11.1170

Boyer M, Haurat J, Samain S et al (2008) Bacteriophage prevalence in the genus *Azospirillum* and analysis of the first genome sequence of an *Azospirillum brasilense* integrative phage. Appl Environ Microbiol 74:861–874. https://doi.org/10.1128/AEM.02099-07

Bryksin AV, Matsumura I (2010) Overlap extension PCR cloning: a simple and reliable way to create recombinant plasmids. Biotechniques 48(6):463–465. https://doi.org/10.2144/000113418

Cassán F, Coniglio A, López G et al (2020) Everything you must know about *Azospirillum* and its impact on agriculture and beyond. Biol Fertil Soils 56:461–479. https://doi.org/10.1007/s00374-020-01463-y

Cruz-Pérez JF, Lara-Oueilhe R, Marcos-Jiménez C et al (2021) Expression and function of the *cdgD* gene, encoding a CHASE–PAS-DGC-EAL domain protein, in *Azospirillum brasilense*. Sci Rep 11(1):520. https://doi.org/10.1038/s41598-020-80125-3

de Zamaroczy M, Paquelin A, Elmerich C (1993) Functional organization of the *glnB-glnA* cluster of *Azospirillum brasilense*. J Bacteriol 175(9):2507–2515. https://doi.org/10.1128/jb.175.9.2507-2515.1993

Del Gallo M, Gratani L, Morpurgo G (1987) *Azospirillum brasilense*, a nitrogen-fixing bacterium which lacks the SOS error-prone repair system. Mutat Res-Fund Mol M 178(2):195–200. https://doi.org/10.1016/0027-5107(87)90269-7

Dennis JJ, Zylstra JG (1998) Plasposons: modular self-cloning minitransposon derivatives for rapid genetic analysis of gram-negative bacterial genomes. Appl Environ Microbiol 64:2710–2715. https://doi.org/10.1128/AEM.64.7.2710-2715.1998

Ditta G, Stanfield S, Corbin D et al (1980) Broad host range DNA cloning system for gram-negative bacteria: construction of a gene bank of *Rhizobium meliloti*. Proc Natl Acad Sci USA 77(12):7347–7351. https://doi.org/10.1073/pnas.77.12.7347

Dubey AP, Pandey P, Singh VS et al (2020) An ECF41 family σ factor controls motility and biogenesis of lateral flagella in *Azospirillum brasilense* Sp245. J Bacteriol 202(16):e00231–e00220. https://doi.org/10.1128/JB.00231-20

Fani R, Bazzicalupo M, Coianiz P et al (1986) Plasmid transformation of *Azospirillum brasilense*. FEMS Microbiol Lett 35:23–27. https://doi.org/10.1111/j.1574-6968.1986.tb01492.x

Fani R, Bazzicalupo M, Ricci F et al (1988) A plasmid vector for the selection and study of transcription promoters in *Azospirillum brasilense*. FEMS Microbiol Lett 50(3):271–276. https://doi.org/10.1111/j.1574-6968.1988.tb02683.x

Figurski DH, Helinski DR (1979) Replication of an origin-containing derivative of plasmid RK2 dependent on a plasmid function provided in *trans*. Proc Natl Acad Sci 76:1648–1652. https://doi.org/10.1073/pnas.76.4.1648

Galimand M, Perroud B, Delorme F et al (1989) Identification of DNA regions homologous to nitrogen fixation genes *nifE*, *nifUS* and *fixABC* in *Azospirillum brasilense* Sp7. Microbiology 135(5):1047–1059. https://doi.org/10.1099/00221287-135-5-1047

Ganusova EE, Rost M, Aksenova A et al (2023) *Azospirillum brasilense* AerC and Tlp4b cytoplasmic chemoreceptors are promiscuous and interact with the two membrane-bound chemotaxis signaling clusters mediating chemotaxis responses. J Bacteriol 205(6):e00484–e00422. https://doi.org/10.1128/jb.00484-22

Gullett J, O'Neal L, Mukherjee T et al (2017) *Azospirillum brasilense*: laboratory maintenance and genetic manipulation. Curr Protoc Microbiol 47:3E.2.1–3E.2.17. https://doi.org/10.1002/cpmc.39

Hallez R, Letesson J-J, Vandenhaute J et al (2007) Gateway-based destination vectors for functional analyses of bacterial ORFeomes: application to the Min system in *Brucella abortus*. Appl Environ Microbiol 73(4):1375–1379. https://doi.org/10.1128/AEM.01873-06

Holguin G, Patten C, Glick B (1999) Genetics and molecular biology of *Azospirillum*. Biol Fertil Soils 29:10–23. https://doi.org/10.1007/s003740050519

Ishikawa M, Yokoe S, Kato S, Hori K (2018) Efficient counterselection for methylococcus capsulatus (bath) by using a mutated *pheS* gene. Appl Environ Microbiol 84:e01875-e01818. https://doi.org/10.1128/AEM.01875-18

Katzen F, Becker A, Ielmini MV, Oddo CG, Ielpi L (1999) New mobilizable vectors suitable for gene replacement in Gram-negative bacteria and their use in mapping of the 3' End of the *Xanthomonas campestris* pv. campestris *gum* Operon. Appl Environ Micro 65(1):278–282. https://doi.org/10.1128/AEM.65.1.278-282

Keen NT, Tamaki S, Kobayashi D, Trollinger D (1988) Improved broad-host-range plasmids for DNA cloning in Gram-negative bacteria. Gene 70(1):191–197. https://doi.org/10.1016/0378-1119(88)90117-5

Kolter R, Helinski DR (1978) Construction of plasmid R6K derivatives in vitro: characterization of the R6K replication region. Plasmid 4:571–580. https://doi.org/10.1016/0147-619x(78)90014-8

Kovach ME, Elzer PH, Hill DS, Robertson GT, Farris MA, Martin RM 2nd, Peterson KM (1995) Four new derivatives of the broad-host-range cloning vector pBBR1MCS, carrying different antibiotic-resistance cassettes. Gene 166(1):175–176. https://doi.org/10.1016/0378-1119(95)00584-1

Kwak Y, Shin JH (2016) First *Azospirillum* genome from aquatic environments: Whole-genome sequence of *Azospirillum thiophilum* BV-S(T), a novel diazotroph harboring a capacity of sulfur-chemolithotrophy from a sulfide spring. Mar Genomics 25:21–24. https://doi.org/10.1016/j.margen.2015.10.002

Labes M, Pühler A, Simon R (1990) A new family of RSF1010-derived expression and lac-fusion broad-host-range vectors for Gram-negative bacteria. Gene 89(1):37–46. https://doi.org/10.1016/0378-1119(90)90203-4

Larsen RA, Wilson MM, Guss AM, Metcalf WW (2002) Genetic analysis of pigment biosynthesis in Xanthobacter autotrophicus Py2 using a new, highly efficient transposon mutagenesis system that is functional in a wide variety of bacteria. Arch Microbiol 178(3):193–201. https://doi.org/10.1007/s00203-002-0442-2

Liang YY, Kaminski PA, Elmerich C (1991) Identification of a nifA-like regulatory gene of Azospirillum brasilense Sp7 expressed under conditions of nitrogen fixation and in the presence of air and ammonia. Mol Microbiol 5(11):2735–2744. https://doi.org/10.1111/j.1365-2958.1991.tb01982.x

Lopez-Reyes L, Soto-Urzua L, Mascarua-Esparza M A, Herrera-Camacho I, Caballero-Mellado J (1989) Antibiotic resistance and β-lactamase activity in *Azospirillum*. Soil Biol Biochem 21: 651–655. https://doi.org/10.1016/0038-0717(89)90059-X

Milcamps A, Van Dommelen A, Stigter J, Vanderleyden J, De Bruijn FJ (1996) The *Azospirillum brasilense* rpoN gene is involved in nitrogen fixation, nitrate assimilation, ammonium uptake, and flagellar biosynthesis. Can J Microbiol 42(5):467–478. https://doi.org/10.1139/m96-064

Mishra S, Singh C, Shanker K et al (2021) Characterization of carotenoids and genes encoding their biosynthetic pathways in *Azospirillum brasilense*. FEMS Microbiol Lett 368(5):fnab025. https://doi.org/10.1093/femsle/fnab025

Mukherjee T, Elmas M, Vo L et al (2019) Multiple CheY homologs control swimming reversals and transient pauses in *Azospirillum brasilense*. Biophys J 116(8):1527–1537. https://doi.org/10.1016/j.bpj.2019.03.006

Nunes CP, Passaglia LMP, Schrank A et al (2000) Directed mutagenesis affects recombination in *Azospirillum brasilense nif* genes. Genet Mol Biol 23:901–905. https://doi.org/10.1590/S1415-47572000000400032

O'Neal L, Akhter S, Alexandre G (2019) A PilZ-containing chemotaxis receptor mediates oxygen and wheat root sensing in *Azospirillum brasilense*. Front Microbiol 10:312. https://doi.org/10.3389/fmicb.2019.00312

Pandey P, Dubey AP, Mishra S et al (2022) β-lactam resistance in *Azospirillum baldaniorum* Sp245 is mediated by lytic transglycosylase and β-lactamase and regulated by a cascade of RpoE7→RpoH3 sigma factors. J Bacteriol 204(4):e00010–e00022. https://doi.org/10.1128/jb.00010-22

Pereg-Gerk L (2004) Expression of flcA, a gene regulating differentation and plant interaction in *Azospirillum*. Soil Biol Biochem 36:1245–1252. https://doi.org/10.1016/j.soilbio.2004.04.005

Prell J, Boesten B, Poole P et al (2002) The *Rhizobium leguminosarum* bv. viciae VF39 gamma-aminobutyrate (GABA) aminotransferase gene (*gabT*) is induced by GABA and highly expressed in bacteroids. Microbiology 148(3):615–623. https://doi.org/10.1099/00221287-148-3-615

Rai S, Singh VS, Gupta P et al (2025) Identification and functional characterization of a fructose-inducible phosphotransferase system *in Azospirillum brasilense* Sp7. Appl Environ Microbiol 91(2):e0082824. https://doi.org/10.1128/aem.00828-24

Ramirez-Mata A, Pacheco MR, Moreno SJ et al (2018) Versatile use of *Azospirillum brasilense* strains tagged with *egfp* and *mCherry* genes for the visualization of biofilms associated with wheat roots. Microbiol Res 215:155–163. https://doi.org/10.1016/j.micres.2018.07.007

Ramos HJ, Roncato-Maccari LD, Souza EM et al (2002) Monitoring *Azospirillum*-wheat interactions using the gfp and gusA genes constitutively expressed from a new broad-host range vector. J Biotechnol 97(3):243–252. https://doi.org/10.1016/s0168-1656(02)00108-6

Rayko M, Kravchenko I, Lapidus A (2024) Complete genome sequence of a novel *Azospirillum* strain TA isolated from Western Siberia Chernevaya Taiga Soil. Microorganisms 12(12):2599. https://doi.org/10.3390/microorganisms12122599

Reeve WG, Tiwari RP, Worsley PS et al (1999) Constructs for insertional mutagenesis, transcriptional signal localization and gene regulation studies in root nodule and other bacteria. Microbiology 145(Pt 6):1307–1316. https://doi.org/10.1099/13500872-145-6-1307

Rivera D, Revale S, Molina R et al (2014) Complete genome sequence of the model rhizosphere strain *Azospirillum brasilense* Az39, successfully applied in agriculture. Genome Announc 2(4):e00683–e00614. https://doi.org/10.1128/genomeA.00683-14

Sant'Anna FH, Andrade DS, Trentini DB et al (2011) Tools for genetic manipulation of the plant growth-promoting bacterium *Azospirillum amazonense*. BMC Microbiol 11(1):1–11. https://doi.org/10.1186/1471-2180-11-107

Schnabel T, Sattely E (2021) Engineering posttranslational regulation of glutamine synthetase for controllable ammonia production in the plant symbiont *Azospirillum brasilense*. Appl Environ Microbiol 87(14):582–603. https://doi.org/10.1128/AEM.00582-21

Simon R, Priefer U, Pühler A (1983) A broad host range mobilization system for in vivo genetic engineering: transposon mutagenesis in gram negative bacteria. Bio/Technology 1(9):784–791. https://doi.org/10.1038/nbt1183-784

Steenhoudt O, Keijers V, Okon Y et al (2001) Identification and characterization of a periplasmic nitrate reductase in *Azospirillum brasilense* Sp245. Arch Microbiol 175(5):344–352. https://doi.org/10.1007/s002030100271

Stephens BB, Loar SN, Alexandre G (2006) Role of CheB and CheR in the complex chemotactic and aerotactic pathway of *Azospirillum brasilense*. J Bacteriol 188:4759–4768. https://doi.org/10.1128/JB.00267-06

Uozumi T, Wang P, Tonouchi N et al (1986) Cloning and expression of the *nifA* gene of *Klebisella oxytoca* in *K. pneumoniae* and *Azospirillum lipoferum*. Agric Biol Chem 50:1539–1544

Vanbleu E, Marchal K, Lambrecht M et al (2004) Annotation of the pRhico plasmid of *Azospirillum brasilense* reveals its role in determining the outer surface composition. FEMS Microbiol Lett 232:165–172. https://doi.org/10.1016/S0378-1097(04)00046-1

Vande Broek A, Vanderleyden J (2000) Introduction of plasmids into *Azospirillum brasilense* by electroporation. In: Eynard N, Teissié J (eds) Electrotransformation of bacteria. Springer, Berlin/Heidelberg, pp 253–256. https://doi.org/10.1007/978-3-662-04305-9_31

Vande Broek A, Van Gool A, Vanderleyden J (1989) Electroporation of *Azospirillum brasilense* with plasmid DNA. FEMS Microbiol Lett 61:177–181. https://doi.org/10.1111/j.1574-6968.1989.tb03574.x

Vanstockem M, Michiels K, Vanderleyden J et al (1987) Transposon mutagenesis of *Azospirillum brasilense and Azospirillum lipoferum*: physical analysis of Tn5 and Tn5-mob insertion mutants. Appl Environ Microbiol 53(2):410–415. https://doi.org/10.1128/aem.53.2.410-415

Vial L, Lavire C, Mavingui P et al (2006) Phase variation and genomic architecture changes in *Azospirillum*. J Bacteriol 188:5364–5373. https://doi.org/10.1128/JB.00521-06

Whitman WB, Woyke T, Klenk HP (2015) Genomic encyclopedia of bacterial and archaeal type strains, phase III: the genomes of soil and plant-associated and newly described type strains. Stand Genomic Sci 10(26). https://doi.org/10.1186/s40793-015-0017-x

Wielbo J, Skorupska A (2001) Construction of improved vectors and cassettes containing gusA and antibiotic resistance genes for studies of transcriptional activity and bacterial localization. J Microbiol Methods 45(3):197–205. https://doi.org/10.1016/s0167-7012(01)00244-5

Wisniewski-Dyé F, Borziak K, Khalsa-Moyers G et al (2011) *Azospirillum* genomes reveal transition of bacteria from aquatic to terrestrial environments. PLoS Genet 7:e1002430. https://doi.org/10.1371/journal.pgen.1002430

Xie Z, Ulrich LE, Zhulin IB et al (2010) PAS domain containing chemoreceptor couples dynamic changes in metabolism with chemotaxis. Proc Natl Acad Sci USA 107(5):2235–2240. https://doi.org/10.1073/pnas.0910055107

Yu Z, Yang G, Liu X et al (2018) Complete genome sequence of the nitrogen-fixing bacterium *Azospirillum humicireducens* type strain SgZ-5(T). Stand Genomic Sci 13(28). https://doi.org/10.1186/s40793-018-0322-2

Part II
Mechanisms Associated with *Azospirillum* spp. Lifestyle and Stress Resistance

Chapter 5
Chemotaxis and Other Forms of Taxis Behaviors

Gladys Alexandre

Abstract Motility and chemotaxis are prerequisites for the ability of *Azospirillum* species to colonize the roots of diverse plants. In addition to chemotaxis, *Azospirillum* species have an oxidative metabolism adapted for microaerobic conditions. Aerotaxis represents a strong behavioral response of *Azospirillum* species that allow them to locate precise low oxygen concentrations in spatial gradients of oxygen/air. Chemotaxis signal transduction has been best characterized in *A. brasilense* strain Sp7. In this strain, two chemotaxis signaling systems and multiple accessory proteins control chemotaxis and aerotaxis. Chemotaxis responses in *A. brasilense* are tightly linked to metabolism and referred to as energy taxis. Both spatial and temporal gradient assays have been developed to measure chemotaxis and aerotaxis in *A. brasilense*, as well as in other *Azospirillum* species. The chemical-in-plug assays and its variations have been used to demonstrate *A. brasilense* chemotaxis attractant responses toward several organic compounds and plant root exudates as well as repellent responses away from redox active quinones or hydrogen peroxide. The root-in-pool assay represents yet another variation of the assay but uses plant roots as the source of chemoeffectors.

Keywords Chemotaxis · Motility · Aerotaxis · Energy taxis · Root colonization

5.1 Motility and Chemotaxis in *Azospirillum* spp.

5.1.1 Motility and Chemotaxis in Azospirillum

The role of motility and chemotaxis in the association of *Azospirillum* with the roots of plants was described soon after the characterization of these bacterial species as motile (Reinhold et al. 1985, Bashan and Holguin 1994, Vande Broek et al. 1998, Greer-Phillips et al. 2004, Mukherjee et al. 2016; O'Neal et al. 2019a, 2020;

G. Alexandre (✉)
Biochemistry & Cellular and Molecular Biology Department, The University of Tennessee, Knoxville, TN, USA
e-mail: galexan2@utk.edu

F. D. Cassán et al. (eds.), *The Definitive Handbook of Azospirillum*,
https://doi.org/10.1007/978-3-032-09222-9_5

Ganusova et al. 2024). Bacteria of the genus *Azospirillum* swim by means of single polar flagella located at one of the cells' poles. The flagellins comprising the polar flagella are glycosylated (Moens et al. 1995a). *Azospirillum* species also swarm using a different set of lateral flagella that are peritrichously organized around the cell body and are not glycosylated (Moens et al. 1995b, 1996). The polar flagellum propels cells under conditions of low viscosity, while the lateral flagella allow cells to move under conditions of increased viscosity (Moens et al. 1995b, 1996). Under laboratory conditions, swimming is typically observed using liquid media or with media solidified with low agar concentrations (0.2–0.3% w/vol), while swarming is observed on the surface of plates solidified with 0.5–0.6% w/vol of agar (Moens et al. 1995b, 1996; Ganusova et al. 2021). The lateral flagella are induced as viscosity increases, and the polar flagellum is not lost, even at high viscosity. The regulatory pathways controlling biosynthesis of lateral flagella have been partly elucidated (Dubey et al. 2020) but the pathways controlling the polar flagellum biosynthesis remain unknown. While swimming and swarming refer to different types of movement powered by different flagellar systems, when observed in soft agar plates, both types of flagella appear to be produced (Ganusova et al. 2021). Bacterial chemotaxis signaling controls the ability of cells to swim in gradients of various chemoeffectors, including those found in the rhizosphere, but its role in modulating swarming has not been investigated.

Motility is often inferred from the ability of inoculated cells (from a suspension or colony) to spread away from the inoculation point within the water-field spaces of soft agar plates. However, spreading in soft agar plates depends on the ability of cells to grow and metabolize the compounds present in the agar plates to produce concentration gradients and then to move along these gradients by chemotaxis. Therefore, the ability to spread on soft agar plates depends on growth, motility, and chemotaxis and are thus indirect indicators of motility and chemotaxis. Microscopy observations are the only direct way to observe motility.

5.1.2 Chemotaxis and Chemosensory Systems in **Azospirillum** *spp. Genomes*

Bacterial chemotaxis depends on dedicated signaling transduction systems that control the probability of changes in the swimming direction powered by the polar flagellum, also referred to as a swimming reversal frequency (Scharf et al. 2016). Chemotaxis signaling pathways are conserved in bacteria and minimally comprise a dedicated histidine kinase, named CheA, and a response regulator, CheY, as well as chemoreceptors (also known as *m*ethyl *a*ccepting *p*roteins or MCP or *t*ransducer-*l*ike *p*roteins or Tlp) and a scaffolding protein, CheW, that couples CheA with chemoreceptors (Wuichet and Zhulin 2010; Colin et al. 2021). Chemoreceptors are assembled as (often, membrane-bound) large arrays of trimers of chemoreceptor dimers that interact with CheA and CheW to form ultrastable higher-order structures that sense environmental signals and modulate the phosphorylation state of CheA (Hazelbauer et al. 2008; Briegel et al. 2009, 2012). Additional proteins that function to modify chemoreceptors and adjust sensory sensitivity (CheB and CheR) and accessory chemotaxis proteins are also encoded in many bacterial genomes (Briegel et al. 2009, 2012; Wuichet and Zhulin 2010; Colin et al. 2021). In response to repellents, CheA becomes autophosphorylated using ATP and then transfers its phosphoryl group to CheY, which, in the phosphorylated state, has a greater affinity for the flagellar motor. Phosphorylated CheY binding to the flagellar motor causes it to switch direction of rotation and leads the cells to reverse swimming direction (Colin et al. 2021). While most chemotaxis signaling pathways regulate changes in the swimming direction, some chemotaxis-like systems, also known as chemosensory signaling pathways, do not possess CheY homologs and regulate other cellular functions, such as type IV pili-mediated motility or twitching, biofilm formation, flagellum biosynthesis, or cyst production (Wuichet and Zhulin 2010; Gumerov et al. 2021).

Species of *Azospirillum* with complete genome sequences available indicate the presence of 3 to 6 chemotaxis and chemosensory systems with many chemoreceptors, ranging from a total of 47 (*A. formorense*) to 88 (*Azospirillum* sp. B510). The chemotaxis signaling systems and chemoreceptor repertoires of *Azospirillum* species are significantly larger than those found in other soil and plant-associated bacteria, which typically range from 20 to 30 chemoreceptors and 2 chemotaxis systems (Buchan et al. 2010; Scharf et al. 2016). Chemotaxis signaling is thus likely to be significant for the lifestyle of all *Azospirillum* species.

Chemotaxis signal transduction has been best characterized in *A. brasilense* strain Sp7. In this strain, two chemotaxis signaling systems (named Che1 and Che4) as well as chemotaxis response regulators (CheY6 and CheY7) encoded outside the genome-encoded chemotaxis signaling pathways control changes in the swimming direction, while the other two are chemosensory systems with functions unrelated to chemotaxis (Che2 and Che3) (Mukherjee et al. 2016, 2019; O'Neal et al. 2019b). Che3 may have a role in controlling flocculation (Bible et al. 2015), while the role of Che2 is unknown. In *A. brasilense* Sp7, Che1 controls the transient increase in swimming speed, while Che4 controls the probability of swimming reversals with both changes contributing to chemotaxis responses (Bible et al. 2008; Bible et al. 2012; Mukherjee et al. 2016). However, the role of Che4 is largely dominant in the control of chemotaxis responses in *A. brasilense* (Mukherjee et al. 2016). Complete genomes of *Azospirillum* species comprise homologs of Che1, Che2, and Che3, which are the likely ancestral chemotaxis and chemosensory signaling systems for the genus. Che4 is also universally present in *Azospirillum* species with 4, 5, and 6 chemotaxis and chemosensory systems (https://mistdb.com/). Chemotaxis and several chemoreceptors characterized to date in *A. brasilense* Sp7 collectively underscore the role of chemotaxis in attraction toward plant roots and subsequent colonization by the bacteria as well as for ensuring nitrogen fixation under free-living conditions (Vande Broek et al. 1998, Xie et al. 2010, Greer-Phillips et al. 2004, Mukherjee et al. 2016; O'Neal et al. 2019a, 2020, Ganusova et al. 2024, Nisha et al. 2024).

5.2 Nature of Chemotaxis Responses in *Azospirillum* spp.

Chemotaxis responses in *A. brasilense* have been well-characterized and shown to be tightly linked to metabolism (Zhulin et al. 1996; Alexandre et al. 2000). The strongest chemotaxis responses of *A. brasilense* are in gradients of air/oxygen, also known as aerotaxis, and toward organic acids, in particular, intermediates of the TCA cycle (Barak et al. 1982; Zhulin et al. 1996; Alexandre et al. 2000). Aerotaxis responses are so strong in *A. brasilense* that they can often mask responses to other chemoeffectors (Alexandre et al. 2000). Organic acids are abundant compounds in diverse plant root exudates and are likely contributing to the attraction of free-swimming *A. brasilense* cells to the roots of plants (O'Neal et al. 2020; Nisha et al. 2024). Starving cells enhance chemotaxis responses to a plethora of chemoeffectors in *A. brasilense* and the strongest attractants for this species are also the best growth substrates (Alexandre et al. 2000). Compounds that affect the activity of the electron transport system, such as electron donors (growth substrates), electron acceptors (oxygen, nitrate) as well as oxidized quinones that can interfere with electron transport through the electron transport chain are strong chemoeffectors for *A. brasilense* (Alexandre et al. 2000; Alexandre 2010). Most chemotaxis responses in *A. brasilense* are tightly coupled to energy metabolism and are collectively referred to as energy taxis responses. The interrelationship between chemotaxis and aerotaxis responses to metabolism is also seen in the sensory specificity and domain architecture of chemoreceptors characterized to date in this species. Several chemoreceptors (Aer and AerC) use FAD as a cofactor to monitor energy metabolism during nitrogen fixation or attraction to plant roots (Xie et al. 2010; O'Neal et al. 2019a). Other chemoreceptors possess a C-terminal PilZ domain that senses c-di-GMP levels, with c-di-GMP binding required for chemoreceptor sensory activity (Russell et al. 2013; O'Neal et al. 2019a). In *A. brasilense*, oxygen is a major regulator of cellular c-di-GMP levels (Russell et al. 2013). Regulating chemotaxis sensory activity of chemoreceptors as a function of intracellular c-di-GMP concentrations would thus optimize motility responses with the preferred microaerobic metabolism of *A. brasilense*.

A. brasilense can sense sugars such as fructose, ribose, arabinose, and galactose (Alexandre et al. 2000; Van Bastelaere et al. 1999; Rai et al. 2025). Fructose is a strong attractant for *A. brasilense,* and chemotaxis to fructose is further enhanced by the induction of a fructose-specific phosphotransferase system (Rai et al. 2025). In contrast to fructose, chemotaxis to galactose is inducible and depends on a periplasmic binding protein named SbpA that functions in both galactose metabolism and chemotaxis (Van Bastelaere et al. 1999; Alexandre et al. 2000).

A. brasilense is chemotactically attracted to amino acids such as glutamate, aspartate, alanine, glutamine, and asparagine, but these elicit only weak chemotaxis responses, while glycine, histidine, isoleucine, threonine, valine, arginine, lysine, and methionine are not chemoeffectors for *A. brasilense* (Alexandre et al. 2000; Okon et al. 1980).

5.3 Assays Suitable for Measuring Chemotaxis, Aerotaxis, and Other Taxis Responses in *Azospirillum* spp.

Chemotaxis is often assessed by inoculating colonies or concentrated cell cultures into medium solidified with low agar concentration (0.2–0.3% agar w/vol), but as indicated above, this is an indirect method to assess chemotaxis responses, and other assays, such as the chemical-in-plug assay and its variations or temporal gradient assays (see below) are much more reliable ways of assessing chemotaxis and aerotaxis responses in *Azospirillum* as well as in most other motile bacterial species.

Chemotaxis and aerotaxis correspond to the responses of motile cells in gradients of chemicals (chemotaxis) or air/oxygen (aerotaxis). Swimming bacteria sense gradients temporally as they move up or down concentration gradients. Therefore, the most direct chemotaxis assays probe the movement of cells toward or away from a chemoeffector using spatial or temporal gradients. In a spatial gradient assay, a chemoeffector is placed at a given concentration as a point source that diffuses in the surrounding medium, creating a spatial gradient of concentration. In a temporal gradient assay, free-swimming cells are challenged by the addition of a chemoeffector, and their swimming reversal frequency before and after the addition of the chemoeffector is determined using computerized motion analysis.

Chemotaxis in many bacterial species is determined semi-quantitatively using a capillary assay in which a capillary tube filled with a chemoeffector is placed in a pool of motile cells for a short time, and the number of cells that migrate up the capillary tube is then determined through cell counts as a proxy for chemotaxis (Adler 1973). The capillary assay cannot be used for *A. brasilense* or any other cells with strong aerotaxis responses because these mask the chemotaxis response measured, i.e., the same number of cells are counted as having moved into the capillaries regardless of chemoeffector or concentration (Barak et al. 1982; Alexandre et al. 2000; Okon et al. 1980).

5.3.1 Spatial Gradient Assays

5.3.1.1 Soft Agar Plate Assay

The soft agar plate assay consists of inoculating a colony or cell suspension into a soft agar plate in which the inoculated cells will grow by metabolizing carbon and nitrogen sources provided as nutrients, thereby depleting the local environment of nutrients and creating a concentration gradient which the cells, if able to chemotax, can sense and pursue by moving outward from the inoculation point. In that case, measuring the chemotaxis swim diameter a few days post inoculation may provide a semi-quantitative measure of the chemotaxis response, as long as there is no growth or motility defect. For *Azospirillum* species, minimal media solidified with 0.2–0.3% agar offer sufficient support for chemotaxis.

It is important to note that in some cases, it may be preferable to use Noble agar as the gelling agent for soft agar plates because commercial agar contains impurities, including unknown nitrogen sources, that can interfere with the chemotaxis responses. Solidifying agar plates with Noble agar produces softer gels compared to regular agar. Last, but not least, briefly boiling soft agar media prior to autoclaving is recommended to fully dissolve agar within the medium. Microwaving the soft agar medium may also produce heterogenous dissolution of the agar. Failure to do so may result in plates that do not solidify evenly.

5.3.1.2 Chemical-in-Plug Assay and Its Variations

This assay produces a spatial gradient of chemoeffectors to be tested by providing a point source of the chemoeffector and placing it within a pool of motile cells, followed by a short incubation that prevents cellular growth to ensure motility responses are observed. The chemoeffector solubilized with buffer (e.g., sterile 10 mM phosphate buffer) may be added to a 2% (w/vol) agar plug that is then placed in a plate containing a dense suspension of motile cells mixed in a soft agar medium (0.3% w/vol), also prepared with the same buffer. Sublethal concentrations of a protein synthesis inhibitor may be added to avoid growth (e.g., 7.5 μg/ml chloramphenicol for *A. brasilense*). Alternatively, the soft agar plate may lack a carbon source to minimize possible growth that would confound the results. The plates are incubated and observed every 20–30 min and to a maximum of a couple of hours until a ring of motile cells forms at some distance from the hard agar plug.

A notable advantage of this method is that both attractant and repellent responses can be detected and measured. By varying the concentrations of chemoeffector, one can establish dose-responses curves that provide an estimate of the strength of the chemotaxis response: strong chemoeffectors elicit chemotaxis responses at lower concentrations compared to weak chemoeffectors. The assay can be miniaturized and performed on a microscope glassslide. There are multiple variations of this method that mainly depend on how the point source of the chemoeffector is prepared and on whether suspension of free-swimming cells or motile cells mixed within a soft agar are used. For example, the chemoeffector may be added to sterile filter paper disks and placed on the surface of a soft agar plate containing motile cells or placed away from motile cells inoculated within a soft agar plate to observe their movement toward or away from the disk containing the chemoeffector to be tested. Variations of this assay consists of inoculating the cells in a soft agar plate lacking carbon and/or nitrogen sources and inoculating a concentrated chemoeffector source as a "spot" at some distance from the inoculated cells. The gradient of the chemoeffector tested will be established by diffusion, and the cells will preferentially move in the direction of the chemoeffector if it functions as an attractant. If it functions as a repellent, the cells will move preferentially away from it.

Chemical-in-plug assays and its variations have been used to demonstrate *A. brasilense* chemotaxis attractant responses toward several organic compounds (Okon et al. 1980; Alexandre et al. 2000; O'Neal et al. 2019a; Rai et al. 2025) and plant root exudates (Okon et al. 1980; O'Neal et al. 2020; Nisha et al. 2024) as well as to demonstrate a repellent response away from redox-active quinones (Alexandre et al. 2000) or hydrogen peroxide (O'Neal et al. 2019a, b).The root-in-pool assay represents yet another variation of the assay but uses plant roots as the source of chemoeffectors exposed to a pool of motile cells (O'Neal et al. 2019a, 2020).

5.3.1.3 Aerotaxis Assay

Aerotaxis describes the movement of motile bacteria in a gradient of air or oxygen. *Azospirillum* species have an oxidative metabolism adapted for microaerobic conditions. Their ability to locate precise low oxygen concentrations in spatial gradients of air is a feature that allows their enrichment during isolation. The precision of aerotaxis in *Azospirillum* results from the ability of the cells to sense both elevated and very low oxygen concentrations as repellents, while a low microaerobic concentration of oxygen preferred by the *Azospirillum* electron transport chain is a strong attractant (Zhulin et al. 1996; Alexandre et al. 2000). The precise aerotaxis response of *Azospirillum* results from the existence of multiple chemoreceptors that sense oxygen directly (e.g., globin-like domain-containing chemoreceptors) and indirectly (e.g., chemoreceptors using cofactors such as FAD that reflect the redox state of the electron transport chain) (https://mistdb.com).

5.3.2 Temporal Gradient Assay

Free-swimming bacteria monitor their environment using chemotaxis signaling by making temporal (and not spatial) comparisons about physicochemical conditions in their surroundings. In the presence of a repellent, the histidine kinase CheA is autophosphorylated from ATP and then transfers a phosphoryl group to CheY, which has a greater affinity for and will bind to the flagellar motor to switch its direction of rotation. In a gradient of an attractant, CheA does not get phosphorylated, so the flagellar motor does not switch direction of rotation. Bacterial chemotaxis is best measured by determining the propensity of cells to change swimming directions when they are challenged by the addition of a repellent or an attractant. The probability of changes in the swimming direction will increase when cells are challenged with a repellent compared to before the addition, and the inverse will occur when cells are challenged with an attractant. Determining the reversal frequency of a population of free-swimming cells requires the ability to record movies and then to analyze bacterial swimming tracks for changes in the swimming directions. Motion tracking analysis is typically developed for particular cell types because a cell's flagellar arrangement affects the turn angles that are detected by computerized motion analysis software and need to be manually adjusted. A computerized tracking software to analyze motility patterns of *Azospirillum* has been validated and published (Mukherjee et al. 2019). The quality of the video recording is paramount to ensure subsequent analysis: the cell suspension should not be too concentrated to visualize single cell tracks and the contrast of the cells relative to the background should be enhanced as best as possible. Computerized motion analysis can yield information on swimming speed, turning angles value and distribution, fraction of motile cells, etc. This information in turn can be used to determine the strength of an attractant or repellent by determining the response or adaptation time, i.e., the time it takes for 50% of cells to return to a pre-stimulus reversal frequency following a challenge with a chemoeffector. The longer it takes for the population to adapt to the stimulus, the stronger the chemoeffector is in eliciting chemotaxis responses.

5.4 Golden Protocols for Measuring Chemotaxis and Aerotaxis in *Azospirillum* spp.

5.4.1 Chemotaxis: Miniaturized Chemical-in-Plug Assay to Analyze Attractant (or Repellent) Responses

A miniaturized chemical-in-plug assay can be used to measure responses to organic compounds, root exudates, etc. (Fig. 5.1). To that end, a concentrated solution of the compound to be tested (e.g., stock solution at 100 mM for malate, succinate, fructose, etc.) is prepared in a buffer (e.g., 10 mM chemotaxis buffer: 1.7 g L^{-1} dipotassium phosphate, 1.36 g L^{-1} monopotassium phosphate, 0.1 mM EDTA) or in sterile water and then dissolved at the final concentration by mixing an aliquot of the stock solution with molten 1% low melting point agarose prepared in buffer or in sterile water. A control plug with buffer or sterile water only should be prepared in a similar manner in a different well. Ten microliter plugs are pipetted out and placed to solidify on a depression slide (microscopic "well" glass slides such as concavity microscope slides from ThermoFisher Scientific (category no. 1518006), and 150 mL of a suspension of motile bacteria are placed within the well of the depression slide and covered with a coverslip. Bacterial motility responses are monitored at low magnification using a phase-contrast microscope. Incubation of 5 min is sufficient to detect a response (Fig. 5.1). The motility responses can also be recorded over time for subsequent analysis. For example, individual frames or images can be extracted from a video using open-source software, such as FIJI (https://imagej.net/) (Schindelin et al. 2012). By using a range of concentrations of the chemoeffector to be tested, a dose-response curve can be used as a semi-quantitative estimation of the chemotaxis response. To increase the chemotaxis response, the cell suspension can be starved in a buffer (e.g., 10 mM phosphate buffer) for up to 48 h.

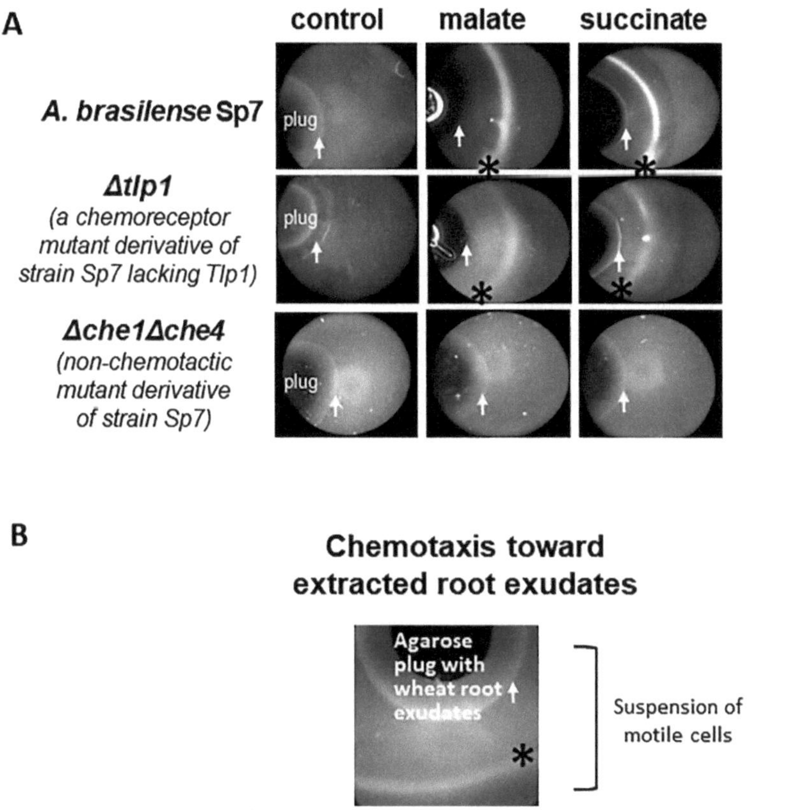

Fig. 5.1 Miniaturized chemical-in-plug assay to assess chemotaxis in *Azospirillum brasilense* Sp7. (**a**) Comparison of the chemotaxis responses of *A. brasilense* strain Sp7, a *tlp1* chemoreceptor mutant derivative (Δ*tlp1*) and a non-chemotactic mutant derivative (Δ*che1*Δ*che4*) to malate and succinate solubilized at 10 mM in chemotaxis buffer in the chemical-in-plug assay. The control corresponds to an agarose plug containing a chemotaxis buffer alone. A chemoreceptor mutant strain has an attenuated response, while a non-chemotactic mutant strain is fully motile but cannot navigate chemical gradients and thus does not form any chemotactic ring. (**b**) The miniaturized chemical-in-plug assay can be adapted to test chemotaxis of *Azospirillum* to plant root exudates, as shown here for wheat root exudates. The white arrow indicates the edge of the agarose plug. The star symbol indicates the position of the chemotactic ring or band that forms at some distance from the plug. See text for details

5.4.2 Aerotaxis

Aerotaxis can be best observed using the capillary assay for aerotaxis where a gradient of air is established by diffusion of air into a suspension of motile cells. In this assay, a culture of *A. brasilense* (typically in minimal medium) grown to mid-log phase is washed in chemotaxis buffer by centrifugation (5000 rpm for 3 min) and then concentrated ten times in the same buffer containing a carbon source and/or nitrogen source. Over 95% of cells remained motile under these conditions. Cells are transferred to an optically flat microcapillary tube (inner dimensions 0.1 × 2 × 50 mm, Vitro Dynamics, Inc., Rockaway, NJ) by immersing a capillary tube into the suspension of motile cells. The formation of the aerotactic band is visualized under a light microscope at low magnification. Formation of a stable aerotactic band is robust and occurs within 2–3 min for wild-type *A. brasilense* Sp7 when using malate and ammonium as carbon and nitrogen sources, respectively. Chemotaxis mutants affected in sensing of oxygen gradients or in control of swimming patterns through chemotaxis signaling may form aerotactic bands within reproducibly longer times, at different distances from the meniscus, and/or with aerotactic bands of different widths (Xie et al. 2010; Mukherjee et al. 2016; O'Neal et al. 2017, 2019a; Elmas et al. 2019; Ganusova et al. 2023). Therefore, measuring the time to stable band formation, the distance of the band relative to the air-liquid meniscus at mid-height of the capillary tube from (the surface of) the meniscus to the center of the band, and band width can provide some information on the strength of the aerotaxis response (Fig. 5.2). The formation of the band should be observed from the side of the capillary that was used to fill the tube. The distance from the meniscus and width of the aerotactic band can be quantified using the measure tool in FIJI in ImageJ (https://imagej.net/). The width of the capillary tube is used as a reference for all measurements. The distribution of bacteria in capillary tubes can also be analyzed using intensity image analysis from digital pictures, using surface plots in Fiji (https://imagej.net/) (Schindelin et al. 2012) after Gaussian adjustment.

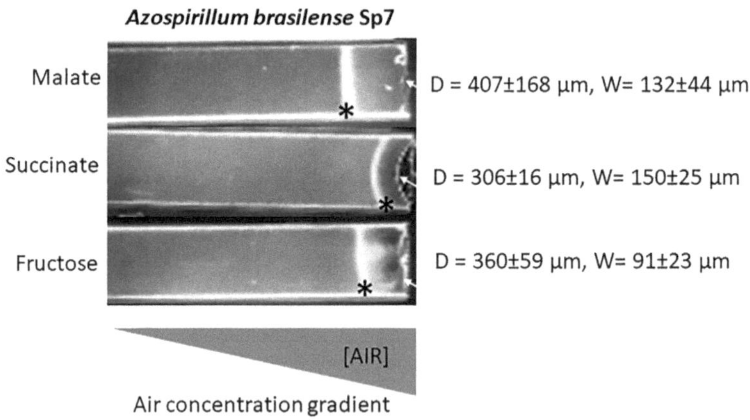

Fig. 5.2 Spatial gradient assay or capillary assay for aerotaxis. These assays were performed using optically flat capillary tubes filled with concentrated suspensions of motile cells to which 10 mM of a carbon source was added. The images were taken 3 min after the capillary tubes were filled with the cell suspensions. The aerotactic bands form at different distances relative to the air-liquid meniscus (indicated in mm as D on the right-hand side) and have different widths (indicated in mm as W on the right-hand side), despite the cell suspensions having similar density. These differences reflect distinct metabolisms on the different substrates tested here. The triangle at the bottom of the figure represents the direction of the air gradient from the air-liquid interface. The white arrow indicates the air-liquid interface. The star symbol indicates the position of the aerotactic band. See text for details

A bespoke gas flow chamber developed for measuring temporal responses of Escherichia coli to changes in ambient oxygen concentrations (Taylor et al. 2007) was used for *A. brasilense* to measure adaptation times to the addition or removal of air using conditions similar to that for *E. coli* and to standardize the diffusion of air into capillary tubes filled with motile cells. To measure free-swimming cells' responses to a temporal change in air, cells are grown, washed, and concentrated as for the capillary assay described above. 50–100 mL of the concentrated cell suspension is placed on a slide inside of the gas flow chamber. Air or humidified N_2 is allowed to flow over the cells for 3 min to let cells adapt to ambient conditions. After equilibration, the atmosphere of the cells is switched to N_2 or air, respectively, thereby producing the stimulatory temporal gradient of oxygen. Under these conditions, free-swimming cells suppress reversals during the transition from N_2 to air or air to N_2, since the lower oxygen conditions experienced by cells during these transitions are attractants for *A. brasilense.* Computerized motion analysis can be used to measure changes in swimming speed and reversal frequency before and after changes in the gas atmosphere of free-swimming cells (Alexandre et al. 2000; O'Neal et al. 2017; Mukherjee et al. 2019).

5.4.3 Root-in-Pool Assay

For the root-in-pool assay, seedlings of plant roots to be tested must be germinated under aseptic conditions. In our experience, germinated seedlings of 3–5 days old are best for the assay, but older plantlets could be used with some adaptation of the basic method (Fig. 5.3). A suspension of motile cells is prepared from cells grown, washed, and concentrated as for the capillary assay and is used as the pool of motile

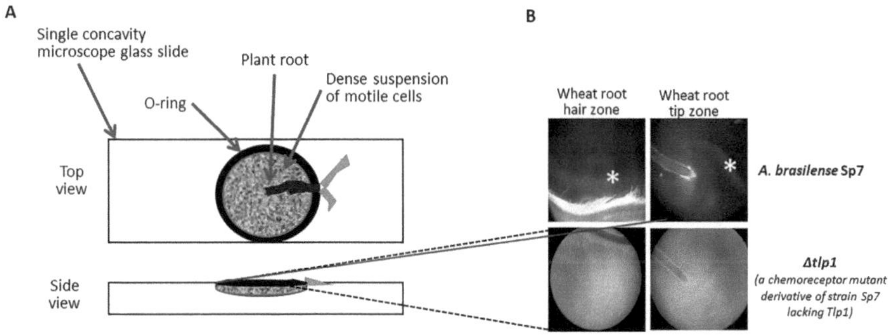

Fig. 5.3 Root in pool assay using 5-day-old wheat root plantlets. (**a**) Schematic representation of the setup for root-in-pool assay using a single concavity glass slide. (**b**) *A. brasilense* Sp7 forms a band at some distance from the wheat root hair zone and the root tip zone after one minute of incubation within the pool of motile cells. A *tlp1* chemoreceptor mutant derivative (Δ*tlp1*) lacks the ability to recognize the different root zones and is also impaired in the ability to colonize root surfaces, underscoring the relationship between chemotaxis toward root surfaces and colonization (Greer et al. 2004; O'Neal et al. 2020). See text for details

cells. A black washer with a slit cut in it is secured to a depression glass slide (e.g., Thermofischer Scientific no. 1518006, which has a concave depression with 5 (bottom)–18 mm (top) diameter well and is 0.6–0.8 mm deep) using vacuum grease. The root(s) of the germinated plantlet is placed in the washer slit, and 400 μL of motile cells is used to fill the chamber created by the dip and washer, and then a coverslip is used to cover the chamber containing motile cells and roots. The motility response of the dense bacterial suspension is monitored through video recording or time lapses for 3–5 min. Image analysis includes converting the root-in-pool assay videos to 8-bit and performing background subtraction applying a Gaussian blur filter using FIJI (https://imagej.net/) (Schindelin et al. 2012). Chemotaxis mutants display an altered response in these assays (Fig. 5.3). Intensity plots can be generated out from the roots to determine bacterial accumulation over time (e.g., every 30 s for 30 min). The average accumulation distance of bacteria exposed to different plant roots can be used to evaluate the chemotaxis responses.

5.5 Conclusions and Outlook

This chapter focused on presenting assays that may be most easily implemented while providing reasonable semi-quantitative estimation of the strengths of the chemotaxis or aerotaxis responses. The temporal assay for chemotaxis or aerotaxis using free-swimming cells and computerized motion analysis is the most specific and precise assay to determine chemotaxis responses in any bacteria. However, the assay is challenging to implement because the computerized motion analysis requires using or adopting algorithms and optimization of codes and criteria for detecting swimming angles. Recent development of microfluidic devices to measure chemotaxis responses in *Azospirillum* is noteworthy (Greipel et al. 2024) because the approach could offer high throughput and quantitative methods to systematically characterize and compare chemotaxis responses of *Azospirillum* species. Similarly, digital holography microscopy is a rising approach to characterize the 3D movement of cells and, as it becomes more widely available, would facilitate the characterization of free-swimming cell tracks in diverse bacterial species (Matthews et al. 2024). Chemotaxis responses of *Azospirillum* are relevant to establishment and maintenance in the rhizosphere but chemotaxis responses have only been characterized in a few strains. The broad range of plant hosts that benefit from *Azospirillum* inoculation suggests that these bacteria also have broad chemotaxis toward a range of plant root exudates. This assumption is supported by recent findings (Nisha et al. 2024) and underscores the need for continuing to characterize the role of chemotaxis in *Azospirillum*-host interactions.

Acknowledgments This research was supported by a National Science Foundation grant NSF-MCB 2130556 (to GA). Any opinions, findings, conclusions, or recommendations expressed in this material are those of the authors and do not necessarily reflect the views of the National Science Foundation.

References

Adler J (1973) A method for measuring chemotaxis and use of the method to determine optimum conditions for chemotaxis by *Escherichia coli*. Microbiology 74(1):77–91. https://doi.org/10.1099/00221287-74-1-77

Alexandre G (2010) Coupling metabolism and chemotaxis-dependent behaviours by energy taxis receptors. Microbiology 156:2283–2293. https://doi.org/10.1099/mic.0.039214-0

Alexandre G, Greer SE, Zhulin IB (2000) Energy taxis is the dominant behavior in *Azospirillum brasilense*. J Bacteriol 182:6042–6048. https://doi.org/10.1128/JB.182.21.6042-6048.2000

Barak R, Nur I, Okon Y et al (1982) Aerotactic response of *Azospirillum brasilense*. J Bacteriol 136(152):643–649. https://doi.org/10.1128/jb.152.2.643-649.1982

Bashan Y, Holguin G (1994) Root-to-root travel of the beneficial bacterium *Azospirillum brasilense*. Appl Environ Microbiol 60:2120–2131

Bible AN, Stephens BB, Ortega DR et al (2008) Function of a chemotaxis-like signal transduction pathway in modulating motility, cell clumping, and cell length in the alphaproteobacterium *Azospirillum brasilense*. J Bacteriol 190:6365–6375. https://doi.org/10.1128/JB.00734-08

Bible A, Russell MH, Alexandre G (2012) The *Azospirillum brasilense* Che1 chemotaxis pathway controls swimming velocity, which affects transient cell-to-cell clumping. J Bacteriol 194: 3343–3355. https://doi.org/10.1128/JB.00310-12

Bible AN, Khalsa-Moyers GK, Mukherjee T et al (2015) Metabolic adaptations of *Azospirillum brasilense* to oxygen stress by cell-to-cell clumping and flocculation. Appl Environ Microbiol 81(24):8346–8357. https://doi.org/10.1128/AEM.02782-15

Briegel A, Ortega DR, Tocheva EI et al (2009) Universal architecture of bacterial chemoreceptor arrays. Proc Natl Acad Sci USA 106:17181–17186. https://doi.org/10.1073/pnas.0905181106

Briegel A, Li X, Bilwes AM et al (2012) Bacterial chemoreceptor arrays are hexagonally packed trimers of receptor dimers networked by rings of kinase and coupling proteins. Proc Natl Acad Sci USA 109:3766–3771. https://doi.org/10.1073/pnas.1115719109

Buchan A, Crombie B, Alexandre GM (2010) Temporal dynamics and genetic diversity of chemotactic-competent microbial populations in the rhizosphere. Environ Microbiol 12(12): 3171–3184. https://doi.org/10.1111/j.1462-2920.2010.02290.x

Colin R, Ni B, Laganenka L et al (2021) Multiple functions of flagellar motility and chemotaxis in bacterial physiology. FEMS Microbiol Rev 45(6):fuab038. https://doi.org/10.1093/femsre/fuab038

Dubey AP, Pandey P, Singh VS et al (2020) An ECF41 family σ factor controls motility and biogenesis of lateral flagella in *Azospirillum brasilense* Sp245. J Bacteriol 202. https://doi.org/10.1128/jb.00231-20

Elmas M, Alexiades V, O'Neal L et al (2019) Modeling aerotaxis band formation in *Azospirillum brasilense*. BMC Microbiol 19(1):101. https://doi.org/10.1186/s12866-019-1468-9

Ganusova EE, Vo LT, Mukherjee T et al (2021) Multiple CheY proteins control surface-associated lifestyles of *Azospirillum brasilense*. Front Microbiol 22(12):664826. https://doi.org/10.3389/fmicb.2021.664826

Ganusova EE, Rost M, Aksenova A et al (2023) *Azospirillum brasilense* AerC and Tlp4b cytoplasmic chemoreceptors are promiscuous and interact with the two membrane-bound chemotaxis signaling clusters mediating chemotaxis responses. J Bacteriol 205(6):e0048422. https://doi.org/10.1128/jb.00484-22

Ganusova EE, Russell MH, Patel S et al (2024) An *Azospirillum brasilense* chemoreceptor that mediates nitrate chemotaxis has conditional roles in the colonization of plant roots. Appl Environ Microbiol 90(6):e0076024. https://doi.org/10.1128/aem.00760-24

Greer-Phillips SE, Stephens BB, Alexandre G (2004) An energy taxis transducer promotes root colonization by *Azospirillum brasilense*. J Bacteriol 186:6595–6604. https://doi.org/10.1128/JB.186.19.6595-6604.2004

Greipel E, Nagy K, Csákvári E et al (2024) Chemotactic interactions of *Scenedesmus* sp. and *Azospirillum brasilense* investigated by microfluidic methods. Microb Ecol 87:52. https://doi.org/10.1007/s00248-024-02366-3
Gumerov VM, Andrianova EP, Zhulin IB (2021) Diversity of bacterial chemosensory systems. Curr Opin Microbiol 61:42–50. https://doi.org/10.1016/j.mib.2021.01.007
Hazelbauer GL, Falke JJ, Parkinson JS (2008) Bacterial chemoreceptors: high-performance signaling in networked arrays. Trends Biochem Sci 33:9–19. https://doi.org/10.1016/j.tibs.2007.11.008
Matthews SA, Coelho C, Rodriguez Salas EE et al (2024) Real-time 3D tracking of swimming microbes using digital holographic microscopy and deep learning. PLoS One 19(4):e0301182. https://doi.org/10.1371/journal.pone.0301182
Moens S, Michiels K, Vanderleyden J (1995a) Glycosylation of the flagellin of the polar flagellum of *Azospirillum brasilense*, a gram-negative nitrogen-fixing bacterium. Microbiology 141(10): 2651–2657. https://doi.org/10.1099/13500872-141-10-2651
Moens S, Michiels K, Keijers V et al (1995b) Cloning, sequencing, and phenotypic analysis of laf1, encoding the flagellin of the lateral flagella of *Azospirillum brasilense* Sp7. J Bacteriol 177: 5419–5426. https://doi.org/10.1128/jb.177.19.5419-5426.1995
Moens S, Schloter M, Vanderleyden J (1996) Expression of the structural gene, laf1, encoding the flagellin of the lateral flagella in *Azospirillum brasilense* Sp7. J Bacteriol 178:5017–5019. https://doi.org/10.1128/jb.178.17.5017-5019.1996
Mukherjee T, Kumar D, Burriss N et al (2016) Azospirillum brasilense chemotaxis depends on two signaling pathways regulating distinct motility parameters. J Bacteriol 198(12):1764–1772. https://doi.org/10.1128/JB.00020-16
Mukherjee T, Elmas M, Vo L et al (2019) Multiple CheY homologs control swimming reversals and transient pauses in *Azospirillum brasilense*. Biophys J 116(8):1527–1537. https://doi.org/10.1016/j.bpj.2019.03.029
Nisha FA, Tagoe JNA, Pease AB et al (2024) Plant seedlings of peas, tomatoes, and cucumbers exude compounds that are needed for growth and chemoattraction of *Rhizobium leguminosarum* bv. *viciae* 3841 and *Azospirillum brasilense* Sp7. Can J Microbiol 70(5):150–162. https://doi.org/10.1139/cjm-2023-0217
Okon Y, Cakmakci L, Nur I et al (1980) Aerotaxis and chemotaxis of *Azospirillum brasilense*: a note. Microb Ecol 6:277–280. https://doi.org/10.1007/BF02010393
O'Neal L, Ryu M, Gomelsky M et al (2017) Optogenetic manipulation of cyclic Di-GMP (c-di-GMP) levels reveals the role of c-di-GMP in regulating aerotaxis receptor activity in *Azospirillum brasilense*. J Bacteriol 199:10. https://doi.org/10.1128/jb.00020-17
O'Neal L, Akhter S, Alexandre G (2019a) A PilZ-containing chemotaxis receptor mediates oxygen and wheat root sensing in *Azospirillum brasilense*. Front Microbiol 10:312. https://doi.org/10.3389/fmicb.2019.00312
O'Neal L, Gullett JM, Aksenova A et al (2019b) Distinct chemotaxis protein paralogs assemble into chemoreceptor signaling arrays to coordinate signaling output. MBio 10(5):e01757–e01719. https://doi.org/10.1128/mBio.01757-19
O'Neal L, Vo L, Alexandre G (2020) Specific root exudate compounds sensed by dedicated chemoreceptors shape *Azospirillum brasilense* chemotaxis in the rhizosphere. Appl Environ Microbiol 86(15):e01026–e01020. https://doi.org/10.1128/AEM.01026-20
Rai S, Singh VS, Gupta P et al (2025) Identification and functional characterization of a fructose-inducible phosphotransferase system in *Azospirillum brasilense* Sp7. Appl Environ Microbiol 91:e00828–e00824. https://doi.org/10.1128/aem.00828-24
Reinhold B, Hurek T, Fendrik I (1985) Strain-specific chemotaxis of *Azospirillum* spp. J Bacteriol 162:190–195. https://doi.org/10.1007/BF00263169
Russell MH, Bible AN, Fang X et al (2013) Integration of the second messenger c-di-GMP into the chemotactic signaling pathway. MBio 4:e00001–e00013. https://doi.org/10.1128/mBio.00001-13

Scharf BE, Hynes MF, Alexandre GM (2016) Chemotaxis signaling systems in model beneficial plant-bacteria associations. Plant Mol Biol 90(6):549–559. https://doi.org/10.1007/s11103-016-0432-4

Schindelin J, Arganda-Carreras I, Frise E et al (2012) Fiji: an open-source platform for biological-image analysis. Nat Methods 9(7):676–682. https://doi.org/10.1038/nmeth.2019

Taylor BL, Watts KJ, Johnson MJ (2007) Oxygen and redox sensing by two-component systems that regulate behavioral responses: behavioral assays and structural studies of Aer using in vivo disulfide cross-linking. Methods Enzymol 422:190–232. https://doi.org/10.1016/S0076-6879(06)22010-X

Van Bastelaere E, Lambrecht M, Vermeiren H et al (1999) Characterization of a sugar-binding protein from *Azospirillum brasilense* mediating chemotaxis to and uptake of sugars. Mol Microbiol 32:703–714. https://doi.org/10.1046/j.1365-2958.1999.01384.x

Vande Broek A, Lambrecht M, Vanderleyden J (1998) Bacterial chemotactic motility is important for the initiation of wheat root colonization by *Azospirillum brasilense*. Microbiology 144:2599–2606. https://doi.org/10.1099/00221287-144-9-2599

Wuichet K, Zhulin IB (2010) Origins and diversification of a complex signal transduction system in prokaryotes. Sci Signal 3ol:ra50. https://doi.org/10.1126/scisignal.2001180

Xie Z, Ulrich LE, Zhulin IB et al (2010) PAS domain containing chemoreceptor couples dynamic changes in metabolism with chemotaxis. Proc Natl Acad Sci USA 107:2235–2240. https://doi.org/10.1073/pnas.0910055107

Zhulin IB, Bespalov VA, Johnson MS et al (1996) Oxygen taxis and proton motive force in *Azospirillum brasilense*. J Bacteriol 178:5199–5204. https://doi.org/10.1128/jb.178.17.5199-5204.1996

Chapter 6
Azospirillum Biofilms, Aggregation, Attachment, and Colonization of Plants

Rebeca Pérez, Alberto Ramírez Mata, Melina Amenta, Luciana Pagnussat, Julia García, Mariana Puente, and Guillermo A. Maroniche

Abstract The success of *Azospirillum* in the rhizosphere is attributed to its chemotactic response to plant signals and nutrients, as well as its capacity to undergo morphological changes, form cellular aggregates, and produce extracellular polymeric substances, which contribute to robust biofilm formation. The biofilm matrix confers *Azospirillum* cells with protection against antibacterial agents released by competing microorganisms and abiotic stress while facilitating root colonization to establish a mutually beneficial association with the host plant. This chapter compiles numerous studies describing the composition of *Azospirillum* biofilms, their regulatory mechanisms, and their role in interactions with other microorganisms. It also addresses the processes of flocculation and aggregation, highlighting their importance in the development of effective microbial inoculants. Furthermore, the chapter provides a detailed overview of attachment and root colonization mechanisms, as well as the role of colonization in plant biostimulation. Finally, standardized and

R. Pérez
Núcleo de Fixação Biológica de Nitrogênio, Departamento de Bioquímica e Biologia Molecular, Universidade Federal do Paraná, Curitiba, Brasil

A. R. Mata
Centro de Investigaciones en Ciencias Microbiológicas, Benemérita Universidad Autónoma de Puebla, Puebla, México

M. Amenta
Facultad de Ciencias Agrarias, Universidad Nacional de Mar del Plata (UNMdP), Balcarce, Buenos Aires, Argentina

L. Pagnussat · G. A. Maroniche (✉)
Facultad de Ciencias Agrarias, Universidad Nacional de Mar del Plata (UNMdP), Balcarce, Buenos Aires, Argentina

Consejo Nacional de Investigaciones Científicas y Técnicas (CONICET), CABA, Buenos Aires, Argentina
e-mail: gmaroniche@mdp.edu.ar

J. García · M. Puente
Instituto de Microbiología y Zoología Agrícola, Instituto Nacional de Tecnología Agropecuaria (INTA), Hurlingham, Buenos Aires, Argentina

F. D. Cassán et al. (eds.), *The Definitive Handbook of Azospirillum*,
https://doi.org/10.1007/978-3-032-09222-9_6

detailed protocols for the assessment of biofilm formation, flocculation, surface attachment, and root colonization by *Azospirillum* are provided.

Keywords Biofilms · Aggregation · Attachment · Colonization · Rhizosphere

6.1 Introduction

The rhizosphere represents a zone of intense microbial activity where critical plant-microbe interactions unfold. This region serves as a selective niche that profoundly influences nutrient acquisition, plant health, and overall soil fertility. For rhizobacteria, including *Azospirillum* species, successful colonization of plant tissues and the subsequent exertion of beneficial effects demand sophisticated physiological adaptations to navigate this competitive environment. A pivotal determinant of their efficacy is the capacity to secrete extracellular polymeric substances (EPS) that not only protect *Azospirillum* against different kinds of stresses but are also the building blocks required to form biofilms. The study of *Azospirillum* interaction with plants has undergone a remarkable evolution, transitioning from initial observations of growth promotion to a detailed understanding of how these bacteria attach, form biofilms, and colonize roots. This journey has been propelled by key contributions from various research groups worldwide.

Early investigations in the late 1970s, pioneered by researchers like Johanna Döbereiner (Embrapa, Brazil) and Yaacov Okon (The Hebrew University of Jerusalem, Israel), focused on *Azospirillum*'s nitrogen-fixing abilities and its association with cereal crops (Döbereiner et al. 1976; Okon et al. 1976). Researchers began to recognize the bacterium's ability to adhere to root surfaces and its tendency to aggregate, or flocculate, as well as the critical role of EPS in *Azospirillum* interactions with plant tissues.

The 1990s witnessed a surge in biofilm research across diverse microbial domains, profoundly impacting *Azospirillum* studies. Scientists began to explicitly investigate biofilm formation on plant roots. For example, the group led by Yoav Bashan at the Center for Biological Research in Mexico utilized advanced microscopy to visualize *Azospirillum* on root surfaces (Bashan and Holguin 1993; Bashan et al. 1991; Burdman et al. 2000); Jos Vanderleyden's team at Katholieke Universiteit Leuven, Belgium, studied the attachment mechanism and specific molecules involved (Croes et al. 1993; Michiels et al. 1991), while Saul Burdman's lab from the Hebrew University of Jerusalem made major contributions to the essential role of EPS in this process (Burdman et al. 1998, 2000, 2001). Later, molecular biology techniques provided deeper insights into the genetic mechanisms underlying *Azospirillum* biofilm formation, such as the identification of genes involved in EPS biosynthesis and cell adherence by the labs led by Lily Pereg at Institut Pasteur, Paris, France, and Edgardo Jofré at Universidad Nacional de Río Cuarto, Argentina (Pereg 2015; Rossi et al. 2016).

In recent years, research has embraced a more systems-level approach, integrating advanced "omics" data to comprehensively understand *Azospirillum* biofilm development, its intricate regulatory networks, and its role in plant colonization and interaction with other microorganisms. Teams from the the Center for Research in Microbiological Sciences at the Benemérita Universidad Autónoma de Puebla in Mexico, along with the Plant and Microbial Biochemistry Laboratory of Mar del Plata National University in Balcarce, Argentina, made significant contributions in this field (Arruebarrena Di Palma et al. 2013; Diaz et al. 2022; Maroniche et al. 2018; Pagnussat et al. 2016; Ramírez-Mata et al. 2016, 2018; Viruega-Góngora et al. 2020), as did De Souza, Pedrosa, and their collaborators from the Department of Biochemistry and Molecular Biology (Universidade Federal do Paraná, Brazil) through their application of NGS techniques (Camilios-Neto et al. 2014).

This chapter aims to provide an updated perspective on *Azospirillum* aggregation, biofilm formation, and colonization of plant tissues, highlighting key advancements that pave the way for future research directions.

6.2 Biofilms

6.2.1 *Overview of* Azospirillum *Biofilm Biology*

The rhizospheric and endophytic lifestyles of *Azospirillum* are governed by multiple mechanisms, which facilitate efficient colonization of the host plant. One of the key processes in this regard is the production of biofilms. Biofilm is the major lifestyle for most bacteria in the environment (Costerton et al. 1995). A biofilm consists of a community of microorganisms, either of a single or multiple species, encased in a self-produced polymeric matrix and adhered to biotic or abiotic surfaces. *Azospirillum* strains have been demonstrated to form biofilms on both abiotic (glass, plastic, bioplastics) and biotic (plant roots) substrates (Fig. 6.1) (Arruebarrena Di Palma et al. 2013; González-Torres et al. 2020; Ramirez-Mata et al. 2018; Shelud'ko et al., 2025; Wang et al. 2017). However, the dynamics of *Azospirillum* biofilm formation have so far focused solely on abiotic substrates using static culture systems. The initial stages of biofilm development (Days 1–2) are characterized by numerous bacterial cells expressing long polar flagella, which interconnect to form intricate networks. This phase is subsequently followed by cellular proliferation and the emergence of microcolonies, culminating in the establishment of a mature, three-dimensional biofilm architecture by Day 4. It is also noteworthy that biofilm dispersal is initiated in response to nutrient depletion (Day 7) to begin the cycle again (Viruega-Góngora et al. 2020).

6.2.2 *Components of* Azospirillum *Biofilm Matrix*

While the exact composition of a biofilm varies, it is predominantly water (approximately 97%) (Rather et al. 2021). The matrix itself is a complex mixture of EPS,

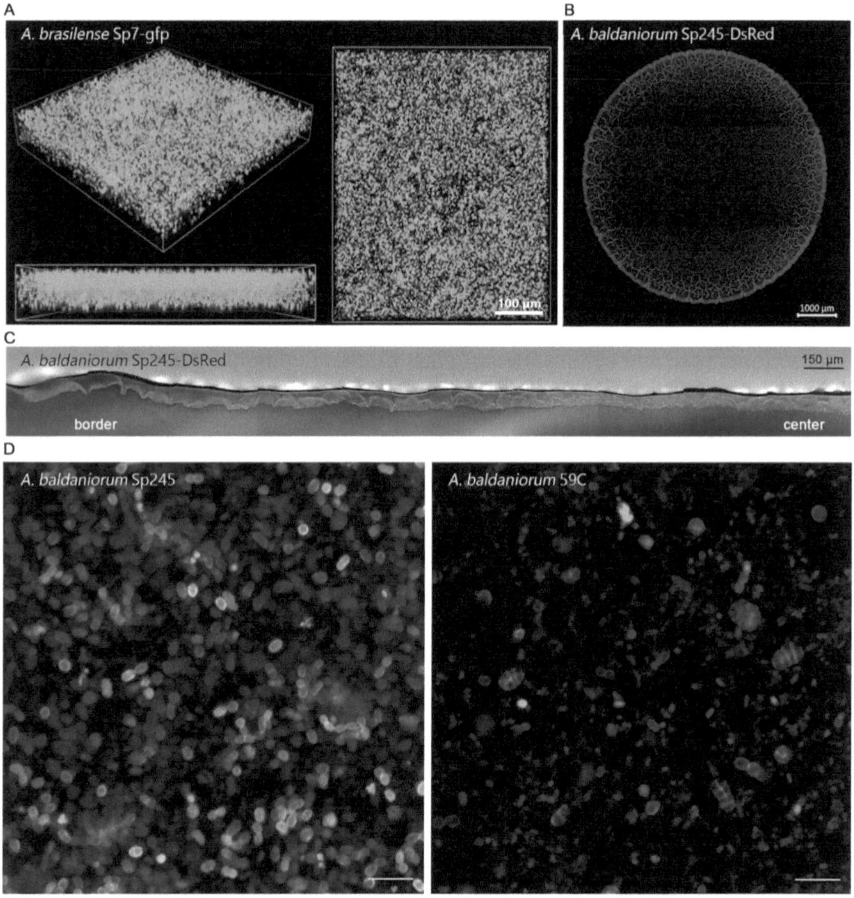

Fig. 6.1 Biofilms formed in vitro by *Azospirillum*. (**a**) Static culture biofilm of *A. brasilense* Sp7 expressing EGFP (Viruega-Góngora et al. 2020). Tangential (right), transversal (left, down), and 3D reconstruction (left, up) are shown. (**b**, **c**) Macrocolony biofilm of *A. baldaniorum* Sp245 expressing DsRed (Diaz et al. 2022). Tangential (**b**) and transversal (**c**) views are shown. (**d**) Static culture biofilms of EGFP-tagged *A. baldaniorum* Sp245 (left) and the isogenic *ΔcdgC* mutant strain 59C (right), dyed with calcofluor CWC (blue). Bars indicate 25 µm

attachment structures like fimbriae and pili, adhesion proteins (amyloids), EPS, extracellular DNA (eDNA), lipids, and cellular debris. This matrix features channels that ensure the transport of essential resources like water, nutrients, and oxygen (O_2) throughout the biofilm, even to its interior region. Consequently, biofilms exhibit diverse microenvironments with varying nutrients, pH, and O_2 levels (Böhning et al. 2024; Branda et al. 2005; Costerton et al. 1995; Davey and O'Toole 2000).

6.2.2.1 Extracellular Polysaccharides

Calcofluor-binding EPS and capsular polysaccharides (CPs) have been observed in flocs, suggesting their involvement in flocculation and biofilm development (Del Gallo et al. 1989; Sadasivan and Neyra 1985). Compositional analyses of *Azospirillum* EPS and CPs have revealed the presence of a diverse range of monosaccharides, including galactose, glucose, fucose, rhamnose, xylose, arabinose, galacturonic acid, galactosamine, and glucosamine. The relative abundance of these monosaccharides varies depending on the species and strain of *Azospirillum*, growth conditions, and physiological state (De Troch and Vanderleyden 1996). For example, *Azospirillum* bacterial cells cultured in media with a high carbon:nitrogen (C:N) ratio exhibit a propensity for biofilm production (Shelud'ko et al. 2020, 2025; Viruega-Góngora et al. 2020; Wang et al. 2017).

6.2.2.2 Extracellular Proteins

Several cell surface proteins have been implicated in *Azospirillum* biofilm formation, including polar flagellin, Tad pilus, OmaA protein, and amyloid proteins. The critical role of the *Azospirillum* polar flagellum in cell adhesion to wheat roots was established in earlier studies (Croes et al. 1993). Furthermore, the polar flagellum significantly contributes to substrate adhesion and cell-cell aggregation during biofilm formation on abiotic surfaces. These observations were supported by immunofluorescence microscopy analyses utilizing flagellin antibodies under biofilm conditions (Viruega-Góngora et al. 2020) and by atomic force microscopy of a non-flagellated *Azospirillum* mutant strain. While this mutant retained the capacity to form biofilms, the resulting structures exhibited a significant reduction in thickness compared to the wild-type strain (Shelud'ko et al. 2015). These findings suggest that EPS and other proteins, such as OmaA, likely play a pivotal role in both the initiation and maintenance of biofilm architecture throughout its developmental stages. OmaA possesses surface-exposed domains that mediate adhesion, which are essential for bacterial attachment and aggregation on plant surfaces (Burdman et al. 2001). Notably, Tad pilus is crucial for biofilm formation in *Azospirillum*. Inactivation of plasmid-encoded genes responsible for Tad pili in *A. baldaniorum* Sp245 results in the suppression of biofilm formation (Wisniewski-Dyé et al. 2011). Finally, the presence of amyloid-like proteins in *Azospirillum* biofilms has been suggested (Shelud'ko et al. 2018). However, to our knowledge, there is no genomic data within the *Azospirillum* genus that supports this hypothesis.

6.2.2.3 Extracellular DNA

eDNA constitutes a significant component of the biofilm matrix, contributing to surface adhesion via acid-base interactions. It plays a crucial role in maintaining biofilm structural integrity and can impede the diffusion of cationic antimicrobial agents. Enzymatic treatment with DNase I, a nuclease that hydrolyzes DNA, has been shown to alter the architecture of biofilms. Using fluorophores (TOTO-1 and propidium iodide) that specifically target DNA, the presence of eDNA has been confirmed within *A. brasilense* Sp7 biofilms grown on glass. These findings reveal an additional matrix constituent in *A. brasilense* biofilms (Ramírez-Mata et al. 2016; Viruega-Góngora et al. 2020). While eDNA is integral to biofilm stability, its presence alone does not fully account for this characteristic. This suggests that other extracellular matrix components in mature biofilms, such as cell-cell-interacting adhesins and EPS, may provide compensatory or synergistic contributions. Alternatively, eDNA within mature biofilms may engage in interactions with other biomolecules, as previously reported (Okshevsky and Meyer 2013).

6.2.3 The Regulation of Biofilm Formation in Azospirillum

Azospirillum strains use the second messenger c-di-GMP to switch between motile and biofilm lifestyles (Carreño-López et al. 2009; Cruz-Pérez et al. 2021; Ramírez-Mata et al. 2016; Russell et al. 2013; Sierra Cacho et al. 2021). This process is regulated by diguanylate cyclases (DGCs, with GGDEF domains) that synthesize c-di-GMP and phosphodiesterases (PDEs, with EAL or HD-GYP domains) that degrade it (Römling et al. 2013). These proteins often have sensory domains (PAS, GAF, REC, CHASE, MHYT) responding to environmental signals crucial for host plant interactions (Randall et al. 2022).

Azospirillum genomes, as *A. baldaniorum* Sp245, typically have around 35 genes for c-di-GMP metabolism, with some proteins likely being membrane-anchored. *A. baldaniorum* Sp245 uses this signaling to form biofilms on plant roots. Its genome includes 25 GGDEF, 5 EAL, and 10 hybrid domain proteins (Mata et al. 2018). Three Cdg proteins have been well characterized in *A. baldaniorum* Sp245. The CdgC, a DGC with REC-GGDEF domains promoting biofilm formation, EPS production, and plant colonization (Sierra Cacho et al. 2021); the CdgD, a hybrid CHASE-PAS-DGC-PDE protein influencing motility, biofilm formation, and plant colonization via its DGC activity (Cruz-Pérez et al. 2021). Reduced c-di-GMP levels due to alterations in such protein impair wheat root internalization (Cruz-Pérez et al. 2021; Sierra Cacho et al. 2021). The third one, CdgB, is a characterized bifunctional (MHYT-PAS-DGC-PDE) enzyme. Overproduction of CdgB in *A. baldaniorum* Sp245 results in phenotypes associated with PDE activity, while in a heterologous host, it primarily shows DGC activity (Viruega-Góngora et al. 2022). CdgB possesses sensory domains, including an MHYT and a PAS domain. The MHYT domain is speculated to be key in controlling CdgB's enzymatic activity, potentially by sensing gases like O_2, CO, or Nitric Oxide (NO), similar to other MHYT-containing proteins like YkoW from *Bacillus subtilis* (Galperin et al. 2001), NbdA/MucR from *Pseudomonas aeruginosa* (which are involved in NO-mediated biofilm dispersion) (Li et al. 2013; Wang et al. 2015), and MucR of *Azotobacter vinelandii* (Martínez-Ortiz et al. 2020). O_2 and NO are critical signals for *A. baldaniorum* Sp245 during interaction with the host plant. The bacterium has chemotaxis receptors (Tlp1, Aer) that sense O_2 and integrate c-di-GMP detection via PilZ domains. CdgB's MHYT and PAS domains (which can bind O_2 or FAD) might allow it to sense similar signals (Arruebarrena Di Palma et al. 2013; O'Neal et al. 2019; Russell et al. 2013). Given that NO promotes root growth (Creus et al. 2005) and biofilm formation in *A. baldaniorum* Sp245 (Arruebarrena Di Palma et al. 2013), and the MHYT domain is proposed to sense NO, it is hypothesized that CdgB could be part of a signaling module controlling biofilm formation in response to NO. This signal might modulate CdgB's DGC and/or PDE activity, though the molecular details require further investigation (Viruega-Góngora et al. 2022).

6.2.4 *Nitrogen Fixation in Biofilms*

Azospirillum is a nitrogen-fixing bacterium that catalyzes the reduction of dinitrogen into ammonia by a nitrogenase enzyme complex, which in turn is inactivated by O_2 (see Chap. 9 for more details). Bacteria have developed several mechanisms to protect nitrogenase from O_2 and increase growth rates, including accelerating the respiratory rate, developing large cells with thicker envelopes (e.g., heterocysts in *Nostoc* sp.), or EPS production as a barrier to O_2 diffusion (Inomura et al. 2017). Studies conducted in the 1980s reported a reduced or absent nitrogenase activity in cyst-like forms of *A. brasilense* and *A. lipoferum* (Bastarrachea et al. 1988; Papen and Werner 1982). Nevertheless, *in situ* hybridization with fluorescent rRNA-targeting oligonucleotide probes showed that cyst-like cells of *A. brasilense* exhibit physiological activity in the rhizosphere of wheat, at least during the early phases of their morphological transformation (Assmus et al. 1995). An actual and detailed study on *A. brasilense* biofilms showed significant nitrogenase activity of cyst-like cells aggregated and surrounded by an EPS envelope into biofilm, implying the formation of a barrier to O_2 diffusion and the generation of an optimal condition for *Azospirillum* free bacteria to perform nitrogen fixation (Wang et al. 2017). Those results were supported by a transcriptomic analysis of cyst-like cells, suggesting an anaerobic/micro-anaerobic environment generated by repression of oxidative phosphorylation pathways and enhanced expression of nitrate/nitrite reduction and nitrogen fixation genes (Malinich and Bauer 2018a).

6.2.5 *The Role of Biofilms in the Interaction with Other Microorganisms*

Biofilms that occur in nature are far more complex because they grow on highly heterogeneous surfaces and as a result of the interaction among different individuals within the bacterial community inhabiting that niche. Significant advancements in the understanding of multispecies biofilms made in recent years have unveiled the complexity of social interactions between bacteria within these structures (Sadiq et al. 2021). Naturally, throughout its lifecycle, *Azospirillum* interacts with the wide diversity of microorganisms that compose the plant microbiome and is expected to engage in multispecies biofilms. In consequence, its beneficial effects on plants may be either hindered or enhanced depending on the outcome. For more information on the interaction of *Azospirillum* with other microbes, see Chap. 16.

Research on *Azospirillum* within multispecies biofilms is scarce. However, some advances have been made by analyzing its interaction with fluorescent *Pseudomonas*, another important group of plant growth-promoting bacteria (PGPB). These two PGPB groups are commonly combined in synthetic consortia for the formulation of mixed inoculants (Valverde et al. 2015). Since fluorescent *Pseudomonas* are known for holding a vast repertoire of metabolites and enzymes with antimicrobial properties (Gross and Loper 2009), appropriate conditions that minimize the production of these factors must be met for *Azospirillum* to successfully engage in dual-species biofilms. For instance, siderophore secretion by *Pseudomonas* hinders the growth of *Azospirillum* within the biofilm (Diaz et al. 2022; Maroniche et al. 2016, 2018). Similarly, the antibiotic 2,4-diacetylphloroglucinol (DAPG), which is produced by several species of fluorescent *Pseudomonas*, can negatively affect *Azospirillum* biofilm formation (Couillerot et al. 2011). However, at the right concentration, DAPG acts as a biofilm-inducing factor and improves *A. baldaniorum* Sp245 phytobeneficial properties (Combes-Meynet et al. 2011).

O_2 is a key player in the regulation of bacterial biofilm shape and functionality, and certainly, *Azospirillum* is not the exception. Due to its microaerophilic nature, biofilms of *A. baldaniorum* Sp245 formed under static culture conditions are formed at the bottom of the well, where O_2 levels are lower (Pagnussat et al. 2016). Similarly, under low N conditions, macrocolony biofilms of Sp245 commonly grow buried in the agar (Fig. 6.1). However, if incubated under a low O_2 atmosphere, *Azospirillum* biofilm growth is enhanced (Diaz et al. 2022). This dependence on low O_2 conditions when forming biofilms is likewise a main factor that governs *Azospirillum*'s capacity to coexist with *Pseudomonas* in dual-species biofilms. When co-cultured with *P. protegens* CHA0, *A. baldaniorum* Sp245 increases biofilm formation and reaches a higher position in the well, suggesting that more favorable conditions for *Azospirillum* might have been created by O_2 consumption due to *Pseudomonas*' accelerated growth and aerobic metabolism. Microscopic observation of mixed biofilms confirms O_2 as a fundamental factor shaping the biofilm, since Sp245 is observed stratified at the bottom and surface of the structure (Pagnussat et al. 2016). Similar results have been observed when *A. baldaniorum* Sp245 interacts with *P. fluorescens* A506 in dual-species macrocolony biofilms. In this case, *Azospirillum* is able to increase its growth considerably within the biofilm, benefiting not only from low O_2 conditions but also from the massive matrix produced by *Pseudomonas*. However, under low O_2 conditions (5%), the relationship between Sp245 and A506 within the biofilm shifts from exploitation to mutual competition, negatively affecting both strains (Diaz et al. 2022). Overall, the microaerophilic behavior of *Azospirillum* strongly influences its interaction with other members of the plant microbiome community in mixed biofilms.

6.3 Aggregation/Flocculation

Cell aggregation or flocculation is the microbial phenomenon by which cells gather together to form stable and multicellular associations under certain physiological conditions. The resulting structures are aggregates, clumps, or flocs (Calleja et al. 1984). The flocculation phenomenon is widespread in the bacterial world, representing a survival strategy under environmental stress conditions. In addition, cell aggregation can contribute to bacterial adhesion and biofilm formation. Several N_2-fixing bacteria, such as *Azotobacter*, *Klebsiella*, and *Azospirillum,* can flocculate, a property that could be beneficial for survival in the soil and success in the colonization of plant roots in the rhizosphere (Burdman et al. 1998; Sadasivan and Neyra 1985).

6.3.1 *Flocculating Cell Morphology*

Azospirillum is characterized by its vibrioid or spirillum shape, possessing long polar and short lateral flagella that contribute to its high motility. However, *Azospirillum* exhibits a notable cellular polymorphism, transitioning from vibrioid-shaped, highly motile vegetative cells to round, non-motile, encapsulated cyst-like forms in response to changes in the growth condition, for example, in aged cultures (Tarrand et al. 1978). This morphological transformation, i.e., flocculation, is not terminal, which means that cells could potentially return to their original shape when environmental conditions change (Pereg 2015).

6.3.2 *Conditions Promoting Flocculation*

Flocculation is influenced by several factors: nitrogen limitation and the presence of an appropriate carbon and energy source, along with a high O_2 tension, such as that found on the surface of agarized media plates, are critical for the transformation of vibroid into cyst-like forms (Bastarrachea et al. 1988). *Azospirillum* strains are highly sensitive to changes in the C:N ratio. A low C:N ratio in the media generally favors dispersive growth, whereas a high C:N ratio induces cell aggregation and flocculation (Burdman et al. 1998; Del Gallo and Haegi 1990; Del Gallo et al. 1989; Sadasivan and Neyra 1985). Under minimal liquid media supplemented with fructose and KNO_3, heavy capsulation gives the cells a particular adhesive nature; therefore, they aggregate in a matrix of polysaccharide material, forming large clumps in the cultures visible to the naked eye (Fig. 6.2) (Pereg-Gerk et al. 1998, 2000; Sadasivan and Neyra 1985).

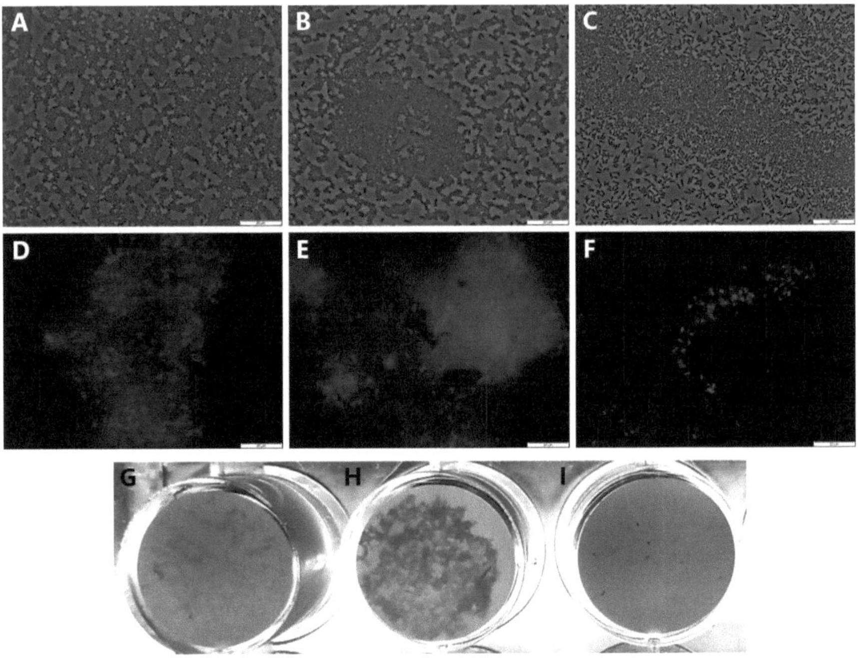

Fig. 6.2 Flocculation time course in *A. brasilense* FP2. Cultures of *wild type* (**a**, **d**, **g**), a super-flocculating mutant (**b**, **e**, **h**), and a non-flocculating mutant (**c**, **f**, **i**) were incubated in flocculation medium and observed after 4 h under the light microscope (**a**–**c**). At 18 h of flocculation, the flocs containing PHB were stained with Nile red and visualized with a green filter (**d**–**f**). Bars depict 20 μm (**a**–**f**). After 24 h of flocculation, the clumps can be seen by the naked eye (**g**–**i**). Perez et al., unpublished results

6.3.3 Physiological Changes over Flocculation

During flocculation, vegetative cells develop a polysaccharide-rich outer layer, lose motility, adopt a larger spherical morphology, and accumulate significant amounts of poly-β-hydroxybutyrate (PHB) granules (Sadasivan and Neyra 1985). The accumulation, degradation, and utilization of PHB by bacteria under stress are thought to represent a mechanism that supports their establishment, growth, survival, and competitive advantage in environments where carbon and energy sources are limited (Okon 1992). The capsule formation and EPS production serve as critical prerequisites for cyst maturation. These cyst-like cells exhibit enhanced resistance to desiccation, osmotic stress, and UV radiation compared to vegetative cells (Bashan et al. 1991; Sadasivan and Neyra 1985).

The structure, composition, and quantity of EPS and CPs produced by *A. brasilense* strains with different patterns of aggregation are different; meanwhile, the outer membrane proteins (OMP) are similar between them, meaning that EPS play a crucial role in flocculation. The composition of EPS and CPs of *A. brasilense* shows a positive correlation between the aggregation and the amount of arabinose. Moreover, a mutant defective in lipopolysaccharide (LPS) synthesis and unable to flocculate produced a glucose-rich EPS, in contrast to mutants with higher aggregation potential, such as *exoB* (UDP-glucose 40′-epimerase), *exoC* (phosphomannomutase), and *phbC* (poly-b-hydroxyburyrate synthase), which produce an arabinose-rich EPS, indicating the importance of EPS composition for flocculation (Bahat-Samet et al. 2004; Burdman et al. 2000). Notably, *A. brasilense* mutant strains devoid of lateral flagella were able to flocculate (Pereg-Gerk et al. 2000) and showed a normal pattern of wheat root colonization at the initial stage of plant-bacteria interaction (Moens et al. 1995). In addition, a mutant lacking both lateral and polar flagella showed the strongest aggregation, indicating that flagella are not involved in flocculation (Burdman et al. 1998). Nevertheless, the Che1 chemotaxis-like pathway from *A. brasilense* may influence the molecular machinery responsible for cell-cell clumping, where motile cells transiently interact by adhering to their non-flagellated poles before dispersing. This clumping event occurs at a prior stage and is essential for flocculation, with both behaviors appearing to be independently regulated (Bible et al. 2012, 2015). Furthermore, cell surface analysis by atomic force microscopy shows that *A. brasilense* cells produce a smooth mucosal extracellular matrix after 24 h of flocculation, while the Che1 mutant exhibits extracellular fibril structures (Bible et al. 2008; Edwards et al. 2011). Interestingly, the *mreB* gene, essential for maintaining the cell rod shape in *E. coli* and *B. subtilis*, was shown to be involved in establishing cell surface properties and in the control of *A. brasilense* cellular differentiation. In consequence, the *mreB-like* mutant has a round shape with a thicker envelope and a high number of PHB granules, resembling the cyst form of *Azospirillum* (Biondi et al. 2004). Additional factors important for the process are proteins containing fasciclin domains (FPD), known to play a critical role in cell-to-cell adhesion, which protect *A. brasilense* Sp7 from photooxidative stress by accumulation of high carotenoid levels and promoting flocculation (Dubey et al. 2021). Also, the presence of multiple lectin-like proteins on the bacterial cell wall surface of *Azospirillum* and their interaction with various carbohydrate residues were demonstrated to be involved in cell aggregation (Burdman et al. 2000; Castellanos et al. 1998).

6.3.4 Genes Directly Involved in Flocculation

The differentiation of vegetative cells of *A. brasilense* into cysts is associated with significant changes in both the transcriptome and proteome, a process that is regulated by the LuxR-type transcription factor, FlcA. The Tn5-mutants *flcA*$^{-}$ do not exhibit flocculation, do not undergo morphological changes, lack EPS material on their surface under all conditions, and show a significant reduction in wheat root colonization, though they have increased nitrogenase activity in hydroponic wheat associations (Katupitiya et al. 1995; Pereg-Gerk et al. 1998). Furthermore, the expression of the *flcA* gene is upregulated during both flocculation and plant root colonization, indicating its significant role in cell adhesion and root colonization (Pereg-Gerk 2004). A study on proteins differentially expressed under flocculation conditions in *A. brasilensis* Sp7 and its knock-out mutant *flcA*$^{-}$ finding proteins involved in stress responses, morphological transformation, and nitrogen fixation, such as the nitrite/nitrate transporter NarK, associated with FlcA (Hou et al. 2014). Notably, the *narL* homologue, potentially involved in regulating the nitrate reductase, is highly expressed in *A. brasilense* FP2 cells on wheat roots (Camilios-Neto et al. 2014). However, the molecular mechanism underlying cell aggregation remains unknown and requires in-depth investigation due to its significance in plant root colonization.

6.3.5 Flocculation as a Base for Inoculant

Since experimental evidence suggests that cell aggregates are more resistant to harsh conditions, plus the fact that flocs can be easily separated from the growth medium, taking advantage of bacterial aggregation turns out to be an important strategy in the production of *Azospirillum* inoculants. In fact, a comparative study of flocculated and standard growth cells of *A. brasilense* revealed that flocculated cells exhibit enhanced survival in alginate beads and disinfected soil after prolonged storage, in addition to higher resistance to dehydration and adhesion efficiency on maize roots when compared to standard-grown cells (Joe et al. 2012). Therefore, the use of *Azospirillum* flocs emerges as an important inoculant formulation tool to promote sustainability by replacing chemical inputs derived from non-renewable sources with bioinputs.

6.4 Attachment and Colonization

6.4.1 Mechanisms of Attachment and Colonization of Roots

Compelling evidence indicates that plant roots can shape the rhizomicrobiome by secreting compounds that recruit specific groups of microorganisms. Bacterial chemotaxis machinery in *Azospirillum* (reviewed in Chap. 5) and other motile rhizobacteria is critical for this attraction process, enabling bacteria to compete for colonization of root surfaces. By reacting to exudate molecules and signals, bacterial chemotaxis and motility contribute decisively to the selection of the initial site of colonization (Fig. 6.3).

Endpoint analyses of *Azospirillum* colonization levels, from a few hours to days after inoculation, have been essential to establish the role of chemotaxis in this process. It is now known that *A. brasilense* chemotaxis enhances the aggregation of motile cells near wheat roots before surface colonization. A key player in this regard

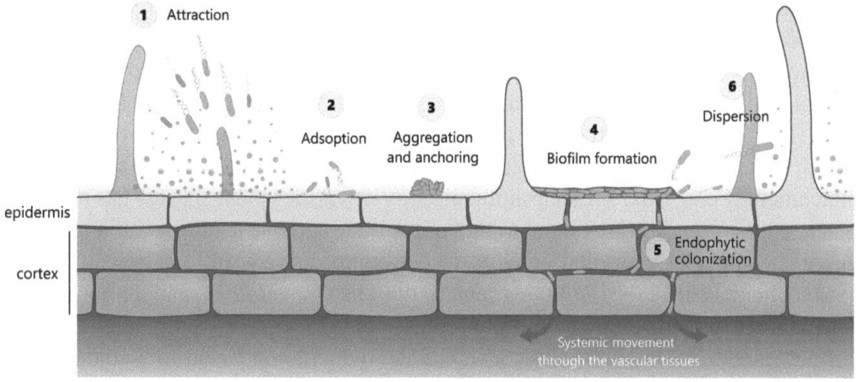

Fig. 6.3 *Azospirillum* root colonization process. The different steps of *Azospirillum* root colonization are schematized: (1) chemotactic attraction of azospirilla present in the rhizosphere to the root by exudates; (2) adsorption of bacteria to the surface of roots; (3) formation of aggregates/flocs to anchor stably to the root; (4) establishment of a biofilm on the surface of the root; (5) occupation of the endosphere through intercellular spaces and subsequent vessel-mediated dissemination to other plant organs; (6) dispersion of azospirilla from mature biofilms to colonize new zones of the root

is the chemoreceptor Tlp1, which is crucial for normal colonization of wheat roots. The absence of Tlp1 in the *Δtlp1* mutants renders cells unable to sense the attractant signal(s) and aggregate in the root hair and elongation zones. Rather, these mutants show an increased attraction to cells of the root tip areas, resulting in discrete bands of bacteria along the surface of the root tip. In contrast to the colonization pattern of the *wild type*, which is characterized by attractant responses to spatial gradients of malate, pyruvate, and succinate, *Δtlp1* cells exhibit diminished responses to spatial gradients of pyruvate and succinate and no reaction at all to malate, indicating that malate is the primary attractant detected by Tlp1 (O'Neal et al. 2020).

Polyamines, including arginine and putrescine, have also been reported as signaling molecules at the root-rhizosphere interface, informing the microbiome about the presence of eukaryotic hosts. The detection of such compounds initiates a lifestyle transition, facilitating adhesion and biofilm development in several bacterial species (Liu et al. 2018). Although the production of polyamines—i.e., cadaverine, putrescine, spermidine, and homospermidine—has been reported in several *Azospirillum* strains, their role in root attachment and colonization remains unclear (Cassan et al. 2009; Hamana and Takeuchi 1998).

In *A. brasilense,* chemotaxis mediates the accumulation of motile cells around wheat roots, which precedes two distinct stages of attachment to the root surface (Fig. 6.3). The first or adsorption phase is rapid yet weak, peaking within 2 h post-inoculation (hpi), and is presumably regulated by bacterial surface proteins. The polar flagellum is an important factor that mediates the adsorption of *A. brasilense* cells to wheat roots. Croes et al. (1993) established that the absence of polar or lateral flagella in the mutant Sp7 p90D084, or the destruction of flagella by heat or acid treatments, inhibits root adsorption. Moreover, polar flagella isolated from *Azospirillum* cells also adhere to wheat roots in vitro (Croes et al. 1993). In addition to its important role in adhesion, this structure has recently emerged as a key *Azospirillum* effector for plant growth promotion (reviewed in Chap. 11). The second or anchoring phase is prolonged, beginning after 8 hpi and peaking at 16 hpi. This phase is more robust and irreversible and is assumed to involve bacterial EPS (Michiels et al. 1991). The aggregation or flocculation of *Azospirillum* cells, which is linked to EPS production, is thought to be essential for stable root surface attachment (Pereg 2015). Electron microscopy of various plant species has revealed that *Azospirillum* cells are bound to the root surface and one another within bacterial aggregates by a large fibrillar network whose composition is still unknown (Bashan 1986; Michiels et al. 1991). In this regard, the Tn5-mutant *flcA*$^{-}$ that do not exhibit flocculation, do not undergo morphological changes, lack EPS material on their surface under all conditions show a significant reduction in wheat root colonization (Katupitiya et al. 1995; Pereg-Gerk et al. 1998). Furthermore, the expression of the *flcA* gene is upregulated during both flocculation and plant root colonization, highlighting its significant role in cell adhesion and root colonization (Pereg-Gerk 2004).

Bacterial EPS are not the only surface component involved in the establishment of *Azospirillum*-plant interaction. Jofre et al. (2004) reported a potential role for LPS in establishing effective associations between *A. brasilense* and maize roots. On the other side, several studies have established that lectins similar to the wheat germ agglutinin (WGA) found on the root surface of various grasses bind to *A. brasilense* and *A. lipoferum* cells. Both the intensity and quantity of binding were shown to vary with culture age, suggesting the presence of specific sugar-bearing receptors for WGA on the cell surface (Burdman et al. 1998; Del Gallo and Haegi 1990; Del Gallo et al. 1989). Thus, it is reasonable to suggest that the interaction between the carbohydrate constituents of the bacterial surface and lectins present on the roots is crucial for bacterial attachment and subsequent host colonization, at least in the *Azospirillum*-Graminaceae systems (Skvortsov and Ignatov 1998).

After stable bacterial attachment during the anchoring phase, the *Azospirillum*-plant association proceeds through the formation of microcolonies that later expand to produce mature **biofilms** on the surface of the root (Camilios-Neto et al. 2014). *Azospirillum* biofilms have been directly visualized on epidermal cells, root hairs, and intercellular spaces of cortical cells (Assmus et al. 1995; Naqqash et al. 2022), from which azospirilla are expected to then *disseminate* to colonize new niches on the root rhizoplane and endosphere, or new roots (Bashan and Holguin 1994). A schematic representation of the *Azospirillum* root association cycle is shown in Fig. 6.3.

The zone of the root that *Azospirillum* colonizes varies along the root axis and between root branches throughout the azospirilla-plant association cycle. The long-term establishment site often differs from the initial point of contact, as bacteria migrate along the roots over a distance range from micrometers to centimeters. To achieve this, bacterial motility must proceed in a context of intense competition with other rhizospheric microorganisms while vying for resources and colonization niches on the developing seedling roots. The root elongation zone and hairs are preferred locations for *Azospirillum* activity (O'Neal et al. 2020). According to

Bashan (1986), the migration of azospirilla from inoculated roots, which continue to exude nutrients, to non-inoculated ones can be explained by at least two mechanisms. (i) The microbial community established in the rhizosphere consumes more nutrients than the roots can provide at any given time. This creates a transient nutrient-deficient microenvironmental gradient that prompts azospirilla migration toward alternate nutrition sources. (ii) *Azospirillum* cells primarily engage in aggregation to achieve colonization, which limits cell mobility due to the synthesis of fibrillar material. These aggregates confer an ecological advantage over individual bacterial cells in the struggle for resources inside the rhizosphere. Consequently, it may be inferred that, to survive, individuals must reach areas devoid of aggregates, which forces them to colonize non-inoculated roots and generate new aggregates (Fig. 6.3).

6.4.2 The Role of Colonization in Plant Biostimulation

The colonization of the roots by *Azospirillum* plays a crucial role in producing plant biostimulation. Camilios-Neto et al. (2014) demonstrated that *A. brasilense* colonization improves nutrient acquisition by not only increasing the root surface area but also regulating the expression of nutrient transporters. Through dual RNA-Seq analysis, the authors identified significant transcriptional changes in wheat roots colonized by *A. brasilense* strain FP2, particularly the expression of 776 plant genes involved in nutrient transport, nitrogen assimilation, cell division, defense mechanisms, and the production of phytohormones and phytochemicals. Additionally, bacterial genes involved in chemotaxis, biofilm formation, and nitrogen fixation were highly expressed, emphasizing the bacteria's active role in root association. In the same sense, another study revealed that the colonization ability of the *A. brasilense* strain CW903 was strongly linked to improved nutrient uptake and plant growth. X-Gal staining of wheat root segments revealed cells of CW903, confirming their colonization and proliferation along the root surface of inoculated plants. Under greenhouse conditions, inoculation with this strain significantly enhanced growth parameters and nutrient absorption in pepper, tomato, and rice (Kim et al. 2010).

Several studies demonstrated that inoculation with *Azospirillum* led to significant improvements in biochemical parameters of plants under abiotic stress conditions such as salinity or drought (García et al. 2017; Gonzalez et al. 2024). Those biochemical parameters included increased levels of pigments, proteins, total sugars, lignin, phenolics, and proline. These molecules are often involved in stress responses, including osmoregulation, cell protection, and overall plant growth under suboptimal conditions. This indicates that the bacteria could be triggering molecular or physiological pathways that help the plant cope better with the stress conditions. The higher levels of proline, for example, suggest improved osmoregulation, which is critical for maintaining cell function under saline or drought conditions.

The colonization of plant roots by *Azospirillum* likely triggers beneficial molecular responses in the plant that help it cope better with salinity and drought, through enhanced nutrient acquisition, production of growth-promoting substances like phytohormones, or other mechanisms such as stress-responsive gene activation (Camilios-Neto et al. 2014; Gonzalez et al. 2024).

6.4.3 Endophytic Colonization by Azospirillum

Understanding the lifestyle of each *Azospirillum* strain (endophytic, rhizospheric, or phyllospheric) and their ability to colonize plant hosts is crucial for selecting the appropriate inoculation and monitoring methods (Nievas et al. 2023). In 1992, Kloepper et al. introduced the term "endophytes" to describe bacteria residing within the internal tissues of the epidermis. Five years later, they proposed classifying these bacteria into two categories: facultative and obligate endophytes. Facultative endophytes are those that can survive in the soil and/or on plant surfaces, in addition to being capable of colonizing the internal tissues of certain plants. Most *Azospirillum* strains are considered facultative endophytes. In contrast, obligate endophytes, such as *Herbaspirillum* spp., *A. diazotrophicus*, and *Burkholderia* spp., are poorly adapted to surviving in the soil and appear to rely on living inside a host plant (Baldani et al. 1997). Among the innovative approaches in agriculture, the use of endophytic microorganisms, beneficial microbes that thrive within plant tissues, holds significant promise for enhancing crop health and productivity.

A well-known case of endophytic behavior is *Azospirillum* sp. B510, which can colonize the inner tissues of rice (*Oryza sativa* cv. Nipponbare). By using a *gus*-tagged variant, it was demonstrated that strain B510 is capable of entering and colonizing the stem of rice. Moreover, re-isolation confirmed its presence in the stems and roots, but not in the leaves. In turn, the endophytic colonization of rice by *Azospirillum* sp. B510 induced resistance against the rice blast disease and bacterial blight disease (Yasuda et al. 2014). A similar response was observed by Fujita et al. (2017), who showed that the endophytic colonization of *Azospirillum* sp. B510 induces disease resistance in tomato plants, protecting them against bacterial leaf spot and gray mold.

Another endophytic strain, *A. brasilense* Yu62, was characterized by tagging with *gfp* to trace its migration within plant systems after inoculating rice and tobacco seeds. Upon germination, Yu62 successfully penetrated the root tissues and ascended through intercellular spaces and the vascular system, reaching the stems and leaves. Some individuals even entered plant cells, indicating a dynamic interaction between *A. brasilense* Yu62 and the host plant (Chi et al. 2004). This behavior underscores the potential of *Azospirillum* endophytes to migrate within plant tissues after root colonization, possibly benefiting different tissues to enhance plant growth and stress tolerance.

Similar to the abovementioned cases, Hong et al. (2019) demonstrated that *A. brasilense* 1224 endophytically colonizes the tissues of onion (*Allium cepa* L.)

seedlings. The tracking of bacterial cells using GUS staining as well as scanning electron microscopy (SEM) showed that, at 3 days post-inoculation (dpi), the 1224 strain colonizes the roots and bulb of the seedlings. However, by 28 dpi, bacterial colonization had spread throughout the entire root system, with a notable increase in bacterial population compared to that at 7 dpi. Additionally, fluorescence microscopy revealed that vegetative cells on the surface of the bulb and roots of onion seedlings undergo differentiation into cyst-form cells as early as 3 dpi, with the conversion process being influenced by the tissue type being colonized.

Fukami et al. (2017) assessed the capacity of *A. brasilense* to colonize maize leaf surfaces using the Ab-V5 and Ab-V6 strains, each carrying reporter plasmids encoding the *egfp* and *eyfp* genes, respectively. The strains were applied via foliar spraying, and colonization was monitored using confocal laser scanning microscopy (CLSM). Fluorescent signals confirmed the presence of bacterial cells on the leaf surface as early as 1 h after inoculation and persisted for up to 24 h; however, no fluorescent cells were detected at 48 h. Quantitative analysis of bacterial populations revealed a progressive decline in cell density, with colony-forming units decreasing after 2 days for both strains, which likely explains the absence of detectable cells in CLSM observations at the final time point. Similarly, regarding the capacity of *Azospirillum* to colonize the aboveground organs of plants, Puente et al. (2020) evaluated the survival and localization of *A. argentinense* Az39 on detached soybean leaves. Though a high concentration of bacteria was localized on the leaf surface using SEM right after bacterization, their number decreased after 24 h, while no cells were detected on the leaves after 48 h. Endophytic colonization of Az39DsRed (fluorescent variant strain) was analyzed using CLSM and the peeling technique, revealing bacterial cells outside the veins, at cell-cell junctions, and around stomata (Fig. 6.4). In addition, cross-sectional analysis of the leaves at 6 h post-bacterization showed the presence of Az39DsRed in the xylem of the treated leaves, suggesting systemic movement within the plant. The endophytic systemic colonization of *A. argentinense* Az39 was later confirmed by Takahashi et al. (2024), who observed

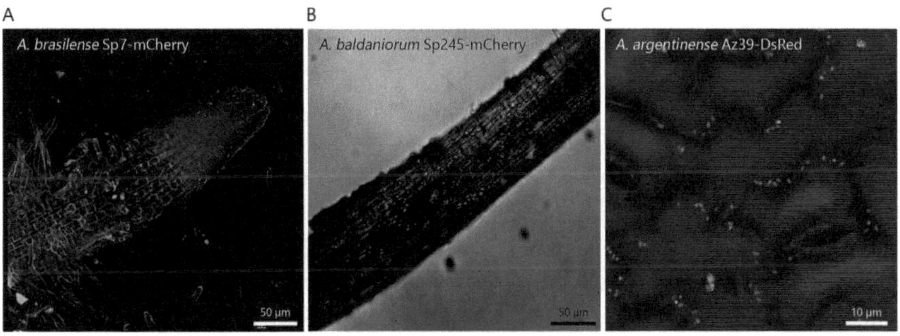

Fig. 6.4 Confocal microscopy of *Azospirillum* colonizing different plant organs. (**a**) *A. brasilense* Sp7-mCherry on the surface of a radish root (green autofluorescence) (Viruega-Góngora et al. 2020). (**b**) Colonization of lettuce roots by *A. baldaniorum* Sp245 (Diaz et al., unpublished results). (**c**) Soybean leaf epidermis colonized by *A. argentinense* Az39-DsRed (Puente et al. 2020)

the bidirectional translocation of this strain between roots and shoots by qPCR, which implies a potential vertical transmission through seeds. In line with this, Malinich and Bauer (2018b) demonstrated that *A. brasilense* represents a significant component of the seed microbiome of *Phaseolus vulgaris*, ranking as the fourth most abundant bacterial genus. Moreover, *A. brasilense* was shown to colonize not only roots and stems but also leaves and developing seeds, with the highest bacterial densities observed in mature seeds. Importantly, this strain is transmitted vertically from the parental generation to the offspring, resulting in increased root fresh mass in F2 plants. These findings reinforce the potential of *Azospirillum* as a growth-promoting endophyte and underscore its relevance as a bioinput for agricultural systems.

6.5 Standard Protocols for Evaluating Biofilm Formation, Aggregation/Flocculation, and Attachment/Colonization of *Azospirillum*

6.5.1 *Quantification of* Azospirillum *Biofilm Production*

6.5.1.1 Biofilms in Static Culture

The determination of in vitro biofilm development by *Azospirillum* is usually done by the classic crystal violet method, which is a crude method that reveals the overall biofilm mass production, including cells and matrix. Initially, *Azospirillum* strains are cultured on Congo red medium (RC) plates for 72 h (Rodrıguez Cáceres 1982). Following incubation, three isolated colonies exhibiting typical morphology are selected and inoculated into 125 ml Erlenmeyer flasks containing 25 ml of LB medium supplemented with 5 mM $MgSO_4$ and 3.3 mM $CaCl_2$ (Jijón-Moreno et al. 2019) or any other suitable growth medium, such as nutrient broth (NB). These pre-cultures are incubated at 30 °C with orbital agitation at 100 rpm for 16 h, until reaching an optical density at 540 nm (OD540) between 1.1 and 1.4. Cells are harvested by centrifugation at 7500× *g* for 10 min, washed with 66 mM phosphate buffer, and resuspended in the same buffer to a final OD540 = 2. Subsequently, the cell suspension is diluted 1:100 in NFb-malic acid medium (Day and Döbereiner 1976) modified to achieve a carbon-to-nitrogen (C: N) ratio of 2. This modification involves the addition of 27.6 mM malic acid and either 13.8 mM NH_4Cl or 13.8 mM KNO_3 as the nitrogen source, or no nitrogen source. Two milliliters per well are transferred to flat-bottom polystyrene 24-well plates (0.2 mL if working with 96-well plates) and incubated statically at 30 °C for 5 days. Biofilm formation is quantified using the crystal violet staining method (O'Toole and Kolter 1998). Briefly, 0.5 mL of 0.5% (w/v) crystal violet solution is added to each well (0.1 mL if working with 96-well plates), and the plates are incubated at room temperature for 30 min and then washed three times with tap water. The crystal violet bound to the biofilm is solubilized with 2 mL of 33 % (v/v) acetic acid (0.2 mL if working with 96-well plates). The absorbance of the resulting solution at 595 nm is measured using a microplate spectrophotometer. Data can be normalized by total protein content, determined using the Bradford assay (Viruega-Góngora et al. 2020), or by total growth estimated by OD540 (Arruebarrena Di Palma et al. 2013).

A complementary protocol was developed by Pagnussat et al. (2016) for *A. baldaniorum* Sp245 by making use of a fluorescent protein (FP) tagged variant of Sp245 to determine, through calibration with UFC count values, the number of cells in the biofilm. For this method, 96-well black microplates are used. Then, planktonic fractions are retrieved to new wells, and the remaining adhered cells are vigorously suspended in 0.2 mL of sterile saline solution (SS). The fluorescence of planktonic and biofilm fractions is then quantified in a microplate fluorescence scanner with suitable filter sets. Control wells with fresh medium have to be added as blanks. In addition, a control with the untagged strain might also be used to rule out its natural fluorescence (but not necessary for *A. baldaniorum* since, in our experience, it shows no autofluorescence whatsoever). If a fluorescence-to-CFU calibration is required, a static culture growth curve for each particular strain (in the same condition as the one evaluated in the experiments) must be conducted in microplates for at least 48 h, and samples must be periodically collected for the simultaneous determination of CFU and fluorescence in each sample. One well is used for each sampling point, so multiple technical repetitions have to be initially seeded (at least one per sampling point). With the registered values, linear regression must be established, and the goodness of fit estimated between fluorescence measurements and UFC.mL^{-1} for each specific strain. The UFC.mL^{-1} values may then be derived from the fluorescence reading of the experiments.

6.5.1.2 Biofilms on Agarized Medium

Macrocolonies can also be used to determine biofilm formation. This biofilm system is not equivalent to static cultures, since the support and aeration conditions are completely different. Bacterial suspensions adjusted to OD600 = 2 are obtained as explained before and used to seed macrocolonies as 10 μl drops on top of the desired agarized medium in Petri dishes or 24-well microplates. If Petri dishes are to be used to seed technical replicates or different treatments, we recommend seeding each macrocolony at the same radial distance from the border of the plate and not too close to each other, as the variation in these parameters might change the growth dynamics of the biofilm. After incubation at 28 °C for 48 h, biofilms can be recollected from the agar (i) with an inoculating loop or (ii) by excising the agar containing the macrocolony and introducing it into a microtube with 1000 mL of SS. Biofilm cells have to be completely resuspended by vigorous vortexing before proceeding to the CFU count in an appropriate medium. This methodology might be tricky when working with *Azospirillum* on low N medium, such as RC medium, because biofilms are buried in the agar and, thus, hard to detach. As an alternative, FP-tagged strains can be used to quantify biofilm growth by fluorescence acquisition and software analysis. This requires specific equipment for imaging (i.e., a transilluminator equipped with filters and a camera or microplate fluorescence reader, depending on the experimental setup), but has more sensitivity to small changes than CFU counts (Diaz et al. 2022). Fluorescence values obtained by this method are a relatively indirect measure of cell number, so appropriate controls must be added to rule out possible interferences or artifacts.

6.5.1.3 Biofilms Imaging

To analyze the structure of biofilms and the distribution of bacteria within them using CLSM, FP-tagged strains are cultured statically, as described previously. In this case, biofilms are formed in chambered coverslips or glass-bottom dishes that allow the direct visualization of the biofilms with inverted microscopes. This system allows the live imaging of biofilms and time-lapse experiments. In addition, matrix components can be visualized by staining with fluorescent dyes such as calcofluor, Congo red (EPS), propidium iodide, TOTO-1 (eDNA), and bromophenol blue (protein fibrils) (Zhang et al. 2020).

Similarly, macrocolony biofilms formed by FP-tagged strains can be imaged directly by photographing the fluorescence with an LED transilluminator and specific emission filter sets, depending on the FP used. If the camera has enough resolution and sensitivity, bacterial distribution patterns can be distinguished within the biofilm. Moreover, a confocal microscope equipped with aerial objectives of low magnification can be used to obtain high-resolution images of the biofilms. This is particularly useful to analyze the dynamics of biofilm formation or to track the spatio-temporal expression of specific functions in the macrocolony if reporter strains are available (Diaz et al. 2022). As in the previous case, matrix components can be stained by including fluorescent dyes in the culture medium. If a lateral view of the biofilm is required, macrocolonies can be transversely sectioned with a scalpel and mounted on a slide for visualization under the confocal microscope (Diaz et al. 2022). Transversal images with single-cell resolution are possible if samples are fixed and sectioned with a cryostat (Serra and Hengge 2017).

6.5.2 *Evaluation of* Azospirillum *Flocculation/Aggregation*

6.5.2.1 Flocculation

This protocol is based on the flocculation conditions described previously by Sadasivan and Neyra (1985). *Azospirillum* is grown on NFb-Lactate medium at 28 °C till late-log phase (OD600 = 0.8–0.9); cells are collected by centrifugation at 5000 rpm for 10 min and then washed twice with NFb minimal medium (without nitrogen source). The resulting suspensions are used to inoculate 10 mL of flocculation medium, i.e., NFb minimal medium supplemented with 8 mM fructose and 0.5 mM KNO_3, to a final OD600 = 0.3–0.4 in a 50 mL Erlenmeyer flask. The flocculation-induced cultures are then incubated at 28 °C with orbital shaking at 200 rpm. Cell aggregation can be observed by light microscopy after 3–4 h, and flocs are visible to the naked eye between 10 and 14 h after the induction. The quantification of flocculation is done according to Burdman et al. (1998): add 1 mL of the flocculation sample to a tube and subject the sample to mild sonication for 20 s on ice, and then measure OD_{600}. The percentage of flocculation can be calculated using the formula:

$$\text{Percentage of flocculation} = (OD_{600A} - OD_{600B}/OD_{600A}) \times 100$$

where OD_{600A} and OD_{600B} are the OD_{600} values after and before sonication, respectively.

6.5.2.2 Sedimentation Rate

This basic protocol, based on the work of Biondi et al. (2004) for analyzing the sedimentation rate of *Azospirillum* cells in suspension, consists of growing the bacteria in nutrient broth (NB) at 28 °C until late-log or stationary phase. Cells are then harvested by centrifugation at 5000 rpm for 10 min and resuspended in the same volume of phosphate-buffered saline (PBS). Sedimentation rate is analyzed by measuring the OD600 at regular intervals (Arunakumari et al. 1992).

6.5.3 Evaluation of the Colonization of Plant Tissues by **Azospirillum**

6.5.3.1 Previous Considerations on *Azospirillum* Cells Quantification

The quantitative analysis of *Azospirillum* colonization requires a methodology to determine the number of cells associated with the plant. The standard method to do this is CFU count on agarized medium, either the classical method or the drop plate method (Di Salvo et al. 2022), which will be described here. However, other methods that take advantage of more modern technologies exist, such as qPCR (Takahashi et al. 2024) or flow cytometry (Valdameri et al. 2015) (reviewed in Chap. 2).

Counting CFU of *Azospirillum* in plant extracts can be really challenging. Since selective media for *Azospirillum* haven't been developed, the RC medium (Rodrıguez Cáceres 1982) is the best option for most of the *Azospirillum* species to facilitate their identification due to their typical scarlet colonies. However, this method is not infallible because azospirilla are not quick growers and tend to be outcompeted by other bacteria. Moreover, some *Azospirillum* strains take longer to get stained by Congo red or do not do it at all, possibly due to the interaction with other microorganisms present in the extract. In consequence, *Azospirillum* CFU counts must rely on additional strategies, such as high-temperature incubation (*Azospirillum* grows well at 37–38 °C while pseudomonads do not) or their natural resistance to ampicillin (typically 200 μg.mL^{-1}) and trimethoprim (20 μg/mL). When a clean count is needed, rifampicin-resistant natural variants can be isolated (Borrajo et al. 2022) and used to avoid the growth of contaminating bacteria. Alternatively, we have used the miniTn7 system to obtain stable gentamicin-resistant strains, which are effective for CFU quantification in root extracts (Maroniche et al. 2018). Plasmids with adequate stability might also be used to obtain antibiotic-resistant variant strains for CFU counting purposes (see Chap. 4). It is essential to consider that these engineered strains are meant to be used under controlled conditions, since they are genetically manipulated and cannot be released into the environment.

Next, typical protocols to analyze *Azospirillum* colonization in different tissues of maize plants are detailed.

6.5.3.2 Colonization of Roots

The following protocol, based on the one published by Balsanelli et al. (2017), is useful for quantifying the adhesion and colonization of roots by *Azospirillum*. It involves the following steps: (a) seed sterilization and germination, (b) preparation of bacterial culture and inoculation, (c) hydroponic plant growth, and (d) CFU quantification.

(a) Seeds must be first surface-sterilized. A typical protocol for maize requires to treat the seeds first with 70% ethanol for 5 min, transfer them to a freshly prepared acidified hypochlorite 1%/tween 0.4% solution (v/v) for 20 min with orbital shaking at low rpm or stirring occasionally by hand, and washing them four times with abundant sterile distilled water. Sterilized seeds are then transferred to water agar plates, wrapped with aluminum foil, and incubated at 30 °C for 2 days.
(b) The inoculum is obtained by culturing *Azospirillum* in an appropriate liquid medium at 30 °C and 120 rpm orbital shaking until an optical density (OD) of 1 (≈10 CFU.mL). Dilute the culture with NFb or plant medium to obtain a bacterial suspension with ≈10 CFU.mL and inoculate the seedlings by complete immersion in 1 mL of bacterial suspension and agitation at 10 rpm for 30 min. As controls, use seedlings treated with NFb or plant medium. Counting of *Azospirillum* CFU in the bacterial suspension (≈10 CFU.ml) should be performed to determine the exact number of CFU in the inoculum.
(c) Sterile 100-mL glass tubes containing 30 ml of C-free plant medium and 10 g of polypropylene beads (2–3 mm diameter) sealed with foam plugs are used to carefully transfer the inoculated seedlings into them. This hydroponic system can be incubated in a plant growth chamber with fixed photoperiod, humidity, and temperature for up to 10 days.

(d) The determination of bacterial adherence to the roots is performed immediately after seedling inoculation. The roots of four or five seedlings (replicates) are cut with a scalpel and, after removing the excess of liquid with filter paper, they are weighed. Then, 1 ml of bacterial culture medium is added to the tube and vortexed for 1 min. The suspension is serially diluted until 10 to quantify the number of *Azospirillum* CFU by the drop method (Di Salvo et al. 2022). The count results are expressed as CFU.g of fresh root.

To determine the kinetics of root colonization, the analysis should be repeated at different time points (typically 1, 3, 7, and 10 days) after inoculation. In this case, the roots are excised, weighed, and homogenized with a previously sterilized mortar and pestle by crushing the roots in 1 ml of bacterial medium. Finally, CFU count is done to obtain total (epiphytic and endophytic) bacteria CFU.g of fresh root.

6.5.3.3 Colonization of Seeds

We describe here a method for analyzing *Azospirillum* colonization of maize seeds, which involves the prior inoculation of the seeds and subsequent recovery and counting of viable culturable cells from the seed surface (REDCAI 2025). Seed disinfection before inoculation was found to be unnecessary, based on an inter-lab experiment involving ten public and private laboratories from different regions of Argentina (Di Salvo et al. 2022).

Inoculate 150 g of seeds as detailed in Sect. 6.5.3.2 and leave them to dry for 2 h. Three biological samples of 50 g of seeds in each are then transferred to sterile Erlenmeyer flasks containing 50 mL of sterile saline solution containing 0.01% (v/v) Tween 80. The flasks are agitated for 15 min; this step is referred to as the direct dilution. After that, serial dilutions are performed, and a 100 μL aliquot from each dilution (direct, 10^{-1} and 10^{-2}) and each replicate is spread into Petri's dish containing RC medium, using a previously sterilized Drigalskyi spatula. For each biological replicate, three technical replicates are performed. Plates are incubated for 4 days at 28–30 °C. After this period, colonies of the expected phenotype are counted on the plates, ranging from 30 to 300 colonies, and the results are expressed as CFU.mL^{-1}.

6.5.3.4 Colonization of Leaves

The colonization of *Azospirillum* on spray-inoculated leaves can be determined by recovering the viable culturable cells from leaves for the CFU count. In a typical assay, maize plants are raised in a growth chamber for 15 days (until stage V2/V3) and then spray-inoculated with 50 mL.plant^{-1} of an *Azospirillum* inoculum containing approximately 10^6 CFU.mL^{-1} (1:500 dilution of the bacterial suspension obtained as explained in Sect. 6.5.3.2), to obtain a final dose of 5.10^7 CFU.plant^{-1}. At a selected time point after inoculation, plant shoots are collected, placed in

100 mL of SDW, and gently agitated for 10 min. If endophytic colonization is to be analyzed, plants are first superficially disinfected with a 15-s rinse with 70% ethanol, followed by a 10-min treatment with 5% sodium hypochlorite and five subsequent washes with SDW, and then mechanically disrupted in 100 mL of SDW with a homogenizer to obtain the initial homogenate (= 10^{-2} dilution). Samples are serially diluted until 10^{-5} and used for CFU count as detailed above. Values obtained are expressed as CFU.plant^{-1}.

References

Arruebarrena Di Palma A, Pereyra CM, Moreno Ramirez L et al (2013) Denitrification-derived nitric oxide modulates biofilm formation in *Azospirillum brasilense*. FEMS Microbiol Lett 338(1):77–85. https://doi.org/10.1111/1574-6968.12030

Arunakumari A, Lamm RB, Neyra-Estens CA (1992) Changes in cell surface properties of azospirilla in relation to cell pleomorphism and aggregation. Symbiosis 13(305):1751–1758

Assmus B, Hutzler P, Kirchhof G et al (1995) *In situ* localization of *Azospirillum brasilense* in the rhizosphere of wheat with fluorescently labeled, rRNA-targeted oligonucleotide probes and scanning confocal laser microscopy. Appl Environ Microbiol 61(3):1013–1019. https://doi.org/10.1128/aem.61.3.1013-1019.1995

Bahat-Samet E, Castrosowinski S, Okon Y (2004) Arabinose content of extracellular polysaccharide plays a role in cell aggregation of *Azospirillum brasilense*. FEMS Microbiol Lett 237(2): 195–203. https://doi.org/10.1016/j.femsle.2004.06.036

Baldani J, Caruso L, Baldani VLD, Goi SR, Döbereiner J (1997) Recent advances in BNF with non-legume plants. Soil Biology and Biochemistry, 29(5-6):911–922. https://doi.org/10.1016/s0038-0717(96)00218-0

Balsanelli E, Pedrosa FO, Souza EM (2017) Quantification of grass colonization by associative bacteria. Curr Protoc Plant Biol 2(2):108–123. https://doi.org/10.1002/9781119077367.2017.2.issue-2, https://doi.org/10.1002/cppb.20047

Bashan Y (1986) Migration of the rhizosphere bacteria *Azospirillum brasilense* and *Pseudomonas fluorescens* towards wheat roots in the soil. J Gen Microbiol 132(12):3407–3414

Bashan Y, Holguin G (1993) Anchoring of *Azospirillum brasilense* to hydrophobic polystyrene and wheat roots. J Gen Microbiol 139(2):379–385. https://doi.org/10.1099/00221287-139-2-379

Bashan Y, Holguin G (1994) Root-to-root travel of the beneficial bacterium *Azospirillum brasilense*. Appl Environ Microbiol 60(6):2120–2131. https://doi.org/10.1128/aem.60.6.2120-2131.1994

Bashan Y, Levanony H, Whitmoyer RE (1991) Root surface colonization of non-cereal crop plants by pleomorphic *Azospirillum brasilense* Cd. Microbiology 137(1):187–196. https://doi.org/10.1099/00221287-137-1-187

Bastarrachea F, Zamudio M, Rivas R (1988) Non-encapsulated mutants of *Azospirillum brasilense* and *Azospirillum lipoferum*. Can J Microbiol 34(1):24–29. https://doi.org/10.1139/m88-005

Bible AN, Stephens BB, Ortega DR et al (2008) Function of a chemotaxis-like signal transduction pathway in modulating motility, cell clumping, and cell length in the Alphaproteobacterium *Azospirillum brasilense*. J Bacteriol 190(19):6365–6375. https://doi.org/10.1128/jb.00734-08

Bible A, Russell MH, Alexandre G (2012) The *Azospirillum brasilense* Che1 chemotaxis pathway controls swimming velocity, which affects transient cell-to-cell clumping. J Bacteriol 194(13): 3343–3355. https://doi.org/10.1128/jb.00310-12

Bible AN, Khalsa-Moyers GK, Mukherjee T et al (2015) Metabolic adaptations of *Azospirillum brasilense* to oxygen stress by cell-to-cell clumping and flocculation. Appl Environ Microbiol 81(24):8346–8357. https://doi.org/10.1128/aem.02782-15

Biondi EG, Marini F, Altieri F et al (2004) Extended phenotype of an *mreB*-like mutant in *Azospirillum brasilense*. Microbiology 150(7):2465–2474. https://doi.org/10.1099/mic.0.26904-0

Böhning J, Tarafder AK, Bharat TAM (2024) The role of filamentous matrix molecules in shaping the architecture and emergent properties of bacterial biofilms. Biochem J 481(4):245–263. https://doi.org/10.1042/bcj20210301

Borrajo MP, Mondino EA, Maroniche GA et al (2022) Potential of rhizobacteria native to Argentina for the control of *Meloidogyne javanica*. Rev Argent Microbiol 54(3):224–232. https://doi.org/10.1016/j.ram.2021.02.010

Branda SS, Vik Å, Friedman L et al (2005) Biofilms: the matrix revisited. Trends Microbiol 13(1):20–26. https://doi.org/10.1016/j.tim.2004.11.006

Burdman S, Jurkevitch E, Schwartsburd B et al (1998) Aggregation in *Azospirillum brasilense*: effects of chemical and physical factors and involvement of extracellular components. Microbiology 144(7):1989–1999. https://doi.org/10.1099/00221287-144-7-1989

Burdman S, Okon Y, Jurkevitch E (2000) Surface characteristics of *Azospirillum brasilense* in relation to cell aggregation and attachment to plant roots. Crit Rev Microbiol 26(2):91–110. https://doi.org/10.1080/10408410091154200

Burdman S, Dulguerova G, Okon Y et al (2001) Purification of the major outer membrane protein of *Azospirillum brasilense*, its affinity to plant roots, and its involvement in cell aggregation. Mol Plant-Microbe Interact 14(4):555–561. https://doi.org/10.1094/mpmi.2001.14.4.555

Calleja GB, Atkinson B, Reichenbach H et al (1984) Aggregation. In: Microbial adhesion and aggregation, pp 302–321. https://doi.org/10.1007/978-3-642-70137-5_20

Camilios-Neto D, Bonato P, Wassem R et al (2014) Dual RNA-seq transcriptional analysis of wheat roots colonized by *Azospirillum brasilense* reveals up-regulation of nutrient acquisition and cell cycle genes. BMC Genomics 15(1). https://doi.org/10.1186/1471-2164-15-378

Carreño-López R, Sánchez A, Camargo N et al (2009) Characterization of *chsA*, a new gene controlling the chemotactic response in *Azospirillum brasilense* Sp7. Arch Microbiol 191(6):501–507. https://doi.org/10.1007/s00203-009-0475-x

Cassan F, Maiale S, Masciarell O et al (2009) Cadaverine production by *Azospirillum brasilense* and its possible role in plant growth promotion and osmotic stress mitigation. Eur J Soil Biol 45(1):12–19. https://doi.org/10.1016/j.ejsobi.2008.08.003

Castellanos T, Ascencio F, Bashan Y (1998) Cell-surface lectins of *Azospirillum* spp. Curr Microbiol 36(4):241–244. https://doi.org/10.1007/s002849900302

Chi F, Shen S-H, Chen S-F et al (2004) Migration of *Azospirillum brasilense* Yu62 from root to stem and leaves inside rice and tobacco plants. J Integr Plant Biol 46(9):1065–1070. https://www.jipb.net/CN/abstract/article_22315.shtml

Combes-Meynet E, Pothier JF, Moenne-Loccoz Y et al (2011) The *Pseudomonas* secondary metabolite 2,4-diacetylphloroglucinol is a signal inducing rhizoplane expression of *Azospirillum* genes involved in plant-growth promotion. Mol Plant-Microbe Interact 24(2):271–284. https://doi.org/10.1094/MPMI-07-10-0148

Costerton JW, Lewandowski Z, Caldwell DE et al (1995) Microbial biofilms. Ann Rev Microbiol 49(1):711–745. https://doi.org/10.1146/annurev.mi.49.100195.003431

Couillerot O, Combes-Meynet E, Pothier JF et al (2011) The role of the antimicrobial compound 2,4-diacetylphloroglucinol in the impact of biocontrol *Pseudomonas fluorescens* F113 on *Azospirillum brasilense* phytostimulators. Microbiology 157(Pt 6):1694–1705. https://doi.org/10.1099/mic.0.043943-0

Creus CM, Graziano M, Casanovas EM et al (2005) Nitric oxide is involved in the *Azospirillum brasilense*-induced lateral root formation in tomato. Planta 221(2):297–303. https://doi.org/10.1007/s00425-005-1523-7

Croes CL, Moens S, van Bastelaere E et al (1993) The polar flagellum mediates *Azospirillum brasilense* adsorption to wheat roots. J Gen Microbiol 139(9):2261–2269. https://doi.org/10.1099/00221287-139-9-2261

Cruz-Pérez JF, Lara-Oueilhe R, Marcos-Jiménez C et al (2021) Expression and function of the cdgD gene, encoding a CHASE–PAS-DGC-EAL domain protein, in *Azospirillum brasilense*. Sci Rep 11(1). https://doi.org/10.1038/s41598-020-80125-3

Davey ME, O'Toole GA (2000) Microbial biofilms: from ecology to molecular genetics. Microbiol Mol Biol Rev 64(4):847–867. https://doi.org/10.1128/mmbr.64.4.847-867.2000

Day JM, Döbereiner J (1976) Physiological aspects of N2-fixation by a *Spirillum* from *Digitaria* roots. Soil Biol Biochem 8(1):45–50. https://doi.org/10.1016/0038-0717(76)90020-1

De Troch P, Vanderleyden J (1996) Surface properties and motility of *Rhizobium* and *Azospirillum* in relation to plant root attachment. Microb Ecol 32(2):149–169. https://doi.org/10.1007/bf00185885

Del Gallo M, Haegi A (1990) Characterization and quantification of exocellular polysaccharides in *Azospirillum brasilense* and *Azospirillum lipoferum*. Symbiosis (Rehovot) 9(1–3):155–161

Del Gallo M, Negi M, Neyra CA (1989) Calcofluor- and lectin-binding exocellular polysaccharides of *Azospirillum brasilense* and *Azospirillum lipoferum*. J Bacteriol 171(6):3504–3510. https://doi.org/10.1128/jb.171.6.3504-3510.1989

Di Salvo LP, Garcia JE, Puente ML et al (2022) The drop plate method as an alternative for *Azospirillum* spp. viable cell enumeration within the consensus protocol of the REDCAI network. Rev Argent Microbiol 54(2):152–157. https://doi.org/10.1016/j.ram.2021.05.002

Diaz PR, Romero M, Pagnussatt L et al (2022) *Azospirillum baldaniorum* Sp245 exploits *Pseudomonas fluorescens* A506 biofilm to overgrow in dual-species macrocolonies. Environ Microbiol 24(12):5707–5720. https://doi.org/10.1111/1462-2920.16195

Döbereiner J, Marriel IE, Nery M (1976) Ecological distribution of *Spirillum lipoferum* Beijerinck. Can J Microbiol 22(10):1464–1473. https://doi.org/10.1139/m76-217

Dubey AP, Pandey P, Mishra S et al (2021) Role of a fasciclin domain protein in photooxidative stress and flocculation in *Azospirillum brasilense* Sp7. Res Microbiol 172(6):103875. https://doi.org/10.1016/j.resmic.2021.103875

Edwards AN, Siuti P, Bible AN et al (2011) Characterization of cell surface and extracellular matrix remodeling of *Azospirillum brasilense* chemotaxis-like 1 signal transduction pathway mutants by atomic force microscopy. FEMS Microbiol Lett 314(2):131–139. https://doi.org/10.1111/j.1574-6968.2010.02156.x

Fujita M, Kusajima M, Okumura Y et al (2017) Effects of colonization of a bacterial endophyte, *Azospirillum* sp. B510, on disease resistance in tomato. Biosci Biotechnol Biochem 81(8):1657–1662. https://doi.org/10.1080/09168451.2017.1329621

Fukami J, Ollero FJ, Megías M et al (2017) Phytohormones and induction of plant-stress tolerance and defense genes by seed and foliar inoculation with *Azospirillum brasilense* cells and metabolites promote maize growth. AMB Express 7(1). https://doi.org/10.1186/s13568-017-0453-7

Galperin MY, Gaidenko TA, Mulkidjanian AY et al (2001) MHYT, a new integral membrane sensor domain. FEMS Microbiol Lett 205(1):17–23. https://doi.org/10.1111/j.1574-6968.2001.tb10919.x

García JE, Maroniche G, Creus C et al (2017) In vitro PGPR properties and osmotic tolerance of different *Azospirillum* native strains and their effects on growth of maize under drought stress. Microbiol Res 202:21–29. https://doi.org/10.1016/j.micres.2017.04.007

Gonzalez AJ, Yarte ME, Llorente BE et al (2024) Biochemical changes by *Azospirillum brasilense* enhance jojoba rooting under salt stress. Plant Cell Tissue Organ Cult (PCTOC) 156(2). https://doi.org/10.1007/s11240-023-02665-6

González-Torres M, Ramírez-Mata A, Melgarejo-Ramírez Y et al (2020) Assessment of biocompatibility and surface topography of poly(ester urethane) – silica nanocomposites reveals multifunctional properties. Mater Lett 276. https://doi.org/10.1016/j.matlet.2020.128269

Gross H, Loper JE (2009) Genomi of secondary metabolite production by *Pseudomonas* spp. Nat Prod Rep 26(11):1408–1446. https://doi.org/10.1039/b817075b

Hamana K, Takeuchi M (1998) Polyamine profiles as chemotaxonomic marker within alpha, beta, gamma, delta, and epsilon subclasses of class *Proteobacteria*: distribution of 2-hydroxyputrescine and homospermidine. Microb Cult Collect 14:1–14. https://cir.nii.ac.jp/crid/1570572701757135360

Hong L, Orikasa Y, Sakamoto H et al (2019) Plant tissue localization and morphological conversion of *Azospirillum brasilense* upon Initial Interaction with Allium cepa L. Microorganisms 7(9). https://doi.org/10.3390/microorganisms7090275

Hou X, McMillan M, Coumans JV et al (2014) Cellular responses during morphological transformation in *Azospirillum brasilense* and its *flcA* knockout mutant. PLoS One 9(12):e114435. https://doi.org/10.1371/journal.pone.0114435

Inomura K, Bragg J, Follows MJ (2017) A quantitative analysis of the direct and indirect costs of nitrogen fixation: a model based on *Azotobacter vinelandii*. ISME J *11*(1):166–175. https://doi.org/10.1038/ismej.2016.97

Jijón-Moreno S, Baca BE, Castro-Fernández DC et al (2019) TyrR is involved in the transcriptional regulation of biofilm formation and D-alanine catabolism in *Azospirillum brasilense* Sp7. PLoS One 14(2). https://doi.org/10.1371/journal.pone.0211904

Joe MM, Karthikeyan B, Chauhan PS et al (2012) Survival of *Azospirillum brasilense* flocculated cells in alginate and its inoculation effect on growth and yield of maize under water deficit conditions. Eur J Soil Biol 50:198–206. https://doi.org/10.1016/j.ejsobi.2012.03.002

Jofre E, Lagares A, Mori G (2004) Disruption of dTDP-rhamnose biosynthesis modifies lipopolysaccharide core, exopolysaccharide production, and root colonization in *Azospirillum brasilense*. FEMS Microbiol Lett 231(2):267–275. https://doi.org/10.1016/S0378-1097(04)00003-5

Katupitiya S, Millet J, Vesk M et al (1995) A mutant of *Azospirillum brasilense* Sp7 impaired in flocculation with a modified colonization pattern and superior nitrogen fixation in association with wheat. Appl Environ Microbiol 61(5):1987–1995. https://doi.org/10.1128/aem.61.5.1987-1995.1995

Kim KY, Deka H, Woo C et al (2010) Isolation and evaluation of inoculation effect of Azospirillum sp. on growth, colonization and nutrient uptake of crops under green house condition. In: Proceedings of the 19th World Congress of Soil Science: Soil solutions for a changing world, Brisbane

Kloepper J, Schippers B, Bakker P (1992) Proposed elimination of the term endorhizosphere. Phytopathology 82(7):726–727

Li Y, Heine S, Entian M et al (2013) NO-induced biofilm dispersion in *Pseudomonas aeruginosa* is mediated by an MHYT domain-coupled phosphodiesterase. J Bacteriol 195(16):3531–3542. https://doi.org/10.1128/jb.01156-12

Liu Z, Beskrovnaya P, Melnyk RA et al (2018) A genome-wide screen identifies genes in rhizosphere-associated *Pseudomonas* required to evade plant defenses. MBio 9(6). https://doi.org/10.1128/mBio.00433-18

Malinich EA, Bauer CE (2018a) Transcriptome analysis of *Azospirillum brasilense* vegetative and cyst states reveals large-scale alterations in metabolic and replicative gene expression. Microb Genom *4*(8). https://doi.org/10.1099/mgen.0.000200

Malinich EA, Bauer CE (2018b) The plant growth promoting bacterium *Azospirillum brasilense* is vertically transmitted in *Phaseolus vulgaris* (common bean). Symbiosis 76(2):97–108. https://doi.org/10.1007/s13199-018-0539-2

Maroniche GA, Rubio EJ, Consiglio A et al (2016) Plant-associated fluorescent *Pseudomonas* from red lateritic soil: beneficial characteristics and their impact on lettuce growth. J Gen Appl Microbiol 62(5):248–257. https://doi.org/10.2323/jgam.2016.04.006

Maroniche GA, Diaz PR, Borrajo MP et al (2018) Friends or foes in the rhizosphere: traits of fluorescent *Pseudomonas* that hinder *Azospirillum brasilense* growth and root colonization. FEMS Microbiol Ecol 94(12):fiy202. https://doi.org/10.1093/femsec/fiy202

Martínez-Ortiz IC, Ahumada-Manuel CL, Hsueh BY et al (2020) Cyclic di-GMP-mediated regulation of extracellular mannuronan c-5 epimerases is essential for cyst formation in *Azotobacter vinelandii*. J Bacteriol 202(24). https://doi.org/10.1128/jb.00135-20

Mata AR, Pacheco CM, Cruz Pérez JF et al (2018) *In silico* comparative analysis of GGDEF and EAL domain signaling proteins from the *Azospirillum* genomes. BMC Microbiol 18(1). https://doi.org/10.1186/s12866-018-1157-0

Michiels KW, Croes CL, Vanderleyden J (1991) Two different modes of attachment of *Azospirillum brasilense* Sp7 to wheat roots. Microbiology 137(9):2241–2246. https://doi.org/10.1099/00221287-137-9-2241

Moens S, Michiels K, Keijers V et al (1995) Cloning, sequencing, and phenotypic analysis of laf1, encoding the flagellin of the lateral flagella of *Azospirillum brasilense* Sp7. J Bacteriol 177(19): 5419–5426. https://doi.org/10.1128/jb.177.19.5419-5426.1995

Naqqash T, Malik KA, Imran A et al (2022) Inoculation with *Azospirillum* spp. acts as the liming source for improving growth and nitrogen use efficiency of potato. Front Plant Sci 13. https://doi.org/10.3389/fpls.2022.929114

Nievas S, Coniglio A, Takahashi WY et al (2023) Unraveling *Azospirillum*'s colonization ability through microbiological and molecular evidence. J Appl Microbiol 134(4). https://doi.org/10.1093/jambio/lxad071

O'Neal L, Akhter S, Alexandre G (2019) A PilZ-containing chemotaxis receptor mediates oxygen and wheat root sensing in *Azospirillum brasilense*. Front Microbiol 10. https://doi.org/10.3389/fmicb.2019.00312

O'Neal L, Vo L, Alexandre G (2020) Specific root exudate compounds sensed by dedicated chemoreceptors shape *Azospirillum brasilense* chemotaxis in the rhizosphere. Appl Environ Microbiol 86(15). https://doi.org/10.1128/aem.01026-20

O'Toole GA, Kolter R (1998) Initiation of biofilm formation in *Pseudomonas fluorescens* WCS365 proceeds via multiple, convergent signalling pathways: a genetic analysis. Mol Microbiol 28(3): 449–461. https://doi.org/10.1046/j.1365-2958.1998.00797.x

Okon Y (1992) Poly-β-hydroxybutyrate metabolism in *Azospirillum brasilense* and the ecological role of PHB in the rhizosphere. FEMS Microbiol Lett 103(2-4):131–139. https://doi.org/10.1016/0378-1097(92)90302-5

Okon Y, Albrecht SL, Burris RH (1976) Factors affecting growth and nitrogen fixation of *Spirillum lipoferum*. J Bacteriol 127(3):1248–1254. https://doi.org/10.1128/jb.127.3.1248-1254.1976

Okshevsky M, Meyer RL (2013) The role of extracellular DNA in the establishment, maintenance and perpetuation of bacterial biofilms. Crit Rev Microbiol 41(3):341–352. https://doi.org/10.3109/1040841x.2013.841639

Pagnussat LA, Salcedo F, Maroniche G et al (2016) Interspecific cooperation: enhanced growth, attachment and strain-specific distribution in biofilms through *Azospirillum brasilense-Pseudomonas protegens* co-cultivation. FEMS Microbiol Lett 363(20):fnw238. https://doi.org/10.1093/femsle/fnw238

Papen H, Werner D (1982) Organic acid utilization, succinate excretion, encystation and oscillating nitrogenase activity in *Azospirillum brasilense* under microaerobic conditions. Arch Microbiol 132(1):57–61. https://doi.org/10.1007/bf00690818

Pereg L (2015) *Azospirillum* cell aggregation, attachment, and plant interaction. In: Handbook for Azospirillum, pp 181–197. https://doi.org/10.1007/978-3-319-06542-7_10

Pereg-Gerk L (2004) Expression of *flcA*, a gene regulating differentiation and plant interaction in *Azospirillum*. Soil Biol Biochem 36(8):1245–1252. https://doi.org/10.1016/j.soilbio.2004.04.005

Pereg-Gerk L, Paquelin A, Gounon P et al (1998) A transcriptional regulator of the LuxR-UhpA family, FlcA, controls flocculation and wheat root surface colonization by *Azospirillum brasilense* Sp7. Mol Plant-Microbe Interact 11(3):177–187. https://doi.org/10.1094/mpmi.1998.11.3.177

Pereg-Gerk L, Gilchrist K, Kennedy IR (2000) Mutants with enhanced nitrogenase activity in hydroponic *Azospirillum brasilense*-wheat associations. Appl Environ Microbiol 66(5): 2175–2184. https://doi.org/10.1128/aem.66.5.2175-2184.2000

Puente ML, Maroniche GA, Panepucci M et al (2020) Localization and survival of *Azospirillum brasilense* Az39 in soybean leaves. *Lett Appl Microbiol*. https://doi.org/10.1111/lam.13444

Ramírez-Mata A, López-Lara LI, Xiqui-Vázquez ML et al (2016) The cyclic-di-GMP diguanylate cyclase CdgA has a role in biofilm formation and exopolysaccharide production in *Azospirillum brasilense*. Res Microbiol 167(3):190–201. https://doi.org/10.1016/j.resmic.2015.12.004

Ramirez-Mata A, Pacheco MR, Moreno SJ et al (2018) Versatile use of *Azospirillum brasilense* strains tagged with *egfp* and *mCherry* genes for the visualization of biofilms associated with wheat roots. Microbiol Res 215:155–163. https://doi.org/10.1016/j.micres.2018.07.007

Randall TE, Eckartt K, Kakumanu S et al (2022) Sensory perception in bacterial cyclic diguanylate signal transduction. J Bacteriol 204(2). https://doi.org/10.1128/jb.00433-21

Rather MA, Gupta K, Mandal M (2021) Microbial biofilm: formation, architecture, antibiotic resistance, and control strategies. Braz J Microbiol 52(4):1701–1718. https://doi.org/10.1007/s42770-021-00624-x

REDCAI (2025) Manual de procedimientos microbiológicos para la evaluación de calidad de inoculantes, bioestimulantes y biocontroladores. Asociación Argentina de Microbiología

Rodrıguez Cáceres EA (1982) Improved medium for isolation of *Azospirillum* spp. Appl Environ Microbiol 44(4):990–991. https://www.ncbi.nlm.nih.gov/pubmed/16346123

Römling U, Galperin MY, Gomelsky M (2013) Cyclic di-GMP: the first 25 years of a universal bacterial second messenger. Microbiol Mol Biol Rev 77(1):1–52. https://doi.org/10.1128/mmbr.00043-12

Rossi FA, Medeot DB, Liaudat JP et al (2016) In *Azospirillum brasilense*, mutations in flmA or flmB genes affect polar flagellum assembly, surface polysaccharides, and attachment to maize roots. Microbiol Res 190:55–62. https://doi.org/10.1016/j.micres.2016.05.006

Russell MH, Bible AN, Fang X et al (2013) Integration of the second messenger c-di-GMP into the chemotactic signaling pathway. MBio 4(2). https://doi.org/10.1128/mBio.00001-13

Sadasivan L, Neyra CA (1985) Flocculation in *Azospirillum brasilense* and *Azospirillum lipoferum*: exopolysaccharides and cyst formation. J Bacteriol 163(2):716–723. https://doi.org/10.1128/jb.163.2.716-723.1985

Sadiq FA, Burmølle M, Heyndrickx M et al (2021) Community-wide changes reflecting bacterial interspecific interactions in multispecies biofilms. Crit Rev Microbiol 47(3):338–358. https://doi.org/10.1080/1040841x.2021.1887079

Serra DO, Hengge R (2017) Experimental detection and visualization of the extracellular matrix in macrocolony biofilms. Methods Mol Biol 1657:133–145. https://doi.org/10.1007/978-1-4939-7240-1_11

Shelud'ko AV, Filip'echeva YA, Shumilova EM et al (2015) Changes in biofilm formation in the nonflagellated flhB1 mutant of *Azospirillum brasilense* Sp245. Microbiology 84(2):144–151. https://doi.org/10.1134/s0026261715010129

Shelud'ko AV, Filip'echeva YA, Telesheva EM et al (2018) Characterization of carbohydrate-containing components of *Azospirillum brasilense* Sp245 biofilms. Microbiology 87(5):610–620. https://doi.org/10.1134/s0026261718050156

Shelud'ko AV, Mokeev DI, Evstigneeva SS et al (2020) Cell ultrastructure in *Azospirillum brasilense* biofilms. Microbiology 89(1):50–63. https://doi.org/10.1134/s0026261720010142

Shelud'ko A, Volokhina I, Mokeev D et al (2025) Multilevel analysis of *Azospirillum* biofilms and cystlike forms and characterization of their possible role in wheat drought tolerance. Rhizosphere 33. https://doi.org/10.1016/j.rhisph.2025.101029

Sierra Cacho D, Zamorano Sánchez DS, Xiqui-Vázquez ML et al (2021) CdgC, a cyclic-di-GMP diguanylate cyclase of *Azospirillum baldaniorum* is involved in internalization to wheat roots. Front Plant Sci 12. https://doi.org/10.3389/fpls.2021.748393

Skvortsov IM, Ignatov VV (1998) Extracellular polysaccharides and polysaccharide-containing biopolymers from *Azospirillum* species: properties and the possible role in interaction with plant roots. FEMS Microbiol Lett 165(2):223–229. https://doi.org/10.1111/j.1574-6968.1998.tb13150.x

Takahashi WY, Galvão CW, Cassán FD et al (2024) Tracking maize colonization and growth promotion by Azospirillum reveals strain-specific behavior and the influence of inoculation method. Plant Physiol Biochem 215. https://doi.org/10.1016/j.plaphy.2024.108979

Tarrand JJ, Krieg NR, Döbereiner J (1978) A taxonomic study of the *Spirillum lipoferum* group, with descriptions of a new genus, *Azospirillum* gen. nov. and two species, *Azospirillum lipoferum* (Beijerinck) comb. nov. and *Azospirillum brasilense* sp. nov. Can J Microbiol 24(8):967–980. https://doi.org/10.1139/m78-160

Valdameri G, Kokot TB, Pedrosa FO et al (2015) Rapid quantification of rice root-associated bacteria by flow cytometry. Lett Appl Microbiol 60(3):237–241. https://doi.org/10.1111/lam.12351

Valverde C, Gonzalez Anta G, Ferraris G (2015) *Pseudomonas* and *Azospirillum*. In: Cassán FD, Okon Y, Creus CM (eds) Handbook for Azospirillum: technical issues and protocols. Springer, Cham, pp 389–409. https://doi.org/10.1007/978-3-319-06542-7_21

Viruega-Góngora VI, Acatitla-Jácome IS, Reyes-Carmona SR et al (2020) Spatio-temporal formation of biofilms and extracellular matrix analysis in *Azospirillum brasilense*. FEMS Microbiol Lett 367(4). https://doi.org/10.1093/femsle/fnaa037

Viruega-Góngora VI, Acatitla-Jácome IS, Zamorano-Sánchez D et al (2022) The GGDEF-EAL protein CdgB from *Azospirillum baldaniorum* Sp245, is a dual function enzyme with potential polar localization. PLoS One 17(11). https://doi.org/10.1371/journal.pone.0278036

Wang Y, Hay ID, Rehman ZU et al (2015) Membrane-anchored MucR mediates nitrate-dependent regulation of alginate production in *Pseudomonas aeruginosa*. Appl Microbiol Biotechnol 99(17):7253–7265. https://doi.org/10.1007/s00253-015-6591-4

Wang D, Xu A, Elmerich C et al (2017) Biofilm formation enables free-living nitrogen-fixing rhizobacteria to fix nitrogen under aerobic conditions. ISME J 11(7):1602–1613. https://doi.org/10.1038/ismej.2017.30

Wisniewski-Dyé F, Borziak K, Khalsa-Moyers G et al (2011) *Azospirillum* genomes reveal transition of bacteria from aquatic to terrestrial environments. PLoS Genet 7(12):e1002430. https://doi.org/10.1371/journal.pgen.1002430

Yasuda M, Isawa T, Shinozak S et al (2014) Effects of colonization of a bacterial endophyte, *Azospirillum* sp. B510, on disease resistance in rice. Biosci Biotechnol Biochem 73(12): 2595–2599. https://doi.org/10.1271/bbb.90402

Zhang L, Li Z, Chen Z (2020) Live cell fluorescent stain of bacterial curli and biofilm through supramolecular recognition between bromophenol blue and CsgA. Chem Commun 56(37): 5014–5017. https://doi.org/10.1039/d0cc01643h

Chapter 7
Cell-Cell Communication in *Azospirillum*

Ludovic Vial, Sofia Nievas, Deniz Sarigol, Anahí Coniglio, and Florence Wisniewski-Dyé

Abstract Cell-cell communication or quorum sensing (QS), a common way of regulating phenotypes at the population level, relies on the self-production of small molecules accumulating extracellularly along growth. In Gram-negative bacteria, the main signaling compounds belong to the *N*-acyl-homoserine lactone (AHL) family and are synthesized by an AHL synthase (LuxI family) and recognized by a transcriptional regulator (LuxR family) that triggers the expression of specific genes. This chapter presents the initial studies of QS that was focused on 40 strains of *Azospirillum* and revealed that only 4 strains isolated from rice were able to synthesize AHLs. A genome mining for QS genes conducted on 43 *Azospirillum* genomes (from 24 validated species) highlights the presence of a complete QS system in only 5 genomes from species belonging to the *lipoferum* clade, while a strain possesses only a *luxI* gene and 12 strains bear orphan *luxR* genes. The potential to interfere with quorum sensing, i.e., quorum quenching, is also surveyed and seems to be widespread, although functional evidence is needed. Finally, methodology for AHL extraction and characterization, as well as for identification of genes mediating AHL-based QS and QS-controlled functions, is provided.

Keywords Quorum sensing · Communication · AHL · LuxR · Regulation

L. Vial · D. Sarigol · F. Wisniewski-Dyé (✉)
Laboratoire d'Ecologie Microbienne -UMR CNRS 5557, UMR INRAE 1418-, Université Claude Bernard Lyon 1, Villeurbanne, France
e-mail: Florence.wisniewski@univ-lyon1.fr

S. Nievas
Laboratorio de Fisiología Vegetal y de la Interacción Planta-Microorganismo, Instituto de Investigaciones Agrobiotecnológicas (INIAB-CONICET), Universidad Nacional de Río Cuarto (UNRC), Río Cuarto, Córdoba, Argentina

A. Coniglio
Laboratorio de Fisiología Vegetal y de la Interacción Planta-Microorganismo, Instituto de Investigaciones Agrobiotecnológicas (INIAB-CONICET), Universidad Nacional de Río Cuarto (UNRC), Río Cuarto, Córdoba, Argentina

School of Biosciences, University of Nottingham, Loughborough, UK

F. D. Cassán et al. (eds.), *The Definitive Handbook of Azospirillum*,
https://doi.org/10.1007/978-3-032-09222-9_7

7.1 Introduction

Among the kingdom Bacteria, cell-cell communication is a widespread way of regulating phenotypes at the population level. Cell-cell communication, also named quorum sensing (QS), relies on the self-production of small molecules (i.e., autoinducers) accumulating extracellularly along growth until a threshold concentration or a quorum of cells is reached (Fuqua and Winans 1994; Soto-Aceves et al. 2023). Detection of the autoinducer triggers the coordinated expression of a given phenotype that is often involved in the interaction between prokaryotic cells and an eukaryotic host. In Gram-negative bacteria, the most studied family of molecules is composed of *N*-acyl-homoserine lactones (AHLs), the first to be described being 3oxo,C_6-HSL produced by the marine bacterium *Aliivibrio fischeri* and coordinating bioluminescence (Eberhard et al. 1981). AHL-based communication mainly relies on the presence of a gene coding an AHL synthase (LuxI family) and a gene encoding a transcriptional regulator able to bind cognate AHL (LuxR family) and to trigger the expression of specific genes. Pioneer studies demonstrated that plant-associated bacteria, whether pathogenic or beneficial, relied on quorum sensing to interact with plants (Jones et al. 1993; Barnard and Salmon 2007; Wisniewski-Dyé and Downie 2002). For example, the synthesis of phenazines, compounds able to protect plants from fungal pathogens, by members of the *Pseudomonas* genus was shown to be under QS control (Pierson et al. 1994). Logically, the question arose whether the plant-associated bacterium belonging to the genus *Azospirillum* used QS regulation or not. This chapter will present the initial studies focusing on QS in *Azospirillum*, as well as a genome mining for QS genes based on the numerous genomes now available and a discussion on putative functions regulated by AHL-based QS. Finally, the ability to interfere with quorum sensing, i.e., quorum quenching (QQ), will also be surveyed and discussed.

7.2 First Discovery of AHL-Based Quorum Sensing Systems in *Azospirillum*

The first study of quorum sensing (QS) in *Azospirillum* dates back to 2006, with an analysis of signal production by 40 strains and gene characterization for one strain (Vial et al. 2006). The screening of AHL production was undertaken by using two biosensors (i.e., bacteria that can no longer make their own AHL but are still able to respond to exogenous AHLs) with different specificity for AHL recognition (Steindler and Venturi 2007) (see Sect. 7.6.1). Four strains isolated from rice could trigger the biosensors: B52, B510, and B518 were isolated from rice disinfected stems in Japan (Elbeltagy et al. 2001), and TVV3 was isolated from the rice rhizosphere in Vietnam (Trân Vân et al. 1997). These strains were all affiliated with the *A. lipoferum* species based on *rrs* sequencing; however, it is likely that they belong to different species, a fact that was later confirmed for B510 when the genome sequence of this strain became available (Kaneko et al. 2010). Thin-layer chromatography (TLC) of AHLs extracted from the supernatant indicated that all four strains were able to produce several AHLs, with two endophytic strains (B518 and B510) displaying very similar AHL patterns. The structural identification was performed by mass spectrometry coupled to liquid chromatography for one Japanese endophyte (B518) and for the strain isolated from Vietnam (TVV3), and the concentration of some AHLs in stationary-phase culture was estimated. TVV3 produced 3O-C_8-HSL (5 pM), C_8-HSL (1 nM), 3O-C_{10}-HSL, 3OH-C_{10}-HSL, and C_{10}-HSL (20 nM), whereas B518 produced 3O-C_6-HSL (50 nM), C_6-HSL, 3O-C_8-HSL, 3OH-C_8-HSL, and C_8-HSL (500 nM). The low quantity of AHLs produced by TVV3 was in accordance with the amount of culture supernatant needed to activate biosensors (i.e., 20 mL, whereas 10 μL was sufficient for B518 and B510). This first study showed that AHL-based QS regulation was rather an exception than a rule within the *Azospirillum* genus (Vial et al. 2006).

Genes involved in AHL production were further characterized for *A. lipoferum* TVV3 by generating a genomic library and complementing a biosensor strain; those genes, designated *alpR* and *alpI* and belonging respectively to the *luxR* and *luxI* families, could direct the synthesis of the five cognate AHLs when cloned in a suitable host (an *Agrobacterium* strain). RT-PCR also indicated that these two genes, although adjacent and in the same orientation, might be transcribed independently. As genomes were not available at that time, Southern hybridization experiments were undertaken using *alpR-alpI* as a probe; these experiments showed that the QS genes were located on an 85-kb plasmid and that the three other AHL-producing strains harbor QS genes that were not closely related to *alpR-alpI* (Vial et al. 2006). Furthermore, several features indicated that *alpR-alpI* might have been acquired by horizontal gene transfer: (i) the *alpR-alpI* genes seemed to be unique to *A. lipoferum* TVV3, (ii) they were located on a small replicon, one of the smallest of the seven plasmids detected for this strain, (iii) the 2600-bp region carrying *alpR-alpI* displayed a G+C content of 59.3 mol% contrasting with the average 69–70 mol% usually described for *Azospirillum* strains, and (iv) prophage genes were found downstream *alpR-alpI*, notably the FETUD operon encoding the major structural components of the P2 contractile tail (Wisniewski-Dyé and Vial 2015; Christie et al. 2002).

When the genome sequence of strain *Azospirillum* sp. B510 was released (Kaneko et al. 2010), the *luxR* and *luxI* homologs were easily identified and appeared to be poorly related to the *alpR-alpI* genes previously characterized. The encoded proteins share only 30% and 20% identity, for respectively LuxI and LuxR (Fig. 7.1). In B510, the genes, distant from 30 kb, are located on the pA chromid, the largest replicon (1.45 Mbp) after the chromosome (3.31 Mbp), the strain bearing 8 replicons. Interestingly, upstream of the *luxI* gene (AZL_a05890) lie several genes encoding transposases, and three genes encoding peptide/polyketide synthase lie between *luxI* and *luxR*; ORFs encoding a DNA invertase-recombinase, a RTX toxin, and a transposase are also located just downstream the *luxR* gene (AZL_a06030) (Wisniewski-Dyé and Vial 2015). These elements support the acquisition by horizontal gene transfer, and this was further strengthened, at least for the *luxI* gene, by

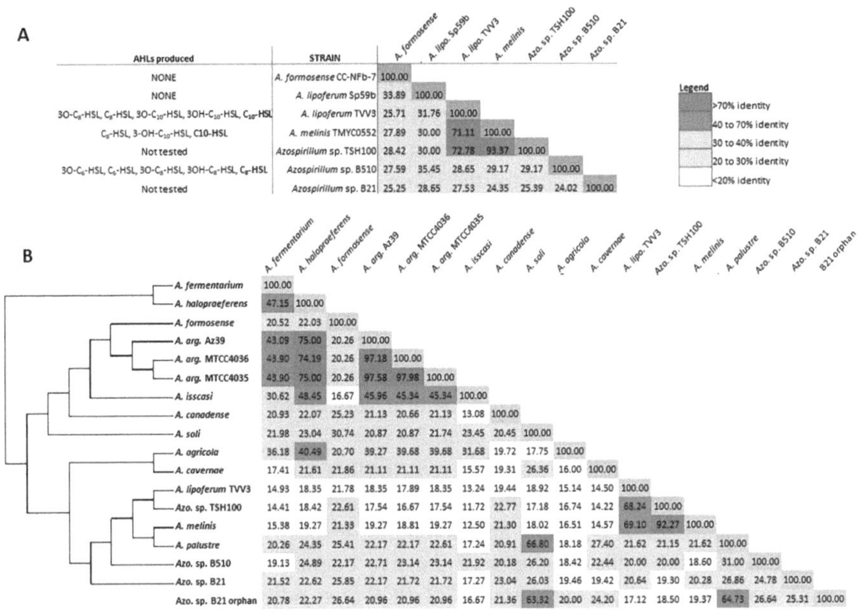

Fig. 7.1 Identity of proteins involved in AHL-mediated quorum sensing in *Azospirillum*. (**a**) All LuxI proteins identified via genome mining were compared with each other and with AlpI of *A. lipoferum* TVV3 (unavailable genome) using ClustalW (https://www.ebi.ac.uk/jdispatcher/msa/clustalo). The profile of AHLs synthesized by the corresponding LuxI was determined as detailed in Sect. 7.6.4. (**b**) All LuxR proteins were compared with each other and with AlpR of *A. lipoferum* TVV3. The phylogenetic tree was inferred from the concatenated taxonomic markers (RpoD-RecA-GyrA) using the neighbor-joining method. As the genome of *A. lipoferum* TVV3 is not available, taxonomic markers of strain *A. lipoferum* Sp59b were used

comparative genomic analyses showing that half of the *Azospirillum* genome has been acquired horizontally from distantly related terrestrial bacteria (Wisniewski-Dyé et al. 2011).

The presence of QS genes was also reported in *A. amazonense* (Sant'Anna et al. 2011), but this species was later reclassified as *Nitrospirillum amazonense* (Lin et al. 2014).

7.3 Occurrence of Uncomplete Quorum Sensing Systems Revealed by Genome Analyses

7.3.1 Choice of Genomes and Methodology

Since these pioneer studies on QS and the update made in the Handbook for *Azospirillum* released in 2015, many new genomes have been released. We thus took the opportunity of this chapter to analyze the occurrence of QS genes among *Azospirillum* genomes. First, a subset of genomes was chosen, based on coverage and completion; incomplete genomes obtained from metagenomes were discarded, and when several sequences were available for one strain, only one sequence was retained. We also tried to obtain a set representative of the diversity of species: to date, 32 *Azospirillum* species have been described, but only 25 are validated (https://lpsn.dsmz.de/). In total, we gathered 43 genomes from 24 validated species (no genome was available for the *Azospirillum largimobile* species), 5 not-yet-validated species, and a few strains whose species remain to be described. LuxI and LuxR homologs were identified by Pfam/INTERPRO domains, as detailed in Sect. 7.6.5, using the MaGe platform (https://mage.genoscope.cns.fr/microscope/home/index.php).

7.3.2 Complete Quorum Sensing Systems

Among the 43 genomes analyzed, genes encoding AHL synthases were identified only in 6 strains (Table 7.1); for 5 strains, the *luxI* gene was linked to a *luxR* gene (*Azospirillum formosense, Azospirillum melinis, Azospirillum* sp. B21, B510, and TSH100). The genome of the sixth strain (*A. lipoferum* Sp59b) possesses only a *luxI* gene. Such a situation is highly unusual, as the only reports of orphan *luxI* (i.e., not linked to a *luxR*) come from marine bacteria having also complete QS systems (*Dinoroseobacter shibae* and *Ruegeria* sp.) (Patzelt et al. 2013; Zan et al. 2015). The orphan LuxI of *A. lipoferum* Sp59b displays ~70% identity with AlpI of TVV3 and LuxI of *A. melinis* and contains the conserved residues of the AHL synthase family (Watson et al. 2002). However, for this strain, under our conditions (growth in AB minimal medium) we were unable to detect the presence of AHLs. Analysis of the region around this orphan *luxI* shows that this gene is located downstream of an ORF with homologies to the *trbC* gene, which is known to be involved in bacterial conjugation. This colocalization of *luxI* and *trbC* is also found in several *Rhizobium* genomes, including *Rhizobium etli*. Finally, over ~5 kb around this *luxI* gene, a GC percentage of 59% was observed, compared to 67% for the entire *A. lipoferum* Sp59b genome.

For LuxI protein associated with a LuxR (i.e., when *luxI* is located near *luxR*), the amino acid sequences of these LuxI proteins were compared with each other and with that of the previously described AlpI protein from *A. lipoferum* TVV3 (whose genome is not available) (Fig. 7.1a). The highest identity (>90%) was observed between AlpI of TVV3 and LuxI of *A. melinis*. For the other AHL synthases, only weak identities were observed and no clear pattern emerged according to phylogeny.

For *A. formosense* CC-Nfb-7, we were unable to detect AHLs in the culture supernatants despite the presence of a complete *luxI-luxR* system (JABEXZ01_12424-JABEXZ01_12425). The LuxI synthase and the LuxR regulator both possess amino acids conserved in LuxI and LuxR proteins. Several ORFs encoding putative transposases and phage proteins are located in the vicinity of the

Table 7.1 List of *Azospirillum* genomes analyzed for the presence of quorum sensing and quorum quenching genes

Azospirillum species[a]	Strain[b]	Origin	Genome size (Mbp)	GC% genome	Number of replicons (or contigs)	Reference GenBank	*luxR* label[c]	Location of *luxR*[d]	GC% of region[e]	Gene encoding acylase[f, g]	Gene encoding amidase[f]
aestuarii	YIM DDC1	Estuarine, China	7.1	68.5	(121)	GCA_016632405.1				**JAEPIV01_10096**	
agricola	**CC-HIH038T**	Cultivated soil, Taiwan	7.8	69.5	8	GCA_020005195.1	AZAGRI_2125	Chr. 3 Mb	67.4	AZAGRI_1635[f]	
argentinense	**Az39**	Surface-disinfected roots of wheat, Argentina	7.4	68.5	6	GCA_000632475.2	ABAZ39_30865	Plas. 641 kb	68.6	**ABAZ39_22635**	ABAZ39_16400
	MTCC4035	Agricultural field, India	7.9	68.5	8	GCA_005222165.1	D3093_27020	Plas. 699 kb	68.6	**D3093_18645**	D3093_17525
	MTCC4036	Agricultural field, India	8.1	68.5	9	GCA_005222185.1	D3867_32520	Plas. 662 kb	68.2	**D3867_22980**	D3867_19790
baldaniorum	**Sp245**	Surface-disinfected roots of wheat, Brazil	7.7	68.5	7	GCA_003119195.2				**AZOBR_p1130068**	
brasilense	**Sp7**	Roots of *Digitaria decumbens*, Brazil	7.0	68.5	(21)	GCA_007827425.1				**D9621_19630**[f]	
	Cd	Agricultural field, USA	7.4	68.5	6	GCA_008274965.1				**D9623_19875**[f]	
	MTCC4038	Agricultural field, India	7.1	68.5	6	GCA_005222145.1				**D3868_17710**	
	MTCC4039	Agricultural field, India	7.2	69.0	6	GCA_005222205.1				**D3869_21380**	D3869_17240
	Ab-V5	Maize, Brazil	6.9	68.5	63	GCA_002940725.1				**POQV01_460185**	
	Ab-V6	Wheat, Brazil	7.2	68.5	63	GCA_002940755.1				**POTD01_450182**	
canadense	**DS2**	Rhizosphere of corn, Canada	9.7	68.0	10	GCA_025961225.1	AZCAN_pf0345	Plas. 499 kb	64.6	AZCAN_pa1121	AZCAN_pb0717
cavernae	**K2W22B-5**	Water from karst cave, China	6.5	66.0	9	GCA_003590795.1	QYUL01_40291	WGS	65.9	QYUL01_10591	
doebereinerae	**GSF71**	Washed roots of *Miscanthus sinensis*, Germany	7.0	69.0	(116)	GCA_003989665.1				RZIJ01_20119	
endophyticum	YIM B02556	Roots of *Paris polyphylla*, China	8.2	68.0	(97)	GCA_016595245.1				JAENHM01_710700	JAENHM01_90063
fermentarium	**CC-LY743**	Fermenter, Taiwan	5.9	68.0	4	GCA_025961205.1	JAOQNH01_20928	WGS	64.6	JAOQNH01_12776	

formosense	**CC-NFb-7**	Agricultural soil, Taiwan	7.0	68.5	7	GCA_019780885.1	**JABEXZ01_12425**	WGS	65.7	**JABEXZ01_20113**	
griseum	**L-25-5 w-1**	Lakewater, China	6.0	66.5	(242)	GCA_003966125.1				RXMA01_1120045	
halopraeferens	**Au4**	Roots of *Leptochloa fusca*, Pakistan	6.5	70.5	(56)	GCA_000429625.1	AUCF_140046	WGS	71.4	–	
humicireducens	**SgZ-5**	Fuel cell inoculated with rice soil, China	6.9	67.5	7	GCA_001639105.2				A6A40_24770	
isscasi	**C340-1**	Paddy soil, China	6.2	69.0	(409)	GCA_030849505.1	JAUJFI01_2610002	WGS	WGS	**JAUJFI01_20038**	
lipoferum	**Sp59b**	Wheat, Brazil	8.0	67.5	(77)	GCA_008364955.1			59.0	AZLIT_0875[f]	AZLIT_pa1289 AZLIT_pb0405
melinis	**TMCY 0552**	*Melinis multiflora*, China	8.0	67.5	(160)	GCA_013340935.1	**AZMEL_pa0036**	Plas. 54 kb	64.1	AZMEL_1213[f]	AZMEL_ph0801
oleiclasticum	RWY-5-1-1	Graminaceous crop in oil field, China	7.8	69.5	(126)	GCA_013423485.1				JABFFR01_410014	
oryzae	**COC8**	*Oryza sativa*, paddy soil, Japan	6.8	67.5	(42)	GCA_008364795.1				VTTM01_110518	
	A2P	Unknown	7.7	67.5	10	GCA_900177515.1				FXAK_51362	FXAK_20015 FXAK_20313
	KACC 14407	Soil, South Korea	6.8	67.5	8	GCA_013347285.1				HUE56_17820[f]	HUE56_07650
	Sh1	Raised bog, Russia	7.3	67.5	(108)	GCA_008365135.1				VTTK01_150130	VTTK01_740025 VTTK01_740137
palustre	**B2**	Sphagnum-dominated peatland, Russia	8.0	68.0	(99)	GCA_002573965.1	PDKW01_410030	WGS	58.8	PDKW01_51173	PDKW01_70361
picis	**IMMIB TAR-3**	Road tar, Taiwan	7.0	68.5	61	GCA_030814895.1				JAGINO01_30066	
ramasamy	**M2T2B2T**	Fermented bovine product, Korea	6.3	68.0	8	GCA_003233655.1				DM194_24440	
rugosum	**IMMIB AFH-6**	Oil-contaminated soil, Taiwan	7.6	69.0	(70)	GCA_030814845.1				JAGINP01_160023	
soli	**CC-LY788T**	Agricultural soil, Taiwan	6.8	67.5	(71)	GCA_017876165.1	AZSOLI_pd0030	Plas. 117 kb	65.8	AZSOLI_pa1500	
tabaci	W712	Rhizosphere of *Nicotiana tabacum*, China	6.3	68.5	(96)	GCA_014596085.1				**JACCJY01_670030**	JACCJY01_340008

(continued)

Table 7.1 (continued)

Azospirillum species[a]	Strain[b]	Origin	Genome size (Mbp)	GC% genome	Number of replicons (or contigs)	Reference GenBank	*luxR* label[c]	Location of *luxR*[d]	GC% of region[e]	Gene encoding acylase[f, g]	Gene encoding amidase[f]
thermophilum	**CFH 70021**	Hot spring, China	6.8	70.0	9	GCA_003130795.1				DEW08_00125[f]	
thiophilum	**BV-S**	Bacterial mat of a sulfide spring, Russia	7.6	68.0	8	GCA_001305595.1				AL072_13025[f]	AL072_30715[f]
zeae	**N7**	Rhizosphere of corn, Canada	7.6	67.5	8	NONE				AZEAE_pe0156	
Undefined	4B	Rice rhizosphere, France	6.8	67.5	7	GCA_000283655.1				AZOLI_p40482	
Undefined	B510	Disinfected stems of rice, Japan	7.6	67.5	7	GCA_000010725.1	**AZL_a06030**	Plas. 1,45 Mb	61.8	AZL_013430[f]	
Undefined	B21	Raised bog, Russia	7.5	67.5	(66)	GCA_008364825.1	VTTO01_100077 **VTTO01_320012**	WGS	63.8 57.1	VTTO01_220538	VTTO01_530029
Undefined	TSA2s	Rice paddy soil, China	8.1	67.0	10	GCA_004923315.1				E6C67_33635[f]	E6C67_13285[f]
Undefined	TSH100	Soil, China	7.2	67.5	8	GCA_004923295.1	**E6C72_31495**	Plas. 54 kb	64.0	E6C72_09925[f]	

[a]In bold are displayed validated species, according to List of Prokaryotic names with Standing in Nomenclature available at https://www.bacterio.net/. Analyses are missing for *A. largimobile* (a validated species with no available genome) and for *Azospirillum himalayense, Aeromicrobium massiliense*, and *Azospirillum palatum* (species not yet validated)

[b]The type strain of each validated species is indicated in bold

[c]When the gene label is in bold, a cognate *luxI* gene was also identified. Strain *A. lipoferum* Sp59b bears only a *luxI* gene (AZLIT_pe0261), located on a 657-kb plasmid

[d]Location of the gene could not be determined in case of genomes obtained using WGS (whole-genome sequencing)

[e]A 5-kb region surrounding the QS gene(s) was analyzed for its GC content

[f]This symbol indicates that the gene is located on the chromosome; in the absence of this symbol, the gene is located on a plasmid or its location could not be determined (in case of WGS)

[g]When the gene label is in bold, the corresponding protein shares more than 90% identity with the acylase of *A. argentinense* Az39

luxI-luxR genes, indicating a possible acquisition by horizontal gene transfer of this region by *A. formosense*.

We were able to detect several AHLs in the supernatant of the *A. melinis* TMCY 0552 culture. This strain produces mainly C_{10}-HSL, but also C_8-HSL and 3OH-C_{10}-HSL. The profile of AHLs is quite similar to that of *A. lipoferum* TVV3 (Vial et al. 2006). It would be interesting to determine the AHL profile of the *Azospirillum* sp. B21 strain (unfortunately not present in our collection). The analysis of these QS systems and the regions carrying these genes reveals some interesting points. Three *luxI-luxR* pairs show strong sequence identity (Fig. 7.1). The LuxI-LuxR proteins of *A. melinis* (encoded by AZMEL_pa0036-AZMEL_pa0035) show strong identity with the LuxI-LuxR pair identified in *Azospirillum* sp. TSH100 (92.3% and 93.4% identity with the LuxR and LuxI of TSH100, respectively) and the AlpI AlpR proteins of *A. lipoferum* TVV3 (68.2% and 71.1% identity). Moreover, for these three genomes, the *luxR* and *luxI* genes are located on small plasmid replicons (54 kb for *A. melinis* and *Azospirillum* sp. TSH100 and 85 kb for *A. lipoferum* TVV3). A GC percentage of 64% is observed for the region surrounding these genes in both *A. melinis* and *Azospirillum* sp. TSH100, while this value is 67.5% for the genome as a whole. Moreover, the synteny around *luxI/luxR* is identical over a region of ~15 kb spanning mainly upstream the *luxI/luxR* pair in these two strains. Upstream and downstream these two *luxI/luxR* pairs lie several ORFs encoding phage proteins, including one encoding lysozyme (located 2 kb downstream *luxI*) (Fig. 7.2). Interestingly, the presence of ORFs encoding phage proteins has also been observed downstream the *alpI/alpR* genes of *A. lipoferum* TVV3 (Wisniewski-Dyé and Vial 2015). However, the products encoded by these ORFs display little or no identity between the three strains, suggesting different horizontal acquisitions.

Finally, the AHL profile was also determined for *Azospirillum* sp. B510 (Fig. 7.1a) and it appears very similar to the one initially published for another rice endophyte, B518 (Vial et al. 2006), suggesting that the two strains might share very similar QS genes.

7.3.3 Orphan luxR *Genes*

Orphan LuxR genes were identified in 12 additional genomes: *Azospirillum isscasi, Azospirillum canadense,* the three *Azospirillum argentinense, Azospirillum soli, Azospirillum palustre, Azospirillum cavernae, Azospirillum agricola, Azospirillum halopraeferens, Azospirillum fermentarium,* and *Azospirillum* sp. TSH100. The latter displays a second *luxR* in addition to the one found in the complete QS system (Table 7.1).

From the few studies available, it appears that orphan *luxR* genes are more abundant and predominant than complete *luxI-luxR* QS systems and that approximately 50% of proteobacterial genomes contain at least one orphan *luxR* gene (Bez et al. 2023). In the *Azospirillum* genomes analyzed, we observed more orphan *luxR*

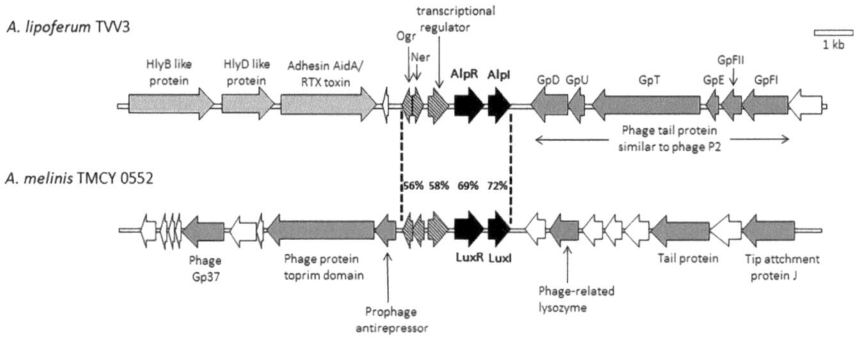

Fig. 7.2 Genetic organization of the region surrounding quorum sensing genes in *Azospirillum lipoferum* TVV3 and *A. melinis* TMCY 0552. The length of each arrow represents the relative ORF size and indicates the direction of transcription. The black arrows indicate *luxI* and *luxR* genes. The dashed arrows indicate ORF sharing sequence homology between *A. lipoferum* TVV3 and *A. melinis* TMCY 0552. The indicated percentage corresponds to the identity between the ORF of *A. lipoferum* TVV3 and that of *A. melinis* TMCY 0552. The gray and black arrows indicate ORFs sharing sequence homology with, respectively, genes encoding putative phage protein and unknown proteins

than complete *luxI-luxR* QS systems. No orphan *luxR* seems to be conserved at the species or genus level. Although the data set is rather small and the location of QS genes is not available for all genomes, the majority of orphan *luxR* in *Azospirillum* appear to be located on plasmidic replicons (with the exception of AZAGRI_2125, which is located on the chromosome in *A. agricola*). The presence of orphan *luxR* on plasmidic replicons has been observed in other alphaproteobacteria such as *Agrobacterium* and *Rhizobium* (Subramoni et al. 2015). The presence of these *luxR* genes on plasmids, which are the main source of genomic plasticity and the main target of gene transfer, certainly partly explains why the percentage of identity between two LuxR does not often match the phylogenetic proximity of the corresponding strains (Fig. 7.1b).

The orphan *luxR* of the type strain *A. argentinense* Az39 has previously been described (Gualpa et al. 2019), and the corresponding protein displays a high identity (>90%) with the orphan LuxR found in the two other *A. argentinense* strains (Fig. 7.1b). This observation, together with the facts that the three *luxR* genes are located on plasmids displaying a high degree of identity with each other, and that the synteny around *luxR* is identical in the three strains, suggests a common origin. The orphan *luxR* from *A. halopraeferens* Au4 shares 75% identity with the *luxR* genes identified in *A. argentinense*, but the corresponding region shows no synteny between the two genomes, with the exception of the gene located in 3′ of the orphan *luxR* (encoding a protein of unknown function).

The protein encoded by the *A. soli luxR* (AZSOLI_pd0030, located on a plasmid of 117 kb) displays a high identity (>60%) with those of *A. palustre* and of *Azospirillum* sp. B21, but the regions surrounding these orphan *luxR* present no synteny.

In *A. canadense* DS2, the LuxR protein (AZCAN_pf0345) shares very little sequence identity with other orphans LuxR found in *Azospirillum* (Fig. 7.1b), but has 67% identity with an orphan LuxR found in *Sphingomonas* sp. 67-36. The presence of an insertion sequence close to AZCAN_pf0345 suggests the possibility of a horizontal transfer of this *luxR* gene and of the upstream *hchA* gene (encoding a chaperone), as the same gene organization is observed in *Sphingomonas* and in some *Stenotrophomonas*. In *A. canadense*, only three nucleotides separate the two ORFs, which are likely to form an operon as observed for *Stenotrophomonas* (whose *luxR* is named *smoR*). It was proposed that these *luxR* genes form a LuxR subfamily named "LuxR regulator chaperone HchA-associated" (Martinez et al. 2015; Subramoni et al. 2015). *Stenotrophomonas maltophilia* SmoR binds exogenous AHL and is involved in swarming motility, although not all key residues of the AHL-binding domain are conserved in *Stenotrophomonas*. Like in SmoR, only four of the six residues involved in AHL binding are conserved in AZCAN_pf0345 (W57, D70, P71, W85, G113 with respect to the TraR amino acid sequence).

A subclass of orphan LuxR (PAB LuxR solo) found only in plant-associated bacteria, both beneficial and pathogenic, responds to plant signals rather than to AHLs (Patel et al. 2013; Bez et al. 2023). Several data indicate that this subclass of LuxR is present in *Azospirillum* sp. B21, a strain that harbors a complete QS system (VTTO01_320012-VTTO01_320015) and an orphan *luxR* (VTTO01_100077). Upstream of this orphan *luxR* gene lies the *pip* gene encoding a proline iminopeptidase, which is known to be located next to PAB LuxR solo and to be regulated by it (Bez et al. 2023). Moreover, a gene annotated as a putative metalloprotease is found just downstream of VTTO01_100077, a feature that is also observed with the PAB *luxR* solo (*psaR2*) of *Pseudomonas syringae* pv. *actinidiae* (Cellini et al. 2022). Finally, two out of the six key residues of the AHL-binding domain are not conserved in this subclass of LuxR, W57, and Y61 (with respect to the TraR amino acid sequence). In VTTO01_100077, only tryptophan 57 is substituted by a phenylalanine. Altogether, these data suggest that VTTO01_100077 encodes a PAB LuxR solo, which is likely able to respond to plant compounds rather than to AHLs, and could be involved in plant-microbe communications like other members of this subclass (Zhang et al. 2023).

From these genome analyses, one can conclude that (i) occurrence of a complete QS system is not common in *Azospirillum;* (ii) all strains bearing a functional QS system (the ones analyzed in Vial et al. 2006, and the ones analyzed here) fall into the *lipoferum* clade as the only strain of the clade *brasilense* that harbors QS genes (*A. formosense*) does not produce AHLs; (iii) the presence of an orphan *luxR* is rather common among *Azospirillum* genomes and these genes may have been acquired by horizontal gene transfer; and (iv) the *A. lipoferum* type strain (Sp59b) possesses an orphan *luxI* gene, a situation that is rather unprecedented.

7.4 Functions Regulated by AHL-Based Quorum Sensing Systems

The most common approach to identify functions that are QS-regulated relies on obtaining a mutant strain deficient in AHL production (*luxI⁻*) or unable to produce the cognate transcriptional regulator (*luxR⁻*). As gene inactivation remains fastidious in some *Azospirillum* strains, especially those of the *lipoferum* species or closed species, another approach has been undertaken (Boyer et al. 2008). This approach relies on the cloning of a gene encoding an enzyme able to inactivate AHLs (see the next section for a focus on these enzymes). By cloning the *attM* gene encoding a lactonase from *Agrobacterium tumefaciens* (Carlier et al. 2003) into strains *A. lipoferum* TVV3 and B518 (one of the three endophytes that produce AHL), it was thus possible to obtain strains with undetectable AHL levels (Boyer et al. 2008). At that time, the choice was arbitrarily put on strain B518, as we were not aware that strain B510 was being sequenced (Kaneko et al. 2010). Several phenotypes shown to be QS-regulated in plant-associated bacteria were then investigated (motility assay, pectinase activity, cellulase activity, siderophore production, production of indole acetic acid). Whereas no phenotype was affected by AHL inactivation in *A. lipoferum* TVV3, the expression of *attM* in the rice endophyte B518 abolished pectinase activity, increased siderophore synthesis, reduced production of indolic compounds (in stationary phase), but had no effect on cellulase activity or on swimming or swarming motilities. The plant growth-promoting ability of these two strains expressing the AHL lactonase was not affected in an in vitro simplified agar system. This targeted analysis was completed by an untargeted proteomic approach (Boyer et al. 2008). No modification was observed for the soluble intracellular proteins for strain TVV3, a result that is consistent with a recent acquisition of the QS genes in this strain. On the contrary, AHL inactivation showed a clear impact (induction or repression) on the expression of numerous soluble intracellular proteins, as well as on some surface-associated and secreted proteins of strain B518, suggesting that QS is a master regulator in this strain. Several induced proteins identified by MS-MS analysis were revealed to be implicated in transport (such as the major outer membrane protein OmaA and components of ABC transporters) and chemotaxis (ChvE), suggesting that QS regulation in *A. lipoferum* B518 is dedicated to control functions linked to root colonization (Boyer et al. 2008).

Such a global approach would be performed nowadays by RNA-seq, but this would require the availability of the B518 genome. B518, together with *Azospirillum* sp. B510 whose genome is available, has been isolated as leaf endophytes of rice, *Oryza sativa* (respectively, from cultivar Kasalath subspecies *indica* and cultivar Nipponbare subspecies *japonica*) (Elbeltagy et al. 2001). The identical AHL profile initially described (Vial et al. 2006) suggested that strains B518 and B510 might harbor very similar QS systems; it would thus be relevant to use the lactonase approach also in strain B510 and to evaluate whether the biocontrol properties of this strain are regulated by QS. Indeed, this strain was shown to enhance rice resistance against diseases caused by the blast fungus *Magnaporthe oryzae* and the bacterial pathogen *Xanthomonas oryzae* (Yasuda et al. 2009) and also tomato resistance against the gray mold *Botrytis cinerea* and the bacterial pathogen *Pseudomonas syringae* pv. *tomato* (Fujita et al. 2017). Finally, the QS genes of B510 are surrounded by genes likely involved in the production of specialized metabolites (Kaneko et al. 2010; Wisniewski-Dyé and Vial 2015) and it is tempting to speculate that QS may regulate the production of these yet-to-be-identified metabolites. A recent genome-mining approach showed that association between biosynthetic gene clusters and *luxR* genes is rather widespread in Pseudomonadota (formerly called Proteobacteria), especially in plant-associated bacteria (Brotherton et al. 2018).

The nature of the signaling molecules that bind to the different LuxR orphans is quite heterogeneous and not always easy to determine. Indeed, orphan LuxRs can bind to self-produced AHLs (by one or more LuxR-LuxI QS systems elsewhere in the genome), to exogenous AHLs, to non-AHL metabolites, or even to signals produced by the plant (Brameyer et al. 2015; Chugani et al. 2001; Zhang et al. 2023). The first step is to determine *in silico* whether the LuxR protein contains both the AHL *N*-terminal AHL-binding domain (PFAM 03472) and the *C*-terminal helix-turn-helix (HTH) DNA-binding domain (PFAM 00196). The next step is to determine whether LuxR can bind to AHLs, usually using an AHL-binding assay. Briefly, a strain overexpressing LuxR was incubated with different AHLs, the culture supernatant was removed after incubation, and the AHLs in the cell pellet were extracted and detected, for example, with a biosensor (Martinez et al. 2015). To identify the genes and functions regulated by the orphan LuxR when it binds to an AHL, it is then possible to compare the transcriptome or proteome of the strain of interest in the presence or absence of the identified AHLs. As previously described, the LuxR PAB family lost the ability to bind to AHLs and respond to low molecular weight compounds produced by plants. Interestingly, the *pipA* gene is always located in close proximity to these *luxR* genes, and LuxR PAB activates the transcription of *pipA* in response to plant compounds. With the help, for example, of *pipA-gfp* transcriptional fusion, it was possible to determine in *Pseudomonas* sp. GM79 that the inducer compound was an ethanolamine derivative present in plant leaf macerates of the *Populus* (Coutinho et al. 2018).

Among plant beneficial Gram-negative bacteria that are known to promote growth of graminaceous plants, several produce AHL. *Gluconacetobacter diazotrophicus* PAL5, an endophytic bacterium isolated from sugar cane stems, can colonize various plants, improve plant growth, and help plants withstand stressful conditions. This strain harbors an active QS system allowing the production of eight different compounds: C_6-HSL, C_8-HSL, C_{10}-HSL, C_{12}-HSL, C_{14}-HSL, and 3O-C_{10}-HSL, 3O-C_{12}-HSL, and 3O-C_{14}-HSL (Nieto-Peñalver et al. 2012). Saline stress (50 mM NaCl) increased by 50% the production of both short and long AHLs (Bertini et al. 2014). To identify QS-regulated genes, a quorum-quenching approach was also undertaken by introducing the *Bacillus aiiA* gene (encoding a lactonase). This strategy allowed to identify a superoxide dismutase, a surface antigen protein, the DnaK chaperone, and several enzymes (6-phosphogluconate dehydrogenase, acetate kinase [AckA]) to be under QS regulation, suggesting that environmental factors can modulate the activity of the QS system (Bertini et al. 2014).

Consequently, *luxI* and *luxR* mutations resulted in a strong reduction of the endophytic colonization of sugar cane seedlings (Hartmann 2020). When inoculated onto rice submitted to drought, the expression of the *luxI* gene of *G. diazotrophicus* PAL5 was shown to be strongly activated in plants submitted to water deficit compared to the control plants. Moreover, plant genes involved in defense pathways (PR1, PR10) and stress (such as those encoding catalase and superoxide dismutase), as well as the accumulation of osmoprotectants, were induced by inoculation with *G. diazotrophicus* PAL5 when plants suffered drought (Filgueiras et al. 2020). These results suggest that AHLs produced by PAL5 may enhance the tolerance to drought in rice plants. However, the direct link of QS regulation and plant protection effect was not demonstrated in this study.

Acidovorax radicis N35, an endophytic bacterium with plant growth-promoting activity on barley and wheat (Li et al. 2011), was shown to produce a unique AHL, C_{10}-HSL. A *luxI* mutant had an altered colonization behavior with less microcolony formation, compared to the wild-type strain, but nevertheless, it retained a significant plant growth-promotion ability. Transcriptomic analyses of axenic barley seedling revealed that the expression of flavonoid biosynthetic genes was increased in plants inoculated with the *luxI* mutant compared to wild-type N35, and this was corroborated with the accumulation of several flavonoid compounds, such as saponarin and lutonarin (Han et al. 2016). It is likely that AHL production in this strain protects it from plant defense compounds, such as flavonoids.

The genus *Herbaspirillum* comprises efficient plant growth-promoting nitrogen-fixing strains, as well as clinical isolates. Notably, two species, *Herbaspirillum seropedicae* and *Herbaspirillum frisingense*, harbor both types of strains. The model strain *Herbaspirillum seropedicae* SRm1, able to colonize endophytically grasses such as rice and sugar cane, does not harbor any QS-related gene (Pedrosa et al. 2011). Among *H. frisingense*, the type strain GSF30^T, isolated from *Miscanthus* in Germany, did not synthesize AHLs, whereas strain Mb11, isolated from the roots of *Pennisetum purpureum* in Brazil, was shown to produce several AHLs (C_{14}-HSL, 3OH-C_{14}-HSL, $C_{14:1}$-HSL) (Rothballer et al. 2008; Hartmann et al. 2024). However, to our knowledge, the role of QS regulation in strain Mb11 has not been further studied and there are no other reports of AHL production in *Herbaspirillum*, suggesting that AHL-based signaling might not be prevalent in this genus.

7.5 Potential for Quorum Quenching Activities

7.5.1 Introduction to Quorum Quenching

Given the importance of QS signaling molecules for the development of bacteria in communities and for host colonization, this is not surprising that a great diversity of organisms have evolved strategies that allow them to interrupt cell-to-cell communication of other microorganisms. These strategies are known as quorum interference or quorum quenching (QQ) (Zhang 2003) and hold great promise for attenuating the virulence of pathogens, whether animal or plant pathogens. The known mechanism of QQ includes inactivation of molecular signal (AHL), inhibition of signal detection, and inhibition of signal biosynthesis (Fetzner 2015). Inactivation of AHL is mediated by AHL-inactivating enzymes, through three distinct enzymatic mechanisms: lactone hydrolysis, amide bond hydrolysis, and acyl chain modification (Utari et al. 2017). Lactone hydrolysis occurs when an AHL lactonase hydrolyzes the homoserine lactone ring (Fetzner 2015). AHL acylases catalyze the complete and irreversible destruction of AHLs through amide bond hydrolysis. AHL oxidases and reductases catalyze a change in the chemical structure of the signals, which impairs AHL signal recognition and interferes with processes regulated by QS (Grandclément et al. 2016). Lactonases and acylases are the most studied system of QQ mechanisms (Fetzner 2015).

7.5.2 *First Evidence of Quorum Quenching in* Azospirillum

The first and unique study dealing on quorum quenching in *Azospirillum* was performed on the model strain *A. argentinense* Az39, a strain widely used as a bioinoculant in Latin America and that does not possess the ability to synthesize AHL (Gualpa et al. 2019). *In silico* analyses failed to identify any lactonase but allowed the identification of several enzymes with potential acylase activity in this strain. Two of them had the strongest potential of having QQ activity, due to their similarity to the main AHL-inactivating enzymes that have been functionally characterized in Gram-negative bacteria: an *N*-acyl-homoserine lactone acylase/penicillin acylase (EC. 3.5.1.11) encoded by ABAZ39_22635 located on plasmid 1 and an aliphatic amidase AmiE (EC. 3.5.1.4) encoded by ABAZ39_16400 located on the 1.9-Mbp megaplasmid. In the case of the AmiE enzyme, predicted MetaCyc pathways suggest its potential involvement in pathways such as acrylonitrile degradation and IAA biosynthesis VI (via indole-3-acetamide). However, AmiE has also been proposed as a novel *N*-acylhomoserine lactone acylase belonging to the amidase family (Ochiai et al. 2014). Although this function remains putative, further studies should be conducted to better understand the role of this enzyme in *Azospirillum*.

When 500 nM of the unsubstituted AHLs (C_4, C_6, C_8, C_{10} , C_{12}, C_{14}) was supplemented in Az39 cultures, AHL levels were lower than in the noninoculated controls and this was assessed using both biosensors and liquid chromatography coupled with mass-mass spectrometry (see Sect. 7.6.4). Similar results were observed upon the addition of AHLs with hydroxy substitutions in C3, showing that this strain is able to inactivate a wide spectrum of AHLs (Gualpa et al. 2019). The enzymatic activity was not released in the supernatant and did not seem to be induced by the presence of AHLs (Gualpa et al. 2019). However, it is not known which of the two enzymes or whether both are involved in AHL inactivation and expression of each enzyme in a heterologous host would be needed to clarify this.

7.5.3 *Genome Mining for Potential Quorum Quenching Activities in* Azospirillum

A first search was performed in the UniProt database using EC numbers of the two main AHL-degrading enzymes found in Gram-negative bacteria, *N*-acyl-homoserine lactonases (EC 3.1.1.81) and *N*-acyl-homoserine lactone acylases (EC 3.5.1.97). A unique lactonase enzyme (RXMA01_1480022) was registered in UniProt for *Azospirillum griseum*, whereas 37 acylases were registered in UniProt across various members of the genus, suggesting that members of this genus are potential quorum quenchers.

Then, using the sequence of enzymes already described (Gualpa et al. 2019) as baits, a BLAST search was carried out in the same genomes as above, using the MaGe platform. Of the 43 strains analyzed, all harbor a protein of the penicillin acylase/ *N*-acyl-homoserine-lactone acylase, with identities varying according to the species (Table 7.1). Not surprisingly, the highest protein identity (>90%) was observed with acylases of strains belonging to the *A. argentinense* species (MTCC4035 and MTCC4038) or to other strains belonging to the *brasilense/ argentinense* clade (*A. baldaniorum* Sp245, *A. brasilense* Sp7/Ab-V5/Ab-V6, *A. formosense* CC-NFb-7, *A. aestuarii* YIM DDC1, *A. tabaci* W712, *A. isscasi* C340-1). Other identified acylases display identities ranging between 60% and 70% with the *A. argentinense* acylase. The region carrying ABAZ39_22635 is well conserved in genomes of phylogenetically closed strains (with up to 230 genes in synteny among genomes of *A. argentinense*) and harbors upstream genes involved in chemotaxis and downstream genes encoding a threonine dehydratase, a flavin reductase, and MotB. In the other genomes, the synteny covers the 15 to 20 downstream genes (such as in *A. isscasi*) or only ABAZ39_22635 and the gene encoding the threonine dehydratase for the genomes of the most phylogenetically distant strains (such as in *A. lipoferum*).

As for the amidase, the BLAST search with ABAZ39_16400 yielded 18 hits whose encoded product is annotated aliphatic amidase or acrylamide amidohydrolase, indicating that the presence of amidase seems to be less common across the genus. High protein identities (>80%) were observed with all the hits, the highest (97%) being observed with amidases of *A. argentinense* MTCC4035 and MTCC4036 and the lowest (84%) being for the amidase found in *A. canadense* D2. Two *A. oryzae* genomes (that of A2P and of Sh1) harbor two copies of this gene. Neighboring genes code for components of ABC transporters, and this region is well conserved among all genomes (from 9 to 37 genes, depending on the genetic proximity).

Thus, the potential for AHL inactivation does not seem to be restricted to the *A. argentinense* species. However, so far, studies aimed at validating this activity have only been conducted in *A. argentinense* Az39, demonstrating that this strain is capable of interfering with quorum sensing in bacteria-bacteria interactions (Gualpa et al. 2019). To ascertain that the other *Azospirillum* strains are truly AHL inactivators, AHL inactivation should be evaluated as described below.

Ecologically, this process could confer competitive advantages in the rhizosphere by inhibiting the communication of other bacteria, thereby promoting its establishment and persistence in specific niches. In bacteria-plant interactions, the QQ mechanism may play a key role in regulating the rhizosphere behavior of certain bacteria. QS signal inactivation in this environment not only modulates root colonization but also influences the production of compounds that affect plant growth (Rodríguez et al. 2020; da Cruz et al. 2024). Furthermore, recent studies have shown that some bacteria can interfere with the QS systems of phytopathogens, reducing their infection capacity and promoting a more favorable environment for plant development (Roca et al. 2024; Warring et al. 2024).

7.6 Methods Related to Quorum Sensing

7.6.1 Detection of AHLs

The discovery that numerous unrelated bacteria have the ability to produce AHL molecules in response to a high cell density has been achieved by using bacterial biosensors (Steindler and Venturi 2007). These biosensors have a functional LuxR homologue but lack the AHL synthase enzyme and consequently do not produce AHLs. Promoter activity of a gene or operon under QS regulation can be induced by the presence of exogenous AHLs. These biosensors have easily detectable phenotypes such as production of pigments, expression of β-galactosidase activity, light emission, or fluorescence. Biosensors are either derivatives of strains that naturally produce AHLs, or entirely engineered plasmids that can be used in several bacterial backgrounds. Among derivatives of natural strains, two are easy to implement and are widely used for AHL detection. The first, *Chromobacterium violaceum* CVO26, harbors a transposon into the *cviI* AHL synthase gene; exposure of this strain to exogenous AHLs with rather short acyl chain AHLs (four to eight carbons) able to interact with CviR (the cognate LuxR) results in production of the pigment violacein (McClean et al. 1997). The second is a derivative *Agrobacterium tumefaciens* strain NT1(pZLR4) that no longer produces its own signal molecule; this strain harbors a *traG-lacZ* fusion plasmid that can be induced when a cognate AHL is recognized by TraR. This second biosensor has the ability to detect a wide range of AHLs, including 3-oxo-HSLs and 3-hydroxy-HSLs (Shaw et al. 1997; Luo et al. 2003).

These biosensors can be used in several ways: (i) by spotting a colony to be tested near a biosensor, i.e., cross-streak assay, that requires that both the sensor and the tested strain can grow on the same medium suitable for AHL detection (i.e., LB for *C. violaceum* CVO26 and AB or AT medium for *A. tumefaciens* strain NT1 (pZLR4)); (ii) by spotting a culture supernatant or an extract of supernatant onto a plate and overlaying with a sensor strain grown in the adequate medium (see Sect. 7.6.7 for a detailed procedure); (iii) by overlaying a thin-layer chromatography where samples have previously been separated (see the next section). After overnight incubation, a colored zone around the site of the spot will appear due to the presence of AHLs.

As one biosensor can only detect a narrow range of AHLs with different sensitivities, more than one type of biosensors is generally needed to test a spectrum of AHLs from one bacterium.

7.6.2 Extraction of AHLs

As biosensors are able to respond to a range of AHLs and as most AHL-producing bacteria produce a set of AHLs in various proportions (Lithgow et al. 2001; Vial et al. 2006), AHLs have to be extracted and separated prior to their identification.

AHLs are extracted from bacterial culture supernatants from the late exponential phase or stationary phase, as AHL accumulation is usually maximal at this growth stage. For *Azospirillum*, the strains have to be grown in 100 mL of LB low salt or AB malate. The cells are eliminated by centrifugation and filtration, and the pH of the supernatant is acidified with HCl 12N down to pH ~3 (as the lactone ring hydrolyzes at pH above 8). Extraction can be performed in two different ways. The classical liquid-liquid extraction can be performed twice with dichloromethane, chloroform, acidified ethyl acetate (0.1 mL glacial acetic acid per liter of ethyl acetate) using one third volume of solvent (Winson et al. 1995), and then the two phases of solvent are pooled prior to evaporation to dryness using Rotavapor. The residue is then dissolved in a small volume of acetonitrile (usually 1/1000th of the initial volume) and stored at −20 °C until analysis. This protocol was used to determine the AHLs produced by *Azospirillum* sp. B510 and *A. melinis* (see Sect. 7.3.2 and Fig. 7.1a).

An alternative is to use solid-phase extraction, which was reported to improve the sensitivity by 2–10-fold compared to liquid-liquid extraction (Li et al. 2006). An optimized protocol has been proposed (Li et al. 2006): the sample (with 25%, v/v ACN) is applied onto a MegaBond Elute cartridge, which is previously conditioned by 2 mL of water and 2 mL of methanol sequentially. The loaded column is then washed with 15/85 (v/v) methanol/water, and the analytes are eluted with 25/75 (v/v) hexane/isopropanol. The elute is then dried and lately resuspended in the appropriate solvent for high-performance liquid chromatography (HPLC) or ultra-high-pressure liquid chromatography (UHPLC).

7.6.3 Separation by Thin-Layer Chromatography and Detection

Thin-layer chromatography (TLC) coupled with biosensors produces a straightforward visual index of the AHLs synthesized by the tested bacteria. AHLs extracted from supernatants (see Sect. 7.6.2) and standards (many of them can now be purchased) are loaded on a C_{18} reverse phase TLC plate. Once the solvent of the samples is evaporated, the plate is placed upright in a glass chamber containing a developing liquid solvent, usually a mixture of methanol and water (for Alugram RP18 plates, we use methanol-water-acetic acid (55:45:0.1, v/v) as the mobile phase). The AHLs migrate at different rates due to their difference in polarity (that is correlated to the length of the acyl chain and the substitution at C3). After migration and drying, the TLC plate is overlaid with soft agar (0.75%) containing the biosensor, i.e., 80 mL of LB inoculated with an overnight 10 mL culture of *C. violaceum* CVO26 or 125 mL AB medium containing 50 $\mu g.mL^{-1}$ X-gal for *A. tumefaciens* strain NT1(pZLR4). After incubation of the TLC overnight at 30 °C, colored halos are visible at the location to which the AHL has migrated onto the plate. The 3-oxo derivatives characteristically produce tailing spots with diffuse edges, whereas the 3-unsubstituted forms produce circular spots with sharp edges. The 3-hydroxy-substituted compounds migrate with the same mobility as their 3-oxo analogues, but the spots do not tail (Shaw et al. 1997). Quantification can be achieved by converting the area of spots into residual concentration of a given AHL via a calibration curve obtained from the standard synthetic AHL.

The absence of a positive response might be due to the presence of a specific acyl chain length or substitution that prevents the detection by the biosensor (Schaefer et al. 2002) or to a concentration of signal molecules below the threshold of sensitivity of the biosensor. In the latter case, an alternative method based on a radioactive incorporation of ^{14}C-carboxy-methionine assay has been developed, allowing notably to detect AHL in complex matrices (Singh et al. 2000). It is noteworthy that the medium used for growing the strain to be tested may influence AHL production (Lithgow et al. 2001; Li et al. 2006). In contrast, compounds present in the supernatant and co-extracted with AHLs can lead to false-positive spots (Wisniewski-Dyé et al. 2002). Finally, if TLC provides a good index of the set of AHLs produced by a strain, one cannot determine the chemical structure of the molecules detected and other more precise and sensitive methods are required to provide a definitive confirmation of their presence and to allow formal structural confirmation and accurate quantification.

7.6.4 Structural Identification of AHLs

Most procedures of AHL characterization have been carried out by using a combination of reverse phase HPLC/UHPLC and mass spectrometry (MS). Because of the very large number of structurally related signaling molecules and the fact that other by-products present in the supernatant are co-extracted, a preliminary separation is indeed required using HPLC/UHPLC. Separation of AHLs by HPLC is performed on a C_8 or C_{18} reverse phase silica column using an isocratic or linear gradient elution of methanol-water or acetonitrile-water as the mobile phase, with a flow of usually 0.4–0.5 mL/min and a monitoring by U.V. at 200 or 210 nm (the only wavelength where AHLs can be hardly traced). A separation test must be carried out with standards in order to determine the optimum gradient, which depends on the type and size of the column. Fractions eluting from the HPLC/UHPLC column with retention times identical to AHL standards or that are able to trigger activation of a biosensor can be rechromatographed; once a single active peak is obtained, it can be further analyzed for structural identification. Alternatively, a single active compound can be obtained via preparative thin-layer chromatography (TLC) and the active spot can be scraped off and extracted with dichloromethane, ethyl acetate, or acetone. After concentration, the active spot can be further analyzed (Shaw et al. 1997; Vial et al. 2006).

Mass spectrometry can make a record of charge/mass ratio of ionized compounds. According to the molecular ion peaks and fragments, the molecular mass and structural features can be estimated, even if picomoles of AHL are present. There are many types of ionization available, the most commonly types used for AHLs are (reviewed in Wang et al. 2010) electron ionization (EI-MS), fast atom bombardment (FAB-MS), chemical ionization (CI-MS), electrospray ionization (ESI-MS), and atmospheric pressure chemical ionization (APCI-MS). Tandem mass spectrometry (MS-MS) is nowadays commonly used, as ions generated in the first MS are separated and fragmented, leading to a more accurate identification of compounds. For quantification of AHLs, the m/z 102 ion, specific for the lactone ring and detected with a good signal-to-noise ratio, allows low detection limits even in complex matrix samples.

7.6.5 Identification of Genes Mediating AHL-Based Quorum Sensing

Detection and characterization of gene mediating AHL-based QS are now greatly facilitated by the availability of complete genome sequences. Rather than BLAST searches or searches by keywords on annotated genes, it is much more reliable to perform searches on protein domains with PF00765/ IPR001690 for LuxI (AHL synthase) and PF03472/IPR005143 for LuxR regulators. It is essential in the case of LuxR homologs as many gene products are annotated as "two component transcriptional regulator, LuxR family" due to the presence of a helix-turn-helix domain of about 65 amino acids, which is the hallmark of the transcriptional regulators of the LuxR/FixJ family. However, only proteins that bear an AHL-binding domain (PF03472) must be considered as true LuxR homologs, and this must be emphasized because some strains previously described as having LuxRs are in fact devoid of them. To illustrate this point, let's put a focus on the largest *Azospirillum* genome, that of *A. canadense* (9.7 Mbp): it contains 13 genes encoding regulators of the LuxR/FixJ family but only one (AUCF_140046) encoding a true LuxR thanks to the presence of the AHL-binding domain (Fig. 7.1).

For strains belonging to the same species, high similarities of LuxI and LuxR proteins have been reported (Wisniewski-Dyé et al. 2002). However, it was also stated that the variation observed among LuxI and LuxR homologues from different strains of the same bacterial species was greater than the variation observed between closely related species (Gray and Garey 2001), emphasizing that BLAST search is not the best method to identify QS genes.

When the genome of the studied strain is not available, identification of QS genes should rely on bacterial genetics. One strategy is to perform a random transposon mutagenesis, and then to screen for a mutant lacking the ability to activate a biosensor; however, this approach is time-consuming, as it relies on the screening of several thousands of mutants. Transposon mutagenesis is not the method of choice on *Azospirillum* strains as a low or very low efficiency of transposition was found in several species (our unpublished results). An alternative is a "gain-of-function" approach that was successfully used to identify the genetic loci involved in AHL production in *Azospirillum* (Vial et al. 2006). This strategy involves the construction of a cosmid library (or sometimes plasmid library) and the complementation of a biosensor strain. In the case of *A. lipoferum* TVV3, a cosmid library was constructed in *E. coli* by inserting DNA fragments of size ranging from 25 to 35 kb. The cosmid library was then transferred by conjugation into the biosensor strain *A. tumefaciens* NTL1(pZLR4), and clones were grown on AB medium containing 50 $\mu g.mL^{-1}$ X-gal. Clones containing a cosmid encoding an AHL synthase gene appear as blue colonies on plates. Complete sequencing of the clone of interest is possible, and gene(s) involved in the production of AHL are then identified. Other biosensors can be used with this strategy, such as *C. violaceum* CV026 (Elasri et al. 2001) or the bioluminescent AHL biosensor pSB403 (Steidle et al. 2002).

7.6.6 Identification of Genes Regulated by AHL-Based Quorum Sensing

Generally, the first step to identify phenotypes or genes regulated by QS relies on the obtention of a strain deficient in signal production or unable to synthesize the cognate transcriptional regulator LuxR. For methods describing the construction of mutants in *Azospirillum*, the reader should refer to Chap. 4. When genetic approaches are fastidious or when several genes encoding synthases are present in the studied strain, an alternative can be contemplated, i.e., the introduction of an enzyme able to inactivate AHLs.

Several AHL degradation enzymes (including lactonases, acylases, oxidoreductases, and paraoxonases) have been identified in a range of living organisms (Chen et al. 2013). The expression of these enzymes in AHL-producing strains efficiently quenches the microbial QS signaling. For example, lactonases cleave the lactone ring of AHLs which are no longer recognized as a signal and incapable of binding to the target transcriptional regulator. The quorum-quenching strategy was initially reported in *Erwinia carotovora* where AiiA, a lactonase from *Bacillus*, inactivates AHL production and strongly attenuates the virulence of this phytopathogen (Dong et al. 2000); this strategy has also been used in *Azospirillum* with the introduction of *attM* (encoding a lactonase from *A. tumefaciens*) in two AHL-producing strains, leading to undetectable levels of AHLs and analysis of regulated phenotypes (Boyer et al. 2008).

The next step is to identify functions that are likely to be regulated by QS. As these functions are often essential for interaction with a eukaryotic host, it is possible to focus on functions essential in plant-bacteria interaction (and/or already shown to be QS-regulated in other plant-associated bacteria), such as motility (swimming and swarming), biofilm formation, root adherence, rhizosphere colonization, phytohormone production, synthesis of antifungal/antimicrobial metabolites, phytostimulatory effects.

However, omics approaches are now the most widely used methods for identifying QS-regulated genes as they allow to gain an overall view of QS-regulated functions in a specific strain. Global changes in gene expression have also been investigated at the level of the transcriptome (microarray or RNA-seq), by comparing the transcriptome of a QS mutant strain with that of a wild-type strain, or with that of the QS mutant complemented with the cognate AHL. RNA-seq was used to decipher the role of two AHL synthases encoding genes (*traI* and *ngrI*) in the symbiotic bacteria *Sinorhizobium fredii* (Krysciak et al. 2014). A common set of 186 genes appears to be regulated by the TraI/R and NgrI/R regulon, with coregulated genes including 42 flagellar biosynthesis genes and 22 genes linked to exopolysaccharide biosynthesis (Krysciak et al. 2014). In the soil saprophyte *Burkholderia thailandensis*, RNA-seq was used to identify the regulons of each of three QS systems and the BtaR1-BtaI1 regulon appeared to be the largest with more than 150 genes regulated (Majerczyk et al. 2014).

A global proteomic approach can also be undertaken, as the one used in *A. lipoferum* TVV3 and B518 using the wild-type strain and a non-AHL-producing strain (containing the *attM* gene) (Boyer et al. 2008).

7.6.7 Methods to Highlight a Quorum-Quenching Activity

7.6.7.1 Evaluation of AHL Inactivation Using Bioassays

To assess the QQ activity, an exponential-phase *Azospirillum* culture (OD_{595} 1.0) obtained in liquid LB medium must be obtained and a volume of 50 μL is used to inoculate a flask containing 50 mL of MMAB medium. When the culture reaches an OD_{595} of 0.8–1.0, corresponding to the exponential growth phase, it is fractionated into 5 mL and treated individually with 10 μM C_6-HSL or C_{10}-HSL. The tubes are then incubated for 12 h at 37 °C with shaking. After incubation, the presence of AHLs is evaluated in the culture medium using bioassays with the reporter strains: *C. violaceum* CV026 and *A. tumefaciens* NTL1(pZLR4). To that end, 500 μL of exponential-phase reporter strain cultures are transferred into glass tubes containing 4.5 mL of semisolid AT medium (0.7% (w/v) agar) containing 50 μg.mL^{-1} X-gal (*A. tumefaciens* NTL1(pZLR4)) or semisolid LB medium (0.7% (w/v) agar)

(*C. violaceum* CV026). The suspensions are plated onto petri dishes containing solidified media, and small holes are created in the petri dishes containing AT- or LB-solidified culture medium using a 5-mm cylindrical punch. A volume of 10 μL of the filtered supernatant to evaluate is dispensed in the holes. The plates are incubated at 30 °C for 24 h to reveal the presence of AHL molecules by a colorimetric reaction. In the presence of AHLs, a violet or light blue halo is formed (depending on the specific AHL and the reporter strain used). AHL inactivation is evidenced by the absence of the colored halo.

7.6.7.2 Evaluation of AHL Inactivation Followed by HPLC-MS/MS

However, the methodology to confirm with accuracy the AHL inactivation is the liquid chromatography. For this, a set of ten flasks containing 20 mL of LB medium is prepared. Seven flasks are inoculated with 20 μL of Az39 culture obtained from liquid LB medium in the late-exponential growth phase (OD_{595} 1.0), and three flasks are left without inoculation (controls). All ten flasks are incubated overnight at 37 °C with shaking for 12 h. About 100 μL methanol is added in one inoculated flask (control treatment to assess potential inhibition of bacterial growth), while all remaining flasks are treated with 100 μL of a methanolic solution containing C_4, C_6, C_8, C_{10}, C_{12}, C_{14}, oxo-C_4, oxo-C_6, oxo-C_8, oxo-C_{10}, oxo-C_{12}, oxo-C_{14}, OH-C_4, OH-C_6, OH-C_8, OH-C_{10}, OH-C_{12}, and OH-C_{14}, each at a concentration of 100 μM, resulting in a final concentration of 500 nM of each AHL molecule. The glass flasks are incubated for 6 h, with sampling intervals at 1, 3, and 6 h. The samples (1 mL) are collected and stored at −20 °C until processing, AHL extraction, and analysis by liquid chromatography-mass spectrometry (LC-MS/MS). AHL inactivation is indicated by a significant reduction in the chromatographic peak area at the different time points in *Azospirillum* cultures with exogenously added AHLs, compared to the uninoculated control samples.

7.7 Conclusion

Investigating cell-cell communication often begins with the evidence of a bacterial signal being produced. In the case of AHLs, various biosensors have been developed for this goal and have been widely used. Identification of genes mediating AHL-based QS is now greatly facilitated by the availability of whole-genome sequences. RNA-seq is now the method of choice to unravel QS-regulated genes. Among the *Azospirillum* genus, AHL-mediated communication seems to be the exception rather than the rule (Vial et al. 2006; Boyer et al. 2008; the genome study carried out for this chapter) as only a few strains have the ability to synthesize these molecules. However, many strains possess orphan LuxR, likely acquired by horizontal gene transfer, and this confers them the ability to pick up signals produced by other soil bacteria. Moreover, given the range of other signaling molecules described so far and the ones that remain uncharacterized, one can assume that some *Azospirillum* might use other types of signals.

The ability to inactivate QS signals, i.e., quorum quenching (QQ), so far experimentally demonstrated for the *A. argentinense* Az39 strain, might be present in other strains. The biotechnological potential of QQ in *Azospirillum* opens new perspectives for the development of agricultural bioinoculants, particularly those aimed at mitigating the impact of crop diseases by inhibiting QS in pathogens.

References

Barnard AM, Salmond GP (2007) Quorum sensing in *Erwinia* species. Anal Bioanal Chem 387: 415–423. https://doi.org/10.1007/s00216-006-0701-1

Bertini EV, Nieto-Peñalver CG, Leguina AC et al (2014) *Gluconacetobacter diazotrophicus* PAL5 possesses an active quorum sensing regulatory system. Antonie Van Leeuwenhoek 106:497–506. https://doi.org/10.1007/s00203-016-1203-y

Bez C, Geller AM, Levy A et al (2023) Cell-cell signaling proteobacterial LuxR solos: a treasure trove of subgroups having different origins, ligands, and ecological roles. mSystems 8: e0103922. https://doi.org/10.1128/msystems.01039-22

Boyer M, Bally R, Perrotto S et al (2008) A quorum-quenching approach to identify quorum-sensing-regulated functions in *Azospirillum lipoferum*. Res Microbiol 159:699–708. https://doi.org/10.1016/j.resmic.2008.08.003

Brameyer S, Kresovic D, Bode HB et al (2015) Dialkylresorcinols as bacterial signaling molecules. Proc Natl Acad Sci USA 112:572–577. https://doi.org/10.1073/pnas.1417685112

Brotherton CA, Medema MH, Greenberg EP (2018) *luxR* homolog-linked biosynthetic gene clusters in proteobacteria. mSystems 3:e00208–e00217. https://doi.org/10.1128/msystems.00208-17

Carlier A, Uroz S, Smadja B et al (2003) The Ti plasmid of *Agrobacterium tumefaciens* harbors an *attM*-paralogous gene, *aiiB*, also encoding N-Acyl homoserine lactonase activity. Appl Environ Microbiol 69:4989e4993. https://doi.org/10.1128/AEM.69.8.4989-4993.2003

Cellini A, Buriani G, Correia C et al (2022) Host-specific signal perception by PsaR2 LuxR solo induces *Pseudomonas syringae* pv. *actinidiae* virulence traits. Microbiol Res 260:127048. https://doi.org/10.1016/j.micres.2022.127048

Chen F, Gao Y, Chen X et al (2013) Quorum quenching enzymes and their application in degrading signal molecules to block quorum sensing-dependent infection. Int J Mol Sci 14:17477–17500. https://doi.org/10.3390/ijms140917477

Christie GE, Temple LM, Bartlett BA et al (2002) Programmed translational frameshift in the bacteriophage P2 FETUD tail gene operon. J Bacteriol 184:6522–6531. https://doi.org/10.1128/jb.184.23.6522-6531.2002

Chugani SA, Whiteley M, Lee KM et al (2001) QscR, a modulator of quorum-sensing signal synthesis and virulence in *Pseudomonas aeruginosa*. Proc Natl Acad Sci USA 98:2752–2757. https://doi.org/10.1073/pnas.051624298

Coutinho BG, Mevers E, Schaefer AL et al (2018) A plant-responsive bacterial-signaling system senses an ethanolamine derivative. Proc Natl Acad Sci USA 115:9785–9790. https://doi.org/10.1073/pnas.1809611115

da Cruz AA, Cabeo M, Durán-Viseras A et al (2024) Interference of AHL signal production in the phytophatogen *Pantoea agglomerans* as a sustainable biological strategy to reduce its virulence. Microbiol Res 285:127781. https://doi.org/10.1016/j.micres.2024.127781

Dong YH, Xu JL, Li XZ et al (2000) AiiA, an enzyme that inactivates the acylhomoserine lactone quorum-sensing signal and attenuates the virulence of *Erwinia carotovora*. Proc Natl Acad Sci USA 97:3526–3531. https://doi.org/10.1073/pnas.97.7.3526

Eberhard A, Burlingame AL, Eberhard C et al (1981) Structural identification of autoinducer of *Photobacterium fischeri* luciferase. Biochemistry 20:2444–2449. https://doi.org/10.1021/bi00512a013

Elasri M, Delorme S, Lemanceau P et al (2001) Acyl-homoserine lactone production is more common among plant-associated *Pseudomonas* spp. than among soilborne *Pseudomonas* spp. Appl Environ Microbiol 67:1198–1209. https://doi.org/10.1128/AEM.67.3.1198-1209.2001

Elbeltagy A, Nishioka K, Sato T et al (2001) Endophytic colonization and in planta nitrogen fixation by *Herbaspirillum* sp. isolated from wild rice species. Appl Environ Microbiol 67:5285–5293. https://doi.org/10.1128/AEM.67.11.5285-5293.2001

Fetzner S (2015) Quorum quenching enzymes. J Biotechnol 201:2–14. https://doi.org/10.1016/j.jbiotec.2014.09.001

Filgueiras L, Silva R, Almeida I et al (2020) *Gluconacetobacter diazotrophicus* mitigates drought stress in *Oryza sativa* L. Plant Soil 451:57–73. https://doi.org/10.1007/s11104-019-04163-1

Fujita M, Kusajima M, Okumura Y et al (2017) Effects of colonization of a bacterial endophyte, *Azospirillum* sp. B510, on disease resistance in tomato. Biosci Biotechnol Biochem 81:1657–1662. https://doi.org/10.1080/09168451.2017.1329621

Fuqua WC, Winans SC (1994) A LuxR-LuxI type regulatory system activates *Agrobacterium* Ti plasmid conjugal transfer in the presence of a plant tumor metabolite. J Bacteriol 176:2796–2806. https://doi.org/10.1128/jb.188.5.1943-1949.2006

Grandclément C, Tannières M, Moréra S et al (2016) Quorum quenching: role in nature and applied developments. FEMS Microbiol Rev 40:86–116. https://doi.org/10.1093/femsre/fuv038

Gray KM, Garey JR (2001) The evolution of bacterial LuxI and LuxR quorum sensing regulators. Microbiology 147:2379–2387. https://doi.org/10.1099/00221287-147-8-2379

Gualpa J, Lopez G, Nievas S et al (2019) *Azospirillum brasilense* Az39, a model rhizobacterium with AHL quorum-quenching capacity. J Appl Microbiol 126:1850–1860. https://doi.org/10.1111/jam.14269

Han S, Li D, Trost E et al (2016) Systemic response of barley to the 3-hydroxy-decanoyl-homoserine lactone producing plant beneficial endophytic *Acidovorax radicis* N35. Front Plant Sci 7:e1868. https://doi.org/10.3389/fpls.2016.01868

Hartmann A (2020) Quorum sensing *N*-acyl-homoserine lactone signal molecules of plant beneficial Gram-negative rhizobacteria support plant growth and resistance to pathogens. Rhizosphere 16:100258. https://doi.org/10.1016/j.rhisph.2020.100258

Hartmann A, Binder T, Rothballer M (2024) Quorum sensing-related activities of beneficial and pathogenic bacteria have important implication for plant and human health. FEMS Microbiol Ecol 100:fiae076. https://doi.org/10.1093/femsec/fiae076

Jones S, Yu B, Bainton NJ et al (1993) The *lux* autoinducer regulates the production of exoenzyme virulence determinants in *Erwinia carotovora* and *Pseudomonas aeruginosa*. EMBO J 12:2477–2482. https://doi.org/10.1002/j.1460-2075.1993.tb05902.x

Kaneko T, Minamisawa K, Isawa T et al (2010) Complete genomic structure of the cultivated rice endophyte *Azospirillum* sp. B510. DNA Res 17:37–50. https://doi.org/10.1093/dnares/dsp026

Krysciak D, Grote J, Rodriguez Orbegoso M et al (2014) RNA Sequencing analysis of the broad-host-range strain *Sinorhizobium fredii* NGR234 identifies a large set of genes linked to quorum sensing-dependent regulation in the background of a *traI* and *ngrI* deletion mutant. Appl Environ Microbiol 80:5655–5671. https://doi.org/10.1128/AEM.01835-14

Li X, Fekete A, Englmann M et al (2006) Development and application of a method for the analysis of *N*-acylhomoserine lactones by solid-phase extraction and ultra high pressure liquid chromatography. J Chromatogr A 1134:186–193. https://doi.org/10.1016/j.chroma.2006.09.047

Li D, Rothballer M, Schmid M et al (2011) *Acidovorax radicis* sp. nov., a wheat-root-colonizing bacterium. Int J Syst Evol Microbiol 61:2589–2594. https://doi.org/10.1099/ijs.0.025296-0

Lin SY, Hameed A, Shen FT et al (2014) Description of *Niveispirillum fermenti* gen. nov., sp. nov., isolated from a fermentor in Taiwan, transfer of *Azospirillum irakense* (1989) as *Niveispirillum irakense* comb. nov., and reclassification of *Azospirillum amazonense* (1983) as *Nitrospirillum amazonense* gen. nov. Antonie Van Leeuwenhoek 105:1149–1162. https://doi.org/10.1007/s10482-014-0176-6

Lithgow JK, Danino VE, Jones J et al (2001) Analysis of *N*-acyl homoserine-lactone quorum-sensing molecules made by different strains and biovars of *Rhizobium leguminosarum* containing different symbiotic plasmids. Plant Soil 232:3–12. https://doi.org/10.1023/A:1010333716277

Luo ZQ, Su S, Farrand SK (2003) In situ activation of the quorum-sensing transcription factor TraR by cognate and noncognate acyl-homoserine lactone ligands: kinetics and consequences. J Bacteriol 185:5665–5672. https://doi.org/10.1128/jb.185.19.5665-5672.2003

Marjeczyk C, Brittnacher M, Jacobs M et al (2014) Global analysis of the *Burkholderia thailandensis* quorum sensing-controlled genes. J Bacteriol 196:1412. https://doi.org/10.1128/jb.01405-13

Martínez P, Huedo P, Martinez-Servat S et al (2015) *Stenotrophomonas maltophilia* responds to exogenous AHL signals through the LuxR solo SmoR (Smlt1839). Front Cell Infect Microbiol 5:41. https://doi.org/10.3389/fcimb.2015.00041

McClean KH, Winson MK, Fish L et al (1997) Quorum sensing and *Chromobacterium violaceum*: exploitation of violacein production and inhibition for the detection of *N*-acylhomoserine lactones. Microbiology 143:3703–3711. https://doi.org/10.1099/00221287-143-12-3703

Nieto-Peñalver CG, Bertini EV, de Figueroa LIC (2012) Identification of *N*-acyl-homoserine lactones produced by *Gluconacetobacter diazotrophicus* PAL5 cultured in complex and synthetic media. Arch Microbiol 194:615–622. https://doi.org/10.1007/s00203-012-0794-1

Ochiai S, Yasumoto S, Morohoshi T et al (2014) AmiE, a novel *N*-acylhomoserine lactone acylase belonging to the amidase family, from the activated-sludge isolate *Acinetobacter* sp. strain Ooi24. Appl Environ Microbiol 80:6919–6925. https://doi.org/10.1128/AEM.02190-14

Patel HK, Suárez-Moreno ZR, Degrassi G et al (2013) Bacterial LuxR solos have evolved to respond to different molecules including signals from plants. Front Plant Sci 4:447. https://doi.org/10.3389/fpls.2013.00447

Patzelt D, Wang H, Buchholz I et al (2013) You are what you talk: quorum sensing induces individual morphologies and cell division modes in *Dinoroseobacter shibae*. ISME J 7:2274–2286. https://doi.org/10.1038/ismej.2013.107

Pedrosa FO, Monteiro RA, Wassem R et al (2011) Genome of *Herbaspirillum seropedicae* strain SmR1, a specialized diazotrophic endophyte of tropical grasses. PLoS Genet 7:1002064. https://doi.org/10.1371/journal.pgen.1002064

Pierson LS III, Keppenne VD, Wood DW (1994) Phenazine antibiotic biosynthesis in *Pseudomonas aureofaciens* 30–84 is regulated by PhzR in response to cell density. J Bacteriol 176:3966–3974. https://doi.org/10.1128/jb.176.13.3966-3974.1994

Roca A, Cabeo M, Enguidanos C et al (2024) Potential of the quorum-quenching and plant-growth promoting halotolerant *Bacillus toyonensis* AA1EC1 as biocontrol agent. Microb Biotechnol 17:1–19. https://doi.org/10.1111/1751-7915.14420

Rodríguez M, Torres M, Blanco L et al (2020) Plant growth-promoting activity and quorum quenching-mediated biocontrol of bacterial phytopathogens by *Pseudomonas segetis* strain P6. Sci Rep 10:4121. https://doi.org/10.1038/s41598-020-61084-1

Rothballer M, Eckert B, Schmid M et al (2008) Endophytic root colonization of gramineous plants by *Herbaspirillum frisingense*. FEMS Microbiol Ecol 66:85–95. https://doi.org/10.1111/j.1574-6941.2008.00582.x

Sant'Anna FH, Almeida LG, Cecagno R et al (2011) Genomic insights into the versatility of the plant growth-promoting bacterium *Azospirillum amazonense*. BMC Genomics 12:409. https://doi.org/10.1186/1471-2164-12-409

Schaefer AL, Taylor TA, Beatty JT et al (2002) Long-chain acyl-homoserine lactone quorum-sensing regulation of *Rhodobacter capsulatus* gene transfer agent production. J Bacteriol 184: 6515–6521. https://doi.org/10.1128/jb.184.23.6515-6521.2002

Shaw PD, Ping G, Daly SL et al (1997) Detecting and characterizing *N*-acyl-homoserine lactone signal molecules by thin-layer chromatography. Proc Natl Acad Sci USA 94:6036–6041. https://doi.org/10.1073/pnas.94.12.6036

Singh PK, Schaefer AL, Parsek MR et al (2000) Quorum sensing signals indicate that cystic fibrosis lungs are infected with bacterial biofilms. Nature 407:762–764. https://doi.org/10.1038/35037627

Soto-Aceves MP, Diggle SP, Greenberg EP (2023) Microbial primer: LuxR-LuxI Quorum Sensing. Microbiol SGM 169:001343. https://doi.org/10.1099/mic.0.001343

Steidle A, Allesen-Holm M, Riedel K et al (2002) Identification and characterization of an *N*-acylhomoserine lactone-dependent quorum-sensing system in *Pseudomonas putida* strain IsoF. Appl Environ Microbiol 68:6371–6382. https://doi.org/10.1016/j.jbiosc.2024.03.001

Steindler L, Venturi V (2007) Detection of quorum-sensing *N*-acyl homoserine lactone signal molecules by bacterial biosensors. FEMS Microbiol Lett 266:1–9. https://doi.org/10.1111/j.1574-6968.2006.00501.x

Subramoni S, Florez Salcedo DV, Suarez-Moreno ZR (2015) A bioinformatic survey of distribution, conservation, and probable functions of LuxR solo regulators in bacteria. Front Cell Infect Microbiol 5:16. https://doi.org/10.3389/fcimb.2015.00016

Trân Vân V, Ngôkê S, Berge O et al (1997) Isolation of *Azospirillum lipoferum* from the rhizosphere of rice by a new, simple method. Can J Microbiol 43:486–490. https://doi.org/10.1139/m97-068

Utari PD, Vogel J, Quax WJ (2017) Deciphering physiological functions of AHL quorum quenching Acylases. Front Microbiol 19:1123. https://doi.org/10.3389/fmicb.2017.01123

Vial L, Cuny C, Gluchoff-Fiasson K et al (2006) *N*-acyl-homoserine lactone-mediated quorum-sensing in *Azospirillum*: an exception rather than a rule. FEMS Microbiol Ecol 58:155–168. https://doi.org/10.1111/j.1574-6941.2006.00153.x

Wang J, Quan C, Wang X et al (2010) Extraction, purification and identification of bacterial signal molecules based on *N*-acyl homoserine lactones. Microb Biotechnol 4:479–490. https://doi.org/10.1111/j.1751-7915.2010.00197.x

Warring SL, Sisson HM, Fineran PC et al (2024) Strategies for the biocontrol *Pseudomonas* infections pre-fruit harvest. Microb Biotechnol 17:e70017. https://doi.org/10.1111/1751-7915.70017

Watson WT, Minogue TD, Val DL et al (2002) Structural basis and specificity of acyl-homoserine lactone signal production in bacterial quorum sensing. Mol Cell 9:685–694. https://doi.org/10.1016/S1097-2765(02)00480-X

Winson MK, Camara M, Latifi A et al (1995) Multiple *N*-acyl-L-homoserine lactone signal molecules regulate production of virulence determinants and secondary metabolites in *Pseudomonas aeruginosa*. Proc Natl Acad Sci USA 92:9427–9431. https://doi.org/10.1073/pnas.92.20.9427

Wisniewski-Dyé F, Downie JA (2002) Quorum-sensing in *Rhizobium*. Antonie Van Leeuwenhoek 81:397–407. https://doi.org/10.1023/A:1020501104051

Wisniewski-Dyé F, Vial L (2015) Cell-cell communication in *Azospirillum* and related PGPR. In: Cassán F, Okon Y, Creus C (eds) Handbook for *Azospirillum*: technical issues and protocols, pp 263–285. https://doi.org/10.1007/978-3-319-06542-7_15

Wisniewski-Dyé F, Jones J, Chhabra SR et al (2002) *railR* genes are part of a quorum-sensing network controlled by *cinI* and *cinR* in *Rhizobium leguminosarum*. J Bacteriol 184:1597–1606. https://doi.org/10.1128/jb.184.6.1597-1606.2002

Wisniewski-Dyé F, Borziak K, Khalsa-Moyers G et al (2011) *Azospirillum* genomes reveal transition of bacteria from aquatic to terrestrial environments. PLoS Genet 7:e1002430. https://doi.org/10.1371/journal.pgen.1002430

Yasuda M, Isawa T, Shinozaki S et al (2009) Effects of colonization of a bacterial endophyte, *Azospirillum* sp. B510, on disease resistance in rice. Biosci Biotechnol Biochem 73:2595–2599. https://doi.org/10.1271/bbb.90402

Zan J, Choi O, Meharena H et al (2015) A solo *luxI*-type gene directs acylhomoserine lactone synthesis and contributes to motility control in the marine sponge symbiont *Ruegeria* sp. KLH11. Microbiology 161:50–56. https://doi.org/10.1099/mic.0.083956-0

Zhang LH (2003) Quorum quenching and proactive host defense. Trends Plant Sci 8:238–244. https://doi.org/10.1016/S1360-1385(03)00063-3

Zhang S, Kan J, Liu X et al (2023) Phytopathogenic bacteria utilize host glucose as a signal to stimulate virulence through LuxR homologues. Mol Plant Pathol 24:359–373. https://doi.org/10.1111/mpp.13302

Chapter 8
Coping with Abiotic and Biotic Stresses in the Rhizosphere by the Plant Growth-Promoting Rhizobacteria *Azospirillum*

Sudhir Singh, Vijay Shankar Singh, and Anil Kumar Tripathi

Abstract Plant growth-promoting bacteria such as *Azospirillum* inhabit the rhizosphere of a large number of C3 and C4 plants. Their association with a wide variety of plants in different types of soils and climatic conditions reflects their resilience and adaptability due to the various mechanisms they possess to cope with the environmental stresses encountered in the rhizosphere or adjacent soils. These stresses include submergence, drought, salinity, thermal stress, oxidative stress, nutrient stress/starvation (carbon, nitrogen, phosphorus, and iron), and competition with other microbes of the rhizosphere. Here, we have discussed how different types of stresses are handled by *A. brasilense* under given stressful conditions and how *A. brasilense* confers growth advantage to its host plants under adverse environmental conditions. In order to survive in fluctuating environmental conditions, *A. brasilense* uses "two-component systems" and "alternative σ factors" to sense and quickly respond to stressful situations. We have also given some protocols that are used for studying the effect of various stresses on *A. brasilense* at the phenotypic as well as protein levels (enzymatic assays and 2D gel electrophoresis).

Keywords Abiotic stress · Drought · Salinity · Oxidative stress · Adaptation

---

S. Singh
Centre for Molecular Biology, Central University of Jammu, Jammu, Jammu and Kashmir, India

V. S. Singh
Department of Microbiology, Sikkim University, Gangtok, Sikkim, India

A. K. Tripathi (✉)
Department of Biological Sciences, Indian Institute of Science Education and Research, Mohali, Punjab, India

F. D. Cassán et al. (eds.), *The Definitive Handbook of Azospirillum*,
https://doi.org/10.1007/978-3-032-09222-9_8

8.1 Introduction

Bacteria live in diverse environments and use mechanisms to adapt to their surrounding environmental conditions. They have mechanisms to sense and respond to the environmental fluctuations (Fig. 8.1). Bacteria inhabiting soil or plant rhizospheres have to cope with fluctuations in their biotic and abiotic environment. High ambient temperatures result in higher rates of transpiration, leading to a reduction in the water content in the soil. This leads to an increase in the concentrations of salts, resulting in a consequent increase in soil salinity and osmolarity in the rhizosphere, which affects the growth and physiology of the rhizosphere bacteria. The increased salinity and osmolarity is sensed by bacteria, which respond by taking up osmoprotectants from their surroundings. But, if the desired osmoprotectants are not available in the rhizosphere, they end up synthesizing them (Sleator and Hill 2002; Zaprasis et al. 2013). But, if they lack the ability to uptake or synthesize osmoprotectants, their survival under salinity stress is compromised. Reduction in the water content in the rhizosphere soil leads to a decrease in the fluidity and an increase in the viscosity of the medium, which can also be sensed by bacteria. When the medium is fluid, bacteria swim with the help of their polar flagellum. But, when the viscosity is high, polar flagellum encounters hindrance in their movement, leading to the production of lateral flagella, which help bacteria in swarming on the surface (Partridge and Harshey 2013). As long as the viscosity of the medium is low, bacteria can run away from a given stressful situation by showing negative chemotaxis. But when running away is difficult or not possible due to the increased

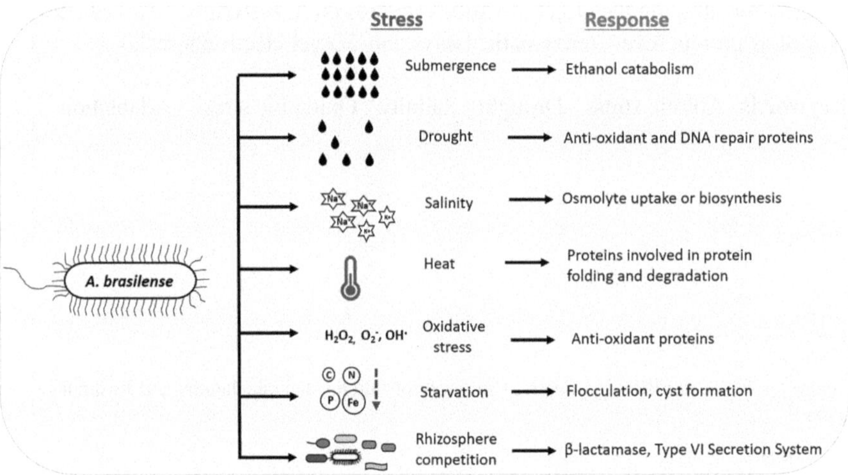

Fig. 8.1 Different types of stresses (submergence, drought, salinity, heat, oxidative stress, starvation, and rhizosphere competition) encountered by plants and the rhizobacterium *A. brasilense,* and the responses invoked by *A. brasilense*

viscosity, they induce the synthesis of proteins and metabolic pathways, which help in coping with the stressful situation (Boric et al. 2012; Diethmaier et al. 2017; Speare et al. 2020; Ribardo et al. 2024).

The bacterial membranes are studded with a diverse array of integral membrane proteins that play crucial roles in cellular functions, including environmental sensing, molecular transport, and host interactions (Jeffery 2016). The cytosolic component of these sensor proteins then transduces the signal/message related to the alteration of environmental condition to their cognate receiver proteins located in the cytosol, which may directly or indirectly involve a few more regulatory proteins to act as a response regulator (RR) (Beier and Gross 2006; Tierney and Rather 2019). The response regulator causes up- or downregulation of pathways that eventually help in coping with the stress caused by the alteration of the ambient environment (Vercruysse et al. 2011; Kaczmarczyk et al. 2014; de Pina et al. 2021). Bacteria often use two-component systems (TCS) consisting of a sensor, which is located in the membrane, and a response regulator, located in the cytosol (Surette et al. 1996). A variety of more complex proteins consisting of multiple sensor and receiver domains also exist, which are involved in a more complex signaling process. The response regulators are often DNA-binding proteins that bind upstream of the promoter elements to interact with the sigma factors to affect initiation of transcription of the genes of the pathways involved in stress regulation (Gao and Stock 2009). Some of the membrane proteins (anti-sigma factors) directly sequester their cognate sigma factors and prevent them from initiating transcription of stress regulatory genes in the absence of stress (Paget 2015). But, under stressful situations, the conformation of the anti-sigma factor changes to set the sigma factor free to initiate transcription of the stress regulatory genes.

8.2 Mechanisms of Coping with Abiotic Stresses in Soil and Rhizosphere

8.2.1 Submergence Stress

Terrestrial plants are often prone to partial to total submergence of their above-ground parts. Submergence of above-ground parts often arises due to flooding, which is one of the major abiotic stresses reported to affect plant growth and crop yield (Bailey-Serres et al. 2012; Voesenek and Bailey-Serres 2015). Submergence of plant parts leads to waterlogging stress, where the leaves and stems of the plant are partially submerged (Nishiuchi et al. 2012). Waterlogging stress leads to a significant decrease in gas diffusion rates, increased vulnerability to fungal diseases, and impaired soil nutrient uptake and light access, thereby inhibiting the ability of plants to carry out photosynthesis and ultimately hindering respiration and photosynthesis (Ram et al. 1999; Nishiuchi et al. 2012). Submerged condition also results in low O_2 availability, leading to a shift from aerobic to anaerobic mode of respiration and ethanol fermentation pathways (Ricard et al. 1994). To minimize the effects of stress due to different abiotic conditions, many studies presented the possibility of using plant growth-promoting rhizobacterium (PGPR) that can improve the ability of plants to grow in different conditions, such as waterlogging (Olanrewaju et al. 2017).

A. brasilense has been shown to enhance plant growth and survival under waterlogged conditions. In radish, *A. brasilense* inoculation improved stomatal conductance, chlorophyll concentration, and photosystem II efficiency under waterlogging stress (Salazar-Garcia et al. 2022). It can survive in soil for extended periods, even at high moisture levels (Belaid et al. 2019). In *Digitaria adscendens*, dual inoculation with *A. brasilense* and *Glomus mossae* maximized growth, nutrient content, and enzyme activity under waterlogged conditions (Rao and Shukla 2002). Similarly, in maize, *A. brasilense* inoculation improved growth, gas exchange, and nutrient uptake under both irrigated and water deficit conditions (Marques et al. 2020). *A. brasilense* has recently been shown to co-metabolize ethanol in the presence of fructose or glycerol, indicating a possibility of mitigating ethanol stress caused by anaerobic conditions by utilizing ethanol as a carbon source in the plant rhizosphere (Singh et al. 2021). These studies demonstrate that *A. brasilense* can mitigate the negative effects of waterlogging on plant physiology and growth, making it a promising tool for sustainable agriculture in the areas prone to waterlogging.

8.2.2 Drought Stress

Drought is one of the major abiotic stress factors that hampers plant growth and development. It affects agricultural productivity and impacts the food and feed supply. It also becomes more intense in recent years, and it is predicted to cause serious destruction on more than half of the world's agricultural land by 2050, thus affecting plant productivity and global food security. It is projected that drought stress reduces cereal, rice, wheat, and maize yields by 10%, 25%, 21%, and 40%, respectively (Anjum et al. 2011). Dry soils and vegetation also increase forest fires (Farooq et al. 2009). During drought, stomata gradually close with a parallel decline in water retention, thus severely affecting water use efficiency, nutrient absorption, and transport from roots to shoots, which alters plant physiological, biochemical, morphological, and molecular traits (Shao et al. 2009).

A. brasilense inoculation enhances crop productivity and stress tolerance in drought-exposed agricultural systems. Studies show that *A. brasilense* inoculation leads to the proliferation of roots, increasing the number of secondary roots and root hairs, which increase root surface area for improved uptake of water and nutrients. This leads to an improved drought tolerance in maize, increasing biomass production, carbon, nitrogen, and chlorophyll levels while reducing stress hormones (Cura et al. 2017). Genetically engineered *A. brasilense* with enhanced trehalose biosynthesis improves drought tolerance and biomass in maize (Rodríguez-Salazar et al. 2009). In *Arabidopsis*, *A. brasilense* inoculation increases ABA levels, leading to the improved drought response, increased biomass, and better reproductive success (Cohen et al. 2015). It alters root architecture, stimulates photosynthetic pigments, and enhances water retention (Cohen et al. 2015). The drought tolerance mechanism involves modulation of plant hormones, with inoculated plants showing lower levels of stress-related hormones like abscisic acid and ethylene (Cohen et al. 2009; Cura et al. 2017). It was also shown that co-application of bentonite and *A. brasilense* in wheat resulted in greater water uptake as compared to the control (Iqbal et al. 2024). It has also been shown that the inoculation of *A. brasilense* in wheat results in significantly diminished inhibitory effects of drought stress on the relative water content of shoots, roots, and leaves, contents of pigments (chlorophyll a and b) and ascorbic acid. Moreover, the bacterial inoculation notably reduced the drought-induced damage indicated by lower leakage of electrolytes and less accumulation of malondialdehyde and hydrogen peroxide (Kasim et al. 2021). These findings demonstrate the potential of *A. brasilense* as a bioinoculant for enhancing crop resilience under drought stress.

8.2.3 Osmotic or Salinity Stress

The excess of salt in soil inhibits the uptake of water from the soil to the plant, causing dehydration, decreased turgor pressure, and finally cell shrinkage, thus affecting the growth of the plant. A limited amount of water inside the plant also leads to the closure of stomata, less amount of CO_2 absorption, and a decrease in the photosynthetic yield. The concentration of solutes (e.g., ions, salts, and metabolites) in soil also plays a vital role in microbial growth and affects their diversity and functions. Excess salt can also affect the growth of microbes due to the difference in osmolarity caused by variation in the optimum concentration of compatible solutes or osmolytes such as sodium chloride, potassium ions, sugars, glycine betaine, proline, trehalose, etc. (Paul 2013). These compatible solutes provide protection against deleterious consequences of water loss from the cell, maintain the appropriate cell volume, and protect intracellular macromolecules (Csonka and Epstein 1996). Microbial membranes are permeable to water but not to most other metabolites. Bacteria accumulate these compatible solutes to cope with salinity or osmotic stress. Proline plays a key role in osmo-adaptation as the dominant osmolyte shifts from glutamate to proline during salinity stress in *A. brasilense* (Riou and Le Rudulier 1990; Tripathi et al. 1998). Different strains of *A. brasilense* exhibit varying levels of salt tolerance, with some strains showing enhanced exopolysaccharide production and cell aggregation as physiological responses to salt stress (Chowdhury et al. 2007). A pair of extracytoplasmic function (ECF) sigma factor and anti-sigma factor (RpoE-ChrR) has been shown to control salt stress tolerance *in A. brasilense* (Mishra et al. 2011). The adaptation of diazotrophs to osmotic stress is of great importance because soil salinity inhibits many of the plant growth promoting activities, such as nitrogen fixation and phytohormone production (Miller and Wood 1996). Two closely related strains of *A. brasilense, viz.*, Sp7 and

Cd, exhibit marked differences in their osmotolerance (Tripathi et al. 1998). In the case of *Azospirillum,* salinity stress has been shown to reduce the attachment of *A. brasilense* Cd to maize and wheat roots (Jofre et al. 2011). Increased salinity was also reported to inhibit nitrogen fixation at the level of *nifH* expression (Jofre et al. 1998) and nitrogenase activity (Tripathi et al. 2002). Comparison of osmotolerance in different species of *Azospirillum* revealed that it increased in the order *A. amazonense*, *A. lipoferum*, *A. brasilense*, and *A. halopraeferens* (Hartmann et al. 1988).

A. brasilense mitigates salinity stress in plants through various mechanisms. It improves plant growth under high salt conditions by regulating the expression of genes involved in salt stress response, abscisic acid signaling, and nutrient transport (Degon et al. 2023). *A. brasilense* enhances antioxidant enzyme activity, reduces proline and malondialdehyde content, and modulates gene expression related to antioxidant activity and pathogenesis in maize (Fukami et al. 2018). In tomato, *A. brasilense* improves root architecture, increases biomass, and enhances antioxidant enzyme activity and proline production under salt stress (Alzate Zuluaga et al. 2022). The bacterium also modulates secondary metabolism and phytohormones to alleviate salinity stress.

8.2.4 Thermal Stress

A. brasilense belongs to a mesophilic class of bacteria whose optimum temperature for growth is 30 °C. Bacteria have developed several mechanisms by which they sense and respond to temperature variations in their surrounding environment. There are various biomolecules that can detect the change in temperature in their surroundings and thus act as thermosensors. These biomolecules include DNA, RNA, and protein and respond appropriately to the requirement (Shapiro and Cowen 2012). Thermal or heat stress is a physical abiotic stress in which microbes are exposed to high temperatures as compared to their ambient growth temperature range. Heat shock response is a universal phenomenon that enables living cells to overcome several environmental stresses, including heat stress, which causes proteins to unfold and aggregate (Baker et al. 1984; Arsène et al. 2000). Most organisms respond to the upshift in temperature by transiently inducing the synthesis of a series of heat shock proteins (HSP), most of which are either chaperones (DnaK, DnaJ, GroEL, and GroES) or ATP-dependent proteases (FtsH, Lon ClpAP, and HslVU), which avoid accumulation of unfolded or misfolded proteins at elevated temperature (Arsene et al. 2000). Numerous HSPs and their homologs have been known to be involved in the nodulation and nitrogen fixation, and their expression is regulated by heat shock sigma factors (RpoH or σ^H).

The genome of *A. brasilense* encodes five paralogs of the RpoH gene (*rpoH1, rpoH2, rpoH3, rpoH4,* and *rpoH5*). In an earlier study, it was shown that out of the 5 RpoH paralogs encoded in *A. brasilense* genome, only RpoH1 is responsible for coping with heat stress, while RpoH3 and RpoH5 are involved in the regulation of expression of catalases (Rai et al. 2018). Heterologous overexpression of all five paralogs of RpoH in a heat-sensitive strain of *E. coli* (CAG 9333) confers improved growth at 42 °C. However, insertional inactivation of all five paralogs of *rpoH* in *A. brasilense* and incubation at higher temperature results in optimal growth of the mutants of all paralogs of RpoHs except RpoH1. While creating *rpoH::km* mutants individually for each *rpoH* paralog in *A. brasilense*, it was expected that it might not be possible to isolate a heat-sensitive *rpoH::km* mutant, as other copies of *rpoH* paralogs might complement for the inactivated copy of *rpoH*. It was presumed that duplication and divergence of *rpoH* in *A. brasilense* and other bacteria seem to have generated multiple copies, of which only one remains committed for the heat stress, while others become specialized for oxidative and photooxidative stress. For example, in an earlier study, it was shown that RpoH2 is involved in coping with photooxidative stress (Kumar et al. 2012). It was also shown that RpoH5 regulates expression of the catalase activity (Rai et al. 2018).

Heat stress significantly impacts *A. brasilense* survival and nitrogen fixation in agricultural soils. Temperatures above 35 °C reduce growth rates of *A. brasilense* strains (da Cunha et al. 2024). However, some *Azospirillum* species, like *A. halopraeferens*, show optimal growth and nitrogen fixation at higher temperatures of 40–41 °C (Tripathi and Klingmuller 1992). Temperature also affects *nif* gene expression and NifA activity, which are crucial for nitrogen fixation (Tripathi and Klingmuller 1992). In field conditions, higher temperatures up to 32 °C enhance acetylene reduction (a proxy for nitrogen fixation) in *Setaria italica* inoculated with *A. brasilense* (Kapulnik et al. 1981).

8.2.5 Oxidative Stress

Bacteria undergo oxidative stress under aerobic conditions by forming reactive oxygen species (ROS) that can have a devastating effect on protein structure and function and can even contribute to cell death. These ROS include superoxide anion ($O_2^{-\cdot}$), hydrogen peroxide (H_2O_2), and the extremely reactive hydroxyl radical ($HO^{\cdot}$) and excess of these ROS leads to oxidative stress, which leads to an upregulation of several genes/proteins during exposures to superoxide/H_2O_2. In *E. coli,* the expression of these genes is controlled by two unique transcriptional activators, OxyR, and SoxRS. The transcription factor OxyR activates and drives the expression of several genes inducible by H_2O_2, while the transcription factors SoxR/SoxS mediate responses to $O_2^{-\cdot}$ (Storz and Imlay 1999; Kiley and Storz 2004). OxyR is a global regulator which is activated by the oxidation of two cysteine residues that form a disulphide bond, altering its structure under oxidative stress (Storz et al. 1990). It regulates the transcription of genes encoding catalase (*katG*), alkylhydroperoxide reductase (*ahpC* and *ahpF*) and several other genes. Alkylhydroperoxide reductase enzyme has been shown to be involved in the detoxification of organic peroxides into corresponding alcohol.

A. brasilense possesses two paralogs of OxyR (*oxyR1* and *oxyR2*), three gene encoding catalases (*katAI*, *katAII*, and *katN*) and one alkylhydroperoxide reductase gene (*ahpC*). The expression of *ahpC* in *A. brasilense* is negatively regulated by the transcriptional regulator (OxyR1) and positively by sigma factor, RpoE2 (Singh et al. 2016). A complex regulatory network involving multiple RpoH sigma factors and OxyR controls the expression of catalases in *A. brasilense*, with RpoH3 and RpoH5 regulating *katN* and *katAII*, respectively (Rai et al. 2018). Under oxidative stress, *A. brasilense* activates the sigma factor (RpoE2), which regulates carotenoid production and enhances resistance to oxidative damage (Gupta et al. 2013).

Azospirillum species play a crucial role in enhancing plant stress tolerance through various molecular mechanisms. Bacterial ability to mitigate oxidative stress extends to its host organisms, as demonstrated in co-cultures with the microalgae *Chlorella sorokiniana*, where *A. brasilense* reduced intracellular ROS accumulation under copper stress and nitrogen limitation (Peng et al. 2021). When inoculated on maize, *A. brasilense* and its metabolites influenced the expression of oxidative stress-related genes (APX1, APX2, CAT1, SOD2, and SOD4) and plant defence genes (PR1, prp2, and prp4) in both leaves and roots (Fukami et al. 2017). It can reduce intracellular ROS accumulation and promote chlorophyll content in plants under stress conditions (Peng et al. 2020). The bacterium's lectins increase peroxidase and superoxide dismutase activities while decreasing catalase activity in wheat seedling roots under short-term thermal stress (Alen'kina et al. 2018).

The influence of *A. brasilense* on plant oxidative stress mechanisms contributes to improved stress tolerance and plant growth promotion. The nitrogen regulatory protein NtrC plays a crucial role in regulating proteins involved in oxidative stress response, such as glutathione S-transferase and hydroperoxide resistance protein (Kukolj et al. 2020). Alkyl hydroperoxide reductase (AhpC) contributes to oxidative stress resistance and modulates cell-surface properties, affecting morphology, aggregation, and flocculation (Wasim et al. 2009).

8.2.6 Nutrient Stress/Starvation (Carbon, Nitrogen, Phosphorus, and Iron)

Starvation is a common experience for bacteria in nature. Under starvation, bacteria enter dormancy characterized by metabolic arrest and slowed or complete inhibition of growth. In some bacteria, this stress triggers an elaborate differentiation response, terminating in the formation of spores that are highly resistant and morphologically distinct structures (Kim et al. 1992; Hoch 1993; Chater 1993). In most bacteria, this differentiation is very subtle and is not accompanied by a comparable morphological change. Since root exudates are rich in carbon and nitrogen, the cells of *Azospirillum* face starvation in the absence of root exudates. Synthesis of PHB granules and production of IAA in *Azospirillum* have been correlated with the onset of stationary phase, i.e., carbon starvation. Members of the genus *Azospirillum* change their metabolic activity according to the changing environment, like the availability of carbon and nitrogen sources, moisture, and oxygen content. Encystation is one such phenomenon, where vegetative cells undergo certain morphological and biological changes to form resting cysts, which enable them to survive under deleterious conditions such as desiccation and enabling long-term survival of these bacteria in soil (Sadoff 1975; Eskew et al. 1977; Sadasivan and Neyra 1985). Cyst formation is favored by fructose as a carbon source. *A. brasilense* can also accumulate PHB up to 70% of the cell dry weight, as PHB-rich cells demonstrate better nitrogen fixation and enzyme activity during starvation (Okon and Itzigsohn 1992). *A. brasilense* also displays phenotypic variation in response to prolonged starvation, altering features such as pigmentation, aggregation, and polysaccharide production (Lerner et al. 2010). PHB degradation and its utilization as the sole source of carbon and energy by *A. brasilense* may be involved in favoring establishment, proliferation, survival, and competition with other organisms in the rhizosphere (Tal et al. 1990).

Phosphorus (P) starvation significantly impacts the growth and nitrogen fixation capabilities of diazotrophic bacteria. Under P deficiency, nitrogen fixation activity is severely repressed (Zhang et al. 2022). Bacteria adapt by synthesizing P-free lipids and upregulating P uptake systems while downregulating nitrogenase and nitrogen fixation-related regulators (Zhang et al. 2022). The application of *A. brasilense* on *Arabidopsis* showed that inoculation alleviated Pi deficiency-induced growth inhibition and anthocyanin accumulation and increased the total P content in *Arabidopsis* plants (Sun et al. 2022). *Azospirillum* species demonstrate remarkable adaptations to phosphorus deficiency, both in their own metabolism and in promoting plant growth. When inoculated with *Arabidopsis*, *A. brasilense* enhances root hair development, acid phosphatase activity, and expression of phosphate transporters, improving P absorption (Sun et al. 2022). In *Brachypodium*, *A. brasilense* initially triggers a defense response but later mitigates P deficiency stress and stimulates root development (Schillaci et al. 2021).

Iron (Fe), an essential element for plant growth, is abundant in soil but with low bioavailability. Thus, plants have developed specialized mechanisms to sequester this element. Iron limitation significantly impacts nitrogen fixation capabilities of *A. brasilense* in agricultural soils. Iron deprivation completely inhibits *nif* gene transcription and nitrogenase activity in *A. brasilense* (Mugnai et al. 1994). However, *A. brasilense* can produce siderophores that sequester iron from soil, potentially improving iron availability for plants (Pii et al. 2015). The ability of this bacterium to alleviate iron deficiency symptoms in plants suggests its potential use in iron-limiting conditions (Pii et al. 2015). Under iron-limiting conditions, *A. brasilense* produces siderophores, which can exhibit antifungal activity against plant pathogens like *Colletotrichum acutatum* (Tortora et al. 2011). These siderophores may also enhance iron availability for plants, as demonstrated in cucumber plants grown on calcareous soils (Pii et al. 2015). Iron-starved *A. brasilense* strains can still promote plant growth and increase iron uptake in host plants, as observed in *Zea mays* (Housh et al. 2021). The ability of *A. brasilense* to adapt to iron-limited environments and its positive effects on plant iron nutrition suggest its potential as a biofertilizer and biocontrol agent in agriculture.

8.2.7 Rhizosphere Competition

The rhizosphere is a zone of intense competition among different groups of microorganisms for carbon and energy source present in the root exudates. To secure a competitive edge, microbial communities use a variety of strategies which include production of specialized metabolites like antibiotics and volatile organic compounds to inhibit their competitors (Chepsergon and Moleleki 2023; Ajinde et al. 2024). Resource depletion occurs through exploitative competition, where bacteria restrict competitors' access to nutrients by secreting siderophores (Chepsergon and Moleleki 2023). Direct antagonism involves competing for space and food with pathogens (Ajinde et al. 2024). *Azospirillum* shows very high levels of resistance to β-lactam antibiotics including ampicillin. In *A. brasilense* Sp245, resistance to antibiotic ampicillin is mediated by an AmpC-type β-lactamase and a lytic transglycosylase, regulated by a cascade of RpoE7 and RpoH3 sigma factors (Pandey et al. 2022). Genome analysis of *A. brasilense* shows that it harbors many copies of genes encoding for enzyme β-lactamase, and expresses a 30 kDa protein that confers high-level of resistance to β-lactam antibiotics (Verreth et al. 1989). When exposed to ampicillin, *A. brasilense* Sp245 which possesses an inducible β-lactamase, will have a competitive edge over other bacteria lacking the β-lactamase.

To restrict proliferation of bacteria in the rhizosphere, plants release different kinds of ROS through a process called oxidative burst, which hampers the growth of microorganisms around roots. These ROS inhibit proliferation of bacteria that lack antioxidant defense mechanisms (Mandon et al. 2021). The genome of *A. brasilense* encodes 23 distinct types of sigma factors (Dubey et al. 2020), along with 2 copies of the oxidative stress response regulator, OxyR1 and OxyR2 (Singh et al. 2016), 3 types of catalases, KatAI, KatN, and KatAII (Rai et al. 2018), and 1 alkylhydroperoxidase, AhpC (Singh et al. 2016). This extensive range of antioxidants which are capable of inactivating different ROS, suggests that *A. brasilense* possesses necessary resilience against elevated levels of ROS released by plants.

In order to outcompete other bacteria in the rhizosphere, *A. brasilense* induces production of a Type 6 Secretion System (T6SS) in the presence of fructose. The T6SS is a common bacterial nanomachine that is vital for pathogenicity and competition among bacteria (Kapitein and Mogk 2013; Coulthurst 2019). Given that the T6SS is well known for its role in releasing effector molecules that suppress the growth of neighboring bacteria, it was demonstrated that *A. brasilense* can inhibit the growth of *E. coli* and *Agrobacterium tumefaciens* (Rai et al. 2025). Earlier also, it was shown that T6SS provides *A. brasilense* antagonistic activities against various plant pathogens such as *Agrobacterium*, *Pectobacterium*, *Dickeya,* and *Ralstonia* species, suggesting a potential biocontrol function (Cassán et al. 2021). This multifaceted system significantly influences bacterial competition and host-microbe interactions across different ecological niches.

8.2.8 Formation of Aggregates and Cysts

The cells of *Azospirillum* tend to aggregate or form visible flocs in the culture medium under diverse stress conditions (Sadasivan and Neyra 1985; Bashan and Holguin 1997). This strategy helps them in mitigating the effects of stressful situations in their surroundings. Transfer of *Azospirillum* from a rich medium to an N-limited medium leads to a switch from a dispersive growth pattern to clumping, aggregation, or flocculation (Burdman et al. 1998). This is characterized by a loss of flagella and motility, the production of thick capsules, and the accumulation of PHB granules. Some of the cells within these aggregates may differentiate into cyst-like dormant cells (Gerk et al. 2000). The ability of *A. brasilense* to aggregate is considered important for its survival in depleted soils. Clumping is a transient cell-to-cell behavior exhibited by motile *Azospirillum* cells where they transiently adhere to each other at the non-flagellated poles. Clumping precedes flocculation, which contributes to the ability of *A. brasilense* to survive in the changing environments due to the presence of carbon storage reserves in the form of intracellular PHB granules and the presence of thick outer walls (Sadasivan and Neyra 1985). The cysts develop in aging cultures of *Azospirillum* grown in minimal media with limited nutrient availability.

The regulation of flocculation in *A. brasilense* was shown to be mediated via a gene named *flcA*, as *flcA* mutants fail to flocculate (Pereg-Gerk et al. 1998). Flocculation contributes to the ability of *A. brasilense* to survive in the soil (Pereg-Gerk et al. 1998; Joe et al. 2010). However, the genes and pathway(s) that control remodeling and changes in cell surface properties during clumping and flocculation are not known yet. Production of EPS and aggregates helps *A. brasilense* in coping with the salt stress (Chowdhury et al. 2007). Identification of salt stress-induced genes in *A. brasilense* Sp7 showed that the genes involved in EPS synthesis and aggregate formation were induced several-fold under salt stress (Chowdhury et al. 2007). The capacity to form aggregates in the presence of NaCl stress can be exploited in the mass production of *A. brasilense* inoculants.

8.3 Mechanisms of Sensing and Responding to Environmental Stresses

Bacteria have the remarkable ability to thrive in nearly every type of environment on the earth. They can grow in diverse habitats such as soil, nuclear waste, hot springs, cold tundras, deep sea, and the human gut (Cox and Battista 2005). Every living organism requires specific optimal conditions for its growth and multiplication; deviations beyond a threshold from these conditions result in stress that negatively impacts their survival and growth. The stresses faced by microorganisms can vary in intensity and consequences (Bogati and Walczak 2022; Sun et al. 2022). Depending on their severity, they can be categorized as mild, moderate, or severe. Under mild stress, cells retain their viability, but their growth rate decreases. In the case of moderate stress, there is a decrease in growth rate along with some loss of viability, while severe stress leads to the loss of the whole population. To protect themselves from such detrimental conditions, bacteria modify their gene expression patterns, leading to the production of proteins that safeguard the bacterial cell against the harmful effects of stress. These protective mechanisms can be classified as adaptive or cross-protective (Rodriguez-Romo and Yousef 2005). An adaptive stress response occurs when bacteria are exposed to non-lethal levels of stress, prompting the expression of proteins that enable them to survive in conditions that would otherwise be lethal. Conversely, in cross-protection, exposure to one type of stress triggers the production of proteins that help bacterial cells endure a different type of stress.

In controlled cultivation conditions, bacteria are subjected to well-defined and relatively stable environments, allowing their growth to occur at a normal rate. In the field conditions, however, bacteria face continually fluctuating environmental conditions such as variations in temperature, salinity, pH, osmotic pressure, nutrient availability, and the presence of excessive levels of ROS, heavy metals, or radiations. To endure and proliferate in such constantly shifting environmental conditions, bacteria must swiftly sense and respond to the altered environmental conditions and adapt by altering protein expression to cope with the new circumstances. Here, we have focused on two distinct mechanisms through which *A. brasilense* is able to sense and respond to the stresses: two-component systems and alternative sigma factors.

8.3.1 Two-Component Systems (TCS)

Two-component systems (TCS) are among the essential mechanisms employed by bacteria to sense environmental changes to modify their behavior in response. TCSs have been implicated in various processes such as virulence, chemotaxis, metabolism, and motility (Gao and Stock 2009). They play a crucial role in relaying changes occurring in the extracellular or intracellular environment to a second component, commonly referred to as the response regulator (RR). The RR is a type of diffusible protein that transmits its signal to downstream signaling proteins or directly interacts with promoter elements of specific genes. In TCSs, the first component is a sensor/kinase protein (HK), which is a transmembrane protein, while the second component is the RR. The HK identifies extracellular signals, leading to a conformational change in the protein that activates its kinase activity. This kinase activity phosphorylates a specific histidine residue, which then transfers its phosphate group to the aspartic acid residue of the RR (Stock et al. 2000).

In certain instances, the HK domain is fused to the RR, which is referred to as a hybrid histidine kinase (HyHK) (Appleby et al. 1996). Phylogenetic studies of the HK and RR domains in *A. lipoferum* 4B indicate that the expansion of this family (HyHK) predominantly occurred through horizontal gene transfer, as well as gene duplications within the *Azospirillum* genus (Borland et al. 2015). The genomes of *Azospirillum* have an abundance of genes encoding TCS, particularly those enriched in hybrid histidine kinases, which may offer regulatory adaptability for responding to changing environmental conditions (Borland et al. 2015). These systems allow *Azospirillum* to react to a variety of stressors, such as osmotic pressure, pesticides, and heavy metals (Gureeva and Gureev 2023).

The NtrY/NtrX system in *A. brasilense* has been recognized and described, displaying similarities to the NtrY/NtrX, a two-component regulatory system found in *Azorhizobium caulinodans* (Ishida et al. 2002). It has been proposed that the NtrY/NtrX system in *A. brasilense* is likely involved in the regulation of nitrate utilization (Ishida et al. 2002). In a separate investigation, the function of two-component systems (FixL/FixJ) has been demonstrated in regulating the genes *nifA* and *fixK*, where FixL serves as an HK and FixJ functions as an RR. It was shown that FixL/FixJ, a two-component system, plays a role in regulating nitrogen fixation genes (*nifA*) under microaerophilic conditions in *A. brasilense* Sp7 (Li et al. 2010). NifA acts as a transcriptional regulator that enhances the expression of the *nifHDK* operon, which encodes for nitrogenase polypeptides (Fischer 1994). Heterologous overexpression of FixL and FixJ from *A. brasilense* Sp7 in the knockout mutants of *fixL*$^-$ and *fixJ* $^-$ in *Sinorhizobium meliloti* resulted in the expression of *nifA*, which was found to be O_2 dependent (Li et al. 2010).

In *A. brasilense*, two separate chemotaxis pathways, Che1 and Che4, are responsible for its movement. The two-component systems (TCSs), such as CheA1/CheA4, allow *Azospirillum* to navigate towards nutrients in the rhizosphere, aiding in plant root colonization. Furthermore, research indicates that the disruption of Che1 and Che4 influences transcription, signaling transport, and cellular metabolism. In particular, it impacts nitrogen metabolism, which includes the processes of nitrate assimilation and nitrogen fixation (Ganusova et al. 2021).

8.3.2 Alternative σ Factors

In the soil or rhizosphere, *A. brasilense* is subjected to varying environmental conditions, including fluctuations in temperature, pH, salinity, moisture levels, and the presence of harmful substances. To adapt to these changing environments, it must produce new sets of proteins. This change in the expression of a new set of genes is brought about by RNA polymerase, which identifies specific promoters of target genes and synthesizes new RNA molecules, which can include mRNA, tRNA, rRNA, or other small regulatory RNAs. The newly formed mRNA is then utilized by ribosomes to create a new set of proteins. The RNA polymerase is composed of five polypeptides: α2, β, β′, and σ, where the combination of four polypeptides (α2, β, β′) constitutes the core RNA polymerase, and the fifth polypeptide, known as the sigma factor (σ), forms a dissociable subunit. The sigma factor binds to specific promoters to initiate gene expression and provides specificity to the core RNA polymerase. The sigma factor, in conjunction with the core RNA polymerase, attaches to the promoter and adds several nucleotides (Feklistov et al. 2014), after which it separates from the core RNA polymerase, leaving the core enzyme to add remaining nucleotides during the elongation phase. Various types of σ factors exist depending on the cellular requirements under different physiological circumstances.

Based on their structure and mechanisms for promoter recognition, these sigma factors can be divided into two families: σ^{54} and σ^{70}. The σ^{54} family, also known as RpoN, is distinct both structurally and functionally from the σ^{70} family. The initiation of transcription by σ^{54} necessitates the involvement of an additional protein partner, known as enhancer binding protein (EBP), which attaches upstream of the promoter (Hopper and Bock 1995), whereas σ^{70} can independently initiate the synthesis of new RNA. The σ^{70} is vital for the expression of most, if not all, genes that are necessary for housekeeping (Davis et al. 2017). The σ^{70} family is further divided into four groups that are phylogenetically and structurally different (Groups 1, 2, 3, and 4). The σ factor RpoD is categorized in Group 1. Group 2 σ factors show close similarities to Group 1 σ factors but are not critical for growth. Group 3 σ factors consist of the heat shock sigma factor, RpoH (Staron et al. 2009), and they regulate cellular processes such as sporulation and flagellum biosynthesis. The largest and most diverse types of σ factors are found in Group 4, which control gene expression in response to external stimuli and are therefore referred to as extracytoplasmic function (ECF) or RpoE σ factors. The activity of these σ factors (Group 4) can be regulated at multiple levels, but their post-translational regulation is especially common due to the direct interaction between σ factors and their corresponding anti-σ factors. This interaction hinders the σ factor's connection with core RNA polymerase (Ho and Ellermeier 2012; Sineva et al. 2017). Anti-σ factors utilize various mechanisms to release the σ factor in response to particular stimuli, which often involves processes that relay extracellular signals to the cytoplasm (Paget 2015).

The genome of *A. brasilense* Sp245 encodes 23 sigma factors, which consist of 1 RpoD (σ^{70}), 1 RpoN (σ^{54}), and 21 alternative sigma factors, including 5 RpoH, 6 RpoI, and 10 RpoE (Fig. 8.2). Among alternative sigma factors of *A. brasilense* Sp7, RpoE1 was the first to be characterized for its role in carotenoid biosynthesis in response to photooxidative stress. RpoE1 is an extracytoplasmic function sigma factor which regulates the expression of genes involved in carotenoid biosynthesis. Under photooxidative stress conditions, singlet oxygen is produced by the interaction of light with the membrane located porphyrins. This singlet oxygen interacts with ChrR1, the cognate anti-sigma factor of RpoE1. The singlet oxygen mediated conformational change in ChrR1 leads to the release of RpoE1, which with the help of core enzyme starts expression of genes involved in the biosynthesis of carotenoids, a known quencher of singlet oxygen (Mishra et al. 2011). The ECF sigma factors, RpoE1 and RpoE2 are cotranscribed with their respective anti-sigma factors, ChrR1 and ChrR2. RpoE1 and RpoE2 have distinct yet overlapping functions in stress response and carotenoid biosynthesis within *A. brasilense*. These ECF sigma factors are inhibited by their corresponding anti-sigma factors (ChrR1 and ChrR2) and become activated during stress conditions (Mishra et al. 2011; Gupta et al. 2013). RpoE1-ChrR1 is associated with tolerance to stresses from salt, ethanol, and

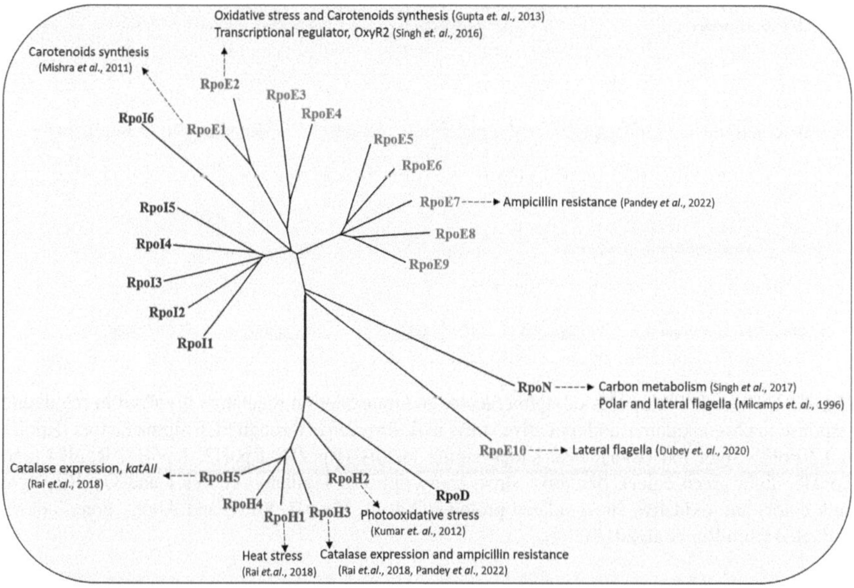

Fig. 8.2 Dendrogram showing evolutionary relationship among 23 different types of sigma factors in *A. brasilense*. Some of the characterized functions of sigma factors have also been mentioned. (Adapted from Dubey et al. 2020)

methylene blue, while also regulating carotenoid production (Mishra et al. 2011). RpoE2-ChrR2 quickly protects against oxidative stress by managing carotenoid synthesis, protein translation, protein folding, and purine biosynthesis (Gupta et al. 2013). Both RpoE-ChrR pairs oversee responses to photooxidative stress, showing evidence of interaction between both cognate and non-cognate pairs, which aids in maintaining redox homeostasis (Gupta et al. 2014). The role of RpoE2 has been demonstrated in the positive regulation of the AhpC (alkylhydroperoxidase) (Singh et al. 2016) and CrtE2 (gene associated with carotenoid synthesis) through RpoH2 (Rai et al. 2016). ECF sigma factor, RpoE7 regulates resistance to β-lactam antibiotics by acting as a negative regulator of the *ampC* expression (β-lactamase) via RpoH3 (Pandey et al. 2022). Further, RpoE10 has been shown to negatively regulate the formation of lateral flagella via a regulatory network involving LafR1 (Dubey et al. 2020). Numerous studies indicate that various ECF sigma factors are involved in controlling different genes to respond to adverse environmental circumstances.

Unlike *E. coli*, the genome of *A. brasilense* Sp7 encodes 5 paralogs of RpoH sigma factors. RpoH1 is crucial for heat tolerance, whereas RpoH3 and RpoH5 play a role in oxidative stress response by regulating the expression of two different types of catalases (Rai et al. 2018). The RpoH2 was shown to be responsible for controlling the photooxidative stress response and managing the expression of chaperones and other proteins related to stress (Kumar et al. 2012). In fact, photooxidative stress response was found to be regulated by a cascade of RpoE1 and RpoH2 sigma factors

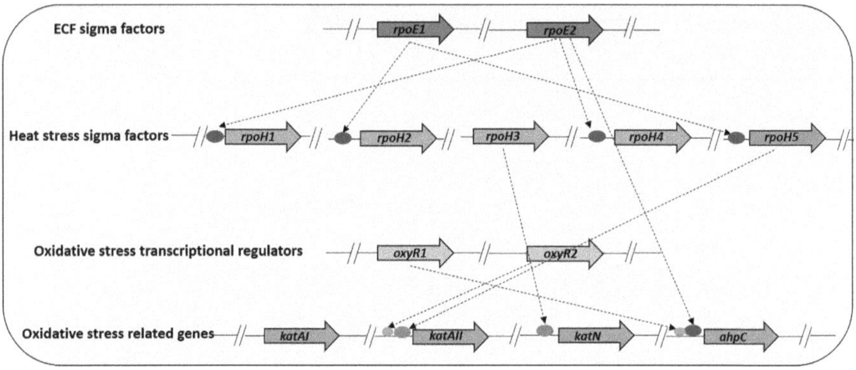

Fig. 8.3 Network of cascades of sigma factors and transcription regulators involved in regulating response to photooxidative and oxidative stress in *A. brasilense* through ECF sigma factors (RpoE1 and RpoE2; light red color), heat shock sigma factors (RpoH1, RpoH2, RpoH3, RpoH4, and RpoH5; light green color), oxidative stress transcriptional regulators (OxyR1 and OxyR2; light pink color) and oxidative stress-related proteins (KatA1, KatAII, KatN, and AhpC; green color). (Adapted from Rai et al. 2018)

(Rai et al. 2016). The expression of these RpoH sigma factors is linked with other regulatory elements, such as OxyR transcription regulators and the RpoE1→RpoH5 pathway (Rai et al. 2018). The availability of a relatively larger number of alternative sigma factors and transcription regulators in *A. brasilense* enables it to fine tune its response to environmental stresses in a more precise way than *E. coli.* It also leads to a more complicated mechanism of stress response using a network of regulatory cascades of alternative sigma factors and transcription regulators (Fig. 8.3). The main reason for the success of *A. brasilense* in the colonization of the rhizosphere of a wide variety of plants is its ability to cope with the diverse types of environmental stresses prevalent in the plant rhizospheres and the surrounding soil. This ability of *A. brasilense* has made it one of the most frequently used PGPR for enhancing crop productivity and will turn out to be important in developing strategies for sustainable agriculture.

8.4 Methodologies

8.4.1 *Effect of Temperature on Growth of* A. brasilense *by Growth Curve*

The effect of temperature on growth can be shown simply by plotting a growth curve at two different temperatures. Here we have given a step by step protocol for comparing growth at two different temperatures.

- Inoculate freshly streaked isolated single colonies into 2 ml minimal malate medium (MMM) (Vanstockem et al. 1987) and incubate at 30 °C with shaking (150 rpm) in an incubator shaker.
- The next day, overnight-grown culture is inoculated into 40 ml of MMM in two duplicate flasks (1% inoculum).
- One flask containing diluted culture is incubated at normal temperature (30 °C) and another at higher temperature (40 °C) under shaking conditions (150 rpm).
- Growth of culture is recorded at $O.D._{600nm}$ at every 2-h interval up to 24 h by transferring a small volume of culture (600 μl) into a cuvette tube.

8.4.2 *Effect of Different Types of Stress by Disc Diffusion Assay*

One can check the effect of different types of stress, such as antibiotics (ampicillin, tetracycline, ciprofloxacin), ROS-inducing agent (H_2O_2, paraquat, cumene hydroperoxide, menadione), and photooxidative stress (methylene blue, rose Bengal, 2,3,5-triphenyl tetrazolium chloride), by disc diffusion assay (Mascher et al. 2007). Here, we have given a step-by-step protocol for carrying out a disc diffusion assay.

1. Freshly streaked isolated single colony is inoculated into 2 ml MMM and grown overnight with shaking.
2. The next day, the overnight-grown culture is reinoculated into 25 ml of MMM (1% inoculum) and grown till 1 $O.D._{600nm}$.
3. To check the effect, an equal number of cells are pelleted by centrifuging 100 μl of overnight-grown culture (corresponding to 0.1 $O.D._{600nm}$) at 12,000 rpm for 10 min.
4. The pelleted cells are resuspended in MMM and spread on the top surface of an already solidified soft MMM agar plate having 1% agar.
5. The sterile disc is soaked in a solution of the test compound (antibiotic/ROS agent), and excess liquid is drained.
6. The soaked disc is placed gently at the center of the agar plate.
7. The agar plate is then incubated at 30 °C for 48 h.
8. A clear zone around the disc will show a zone of inhibition (absence of growth), and the diameter of the zone of inhibition is measured.

8.4.3 *Determination of the Flocculation Test*

A. brasilense under conditions of high carbon and low nitrogen promotes cell aggregation properties accompanied by the production of EPS and cytoplasmic PHB granules. *A. brasilense* Sp7 does not flocculate in medium with a low C:N ratio (40 mM malate and 10 mM NH_4Cl), but it shows extensive flocculation in medium with a high C:N ratio (40 mM malate and 4 mM NH_4Cl). Here we have given a protocol for quantitative estimation of flocculation in *A. brasilense*.

1. Inoculate a freshly streaked isolated single colony into 2 ml MMM containing 10 mM NH_4Cl and incubate at 28 °C under shaking condition (150 rpm).
2. The next day, an equal number of cells are harvested by centrifugation at 12,000 rpm for 10 min and washed twice with nitrogen-free MMM.
3. Washed cells are reinoculated into fresh MMM containing 4 mM NH_4Cl, and the culture grown at 30 °C at 150 rpm for 24 h.
4. Liquid culture containing cell aggregates is transferred into a 50 ml Falcon tube and allowed to stand for 30 min.
5. Once the complete aggregate gets settled and the suspension consists of free cells, the turbidity of each suspension is measured at 540 nm.
6. The aggregate (settled at the bottom of the Falcon tube, step 4) is then dispersed by sonication for 1 min, and total turbidity was measured at 540 nm (OD_t).
7. The percentage of cell aggregation is quantified as follows:

$$\%\text{aggregation} = \text{OD}_\text{t} - \text{OD}_\text{s} \times 100 \div \text{OD}_\text{t}$$

Where (OD_t) = O.D. at 540 nm for total turbidity after sonication
Where (OD_s) = O.D. at 540 nm for cell suspension

8.4.4 In-Gel Catalase Activity

Catalase is an enzyme which converts H_2O_2 into H_2O and O_2. Here we have given step by step method to show in-gel activity of catalase using crude extract.

1. Inoculate freshly streaked, isolated single colonies of the parent and mutant strain into 2 mL MMM.
2. The next day, the overnight-grown culture is reinoculated into 50 ml of MMM (1% inoculum) and allowed to grow up to the mid-log phase.
3. Once culture O.D. reached mid-log phase, half the volume of the grown culture (25 ml) is transferred into another flask, and this culture treated with 1 mM H_2O_2.
4. Crude lysate is prepared from treated and untreated cultures by sonication, and the total protein quantified using the Bradford assay.
5. Equal amounts of protein (100 μg) is loaded onto 7% native PAGE and resolved by electrophoresis at 4 °C.
6. The gel is washed 3× with distilled water for 10 min each at normal RT.
7. The washed gel is then dipped into a 0.003% (V/V) solution of H_2O_2 for 10 min and again rinsed with distilled water two times for 5 min each.
8. The top of the gel is poured with equal volumes of 2% ferric chloride and 2% potassium ferricyanide for staining.

9. An intense band will be visible near the well of the gel, showing catalase activity. Once the visible solution of ferric chloride and potassium ferricyanide is poured off, the gel is washed with water 3× for 3 min each.

8.4.5 *In Vitro Alkylhydroperoxidase Activity*

Enzyme AhpC converts organic peroxide into simple alcohol and during conversion it itself gets oxidized. The oxidized form of AhpC is reduced by the reducing agent, DTT. Transfer of electrons from DTT to AhpC enzyme in turn converts DTT to its oxidised form, which will give absorption maxima at 310 nm. Reduction in absorption at 310 nm with time reflects the activity of the enzyme AhpC (Iyer and Klee 1973). Here we have given step by step protocol for measuring the in vitro catalytic activity of enzyme AhpC using crude protein.

1. Inoculate freshly streaked, isolated single colonies into 2 ml MMM.
2. The next day, the overnight-grown culture is reinoculated into 50 ml of MMM (1% inoculum) and allowed to grow up to the mid-log phase.
3. Mid-log-phase culture is then pelleted by centrifugation and washed twice with 1X-PBS and once with 25 mM Tris/HCl (pH 8.0) containing 1 mM protease inhibitor PMSF.
4. Pelleted and washed cells are resuspended in the same buffer, and cells lysed by sonication using 5 s ON and 10 s OFF for four cycles.
5. Cell debris are removed by centrifugation, and clear lysate transferred into a separate Eppendorf tube.
6. Protein content of the prepared crude lysate is quantified using Bradford assay.
7. To measure the activity of AhpC protein, 50 μg of crude lysate is incubated in 100 mM potassium phosphate buffer (pH 7.0), 1 mM EDTA, 10 mM DTT, and 50 μM substrate tert-butylhydroperoxide. Reaction is performed in 1 ml quartz cuvette of 1 ml reaction volume at 25 °C.
8. Decreased absorbance at 310 nm with corresponding time reflects the activity of enzyme AhpC.

8.4.6 *Protein Isolation and Two-Dimensional Gel Electrophoresis*

To identify the proteins that are involved in mediating a specific stress response, 2D gel electrophoresis can be used for comparing the proteome of *A. brasilense* grown under normal conditions with that grown in stressed conditions. Here we have given a step-by-step protocol for comparing the proteomes by 2D gel electrophoresis.

1. Inoculate freshly streaked isolated single colonies into 2 ml Luria-Bertani (LB) broth.

2. The next day, the overnight-grown cultures are reinoculated into 25 ml of MMM broth (1% inoculum) with and without stress and allowed to grow up to the mid-log phase (~0.8–1 $O.D._{600nm}$).
3. Grown cells are pelleted by centrifugation at 10,000 rpm for 2 min at 4 °C.
4. Pelleted cells are washed twice with Tris-Cl buffer (pH 8.0) and resuspended in the same buffer containing 1 mM PMSF (protease inhibitor).
5. Resuspended cells are lysed by sonication using five cycles of 15 s with setting intervals of 10 s under cooling conditions.
6. Cell debris and intact cells are sorted out by centrifugation at 13,000 rpm for 15 min at 4 °C.
7. Supernatant containing soluble protein is transferred into another microtube, and protein is quantified using the Bradford assay.
8. An equal amount of protein (1 mg) is precipitated from the stressed and unstressed cultures by adding four volumes of chilled acetone and shifting the microtube at −20 °C for 1 h.
9. Precipitated proteins are separated from the rest of the organic solvent, acetone, by centrifugation at 13,000 rpm for 5 min at 4 °C.
10. Precipitated proteins are washed twice with chilled acetone.
11. The protein pellet is dried out and resuspended into solubilization buffer having 8 M urea, 50 mM DTT, 2% CHAPS (W/V), 0.2% carrier ampholites, and a trace of bromophenol blue as an indicator dye.
12. Subsequent steps are performed as suggested by the manufacturing company, depending on the required length of the pH strip used for protein loading and machine programming.

8.5 Future Perspective

A. brasilense, a plant growth-promoting rhizobacterium (PGPR) shows great promise for sustainable agriculture due to its excellent ability to endure various environmental stresses while providing benefits to host plants. As the agricultural sector faces mounting challenges from global warming, climate change and environmental degradation, it becomes increasingly vital to understand how this beneficial bacterium handles stress during interactions with plants. Heavy dependence of global agriculture on the Haber Bosch process for producing nitrogen fertilizers is leading to catastrophic consequences of global warming. There is an urgent need to reduce our dependence on chemical fertilizers. Under such challenging circumstances *A. brasilense* provides hope for reducing the use of nitrogen fertilizers as it possesses mechanisms to cope with a variety of environmental stresses. Hence, the prospects of using *A. brasilense* as biofertilizer is encouraging, especially as we confront growing environmental issues. By gaining insights into how this bacterium reacts to different stresses in plant interactions, researchers can create more efficient inoculation techniques, select suitable strains, and refine application methods. *A. brasilense* holds significant potential for reducing dependence on chemical inputs while aiding plants in coping with environmental stresses. Additional research focusing on the molecular mechanisms of stress responses will enable us to harness the full potential of this plant growth-promoting rhizobacterium for sustainable agriculture.

References

Ajinde AO, Dayo-Olagbende OG, Akpor OB (2024) Direct and indirect mechanisms of growth promotion by plant growth promoting rhizobacteria. In: International conference on science, engineering and business for driving sustainable development goals, Omu-Aran, Nigeria, pp 1–16. https://doi.org/10.1109/SEB4SDG60871.2024.10630032

Alen'kina SA, Romanov NI, Nikitina VE (2018) Regulation by *Azospirillum* lectins of the activity of antioxidant enzymes in wheat seedling roots under short-term stresses. Rev Bras Bot 41:579–587. https://doi.org/10.1007/S40415-018-0489-1/METRICS

Alzate Zuluaga MY, Miras-Moreno B, Monterisi S et al (2022) Integrated metabolomics and morpho-biochemical analyses reveal a better performance of *Azospirillum brasilense* over plant-derived biostimulants in counteracting salt stress in tomato. Int J Mol Sci 23:142216. https://doi.org/10.3390/IJMS232214216

Anjum SA, Xie X-Y, Wang L-C et al (2011) Morphological, physiological and biochemical responses of plants to drought stress. Afr J Agric Res 6:2026–2032. https://doi.org/10.5897/AJAR10.027

Appleby JL, Parkinson JS, Bourret RB (1996) Signal transduction via the multi-step phosphorelay: not necessarily a road less traveled. Cell 86:845–848. https://doi.org/10.1016/S0092-8674(00)80158-0

Arsene F, Tomoyasu T, Bukau B (2000) The heat shock response of *Escherichia coli*. Int J Food Microbiol 55:3–9. https://doi.org/10.1016/S0168-1605(00)00206-3

Bailey-Serres J, Lee SC, Brinton E (2012) Waterproofing crops: effective flooding survival strategies. Plant Physiol 160:1698–1709. https://doi.org/10.1104/PP.112.208173

Baker TA, Grossman AD, Gross CA (1984) A gene regulating the heat shock response in *Escherichia coli* also affects proteolysis. Proc Natl Acad Sci USA 81:6779–6783. https://doi.org/10.1073/PNAS.81.21.6779

Bashan Y, Holguin G (1997) *Azospirillum*-plant relationships: environmental and physiological advances (1990-1996). Can J Microbiol 43:103–121. https://doi.org/10.1139/M97-015

Beier D, Gross R (2006) Regulation of bacterial virulence by two-component systems. Curr Opin Microbiol 9:143–152. https://doi.org/10.1016/j.mib.2006.01.005

Belaid FM, Abubaker NS, El-Komy HMA (2019) Survival of *Azospirillum brasilense* under different moisture content levels. EPH-Int J Appl Sci 5:27–31. https://doi.org/10.53555/EIJAS.V5I4.111

Bogati K, Walczak M (2022) The impact of drought stress on soil microbial community, enzyme activities and plants. Agronomy 12:189. https://doi.org/10.3390/AGRONOMY12010189

Boric M, Danevcic T, Stopar D (2012) Viscosity dictates metabolic activity of *Vibrio ruber*. Front Microbiol 3:255. https://doi.org/10.3389/FMICB.2012.00255

Borland S, Oudart A, Prigent-Combaret C et al (2015) Genome-wide survey of two-component signal transduction systems in the plant growth-promoting bacterium *Azospirillum*. BMC Genomics 16:1–17. https://doi.org/10.1186/S12864-015-1962-X

Burdman S, Vedder D, German M et al (1998) Legume crop yield promotion by inoculation with *Azospirillum*. In: Elmerich C, Kondorosi A, Newton WE (eds) Biological nitrogen fixation for 21st century. Current plant science and biotechnology in agriculture. Springer, Dordrecht, pp 609–612. https://doi.org/10.1007/978-94-011-5159-7_381

Cassán FD, Coniglio A, Amavizca E et al (2021) The *Azospirillum brasilense* type VI secretion system promotes cell aggregation, biocontrol protection against phytopathogens and attachment to the microalgae *Chlorella sorokiniana*. Environ Microbiol 23:6257–6274. https://doi.org/10.1111/1462-2920.15749

Chater KF (1993) Genetics of differentiation in *Streptomyces*. Ann Rev Microbiol 47:685–711. https://doi.org/10.1146/ANNUREV.MI.47.100193.003345

Chepsergon J, Moleleki LN (2023) Rhizosphere bacterial interactions and impact on plant health. Curr Opin Microbiol 73:102297. https://doi.org/10.1016/J.MIB.2023.102297

Chowdhury SP, Nagarajan T, Tripathi R et al (2007) Strain-specific salt tolerance and osmoregulatory mechanisms in *Azospirillum brasilense*. FEMS Microbiol Lett 267:72–79. https://doi.org/10.1111/J.1574-6968.2006.00540.X

Cohen AC, Travaglia CN, Bottini R et al (2009) Participation of abscisic acid and gibberellins produced by endophytic *Azospirillum* in the alleviation of drought effects in maize. Botany 87:455–462. https://doi.org/10.1139/B09-023

Cohen AC, Bottini R, Pontin M et al (2015) *Azospirillum brasilense* ameliorates the response of Arabidopsis thaliana to drought mainly via enhancement of ABA levels. Physiol Plant 153:79–90. https://doi.org/10.1111/PPL.12221

Coulthurst S (2019) The type VI secretion system: a versatile bacterial weapon. Microbiology (UK) 165:503–515. https://doi.org/10.1099/MIC.0.000789

Cox MM, Battista JR (2005) *Deinococcus radiodurans*—the consummate survivor. Nat Rev Microbiol 3:882–892. https://doi.org/10.1038/NRMICRO1264

Csonka LN, Epstein W (1996) Osmoregulation. In Neidhardt FC (Ed) *Escherichia coli* and *Salmonella* Cellular and molecular biology. ASM Press, Washington, DC, pp 1210–1223. https://www.scirp.org/reference/referencespapers?referenceid=1167700

Cura JA, Franz DR, Filosofía JE et al (2017) Inoculation with *azospirillum* sp. and *herbaspirillum* sp. bacteria increases the tolerance of maize to drought stress. Microorganisms 5:41 10.3390/MICROORGANISMS5030041

da Cunha ET, Pedrolo AM, Arisi ACM (2024) Thermal and salt stress effects on the survival of plant growth-promoting bacteria *Azospirillum brasilense* in inoculants for maize cultivation. J Sci Food Agric 104:5360–5367. https://doi.org/10.1002/JSFA.13366

Davis MC, Kesthely CA, Franklin EA et al (2017) The essential activities of the bacterial sigma factor. Can J Microbiol 63:89–99. https://doi.org/10.1139/CJM-2016-0576

de Pina LC, da Silva FSH, Galvão TC et al (2021) The role of two-component regulatory systems in environmental sensing and virulence in *Salmonella*. Crit Rev Microbiol 47:397–434. https://doi.org/10.1080/1040841X.2021.1895067

Degon Z, Dixon S, Rahmatallah Y et al (2023) *Azospirillum brasilense* improves rice growth under salt stress by regulating the expression of key genes involved in salt stress response, abscisic acid signalling, and nutrient transport, among others. Front Agron 5:12165003. https://doi.org/10.3389/FAGRO.2023.1216503

Diethmaier C, Chawla R, Canzoneri A et al (2017) Viscous drag on the flagellum activates *Bacillus subtilis* entry into the K-state. Mol Microbiol 106:367–380. https://doi.org/10.1111/MMI.13770

Dubey AP, Pandey P, Singh VS et al (2020) An ECF41 family σ factor controls motility and biogenesis of lateral flagella in *azospirillum brasilense* Sp245. J Bacteriol 202:10 1128. https://doi.org/10.1128/JB.00231-20

Eskew DL, Focht DD, Ting IP (1977) Nitrogen fixation, denitrification, and pleomorphic growth in a highly pigmented *Spirillum lipoferum*. Appl Environ Microbiol 34:582–585. https://doi.org/10.1128/AEM.34.5.582-585.1977

Farooq M, Wahid A, Kobayashi N et al (2009) Plant drought stress: effects, mechanisms and management. Agron Sustain Dev 29:185–212. https://doi.org/10.1051/AGRO:2008021

Feklistov A, Sharon BD, Darst SA et al (2014) Bacterial sigma factors: a historical, structural, and genomic perspective. Ann Rev Microbiol 68:357–376. https://doi.org/10.1146/ANNUREV-MICRO-092412-155737

Fischer HM (1994) Genetic regulation of nitrogen fixation in rhizobia. Microbiol Rev 58:352–386. https://doi.org/10.1128/MR.58.3.352-386.1994

Fukami J, Ollero FJ, Megias M et al (2017) Phytohormones and induction of plant-stress tolerance and defense genes by seed and foliar inoculation with *Azospirillum brasilense* cells and metabolites promote maize growth. AMB Express 7:1–13. https://doi.org/10.1186/S13568-017-0453-7

Fukami J, De La Osa C, Ollero FJ et al (2018) Co-inoculation of maize with *Azospirillum brasilense* and *Rhizobium tropici* as a strategy to mitigate salinity stress. Funct Plant Biol 45:328–339. https://doi.org/10.1071/FP17167

Ganusova EE, Vo LT, Abraham PE et al (2021) The *Azospirillum brasilense* core chemotaxis proteins CheA1 and CheA4 link chemotaxis signaling with nitrogen metabolism. mSystems 6: 10–1128. https://doi.org/10.1128/MSYSTEMS.01354-20

Gao R, Stock AM (2009) Biological insights from structures of two-component proteins. Ann Rev Microbiol 63:133–154. https://doi.org/10.1146/ANNUREV.MICRO.091208.073214

Gerk LP, Gilchrist K, Kennedy IR (2000) Mutants with enhanced nitrogenase activity in hydroponic *Azospirillum brasilense* wheat associations. Appl Environ Microbiol 66:2175–2184. https://doi.org/10.1128/AEM.66.5.2175-2184.2000

Gupta N, Kumar S, Mishra MN et al (2013) A constitutively expressed pair of *rpoE2-chrR2* in *Azospirillum brasilense* Sp7 is required for survival under antibiotic and oxidative stress. Microbiology (UK) 159:205–218. https://doi.org/10.1099/MIC.0.061937-0

Gupta N, Gupta A, Kumar S et al (2014) Cross-talk between cognate and noncognate *rpoE* sigma factors and Zn^{+2} binding anti-sigma factors regulates photooxidative stress response in *Azospirillum brasilense*. Antioxid Redox Signal 20:42–59. https://doi.org/10.1089/ARS.2013.5314

Gureeva MV, Gureev AP (2023) Molecular mechanisms determining the role of bacteria from the genus *Azospirillum* in plant adaptation to damaging environmental factors. Int J Mol Sci 24: 9122. https://doi.org/10.3390/IJMS24119122

Hartmann A, Fu HA, Burris RH (1988) Influence of amino acids on nitrogen fixation ability and growth of *Azospirillum* spp. Appl Environ Microbiol 54:87–93. https://doi.org/10.1128/AEM.54.1.87-93.1988

Ho TD, Ellermeier CD (2012) Extra cytoplasmic function σ factor activation. Curr Opin Microbiol 15:182–188. https://doi.org/10.1016/J.MIB.2012.01.001

Hoch JA (1993) Regulation of the phosphorelay and the initiation of sporulation in *Bacillus subtilis*. Ann Rev Microbiol 47:441–465. https://doi.org/10.1146/ANNUREV.MI.47.100193.002301

Hopper S, Bock A (1995) Effector-mediated stimulation of ATPase activity by the σ54-dependent transcriptional activator FHLA from *Escherichia coli*. J Bacteriol 177:2798–2803. https://doi.org/10.1128/JB.177.10.2798-2803.1995

Housh AB, Powell G, Scott S et al (2021) Functional mutants of *Azospirillum brasilense* elicit beneficial physiological and metabolic responses in Zea mays contributing to increased host iron assimilation. ISME J 15:1505–1522. https://doi.org/10.1038/s41396-020-00866-x

Iqbal R, Valipour M, Ali B (2024) Maximizing wheat yield through soil quality enhancement: a combined approach with *Azospirillum brasilense* and bentonite. Plant Stress 11:100321. https://doi.org/10.1016/J.STRESS.2023.100321

Ishida ML, Assumpção MC, Machado HB et al (2002) Identification and characterization of the two-component NtrY/NtrX regulatory system in *Azospirillum brasilense*. Braz J Med Biol Res 35:651–661. https://doi.org/10.1590/S0100-879X2002000600004

Iyer KS, Klee WA (1973) Direct spectrophotometric measurement of the rate of reduction of disulfide bonds: The Reactivity of the disulfide bonds of bovine α-lactalbumin. J Biol Chem 248:707–710. https://doi.org/10.1016/s0021-9258(19)44430-x

Jeffery CJ (2016) Expression, solubilization, and purification of bacterial membrane proteins. Curr Protoc Protein Sci 83:29–15. https://doi.org/10.1002/0471140864.PS2915S83

Joe M, Karthikeyan MB, Sekar C et al (2010) Optimization of biofloc production in *Azospirillum brasilense* (MTCC-125) and evaluation of its adherence with the roots of certain crops. Indian J Microbiol 50:21–25. https://doi.org/10.1007/S12088-010-0050-9

Jofré E, Rivarola V, Balegno H et al (1998) Differential gene expression in *Azospirillum brasilense* Cd under saline stress. Can J Microbiol 44:929–936. https://doi.org/10.1139/cjm-44-10-929

Jofré E, Fischer S, Rivarola V et al (2011) Saline stress affects the attachment of *Azospirillum brasilense* Cd to maize and wheat roots. Can J Microbiol 44:416–422. https://doi.org/10.1139/W98-024

Kaczmarczyk A, Hochstrasser R, Vorholt JA et al (2014) Complex two-component signaling regulates the general stress response in Alphaproteobacteria. Proc Natl Acad Sci USA 111: E5196–E5204. https://doi.org/10.1073/PNAS.1410095111

Kapitein N, Mogk A (2013) Deadly syringes: type VI secretion system activities in pathogenicity and interbacterial competition. Curr Opin Microbiol 16:52–58. https://doi.org/10.1016/j.mib.2012.11.009

Kapulnik Y, Okon Y, Kigel J et al (1981) Effects of temperature, nitrogen fertilization, and plant age on nitrogen fixation by *Setaria italica* Inoculated with *Azospirillum brasilense* (strain cd). Plant Physiol 68:340–343. https://doi.org/10.1104/PP.68.2.340

Kasim WA, Osman MEH, Omar MN et al (2021) Enhancement of drought tolerance in Triticum aestivum L. seedlings using *Azospirillum brasilense* NO40 and Stenotrophomonas maltophilia B11. Bull Natl Res Cent 45:95. https://doi.org/10.1186/S42269-021-00546-6

Kiley PJ, Storz G (2004) Exploiting thiol modifications. PLoS Biol 2:e400. https://doi.org/10.1371/JOURNAL.PBIO.0020400

Kim SK, Kaiser D, Kuspa A (1992) Control of cell density and pattern by intercellular signaling in *Myxococcus* development. Ann Rev Microbiol 46:117–139. https://doi.org/10.1146/ANNUREV.MI.46.100192.001001

Kukolj C, Pedrosa FO, De Souza GA et al (2020) Proteomic and metabolomic analysis of *Azospirillum brasilense ntrC* mutant under high and low nitrogen conditions. J Proteome Res 19:92–105. https://doi.org/10.1021/ACS.JPROTEOME.9B00397

Kumar S, Rai AK, Mishra MN et al (2012) RpoH2 sigma factor controls the photooxidative stress response in a non-photosynthetic rhizobacterium, *Azospirillum brasilense* Sp7. Microbiology (UK) 158:2891–2902. https://doi.org/10.1099/MIC.0.062380-0

Lerner A, Valverde A, Castro-Sowinski S et al (2010) Phenotypic variation in *Azospirillum brasilense* exposed to starvation. Environ Microbiol Rep 2:577–586. https://doi.org/10.1111/J.1758-2229.2010.00149.X

Li H, Xu F, Ren X et al (2010) Functional analysis of the *fixL/fixJ* and *fixK* genes in *Azospirillum brasilense* Sp7. Ann Microbiol 60:469–480. https://doi.org/10.1007/S13213-010-0065-9/FIGURES/4

Mandon K, Nazaret F, Farajzadeh D et al (2021) Redox regulation in diazotrophic bacteria in interaction with plants. Antioxidants 10:880. https://doi.org/10.3390/ANTIOX10060880

Marques DM, Magalhães PC, Marriel IE et al (2020) *Azospirillum brasilense* favors morphophysiological characteristics and nutrient accumulation in maize cultivated under two water regimes. Rev Bras Milho Sorgo 19:e1152. https://doi.org/10.18512/1980

Mascher T, Hachmann AB, Helmann JD (2007) Regulatory overlap and functional redundancy among *Bacillus subtilis* extracytoplasmic function σ factors. J Bacteriol 189:6919–6927. https://doi.org/10.1128/JB.00904-07

Miller KJ, Wood JM (1996) Osmoadaptation by rhizosphere bacteria. Ann Rev Microbiol 50:101–136. https://doi.org/10.1146/ANNUREV.MICRO.50.1.101

Mishra MN, Kumar S, Gupta N et al (2011) An extracytoplasmic function sigma factor cotranscribed with its cognate anti-sigma factor confers tolerance to NaCl, ethanol and methylene blue in *Azospirillum brasilense* Sp7. Microbiology (N Y) 157:988–999. https://doi.org/10.1099/MIC.0.046672-0

Mugnai M, Bazzicalupo M, Fani R et al (1994) Factors affecting nitrogen fixation and *nif* gene transcription in *Azospirillum brasilense*. FEMS Microbiol Lett 120:133–136. https://doi.org/10.1111/J.1574-6968.1994.TB07019.X

Nishiuchi S, Yamauchi T, Takahashi H et al (2012) Mechanisms for coping with submergence and waterlogging in rice. Rice 5:1–14. https://doi.org/10.1186/1939-8433-5-2

Okon Y, Itzigsohn R (1992) Poly-β-hydroxybutyrate metabolism in *Azospirillum brasilense* and the ecological role of PHB in the rhizosphere. FEMS Microbiol Lett 103:131–139. https://doi.org/10.1016/0378-1097(92)90302-5

Olanrewaju OS, Glick BR, Babalola OO (2017) Mechanisms of action of plant growth promoting bacteria. World J Microbiol Biotechnol 33:1–16. https://doi.org/10.1007/S11274-017-2364-9

Paget MS (2015) Bacterial sigma factors and anti-sigma factors: structure, function and distribution. Biomolecules 5:1245–1265. https://doi.org/10.3390/BIOM5031245

Pandey P, Dubey AP, Mishra S et al (2022) β-lactam resistance in *Azospirillum baldaniorum* Sp245 is mediated by lytic transglycosylase and β-lactamase and regulated by a cascade of RpoE7-RpoH3 sigma factors. J Bacteriol 204:e00010–e00022. https://doi.org/10.1128/JB.00010-22

Partridge JD, Harshey RM (2013) Swarming: flexible roaming plans. J Bacteriol 195:909–918. https://doi.org/10.1128/JB.02063-12

Paul D (2013) Osmotic stress adaptations in rhizobacteria. J Basic Microbiol 53:101–110. https://doi.org/10.1002/JOBM.201100288

Peng H, de-Bashan LE, Bashan Y et al (2020) Indole-3-acetic acid from *Azosprillum brasilense* promotes growth in green algae at the expense of energy storage products. Algal Res 47:101845. https://doi.org/10.1016/J.ALGAL.2020.101845

Peng H, de-Bashan LE, Higgins BT (2021) *Azospirillum brasilense* reduces oxidative stress in the green microalgae *Chlorella sorokiniana* under different stressors. J Biotechnol 325:179–185. https://doi.org/10.1016/j.jbiotec.2020.10.029

Pereg-Gerk L, Paquelin A, Gounon P et al (1998) A transcriptional regulator of the LuxR-UhpA family, FlcA, controls flocculation and wheat root surface colonization by *Azospirillum brasilense* Sp7. Mol Plant-Microbe Interact 11:177–187. https://doi.org/10.1094/MPMI.1998.11.3.177

Pii Y, Penn A, Terzano R et al (2015) Plant-microorganism-soil interactions influence the Fe availability in the rhizosphere of cucumber plants. Plant Physiol Biochem 87:45–52. https://doi.org/10.1016/J.PLAPHY.2014.12.014

Rai AK, Dubey AP, Kumar S et al (2016) Carotenoid biosynthetic pathways are regulated by a network of multiple cascades of alternative sigma factors in *Azospirillum brasilense* Sp7. J Bacteriol 198:2955–2964. https://doi.org/10.1128/JB.00460-16

Rai AK, Singh S, Dwivedi SK et al (2018) Catalase expression in *Azospirillum brasilense* Sp7 is regulated by a network consisting of OxyR and two RpoH paralogs and including an RpoE1→RpoH5 regulatory cascade. Appl Environ Microbiol 84:e01787–e01718. https://doi.org/10.1128/AEM.01787-18

Rai S, Singh VS, Gupta P et al (2025) Identification and functional characterization of a fructose-inducible phosphotransferase system in *Azospirillum brasilense* Sp7. Appl Environ Microbiol 91:e00828–24. https://doi.org/10.1128/AEM.00828-24

Ram PC, Singh AK, Singh BB et al (1999) Environmental characterization of floodwater in eastern india: relvance to submergence tolerance of lowland rice. Exp Agric 35:141–152. https://doi.org/10.1017/S0014479799002057

Rao VM, Shukla R (2002) Effect of moisture on growth and N_2 fixation in *Digitaria adscendens* inoculated with *Azospirillum brasilense* and *Glomus mossae*. Indian J Microbiol 42:299–301

Ribardo DA, Johnson JJ, Hendrixson DR (2024) Viscosity-dependent determinants of *Campylobacter jejuni* impacting the velocity of flagellar motility. MBio 15:e02544–23. https://doi.org/10.1128/MBIO.02544-23

Ricard B, Couée I, Raymond P et al (1994) Plant metabolism under hypoxia and anoxia. Plant Physiol Biochem 32:1. https://surli.cc/cratkf

Riou N, Le Rudulier D (1990) Osmoregulation in *Azospirillum brasilense*: glycine betaine transport enhances growth and nitrogen fixation under salt stress. J Gen Microbiol 136:1455–1461. https://doi.org/10.1099/00221287-136-8-1455
Rodriguez-Romo LA, Yousef AE (2005) Cross-protective effects of bacterial stress. In: Understanding pathogen behaviour, pp 128–151. https://doi.org/10.1533/9781845690229.1.128
Rodriguez-Salazar J, Suarez R, Caballero-Mellado J et al (2009) Trehalose accumulation in *Azospirillum brasilense* improves drought tolerance and biomass in maize plants. FEMS Microbiol Lett 296:52–59. https://doi.org/10.1111/J.1574-6968.2009.01614.X
Sadasivan L, Neyra CA (1985) Flocculation in *Azospirillum brasilense* and *Azospirillum lipoferum*: exopolysaccharides and cyst formation. J Bacteriol 163:716–723. https://doi.org/10.1128/JB.163.2.716-723.1985
Sadoff HL (1975) Encystment and germination in *Azotobacter vinelandii*. Bacteriol Rev 39:516–539. https://doi.org/10.1128/BR.39.4.516-539.1975
Salazar-Garcia G, Balaguera-Lopez HE, Hernandez JP (2022) Effect of plant growth-promoting bacteria *Azospirillum brasilense* on the physiology of radish (*Raphanus sativus* L.) under waterlogging stress. Agronomy 12:726. https://doi.org/10.3390/AGRONOMY12030726
Schillaci M, Kehelpannala C, Martinez-Seidel F et al (2021) The metabolic response of *Brachypodium* roots to the interaction with beneficial bacteria is affected by the plant nutritional status. Metabolites 11:358. https://doi.org/10.3390/METABO11060358
Shao HB, Chu LY, Jaleel CA et al (2009) Understanding water deficit stress-induced changes in the basic metabolism of higher plants-biotechnologically and sustainably improving agriculture and the eco environment in arid regions of the globe. Crit Rev Biotechnol 29:131–151. https://doi.org/10.1080/07388550902869792
Shapiro RS, Cowen LE (2012) Thermal control of microbial development and virulence: Molecular mechanisms of microbial temperature sensing. MBio 3:10–1128. https://doi.org/10.1128/MBIO.00238-12
Sineva E, Savkina M, Ades SE (2017) Themes and variations in gene regulation by extracytoplasmic function (ECF) sigma factors. Curr Opin Microbiol 36:128–137. https://doi.org/10.1016/J.MIB.2017.05.004
Singh S, Dwivedi SK, Singh VS et al (2016) Expression of alkyl hydroperoxide reductase is regulated negatively by OxyR1 and positively by RpoE2 sigma factor in *Azospirillum brasilense* Sp7. Microbiology (UK) 162:1868–1881. https://doi.org/10.1099/MIC.0.000363
Singh VS, Dubey BK, Pandey P et al (2021) Cometabolism of ethanol in *Azospirillum brasilense* Sp7 Is mediated by fructose and glycerol and regulated negatively by an alternative sigma factor RpoH2. J Bacteriol 203:10–1128. https://doi.org/10.1128/JB.00269-21
Sleator RD, Hill C (2002) Bacterial osmoadaptation: the role of osmolytes in bacterial stress and virulence. FEMS Microbiol Rev 26:49–71. https://doi.org/10.1111/J.1574-6976.2002.TB00598.X
Speare L, Smith S, Salvato F et al (2020) Environmental viscosity modulates interbacterial killing during habitat transition. MBio 11:e03060–e03019. https://doi.org/10.1128/MBIO.03060-19
Staron A, Sofia HJ, Dietrich S et al (2009) The third pillar of bacterial signal transduction: classification of the extracytoplasmic function (ECF) σ factor protein family. Mol Microbiol 74:557–581. https://doi.org/10.1111/J.1365-2958.2009.06870.X
Stock AM, Robinson VL, Goudreau PN (2000) Two-component signal transduction. Annu Rev Biochem 69:183–215. https://doi.org/10.1146/ANNUREV.BIOCHEM.69.1.183
Storz G, Imlay JA (1999) Oxidative stress. Curr Opin Microbiol 2:188–194. https://doi.org/10.1016/S1369-5274(99)80033-2
Storz G, Tartaglia LA, Ames BN (1990) Transcriptional regulator of oxidative stress-inducible genes: direct activation by oxidation. Science (1979) 248:189–194. https://doi.org/10.1126/SCIENCE.2183352
Sun N, Huang L, Zhao H et al (2022) Beneficial bacterium *Azospirillum brasilense* induces morphological, physiological and molecular adaptation to phosphorus deficiency in *Arabidopsis*. Plant Cell Physiol 63:1273–1284. https://doi.org/10.1093/PCP/PCAC101

Surette MG, Leviti M, Liu Y et al (1996) Dimerization is required for the activity of the protein histidine kinase CheA that mediates signal transduction in bacterial chemotaxis. J Biol Chem 271:939–945. https://doi.org/10.1074/JBC.271.2.939

Tal S, Smirnoff P, Okon Y (1990) The regulation of poly-β-hydroxybutyrate metabolism in *Azospirillum brasilense* during balanced growth and starvation. J Gen Microbiol 136:1191–1196. https://doi.org/10.1099/00221287-136-7-1191

Tierney ARP, Rather PN (2019) Roles of two-component regulatory systems in antibiotic resistance. Future Microbiol 14:533–552. https://doi.org/10.2217/FMB-2019-0002

Tortora ML, Díaz-Ricci JC, Pedraza RO (2011) *Azospirillum brasilense* siderophores with antifungal activity against *Colletotrichum acutatum*. Arch Microbiol 193:275–286. https://doi.org/10.1007/S00203-010-0672-7

Tripathi AK, Klingmuller W (1992) Temperature sensitivity of nitrogen fixation in *Azospirillum* spp. Can J Microbiol 38:1238–1241. https://doi.org/10.1139/M92-204

Tripathi AK, Mishra BM, Tripathi P (1998) Salinity stress responses in the plant growth promoting rhizobacteria, *Azospirillum* spp. J Biosci 23:463–471. https://doi.org/10.1007/BF02936140

Tripathi AK, Nagarajan T, Verma SC et al (2002) Inhibition of biosynthesis and activity of nitrogenase in *Azospirillum brasilense* Sp7 under salinity stress. Curr Microbiol 44:363–367. https://doi.org/10.1007/S00284-001-0022-8

Vanstockem M, Michiels K, Vanderleyden J et al (1987) Transposon mutagenesis of *Azospirillum brasilense* and *Azospirillum lipoferum*: physical analysis of Tn5 and Tn5-Mob insertion mutants. Appl Environ Microbiol 53:410–415. https://doi.org/10.1128/AEM.53.2.410-415.1987;WGROUP:STRING:PUBLICATION

Vercruysse M, Fauvart M, Jans A et al (2011) Stress response regulators identified through genome-wide transcriptome analysis of the (p)ppGpp-dependent response in *Rhizobium etli*. Genome Biol 12:1–9. https://doi.org/10.1186/GB-2011-12-2-R17

Verreth C, Cammue B, Swinnen P et al (1989) Cloning and expression in *Escherichia coli* of the *Azospirillum brasilense* Sp7 gene encoding ampicillin resistance. Appl Environ Microbiol 55:2056–2060. https://doi.org/10.1128/AEM.55.8.2056-2060.1989

Voesenek LACJ, Bailey-Serres J (2015) Flood adaptive traits and processes: an overview. New Phytol 206:57–73. https://doi.org/10.1111/NPH.13209

Wasim M, Bible AN, Xie Z et al (2009) Alkyl hydroperoxide reductase has a role in oxidative stress resistance and in modulating changes in cell-surface properties in *Azospirillum brasilense* Sp245. Microbiology (NY) 155:1192–1202. https://doi.org/10.1099/MIC.0.022541-0

Zaprasis A, Brill J, Thüring M et al (2013) Osmoprotection of *Bacillus subtilis* through import and proteolysis of proline-containing peptides. Appl Environ Microbiol 79:576–587. https://doi.org/10.1128/AEM.01934-12

Zhang LM, Silvano E, Rihtman B et al (2022) Biochemical mechanism of phosphorus limitation impairing nitrogen fixation in diazotrophic bacterium *Klebsiella variicola* W12. J Sustain Agric Environ 1:108–117. https://doi.org/10.1002/SAE2.12024

Part III
Plant Growth Promotion and Pathogen Biocontrol Mechanisms of *Azospirillum* spp.

Chapter 9
Nitrogen Metabolism and Genome Tailoring for Improving Ammonium Excretion in *Azospirillum* spp. and Other Free-Living Diazotrophs

Adriano Alves Stefanello, Lara Sanchez Rizza, Thiago Moia Apolonio, Fábio de Oliveira Pedrosa, and Emanuel Maltempi de Souza

Abstract *Azospirillum* spp. is a diazotrophic bacterium widely recognized for its plant growth-promoting abilities, attributed to both phytohormone production and nitrogen fixation. However, under natural conditions, fixed nitrogen is typically released only after bacterial death. To enhance its biofertilizer potential, efforts have focused on engineering strains capable of excreting ammonium during their active lifespan. This involves decoupling nitrogen fixation from intracellular assimilation through targeted genetic modifications. Key regulatory nodes include NifA, GlnE, GlnD, GlnB, and GlnZ, whose alterations can enable constitutive nitrogenase activity and ammonium excretion even under repressing conditions. Ammonium-excreting strains can be obtained either by directed mutagenesis or by selecting spontaneous mutants resistant to ethylenediamine (EDA), a compound toxic to glutamine synthetase-active cells. Selected strains are screened using nitrogen-free semi-solid media to assess constitutive nitrogenase activity under both nitrogen-limiting and nitrogen-rich conditions. Ammonium release is quantified through the Berthelot indophenol colorimetric method. Future prospects include integrating precision genome-editing tools such as CRISPR/Cas with synthetic circuits responsive to plant-derived rhizospheric signals, allowing context-dependent control of nitrogen metabolism. Such innovations could simulate persistent nitrogen-starvation states while maintaining symbiotic compatibility. These engineered strains hold promise for low impact, sustainable agriculture by improving nitrogen availability directly in the rhizosphere. Nevertheless, challenges remain, particularly regarding

A. A. Stefanello · T. M. Apolonio · F. de Oliveira Pedrosa (✉) · E. M. de Souza (✉)
Departamento de Bioquímica e Biologia Molecular, Universidade Federal do Paraná, Curitiba, Brasil
e-mail: fpedrosa@ufpr.br; souzaem@ufpr.br

L. S. Rizza
INBIOTEC-CONICET y FIBA, Mar del Plata, Argentina

F. D. Cassán et al. (eds.), *The Definitive Handbook of Azospirillum*,
https://doi.org/10.1007/978-3-032-09222-9_9

genetic stability, environmental performance, and plant interaction dynamics. Field trials and ecological risk assessments will be essential to transition these strains from laboratory models to viable agricultural technologies.

Keywords Nitrogen metabolism · Ammonium excretion · Fertilization · Nitrogenase · Biofertilizer

9.1 Introduction

Azospirillum is a well-known plant-growth promoter, a characteristic often attributed to both plant hormone production and its ability to fix atmospheric nitrogen (Dobereiner and Pedrosa 1987). The close association between certain *Azospirillum* species and nonleguminous plants makes them promising for biofertilizer development, with many commercial formulations already available.

Currently, *Azospirillum* inoculants are based on selected wild-type strains that have shown remarkable plant growth promotion in controlled studies. However, with the steady growth in knowledge of bacterial metabolism, engineering improved strains for agricultural applications is becoming more feasible.

Particularly for *Azospirillum*, there is much to be gained by enhancing the nitrogen fixation rate and, perhaps more crucially, the mobilization rate of fixed nitrogen to the host plant. *Azospirillum* spp. only fix atmospheric nitrogen when more energetically cost-effective alternatives are absent, and most fixed nitrogen becomes available to plants only after bacterial death. Engineering an early release of fixed nitrogen, as ammonium excreted from the bacterial cell, could lead to a much larger amount of this nutrient being made available for plant consumption over the bacterial cell's lifetime. Indeed, the agronomic potential of such a strain is evident in recent results obtained with *A. brasilense* HM053, an ammonium-excreting derivative of FP2 (Pedrosa et al. 2020).

9.1.1 Nitrogen Metabolism in Azospirillum *spp.*

The best-studied species in the genus are *A. brasilense, A. baldaniorum, A. argentinense,* and *A. lipoferum*. In these organisms, all nitrogen cycle pathways, with the exception of nitrification and anammox, have been described in these species.

A. baldaniorum, formerly classified as *A. brasilense* Sp245 (Santos et al. 2020), has been used to identify various nitrogen cycle pathways, and its genome was the first to be completely sequenced. The strain is well characterized physiologically. Genes encoding the ABC-type nitrate/nitrite transport system (*nasFED*), as well as those for assimilatory nitrite reductase (*nirBD*), assimilatory nitrate reductase (*nasA*), and uroporphyrin III C-methyltransferase (*nasG* or *cysG*), are located on plasmid 1. Homologous genes are also found in the genomes of *A. brasilense*, *A. lipoferum*, *A. halopraeferens*, and *A. irakense*. Downstream of *nasFED*, the *nasT* gene, encoding an antitermination protein, is present.

In plasmid 2 of *A. baldaniorum* Sp245, genes for the periplasmic nitrate reductase (*napEFDABC*) are found. Another set of *nap* genes (*napEDABC*) is located on plasmid 4. The gene for the copper-containing nitrite reductase (*nirK*, also known as *aniA*) is located on plasmid 3. These genes are also conserved in *A. brasilense*, *A. lipoferum*, *A. halopraeferens*, and *A. irakense*.

Genes encoding nitric oxide reductase (*norBCQD*) and nitrous oxide reductase (*nosZ*) have been identified in *A. brasilense*, *A. argentinense*, and *A. baldaniorum*. The regulation of nitrate metabolism in *Azospirillum* appears to be complex and is influenced by multiple factors, including nitrate, nitrite, and oxygen levels (Döbereiner and Pedrosa 1987; Steenhoudt et al. 2001; Pedrosa and Elmerich 2007).

A *nif* gene cluster is present in most sequenced species. As in other diazotrophs, nitrogenase synthesis and activity are tightly regulated by ammonium and oxygen. This regulation occurs both at the transcription and translation levels (Pedrosa and Elmerich 2007).

The next section briefly describes the regulation of ammonium assimilation in *A. brasilense*, with a focus on the key steps involved in ammonium release. This is followed by protocols for selecting *Azospirillum* strains with enhanced ammonium excretion capacity. Although these tools were primarily developed for *A. brasilense*, they can be applied to other species within the genus.

9.1.2 Regulation of Ammonium Assimilation and Biological Nitrogen Fixation

Nitrogen assimilation in *Azospirillum* is controlled by a complex network of transcriptional and posttranslational mechanisms. These mechanisms are crucial to ensure that the cell's nitrogen uptake does not overload its carbon and energy metabolism, which could indirectly hinder cell growth and reproduction. Control is also exerted to prioritize the acquisition of nitrogen sources requiring the least energy expenditure. For instance, ammonium is preferred over nitrate and nitrite, and these are preferred over atmospheric nitrogen.

As in other proteobacteria, nitrogen metabolism regulation in *Azospirillum* primarily relies on monitoring the intracellular levels of glutamine and its equivalent α-ketoacid, 2-oxoglutarate (2OG). Under high ammonium conditions, increased intracellular glutamine concentrations coupled with low 2OG concentrations signal nitrogen sufficiency. This leads to the downregulation of the main nitrogen assimilation pathway and its shutdown. Conversely, under low ammonium conditions, glutamine levels decrease with a concomitant increase of 2OG concentrations, signaling nitrogen starvation.

Ammonium assimilation in *Azospirillum* spp. occurs via two pathways, one involving glutamate dehydrogenase (encoded by *gdhA*) under high NH_4^+ concentration and the other involving glutamine synthetase (GS) (encoded by *glnA*) and glutamate synthase (encoded by *gltBgltD*) under limiting NH_4^+. Genes for these pathways are present in all species so far analyzed, but *A. brasilense* is the species most studied in this respect.

The GS of *A. brasilense* shares structural and functional characteristics with those of other well-studied Proteobacteria, such as *Escherichia coli* and *Salmonella typhimurium*. It is a dodecameric enzyme composed of 12 identical subunits, with an approximate molecular mass of 630 kDa, similar to that observed in *E. coli* (Pirola et al. 1992). GS catalyzes the ATP-dependent synthesis of glutamine from glutamate and ammonium, and its activity is regulated both transcriptionally and posttranslationally to respond to variations in nitrogen availability. Posttranslational regulation involves the bifunctional adenylyltransferase/deadenylylase, GlnE, that acts based on the internal carbon and nitrogen status (Van Heeswijk et al. 2013). In *E. coli*, GlnE activity is regulated directly by the concentration of glutamine and the interaction with PII proteins and indirectly by the concentration of 2-oxoglutarate, which binds to PII and has an influence on its uridylylation status. When intracellular glutamine is high and PII is found in its non-uridylylated state, GlnE adenylylates individual subunits of GS, rendering them inactive (Jiang et al. 1998). Conversely, under low glutamine concentration and in the presence of uridylylated PII, GlnE deadenylates GS, restoring its activity. Evidence indicates that *A. brasilense* adopts similar regulatory mechanisms, with specific adaptations to its diazotrophic lifestyle. Van Dommelen et al. (2009) isolated the *glnE* gene from *A. brasilense* and demonstrated that posttranslational regulation of GS is also mediated by GlnE, since a *glnE* mutant lost the capacity to adenylate GS. The role of PII proteins in this circuit is somewhat different when compared to the classical description in *E. coli*. *Azospirillum brasilense* harbors two PII paralogs, GlnB and GlnZ, which possess distinct cellular functions (de Zamaroczy 1998). GlnB is expressed by the *glnBA* operon, which also encodes glutamine synthetase (GS) and is subject to complex regulation. In contrast, GlnZ is expressed monocistronically, a characteristic that differentiates it from its ortholog GlnK in *Escherichia coli*, which is encoded by the *glnKamtB* operon (Moure et al. 2015; Huergo et al. 2013).

PII proteins are ubiquitous signal transduction molecules found in Bacteria, Archaea, and the chloroplasts of plants, where they integrate information on cellular nitrogen, carbon, and energy status to regulate metabolic pathways and, indirectly, gene expression, thereby serving as key indicators of nitrogen availability. Proteins of this family form homotrimers (~36 kDa), composed of ~12 kDa subunits, featuring a loop (the T-loop) responsible for interaction with most of its target proteins. Their activity is modulated by binding ATP, ADP, and 2-oxoglutarate, which alter their functional conformation (Moure et al. 2015; Huergo and Dixon 2015). In addition, PII protein activity is modified by uridylylation catalyzed by GlnD. In *A. brasilense*, GlnD, a bifunctional enzyme, modulates GlnB and GlnZ activity in response to the intracellular levels of glutamine and 2-oxoglutarate by uridylylation in the presence of ATP and 2-oxoglutarate under low ammonium conditions (Araujo et al. 2004; Araujo et al. 2008). Conversely, GlnD deuridylylates the PII proteins in response to elevated intracellular glutamine, as occurs when the ammonium levels are high (Araujo et al. 2008; Huergo and Dixon 2015). Uridylylation regulates PII interaction with a host of target proteins, such as the aforementioned GlnE and the *nif*-specific transcriptional activator NifA, thereby modulating nitrogen assimilation and fixation (Huergo et al. 2013). Most importantly, in *A. brasilense*, uridylylation of GlnB prevents its interaction with NtrB, allowing its kinase activity to phosphorylate the transcription factor NtrC, which in turn activates the σ^{54}-dependent transcription of ammonium uptake and assimilation genes. The NtrC regulon includes the *glnBA* operon, the *glnZ* gene (which encodes the PII paralog GlnZ), and the *amtB* gene (which encodes a passive ammonium channel). Deuridylation of GlnB under ammonium excess reverses these effects by stimulating NtrB phosphatase activity, leading to the dephosphorylation of NtrC and consequently repressing NtrC-dependent genes. The concomitant deuridylylation of GlnZ inhibits ammonium uptake by allowing it to interact with the ammonium-transporting channel AmtB, blocking it (Huergo et al. 2007).

In *A. brasilense*, the *glnA* (GS) gene is co-transcribed with *glnB* from the *glnBp1*-σ^{70} promoter (active under nitrogen sufficiency) and *glnBp2*-σ^{54} (active under nitrogen limitation), both regulated by NtrC (de Zamaroczy et al. 1993; Huergo et al. 2003). In *Azospirillum brasilense*, a distinct promoter for *glnA*, located within the *glnB-glnA* intergenic region, allows for its expression under conditions of excess ammonium, independent of NtrC. This ensures a basal level of glutamine synthetase even when nitrogen is abundant. The *glnZ* gene is independently transcribed from a σ^{54}-dependent promoter activated by NtrC under both nitrogen limitation and excess. The *amtB* gene expression is also activated by NtrC under nitrogen limitation (Van Dommelen et al. 1998). The *ntrBC* genes are organized in an operon with *nifR3* (Liang et al. 1993; Machado et al. 1995).

Despite its central role in controlling the nitrogen metabolism, NtrC is not essential for nitrogen fixation (Liang et al. 1993; Machado et al. 1995; Fadel-Picheth et al. 1999). The genes encoding the nitrogenase complex and its many associated proteins are under the control of a specific σ^{54}-dependent transcriptional activator, NifA. In *A. brasilense*, *nifA* expression is independent of NtrC and only partially regulated by nitrogen and oxygen (Liang et al. 1991; Fadel-Picheth et al. 1999). *A. brasilense* NifA has a typical three-domain structure in which the N-terminal domain contains a GAF motif that responds to ammonium (Arsene et al. 1996); the central domain is responsible for ATP hydrolysis and interaction with σ^{54}, while the C-terminal domain binds to upstream activating sequences in the promoter region of *nif* genes (Dixon and Kahn 2004). An interdomain linker located between the central and C-terminal domains contains a four-cysteine motif, which is believed to inactivate the protein under high oxygen conditions. NifA activity is dependent on the presence of GlnB (de Zamaroczy et al. 1993; Araujo et al. 2004), which counteracts the inhibitory effect of the N-terminal GAF domain over the protein: deletions in the GAF domain result in constitutively active, GlnB-independent NifA (Nishikawa et al. 2012). This is similar to what was observed in other diazotrophic Proteobacteria: *Rhodospirillum rubrum* NifA also requires GlnB-

UMP (Zou et al. 2008); in *Herbaspirillum seropedicae*, mutations in the GAF domain also yield constitutively active, PII-independent NifA (Aquino et al. 2015); *Rhodopseudomonas palustris* shows constitutive NifA activation with mutations in the GAF domain between residues 202 and 217 (McKinlay and Harwood 2010). This is in contrast to the well-studied *Klebsiella pneumoniae* and *Azotobacter vinelandii* where the regulation of NifA activity requires the NifL protein, which inhibits NifA in the presence of oxygen and/or ammonium (Dixon and Khan 2004).

In *A. brasilense*, nitrogenase activity is also regulated by the reversible ADP-ribosylation of the dinitrogenase reductase (NifH), a modification catalyzed by DraT (dinitrogenase reductase ADP-ribosyl transferase) and reversed by DraG (dinitrogenase reductase-activating glycohydrolase). *draT* mutants retain nitrogenase activity even when exposed to a sudden increase in ammonium availability, whereas *draG* mutants are unable to reactivate the enzyme following its inactivation due to increased ammonium levels (Zhang et al. 1992). PII proteins, GlnB and GlnZ, participate in the regulation of DraG and DraT. Under nitrogen excess, non-uridylylated ADP-bound GlnB interacts with DraT, activating it and promoting ADP-ribosylation in NifH. Deuridylylated GlnZ interacts with DraG, targeting it to membrane-bound AmtB, thus inhibiting DraG and promoting nitrogenase switch-off. Under nitrogen limitation, PII proteins are uridylylated and saturated with ATP and 2-oxoglutarate, preventing their interactions with DraT and DraG, thus allowing continued nitrogenase activity (Klassen et al. 2001, 2005; Huergo et al. 2007; Moure et al. 2013, 2019). Figure 9.1 summarizes the current model of nitrogen metabolism and nitrogen fixation regulation in *A. brasilense*.

9.2 Decoupling Nitrogen Sensing from Nitrogen Fixation Repression

In wild-type strains, N_2 fixation is tightly coupled to intracellular ammonium assimilation, minimizing energy loss and preventing nitrogen release to the environment. For *A. brasilense* to be used as an ammonium-excreting biofertilizer, it is necessary to modulate or disrupt key nodes of this regulatory network. Based on this model, in vitro strain selection, genetic engineering, and genome editing can be used to enable constitutive nitrogen fixation and ammonium excretion, overcoming the physiological limitations of wild strains.

A central strategy involves modifying the NifA protein, the essential transcriptional activator of *nif* genes. Alterations in the N-terminal domain or Q-linker region can yield constitutively active NifA variants, insensitive to repression mediated by GlnB or induced by ammonium. These changes support sustained nitrogenase gene activation even under inhibitory conditions.

In this context, Arsene et al. (1996) and Souza et al. (1999) showed that *A. brasilense* strains expressing NifA variants lacking the N-terminal domain exhibit constitutive nitrogenase activity, although without detectable ammonium excretion

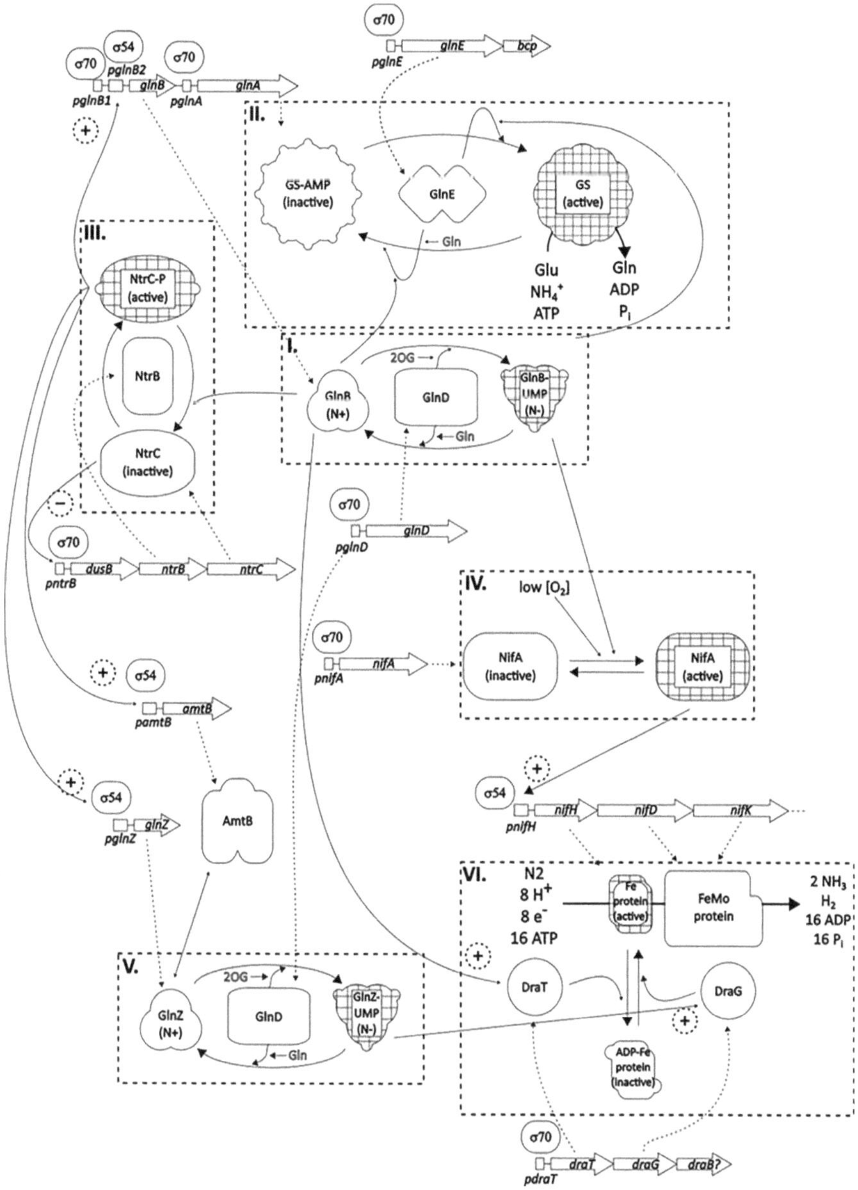

Fig. 9.1 Schematic representation of genes and proteins involved in nitrogen fixation and ammonium assimilation in *A. brasilense.* **References: I. GlnD and PII Regulation:** The *glnD* gene is constitutively expressed from a σ70 promoter. Its product, the bifunctional GlnD protein (PII-uridylyl transferase/uridylyl removing enzyme; EC: 2.7.7.59), is sensitive to intracellular glutamine concentration. GlnD uridylylates the PII protein GlnB at residue Y51 when glutamine is insufficient and removes this uridylyl group when glutamine becomes abundant. The uridylylation state of GlnB then further regulates the activity of other proteins in nitrogen assimilation pathways. **II. Glutamine Synthetase (GS) Activity Regulation:** GS (glutamine synthetase; EC: 6.3.1.2), the product of the *glnA* gene, is responsible for most ammonium assimilation in *A. brasilense*. Its activity is controlled by the bifunctional protein GlnE (glutamine

(Souza et al. 1999). Posttranslational regulation of nitrogenase can also be targeted by deleting the *draT* gene, which encodes the ADP-ribosylation enzyme that inactivates NifH under high ammonium levels. With this modification, nitrogenase remains active even when ammonium ion levels increase. Additionally, engineering

Fig. 9.1 (continued) synthetase adenylyltransferase/glutamine synthetase-adenylyl-L-tyrosine phosphorylase; EC: 2.7.7.42; EC: 2.7.7.89), which is constitutively expressed. When intracellular glutamine concentration is high, the non-uridylylated form of GlnB predominates in the cytosol. This form can interact with GlnE to stimulate the adenylylation of residue Y397 of GS, a modification that renders GS inactive. Conversely, when intracellular glutamine concentration is low, GlnB becomes uridylylated by GlnD, thereby signaling to GlnE to remove the adenylyl group from GS, activating it. The expression of the *glnBA* operon is complex: there are two constitutive promoters (one upstream of *glnB* and one upstream of *glnA*), ensuring the presence of some GlnB and GlnA regardless of the cell's nitrogen status. Additionally, a stronger σ^{54}-dependent promoter is active only under low glutamine concentrations. **III. NtrB-NtrC Two-Component System:** GlnB regulates the activity of the NtrB-NtrC two-component system, which is crucial for the expression of many σ^{54}-dependent genes. The interaction between non-uridylylated GlnB (the form predominant when glutamine is abundant) and the NtrB histidine kinase (also called GlnL; EC: 2.7.13.3) prevents NtrB's autophosphorylation and stimulates its phosphoryl-removing activity, keeping the transcription factor NtrC (also called GlnG) in its inactive form. In contrast, uridylylated GlnB does not interact with NtrB, allowing NtrB to autophosphorylate and transfer the phosphoryl group to NtrC, rendering it active. Phosphorylated NtrC is a positive regulator for the σ^{54}-dependent *glnBA*, *glnZ*, and *amtB* operons, in addition to many other genes required for the acquisition of nitrogen from sources other than ammonium. Non-phosphorylated NtrC also downregulates its own synthesis by binding to the promoter region of the *nifR3ntrBC* operon. **IV. NifA Regulation:** NifA is the sole transcriptional activator for the expression of the *nif* genes, which encode the nitrogenase complex. A σ^{70} promoter causes constitutive low expression of the *nifA* gene; its product, NifA, is an oxygen- and ammonium-sensitive, uridylylated-GlnB-dependent enhancer-binding protein. **V. GlnZ and AmtB Regulation:** Like GlnB, GlnZ is a PII protein that can be uridylylated and deuridylylated by GlnD. GlnZ expression is dependent on σ^{54} and phosphorylated NtrC, and the protein targets distinct partners from GlnB. Notably, GlnZ is the main interaction partner for AmtB, an ammonia transporter that is also expressed from a $\sigma54$ and NtrC-dependent promoter. When intracellular glutamine concentration is low, both GlnZ and AmtB are expressed; GlnZ is uridylylated by GlnD, preventing it from interacting with AmtB and allowing rapid internalization of ammonium if present in the extracellular medium. Once glutamine levels increase, GlnZ becomes deuridylylated by GlnD, enabling it to interact with AmtB, thereby blocking it. **VI. Molybdenum-Dependent Nitrogenase Complex and DraT/DraG Regulation:** The molybdenum-dependent nitrogenase complex (EC: 1.18.6.1) is responsible for reducing N_2 to ammonium, which is then assimilated by GS. The dimeric Fe-protein (also called NifH or dinitrogenase reductase) acquires electrons from cellular ferredoxins and channels them to the FeMo-protein (dinitrogenase), a NifDK heterotetramer, which transfers them to N_2 to produce NH_3 and H_2. The FeMo-protein is active under low oxygen concentration. In *A. brasilense*, the Fe-protein can be inactivated and reactivated by the DraT/DraG system. DraT (NAD^+/dinitrogen-reductase ADP ribosyltransferase; EC:2.4.2.37) interacts with non-uridylylated GlnB, stimulating its activity, allowing the transfer of an ADP-ribosyl group from NAD^+ to residue R101 of NifH, inactivating it. Simultaneously, DraG (ADP-ribosyl-dinitrogen-reductase hydrolase; EC: 3.2.2.24) is sequestered to the membrane-bound AmtB through interaction with non-uridylylated GlnZ, rendering it inactive. Once intracellular glutamine decreases, GlnB becomes uridylylated and can no longer interact with DraT, which becomes inactive. DraG, freed from its inhibitory interaction with GlnZ since it cannot form a complex with uridylylated GlnZ, removes the ADP-ribosyl from NifH, restoring nitrogenase activity ammonium assimilation pathways is critical: mutations that reduce GS activity, such as those decreasing its affinity for ammonium or ATP, can limit nitrogen incorporation and favor its excretion. Likewise, mutations in *glnD* that impair PII deuridylylation but maintain GlnD uridylylation activity disrupt nitrogen sensing and signal a persistent ammonium ion deficiency state even under ammonium excess, promoting constitutive nitrogen fixation. These modifications, whether applied individually or synergistically, can decouple nitrogen sensing, nitrogenase activity, and ammonium assimilation and conservation. This capability enables the development of high-potential biotechnological strains.

Schnabel and Sattely (2021) demonstrated that a Δ*glnE* mutant promoted millimolar-level ammonium excretion. *A. brasilense* GS-deficient mutants have been described by several authors (Machado et al. 1991; Gauthier and Elmerich 1977; Fischer et al. 1986; Christiansen-Weniger and Van Veen 1991; Van Dommelen et al. 2003). These strains fix nitrogen constitutively and excrete ammonium in nitrate-containing or diazotrophic media. Some have been tested as inoculants under controlled conditions (Christiansen-Weniger and Van Veen 1991; Pankievicz et al. 2015; Santos et al. 2017) and in the field (Pedrosa et al. 2020), demonstrating that the high root colonization capacity of *A. brasilense*, combined with ammonium release, makes it a promising plant growth-promoting strategy.

9.3 Tools to Study N Metabolism in *Azospirillum:* Selection of Ammonium-Excreting Strains

The following protocols describe experimental approaches for developing and characterizing ammonium-excreting mutants of *Azospirillum*, based on our experience in working mostly with *A. brasilense*. These strategies should roughly work for other species in the genus, but some optimization might be needed to account for the physiological uniqueness of each.

9.3.1 General Methods

9.3.1.1 Cultivation and Storage of *Azospirillum*

Azospirillum is routinely grown in the NFb medium (Okon et al. 1977; Dobereiner 1995; Baldani et al. 2014). Pedrosa and Yates (1984) changed the composition to incorporate a higher concentration of phosphate to improve pH control and renamed it NFbHP, which has the following composition per liter (described in Machado et al. 1991): KH_2PO_4 4 g, K_2HPO_4 6 g, $MgSO_4.7H_2O$ 0.2 g, NaCl 0.1 g, $CaCl_2$ 0.02 g, nitrilotriacetic acid (NTA) 0.056 g, $FeSO_4.7H_2O$ 0.02 g, biotin 100 μg, trace elements solution 10 mL, carbon source (lactic acid or malic acid) 5 g. The trace element solution contains, per liter: $NaMoO_4.2H_2O$ 200 mg, $MnSO_4.H_2O$ 235 mg,

H_3BO_3, 280 mg, $CuSO_4.5\ H_2O$ 8 mg, $ZnSO_4.7H_2O$ 24 mg; its pH is adjusted to 6.8 with 1 M KOH. Either lactic acid or malic acid can be used as a carbon source in the experiments described here. The medium is usually prepared in distilled water without the phosphate content and sterilized by autoclaving. The phosphate stock solution is prepared 20-fold concentrated and sterilized separately and added to the culture medium as needed. This may be handy when dealing with species other than *A. brasilense*; for example, *A. lipoferum* requires 1/5 of the phosphate concentration for optimal growth. Nitrogen source is usually 20 mM NH_4Cl, and the medium is now named NFbHPN ("high phosphate and nitrogen"). For liquid medium, the culture is incubated in a rotary shaker at 120 rpm and 30 °C. Solid NFbHPN plates are incubated for 24 h at 30 °C. The cultures are viable for weeks, provided the medium does not dehydrate. *A. brasilense* can be stored at 8 °C for longer periods on solid NFbHPN. *Azospirillum* can also be stored at −20 °C in 25–50% glycerol.

9.3.1.2 Bacterial Counting

Viable bacterial counts per mL can be determined via serial dilution (10^{-1} to 10^{-8}) of the original culture, followed by spotting three drops of 10 μL from each dilution on NFbHPN plates. After 24 h at 30 °C, the small colonies are counted using a magnifying glass.

9.3.1.3 Determination of Constitutive Nitrogenase Activity

9.3.1.3.1 Growth in Semi-solid Medium for Nitrogenase Activity Detection

Azospirillum spp. are aerobic but will only fix nitrogen in the presence of a constant supply of low oxygen (micro-aerobic) concentrations; higher pO_2 irreversibly inactivates the nitrogenase and suppresses *nif* gene transcription, while the total lack of oxygen causes the cell to lack sufficient energy for nitrogen fixation. Johanna Döbereiner's group introduced the N-free semi-solid media for the study of nitrogen fixation in *Azospirillum* and other motile aerotactic diazotrophs in the mid-1970s (Baldani et al. 2014). This is the simplest approach to measure in vivo nitrogenase activity in *Azospirillum*. The bacteria are inoculated in N-free minimum growth media containing low agar concentration and incubated under static conditions; as the cell population increases, the culture grows a veil-like pellicle that continuously migrates vertically to regions where the oxygen concentration is ideal for the expression and activity of nitrogenase. The ideal agar concentration is 1.65 g/L (it may vary depending on the agar supplier).

For the nitrogenase assay, cells are initially cultivated at 30 °C and 120 rpm in liquid NFbHP medium supplemented with 5 mM of NH_4Cl, which will be completely consumed as the culture approaches saturation. Cultures deficient in GS can be supplemented with 2 mM glutamine to allow for faster growth. A small aliquot (10 mL) of each *Azospirillum* culture is then inoculated into 10 mL bijou bottles containing 4 mL of semi-solid NFbHP medium. Inoculation is carried out with a micropipette by dispensing the cells close to the flask bottom, through the middle of the semi-solid medium column. The flask is then incubated at 30 °C for the duration of the experiment.

Nitrogenase activity should be evaluated both in nitrogen-limiting and nitrogen-excess conditions (5–10 mM NH_4Cl). Wild-type *Azospirillum* will only show activity in the former; nitrogenase constitutive strains, on the other hand, will show nitrogenase activity in both conditions, at varying levels. Semi-solid NFbHP medium may contain a small amount of glutamate or glutamine (up to 0.5 mM) to accelerate the initial growth, allowing for nitrogenase activity to be quantifiable after about 24 h from the start. Glutamine is preferable when dealing with GS-deficient strains.

9.3.1.3.2 Time Course of Nitrogenase Activity

Nitrogenase activity in semi-solid medium is determined by the reduction of acetylene, an alternative substrate for nitrogenase, to ethylene. The two gases (ethylene and acetylene) can be separated by gas chromatography.

In order to detect maximum nitrogenase activity during semi-solid growth, it is often necessary to make time-course measurements, especially when working with mutant strains that grow more slowly due to the energy spent excreting ammonium. At the early stages of growth, the bacterial population can be too small for nitrogenase detection, while at the later stages, the number of cells is too large and the oxygen supply falls below what is needed to sustain energy production and full nitrogen fixation. *A. brasilense* usually reaches peak nitrogenase activity 24–48 h after inoculation in the semi-solid medium. To determine the optimum growth period, it is recommended to perform measurements after 12 h, 24 h, 36 h, 48 h, 72 h, and 96 h. When fixing nitrogen, a veil-like bacterial growth should be visible just below the medium surface.

For nitrogenase measurement at each time point, the bottles are tightly sealed with subaseals or rubber stopper with aluminum seals, and 0.6 ml of acetylene (10% of the gas phase) is injected into each bottle. The cultures are further incubated at the same temperature for 60 min, and 0.5 ml of the gas phase is analyzed by gas chromatography to determine the amount of ethylene formed. The gas chromatograph is equipped with a flame ionization detector (FID) and a Porapak N (100–200 mesh) column (2 m, 1 mm diameter) operated at 110 °C. Ethylene (100 ppm in N_2) is used as a standard, and 0.5 ml contains 2.232 nmol of ethylene at 25 °C and 1 atm.

To calculate the specific activity of nitrogenase, it is necessary to determine the culture protein concentration. For this, the cultures in semi-solid medium are quickly heated in boiling water to melt the agar and vigorously vortexed. Melted culture samples are then lysed in NaOH 0.1 M (final concentration), and protein concentration is determined by the standard Bradford method, following the same procedure used for liquid cultures.

Nitrogenase-specific activity is calculated by the following equation:

$$N_2ase = (\text{nmol } C_2H_4 \text{ in } 0.5 \text{ ml}) \times 12/(\text{mg protein} \times 60 \text{ min})$$

where 12 is the correction factor for the gas volume (considering a 10 mL bottle with 4 mL of semi-solid medium, and mg protein is the total protein content of the culture in milligrams. Once an optimum of nitrogenase activity is found, the subsequent assays can be done solely at this time point.

9.3.1.3.3 Effect of Inoculum Size

It is possible to determine the maximum nitrogenase activity using a different strategy, namely varying the inoculum concentration and measuring the activity at a fixed time point. The *Azospirilla* inocula are grown as described above and inoculated in increasing amounts (e.g., 1, 2, 4, 8, 16, 32, and 64 microliters) into 4 ml of N-free semi-solid medium in 10 ml bijou or penicillin bottles. The cultures are incubated statically at 30 °C for ~24 h until a veil-like bacterial growth is visible just below the medium surface and assayed for nitrogenase activity. For nitrogenase measurement at each point, the bottles are tightly sealed with subaseals or rubber stoppers with aluminum seals, and 0.6 ml of acetylene (10% of the gas phase) is injected into each bottle. The cultures are further incubated at the same temperature for 60 min, and the ethylene formed is determined by gas chromatography as before. Strains capable of reducing acetylene in the presence of ammonium (constitutive nitrogenase activity) are selected for further assays.

9.3.1.3.4 Protein Quantification

Protein concentrations are measured using the Bradford method (Kruger 2002), with bovine serum albumin (0.2 $g \cdot L^{-1}$) as the standard. Protein assays in cell suspensions are performed after alkaline lysis in 0.1 M NaOH at 80 °C for 30 min.

9.3.2 Determination of Ammonium Excretion Ability

Ammonium excretion by *A. brasilense* is measured using a time-course approach in semi-solid, N-free NFbHP medium at 30 °C. Cultures, grown without agitation, are thoroughly mixed and centrifuged, and the cell- and agar-free supernatant is used for NH_4^+ determination. The method, adapted from nitrogenase activity assays, requires an ammonium-free environment, though 0.5 mM glutamate or glutamine may be added for GS-deficient strains. Cultures are sampled periodically (24, 48, 72, 96 h), as ammonium is typically detectable on days 2–3, after which increased cell density can lead to re-assimilation. For semiquantitative analysis, 50 μL aliquots are taken below the pellicle every 24 h. While ammonium concentration can be normalized to total protein, this requires sacrificing cultures for each time point. A high-throughput method using 96-well deep-well plates (650 μL NFbHP medium/well, 10 μL inoculum) with punctured plastic seals is also viable, otherwise following the same experimental procedure.

9.3.2.1 Ammonium Detection and Quantification

Among the various methods available for ammonium detection and quantification (ninhydrin, Nessler, and Berthelot), the Berthelot indophenol method is preferred due to its high sensitivity (<50 μM) and excellent linear response across the range of 50 μM to 2 mM (Chaney and Marbach 1962). However, fluctuations in sodium hypochlorite concentration may cause inconsistent results. Therefore, it is essential to source reagents from reliable suppliers and verify their quality before use.

Prepare the ammonium chloride standard solutions in advance using NH_4Cl at concentrations of 0, 50 μM, 100 μM, 150 μM, 200 μM, and 300 μM. Store these standards in capped plastic tubes at room temperature for convenience.

Next, prepare the stock solutions S0 (5% w/v phenol in ultrapure water), S1 (2.5% w/v NaOH in ultrapure water), and S2 (1% w/v sodium nitroprusside in ultrapure water). Phenol is toxic and carcinogenic; it must be handled only in a fume hood. Use only colorless phenol, as it turns pink or yellow upon oxidation and becomes ineffective. Stock solutions S0 and S1 can be stored at room temperature for several months. Solution S2 should be protected from light and preferably refrigerated; it is also stable for months under these conditions.

From these stock solutions, prepare fresh working solutions as needed for daily assays:

1. Prepare working solution W1 by mixing equal volumes of S0 and S1.
2. Prepare working solution W2 by diluting 1 milliliter of S2 into 49 milliliters of ultrapure water.
3. Prepare working solution W3 by diluting sodium hypochlorite in water to a final concentration of 1.0–1.25%. High-purity hypochlorite is unnecessary; standard household bleach is acceptable if it is freshly prepared and within the correct concentration range.

W3 is the most critical reagent in this assay. Hypochlorite concentration must be between 1.0% and 1.25%; deviations from this range can lead to failed reactions. Since hypochlorite degrades over time (releasing Cl_2 gas), W3 should be prepared fresh and used within 2 weeks. When troubleshooting failed assays, always begin by checking the quality and concentration of the hypochlorite.

Before applying the assay to valuable samples, validate reagent performance. On a 96-well clear microplate, pipette 125 μL of ultrapure water (blank) and 50 and 300 μM NH_4Cl standard solutions into separate wells. Then, sequentially add 25 μL of W1, 37.5 μL of W2, and 37.5 μL of W3. Incubate at room temperature for 40 min. The water control should remain colorless or faintly yellow, while the ammonium standards should develop a distinct blue color of varying intensities. If this test is successful, proceed with the actual assay.

For sample analysis, pipette 125 μL of each ammonium standard or sample into two wells of a 96-well clear microplate. Add 25 μL of each sample to individual wells, in triplicate, followed by 100 μL of ultrapure water. As before, to each well add sequentially: 25 μL of W1, 37.5 μL of W2, and 37.5 μL of W3.

Incubate for 40 min at room temperature and measure absorbance at 625 nm using a microplate reader. Plot a standard curve using your preferred software and determine sample concentrations by linear regression. Finally, multiply the measured ammonium concentration by five to account for sample dilution during the assay.

9.3.3 *Strategies for Obtaining Ammonium-Excreting Mutants*

9.3.3.1 Selection of *Azospirillum* Resistance to Ethylenediamine

The following protocols describe experimental approaches for selecting ammonium-excreting mutants of *A. brasilense* through spontaneous resistance to ethylenediamine (EDA) and for quantifying ammonium excretion in nitrogen-free semi-solid medium. The selection protocol relies on the toxic incorporation of EDA via glutamine synthetase activity, favoring the survival of mutants with impaired ammonium assimilation. Thus, this selection aims to seek strains with decoupled ammonium assimilation from nitrogen fixation, allowing excess fixed nitrogen to leak to the environment. The ammonium released is quantified by the method based on the indophenol blue colorimetric assay and provides a reliable estimate of nitrogen excretion by individual strains. Together, these protocols allow the identification and characterization of mutants with potential applications in plant growth promotion and ecology.

The ethylenediamine (EDA) solution is prepared in water (10% v/v), neutralized to pH 7.0 with HCl, and sterilized by filtration using a Millipore HAWP membrane (0.45 μm). Two hundred microliters (OD600 = 1.5) of an *Azospirillum brasilense* strain culture grown in liquid NFbHPG (the nitrogen source is glutamate 5 mM) medium are spread using sterile spatulas on solid NFbHPG plates supplemented with EDA at different concentrations (0, 0.01, 0.02, 0.05, 0.1, 0.2, 0.5, and 1%). The number of colony-forming units (CFU) is also determined (see below). After 7 days of incubation at 30 °C, colonies are counted to determine the minimum inhibitory concentration (MIC) and the frequency of emergence of EDA-resistant mutants (Machado et al. 1991). The MIC is defined as the lowest concentration of the inhibitor that completely prevents colony formation. The mutation frequency is calculated as the ratio between the number of resistant colonies and the total number of bacterial colonies.

EDA-resistant colonies are isolated, purified by streaking in fresh plates, cultivated in liquid NFbHPG medium, and stored in 50% glycerol at −20 °C for further characterization, including nitrogenase activity in the presence or absence of NH_4Cl

(5, 10, and 20 mM), ammonium excretion in N-free semi-solid medium, and growth in different nitrogen sources.

9.3.3.2 Directed Mutagenesis in *Azospirillum*

While the EDA selection method remains a rich source of novel mutations, it is also useful to have the toolset for introducing specific changes in the bacterial genome. This allows, for example, for spontaneous mutations obtained in one strain background to be readily transferred to other Azospirilla; and, as knowledge about the control of nitrogen metabolism progresses in this genus and other Proteobacteria, directed mutagenesis can play an ever-larger role in the engineering of ammonium-excreting strains.

This section describes a method for introducing desired substitutions, deletions, or insertions in the *Azospirillum* genome for achieving ammonium excretion. All the steps described here rely on standard molecular biology and microbiology techniques.

9.3.3.2.1 Directed Mutagenesis Rationale

A comprehensive set of strategies for modifying bacterial genomes and tagging can be found in Ramos et al. (2011). Directed mutagenesis in *Azospirillum* is based on homologous recombination between the bacterial chromosome and a suicidal plasmid. The mutagenic plasmid carries the desired target modification flanked by sequences identical to the genome sequence, plus a marker for positive selection (allowing easy detection of the first recombination event) and a marker for negative selection (allowing detection of the second recombination event, which removes the plasmid from the genome). Together, these markers allow for the two sequential recombination events, leading to precision base exchange or genomic deletion of selected targets. For the first recombination event, mutant cells are usually selected for antibiotic resistance, indicating integration of the suicidal vector in the bacterial chromosome. The recombinant colonies obtained are then exposed to culture conditions that select against cells with the negative selection gene marker, thereby isolating double recombinant strains that have excised the integrated suicidal vector from their genome; these strains are subsequently screened for the presence of the desired mutation. This scarless procedure permits DNA editing, with no leftover DNA from the plasmidial backbone. Figure 9.2 summarizes the process.

9.3.3.2.2 Design of the Mutagenic DNA Fragment

The following detailed protocol has been used for the genome editing of *A. brasilense* and other Gram-negative bacteria. The mutagenic DNA fragment should contain the desired DNA alteration flanked by around 800–1000 bp of the

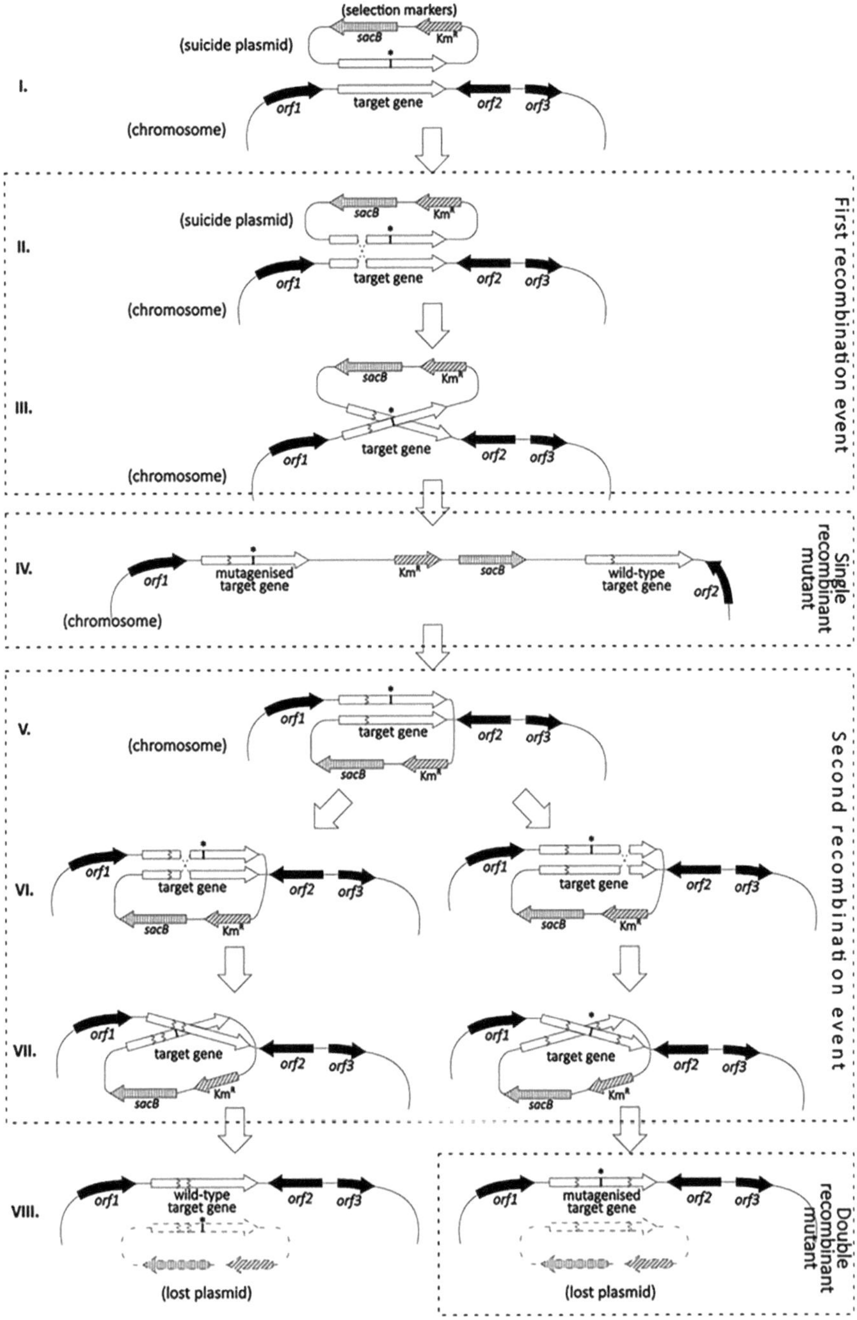

Fig. 9.2 Schematic representation of the events leading to the establishment of a scarless mutation in the genome of *Azospirillum*. **References: I. Plasmid Introduction and Integration Preparation:** A suicide plasmid is introduced into the bacterial cell, typically by biparental conjugation. This plasmid cannot replicate independently within the cell; therefore, it will only be passed down after **mitosis** if it integrates into the bacterial chromosome. The plasmid employed is a pK18

immediately upstream and immediately downstream sequences, as found in the genome. For substitutions, the standard method for generating this fragment is overlap-extension PCR (OE-PCR), for which four construction primers are needed. A visual representation of this procedure is shown in Fig. 9.3. Primer design should follow the general guidelines, with primers 1 and 4 annealing to the sequences upstream and downstream of the target DNA, respectively. For easy cloning, they should contain suitable restriction sites, and care should be taken to avoid choosing a site already present anywhere in the target DNA.

Primers 2 and 3 are employed to introduce the desired substitution or modification in the genome sequence. They are usually complementary to each other, though this is not strictly required, as long as they have partial overlap (at least 10 bases).

Fig. 9.2 (continued) derivative, containing a gene for positive selection (KmR, conferring kanamycin resistance), a gene for negative selection (*sacB* from *Bacillus*, which confers sensitivity to sucrose in Gram-negative bacteria), and a region with high homology to the *Azospirillum* genome. This homologous region will harbor the desired mutation. For simplicity, the figure shows a homologous region covering an entire **ORF** with one mutation close to its center, but this is not always the case. For small genes, it is necessary to clone the regions upstream and downstream of the **ORF** to ensure a sufficiently long homologous tract flanking each side of the mutation. Likewise, for mutations inserted too close to either end of the **ORF**, the adjacent noncoding region must be included in the construct. **II. and III. First Recombination Event:** The suicide plasmid integrates into the genome following a double-strand break in the bacterial chromosome within the target region, which is the same region cloned into the suicide plasmid. A second double-stranded break must then occur in the suicide plasmid, allowing the complementary sequences in both molecules to align and join after cellular DNA-repairing enzymes digest the 3' ends in both. The model shows the double-strand break occurring upstream of the desired mutation, but it is equally likely to occur downstream of it. **IV. Selection of Single Recombinant Mutants:** The single recombinant mutant contains two copies of the target region: one identical to the preexisting wild-type genotype, and the other harboring the introduced mutation. These copies flank the remainder of the plasmid. This mutant can be selected from the mix of transconjugants by plating on medium containing kanamycin; the positive selection marker confers resistance to this antibiotic and will only be expressed in cells that have successfully integrated the plasmid into the chromosome. **V., VI., and VII. Second Recombination Event (Excision):** The suicide plasmid is excised from the genome following double-strand breaks in both copies of the target region, followed by digestion of the 3' ends, pairing of homologous sequences, and DNA ligation. The position of this second recombination event determines the final genotype: if the first recombination occurred upstream of the introduced mutation, then a second recombination event also taking place upstream of it will lead to the excision of a plasmid containing the desired mutation and the consequent reversion of the bacterial genome to its original sequence. However, if the second recombination event occurs downstream of the introduced mutation, then the excised plasmid will contain the wild-type copy of the target region, and the genome will retain the mutagenized copy. **VIII. Selection and Verification of Double Recombinants:** The double recombinants carry only one copy of the target region, which is either wild-type or mutagenized. They are selected from the single recombinant mutants by plating on medium containing sucrose, as the negative selection marker *sacB* is lethal to Gram-negative cells under these conditions. The presence of the desired mutation can be verified by PCR, followed by RFLP analysis or DNA sequencing. In the case of putative ammonium excreters, a phenotypic selection is sometimes visible directly on the selection plate: ammonium excreters sometimes grow at a slower pace than the parental strain when kept on solid medium without glutamine

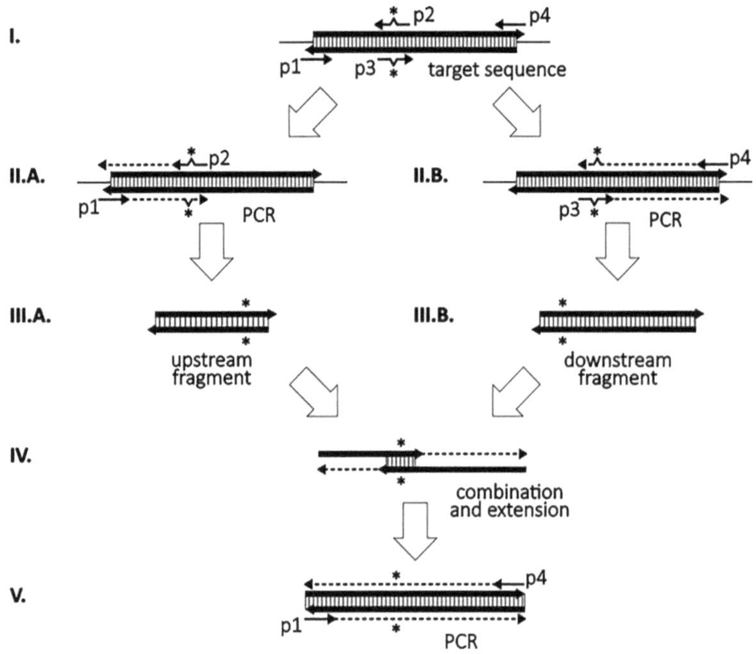

Fig. 9.3 Overlap-extension PCR for the introduction of mutations in a target region. **References: I. Primer Design:** A target genomic region is chosen, and two pairs of primers are designed for mutagenesis. Primer 1 (the upstream external primer) and primer 4 (the downstream external primer) flank the entire desired sequence and usually harbor restriction sites for subsequent cloning. Primer 2 (the downstream mutagenic primer) and primer 3 (the upstream mutagenic primer) align to the site where the mutation(s) will be introduced and contain the necessary base pair mismatches to achieve this. **II.A. and III.A. Amplification of the Upstream Fragment:** The upstream mutagenic fragment is amplified from the genome using primers 1 and 2. **II.B. and III.B. Amplification of the Downstream Fragment:** In a separate PCR, the downstream mutagenic fragment is amplified from the genome using primers 3 and 4. **IV. Extension Reaction:** The two mutagenic fragments are purified and mixed, allowing the pairing of complementary regions and the synthesis of the full mutagenized target region by extension with a DNA polymerase. **V. Final Amplification and Cloning Preparation:** The full mutagenized region is then amplified with primers 1 and 4, producing a sufficient amount of DNA for subsequent cloning into the suicide vector

They may also contain silent substitutions to create or destroy a restriction site, which can be useful to confirm the presence of the desired mutation in the target gene.

A second set of primers might be convenient to facilitate the screening of the target region to confirm the presence of the intended mutation by PCR. These are primers that bind to sequences closer to the desired mutation, allowing for the PCR amplification of a smaller fragment than that produced by primers 1 and 4.

OE-PCR can also be used for constructing deleted strains. In this case, primers 2 and 3 must have their sequences corresponding to the bases immediately upstream of the desired deletion target sequence, followed by the bases immediately downstream of the deletion point.

Insertions can also be done by OE-PCR, though this tends to be practical only for smaller fragments. In the case of insertions of a few codons, this can be done directly by adding the desired bases to primers 2 and 3. Medium-sized fragments (generally up to 500 bp) require an additional set of primers and one more iteration of PCR.

9.3.3.2.3 Production of the Mutagenic DNA Fragment

Template DNA can be readily obtained from *Azospirillum* cells by heat lysis. Two hundred microliters of a saturated liquid culture are centrifuged for 1 min at 13,000× *g*, resuspended in ultrapure water, and heated at 90 °C for 20 min; the lysed cell suspension is then centrifuged for another minute at 13,000× *g*. Alternatively, a colony growing on solid medium can be resuspended in 100 μL of ultrapure water and treated the same as described for the liquid culture. Usually, 0.5 microliters of lysate prepared this way will be used as a template in a 10–25 μL PCR reaction. Finally, when working with robust DNA polymerases, a very small quantity of an *Azospirillum* colony can be collected by pricking it with an entomological pin and then adding it directly to the PCR reaction mix. Fresh colonies (<1 week old) tend to give more consistent results than older colonies.

For more challenging PCR (due to either suboptimal primers or difficult genomic context), it is recommended to use purified genomic DNA. The standard SDS lysis method followed by phenol-chloroform extractions works well for *Azospirillum*.

To produce the mutagenized fragment shown in Fig. 9.3, *Azospirillum* DNA is first amplified in two separate reactions with primers 1 and 2 and primers 3 and 4; the products are then recovered by agarose gel after electrophoresis, if necessary purified, and combined to act as the template for a new PCR reaction with primers 1 and 4, yielding the mutagenized fragment. DNA purification of the two initial PCR products is not strictly required, but is recommended in order to remove the original template DNA, which, when amplified by primers 1 and 4, yields the original, non-mutagenized sequence of the target region. If a restriction site was added close to the desired substitution, it is possible to confirm the exchange by restriction analysis.

9.3.3.2.4 Production of the Mutagenic Plasmid

Plasmid pK18mobsacB (Schafer et al. 1994), a pUC18 derivative, cannot replicate in *Azospirillum* and will only be passed down to all daughter cells if it integrates into the genome, which requires a region (typically 500–1000 bp) homologous between the two DNA molecules. This vector contains one positive selection marker, *kanR* (which confers resistance to kanamycin), and one negative selection marker, *sacB* (which confers sensitivity to sucrose in Gram-negative bacteria). It also contains a *mobile* site, allowing transfer between bacteria by conjugation.

To produce the desired mutagenic plasmid, the final PCR product described in the previous section can be digested with the appropriate restriction enzymes and ligated to pK18mobsacB, similarly digested; the ligation product is transformed into *Escherichia coli*, and transformants are selected on LA plates containing kanamycin 50 μg/mL. When working with pK18mobsacB, it is important to make sure that the culture media used to grow the cells harboring it is devoid of sucrose to avoid cell death; not every grade of tryptone or yeast extract will be appropriate.

The identity and integrity of the mutagenized DNA should be confirmed after cloning, preferably by DNA sequencing, to prevent unwanted mutations from being carried to the chromosome. Low-error-rate DNA polymerases (e.g., Pfu) are recommended to minimize this problem.

9.3.3.2.5 Plasmid Transfer by Conjugation

The biparental conjugation method outlined below has given satisfactory results in our experience.

First, transform the pK18mobsacB-derived plasmid in *E. coli* S17-1 (Simon et al. 1983), and plate in sucrose-free LA medium with 50 μg/mL of kanamycin; incubate at 37 °C overnight. Pick one colony with a sterile toothpick, and inoculate 3 mL of LB medium with 50 μg/mL kanamycin; grow overnight at 37 °C, at 120 rpm. On the same day, inoculate a colony of *A. brasilense* in 20 mL of NFbHPN liquid medium, and incubate at 30 °C at 120 rpm for approximately 24 h. Once both cultures reach saturation, use 60 μL of saturated *E. coli* S17.1 carrying the mutagenic plasmid culture to inoculate 3 mL of fresh LB medium with kanamycin; grow for approximately 3 h at 37 °C, 120 rpm. Centrifuge 1 mL of the (fresh) *E. coli* culture and 4 mL of a saturated *A. brasilense* culture for 3 min at ~1700× *g*, separately; discard the liquid, and resuspend the cell pellets in 1 mL of sterile 0.8% KCl. Repeat this process twice; remove the remaining supernatant and resuspend each cell pellet in 100 μL of 0.8% KCl. Mix 50 μL of each cell suspension in a new sterile tube, and plate ten 10 μL drops on a plate containing a mixture of NFbHPN and LA at the ratio of 3:1 (Pedrosa and Yates 1984), without any antibiotics. Do not spread the drops; let them dry on the medium, then close the plates and incubate at 30 °C for 24 h.

9.3.3.2.6 First Recombination Event

After 24 h, scrape the cell lawn with a sterile toothpick and resuspend the paste in 1 mL of sterile 0.8% KCl. Centrifuge for 3 min at ~1700× *g*; discard the liquid, and resuspend the cell pellets in 600 μL of sterile 0.8% KCl. Dilute the cell suspension 10-fold and a 100-fold in 0.8% KCl to a final volume of 1 mL for each dilution. Plate 100 μL aliquots of undiluted cell suspension and of each dilution on NFbHPN containing 100 μg/mL kanamycin. Allow it to grow at 30 °C; colonies usually appear by the third day. These kanamycin-resistant colonies have integrated the mutagenic plasmid into their genome and now contain both the wild and mutagenized versions of the target sequence. Figure 9.2 shows the integration of the plasmid by the homologous recombination apparatus to obtain a strain with the mutagenic plasmid integrated in the genome. The integration of the mutagenic plasmid into the genome can be confirmed by PCR followed by digestion at the restriction site linked to the introduced mutation.

9.3.3.2.7 Second Recombination Event

The negative selection marker in pK18mobsacB is used to eliminate the plasmid backbone and the wild-type copy of the target region, retaining only the mutagenized version. The *sacB* gene from *Bacillus subtilis* encodes an enzyme that synthesizes (2 → 6)-β-D-fructan from sucrose; accumulation of this polymer is lethal to *Azospirillum* and other Gram-negative bacteria. As a result, *A. brasilense* cells carrying *sacB* are not viable in the presence of sucrose, enabling the selection of clones that have undergone a second homologous recombination event. This event leads to the excising of the *sacB* gene from the genome, leaving behind either the wild-type or the mutagenized version of the target region, depending on the recombination breakpoint (Fig. 9.2).

For the selection of strains resulting from the second recombination event, inoculate kanamycin-resistant colonies, obtained during the first recombination event, in 3 mL of NFbHPN liquid medium, and grow at 30 °C, 120 rpm overnight until culture saturation. Transfer 100 μL of the saturated culture to a fresh 3 mL aliquot of liquid medium, and grow as before. Repeat this process three times. Then, dilute the final saturated liquid culture 10-, 100-, and 1,000-fold in sterile 0.8% KCl to a final volume of 1 mL for each dilution. Plate 100 μL of the undiluted culture and each dilution onto NFbHPN plates containing 10% (w/v) sucrose. Incubate at 30 °C for up to 3–4 days. Streak resulting colonies onto fresh NFbHPN plates with 10% sucrose to ensure that no sucrose-sensitive cells are carried over.

Sucrose is prepared as a 50% (w/v) solution in distilled water. Sterilization is done by filtering through a 0.22 μm filter, as the high temperature in the autoclave would cause it to partially hydrolyze.

As mutants with reduced GS activity are expected, the addition of 2 mM glutamine may support the growth of recombinant cells carrying the mutagenized region. Severe GS deficiency can substantially delay growth, causing these colonies to appear several days later than wild-type revertants. Additional supplements may be included in the culture medium as needed, depending on the expected phenotypes of the edited strains.

9.3.3.2.8 Screening for Mutagenized Double Recombinants

Since sucrose-resistant double recombinants may carry either the wild-type or the mutagenized target region (Fig. 9.2 (VIII)), it is necessary to screen the colonies to identify the desired mutants. The preferred method is PCR screening, followed by restriction enzyme digestion.

To perform the screening, the PCR battery should include one negative control (no bacterial template), one wild-type control (parental *Azospirillum* strain), and one positive control (a dilution of the mutagenic plasmid). Colonies of *Azospirillum* can be picked with a sterile toothpick or the tip of an entomological pin, which is then immersed in the PCR reaction mixture. The toothpick is removed, the tubes sealed, and the PCR reaction may proceed.

PCR can be performed using the outer primers (primers 1 and 4, as shown in Fig. 9.3), but it is usually faster and more reliable to use primers that flank the mutation site more closely. For large deletions in the target sequence, simple size-based separation by gel electrophoresis is sufficient to distinguish colonies carrying only the mutagenized fragment. For small deletions, insertions, or substitutions, however, the PCR products from wild-type and mutant templates typically co-migrate, making restriction digestion at a site introduced or removed by the mutation necessary for differentiation.

9.3.3.2.9 Ammonium Excretion by Nitrogen-Fixing Strains

Strains in which nitrogen fixation and ammonium assimilation are decoupled may release ammonium into the culture medium. Higher nitrogenase activity or lower GS activity leads to greater ammonium excretion. To evaluate this potential, strains should be tested for nitrogenase activity and for ammonium detection as described above.

9.4 Future Perspectives

Engineering *A. brasilense* strains with the capacity for constitutive nitrogen fixation and ammonium excretion represents a strategic step toward more efficient and sustainable biofertilizers. However, the genetic stability of these modifications, their performance under diverse environmental conditions, and the maintenance of beneficial plant interactions remain major challenges.

Future approaches may include precision genome editing (e.g., CRISPR/Cas) combined with plant-signal-sensitive regulatory promoters, enabling context-dependent expression of fixation and excretion genes. Synthetic circuits simulating persistent nitrogen deficiency states, while remaining responsive to rhizosphere cues, also represent a promising alternative for fine-tuned diazotrophic metabolism control. Altogether, these strategies may consolidate *A. brasilense* as an optimized microbial platform for low impact, sustainable agriculture. Field trials and ecological assessments of excreting strain release will be essential to advance from lab-scale to agricultural implementation.

References

Aquino B, Stefanello AA, Oliveira MA et al (2015) Effect of point mutations on Herbaspirillum seropedicae NifA activity. Braz J Med Biol Res 48:683–690. https://doi.org/10.1590/1414-431X20154522

Araujo MS, Baura VA, Souza EM et al (2004) In vitro uridylylation of the Azospirillum brasilense N-signal transducing GlnZ protein. Protein Expr Purif 33:19–24. https://doi.org/10.1016/j.pep.2003.08.024

Araújo LM, Huergo LF, Invitti AL et al (2008) Different responses of the GlnB and GlnZ proteins upon in vitro uridylylation by the Azospirillum brasilense GlnD protein. Braz J Med Biol Res 41:289–294. https://doi.org/10.1590/s0100-879x2008000400006

Arsène F, Kaminski PA, Elmerich C (1996) Modulation of NifA activity by PII in Azospirillium brasilense: evidence for a regulatory role of the NifA N-terminal domain. J Bacteriol 178:4830–4838. https://doi.org/10.1128/jb.178.16.4830-4838.1996

Baldani JI, Reis VM, Videira SS et al (2014) The art of isolating nitrogen-fixing bacteria from non-leguminous plants using N-free semi-solid media: a practical guide for microbiologists. Plant Soil 384:413–431. https://doi.org/10.1007/s11104-014-2186-6

Chaney AL, Marbach EP (1962) Modified reagents for determination of urea and ammonia. Clin Chem 8:130–132

Christiansen-Weniger C, Van Veen JA (1991) NH4+-excreting Azospirillum brasilense mutants enhance the nitrogen supply of a wheat host. Appl Environ Microbiol 57:3006–3012. https://doi.org/10.1128/aem.57.10.3006-3012.1991

de Zamaroczy M (1998) Structural homologues PII and Pz of Azospirillum brasilense provide intracellular signalling for selective regulation of various nitrogen-dependent functions. Mol Microbiol 29:449–463. https://doi.org/10.1046/j.1365-2958.1998.00938.x

de Zamaroczy M, Paquelin A, Elmerich C (1993) Functional organization of the glnB-glnA cluster of Azospirillum brasilense. J Bacteriol 175:2507–2515. https://doi.org/10.1128/jb.175.9.2507-2515.1993

Dixon R, Kahn D (2004) Genetic regulation of biological nitrogen fixation. Nat Rev Microbiol 2: 621–631. https://doi.org/10.1038/nrmicro954

Döbereiner J (1995) Isolation and identification of aerobic nitrogen-fixing bacteria from soil and plants. In: Alef K, Nannipieri P (eds) Methods in applied soil microbiology and biochemistry. Academic, San Diego, pp 134–141

Döbereiner J, Pedrosa FO (1987) Nitrogen-fixing bacteria in nonleguminous crop plants. Science Tech/Springer, Madison

Fadel-Picheth CMT, Souza EM, Wassem R et al (1999) Expression of the nifA gene of Azospirillum brasilense is regulated at the transcriptional level by ammonium. FEMS Microbiol Lett 170:185–190

Fischer M, Levy E, Geller T (1986) Regulatory mutation that controls nif expression and histidine transport in Azospirillum brasilense. J Bacteriol 167:423–426. https://doi.org/10.1128/jb.167.1.423-426.1986

Gauthier D, Elmerich C (1977) Relationship between glutamine synthetase and nitrogenase in Spirillum lipoferum. FEMS Microbiol Lett 2:101–104

Huergo LF, Dixon R (2015) The emergence of 2-oxoglutarate as a master regulator metabolite. Microbiol Mol Biol Rev 79:419–435. https://doi.org/10.1128/MMBR.00038-15

Huergo LF, Souza EM, Steffens MBR et al (2003) Regulation of glnB gene promoter expression in Azospirillum brasilense by the NtrC protein. FEMS Microbiol Lett 223:33–40. https://doi.org/10.1016/S0378-1097(03)00346-X

Huergo LF, Merrick M, Pedrosa FO et al (2007) Ternary complex formation between AmtB, GlnZ and the nitrogenase regulatory enzyme DraG reveals a novel facet of nitrogen regulation in bacteria. Mol Microbiol 66:1523–1535

Huergo LF, Chandra G, Merrick M (2013) PII signal transduction proteins: nitrogen regulation and beyond. FEMS Microbiol Rev 37:251–283

Jiang P, Peliska JA, Ninfa AJ (1998) The regulation of Escherichia coli glutamine synthetase revisited: role of 2-ketoglutarate in the regulation of glutamine synthetase adenylylation state. Biochemistry 37:12782–12789. https://doi.org/10.1021/bi980666u

Klassen G, Souza EM, Yates MG et al (2001) Control of nitrogenase reactivation by the GlnZ protein in Azospirillum brasilense. J Bacteriol 183. https://doi.org/10.1128/jb.183.22.6710-6713.2001

Klassen G, Souza EM, Yates MG et al (2005) Nitrogenase switch-off by ammonium ions in Azospirillum brasilense requires the GlnB nitrogen signal-transducing protein. Appl Environ Microbiol 71:5637–5641. https://doi.org/10.1128/AEM.71.9.5637-5641.2005

Kruger NJ (2002) The Bradford method for protein quantitation. In: The protein protocols handbook, pp 15–21

Liang YY, Kaminski PA, Elmerich C (1991) Identification of a nifA-like regulatory gene of Azospirillum brasilense Sp7 expressed under conditions of nitrogen fixation and in the presence of air and ammonia. Mol Microbiol 5:2735–2744. https://doi.org/10.1111/j.1365-2958.1991.tb01982.x

Liang CM, Arsene F, Elmerich C (1993) The ntrBC operon of Azospirillum brasilense is not required for nitrogen fixation. Mol Gen Genet 241:129–136

Machado HB, Funayama S, Rigo LU et al (1991) Excretion of ammonium by Azospirillum brasilense mutants resistant to ethylenediamine. Can J Microbiol 37. https://doi.org/10.1139/m91-092

Machado B, Funayama S, Rigo LU et al (1995) The ntrBC genes of Azospirillum brasilense are part of a nifR3-like–ntrB–ntrC operon and are negatively regulated. Can J Microbiol 41. https://doi.org/10.1139/m95-093

McKinlay JB, Harwood CS (2010) Carbon dioxide fixation as a central redox cofactor recycling mechanism in bacteria. Proc Natl Acad Sci USA 107:11669–11675

Moure VR, Danyal K, Yang Z et al (2013) The nitrogenase regulatory enzyme dinitrogenase reductase ADP-ribosyltransferase (DraT) is activated by direct interaction with the signal transduction protein GlnB. J Bacteriol 195. https://doi.org/10.1128/jb.01517-12

Moure VR, Costa FF, Cruz LM et al (2015) Regulation of nitrogenase by reversible mono-ADP-ribosylation. Curr Top Microbiol Immunol 384:89–106

Moure VR, Siöberg CLB, Valdameri G et al (2019) The ammonium transporter AmtB and the PII signal transduction protein GlnZ are required to inhibit DraG in Azospirillum brasilense. FEBS J 286:1214–1229. https://doi.org/10.1111/febs.14745

Nishikawa CY, Araújo LM, Kadowaki MA et al (2012) Expression and characterization of an N-truncated form of the NifA protein of Azospirillum brasilense. Braz J Med Biol Res 45(2): 113–117. https://doi.org/10.1590/s0100-879x2012007500006

Okon Y, Albrecht SL, Burris RH (1977) Methods for growing Spirillum lipoferum and for counting it in pure culture and in association with plants. Appl Environ Microbiol 33:85–88

Pankievicz VC, do Amaral FP, Santos KF et al (2015) Robust biological nitrogen fixation in a model grass-bacterial association. Plant J 81(6):907–919. https://doi.org/10.1111/tpj.12777

Pedrosa FO, Elmerich C (2007) Regulation of nitrogen fixation and ammonium assimilation in associative and endophytic nitrogen fixing bacteria. In: Elmerich C, Newton WE (eds) Associative and endophytic nitrogen-fixing bacteria and cyanobacterial associations. Springer, Dordrecht, The Nederlands, pp 41–71

Pedrosa FO, Yates MG (1984) Regulation of nitrogen fixation *(nif)* genes of *Azospirillum brasilense*by *nifA* and *ntr (gln)* type gene products. FEMS Microbiol Lett 23:95–101. https://doi.org/10.1111/j.1574-6968.1984.tb01042.x

Pedrosa FO, Oliveira ALM, Guimarães VF et al (2020) The ammonium excreting Azospirillum brasilense strain HM053: a new alternative inoculant for maize. Plant Soil 451:45–56. https://doi.org/10.1007/s11104-019-04124-8

Pirola MC, Monopoli R, Aliverti A et al (1992) Isolation and characterization of glutamine synthetase from the diazotroph Azospirillum brasilense. Int J Biochem 24:1749–1754

Ramos HJO, Yates MG, Pedrosa FO et al (2011) Strategies for bacterial tagging and gene expression in plant-host colonization studies. Soil Biol Biochem 43:1626–1638. https://doi.org/10.1016/j.soilbio.2011.03.029

Santos Ferreira N, Hayashi Sant' Anna F, Massena Reis V et al (2020) Genome-based reclassification of Azospirillum brasilense Sp245 as the type strain of Azospirillum baldaniorum sp. nov. Int J Syst Evol Microbiol 70:6203–6212. https://doi.org/10.1099/ijsem.0.004517

Santos KFDN, Moure VR, Hauer V et al (2017) Wheat colonization by an Azospirillum brasilense ammonium-excreting strain reveals upregulation of nitrogenase and superior plant growth promotion. Plant Soil 415:245–255. https://doi.org/10.1007/s11104-016-3140-6

Schäfer A, Tauch A, Jäger W et al (1994) Small mobilizable multi-purpose cloning vectors derived from the Escherichia coli plasmids pK18 and pK19: selection of defined deletions in the chromosome of Corynebacterium glutamicum. Gene 145:69–73. https://doi.org/10.1016/0378-1119(94)90324-7

Schnabel T, Sattely E (2021) Engineering posttranslational regulation of glutamine synthetase for controllable ammonia production in the plant symbiont Azospirillum brasilense. Appl Environ Microbiol 87(14):e0058221. https://doi.org/10.1128/AEM.00582-21

Simon R, Priefer U, Pühler A (1983) A broad host range mobilization system for in vivo genetic engineering: transposon mutagenesis in gram negative bacteria. Biotechnology 1:784–791

Souza EM, Pedrosa FO, Drummond M et al (1999) Control of Herbaspirillum seropedicae NifA activity by ammonium ions and oxygen. J Bacteriol 181:681–684. https://doi.org/10.1128/jb.181.2.681-684.1999

Steenhoudt O, Keijers V, Okon Y et al (2001) Identification and characterization of a periplasmic nitrate reductase in Azospirillum brasilense Sp245. Arch Microbiol 175:344–352

Van Dommelen A, Keijers V, Vanderleyden J et al (1998) (Methyl)ammonium transport in the nitrogen-fixing bacterium Azospirillum brasilense. J Bacteriol 180:2652–2659

Van Dommelen A, Keijers V, Wollebrants A et al (2003) Phenotypic changes resulting from distinct point mutations in the Azospirillum brasilense glnA gene, encoding glutamine synthetase. Appl Environ Microbiol 69:5699–5701

Van Dommelen A, Spaepen S, Vanderleyden J (2009) Identification of the glutamine synthetase adenylyltransferase of Azospirillum brasilense. Res Microbiol 160:333–339. https://doi.org/10.1016/j.resmic.2009.03.005

Van Heeswijk WC, Westerhoff HV, Boogerd FC (2013) Nitrogen assimilation in Escherichia coli: putting molecular data into a systems perspective. Microbiol Mol Biol Rev 77:628–695. https://doi.org/10.1128/MMBR.00025-13

Zhang YM, Burris RH, Roberts GP (1992) Cloning and characterization of the draT and draG genes from Azospirillum brasilense. J Bacteriol 174:7117–7124

Zou X, Zhu Y, Pohlmann EL et al (2008) Identification and functional characterization of NifA variants that are independent of GlnB activation in the photosynthetic bacterium Rhodospirillum rubrum. Microbiology (Reading) 154:2689–2699. https://doi.org/10.1099/mic.0.2008/019406-0

Chapter 10
Phytohormones, Plant Growth Regulators, and Other Related Metabolites Produced by *Azospirillum*

Daniela Torres, Verónica Mora, Sanja Ćavar Zeljković, Nuria De Diego, and Fabricio Dario Cassán

Abstract One of the most studied mechanisms proposed to explain plant growth promotion due to *Azospirillum* inoculation is the production of bacterial phytohormones, plant growth regulators, and other related metabolites. Nowadays, we know that some beneficial effects of inoculation with rhizobacteria, particularly *Azospirillum,* can be correlated with the production and metabolism of chemically defined compounds with capacity to modify plant physiology. These include plant growth regulators such as auxins, cytokinins, gibberellins, abscisic acid, and nitric oxide, as well as cell structural molecules such as lipopolysaccharides, exopolysaccharides, and flagellin. In this chapter, we highlight the analytical and biological methods used to identify and quantify the most relevant phytohormones, plant growth regulators, and other related molecules produced by *Azospirillum.*

Keywords Phytohormones · Plant growth regulators · Auxins · Cytokinins · Gibberellins · Nitric oxide

10.1 Introduction

Plant growth and development are processes defined by the interactions among their genomes, microbiomes (referred to as the second genome), and the environment. Consequently, they evolved a multitude of complex signaling systems to respond to external and internal cues to regulate these processes. Plant hormones or

D. Torres · V. Mora · F. D. Cassán (✉)
Instituto de Investigaciones Agrobiotecnológicas (INIAB-CONICET), Universidad Nacional de Río Cuarto, Río Cuarto, Córdoba, Argentina
e-mail: fcassan@exa.unrc.edu.ar

S. Ć. Zeljković · N. De Diego
Czech Agrifood Research Center (CARC), Olomouc, Czech Republic

Czech Advanced Technology and Research Institute (CATRIN), Palacký University Olomouc, Olomouc, Czech Republic

F. D. Cassán et al. (eds.), *The Definitive Handbook of Azospirillum*,
https://doi.org/10.1007/978-3-032-09222-9_10

phytohormones play critical regulatory roles in this process. Various molecules or molecule families have been discovered as those belonging to the auxins (AUXs), gibberellins (GAs), cytokinins (CKs), brassinosteroids (BRs), and strigolactones (SLs), as well as the abscisic acid (ABA), ethylene (ET), jasmonic acid (JA), salicylic acid (SA), and nitric oxide (NO), among others (Santner and Estelle 2009). These compounds control essentially the plant life cycle, cell identity and differentiation, and plant responses to the environment or stress conditions (Feussner and Wasternack 2002). The phytohormone activity in a biological system is determined by its availability, metabolism (i.e., biosynthesis and catabolism), distribution, and signal perception (receptors). Furthermore, an additional level of complexity in the regulation of developmental processes took place when interactions occur between phytohormones in the same biological system or between phytohormone-producer organisms (i.e., different biological systems) as in the case of plant-microbe holobionts. Besides the plants, many genera of fungi and bacteria are able to produce phytohormones. Endophytic fungi such as those of the genera *Penicillium*, *Trichoderma*, *Aspergillus*, *Cladosporium*, and *Purpureocillium* produce hormones such as JA, indole-3-acetic acid (IAA), CKs, GAs, ET, ABA, and SA (Baron and Rigobelo 2022; Badawy et al. 2021; Illescas et al. 2021). Bacteria of the genus *Azospirillum*, *Bacillus*, *Achromobacter*, *Acetobacter*, *Arthrobacter*, *Azotobacter*, *Burkholderia*, *Enterobacter*, *Halobacillus*, *Herbaspirillum*, *Micrococcus*, *Moraxella*, *Pseudomonas*, *Serratia*, *Staphylococcus*, *Xanthomonas*, *Klebsiella*, *Agrobacterium*, *Enterococcus*, *Enterobacter*, *Mesorhizobium*, *Rhizobium*, *Streptomyces*, and *Ensifer* (formerly *Sinorhizobium*) (Tsavkelova et al. 2006; Etesami and Glick 2020) are producers of phytohormones as AUXs, CKs, GAs, ET, ABA, and SA among others (Ping and Boland 2004). *Azospirillum* has been one of the most studied bacteria in the last 50 years due its capacity to promote plant growth (Cassán et al. 2020). One of the most commonly cited bacterial mechanisms used to explain plant growth promotion due to inoculation with *Azospirillum* is its capacity to produce active phytohormones and plant growth regulators (Table 10.1). Based on the available data, *Azospirillum* can be linked to the production of several active molecules as AUXs, but mainly IAA (Reynders and Vlassak 1979), GAs (Bottini et al. 1989), CKs (Tien et al. 1979), ET (Strzelczyk et al. 1994), ABA (Kolb and Martin 1985), NO (Creus et al. 2005), and several polyamines, such as cadaverine (Cad) (Cassán et al. 2009).

The aim of this chapter is to present the current knowledge on phytohormone production in bacteria belonging to genus *Azospirillum* as well as different extraction, purification, and quantification protocols for the most important groups of molecules produced by these bacteria.

10.1.1 Auxins

Auxin is a generic term representing a group of chemical compounds characterized by their ability to induce cell elongation in the subapical region of the stem and to reproduce the physiological effect of the most abundant, naturally occurring auxin

Table 10.1 Biologically active phytohormones and plant growth regulators produced by *Azospirillum* sp. and its hierarchy (i.e., related to the importance of its interaction with plants)

Class	Hierarchy	Molecules	References
Auxins	First[t]	IAA, PAA, IBA	Reynders and Vlassak (1979), Tien et al. (1979), Prinsen et al. (1993), Somers et al. (2005), and Mora et al. (2023)
Gibberellins	Fourth	GA_3, GA_1	Bottini et al. (1989) and Piccoli and Bottini (1994)
Cytokinins	Third	iP, iPr, Z, *t*-Zr	Horemans (1986), Esquivel-Cote et al. (2010), and Tien et al. (1979)
Ethylene	Fifth	Et	Strzelczyk et al. (1994) and Ribaudo et al. (2006)
ABA	Sixth	ABA	Kolb and Martin (1985) and Cohen et al. (2008)
Nitric oxide	Second	NO	Creus et al. (2005)
Polyamines	Seventh	Cad, Spm, Spd, Put	Thuler et al. (2003) and Cassán et al. (2009)

Modified from Cassán et al. (2014)
IAA indole-3-acetic acid, *PAA* phenylacetic acid, *IBA* indole-3-butyric acid, *NO* nitric oxide, *iP* isopentenyl adenine, *iPr* isopentenyl adenine riboside, *Z* zeatin, *t-Zr* trans-zeatin riboside, $GA_{3,1}$ gibberellins n, *ET* ethylene, *ABA* abscicic acid, *Cad* cadaverine, *Spm* spermine, *Spd* spermidine, *Put* putrescine

molecule, IAA. These compounds have been associated with different plant processes such as gravitropism and phototropism, vascular tissue differentiation, apical dominance, lateral and adventitious root initiation, together with CKs, cell division stimulation and cell elongation, and stem and root elongation (Teale et al. 2006). IAA is structurally related to the amino acid tryptophan (Trp), and the first studies on auxin biosynthesis hundreds of organisms proposed this compound as the main precursor in nature. Plants convert Trp to IAA by several pathways including indole-3-pyruvic acid (IPA), tryptamine (TAM), and indole-3-acetonitrile (IAN) pathway; however, there is an amino-acid-independent pathway from indole or indole-3-glycerol phosphate. The first reports of auxin biosynthesis in the genus *Azospirillum* were simultaneously published by Reynders and Vlassak (1979) and Tien et al. (1979). In the first paper, the authors reported the conversion of Trp to auxins in pure cultures of several strains of *A. brasilense* and confirmed the role of this amino acid as a precursor of IAA by thin-layer chromatography (TLC) and plant bioassay. In the second report, the authors reported the *A. brasilense* capacity to produce IAA from L-Trp and the increase of the hormone concentration at the stationary growth phase in liquid culture medium. The methodology used for the determination of auxins was TLC, high-pressure liquid chromatography (HPLC), and plant bioassays. At least six metabolic routes for IAA biosynthesis have been proposed in bacteria and, most of them use Trp as a precursor and as the IPyA, TAM, IAN, indole acetamide (IAM), and tryptophan side-chain oxidase (TSO) pathways, besides a tryptophan-independent pathway. Despite this diversity, prokaryotic IAA biosynthesis seems to follow predominantly two major routes: the IAM and the IPA (Lambrecht et al. 2000), but at least four different pathways have been described for *Azospirillum*, three Trp-dependent (via IPyA, IAM, and TAM) and one Trp-independent one (Cassán et al. 2014). *A. baldaniorum* Sp245 is one of the most studied strains of *Azospirillum* and is considered as a type strain for auxin biosynthesis world-wide. The IAA production by *A. baldaniorum* Sp245 has been extensively studied with the IPA pathway, the main route for IAA biosynthesis. All genes involved in this pathway were found in the genome sequence of Sp245, but additionally, nitrilase was also identified in the Sp245 genome sequence, which can catalyze the conversion of IAN to IAA. Studies on the biosynthesis of IAA have been extended to the complete metabolism of auxins in some strains of *Azospirillum*. Rivera et al. (2018) found that some amino acids, such as L-met, L-val, L-cys, and L-ser, inhibit bacterial growth and reduce IAA biosynthesis, while the expression of *ipdC* and IAA biosynthesis, but not bacterial growth, are affected by L-leu, L-phe, L-ala, L-ile, and L-pro. Furthermore, L-arg, L-glu, L-his, L-lys, L-asp, and L-thr do not affect bacterial growth, IAA biosynthesis, or *ipd*C gene expression; this fact should have some impact on the rhizosphere during plant-microbe interactions (Coniglio et al. 2022). It was also confirmed that the *A. brasilense* Cd, *A. baldaniorum* Sp245 (formerly *A. brasilense*), and *A. argentinense* Az39 (formerly *A. brasilense*) only produce IAA in the presence of L-Trp (biosynthesis); these strains are unable to degrade auxins (catabolism), conjugate IAA with sugars and/or L-amino acids (conjugation), or hydrolyze conjugates to release free IAA (hydrolysis). The IAA or IAA-conjugate content was quantified by HPLC. Under biotic and abiotic stress conditions, IAA biosynthesis by Sp245 and Az39 is affected. IAA production increases with daylight or in the presence of polyethylene glycol (PEG_{6000}), ABA, salicylic acid, chitosan, and *Fusarium oxysporum*. In contrast, exposure to 45°C or H_2O_2, NaCl, Na_2SO_4, 1-aminocyclopropane-1-carboxylic acid (ACC), methyl jasmonic acid (MeJA), or *Pseudomonas savastanoi* supernatants decreased IAA biosynthesis (Molina et al. 2018). The colorimetric technique of the Salkowski reaction (Pilet and Chollet 1970) was used and then confirmed by HPLC according to Jensen et al. (1995). Several studies have demonstrated that *Azospirillum* modifies the root architecture by promoting the development and elongation of lateral roots (LR) and root hairs (RH) and inhibiting the elongation of the primary root (PR). These morphological changes have been mostly related to the bacterial capacity to produce and release IAA (Prinsen et al. 1993; Mora et al. 2023).

10.1.2 Gibberellins

Gibberellins are a large family of tetracyclic diterpenoid plant hormones that regulate many aspects of both vegetative and reproductive growth in plants. These include seed germination, stem elongation, leaf expansion, transitions from meristematic to shoot growth, the shift from vegetative growth to flowering, promotion of fruit set and subsequent growth, and disease resistance in rice in response to different abiotic stresses (Seo et al. 2009; Yang et al. 2013; Colebrook et al. 2014). Currently, more than 130 GAs have been identified in plants, fungi, and bacteria; however, only a small number of GAs is biologically active in plants: GA_1, GA_3, GA_4, and GA_7. Many GAs are biosynthetic intermediates or catabolites of bioactive molecules. They are biosynthesized from geranylgeranyl diphosphate (GGDP), a common C-20 precursor for diterpenoid, and the plant biosynthesis can be divided into three steps according to the characteristics of the biosynthetic enzymes. In the first step, *ent*-kaurene, a tetracyclic hydrocarbon intermediate, is synthesized from GGDP by the *ent*-copalyl diphosphate synthase (CPS) and *ent*-kaurene synthase (KS) in proplastid. In the second step, *ent*-kaurene is converted to GA_{12} by two microsomal cytochrome P450-monooxygenase in the endoplasmic reticulum. The *ent*-kaurene oxidase (KO) catalyzes the sequential oxidation on C-19 to produce *ent*-kaurenoic acid, which is subsequently converted to GA_{12} by the *ent*-kaurenoic acid oxidase (KAO). The reactions in the final step involved the formation of C-20 and C-19 GAs in the cytoplasm. Thereby, GA_{12} is converted to GA_4 (13-non-hydroxylated pathway) through oxidations on C-20 and C-3 by GA 20-oxidase (GA20ox) and GA 3-oxidase (GA3ox), respectively. GA_{12} is also a substrate for GA13ox for the production of GA_{53}, which is a precursor to GA_1 in the 13-hydroxylated pathway. GA20ox catalyzes the sequential oxidation of C-20, including the loss of C-20 as CO_2 and the formation of γ-lactone; thereby, C19-GAs are produced. The introduction of a 3β-hydroxyl group in the inactive precursors results in the formation of the bioactive GAs. Bottini et al. (1989) were the first to confirm the ability of *Azospirillum* sp. to produce GAs in chemically defined culture medium. They reported the production of GA_1 and GA_3 in a free nitrogen culture of *A. lipoferum* by gas chromatography-mass spectrometry (GC-MS). Similar results were reported in *A. brasilense* by Janzen et al. (1992) and in *A. lipoferum* and *A. brasilense* by Esquivel-Cote et al. (2010). Additionally, the production of GA_{19} and GA_9 in a chemically defined medium of *A. lipoferum* was reported, suggesting the existence of different routes of synthesis in those bacteria. The first pathways presumably include the metabolism of GA_{19} (and its metabolite GA_{20}) to GA_1. On the other hand, GA_9 is (presumably) a precursor of GA_3. These pathways were confirmed in subsequent reports for *A. lipoferum* using minimal medium supplemented with deuterated precursors GA_{20} and GA_9 and subsequent identification as GA_1 and GA_3 by GC-MS. On the other hand, *A. lipoferum* can produce GA_{20} and GA_5, suggesting the existence of a second 13α-hydroxylation pathway for the metabolism of GA_{20} to GA_5 and GA_3 3ß-hydroxylation; however, this route has not been confirmed unequivocally. GAs production, metabolism, and hydrolysis of conjugates by *Azospirillum* sp. were summarized by Bottini et al. (2004). Perrig et al. (2007) evaluated the production of gibberellin in the agronomically used *A. brasilense* Cd and *A. argentinense* Az39 and found that both strains produced GA_3 in the supernatants of a chemically defined medium. However, GA_3 levels were significantly higher in the Cd strain. Gibberellic acid GA_3 production was analyzed by GC-MS.

10.1.3 Cytokinins

Cytokinins play an important role in all phases of plant development, from seed germination to senescence, such as maintenance of stem cell systems in shoots and roots, nodule organogenesis, leaf senescence, root vascular development, control of shoot branching in interaction with auxin, and in response to abiotic and biotic stress (Kouchi et al. 2010; O'Brien and Benková 2013). Structurally, CKs are adenine derivatives substituted in the N6 position, and they can be divided into two groups based on their side chain: those with isoprene-derived side chains, which are predominant in plants; and those with aromatic side chains that are found in plants at a lower abundance. In higher plants, the predominant CKs are isopentenyladenine (iP), trans-zeatin (tZ)-type, cis-zeatin (cZ)-type, or dihydrozeatin-type (dZ) derivatives, and they can exist as free bases and also in the form of nucleosides and nucleotides. CKs biosynthesis has progressed greatly due in large part to the identification of key pathways. A key enzyme in isoprenoid CKs biosynthesis of plants is adenosine phosphate-isopentenyl transferase (IPT), which catalyzes N-prenylation of the adenosine 5′-phosphates Adenosine Triphosphate (ATP), Adenosine Diphosphate (ADP) and Adenosine Monophosphate (AMP). The preferred donor of the isoprenoid side chain is dimethylallyl pyrophosphate (DMAPP), and the preferred acceptors are ATP and ADP. Therefore, the primary products of synthesis are either iP riboside 5′-triphosphate (iPRTP) or iP ribose 5′-diphosphate. The iP nucleotide is converted into tZ nucleotide by two cytochrome P450 monooxygenases. To become biologically active, iP and tZ nucleotides are converted to nucleobase form via dephosphorylation and deribosylation. An indirect pathway of CKs production involves the release of CKs by turnover of tRNA (Hwang and Sakakibara 2006; Miyawaki et al. 2006; Sugawara et al. 2008). Barea et al. (1976) found that at least 90% of the bacteria isolated from the rhizosphere of crops of agricultural interest were able to produce CK-type compounds in a defined culture medium. Horemans (1986) modified the analytical procedure and demonstrated that *A. brasilense* produces isopentenyl adenine (iP), isopentenyl adenine riboside (iPR), and zeatin (Z) in a chemically defined culture medium. A reference on CKs production by *Azospirillum* was published by Strzelczyk et al. (1994), using a culture medium supplemented with different sources of carbon. They reported the production of iP, iPR, trans-zeatin riboside (t-ZR), and Z in three strains of *Azospirillum* sp. isolatcd from sporocarps of the ectomycorrhizal fungi *Rhizopogon vinicolor*, *Laccaria laccata*, and *Hebeloma crustuliniforme*; however, they could only confirm iPR production in one of the three strains by gas chromatography (GC). An interesting case of synergism has been described for a mixed culture of *A. brasilense* and *Arthrobacter giacomello* showing a higher content of CKs, compared to those found in individual cultures of each microorganism (Cacciari et al. 1989). Tien et al. (1979) were the first to demonstrate the ability of *A. brasilense* to produce molecules of type-CK using different types of chromatography (HPLC and TLC) and a bioassay inoculation of pearl millet; however, the partially purified compounds were not characterized. It is reported that inoculation caused significant changes in the morphology of the root by increasing the number of lateral roots, and the root hairs density was similar to that obtained by the exogenous application of CKs. Similar results were found in *A. lipoferum* by Muralidhara and Rai (1986).

10.1.4 Abscisic Acid

Abscisic acid is an important phytohormone that plays many physiological roles in plants, including seed dormancy, stomatal movement, growth, and various environmental stresses (abiotic and biotic) (Jiang and Hartung 2008; Sreenivasulu et al. 2010). It is produced primarily in the vascular tissues of roots and leaves, in guard cells, and in seeds. There is an indirect pathway for ABA biosynthesis in higher plants, which consists of the cleavage of a C_{40} carotenoid precursor followed by a two-step conversion of the intermediate xanthoxin to ABA via ABA aldehyde (Taylor et al. 2000; Finkelstein et al. 2002; Seo and Koshiba 2002; Schwartz et al. 2003). The ABA biosynthesis pathway takes place in plastids, while the last two steps occur in the cytosol. ABA is an apocarotenoid synthesized via the oxidative cleavage of epoxy-carotenoids in plants (Zeevaart and Creelman 1988). Zeaxanthin and antheraxanthin are epoxidized to form violaxanthin and neoxanthin, a process catalyzed by zeaxanthin epoxidase (ZEP/AtABA1) (Marin et al. 1996). The products are isomerized to produce 9-cis isomers, which are cleaved by nine-*cis*-epoxycarotenoid dioxygenase (NCED) to form xanthoxin (Schwartz et al. 1997). Xanthoxin is then transported to the cytosol, where it is converted by the action of the enzyme ABA-aldehyde to ABA. Short-chain dehydrogenase/reductase (SDR) generates ABA aldehyde, and then aldehyde oxidase 3 (AAO3) catalyzes the last step in ABA biosynthesis (Xiong and Zhu 2003). In certain bacteria, ABA is synthesized as a C-15 terpenoid (sesquiterpene); however, nothing is known about the biochemistry of ABA (biosynthesis and metabolism) or about a possible function of this molecule in prokaryotes. There are only a few reports on ABA production by *Azospirillum* in a chemically defined culture medium and in inoculated plants. Kolb and Martin (1985) were the first to report on ABA production by *A. brasilense* in a defined culture medium. *A. argentinense* Az39 and *A. brasilense* have the capacity to produce ABA (75 and 6.5 $ng \cdot mL^{-1}$, respectively) in a chemically defined medium as identified by Perrig et al. (2007) by gas chromatography-mass spectrometry. ABA was also characterized by gas chromatography with electron impact mass spectrometry (GC-EI-MS) in the supernatant of the model strain *A. baldaniorum* Sp245 from the chemically defined media by Cohen et al. (2008, 2015).

10.1.5 Salicylic Acid

Salicylic acid is a phenolic compound produced by a wide range of prokaryotic and eukaryotic organisms. Indeed, SA has been found to play a key role in various physiological and biochemical activities of plants such as thermogenesis, photosynthesis, stomatal closure, seed germination, plant growth (Abreu and Munné-Bosch 2009), response to local and systemic against pathogens, and response to abiotic stresses such as drought, chilling, heavy metal toxicity, heat, and osmotic stress (Ng et al. 2011). SA biosynthesis in bacteria is associated with the function of the siderophore and antibacterial agents (Grobelak and Hiller 2017; Maruri-López et al. 2019; Kaduskar et al. 2017). For SA biosynthesis, both plants and bacteria require chorismate as an initial precursor (Mishra and Baek 2021). SA is synthesized in plants through two distinct enzymatic pathways: the phenylalanine ammonia lyase (PAL)-mediated phenylalanine pathway in the cytoplasm and the isochorismate synthase (ICS)-mediated isochorismate pathway in the chloroplast. Despite having two pathways, the main pathway in plants is the ICS, contributing to more than 90% of SA synthesis (Chen et al. 2009; Rekhter et al. 2019). Both of these pathways originate in plastids from chorismate, which is an intermediate of plant phenylpropanoid pathway. The common and initial enzymatic step of PAL pathway is the conversion of chorismate-derived L-phenylalanine to *trans*-cinnamic acid by phenylalanine ammonia-lyase. Subsequently, *trans*-cinnamic acid is hydroxylated to form *O*-coumarate followed by oxidation of the side chain to yield SA. Alternatively, the side chain of *trans*-cinnamic acid can be initially oxidized to give benzoic acid, which is then hydroxylated to produce SA. Thus, the difference between the two routes is the hydroxylation of the aromatic ring before or after the chain-shortening reactions. Several genera of bacteria such as *Pseudomonas aeruginosa* and *P. fluorescens* can synthesize SA from chorismate via two reactions catalyzed by isochorismate synthase (ICS) and isochorismate pyruvate lyase (IPL). ICS catalyzes the synthesis of isochorismate from chorismate, and IPL catalyzes the conversion of SA from isochorismate (Mercado-Blanco et al. 2001). No gene encoding IPL has been cloned from plant species; in plants, the conversion from IC to SA may be mediated by a route different from that characterized in bacteria. By contrast, SA synthesis in *Yersinia enterocolitica* and *Mycobacterium tuberculosis* is mediated by a single, bifunctional enzyme SA synthase (SAS) that directly converts chorismate to SA via an isochorismate intermediate (Harrison et al. 2006). One of the first reports of SA synthesis in the genus *Azospirillum* was by Saxena et al. (1986), in which they identified SA in *A. lipoferum* using the TLC technique on silica gel. Another report of the presence of SA in *Azospirillum* cultures was published by Shah et al. (1992) which is also by silica TLS. Note that these methodologies were not fully suitable for the determination of SA. Tortora et al. (2011) confirmed SA production by *A. brasilense* REC2 and REC3 under iron-starved conditions, and isolated this compound from other catechols using silica TLC coupled with fluorescence spectroscopy and GC-MS analysis. The siderophores produced by them showed in vitro antifungal activity against *Colleotrhicum acutatum* M11. This fact was coincident with the results obtained from phytopathological tests performed *in planta*, where a reduction of anthracnose symptoms on strawberry plants previously inoculated with *A. brasilense* was observed (Tortora et al. 2011).

10.1.6 Ethylene

Ethylene is an important phytohormone related to plant growth and development (Burg 1962). Because of its gaseous state under physiological conditions, for a long time, it was not considered a phytohormone, but various studies showed that its synthesis and action are critical for certain physiological processes such as leaf abscission and senescence, seed germination, response to stress, rhizobia nodule formation, root initiation, elongation, and branching, flowering and flower wilting, fruit ripening, as well as a plant's interaction with beneficial mycorrhizal fungi (Glick 2014). ET is a simple and symmetrical molecule composed of two carbon atoms (joined by a double bond) and four H atoms. ET can exert its physiological effects at very low concentrations in plant tissues (almost 0.1 ppm). In higher plants, all tissues have the capacity to synthesize this hormone, but in general, ET concentration is associated with the growth state and developmental phase of the plant, with a higher concentration in those tissues involved in active cell division, which are under stressful conditions or in a senescence stage (Burg and Burg 1968). There is little published information on ET production by plant growth-promoting rhizobacteria (PGPR) and its effect on plant growth. The ability of the free-living bacteria *Azotobacter* sp. and *Bacillus* sp. to produce ET in chemically defined medium was assessed in the early days (Primrose and Dilworth 1976). Later, it was shown that *Azospirillum* sp. produces ET in culture media with malate, succinate, or pyruvate as a carbon source, reaching a maximal production of 0.17 μmol g^{-1} dry weight (DW) when L-methionine is added. ET determination was performed by chromatographic analysis (Strzelczyk et al. 1994). Ribaudo et al. (2006) showed that inoculation of tomato seedlings with *A. brasilense* promotes plant development by increasing shoot and root biomass, shoot height, and total root surface area. Similar results were previously reported by Bashan (1998). For the evaluation of ET produced by *A. brasilense*, 0.1 mL of nitrogen-free medium (NFb) broth culture (10^7 CFU·mL^{-1}) is aseptically injected into 5 mL of semisolid NFb medium contained in a 10 mL flask. After sealing, the flasks were incubated in the dark for 24 h at 30 °C; 1 mL of headspace gas was withdrawn from all samples and ET was quantified on a gas chromatograph equipped with a flame ionization detector (FID). Not only can bacterial production of ET modify plant growth and hormonal status, other hormones (e.g., IAA) or bacterial enzymes (e.g., ACC deaminase) can alter the level of ET in colonized plants. Tomato seedlings inoculated with the auxin-producing *A. brasilense* showed a positive correlation between IAA concentration, number and length of roots, and plant ET production (up to ten times higher than the controls due to the increase in ACC synthase activity) (Ribaudo et al. 2006). This indicates that the plant ET increase is at least partly due to crosstalk between the bacterium-produced IAA and the plant ET biosynthesis, as previously proposed by Rahman et al. (2002). The IAA and ET content in plants increased strongly in response to inoculation with *A. brasilense* (Ribaudo et al. 2006). *Azospirillum* is an IAA producer, so the measured hormone level will be the contribution of unknown proportions from the bacteria and the plant. An increase in bacterial hormone synthesis has important consequences on plant physiology. The IAA can activate biochemical signals leading to enhanced plant growth and, on the other hand, IAA can stimulate ET production through the activation of 1-aminocyclopropane-1-carboxylate synthase (ACS) (Abeles 1992; Kende 1993). An increase in ACS expression could be detected in tomato shoots, strongly suggesting that ACS expression and/or activity are modulated in the process (Ribaudo et al. 2006). In addition, experiments performed with 1-methylcyclopropene (1-MCP), an ETe antagonist, demonstrated that ET is a positive regulator of root hair development and, consequently, of the increase in root surface area (Ribaudo et al. 2006). Inoculation with *A. brasilense* or *A. argentinense* results in increased root hair density (Ribaudo et al. 2006; Mora et al. 2023) and shoot development in plants, which is correlated with an increase in ET levels (Ribaudo et al. 2006). Inoculation with *A. brasilense* provides a concentration of ET sufficient to promote root hair development but not to inhibit main root development. This hypothesis would explain why, although *Azospirillum* lacks ACC deaminase activity, it behaves as a PGPR. However, some studies have shown a decrease or inhibition of main root development in plants inoculated with *Azospirillum* as those published by Mora et al. (2023). Plant-*Azospirillum* interaction studies provide strong evidence that ET is an intermediary in a signaling pathway that leads to improved root hair density in inoculated tomato plants. However, further studies are needed to obtain information on the biochemical mechanisms involved in the interaction.

10.1.7 Nitric Oxide

Nitric oxide is a volatile, lipophilic-free radical that participates in metabolic, signaling, defense, and developmental pathways in plants (Lamattina and Polacco 2007). NO biosynthesis in plants can occur in two pathways: oxidative and reductive (Gupta et al. 2011; Kolbert et al. 2019). First, a nitrous oxide synthase (NOS)-like enzyme together with several cofactors converts L-arginine into L-citrulline and NO. Although several studies have proposed the occurrence of oxidative NO production, only reductive routes for NO production have been evidenced to date in land plants, as no canonical NOS in plants were encountered (Kolbert et al. 2019). In the reductive synthesis pathway, there are four different ways to synthesize NO: (1) The enzyme xanthine oxidoreductase (XOR) catalyzes the reduction of nitrite to NO using NADH or xanthine as a reducing substrate. (2) The enzyme nitrate reductase (NR) catalyzes the reduction of nitrite (NO_2) to NO. The activity of this enzyme is regulated by cytoplasmic nitrate (NO_3), which competitively inhibits the reduction of nitrite under aerobic conditions. However, when the oxygen concentration and pH are low, nitrite reductase (NIR) activity is inhibited, resulting in increased nitrite concentration and subsequent NO production. (3) Under hypoxic conditions, the generated NO_2 is transported to mitochondria via a putative nitrite transporter and nitrite reduction to NO takes place in complexes III and IV, and possibly in AOX (alternative oxidase). (4) By the combined action of two enzymes, the plasma membrane-bound NR enzyme (PM-NR) reduces NO_3 to NO_2, then plasma membrane-bound nitrite-NO reductase (PM-NiNOR) reduces NO_2 to NO (Gupta et al. 2022). NO plays a major role in the IAA signaling pathway, and its participation leads to lateral and adventitious root formation, wherein NO acts as an intermediate in IAA-induced root development (Correa-Aragunde et al. 2006). *A. baldaniorum* can produce NO in vitro under aerobic culture conditions. This was detected by electron paramagnetic resonance (EPR) and confirmed by the NO-specific fluorescent probe 4,5-diaminofluorescein diacetate (DAF-2 DA) (Creus et al. 2005). In this bacterium, NO can probably be synthesized via multiple pathways such as aerobic denitrification and heterotrophic nitrification. NO is produced during the middle and late logarithmic phases of growth (Molina-Favero et al. 2007, 2008). NO production in *A. baldaniorum* induces morphological changes in lateral and adventitious tomato roots regardless of the full bacterial capacity to synthesize IAA. An IAA-attenuated mutant of this strain induces the same physiological changes with slightly less effect on root development as observed for the wild-type strain (Molina-Favero et al. 2008). Arruebarrena Di Palma et al. (2013) carried out mutational studies to establish that NO is the signaling molecule influencing biofilm formation in *A. baldaniorum* Sp245 in a dose-dependent manner. The interplay between IAA and NO production in *A. brasilense* SM was also explored (Koul et al. 2015). Production of NO and the presence of NO metabolism genes, nitrous oxide reductase (*nosZ*), nitrous oxide reductase regulator (*nosR*), and nitric oxide reductase (*norB*), were identified in this strain. Quantitative IAA estimation highlighted a crucial role of nitrous oxide reductase regulator (*nosR*) in regulating IAA biosynthesis. By using the NO quencher and donor, they showed reduced IAA biosynthesis, indicating their common regulatory role in IAA biosynthesis. Tryptophan, substrate for IAA biosynthesis, showed higher expression of NO metabolism genes, while in the case of *ipd*C, only Trp and IAA increased expression, while arginine had no significant effect (Koul et al. 2015). These results indicate a strong correlation between IAA and NO in *A. brasilense* SM and suggest the existence of crosstalk or shared signaling mechanisms in these two growth regulators, in bacteria as well as it occurs in plants.

10.1.8 Polyamines

Polyamines are low-molecular-weight organic compounds with two or more primary amino groups. They are ubiquitous in plant, animal, and microbial cells (Davies and Davies 1995). Polyamines serve as growth-regulating compounds in plants and, like phytohormones, when they are present at the appropriate level, they display biological activity in processes such as plant growth, development, and stress mitigation (Kuznetsov et al. 2006). In this sense, one of the best documented molecules is diamine cadaverine (1,5-diaminopentane), which has been correlated with root growth promotion in pine and soybean (Niemi et al. 2002), response to osmotic stress in turnip (Aziz et al. 1997), and control of stomatal activity in *Vicia faba* beans

(Liu et al. 2000). The production of putrescine (Put), spermidine (Spd), and spermine (Spm) in the chemically defined medium of *Azospirillum* isolated from manioc roots was reported by Thuler et al. (2003). The production of cadaverine (Cad) and other polyamines was first reported by Hamana et al. (1988, 1990) for a wide group of α-proteobacteria belonging to the order Rhizobiales, but later proposed for *A. argentinense* by Cassán et al. (2009). The diamine Cad has been correlated with root growth promotion (Gamarnik and Frydman 1991; Niemi et al. 2002) and osmotic stress mitigation (Aziz et al. 1997; Liu et al. 2000) in plants. *A. argentinense* has the capability to produce Cad in a chemically defined medium modified by addition of L-ornithine as precursor, and as an endophytic state in rice seedlings. Cadaverine was analyzed by reverse-phase HPLC as described in Pieckenstain et al. (2001). This capacity has been correlated with root growth promotion and osmotic stress mitigation under hydroponic conditions (Cassán et al. 2009).

The following section describes the analytical and biological standardized methods used for the extraction, purification, identification, and quantification of the most important phytohormones, plant growth regulators, and other related molecules produced by *Azospirillum*.

10.2 Methods

10.2.1 Phytohormones Extraction and Purification

10.2.1.1 Samples Assessment for Simultaneous Determinations

The choice of extraction solvent and conditions plays a crucial role in determining which analytes are effectively extracted from a sample matrix. Currently, no analytical method provides a universal solution, and each introduces a degree of bias. Phytohormones and other plant growth-related compounds have diverse physicochemical properties, stability, and volatility, making extraction particularly challenging. Common practice involves using aqueous mixtures with high concentrations of organic solvents such as methanol (MeOH), acetonitrile (ACN), and isopropanol (IPA). At appropriate concentrations, organic solvents can also reduce enzymatic activity, thereby minimizing chemical changes during extraction. Proper sampling and quenching the metabolism are critical to prevent undesirable chemical changes, while factors such as mechanical stress and circadian rhythms can significantly influence levels of analytes of interest. Thus, careful planning and precise sampling are essential steps in trace analysis (Vrobel and Tarkowski 2023). The most commonly used solvents for extraction include modified Bieleski solvent (15:4:1; MeOH:H_2O:formic acid) and aqueous mixtures of MeOH or ACN. Originally developed to prevent phosphatase activity during CK extraction, the modified Bieleski solvent omits chloroform and provides comparable or superior extraction efficiency. In addition to CKs, it has been successfully used for extracting other hormone classes such as ABAs, JAs, SAs, GAs, AUXs, and 1-aminocyclopropane-1-carboxylic acid (ACC) (Kojima et al. 2009; Wiszniewska et al. 2019). MeOH and ACN mixtures are suitable for milder extraction conditions, allowing for the recovery of analytes with a wide range of polarities. For specific hormone classes, alternative solvents are often required to accommodate the unique stability and chemical properties of certain compounds. Auxin analysis, for example, is complicated by the instability of biosynthetic precursors, which can degrade during extraction and lead to overestimation of IAA content. Strigolactones also present challenges due to their limited stability, which is highly dependent on pH, the presence of nucleophiles, and the choice of solvent (Halouzka et al. 2020). Therefore, selecting appropriate internal standards and minimizing enzymatic activity, hydrolysis, or other induced changes during extraction is essential to ensure reliable results and high signal intensity. Next crucial step is sample purification, primarily aimed at removing matrix components while retaining analytes to generate cleaner samples and reduce matrix effects (ME). Although recent reviews have comprehensively covered this topic, it is worth noting that solid-phase extraction (SPE) and liquid-liquid extraction (LLE), along with their miniaturized variants, are the most widely used methods. Typically, purification exploits common features of analytes, such as polarity or basicity, to separate them from matrix components. For instance, lipophilic analytes can be retained on a C_{18} or hydrophilic-lipophilic balance (HLB) sorbent during SPE, or compounds may be transferred from a polar solvent to a nonpolar solvent during LLE, thereby removing polar and mid-polar compounds from the sample. CKs purification leverages their basicity, with ion-exchange or mixed-mode SPE achieving high purification efficiency by removing neutral, acidic, polar, and nonpolar compounds. Analogous methods can be applied to acidic hormones such as ABA, JA, GAs, and AUXs (Liu et al. 2019; Wang et al. 2020). In recent years, there has been a clear trend toward abandoning labor-intensive purification protocols in favor of faster, higher-throughput methods that include a broader range of analytes. These simplified approaches reduce preparation time and increase efficiency but often involve less specific purification steps. While plant metabolomics aim to minimize purification by excluding matrix components based on their differing chemical properties, the diversity of analytes poses a challenge. Some analytes may be lost during purification, forcing researchers to prioritize either simple purification that risks undetectable signals or more extensive purification that reduces matrix effects but may remove key analytes (Vrobel and Tarkowski 2023).

10.2.1.1.1 For Simultaneous Determination of ABA, IAA, and SA

This methodology is described by Durgbanshi et al. (2005). For that, centrifuge 20 mL of defined bacterial cultures, in exponential growth phase, separate at 8000 rpm, 4 °C, for 15 min and add 50 ng (5 μl) of the corresponding deuterated standard (ABA, SA, GA_3, JA, and IAA). Acidify the supernatants at pH 2.5–2.8 with acetic acid solution (15% v/v), and extract twice with 20 mL of ethyl ether.

Evaporate ethyl ether to dryness at 36 °C and resuspend dried samples in 1500 μL methanol. Filter through a 0.22 μm cellulose acetate filter and dry methanol in a speed vacuum. Resuspend in 5 μL methanol (100%) and injected into the analytical system.

Note: This methodology can be used for simultaneous determination of ABA, IAA, and SA from bacterial culture medium or inoculated plant tissues. In the second case, approximately 200 mg of dry weight plant material must be ground in a mortar with liquid nitrogen, then add 5 mL ultra-pure water to the ground material.

10.2.1.1.2 For Simultaneous Determination of IAA, ABA, GA_3, and Z

This methodology is described by Perrig et al. (2007). For that, centrifuge 20 mL of defined bacterial cultures in the exponential growth phase, separated at 8000 rpm, 4 °C, for 15 min. Add 100 ng of the corresponding deuterated standard (ABA, SA, GA_3, JA, and IAA) and acidify the supernatants at pH 2.5–2.8 with acetic acid solution (15%, v/v). Keep the sample at 4 °C for 2 h and partition four times with the same volume of acetic-acid-saturated ethyl acetate (1%, v/v). Evaporate ethyl ether to dryness at 36 °C, and dried samples are diluted in a 100 μL acetic acid/methanol/water solution (1:30:70, v/v) for ABA determination, acetic acid/acetonitrile/water solution (1:15:85, v/v) for IAA determination, and a methanol/water solution (30:70, v/v) for GA_3. Inject into a reverse-phase octadecyl carbon chain (C_{18}) column (300 × 3.9 mm) in an HPLC system coupled to a UV-Vis diode-array spectrometer.

Note: This methodology can be used for simultaneous determination of IAA, ABA, GA_3, and Z from bacterial culture medium or inoculated plant tissues. In the second case, approximately 200 mg of dry weight plant material must be ground in a mortar with liquid nitrogen, then add 5 mL ultra-pure water to the ground material.

10.2.1.2 Samples Assessment for Simple Determination

10.2.1.2.1 For ABA Determination by Minicolumn Amino Anion Exchange (NH_2)

This methodology is described by Luna et al. (1992). For that, homogenize the lyophilized culture medium (approximately 0.2 g dry weight, DW) with 20 mL imidazole buffer (pH 7.0) plus 2,6-di-tert-butyl-p-cresol as an antioxidant and add 50 ng (5 μL) of the corresponding deuterated standard ABA. Incubate each sample overnight at 4 °C to allow extraction and standard equilibration. Centrifuge the sample for 15 min at 1530 *g*, then collect the supernatant and evaporate it under vacuum until eliminating the isopropanol. An amino anion exchange minicolumn (NH_2) or similar is conditioned with 100% methanol, deionized water, and imidazole buffer 0.02 M. The aqueous fraction is loaded onto the minicolumn. Wash sequentially the minicolumn with 6 mL each of hexane, ethyl acetate, and acetonitrile and discard these fractions. ABA is eluted with 6 mL of methanol:acetic acid (95:5 v/v), and the eluate fraction is evaporated to dryness under vacuum in a rotary evaporator. For further ABA identification and quantification, inject 15 μL into high-performance liquid chromatography coupled to mass spectrometers (LC-MS/MS).

Note: This methodology can be used for determination of ABA from bacterial culture medium or inoculated plant tissues. In the second case, approximately 200 mg of dry weight plant material must be ground in a mortar with liquid nitrogen and add 20 mL imidazole buffer (pH 7.0) plus 2,6-di-tert-butyl-p-cresol as antioxidant to the ground.

10.2.1.2.2 Direct Use of the Bacterial Culture for IAA Determination

Centrifuge 1.5 mL of bacterial culture at 10,000 rpm for 15 min at room temperature and filter the supernatant using 0.22 μm cellulose acetate filters. Keep the sample in dark conditions at 4 °C and use the clarified supernatant for the next steps.

Note: This methodology can be used for the determination of IAA and other related auxins from bacterial culture medium.

10.2.2 Phytohormone Identification and Quantification

10.2.2.1 Biological Methods

The biological activity of microbial phytohormones depends on their concentrations in culture supernatants and the microbial ability to produce these molecules (Cassán et al. 2009). Phytohormones may influence early seed germination, early seedling growth, plant colonization, and bacterial establishment. The methodology proposed in this section allows you to estimate the bacterial ability to produce GAs, AUXs, and/or CKs in inoculated plants or in chemically defined culture medium.

10.2.2.1.1 Quantification of Biologically Active Gibberellins

A sensitive, easy, and fast bioassay is described by Cassán et al. (2009) for detection and quantification of active GAs from bacterial solutions. For that, maize (*Zea mays* L.) or rice (*Oryza sativa* L.) seeds are surface disinfected by soaking for 3 min in 1% NaClO and then washed with sterile distilled water to eliminate traces of this compound. Seeds are pregerminated for 48 h at 30 °C and 80 rpm shaking in an Erlenmeyer flask containing sterile 80 μM uniconazole [S-3307, (±) (E)-1-(4-clorophenyl)-4,4-dimethyl-2-(1,2,4-triazol-1-yl)-1-penten-3-ol]. Almost five seeds are planted at 1 cm depth in individual pots, each one containing 200 mL of sterile perlite-sand (1:1) mixture. After full emergence, seedlings are thinned to three plants per pot. Plant pots are kept at field capacity by daily irrigation with sterile distilled water, and once a week fertilized with 50% of Hoagland´s solution

(Hoagland and Boyer 1936). Plants (three plants per pot, five replicates per treatment) are cultured in a growth chamber with daily cycles of 16 h at 25–30 °C with a light intensity of 450 μmol·m^{-2}·s^{-1} and 8 h at 20 °C in dark for 7 days. Plants are grouped in the following treatments—(a) cell inoculated: each seedling is inoculated with 1 mL of bacterial culture medium with up to 10^9 CFU·mL^{-1}; (b) supernatant inoculated: each seedling is inoculated with 1 mL of supernatant of culture medium (10,000 rpm × 10 min at room temperature); (c) GA_3 treated: GA_3 is dissolved in pure ethanol to obtain 0.1, 0.5, 1.0, 5.0, and 10.0 μg·mL^{-1} solutions, and then 10 μL are applied individually at the first leaf with a microsyringe for every seedling for the same treatment; and (d) control: seedlings individually treated with pure ethanol. At the end of the experiment, the plants are carefully pulled out from the pots and rinsed for 5 s in the potassium phosphate buffer. The following parameters are measured as a plant basis: first internodes and total shoot length.

Note: There are some *dwarf* mutants deficient in the production of physiologically active gibberellins, like the *dwarf-1* (*d1*) described in maize (*Zea mays* L.) by Fujioka et al. (1988) and the *dwarf-x* (dx or cv. Tan-ginbozu) described in rice (*Oryza sativa* L.) by Kobayashi et al. (1989). These mutants express the dwarf phenotype from the young seedling stage, but the exogenous application of GA_3 allows expression of the normal phenotype. The use of these mutants is strongly recommended.

10.2.2.1.2 Quantification of Biologically Active Auxins

A sensitive, easy, and fast bioassay is described by Epel et al. (1987) for the detection and quantification of active auxins in bacterial cultures. The bioassay consists of measuring the weight increase of hypocotyl sections from etiolated cucumber seedlings incubated in a simple growth medium. The sensitivity of the test is high with a significant response at an IAA concentration as low as 1 ng·mL^{-1}. The bioassay requires at least 3 h to evaluate and is easy to perform. Some advantages of this methodology are listed below: (1) the slope of the log-linear concentration-response curve is moderately steep; (2) the assay is insensitive to the pH of the media between 4.0 and 7.0 and can be performed in the presence or absence of buffer; and (3) the growth of the etiolated hypocotyl sections is insensitive to GA and to kinetin. Cucumber seeds (*Cucumis sativus* L.) are germinated and grown in vermiculite substrate under dark and watery irrigation conditions at 25 °C. When the etiolated seedlings were 5–8 cm in length (generally 4–5 days, depending on the freshness of seeds), hypocotyl sections 10 mm in length were excised from the hypocotyl, starting about 2 mm below the hook, unless stated otherwise. The excised sections are immediately placed in tap water until sufficient sections have been cut. The sections are then washed in running tap water for about 30 min. The seedlings should be protected from direct light during this procedure. Washed sections, gently blotted with filter paper, are weighed in groups of 10–12 with a tared electronic balance with mg precision (typical weight, about 250 mg) and placed in glass scintillation vials containing 3 mL standard growth medium (pH 7.0), containing

2 mM KCl, 0.1 mM CaCl, and 10 mg·L^{-1} chloramphenicol. IAA solutions are added individually into scintillation vials to produce a typical dose-response calibration curve. These solutions should contain a concentration of pure IAA between 0.1 and 100 μg·mL^{-1}. The bacterial supernatants are also added in specific vials as problem treatment. After the treatment, the covered vials are gently shaken (80 rpm·min^{-1}) at 30 °C between 3 and 12 h (linearity increases with the time of exposition) and sections are removed, blotted, and weighed in groups. Each point should represent three replicates of 10–12 sections each. The growth response of IAA solutions is expressed as a percentage of increase in weight (%) versus log μg·mL^{-1} IAA. The supernatant IAA concentration calculation is derived from that dose-response curve.

10.2.2.1.3 Quantification of Biologically Active Cytokinins

The cucumber cotyledon greening bioassay reported by Fletcher et al. (1982) with modifications is used for detecting CKs from several natural sources, including bacterial cultures. The sensitivity of the methodology is extremely low and useful to detect a minimum of 0.0001 mg·L^{-1} of N′-benzyladenine, Z, kinetin, or Z riboside in solution. Of the more frequent CKs tested, kinetin appeared to be the least active. Cucumber seeds are planted in vermiculite, in peat flats (20 × 15 × 7 cm) and germinated in the dark at 28 °C. The cotyledons from 5-d old plants are excised in absence of green light, making certain that the hypocotyl hook is removed. The cotyledons are placed in 5-cm Petri dishes containing 3 mL of test solution, which consists of distilled H_2O, CKs at various concentrations to generate a dose-response curve or bacterial supernatant, and 40 mM KCl. The dishes are returned to the dark at 28 °C for 20 h. After the incubation period, they are exposed to fluorescent light with an intensity of 12.9 w·m^2. After 3.5 h, cotyledons are homogenized, and the chlorophyll is extracted in an 8 mL 80% acetone solution. The volume is brought up to 10 mL with acetone and then centrifuged at 2500 *g* for 10 min. The absorbance of the supernatants are read at 663 (chlorophyll A) and 645 nm (chlorophyll B) to obtain the partial and total concentrations according to Arnon (1949).

10.2.2.1.4 Evaluation of *Azospirillum*'s Phytohormone Production *in Planta*

Using Arabidopsis *as a Model Plant*

To check the effects of *Azospirillum* inoculation on *Arabidopsis* root morphology, *Arabidopsis thaliana* (L.) ecotype Columbia 0 (Col-0, wild type) are used. *Arabidopsis* seeds are surface sterilized with 95% (v/v) ethanol (5 min), followed by an aqueous 2.5% (v/v) sodium hypochlorite solution containing 0.1% (v/v) Triton X-100 (5 min). After being washed four times with sterile distilled water, the seeds are stratified in darkness at 4 °C in plastic tubes containing sterile distilled water. Two days later, the seeds are sown in Petri dishes containing Murashige and Skoog (MS) medium (Murashige and Skoog 1962) supplemented with 0.8% (w/v) agar and

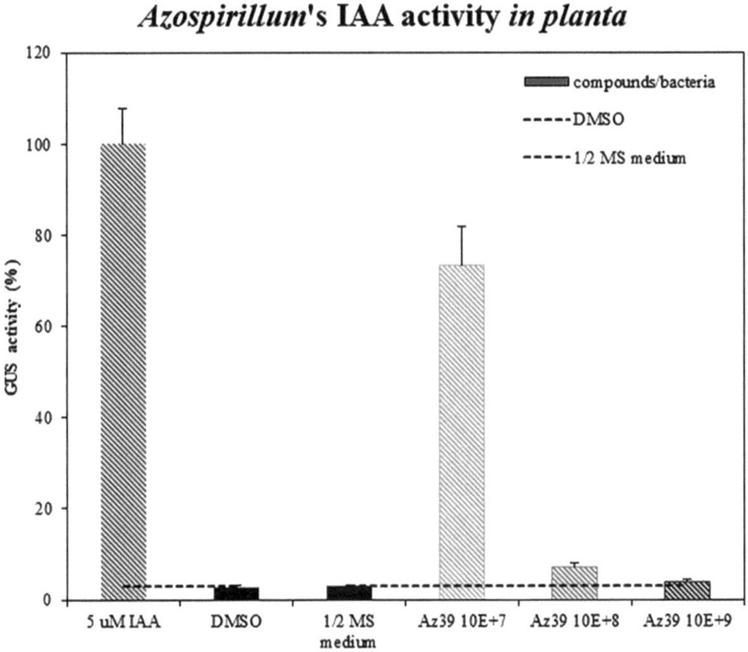

Fig. 10.1 Expression of DR5::GUS in *A. thaliana* Col-0 seedlings inoculated with different titers of *A. argentinense* Az39. Seedlings were exposed for 17 h to 10^7 to 10^9 CFU·mL^{-1} of this strain. To obtain a positive control, seedlings were exposed for the same time to 5 μM IAA dissolved in 0.1% acetic acid (ac). Seedlings exposed to Murashige and Skoog (MS) medium with or without 0.1% (w/v) Dimethyl Sulfoxide (DMSO) were used as negative controls

a final pH of 5.7. Subsequently, plates are sealed with micropore tape, and seeds are allowed to germinate and grow vertically for 7 days in a growth chamber at 22 °C, under a 16–8-h light-dark photoperiod (intensity of 60 μmol·m^{-1}·s^{-1}). Seven days after sowing, five *Arabidopsis* seedlings per plate are aseptically transferred to new Petri dishes containing MS medium supplemented with 0.8% (w/v) agar and subjected to the interest treatments (in different sets of experiments): inoculation between 10^3 and 10^8 CFU·mL^{-1} of *Azospirillum* and addition of 0.01–25.0 μg·mL^{-1} IAA. In this example, we are focused to evaluate the capacity of *Azospirillum* to produce IAA and its effects on *Arabidopsis* root morphology (Fig. 10.1). For inoculation procedure, bacteria from the late exponential growth phase in Minimal Medium for Azospirillum brasilense (MMAB) medium are obtained through centrifugation (3900× *g*, 12 min, 16 °C), washed with 0.85% (w/v) sterile saline, and resuspended in 0.01 M $MgSO_4$. Subsequently, variable volumes (0.02–2.00 mL) of bacterial suspensions (about 10^9 CFU·mL^{-1}) are mixed with variable volumes (19.98–18.00 mL) of molten MS medium (below 40 °C) to ensure a final titer between 10^3 and 10^8 CFU·mL^{-1}, homogeneously distributed in the growing medium. In the case of IAA, dissolved in MS medium to reach final concentrations of 0.01; 0.1; 1.0, 5.0; 10.0; 15.0, and 25.0 μg IAA mL^{-1}.

Using Arabidopsis *Reporter Lines*

Using *Arabidopsis* reporter lines is a simple and speedy technique to assess the phytohormone production and in planta activity. An example is the use of *Arabidopsis* DR5::GUS reporter line to evaluate the capacity of *Azospirillum* to produce IAA and the capacity to activate the receptors. In the assay, different concentrations of bacteria are compared with the supplementation of the phytohormone (Fig. 10.2). The procedure is as follows. The seeds of the *Arabidopsis* reporter lines are placed separately in 96-well microplates containing half-strength MS medium and stratified at 4 °C in the darkness for 4 days. Subsequently, the microplates are placed into a phytotron under controlled conditions (75 rpm, 21 °C, 16/8 h light/dark, light intensity 100 μmol·m^{-2}·s^{-1}) for 7 days, after which 2 μL of the *Azospirillum* suspension containing different CFU·mL^{-1} is added, in this case from 10^7 to 10^9. An equivalent volume of a pure phytohormone solution (5 μM IAA, dissolved in 0.1% acetic acid) is added to a different set of *A. thaliana* seedlings to obtain a positive control. Finally, a negative control is achieved by applying a half-strength MS medium with 0.1% acetic acid in a third set of seedlings. After 17 h of incubation, the medium is discarded, and the seedlings are directly treated with 150 μL lysis buffer (50 mM sodium phosphate, pH 7.0, 10 mM EDTA, 0.1% Triton X-100) containing 1 mM of 4-methylumbelliferyl glucuronide (4-MUG), and incubated at 37 °C for 90 min. At the end of the incubation period, 50 μL of 1 M Na_2CO_3 (stop solution) is added to each well, and the fluorescence due to 4-methylumbelliferone (4-MU) formation is measured in a microplate reader device (excitation/emission wavelengths 365/460 nm).

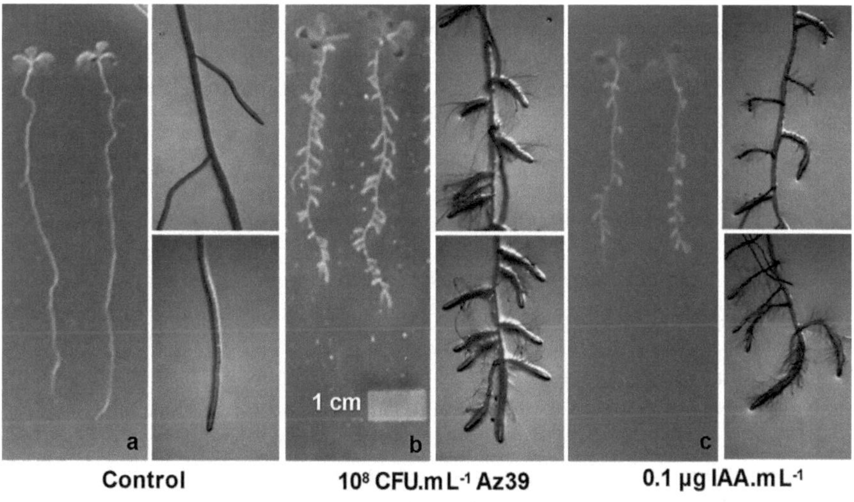

Fig. 10.2 *Arabidopsis thaliana* (L.) ecotype Columbia 0 (Col-0, wild type) corresponding to (**a**) untreated control seedlings; (**b**) seedlings inoculated with 10^8 CFU·mL^{-1} of *A. argentinense* Az39; and (**c**) seedlings treated with final concentrations of 0.1 μg IAA·mL^{-1}. The right-side images correspond to the middle and apical zones of the roots of *A. thaliana* seedlings of each treatment. (The figure was modified from Mora et al. 2023)

10.2.2.2 Analytical Methods

10.2.2.2.1 Quantification of Indole-3-Acetic Acid by Light Spectrophotometry

The precise identification of auxin and related molecules can be obtained according to previous extraction, concentration, separation, and identification of the compounds by complex methodologies such as high-performance liquid chromatography or gas chromatography-mass spectrometry. Such methods, however, are time consuming and cannot be used as routine assays. To solve these problems, various authors have used a colorimetric technique derived from that of Salkowski (1885) for indole detection. This method has been used for years because it is simple, rapid, and cheap and allows the daily analysis of numerous bacterial supernatants. Glickman and Dessaux (1995) examined the sensitivity and the specificity of three versions of the Salkowski colorimetric technique. Two of these allowed the detection of IAA over a low range of concentrations (0.5–20 $mg{\cdot}mL^{-1}$), while the third permitted the detection of IAA over a range of higher concentrations (5–200 $mg{\cdot}mL^{-1}$). These methodologies are shown to be specific for IAA, IPA, and IAM. The first colorimetric technique was performed according to Pilet and Chollet (1970), using reagent R1, which consisted of 12 g of $FeCl_3$ per liter in 7.9 M H_2SO_4. One milliliter of reagent R1 will be added to 1 mL of the sample solution, well mixed, deposited in a 3 mL spectrophotometer cuvette, and kept in the darkness conditions for 30 min at room temperature. The second and third colorimetric methods, termed, were derived from that of Tang and Bonner (1947). They used reagent R2, which consisted of 4.5 g of $FeCl_3$ per liter in 10.8 M H_2SO_4. An aliquot of 1 or 2 mL of reagent R2 will be added to 1 mL of sample solution, and the mixture will be processed as indicated for the first technique. In all cases, mixtures will be examined at OD_{530} for determination of indolic compounds. The quantification of indole-3-acetic acid will be performed by the use of a calibration curve of pure IAA solution.

10.2.2.2.2 Liquid Chromatography

For complex matrices, such as plant tissue or bacterial cultures including cells, efficient chromatographic separation is critical to avoid coelution and enhance sensitivity, especially as many analytes exist in trace amounts. Reversed-phase liquid chromatography (RP-LC) is one of the most commonly used techniques for analyzing plant tissues due to its reliability and versatility. Its dominant position stems from its ability to separate a wide range of compounds with high predictability, driven primarily by the physicochemical properties of analytes such as LogP, LogD, and pKa. The availability of a broad range of analytical columns with various bonded phases, particle sizes, and column dimensions allows for the separation of numerous compounds. Among these, the octadecyl C_{18} phase is considered the workhorse due to its high stability and retention capacity. Specialized columns, such as the Atlantis T3 and Acquity BEH C_{18}, offer distinct advantages. The Atlantis T3 column allows for the use of a 100% aqueous mobile phase, making it suitable for retaining polar compounds, while the BEH C_{18} column is known for its superior mechanical and chemical stability. Although RP-LC is widely used, hydrophilic interaction chromatography (HILIC) is gaining popularity as a complementary technique. HILIC offers an orthogonal separation mechanism and better sensitivity for mid-polar and highly polar compounds like ACC. As technological advances continue, new separation strategies will likely play an increasingly important role in phytohormone analysis, pushing the boundaries of what chromatographic techniques can achieve (Vrobel and Tarkowski 2023; Vrobel et al. 2024; Ćavar Zeljković et al. 2024).

Simultaneous Quantification of IAA, ABA, and GA_3 by HPLC

The chromatographic analysis is performed according to Kelen et al. (2004). For that, the chromatography system consists of a pump with an autoinjector and diode-array detector (DAD), a column oven, and a degasser system. The column used is C_{18} (250 mm × 4.6 ID stainless steel analytical column with 5 μm particle size). The electromotive force (e.m.f.) values used to evaluate the pH of the mobile phase are measured with a pH/ion analysis apparatus using an Ag/AgCl combination pH electrode. All solutions are externally thermostated at 25 °C ± 0.1 °C. Stabilize the electrode in appropriate acetonitrile-water mixtures before the e.m.f. measurements. The mobile phases are acetonitrile-water mixture (26:74 v/v), containing 30 mM p phosphoric acid with sodium hydroxide at pH 4.0. Equilibrate the C_{18} column for each mobile phase condition with a time limit of 30 min. Maintain the column temperature at constant 25 °C ± 0.1 °C. The separation is carried out by isocratic elution at 0.8 $mL{\cdot}min^{-1}$ of flow rate. An injection volume of 10 μL is used for each analysis. The standard solution of the individual acids is prepared in the mobile phase and chromatographed separately to determine the retention time for each molecule. The signal of the compounds is monitored at 208, 265, and 280 nm for GA_3, ABA, and IAA, respectively. Calculate the capacity factors from $k = (tR - to)/to$, where t is the hold-up time, and tR is the retention time of each hormone for each mobile phase. In this equation, the hold-up time, to, is established for every mobile phase composition using potassium bromide solution (0.01% (w/v)) in water (μmax = 200 nm). Determinate the retention time and capacity factor of solutes from three different injections. The identification of different peaks is based on retention time and spiking of the sample.

10.2.2.2.3 Mass Spectrometry

The first step in mass spectrometry (MS) detection is ionization, which occurs in the ion source. For the analysis of small molecules, several ionization techniques can be employed, including electrospray ionization (ESI), atmospheric pressure chemical ionization (APCI), and atmospheric photoionization (APPI). The efficiency of ionization is often influenced by coeluting compounds, which contribute to matrix effects (ME). Among these methods, ESI is the most commonly used due to its broad applicability to a wide range of analytes (Konermann et al. 2013). However, it is also the most prone to matrix effects, likely because analytes within the ion source compete for limited charges in a process similar to an electrolytic cell operating at a constant current. By contrast, APCI and APPI are less susceptible to matrix effects because they rely on different ionization mechanisms. In APCI, ionization occurs in the reaction zone of a corona discharge needle, while in APPI, ionization results from an 8–12 eV photon that ejects an electron from the molecule. These differences offer distinct selectivity advantages: APCI is more effective for ionizing nonpolar compounds that are challenging to ionize with ESI, making it particularly useful for analyzing oxylipins, brassinosteroids, and various acidic hormones following derivatization. Despite its higher susceptibility to matrix effects, ESI remains the most popular choice due to its capacity to ionize a wider array of compounds. For plant metabolomic analysis, tandem mass spectrometry (MS/MS) is often used to ensure high selectivity and sensitivity, with triple quadrupole (QqQ) mass analyzers being particularly favored for their quantitative accuracy and efficiency in multiple reaction monitoring (MRM) mode. However, QqQ instruments face some limitations, as their acquisition speed restricts the number of analytes that can be monitored during a chromatographic run. While monitoring a single MRM transition may be sufficient for quantification, it can compromise selectivity and increase the risk of false identification. Monitoring multiple transitions improves selectivity but is not always feasible for small molecules, as their fragmentation spectra are often too simple to generate multiple strong signals. The ongoing development of high-resolution mass spectrometry (HRMS) has opened new possibilities in plant research, especially in the analysis of high-molecular weight targets such as signaling proteins (Vrobel and Tarkowski 2023).

Simultaneous Quantification of Phytohormones

Novel ultra-high-performance liquid chromatography coupled with heated electrospray tandem mass spectrometry (UHPLC-HESI-MS/MS) method described by Mogal et al. (2022) emphasizes significance in accurately quantifying phytohormones. The method is optimized for each phytohormone to ensure high sensitivity and resolution. This advancement allows for a deeper understanding of how specific bacteria influence phytohormonal profiles, providing insights that could lead to more effective agricultural strategies. The sample extraction and cleanup process involve meticulous preparation to isolate phytohormones from root tissues. Initially, root samples are homogenized using a heavy-duty variable speed homogenizer to create a uniform mixture. A representative sample weighing approximately 15 ± 0.1 g is placed in 50 mL polypropylene centrifuge tubes. To extract the target phytohormones, 15 mL of 1% acetic acid in acetonitrile is added as the extracting solvent. The mixture is then subjected to freezing for 20–30 min to facilitate the extraction process. After freezing, 6 g of magnesium sulfate ($MgSO_4$) and 1.5 g of sodium acetate are introduced to the sample, which is then shaken vigorously for 1 min. Following this, the mixture is centrifuged at 2205 *g* for 2 min, and the supernatant is carefully transferred to new tubes containing anhydrous $MgSO_4$ (0.9 g) and primary secondary amines (PSA, 0.3 g) for further cleanup. This step is crucial for removing any remaining impurities and ensuring the clarity of the extract. The mixture is vortexed for 1 min and centrifuged again at 1125 *g* for 2 min. For analysis, an aliquot of 2.0 mL of the cleaned extract is taken and evaporated to dryness under nitrogen gas using a TurboVap system. Before injection into the UHPLC system, the dried samples are reconstituted in a suitable solvent and filtered through 0.22 μm Polyvinylidene Difluoride (PVDF) syringe filters, ensuring that only clean, particle-free extracts are analyzed. The mobile phase consists of water with 0.1% formic acid and methanol with 0.1% formic acid. Specific parameters, such as flow rate, column type, and gradient profiles, are established to maximize sensitivity and accuracy (Mogal et al. 2022). Recently, Vrobel et al. (2024) published a novel analytical method utilizing hydrophilic interaction liquid chromatography coupled with tandem mass spectrometry (HILIC-MS/MS) for the simultaneous quantification of multiple classes of plant hormones, including cytokinins, auxins, jasmonates, and abscisic acid, without the need for complex sample derivatization. The authors developed a rapid sample preparation protocol and a robust HILIC method that improves sensitivity and selectivity over conventional reversed-phase chromatography. This integrated approach allowed for a deeper understanding of plant hormone dynamics and their interactions under multifactorial stress conditions, highlighting the potential for further research into plant-environment interactions. The authors optimized the extraction procedure using 80 mg of plant tissue (or its lyophilized equivalent) and a 7:2:1 solvent mixture of ACN, water, and isopropanol, which effectively extracts a wide range of polar and moderately polar plant hormones while minimizing losses. The extraction process is performed in an ice-cold ultrasonic bath for 30 min to enhance the solubilization of analytes, followed by centrifugation at 20,000 *g* for 15 min at 4 °C to separate the solid residue from the liquid phase. The supernatant, which contains the extracted hormones, is then transferred to a new tube and evaporated under reduced pressure to remove solvents, yielding a concentrated residue. Next, the dried residues are reconstituted in a small volume of methanol and treated with a strong acid (1 M formic acid) to facilitate the dissociation of analytes and improve their ionization during mass spectrometry. Following this, the samples undergo solid-phase extraction (SPE) using Oasis MCX cartridges, which are activated and equilibrated prior to loading. This SPE step is critical for removing matrix components and concentrating the target analytes. The cartridges are washed with water to eliminate polar contaminants, then eluted with a solution containing ammonium hydroxide in methanol to recover the retained hormones. Finally, the eluted samples are evaporated again to dryness, and the residues are reconstituted in a specific sample diluent (a mixture of 30 mM ammonium formate buffer at pH 3.0, isopropanol, and acetonitrile) for analysis. Subsequently, purified samples underwent HILIC-MS/MS analysis, with the method validated for 45 analytes, including cytokinins, auxins, jasmonates, salicylic acid, and abscisic acid. Two distinct chromatographic methods (Method A5.5 and Method B3.0) were developed, each utilizing different pH conditions to optimize the retention and separation of various hormone classes. The different pH conditions in Method A5.5 and Method B3.0 significantly influence the retention and separation of various hormone classes in HILIC-MS/MS analysis due to their effects on the ionization state of the analytes. Method A5.5 is generally favorable for retaining acidic and neutral compounds. At pH 5.5, many acidic hormones, such as auxins and jasmonates, remain largely deprotonated, which enhances their hydrophobic interactions with the stationary phase and leads to improved retention. Conversely, Method B3.0 operates at a lower pH of 3.0, which promotes the protonation of certain analytes, particularly weakly basic compounds like CKs and 1-aminocyclopropane-1-carboxylic acid (ACC). At this lower pH, these compounds are more positively charged, increasing their electrostatic interactions with the stationary phase, thereby enhancing their retention and separation.

Simultaneous Quantification of Amino Acids and Polyamines

Free amino acids are generally analyzed using (Kawasaki et al. 2016; Kolega et al. 2020) AccQ-Tag Chemistry Kit (Waters Corporation) method which include precolumn derivatization to obtain fluorescent amino acid derivatives and HPLC separation on C_{18} column. The amino acids were quantified based on retention times compared to known standards, ensuring accurate measurement of their concentrations in the exudate samples. This comprehensive approach allows for precise identification and quantification of amino acids, facilitating insights into their role in plant-microbe interactions in the rhizosphere. In addition, free polyamines are generally extracted with perchloric or trichloroacetic acid, and then derivatized by dansyl or benzoyl chloride and analyzed by HPLC on C_{18} column (Zahedi and Abassi 2015; Pereyra et al. 2023). However, both of these methods require significant amounts of either plant tissue or bacterial or plant exudates and, due to the derivatization, are quite laborious. However, recently Ćavar Zeljković et al. (2024) published a novel method for the comprehensive analysis of nitrogen-related metabolites in plants using liquid chromatography-tandem mass spectrometry (LC-MS/MS). This method allows for the simultaneous quantification of amino acids and biogenic amines, and their derivatives, which are crucial for understanding plant metabolism and responses to environmental stress. Authors developed a one-step extraction process from lyophilized plant material, which significantly enhances the efficiency and reduces the cost of analysis. The method was validated across various plant species, including *Arabidopsis*, maize, and barley, revealing significant variations in metabolite profiles due to species differences, developmental stages, and growth conditions. Moreover, this methodology enables quantification of different forms of biogenic mines, i.e., their free, conjugated, and bounded forms. Briefly, lyophilized tissue (2–5 mg) is extracted with 1 mL of 50% EtOH. After sonication for 15 min, each extract was centrifuged for 5 min at 25 °C at 14,500 *g*, then three 250 μL samples of the supernatant were placed in different vials. Sample I was used for analysis of free amino acids and methylated and acetylated metabolites, while samples II and III were used for analysis of free and conjugated biogenic amines.

Sample I was evaporated to dryness and redissolved in 50 μL of mobile phase for HILIC separation on the BEH AMIDE column to quantify free amino acids and biogenic amines and their acetylated and methylated derivatives, as well as selected free-nitrogen-containing compounds (i.e., allantoin, allantoic acid, etc.). Sample II was subjected to the derivatization protocol of Taibi et al. (2000) with slight modifications. After evaporation to dryness, the sample was dissolved in a mobile phase for RP separation on the BEH C_{18} column to quantify free biogenic amines. Sample III was subjected to acid hydrolysis (100 μL of 12 M HCl) for 16 h at 40 °C to liberate conjugated biogenic amines (Piqueras et al. 2002). After derivatization, as described below, the sample was prepared for analysis on the RP column in the same way as sample II. As a result, sample III contained both free and conjugated biogenic amines whereas sample II contained only free biogenic amines. The concentrations of conjugated amines were thus determined by subtracting the values obtained for sample II from those for sample III. The pellet that remained after extraction was hydrolyzed by treatment with 200 μL of 6 M NaOH under sonication for 15 min, followed by heating at 40 °C in 200 μL of 12 M HCl to release macromolecule-bound biogenic amines. After derivatization, the resulting sample IV was prepared for analysis on the RP column.

Other Related Compounds

Beside phytohormones and nitrogen-containing compounds, such as amino acids and polyamines and their derivatives, several studies pointed that analysis of small organic acids (Tahir et al. 2015; Greipel et al. 2024), phenolic acids (Dal Lago et al. 2015; Kolega et al. 2020; Pereyra et al. 2023), as well as sugars (Kawasaki et al. 2016; Brusamarello-Santos et al. 2017; Schillaci et al. 2021) released in the rhizosphere or from bacterial cultures are also important for understanding the plant-PGPR interactions. Small organic acids are analyzed using high-performance liquid chromatography (HPLC) coupled with diode-array detectors (DAD) is used at 210 nm for detection and UV spectra are recorded from 190 to 400 nm. Organic acids are separated using a Microguard Cation-H precolumn and an Aminex HPx-87H analytical column. Sulfuric acid (0.001 N) is used as the mobile phase at a flow rate of 0.6 mL·min^{-1}. Standard solutions (100 μg·L^{-1}) of citric acid, malic acid, acetic acid, succinic acid, gluconic acid, lactic acid, and oxalic acid are used for calibration (Tahir et al. 2015). For sugar analysis, the same mobile phase and flow rate are used as for organic acids. Sugars are detected using a refractive index detector (Shimadzu Co., Japan) with the column temperature set at 85 °C. Standard solutions (100 μg·L^{-1}) of sucrose, maltose, glucose, galactose, and xylose are used for calibration. Quantification of organic acids and sugars is performed by comparing the peak areas and retention times of the samples with those of the standards. Free and cell-wall-bound phenolics are extracted with methanol/distilled water (80:20, v/v) and absolute ethanol, and then centrifuged at 930× *g* for 10 min. The supernatant contains free phenolics, and the remaining residue, representing the cell-wall-bound phenolic fraction, is subjected to alkaline hydrolysis with 1 M sodium hydroxide for 18 h in the dark. The saponified extract is acidified to pH

1–2 using hydrochloric acid and extract with 7 mL of ethyl acetate. The organic phase is evaporated to dryness under a nitrogen stream. Known quantities of trans-cinnamic acid are added prior to saponification for recovery assays. High-performance liquid chromatography (HPLC) analysis of ferulic acid (FA), diferulic acid (DFA), and *p*-coumaric acid is performed using a Hewlett Packard 1100 HPLC system with a binary pump, a Phenomenex reverse-phase LUNA C-18 column (250 mm × 4.6 mm, 5 μm), a temperature control, and a photodiode array UV detector (Dal Lago et al. 2015; Pereyra et al. 2023).

10.2.2.2.4 Methods for NO Measuring

The experiments must create conditions that favor NO production by *Azospirillum* spp. Thus, the composition of the growth media is determinant for NO production, and the media must provide the precursor for its synthesis. In denitrifying bacteria, media must contain NO_3^- and low O_2 tension must be supplied. If the bacterium harbors a periplasmic nitrate reductase, an enzyme neither repressed nor inhibited by O_2, then high agitation can be done. Besides, the bacterial growth phase must be taken into consideration as it represents a particular physiological stage in which some metabolic pathways are activated and other repressed. It is suggested that NO production be investigated in each phase to acquire a complete picture of the NO metabolism in the studied bacteria. The inhibition of key enzymes from metabolic pathways leading to NO production could help in the study of the biogenic origin of NO. Tungstate is widely used as a NR inhibitor both in plant research (Mendel 2007) and in bacteria culture (Betlach and Tiedje 1981). Tungstate can compete with molybdate for incorporation into the enzyme complex and results in the inactivation of molybdate-dependent enzymes (Hille 2002). However, it was demonstrated that tungstate not only inactivates NR but also inhibits other molybdate-dependent enzymes (Xiong et al. 2012) and root growth (Adamakis et al. 2008). Therefore, this inhibitor must be used with caution in analyzing plant-microbe interaction experiments. Dissimilatory NIR is considered the major source of NO in bacteria. Its localization is periplasmic, and it can be inhibited using KCN. This approach is appropriate when testing purified enzymes, but not when pharmacological experiments are done with living bacteria due to the blocking of other vital processes yielding misinterpretation of results. A combination of pharmacological and biochemical analysis with a genetic approach will be necessary to investigate the roles of NO in plants (Xiong et al. 2012), bacteria, and their interaction. Here, we describe some of the most used methods for NO determination in bacteria.

Griess Assay

Since NO is not stable and persistent in stored biological samples, its production in bacteria is often measured indirectly by quantifying the stable end products of its metabolism. The stable oxidation products of NO metabolism are inorganic nitrite and nitrate (Archer 1993). Due to its simplicity and commercial availability, the colorimetric Griess assay is one of the most used techniques today for indirect quantify NO production and was suffering from many improvements that better its sensibility (Amenta et al. 2015). In this technique, nitrite is first treated with sulfanilamide, a diazotizing reagent, in acidic media. The transient diazonium salt formed is then allowed to react with a coupling reagent, N-naphthylethylenediamine (NED), to form a stable azo compound. The intense purple color of the product allows assaying nitrite with high sensitivity and can be used to measure nitrite concentration as low as ~0.5 μM (Amenta et al. 2015). For a more accurate measurement of NO produced in a sample, the nitrate formed via oxidation of nitrite must also be measured. This is often accomplished by reducing nitrate to nitrite immediately prior to the addition of the Griess reagents to the initial sample. To perform the assay, transfer 100 μL of the cell-free culture or standard nitrite solution to each well of a microtiter plate. It is recommended to prepare triplicate replicas. Then, add 50 μL of reagent A (which is a sulfanilamide solution composed of 1% (w/v) sulfanilamide in 5% (v/v) phosphoric acid) to each well and incubate for 10 min at 20–25 °C, in darkness. Subsequently, add 50 μL of reagent B (a NED solution containing 0.1% (w/v) N-1-napthylethylenediamine dihydrochloride) to the wells and incubate at 20–25 °C for 10 min in darkness. After the incubation period, record the absorbance at 540 nm with a microtiter plate reader and compare the obtained values with a standard curve prepared with 0.1 M sodium nitrite (linear between 0 and 100 μM). Calculate the concentration from the curve.

Electron Paramagnetic Resonance

Electron paramagnetic resonance (EPR) is a versatile technology specific to detect atoms, molecules, and complexes with unpaired electrons, as in free radicals. This technology allows us to identify and quantify a paramagnetic molecule as NO with high specificity and sensibility (detection limit ~10^{-9} M). It is the only method that unequivocally discriminates NO from other related nonfree radicals, as nitrate and nitrite. The main drawback of EPR is that specialized, expensive equipment, and technical expertise are needed. Due to the transient nature of NO in biological samples, NO must be trapped to form stable paramagnetic adducts before being quantified. Several diamagnetic compounds have been used as "spin trap" of NO as the diethyldithiocarbamate (DETC), N-Methyl-D-glucamine dithiocarbamate (MGD), and dithiocarboxy sarcosine (DTCS) (Puntarulo et al. 2009). For the EPR assay, centrifuge bacterial cultures and weigh them as a minimum of approximately 0.15 g of fresh cells harvested by centrifugation. Immediately add 220 μL MGD solution (solution of 10 mM sodium N-methyl-D-glucamine dithiocarbamate [MGD] in 1 mM $FeSO_4$) to the bacterial sample. Then, transfer the sample with the spin trap to a bottom-sealed Pasteur pipette and introduce it into the spectrometer cavity for measurements. Record the spectra at room temperature (~20 °C) in the EPR spectrometer operating at 9.5 GHz (X-band). Set the instrument at 200 G field scan, 83.886 s sweep time, 327.68 ms time constant, 5.983 G modulation amplitude, 50 kHz modulation frequency, and 20 mW microwave power. Different modulation amplitudes and microwave powers can be selected to improve the resolution of the spectra depending on the spectrometer used. Record the spectra of the TEMPOL solution (which is an aqueous solution of 4-hydroxy-2,2,6,6-tetramethyl piperidine 1-oxyl, TEMPOL, a stable free radical used as standard to obtain the concentration of free radical adducts other than NO) with the same spectrometer setting. The NO concentration is quantified by double integration of the three-line spectra (nitrite, nitrate, and NO) and is referenced to TEMPOL spectra, which has a known area to concentration ratio.

Fluorescent Dye Probes

Diamine derivatives of fluorescein (DAF) are the most widely used NO fluorescent probes. Diaminofluorescein-2 (DAF-2) was the first product to become commercially available. It reacts with NO, in the presence of oxygen, resulting in the formation of a triazolo-fluorescein analogue (DAF-2T) that exhibits a strong green fluorescence (Kojima et al. 1998). DAF-2 DA, a diacetate analogue of DAF-2, was designed for imaging of NO produced in living cells (Kojima et al. 1998). Because of the membrane permeability of DAF-2 DA, it is possible to monitor endogenous NO production in a real-time fashion (Arita et al. 2007). DAF-FM has been developed as a more sensitive NO sensor than DAF-2 DA, ~3 nM and ~5 nM, respectively (Murad 1999). DAF-2 DA was employed to detect NO production by *A. baldanorium* Sp245 growing under aerobic conditions (Creus et al. 2005; Molina-Favero et al. 2008) and to establish the involvement of NO in the formation of lateral roots by *Azospirillum* (Creus et al. 2005).

Reagent Preparation and Assay

For reagent preparation and assay, prepare a 1× working solution of NO-specific probes DAF-2 DA or DAF-FM DA (e.g., from Calbiochem, La Jolla, CA, USA) and immediately prior to use, because DAF-probes are very light sensitive. A 1:500 dilution in 0.1 M phosphate buffer (pH 7.4) is generally required. Do not save 1× solution for future use. For the assay, transfer 100 μL of each treatment of *Azospirillum* culture to a 96-well microtiter plate. It is recommended to perform triplicate replicas. Add 1× working solution of DAF-probe at 10 μM final concentration. Then, incubate for 20 min at room temperature, in darkness and with gentle agitation. Place the 96-well microtiter plate into a fluorescence plate reader (e.g., Fluoroskan Ascent; Thermo Electron, with the appropriate filters [excitation 480 nm; emission 515 nm]). The fluorescence intensity must be measured every 2–4 min over a period of at least 2 h. Determinate the NO concentration comparing it to DAF-2T or DAF-FMT standards, such as from Calbiochem, La Jolla, CA, USA, and referencing a linear calibration curve between 0 and 1000 nmol.

References

Abeles FW (1992) Roles and physiological effects of ethylene in plant physiology: dormancy, growth, and development. In: Ethylene in plant biology

Abreu ME, Munne-Bosch S (2009) Salicylic acid deficiency in NahG transgenic lines and sid2 mutant's increases seed yield in the annual plant *Arabidopsis thaliana*. J Exp Bot 60(4): 1261–1271

Adamakis IDS, Eleftheriou EP, Rost TL (2008) Effects of sodium tungstate on the ultrastructure and growth of pea (*Pisum sativum*) and cotton (*Gossypium hirsutum*) seedlings. Environ Exp Bot 63(1–3):416–425

Amenta M, Molina-Favero C, Creus CM et al (2015) Nitric oxide in Azospirillum and related bacteria: production and effects. In: Handbook for Azospirillum: technical issues and protocols. Springer, Cham, pp 155–180

Archer S (1993) Measurement of NO in biological models. FASEB J 7(2):349–360

Arita NO, Cohen MF, Tokuda G et al (2007) Fluorometric detection of nitric oxide with diaminofluoresceins (DAFs): applications and limitations for plant NO research. In: Nitric oxide in plant growth, development and stress physiology, pp 269–280

Arnon DI (1949) Copper enzymes in isolated chloroplasts. Polyphenoloxidase in Beta vulgaris. Plant Physiol 24(1):1

Arruebarrena Di Palma A, Pereyra CM, Moreno Ramirez L et al (2013) Denitrification-derived nitric oxide modulates biofilm formation in *Azospirillum brasilense*. FEMS Microbiol Lett 338(1):77–85

Aziz A, Martin-Tanguy J, Larher F (1997) Plasticity of polyamine metabolism associated with high osmotic stress in rape leaf discs and with ethylene treatment. Plant Growth Regul 21:153–163

Badawy AA, Alotaibi MO, Abdelaziz AM et al (2021) Enhancement of seawater stress tolerance in barley by the endophytic fungus *Aspergillus ochraceus*. Metabolites 11(7):428

Barea JM, Navarro E, Montoya E (1976) Production of plant growth regulators by rhizosphere phosphate-solubilizing bacteria. J Appl Bacteriol 40(2):129–134

Baron NC, Rigobelo EC (2022) Endophytic fungi: a tool for plant growth promotion and sustainable agriculture. Mycology 13(1):39–55

Bashan Y (1998) Azospirillum plant growth-promoting strains are nonpathogenic on tomato, pepper, cotton, and wheat. Canadian Journal of Microbiology, 44(2), 168–174

Betlach MR, Tiedje JM (1981) Kinetic explanation for accumulation of nitrite, nitric oxide, and nitrous oxide during bacterial denitrification. Appl Environ Microbiol 42(6):1074–1084

Bottini R, Fulchieri M, Pearce D et al (1989) Identification of gibberellins A1, A3, and iso-A3 in cultures of *Azospirillum lipoferum*. Plant Physiol 90(1):45–47

Bottini R, Cassán F, Piccoli P (2004) Gibberellin production by bacteria and its involvement in plant growth promotion and yield increase. Appl Microbiol Biotechnol 65:497–503

Brusamarello-Santos LC, Gilard F, Brulé L et al (2017) Metabolic profiling of two maize (Zea mays L.) inbred lines inoculated with the nitrogen fixing plant-interacting bacteria *Herbaspirillum seropedicae* and *Azospirillum brasilense*. PLoS One 12(3):e0174576

Burg SP (1962) The physiology of ethylene formation. Annu Rev Plant Physiol 13(1):265–302

Burg SP, Burg EA (1968) Ethylene formation in pea seedlings; its relation to the inhibition of bud growth caused by indole-3-acetic acid. Plant Physiol 43(7):1069–1074

Cacciari I, Lippi D, Pietrosanti T et al (1989) Phytohormone-like substances produced by single and mixed diazotrophic cultures of Azospirillum and Arthrobacter. Plant Soil 115:151–153

Cassán F, Maiale S, Masciarelli O et al (2009) Cadaverine production by Azospirillum brasilense and its possible role in plant growth promotion and osmotic stress mitigation. Eur J Soil Biol 45(1):12–19

Cassán F, Vanderleyden J, Spaepen S (2014) Physiological and agronomical aspects of phytohormone production by model plant-growth-promoting rhizobacteria (PGPR) belonging to the genus Azospirillum. J Plant Growth Regul 33:440–459

Cassán F, Coniglio A, López G et al (2020) Everything you must know about Azospirillum and its impact on agriculture and beyond. Biol Fertil Soils 56:461–479

Ćavar Zeljković S, De Diego N, Drašar L et al (2024) Comprehensive LC-MS/MS analysis of nitrogen-related plant metabolites. J Exp Bot 75(17):5390–5411

Chen Z, Zheng Z, Huang J et al (2009) Biosynthesis of salicylic acid in plants. Plant Signal Behav 4(6):493–496

Cohen AC, Bottini R, Piccoli PN (2008) *Azospirillum brasilense* Sp245 produces ABA in chemically-defined culture medium and increases ABA content in Arabidopsis plants. Plant Growth Regul 54:97–103

Cohen AC, Bottini R, Pontin M et al (2015) *Azospirillum brasilense* ameliorates the response of *Arabidopsis thaliana* to drought mainly via enhancement of ABA levels. Physiol Plant 153(1): 79–90

Colebrook EH, Thomas SG, Phillips AL et al (2014) The role of gibberellin signalling in plant responses to abiotic stress. J Exp Biol 217(1):67–75

Coniglio A, Larama G, Molina R et al (2022) Modulation of maize rhizosphere microbiota composition by inoculation with *Azospirillum argentinense* Az39 (formerly A. brasilense Az39). J Soil Sci Plant Nutr 22(3):3553–3567

Correa-Aragunde N, Graziano M, Chevalier C et al (2006) Nitric oxide modulates the expression of cell cycle regulatory genes during lateral root formation in tomato. J Exp Bot 57(3):581–588

Creus CM, Graziano M, Casanovas EM et al (2005) Nitric oxide is involved in the *Azospirillum brasilense*-induced lateral root formation in tomato. Planta 221:297–303

Dal Lago CC, Creus CM, Pereyra MA (2015) *Azospirillum* growth promotion is related to changes in ferulate and dehydrodiferulate contents in cell wall of inoculated cucumber seedlings hypocotyls. Am J Plant Sci 6(10):1650–1661

Davies PJ, Davies PJ (1995) The plant hormones: their nature, occurrence and functions. In: Plant hormones: physiology, biochemistry and molecular biology. Kluwer Academic Publishers, Dordrecht, pp 1–12

Durgbanshi A, Arbona V, Pozo O et al (2005) Simultaneous determination of multiple phytohormones in plant extracts by liquid chromatography–electrospray tandem mass spectrometry. J Agric Food Chem 53(22):8437–8442

Epel BL, Erlanger M, Yahalom A (1987) The etiolated cucumber hypocotyl weight-growth test: a sensitive, easy and ultrafast bioassay for IAA. Plant Growth Regul 5:3–14

Esquivel-Cote R, Ramírez-Gama RM, Tsuzuki-Reyes G et al (2010) *Azospirillum lipoferum* strain AZm5 containing 1-aminocyclopropane-1-carboxylic acid deaminase improves early growth of tomato seedlings under nitrogen deficiency. Plant Soil 337:65–75

Etesami H, Glick BR (2020) Halotolerant plant growth–promoting bacteria: Prospects for alleviating salinity stress in plants. Environ Exp Bot 178:104124

Feussner I, Wasternack C (2002) The lipoxygenase pathway. Annu Rev Plant Biol 53(1):275–297

Finkelstein RR, Gampala SS, Rock CD (2002) Abscisic acid signaling in seeds and seedlings. Plant Cell 14(suppl_1):S15–S45

Fletcher RA, Kallidumbil V, Steele P (1982) An improved bioassay for cytokinins using cucumber cotyledons. Plant Physiol 69(3):675–677

Fujioka S, Yamane H, Spray CR et al (1988) Qualitative and quantitative analyses of gibberellins in vegetative shoots of normal, dwarf-1, dwarf-2, dwarf-3, and dwarf-5 seedlings of Zea mays L. Plant Physiol 88(4):1367–1372

Gamarnik A, Frydman RB (1991) Cadaverine, an essential diamine for the normal root development of germinating soybean (Glycine max) seeds. Plant Physiol 97(2):778–785

Glick BR (2014) Bacteria with ACC deaminase can promote plant growth and help to feed the world. Microbiol Res 169(1):30–39

Glickmann E, Dessaux Y (1995) A critical examination of the specificity of the Salkowski reagent for indolic compounds produced by phytopathogenic bacteria. Appl Environ Microbiol 61(2): 793–796

Greipel E, Nagy K, Csákvári E et al (2024) Chemotactic interactions of Scenedesmus sp. and *Azospirillum brasilense* investigated by microfluidic methods. Microb Ecol 87(1):52

Grobelak A, Hiller J (2017) Bacterial siderophores promote plant growth: screening of catechol and hydroxamate siderophores. Int J Phytoremed 19(9):825–833

Gupta KJ, Fernie AR, Kaiser WM et al (2011) On the origins of nitric oxide. Trends Plant Sci 16(3): 160–168

Gupta KJ, Kaladhar VC, Fitzpatrick TB et al (2022) Nitric oxide regulation of plant metabolism. Mol Plant 15(2):228–242

Halouzka R, Zeljković SĆ, Klejdus B et al (2020) Analytical methods in strigolactone research. Plant Methods 16:1–13

Hamana K, Matsuzaki S (1990) Occurrence of homospermidine as a major polyamine in the authentic genus Flavobacterium. Can J Microbiol 36(3):228–231

Hamana K, Niitsu M, Samejima K et al (1988) Occurrence of aminopropylcadaverine and its aminopropyl derivatives aminopentylnorspermidine and N,N′-bis(3-aminopropyl)cadaverine in *Halococcus acetoinfaciens*. FEMS Microbiol Lett 50(1):79–83

Harrison AJ, Yu M, Gårdenborg T et al (2006) The structure of MbtI from Mycobacterium tuberculosis, the first enzyme in the biosynthesis of the siderophore mycobactin, reveals it to be a salicylate synthase. J Bacteriol 188(17):6081–6091

Hille R (2002) Molybdenum and tungsten in biology. Trends Biochem Sci 27(7):360–367

Hoagland DR, Broyer TC (1936) General nature of the process of salt accumulation by roots with description of experimental methods. Plant Physiol 11(3):471

Horemans S (1986) Production of plant growth substances by *Azospirillum* sp. and other rhizosphere bacteria. Symbiosis 2:341–346

Hwang I, Sakakibara H (2006) Cytokinin biosynthesis and perception. Physiol Plant 126(4): 528–538

Illescas M, Pedrero-Méndez A, Pitorini-Bovolini M et al (2021) Phytohormone production profiles in Trichoderma species and their relationship to wheat plant responses to water stress. Pathogens 10(8):991

Janzen RA, Rood SB, Dormaar JF et al (1992) *Azospirillum brasilense* produces gibberellin in pure culture on chemically-defined medium and in co-culture on straw. Soil Biol Biochem 24(10): 1061–1064

Jensen JB, Egsgaard H, Van Onckelen H et al (1995) Catabolism of indole-3-acetic acid and 4- and 5-chloroindole-3-acetic acid in *Bradyrhizobium japonicum*. J Bacteriol 177(20):5762–5766

Jiang F, Hartung W (2008) Long-distance signalling of abscisic acid (ABA): the factors regulating the intensity of the ABA signal. J Exp Bot 59(1):37–43

Kaduskar RD, Scala GD, Al Jabri ZJ et al (2017) Promysalin is a salicylate-containing antimicrobial with a cell-membrane-disrupting mechanism of action on Gram-positive bacteria. Sci Rep 7(1): 8861

Kawasaki A, Donn S, Ryan PR et al (2016) Microbiome and exudates of the root and rhizosphere of *Brachypodium distachyon*, a model for wheat. PLoS One 11(10):e0164533

Kelen M, Demiralay EC, Şen S et al (2004) Separation of abscisic acid, indole-3-acetic acid, gibberellic acid in 99 R (*Vitis berlandieri* x *Vitis rupestris*) and rose oil (*Rosa damascena* Mill.) by reversed phase liquid chromatography. Turk J Chem 28(5):603–610

Kende H (1993) Ethylene biosynthesis. Annu Rev Plant Physiol Plant Mol Biol 44:283–307

Kobayashi M, Sakurai A, Saka H et al (1989) Quantitative analysis of endogenous gibberellins in normal and dwarf cultivars of rice. Plant Cell Physiol 30(7):963–969

Kojima H, Sakurai K, Kikuchi K et al (1998) Development of a fluorescent indicator for nitric oxide based on the fluorescein chromophore. Chem Pharm Bull 46(2):373–375

Kojima M, Kamada-Nobusada T, Komatsu H et al (2009) Highly sensitive and high-throughput analysis of plant hormones using MS-probe modification and liquid chromatography–tandem mass spectrometry: an application for hormone profiling in *Oryza sativa*. Plant Cell Physiol 50(7):1201–1214

Kolb W, Martin P (1985) Response of plant roots to inoculation with Azospirillum brasilense and to application of indole acetic acid. In: Azospirillum III: genetics physiology ecology. Springer, Berlin, pp 215–221

Kolbert ZS, Barroso JB, Brouquisse R et al (2019) A forty-year journey: the generation and roles of NO in plants. Nitric Oxide 93:53–70

Kolega S, Miras-Moreno B, Buffagni V et al (2020) Nutraceutical profiles of two hydroponically grown sweet basil cultivars as affected by the composition of the nutrient solution and the inoculation with *Azospirillum brasilense*. Front Plant Sci 11:596000

Konermann L, Ahadi E, Rodriguez AD et al (2013) Unraveling the mechanism of electrospray ionization. Anal Chem 85(1):2–9

Kouchi H, Imaizumi-Anraku H, Hayashi M et al (2010) How many peas in a pod? Legume genes responsible for mutualistic symbioses underground. Plant Cell Physiol 51(9):1381–1397

Koul V, Tripathi C, Adholeya A et al (2015) Nitric oxide metabolism and indole acetic acid biosynthesis cross-talk in *Azospirillum brasilense* SM. Res Microbiol 166(3):174–185

Kuznetsov VV, Radyukina NL, Shevyakova NI (2006) Polyamines and stress: biological role, metabolism, and regulation. Russ J Plant Physiol 53:583–604

Lamattina L, Polacco JC (2007) Nitric oxide in plant growth, development and stress physiology. In: Lamattina L, Polacco JC (eds) Plant cell monographs. Springer, Dordrecht

Lambrecht M, Okon Y, Van den Broek A et al (2000) Indole-3-acetic acid: a reciprocal signalling molecule in bacteria–plant interactions. Trends Microbiol 8(7):298–300

Liu K, Fu H, Bei Q et al (2000) Inward potassium channel in guard cells as a target for polyamine regulation of stomatal movements. Plant Physiol 124(3):1315–1326

Liu Y, Fang XA, Chen G et al (2019) Recent development in sample preparation techniques for plant hormone analysis. TrAC Trends Anal Chem 113:224–233

Luna V, Soriano MD, Bottini R et al (1992) Levels of endogenous gibberellins, abscisic acid, indol 3 acetic acid, and naringenin during dormancy of peach flower buds. In: VII international symposium on plant growth regulators in fruit production, vol 329, pp 265–267

Marin E, Nussaume L, Quesada A et al (1996) Molecular identification of zeaxanthin epoxidase of Nicotiana plumbaginifolia, a gene involved in abscisic acid biosynthesis and corresponding to the ABA locus of Arabidopsis thaliana. EMBO J 15(10):2331–2342

Maruri-López I, Aviles-Baltazar NY, Buchala A et al (2019) Intra and extracellular journey of the phytohormone salicylic acid. Front Plant Sci 10:423

Mendel RR (2007) Biology of the molybdenum cofactor. J Exp Bot 58(9):2289–2296

Mercado-Blanco J, Rodríguez-Jurado D, Pérez-Artés E et al (2001) Detection of the nondefoliating pathotype of Verticillium dahliae in infected olive plants by nested PCR. Plant Pathol 50(5): 609–619

Mishra AK, Baek KH (2021) Salicylic acid biosynthesis and metabolism: a divergent pathway for plants and bacteria. Biomolecules 11(5):705

Miyawaki K, Tarkowski P, Matsumoto-Kitano M et al (2006) Roles of Arabidopsis ATP/ADP isopentenyltransferases and tRNA isopentenyltransferases in cytokinin biosynthesis. Proc Natl Acad Sci 103(44):16598–16603

Mogal CS, Solanki VH, Kansara RV et al (2022) UHPLC-MS/MS and QRT-PCR profiling of PGP agents and Rhizobium spp. of induced phytohormones for growth promotion in mungbean (var. Co4). Heliyon 8(5)

Molina R, Rivera D, Mora V et al (2018) Regulation of IAA biosynthesis in Azospirillum brasilense under environmental stress conditions. Curr Microbiol 75:1408–1418

Molina-Favero C, Creus CM, Lanteri ML et al (2007) Nitric oxide and plant growth promoting rhizobacteria: common features influencing root growth and development. Adv Bot Res 46: 1–33

Molina-Favero C, Creus CM, Simontacchi M, Puntarulo S, Lamattina L (2008) Aerobic nitric oxide production by Azospirillum brasilense Sp245 and its influence on root architecture in tomato. Molecular plantmicrobe interactions, 21(7), 1001–1009.

Mora V, López G, Molina R et al (2023) *Azospirillum argentinense* modifies Arabidopsis root architecture through auxin-dependent pathway and flagellin. J Soil Sci Plant Nutr 23(3): 4543–4557

Murad F (1999) Discovery of some of the biological effects of nitric oxide and its role in cell signaling (Nobel lecture). Angew Chem Int Ed 38(13–14):1856–1868

Muralidhara R, Rai P (1986) Plant growth regulators produced by diazotrophic bacteria. In: National seminar on microbial ecology, pp 18–23

Murashige T, Skoog F (1962) A revised medium for rapid growth and bio assays with tobacco tissue cultures. Physiologia plantarum, 15(3)

Ng G, Seabolt S, Zhang C et al (2011) Genetic dissection of salicylic acid-mediated defense signaling networks in Arabidopsis. Genetics 189(3):851–859

Niemi K, Häggman H, Sarjala T (2002) Effects of exogenous diamines on the interaction between ectomycorrhizal fungi and adventitious root formation in Scots pine in vitro. Tree Physiol 22(6): 373–381

O'Brien JA, Benková E (2013) Cytokinin cross-talking during biotic and abiotic stress responses. Front Plant Sci 4:451

Pereyra CM, Dal Lago CC, Creus CM et al (2023) *Azospirillum baldaniorum* Sp 245 inoculation affects cell wall and polyamines metabolisms in cucumber seedling roots. FEMS Microbiol Lett 370:fnad005

Perrig D, Boiero ML, Masciarelli OA et al (2007) Plant-growth-promoting compounds produced by two agronomically important strains of *Azospirillum brasilense*, and implications for inoculant formulation. Appl Microbiol Biotechnol 75:1143–1150

Piccoli P, Bottini R (1994) Metabolismo de 17, 17-[$^2H_2$] giberelina A_{20} a 17, 17-[$^2H_2$] giberelina A_1 por cultivos de Azospirillum tipoferum. AgriScientia, 11, 13–15

Pieckenstain FL, Gárriz A, Chornomaz EM et al (2001) The effect of polyamine biosynthesis inhibition on growth and differentiation of the phytopathogenic fungus *Sclerotinia sclerotiorum*. Antonie Van Leeuwenhoek 80:245–253

Pilet PE, Chollet R (1970) Sur le dosage colorimétrique de l'acide indolylacétique. CR Acad Sci Ser D 271:1675–1678

Ping L, Boland W (2004) Signals from the underground: bacterial volatiles promote growth in Arabidopsis. Trends Plant Sci 9(6):263–266

Piqueras A, Cortina M, Serna MD et al (2002) Polyamines and hyperhydricity in micropropagated carnation plants. Plant Sci 162(5):671–678

Primrose SB, Dilworth MJ (1976) Ethylene production by bacteria. J Gen Microbiol 93:177–181

Prinsen E, Costacurta A, Michiels K et al (1993) *Azospirillum brasilense* indole-3-acetic acid biosynthesis: evidence for a non-tryptophan dependent pathway. Mol Plant-Microbe Interact 6:609–609

Puntarulo S, Jasid S, Boveris AD et al (2009) Electron paramagnetic resonance as a tool to study nitric oxide generation in plants. Nitric Oxide 21(1):1–13

Rahman A, Hosokawa S, Oono Y et al (2002) Auxin and ethylene response interactions during Arabidopsis root hair development dissected by auxin influx modulators. Plant Physiol 130(4): 1908–1917

Rekhter D, Lüdke D, Ding Y et al (2019) Isochorismate-derived biosynthesis of the plant stress hormone salicylic acid. Science 365(6452):498–502

Reynders L, Vlassak K (1979) Conversion of tryptophan to indoleacetic acid by Azospirillum brasilense. Soil Biol Biochem 11:547–548

Ribaudo CM, Krumpholz EM, Cassán FD et al (2006) *Azospirillum* sp. promotes root hair development in tomato plants through a mechanism that involves ethylene. J Plant Growth Regul 25:175–185

Rivera D, Mora V, Lopez G et al (2018) New insights into indole-3-acetic acid metabolism in *Azospirillum brasilense*. J Appl Microbiol 125(6):1774–1785

Salkowski E (1885) Ueber das verhalten der skatolcarbonsäure im organismus. Ber Dtsch Chem Ges 18:145–147

Santner A, Estelle M (2009) Recent advances and emerging trends in plant hormone signalling. Nature 459(7250):1071–1078
Saxena B, Modi M, Modi VV (1986) Isolation and characterization of siderophores from Azospirillum lipoferum D-2. J Gen Microbiol 132(8):2219–2224
Schillaci M, Kehelpannala C, Martinez-Seidel F et al (2021) The metabolic response of Brachypodium roots to the interaction with beneficial bacteria is affected by the plant nutritional status. Metabolites 11(6):358
Schwartz SH, Tan BC, Gage DA et al (1997) Specific oxidative cleavage of carotenoids by VP14 of maize. Science 276(5320):1872–1874
Schwartz SH, Qin X, Zeevaart JA (2003) Elucidation of the indirect pathway of abscisic acid biosynthesis by mutants, genes, and enzymes. Plant Physiol 131(4):1591–1601
Seo M, Koshiba T (2002) Complex regulation of ABA biosynthesis in plants. Trends Plant Sci 7(1): 41–48
Seo M, Nambara E, Choi G et al (2009) Interaction of light and hormone signals in germinating seeds. Plant Mol Biol 69:463–472
Shah S, Karkhanis V, Desai A (1992) Isolation and characterization of siderophore, with antimicrobial activity, from *Azospirillum lipoferum* M. Curr Microbiol 25:347–351
Somers E, Ptacek D, Gysegom P, Srinivasan M, Vanderleyden J (2005) Azospirillum brasilense produces the auxin-like phenylacetic acid by using the key enzyme for indole-3-acetic acid biosynthesis. Applied and environmental microbiology, 71(4):1803–1810
Sreenivasulu N, Radchuk V, Alawady A et al (2010) De-regulation of abscisic acid contents causes abnormal endosperm development in the barley mutant seg8. Plant J 64(4):589–603
Strzelczyk E, Kampert M, Li CY (1994) Cytokinin-like substances and ethylene production by Azospirillum in media with different carbon sources. Microbiol Res 149(1):55–60
Sugawara H, Ueda N, Kojima M et al (2008) Structural insight into the reaction mechanism and evolution of cytokinin biosynthesis. Proc Natl Acad Sci 105(7):2734–2739
Tahir M, Mirza MS, Hameed S et al (2015) Cultivation-based and molecular assessment of bacterial diversity in the rhizosheath of wheat under different crop rotations. PLoS One 10(6):e0130030
Taibi G, Schiavo MR, Gueli MC et al (2000) Rapid and simultaneous high-performance liquid chromatography assay of polyamines and monoacetylpolyamines in biological specimens. J Chromatogr B Biomed Sci Appl 745(2):431–437
Tang YW, Bonner J (1947) The enzymatic inactivation of indoleacetic acid; some characteristics of the enzyme contained in pea seedlings. Arch Biochem 13(1):11–25
Taylor IB, Burbidge A, Thompson AJ (2000) Control of abscisic acid synthesis. Journal of experimental Botany, 51(350):1563–1574
Teale WD, Paponov IA, Palme K (2006) Auxin in action: signalling, transport and the control of plant growth and development. Nat Rev Mol Cell Biol 7(11):847–859
Thuler DS, Floh EIS, Handro W et al (2003) Plant growth regulators and amino acids released by Azospirillum sp. in chemically defined media. Lett Appl Microbiol 37(2):174–178
Tien TM, Gaskins MH, Hubbell D (1979) Plant growth substances produced by *Azospirillum brasilense* and their effect on the growth of pearl millet (*Pennisetum americanum* L.). Appl Environ Microbiol 37(5):1016–1024
Tortora ML, Díaz-Ricci JC, Pedraza RO (2011) *Azospirillum brasilense* siderophores with antifungal activity against *Colletotrichum acutatum*. Arch Microbiol 193:275–286
Tsavkelova EA, Klimova SY, Cherdyntseva TA et al (2006) Microbial producers of plant growth stimulators and their practical use: a review. Appl Biochem Microbiol 42:117–126
Vrobel O, Tarkowski P (2023) Can plant hormonomics be built on simple analysis? A review. Plant Methods 19(1):107
Vrobel O, Ćavar Zeljković S, Dehner J et al (2024) Multi-class plant hormone HILIC-MS/MS analysis coupled with high-throughput phenotyping to investigate plant–environment interactions. Plant J 120(2):818–832
Wang L, Zou Y, Kaw HY et al (2020) Recent developments and emerging trends of mass spectrometric methods in plant hormone analysis: a review. Plant Methods 16:1–17

Wiszniewska A, Koźmińska A, Hanus-Fajerska E et al (2019) Insight into mechanisms of multiple stresses tolerance in a halophyte *Aster tripolium* subjected to salinity and heavy metal stress. Ecotoxicol Environ Saf 180:12–22
Xiong L, Zhu JK (2003) Regulation of abscisic acid biosynthesis. Plant Physiol 133(1):29–36
Xiong J, Fu G, Yang Y et al (2012) Tungstate: is it really a specific nitrate reductase inhibitor in plant nitric oxide research? J Exp Bot 63(1):33–41
Yang DL, Dong WX, Zhang YY et al (2013) Gibberellins modulate abiotic stress tolerance in plants. Sci Sin Vitae 43:1119–1126
Zahedi H, Abbasi S (2015) Effect of plant growth promoting rhizobacteria (PGPR) and water stress on phytohormones and polyamines of soybean. Indian J Agric Res 49(5):427–431
Zeevaart JA, Creelman RA (1988) Metabolism and physiology of abscisic acid. Annu Rev Plant Physiol Plant Mol Biol 39(1):439–473

Chapter 11
Role of Bacterial Surface Components in the *Azospirillum*-Plant Interaction

Alma Alejandra Hernández-Esquivel, Manuel Méndez-Gómez, Amanda Ortíz-Calderón, Elda Castro-Mercado, and Ernesto García-Pineda

Abstract The interaction between *Azospirillum* and plant roots is complex, involving bacterial derived molecules whose precise functions remain largely unknown. However, the role of some of these molecules in establishing root associations and stimulating biochemical responses in plant-root cells is gradually being uncovered. Bacterial molecules involved in root interaction include structural components of the outer membrane and flagellum, such as lipopolysaccharides, lectins, and flagellin, as well as excreted compounds like exopolysaccharides. These molecules actively promote plant growth and elicit defense responses. Current research suggests that these bacterial molecules, together with phytohormones, work synergistically to enhance the biological activity of *Azospirillum* in plants. Future studies will be essential to elucidate the specific contributions of each component to the beneficial effects of the rhizobacteria on plant health and development.

Keywords Surface components · Lipopolysaccharides · Flagellin · Exopolysaccharides · Plant interaction

11.1 *Azospirillum*-Derived Proteins, Lipids, and Carbohydrates as Root Colonization Mediators

Azospirillum bacteria are recognized as colonizers of the root tissue of various plant species, primarily cereals and grasses such as wheat and sorghum. They can attach to different areas of the root, including intercellular spaces of root hairs, the meristematic zone, and the elongation zone, exhibiting species-specific attachment

A. A. Hernández-Esquivel · A. Ortíz-Calderón · E. Castro-Mercado · E. García-Pineda (✉)
Instituto de Investigaciones Químico Biológicas, Universidad Michoacana de San Nicolás de Hidalgo, Morelia, Michoacán, México
e-mail: egpineda@umich.mx

M. Méndez-Gómez
Centro de Investigación y de Estudios Avanzados, I.P.N., Irapuato, Guanajuato, México

F. D. Cassán et al. (eds.), *The Definitive Handbook of Azospirillum*,
https://doi.org/10.1007/978-3-032-09222-9_11

preferences. However, only a limited number of *Azospirillum* strains have been identified as capable of colonizing plant tissues (Yegorenkova et al. 2001; Zhu et al. 2002).

The initial stage of the *Azospirillum*-plant interaction is influenced by the metabolic state of the bacteria (Borland et al. 2015). Through metabolic processes such as motility and chemotaxis, *Azospirillum* can assess its energy levels and migrate toward nutrient-rich niches. These attractants, secreted by plants, guide bacterial movement and correlate with the specificity of the chemotactic response, integrating a signal transduction pathway. This ultimately leads to the attachment and anchoring of the bacteria to the root (Wisniewski-Dyé et al. 2012).

Several plant-secreted metabolites stimulate metabolic changes in bacteria. Root exudates from various plants, including wheat, maize, bean, and alfalfa, induce protein secretion in *Azospirillum brasilense* Sp7. Notably, a 40-kDa acidic protein is highly induced in response to these exudates. This protein named SbpA and likely to be preilplasmic, mediates the chemotactic response to saccharides such as D-fucose, L-arabinose, and D-galactose, facilitating root colonization (Van Bastelaere et al. 1993). Cell surface and secreted proteins of *Azospirillum* have been poorly studied as potential factors in bacterial interactions with plant roots. Their role as integral components of the bacteria-plant interaction complex or as elicitors of plant responses remains a promising area of research.

A two-step mechanism has been proposed for the adhesion of *Azospirillum* with plant roots. The first step involves a fast and reversible binding mediated by proteins (Michiels et al. 1991). Notably, the Fla1 protein, a component of the polar flagellum, enables bacterial adsorption to plant roots (Croes et al. 1993). Studies on *Azospirillum brasilense* Sp7 have shown that a mutant lacking the Tlp1 protein (transducer-like protein 1), a chemoreceptor involved in bacterial chemotaxis toward organic acids, exhibits deficiencies in wheat root colonization (Greer-Phillips et al. 2004). Similarly, the Tlp2 protein (transducer-like protein 2) plays a role in the efficient colonization of red clover (*Trifolium pratense*) and bean (*Vigna unguiculata*), but not wheat, suggesting that these proteins may regulate host specificity in a selective manner (Ganusova et al. 2024).

Likewise, in *A. brasilense*, the CheA1 and CheA4 proteins (chemoreceptor-associated histidine kinases) regulate key aspects of bacterial metabolism, including motility based on chemotactic responses to various molecules, recognition of root surface regions, and colonization (Ganusova et al. 2021).

The second step involves strong and irreversible anchoring to the root tissue, mediated by the bacterial production of various extracellular polysaccharides. In this process, polysaccharide-derived microfibrils contribute to the attachment of *A. brasilense* Sp7 to wheat roots (Michiels et al. 1991). In the same strain, specific polysaccharides constituted by N-acetyl-D-glucosamine, N-acetyl-D-galactosamine, galactose, glucose, and mannose can bind to lectins found in plant roots (Skvortsov and Ignatov 1998).

Lectins are glycoproteins that play a crucial role in bacterial aggregation and subsequent root colonization (Wisniewski-Dyé et al. 2013). The Wheat germ agglutinin (WGA) lectin can interact with certain *Azospirillum* species due to the presence of N-acetyl-D-glucosamine-containing polysaccharide complexes on the bacterial cell surface. As a result, WGA has been suggested to function as an acceptor in *Azospirillum* root colonization (Yegorenkova et al. 2001; Skvortsov and Ignatov 1998). *Azospirillum brasilense* strains Sp7, Sp107, S17, and *A. baldaniorum* Sp245 produce at least two polysaccharide-containing complexes, which have been shown to bind wheat germ agglutinin. The polysaccharides from all strains contain rhamnose, galacturonic acid, and glucosamine. However, strains Sp7 and S17 also include fucose and mannose, respectively, and both contain galactose. These polysaccharide-containing complexes may contribute to the interaction between *Azospirillum* and wheat root surfaces (Konnova et al. 1994).

Lipopolysaccharides (LPSs) are molecules found in the outer membrane of Gram-negative bacteria, including the *Azospirillum* genus and are composed of a lipid component (lipid A) and a glycosidic component (core and O-antigen). They play a crucial role in plant recognition and subsequent root anchoring, as observed in interactions between *Pseudomonas putida* and potato (Mazgaeen and Gurung 2020). In wheat, *Azospirillum* LPS is essential for root hair deformation and colonization. Similarly, *A. brasilense* LPS mutants exhibit reduced colonization due to modifications in their glycosidic structures (Yegorenkova et al. 2001; Jofré et al. 2004). Whether plant roots have receptors for LPS remains unknown. Although receptors, such as lipooligosaccharide-specific reduced elicitation (LORE), have been identified in *Arabidopsis thaliana* as a sensor of LPS, their precise mode of action is still unclear (Shang-Guan et al. 2018).

Identifying the molecules from *Azospirillum* involved in the sequential events leading to successful root colonization would provide insights into the precise role of each player in the *Azospirillum*-plant-root interaction.

11.2 *Azospirillum* Exopolysaccharides and Lectins in Plant Growth Stimulation

Bacterial exopolysaccharides (EPS) are high-molecular-weight carbohydrates present in the cell walls of bacteria. Unlike capsular polysaccharides, EPS are easily released into the environment (Netrusov et al. 2023). In *Azospirillum*, several genes involved in EPS biosynthesis have been identified, including *exoB* (UDP-glucose 4′-epimerase), *exoC* (phosphomannomutase), *rmlD* (dTDP 4-rhamnose reductase), *phbC* (poly-b-hydroxyburyrate synthase), *noeJ* (mannose-6-phosphate isomerase), and *noeL* (GDP-mannose 4,6-dehydratase) (Jofré et al. 2004; Lerner et al. 2009; Michiels et al. 1988). Although the structural characteristics of *Azospirillum* EPS remain unknown, staining with calcofluor and Congo red suggests the presence of β-1,4- and β-1,3-linked glucans. These EPS are heteropolymers primarily composed of glucose, galactose, fucose, rhamnose, and arabinose, with smaller amounts of xylose and allose. However, their composition can vary depending on the species, strain, growth phase, and environmental conditions (Bahat-Samet et al. 2004; Burdman et al. 2000; Jofré et al. 2004; Lerner et al. 2009; Skvortsov and Ignatov 1998). For instance, under salt stress, *A. brasilense* Cd produces EPS rich in galactose, whereas under normal conditions, the EPS primarily consists of glucose (Fischer et al. 2003). Additionally, root exudates stimulate the production of arabinose and xylose in the EPS of *A. brasilense* Cd (Fischer et al. 2003). Furthermore, genome sequencing of *Azospirillum* strains has revealed the presence of unique genes, suggesting possible differences in biosynthetic pathways, as well as variations in the quantity and composition of these EPS (Wisniewski-Dyé et al. 2011, 2012).

Recent reviews suggest that EPS from rhizobacteria enhance plant growth through various mechanisms (Carezzano et al. 2023), and those produced by *Azospirillum* are no exception. For instance, maize plants inoculated with *A. brasilense* Ab-V5, a high-EPS-producing strain, exhibited growth levels comparable to fully fertilized plants, even when receiving only 20% of the usual fertilizer (Oliveira et al. 2017). Moreover, EPS alone has biological activity that promotes plant growth. Sankari et al. (2011) reported that EPS application from *Azospirillum* isolates A-17, A-18, A-26, and A-37 enhances rice growth and induces systemic resistance (ISR), which protects plants against *Pyricularia oryzae* infection, similar to findings reported by Park et al. (2008). ISR is well known to involve the hormones jasmonic acid (JA) and ethylene (ET), as well as the transcription factors MYB and MYC, which may be stimulated by EPS (Vlot et al. 2021). These molecules can also contribute to plant tolerance to metal toxicity. For example, *A. argentinense* Az39 has been identified as a potential bioinoculant for arsenic-contaminated soils due to its high-EPS and biofilm production in the presence of arsenic (Armendariz et al. 2015). Additionally, EPS and other bacterial components improve biochemical properties in plants under abiotic stress. Co-inoculation with *Bacillus subtilis* and *A. brasilense*, both high-EPS producers, enhances wheat growth, osmolyte accumulation (proline), and antioxidant activity (superoxide dismutase [SOD], catalase [CAT], and peroxidase [POD]) under drought conditions (Ilyas et al. 2020). Furthermore, increased EPS production enhances bacterial viability, protects inoculants, and improves plant growth promotion. This is particularly relevant because *Azospirillum* mutants deficient in EPS are more susceptible to environmental stressors such as salinity, high temperatures, antibiotics, and hydrogen peroxide (Fischer et al. 2003; Jofré et al. 1998; Lerner et al. 2009).

Although the contribution of EPS to plant growth promotion is well documented, the molecular mechanisms behind this effect remain unclear. Exopolysaccharide receptors in plant cells may play a role in recognizing these molecules. One example is Epr3 (*exopolysaccharide receptor 3*), which encodes a LysM receptor-like kinase expressed in the root epidermis of *Lotus japonicus* (Kawaharada et al. 2015).

Additionally, EPS can associate with proteins found in bacterial cell walls (Ilyas et al. 2020).

Lectins are glycoproteins that bind specifically to carbohydrates on cell surfaces through carbohydrate-recognition domains (CRDs). This interaction offers microbes several advantages, including enhanced adhesion, facilitation of infection, and inhibition of competing microorganisms (Fares et al. 2025; Kotecha and Poduval 2019). *A. brasilense* and *A. lipoferum* have been reported to contain lectins of approximately 43 kDa in their cell walls (Castellanos et al. 1998). Neoglycoprotein-labeling assays indicate that these proteins interact with galactosamide, glucosamide, mannopyranoside, and fucosylamide moieties (Castellanos et al. 1998). Similarly, *A. brasilense* Sp7 produces a 67 kDa lectin that recognizes the EPS of the same bacterium. Analysis of lectin-polysaccharide interactions using EPS, D-mannose, D-glucose, and L-arabinose immobilized on Sepharose suggests that this lectin binds to these polysaccharides and monosaccharides to a lesser extent, indicating that EPS composition may influence lectin-EPS interactions (Mora et al. 2008). Additionally, the 36 kDa lectins of *A. brasilense* Sp7 interact with L-fucose and D-galactose monomers (Alen'kina and Nikitina 2010).

The role of *Azospirillum* lectins in plant-microorganism interactions remains relatively understudied, though they are known to contribute to plant colonization and growth stimulation. It has been suggested that *Azospirillum* lectins act as signaling molecules that enhance the activity of superoxide dismutase, peroxidases, oxalate oxidase, and lipoxygenase, as well as increase nitric oxide levels and salicylic acid content in wheat roots (Alen'kina et al. 2014; Alen'kina and Nikitina 2010, 2015). However, plant responses differ when treated with EPS from *A. brasilense* Sp7.2.3 mutants deficient in lectin activity (Alen'kina and Nikitina 2010). In addition, *Azospirillum* lectins appear to play a role in seed germination. Lectins from the endophytic strain *A. baldaniorum* Sp245 are more effective in promoting wheat seed germination under various abiotic stress conditions than those from the epiphytic strain *A. brasilense* Sp7 (Alen'kina and Nikitina 2021). Therefore, differences in the EPS and lectin structures of *Azospirillum*, as well as their interactions with plant cell components, may account for variations in colonization capacity and plant growth promotion. A deeper understanding of how plants perceive *Azospirillum* EPS, interact with bacterial lectins, and activate the signaling pathways involved in growth is essential for the effective application of these bacteria in crop management.

11.3 *Azospirillum* Lipopolysaccharides as Plant Growth Modulators

Signaling molecules and surface components from both plant and bacterial cells play a crucial role in establishing plant-bacterium associations (Sigida et al. 2019a). Among these components, lipopolysaccharide (LPS), a major constituent of the

Gram-negative bacterial outer membrane, is particularly significant. LPS was discovered in the late nineteenth century and initially termed as endotoxin (Knirel and Valvano 2011). The first isolation of LPS was performed on strain *A. lipoferum* SpBr17 (Choma et al. 1984).

In *Azospirillum* cells, LPS constitutes approximately 70% of the cell surface and is considered essential for plant-bacterium interactions (Fedonenko et al. 2008a; Fedonenko et al. 2015).

Structurally, LPSs are amphiphilic molecules consisting of a hydrophobic lipid A component, which is linked to a core oligosaccharide (core). This core, in turn, is connected to an O-specific polysaccharide (O antigen, O chain, or OPS) composed of repeating oligosaccharide units. The S (smooth) form of LPS includes all three components (lipid A, the core oligosaccharide, and the O-specific polysaccharide), whereas the R (rough) form lacks the O-specific polysaccharide and consists only of lipid A and the core oligosaccharide (Sigida et al. 2013a, b).

The analysis of LPS from *A. lipoferum* SpBr17 revealed the neutral sugar and fatty acid composition of the LPS, identifying rhamnose and glucose as the primary components of the O-side chains, which together accounted for more than 95% of the neutral sugars. Additionally, a high concentration of two distinct 3-hydroxylated fatty acids and the presence of D-glycero-D-mannoheptose were detected, suggesting that the LPS of *A. lipoferum* SpBr17 belongs to the S type (Choma et al. 1984). Shortly afterward, the LPS of ten different *Azospirillum* strains, five from *A. brasilense* and five from *A. lipoferum*, was analyzed. The LPS composition was easily distinguishable between the species, with *A. brasilense* exhibiting a more complex sugar profile, including rhamnose, fucose, galactose, mannose, and glucose in varying proportions (Choma et al. 1987).

The characterization of the O-polysaccharides of LPS from various *Azospirillum* strains (Table 11.1) revealed a high degree of similarity among them, attributed to the presence of repeating structural units in their O-antigens. The serological properties and chemotypes of O-antigens have led to the classification of these strains into three serogroups. Serogroup I is characterized by linear D-rhamnose. These bacteria are widely distributed across different climate zones and are commonly associated with wheat. A representative species of this group is *A. baldaniorum* Sp245 (Fedonenko et al. 2011). Serogroup II encompasses the largest number of strains and exhibits the greatest diversity in the OPS of their LPS. Classification into this serogroup is based on serological cross-reactions with the type strain *A. brasilense* Sp7 (Konnova et al. 2008; Fedonenko et al. 2015). The host range for Serogroup II Azospirilla includes various gramineous plants, such as crabgrass, Cynodon, millet, rice, and wheat. Serogroup III is distinguished by the presence of L-rhamnose in the main chain and D-glucose in the side chain. A representative strain of this group is *A. lipoferum* 59b (Sigida et al. 2014a). The structure of lipid A in *Azospirillum* has only been reported for *A. lipoferum* SpBr17. Its backbone consists of a β-(1,6)-linked D-glucosamine disaccharide but notably lacks phosphate residues, in contrast to enterobacterial lipid A, which contains phosphate residues and is strongly associated with the endotoxic activity of LPS molecules. This

Table 11.1 Structures of repeating units of LPS from *Azospirillum* spp.

	Strain	Structure of repeating units	References
Azospirillum lipoferum	SpBr17	→3)-2-O-Ac-α-L-Rhap-(1→2)-α-L-Rhap-(1→3)-α-L-Rhap-(1→ 2 (O-Ac); 3 ↑ 1 β-D-Glcp	Choma et al. (1984, 2009)
	Sp59b	α-L-Rhap-(1→3)-α-L-Rhap-(1→2)-α-L-Rhap-(1→3)-β-D-Manp-(1→4) 1 ↓ 4 →3)-β-D-Galp-(1→3)-α-D-Galp-(1→	Fedonenko et al. (2005)
	Sp59b SR66	β-D-Glcp-(1→3)┐ O-Ac 2 →2)-α-L-Rhap-(1→3)-α-L-Rhap-(1→3)-α-L-Rhap-(1→	Smol'kina et al. (2010) and Fedonenko et al. (2011)
	SR65 SR85	β-D-Glcp-(1→3)┐ →2)-α-L-Rhap-(1→3)-α-L-Rhap-(1→3)-α-L-Rhap-(1→	Fedonenko et al. (2008b, 2011)
	SR5 SR16 SR38 SR61	β-D-Glcp-(1→3)┐ 2-O-Ac →2)-α-L-Rhap-(1→3)-α-L-Rhap-(1→3)-α-L-Rhap-(1→	Fedonenko et al. (2015)
Azospirillum brasilense	SR75 Sp107 RG20a	→2)-β-D-Rhap-(1→3)-α-D-Rhap-(1→3)-α-D-Rhap-(1→2)-α -D-Rhap-(1→2)-α-D-Rhap-(1→	Fedonenko et al. (2005) and Boiko et al. (2010)
	S17 S27	β-D-GlcpNAc(1→4)┐ →4)-α-L-Rhap2Me-(1→3)-α-D-ManpN(SHb)-(1→ **1** β-D-Glcp-(1→2)┐ →3)-α-L-Rhap-(1→3)-α-L-Rhap-(1→2)-α-L-Rhap-(1→ **2**	Fedonenko et al., (2008a, 2015)
	R15	→2)-α-D-Rhap-(1→ 2)-β-D-Rhap-(1→3)-α-D-Rhap-(1→2)-α -D-Rhap-(1→	Boiko et al. (2010)

(continued)

Table 11.1 (continued)

	Strain	Structure of repeating units	References
	54 SR8	β-D-Glc*p*-(1→4)┐ →3)-α-L-Rha*p*-(1→3)-α-L-Rha*p*-(1→2)-α-L-Rha*p*-(1→	Fedonenko et al. (2011, 2015)
	R55	→3)-β-L-Rha*p*-(1→4)-α-L-Rha*p*-(1→3)-α-D-Gal*p*-(1→3)-β-L-Rha*p*-(1→4)-β-L- α-D-Glc*p*A-(1→2)┘ Rha*p*3OMe-(1→4)-α-L-Rha*p*-(1→3)-α-D-Gal*p*-(1→	Boyko et al. (2011)
	Jm6B2	α-D-Rha*p*3OMe-(1→3)┐ →4)-α-L-Fuc*p*-(1→4)-β-D-Xyl*p*-(1→	Boyko et al. (2012)
	SR80 SR50	α-L-Fuc*p*-(1→2)┐ →3)-β-D-Gal*p*NAc-(1→3)-α-D-Gal*p*-(1→3)-β-D-Gal*p*-(1→	Sigida et al. (2013a, 2014a)
	Sp7	→3)-α-L-Rha*p*-(1→3)-β-D-Gal*p*-(1→3)-β-D-Glc*p*NAc-(1→ MeO2 ┘ α-L-Fuc*p*-(1→4) ┘	Sigida et al. (2013b)
	SR88 SR109 SR111 SR115	α-D-Rha*p*-(1→3)┐ →4)-α-L-Fuc*p*-(1→4)-β-D-Xyl*p*-(1→	Sigida et al. (2014a)
	Sp246 SpBr14	β-D-Glc*p*-(1→4)┐ →2)-α-L-Rha*p*-(1→3)-α-L-Rha*p*-(1→3)-α-L-Rha*p*-(1→ **1** α-D-Fuc*p*3NAc-(1→2)┐ →4)-α-L-Rha*p*-(1→3)-β-D-Man*p*NAc-(1→ **2**	Sigida et al. (2014b)
	Jm125A2	β-D-Glc*p*-(1→2)┐ →3)-α-L-Rha*p*-(1→3)-α-L-Rha*p*-(1→2)-α-L-Rha*p*-(1→ **1** α-D-Fuc*p*3NAc-(1→2)┐ →3)-α-L-Rha*p*-(1→3)-β-D-Glc*p*NAc-(1→ **2**	Sigida et al. (2015)

	SR8	β-D-Glc*p*-(1→4)┐ →2)-α-L-Rha*p*-(1→3)-α-L-Rha*p*-(1→3)-α-L-Rha*p*-(1→ **1** β-D-Gal*p*NAc4Me-(1→4)┐ →4)-α-L-Rha*p*-(1→3)-β-D-Man*p*NAc-(1→ **2**	Sigida et al. (2019a)
Azospirillum	*baldaniorum* Sp245	→2)-β-D-Rha*p*-(1→3)-α-D-Rha*p*-(1→3)-α-D-Rha*p*-(1→2)-α -D-Rha*p*-(1→2)-α-D-Rha*p*-(1→	Fedonenko et al. (2002)
	fermentarium CC-LY743^T	α-D-Man*p*-(1→3)-α-L-Fuc*p*-(1→3)-α-D-Man*p*-(1→3)-α-L-Fuc*p*-(1→	Sigida et al. (2018)
	doebereinerae GSF-71^T	→3)-α-L-Rha*p*-(1→3)-α-L-Rha*p*-(1→3)-α-L-Rha*p*-(1→3)-β-D-Glc*p*NAc-(1→ └ 2-O-Ac └ 6-O-Ac	Sigida et al. (2019b)
	formosense CC-Nfb-7^T	α-D-Man*p*3Me-(1→3)┐ →4)-α-L-Fuc*p*-(1→4)-β-D-Xyl*p*-(1→	Sigida et al. (2020)
	zea N7^T *melinis* TMCY 0552^T *palustre* B2^T	β-D-Glc*p*-(1→3)┐ 2-O-Ac →2)-α-L-Rha*p*-(1→3)-α-L-Rha*p*-(1→3)-α-L-Rha*p*-(1→	Sigida et al. (2022)
	agricola CC-HIH038	→3)-α-L-Rha*p*2Ac-(1→3)-α-L-Rha*p*-(1→3)-α-L-Rha*p*-(1→3)-β-D-Glc*p*NAc6Ac-(1→	Sigida et al. (2024)

structural difference suggests that *Azospirillum* lipid A may exhibit lower toxicity toward eukaryotic cells (Choma et al. 2009).

Research on these macromolecules, particularly their chemical structure, is crucial for advancing our molecular-level understanding of the mechanisms that drive effective plant-bacteria symbiosis (Fedonenko et al. 2005). Mutants with impaired LPS biosynthesis show reduced adsorption efficiency to wheat roots and a diminished capacity to colonize maize roots compared to their wild-type counterparts (Fedonenko et al. 2001; Jofré et al. 2004; Fedonenko et al. 2008b).

Lipopolysaccharides of plant pathogenic bacteria are classified as microbe-associated molecular patterns (MAMPs), which are recognized by the plant innate immune system. This recognition triggers a range of defense responses, including the production of reactive oxygen species (ROS), nitric oxide (NO), and pathogenesis-related (PR) proteins (Newman et al. 2013). The recognition of MAMPs involves pattern recognition receptors (PRRs), which can be kinase-type receptors featuring an extracellular domain for ligand perception, a transmembrane domain, and an intracellular kinase domain (Boutrot and Zipfel 2017).

Recent studies have focused on elucidating the role of specific LPS components in plant-microbe interactions, with evidence suggesting that the OPS moiety may act as an elicitor, triggering a cascade of plant defense responses (Sigida et al. 2015).

Plants can use MAMP recognition not only to mount immune responses but also to establish beneficial symbiotic relationships and improve plant growth. The application of LPS from *Azospirillum* strains to wheat (*Triticum aestivum*), cucumber (*Cucumis sativus*), potato, and *Arabidopsis* has been shown to induce early growth responses in seedlings. However, to date, no specific receptors for *Azospirillum* LPS have been identified. LPS from *A. baldaniorum* Sp245 is also able to activate the mitotic activity of meristematic cells in wheat roots, which was comparable to the whole bacterial cells (Evseeva et al. 2011).

Plant growth parameters such as enhanced leaf elongation, dry weight, and spike formation were indeed improved with the exogenous application of *A. baldaniorum* Sp245 LPS on wheat. However, they also accelerated plant aging, suggesting that LPS influences developmental processes in wheat (Chávez-Herrera et al. 2018).

LPS from Sp245 has been shown to stimulate biochemical responses in wheat and cucumber plants, including the production of reactive oxygen species (ROS) (Vallejo-Ochoa et al. 2017), elevated peroxidase activity, increased Ca^{2+} availability (Hernández-Esquivel et al. 2020), and total phenolic content (Morales-Ascencio et al. 2019).

Interestingly, LPS from *Azospirillum* significantly stimulates morphogenetic activity of wheat somatic tissues in comparison with LPS from *Enterobacteriaceae*, in addition to the increase in plant size and weight (Evseeva et al. 2018). Furthermore, the O-polysaccharide appears to be important in the morphogenetic activity of *Azospirillum* LPS. A comparative study of LPS from three *Azospirillum* strains (*A. brasilense* SR55, *A. brasilense* SR75, and *A. lipoferum* SR65) on calluses derived from two wheat (*Triticum aestivum*) lines (LRht-B1c and LRht-B1a) differing in their morphogenetic activity shows that LPS from *A. lipoferum* SR65 significantly promoted callus morphogenesis and regenerative development in both wheat lines. Meanwhile, LPS from *A. brasilense* SR55 and SR75 selectively enhanced either the yield of morphogenetic calluses or the yield of regenerated plants, respectively. Overall, *Azospirillum* LPS had a stronger effect on the weakly morphogenetic LRht-B1a line than on the highly morphogenetic LRht-B1c line (Tkachenko et al. 2021).

The monosaccharide residues of the O-polysaccharide play a major role in the stimulatory effects of LPS on plant growth promotion. In a study analyzing the effects of LPS from six bacterial strains belonging to different taxonomic groups (*Ochrobactrum quorumnocens* T1Kr02, *Pseudomonas chlororaphis* K3, *Azospirillum brasilense* SR80, *Azospirillum brasilense* SR88, *Azospirillum thiophilum* BV-S, and *Azospirillum zeae* N7) on potato microplants, growth parameters such as shoot length, total root length, and dry weight of shoots and roots were evaluated. The LPS from *O. quorumnocens* T1Kr02 and *A. thiophilum* BV-S exhibited the most significant positive effects. Notably, the O-polysaccharides of these strains consist predominantly or entirely of 6-deoxyhexose residues, specifically fucose and rhamnose (Ivanova et al. 2024).

The molecular mechanism by which LPS influences plant growth remains under investigation. However, the target of rapamycin (TOR) signaling pathway has been shown to play a role in the cellular response to LPS. *Arabidopsis thaliana* plants sprayed with two different concentrations of *A. baldaniorum* Sp245 LPS (10 and 100 μg/mL) exhibited enhanced growth and development. This effect correlated with increased expression and activity of the TOR kinase, as well as phosphorylation of its downstream target, the S6K protein. These changes also promoted cell proliferation, suggesting that *A. baldaniorum* LPS may modulate TOR-mediated processes (Méndez-Gómez et al. 2020). While these findings provide valuable insight into the molecular components of LPS activity, many questions remain. The upstream elements that activate this pathway, potential crosstalk with phytohormone signaling, and the precise mechanisms by which plants recognize *Azospirillum* LPS are still unknown. Further research is needed to fully elucidate these complex interactions.

11.4 Function of Flagellin in the Interaction Between Plants and *Azospirillum*

A. brasilense is a Gram-negative, nitrogen-fixing rhizobacterium distinguished by a mixed flagellation pattern. It has a single polar flagellum, which facilitates swimming (Tarrand et al. 1978; Zhulin and Armitage 1993), and multiple lateral flagella that enable swarming on solid surfaces (Hall and Krieg 1983).

Motility and chemotaxis allow *A. brasilense* to detect and move toward energy sources. The bacterium exhibits chemotactic responses to various amino acids, organic acids, and sugars (Zhulin and Armitage 1993). Among these, oxalate secreted by rice and malate exuded by maize play crucial roles in determining host specificity during colonization (Reinhold et al. 1985).

The initial stage of root colonization, involving adsorption to plant roots, is facilitated by the polar flagellum (Croes et al. 1993). This structure is critical for early colonization, as mutants lacking both polar and lateral flagella show significantly reduced adsorption capacity (Tarrand et al. 1978; Croes et al. 1993). Supporting this, studies have demonstrated that purified polar flagella effectively bind to wheat roots, whereas lateral flagella do not exhibit this binding ability (Croes et al. 1993).

Flagellin, the primary structural protein of the bacterial flagellum filament, is one of the most abundant and widely distributed proteins in motile bacteria (Chaban et al. 2015). The flagellar filament, a tubular structure extending up to 15 μm in length, is assembled through the distal polymerization of up to 30,000 flagellin (FliC) subunits (Paradis et al. 2017).

The role of flagellins from plant-growth-promoting rhizobacteria (PGPR) in triggering plant responses remains largely unexplored (Sumayo et al. 2013), though progress has been made in their characterization. Glycosylation of the flagellin in the polar flagellum was first identified in *A. brasilense* type strain Sp7 (Moens et al. 1995). Subsequent studies confirmed the presence of an O-linked polysaccharide in the glycosylated flagellin of Sp7 and characterized the structure of its tetrasaccharide repeating unit (Belyakov et al. 2012). More recently, research has shown that flagellin glycosylation is widespread among Gram-negative bacteria (Merino and Tomás 2014). However, the biological significance of flagellin glycosylation in plant-bacteria interactions remains unknown.

Research has identified a conserved 22-amino acid peptide, Flg22, within the flagellin protein of *Pseudomonas aeruginosa*, as the key elicitor responsible for these immune responses. Flg22 is embedded within the bacterial flagellum and is released by plant-secreted proteinases in its biologically active form (Sanguankiattichai et al. 2022). Whether a similar mechanism exists for generating Flg22 during interactions between beneficial bacteria and plants remains unknown and warrants further investigation. To date, no studies have reported the biological activity of *Azospirillum* Flg22 in plants. It remains unclear whether this peptide plays a role in the plant growth-promoting effects of *Azospirillum* or whether plants associated with *Azospirillum* possess specific receptors for its Flg22.

Flagellin is perceived by FLS2, a receptor-like kinase (RLK) composed of an extracellular leucine-rich repeat (LRR) domain, a single membrane-spanning domain, and an intracellular serine/threonine protein kinase domain (Gómez-Gómez and Boller 2000; Shiu and Bleecker 2001; Song et al. 1995). The LRR domain mediates interactions with ligands, which suggests that the LRR domain of FLS2 may serve as a receptor for flg22 and flagellin (Gómez-Gómez and Boller 2000). In addition, flagellin can be detected by different plant receptors, like FLS3, as was reported in Solanaceous plants (Hind et al. 2016). Whether FLS receptors perceive flagellin from pathogenic and beneficial bacteria in the same manner remains an open question.

Although the flagellum plays a key role in the adsorption of *A. brasilense* to plant roots (Croes et al. 1993), the specific contributions of flagellin and lipopolysaccharides (LPSs) in this process remain unclear. Notably, LPS appears to form a sheath around the flagellar filament (Burygin et al. 2007), which may influence its interaction with plant surfaces.

Moens et al. (1995) demonstrated that the flagellin of the polar flagellum in *A. brasilense* Sp7 is glycosylated, while Belyakov et al. (2012) further characterized these glycan chains as a polysaccharide with a branched tetrasaccharide repeating unit composed of rhamnose (Rha), fucose (Fuc), galactose (Gal), and *N*-acetylglucosamine (GlcNAc). Notably, this tetrasaccharide shares the same monosaccharide residue ratio as the O-specific polysaccharide chain of the lipopolysaccharide (LPS) from the same strain, highlighting a close structural relationship between these two glycopolymers.

Flagellin glycosylation may influence the interaction of *Azospirillum* with plant hosts. While direct studies on *Azospirillum* flagellin glycosylation and immune evasion remain limited, research on other bacteria suggests that glycosylation enhances the structural stability of flagella, potentially reducing the release of flagellin monomers and, consequently, minimizing recognition by plant immune receptors (Rossez et al. 2015). This raises the possibility that a key distinction between plant-beneficial and pathogenic bacteria in plant recognition and defense activation could lie in flagellin glycosylation.

Beyond its structural role, the flagellin of *A. baldaniorum* Sp245 functions as an elicitor of molecular changes associated with plant growth promotion. Remarkably, it can independently stimulate mitosis in wheat roots (Shirokov et al. 2020), demonstrating that bacterial structural components play a crucial role in activating plant growth responses. These effects, combined with bacterial phytohormone production, contribute to the mechanisms by which plant growth-promoting rhizobacteria (PGPR) enhance plant development.

Additionally, *Azospirillum* flagellin (from *A. brasilense* REC3) has been shown to act as a microbe-associated molecular pattern (MAMP), triggering plant defense responses. These include reactive oxygen species (ROS) accumulation, salicylic acid production, callose and lignin deposition, increased biofilm formation on leaves, and stomatal closure, ultimately providing strawberry plants with protection against the pathogenic fungus *Macrophomina phaseolina* (Elías et al. 2022). This property of flagellin suggests that it may play a role in activating plant responses involved in *Azospirillum*-mediated biological control.

11.5 In Vitro Assay of *Arabidopsis* Biostimulation by Lipopolysaccharides from *Azospirillum*

Lipopolysaccharide (LPS) is a key component of the outer membrane of *Azospirillum*. Structurally, LPS consists of three distinct regions: lipid A, which anchors it to the bacterial membrane; a core oligosaccharide; and repeating O-antigen units that extend outward from the cell surface. Lipid A, characterized by multiple fatty acids, plays a crucial role in embedding LPS within the membrane. Here, we describe an *in vitro* bioassay to evaluate the biostimulatory effects of *Azospirillum*-derived LPS on *Arabidopsis* plants.

11.5.1 Lipopolysaccharides Extraction (Adapted from Renukadevi et al. 2012 and Nguyen et al. 2019)

1. Pre-inoculum: In a glass tube, add 3 mL of LB minimum (LBm) medium. Inoculate with a colony of *Azospirillum*.
2. Incubate at 30 °C with constant agitation for 16 h. Include a negative control (a glass tube containing only LBm, without bacterial inoculation).
3. In an Erlenmeyer flask, add 200 mL of LBm, then inoculate with 300 μL of the pre-inoculum.
4. Incubate at 30 °C with constant agitation for 24 h. Include a negative control (an Erlenmeyer flask containing only 100 mL of LBm, without pre-inoculum). Ensure that the OD_{600} nm ≥ 0.9.
5. Distribute the culture into four 50 mL Falcon tubes (50 mL per tube).
6. Centrifuge at 3200 rpm for 10 min.
7. Remove 40 mL of the supernatant from each tube, leaving 10 mL of culture medium in each. Resuspend the bacterial pellet using a vortex mixer.
8. Combine the resuspended pellets into a single Falcon tube (total volume: 40 mL).
9. Centrifuge again at 3200 rpm for 5 min.
10. Resuspend the pellet in 5 mL of sterile distilled water using a vortex mixer.
11. Add 5 mL of ice-cold Tris-saturated phenol and mix thoroughly.
12. Incubate at 65 °C for 20 min.
13. Store at 4 °C overnight.
14. Centrifuge at 3800 rpm for 10 min.
15. Transfer the aqueous phase into a dialysis membrane, sealing the bottom with a cotton thread (Before adding the sample, test the dialysis membrane by filling it with sterile distilled water to ensure it is securely sealed).
16. Seal both ends of the dialysis membrane (forming a "candy" shape) and place it in a beaker containing sterile distilled water.
17. Dialyze for 3–4 days, replacing the sterile distilled water every 24 h to remove phenol traces. The resulting liquid is referred to as the "dialysate."
18. Transfer the dialysate into Eppendorf tubes.
19. Centrifuge at 12,000 rpm for 10 min.
20. Weigh sterile Eppendorf tubes before use.
21. Transfer the aqueous phase into a pre-weighed Eppendorf tube.
22. Add 0.15 M sodium acetate (pH 8.0) to the dialysate.

23. Add absolute ethanol in a 1:4 ratio (v/v).
24. Incubate at −20 °C for 24 h.
25. Centrifuge at 4200 rpm for 20 min.
26. Remove the supernatant and air-dry the pellet at room temperature for 20 min. To dry, invert the Eppendorf tube on absorbent paper, ensuring the pellet remains intact.
27. Resuspend the pellet in 200 μL of sterile distilled water and transfer it to a pre-weighed Eppendorf tube. Ensure complete dissolution by pipetting.
28. Add 5 μL of DNase I and 5 μL of RNase from a 10 mg/mL stock solution (final concentration: 0.25 μg/μL). Incubate at 37 °C for 30 min.
29. Add 10 μL of Proteinase K from a 10 mg.mL^{-1} stock solution (final concentration: 0.50 μg.μL^{-1}). Incubate at 59 °C for 1 h.
30. Precipitate the lipopolysaccharides (LPSs) by adding ethanol in a 3:1 ratio (v/v). Store at −20 °C overnight.
31. Centrifuge at 4200 rpm for 20 min and remove the supernatant.
32. Weigh the tube containing the dried pellet to calculate the LPS yield by subtracting the pre-weighed tube mass.
33. Resuspend the LPS pellet in sterile distilled water.
34. Store at 4 °C until use.

11.5.2 *In Vitro Bioassay*

1. Wash and disinfect *Arabidopsis thaliana* seeds as follows:
   - Add 96% (v/v) ethanol for 5 min.
   - Remove the ethanol and add 20% (v/v) chlorine for 5 min.
   - Remove the chlorine and wash the seeds five times with sterile deionized water, each wash lasting 2 min.
2. Store the washed seeds at 4 °C for 48 h.
3. Germinate the seeds on Petri dishes containing 0.2 MS medium for 4 days until they reach approximately 1 cm in length.
4. Prepare 0.2× MS culture medium, sterilize it, and allow it to cool to a warm temperature. Add lipopolysaccharides to a final concentration of 100 μg.mL^{-1}.
5. Pour the prepared culture medium into Petri dishes and allow it to solidify.
6. Using 1000 μL pipette tips, transfer 4-day-old germinated seedlings to the Petri dishes containing MS culture medium with lipopolysaccharides, ensuring a 5 mm spacing between plants. Place 10 plants per Petri dish.
7. Seal the Petri dishes with Parafilm and place them horizontally at a 65° angle in a plant growth chamber for 7 days or the duration required for the experiment (Fig. 11.1).

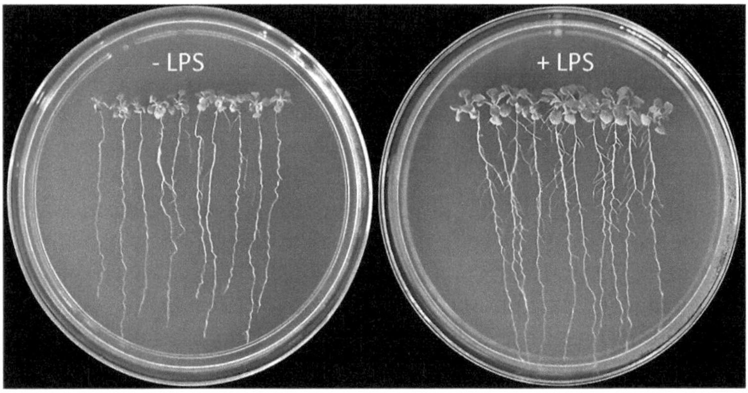

Fig. 11.1 *Arabidopsis* seedlings grown for 10 days in the presence of *Azospirillum baldaniorum* Sp245 LPS (100 μg/ml). (Image courtesy of Eusebio Álvarez-Santamaría)

11.6 Conclusion and Perspectives

The colonization of plant roots by *Azospirillum* is a multi-stage process orchestrated by a complex array of bacterial surface molecules, including proteins, lipids, and carbohydrates. The fact that some components are involved in the colonization of different plant species suggests that *Azospirillum* has evolved sophisticated mechanisms for host-specific recognition and attachment.

Next to root colonization, some surface components act as elicitors, triggering specific responses in plants. *Azospirillum* EPS promotes plant development, enhances nutrient uptake efficiency, protects plants against abiotic stresses, and activates plant defense systems. With regard to LPS, these can trigger beneficial plant responses such as enhanced cell proliferation, improved morphogenetic activity, and activation of the TOR signaling pathway.

Finally, flagellin, beyond its structural role in motility, chemotaxis, and adhesion to roots, can trigger molecular modifications associated with plant growth response and provide protection against pathogens.

Significant knowledge gaps remain, particularly concerning how these bacterial elicitors influence the biochemical pathways that regulate plant morphology and growth, as well as the identification of their corresponding plant receptors.

References

Alen'kina SA, Nikitina VE (2010) Effect of lectins from *Azospirillum brasilense* to peroxidase and oxalate oxidase activity regulation in wheat roots. Biol Bull 37:89–92. https://doi.org/10.1134/S106235901001012

Alen'kina SA, Nikitina VE (2015) Effect of *Azospirillum* lectins on the activity of proteolytic enzymes and their inhibitors in wheat seedling roots. Microbiol (Russian Federation) 84:630–635. https://doi.org/10.1134/S0026261715050021

Alen'kina SA, Nikitina VE (2021) Stimulating effect from lectins of associative bacteria of the genus *Azospirillum* on the germination and morphometric characteristics of spring wheat sprouts in simulated abiotic stress. Russ J Plant Physiol 68:315–321. https://doi.org/10.1134/S1021443721010027

Alen'kina SA, Bogatyrev VA, Matora LY et al (2014) Signal effects of the lectin from the associative nitrogen-fixing bacterium *Azospirillum brasilense* Sp7 in bacterial-plant root interactions. Plant Soil 381:337–349. https://doi.org/10.1007/s11104-014-2125-6

Armendariz AL, Talano MA, Wevar Oller AL et al (2015) Effect of arsenic on tolerance mechanisms of two plant growth-promoting bacteria used as biological inoculants. J Environ Sci 33: 203–210. https://doi.org/10.1016/j.jes.2014.12.024

Bahat-Samet E, Castro-Sowinski S, Okon Y (2004) Arabinose content of extracellular polysaccharide plays a role in cell aggregation of *Azospirillum brasilense*. FEMS Microbiol Lett 237:195–203. https://doi.org/10.1016/j.femsle.2004.06.036

Belyakov AY, Burygin GL, Arbatsky NP et al (2012) Identification of an O-linked repetitive glycan chain of the polar flagellum flagellin of *Azospirillum brasilense* Sp7. Carbohydr Res 361:127–132. https://doi.org/10.1016/j.carres.2012.08.019

Boiko AS, Smol'kina ON, Fedonenko Y et al (2010) O-polysaccharide structure in serogroup I azospirilla. Microbiology 79:197–205. https://doi.org/10.1134/S0026261710020116

Borland S, Oudart A, Prigent-Combaret C et al (2015) Genome-wide survey of two-component signal transduction systems in the plant growth-promoting bacterium *Azospirillum*. BMC Genomics 16:833. https://doi.org/10.1186/s12864-015-1962-x

Boutrot A, Zipfel C (2017) Function, discovery, and exploitation of plant pattern recognition receptors for broad-spectrum disease resistance. Annu Rev Phytopathol 55:257–286. https://doi.org/10.1146/annurev-phyto-080614-120106

Boyko AS, Konnova SA, Fedonenko YP et al (2011) Structural and functional peculiarities of the lipopolysaccharide of *Azospirillum brasilense* SR55, isolated from the roots of *Triticum durum*. Microbiol Res 166:585–593. https://doi.org/10.1016/j.micres.2011.01.002

Boyko AS, Dmitrenok AS, Fedonenko YP et al (2012) Structural analysis of the O-polysaccharide of the lipopolysaccharide from *Azospirillum brasilense* Jm6B2 containing 3-O-methyl-D-rhamnose (D-acofriose). Carbohydr Res 355:92–95. https://doi.org/10.1016/j.carres.2012.04.006

Burdman S, Jurkevitch E, Soria-Díaz ME et al (2000) Extracellular polysaccharide composition of *Azospirillum brasilense* and its relation with cell aggregation. FEMS Microbiol Lett 189:259–264. https://doi.org/10.1111/j.1574-6968.2000.tb09240.x

Burygin GL, Shirokov AA, Shelud'ko AV (2007) Detection of a sheath on *Azospirillum brasilense* polar flagellum. Microbiology 76:728–734. https://doi.org/10.1134/S0026261707060124

Carezzano ME, Alvarez Strazzi FB, Pérez V et al (2023) Exopolysaccharides synthesized by rhizospheric bacteria: a review focused on their roles in protecting plants against stress. Appl Microbiol 3:1249–1261. https://doi.org/10.3390/applmicrobiol3040086

Castellanos T, Ascencio F, Bashan Y (1998) Cell-surface lectins of *Azospirillum* spp. Curr Microbiol 36:241–244. https://doi.org/10.1007/s002849900302

Chaban B, Hughes HV, Beeby M (2015) The flagellum in bacterial pathogens: for motility and a whole lot more. Semin Cell Dev Biol 46:91–103. https://doi.org/10.1016/j.semcdb.2015.10.032

Chávez-Herrera E, Hernández-Esquivel AA, Castro-Mercado E et al (2018) Effect of *Azospirillum brasilense* Sp245 lipopolysaccharides on wheat plant development. J Plant Growth Regul 37: 859–866. https://doi.org/10.1007/s00344-018-9782-2

Choma A, Russa R, Lorkiewicz Z (1984) Chemical composition of lipopolysaccharide from *Azospirillum lipoferum*. Microbiol Lett 22:245–248. https://doi.org/10.1111/j.1574-6968.1984.tb00735.x

Choma A, Russa R, Meyer H et al (1987) Chemical analysis of *Azospirillum* lipopolysaccharides. Arch Microbiol 146:341–345. https://doi.org/10.1007/BF00410933

Choma A, Komaniecka I, Sowinski P (2009) Revised structure of the repeating unit of the O-specific polysaccharide from *Azospirillum lipoferum* strain SpBr17. Carbohydr Res 344: 936–939. https://doi.org/10.1016/j.carres.2009.02.021

Croes CL, Moens S, Van Bastelaere E et al (1993) The polar flagellum mediates *Azospirillum brasilense* adsorption to wheat roots. J Gen Microbiol 139:2261–2269. https://doi.org/10.1099/00221287-139-9-2261

Elías JM, Ramírez-Mata A, Albornóz PL et al (2022) The polar flagellin of *Azospirillum brasilense* REC3 induces a defense response in strawberry plants against the fungus *Macrophomina phaseolina*. J Plant Growth Regul 41:2992–3008. https://doi.org/10.1007/s00344-021-10490-4

Evseeva NV, Matora LY, Burygin GL et al (2011) Effect of *Azospirillum brasilense* Sp245 lipopolysaccharide on the functional activity of wheat root meristematic cells. Plant Soil 346: 181–188. https://doi.org/10.1007/s11104-011-0808-9

Evseeva NV, Tkachenko OV, Burygin GL et al (2018) Effect of bacterial lipopolysaccharides on morphogenetic activity in wheat somatic calluses. World J Microbiol Biotechnol. https://doi.org/10.1007/s11274-017-2386-3

Fares M, Imberty A, Titz A (2025) Bacterial lectins: multifunctional tools in pathogenesis and possible drug targets. Trends Microbiol. https://doi.org/10.1016/j.tim.2025.03.007

Fedonenko YP, Egorenkova IV, Konnova SA et al (2001) Involvement of the lipopolysaccharides of *Azospirilla* in the interaction with wheat seedling roots. Microbiology 70:329–334. https://doi.org/10.1023/A:1010411629428

Fedonenko YP, Zatonsky GV, Konnova SA et al (2002) Structure of the O-specific polysaccharide of the lipopolysaccharide of *Azospirillum brasilense* Sp245. Carbohydr Res 337:9869–9872. https://doi.org/10.1016/S0008-6215(02)00061-7

Fedonenko YP, Borisov IV, Konnova ON et al (2005) Determination of the structure of the repeated unit of the *Azospirillum brasilense* SR75 O-specific polysaccharide and homology of the lps loci in the plasmids of *Azospirillum brasilense* strains SR75 and Sp245. Microbiology 74:542–548. https://doi.org/10.1007/s11021-005-0101-0

Fedonenko P, Konnova ON, Zdorovenko EL et al (2008a) Structural analysis of the O-polysaccharide from the lipopolysaccharide of *Azospirillum brasilense* S17. Carbohydr Res 343:810–816. https://doi.org/10.1016/j.carres.2007.12.013

Fedonenko P, Zdorovenko EL, Konnova SA et al (2008b) Structural analysis of the O-antigen of the lipopolysaccharide from *Azospirillum lipoferum* SR65. Carbohydr Res 343:2841–2844. https://doi.org/10.1016/j.carres.2008.05.022

Fedonenko P, Boiko AS, Zdorovenko EL et al (2011) Structural peculiarities of the O-specific polysaccharides of *Azospirillum* bacteria of serogroup III. Biochem Mosc 76:797–802. https://doi.org/10.1134/S0006297911070108

Fedonenko YP, Sigida EN, Konnova SA et al (2015) Structure and serology of O-antigens of nitrogen-fixing rhizobacteria of the genus *Azospirillum*. Russ Chem Bull 64:1024–1031. https://doi.org/10.1007/s11172-015-0971-x

Fischer SE, Miguel MJ, Mori GB (2003) Effect of root exudates on the exopolysaccharide composition and the lipopolysaccharide profile of *Azospirillum brasilense* Cd under saline stress. FEMS Microbiol Lett 219:53–62. https://doi.org/10.1016/S0378-1097(02)01194-1

Ganusova EE, Vo LT, Abraham PE et al (2021) The *Azospirillum brasilense* core chemotaxis proteins CheA1 and CheA4 link chemotaxis signaling with nitrogen metabolism. mSystems 6. https://doi.org/10.1128/msystems.01354-20

Ganusova EE, Russell MH, Patel S et al (2024) An *Azospirillum brasilense* chemoreceptor that mediates nitrate chemotaxis has conditional roles in the colonization of plant roots. Appl Environ Microbiol 90:e00760–e00724. https://doi.org/10.1128/aem.00760-24

Gómez-Gómez L, Boller T (2000) FLS2: an LRR receptor-like kinase involved in the perception of the bacterial elicitor flagellin in *Arabidopsis*. Mol Cell 5:1003–1011. https://doi.org/10.1016/s1097-2765(00)80265-8

Greer-Phillips SE, Stephens BB, Alexandre G (2004) An energy taxis transducer promotes root colonization by *Azospirillum brasilense*. J Bacteriol 186:6595–6604. https://doi.org/10.1128/JB.186.19.6595-6604.2004

Hall PG, Krieg NR (1983) Swarming of *Azospirillum brasilense* on solid media. Can J Microbiol 29:1592–1594. https://doi.org/10.1139/m83-243

Hernández-Esquivel AA, Castro-Mercado E, García-Pineda E (2020) Comparative effects of *Azospirillum brasilense* Sp245 and *Pseudomonas aeruginosa* PA01 lipopolysaccharides on wheat seedling growth and peroxidase activity. J Plant Growth Regul 40:1903–1911. https://doi.org/10.1007/s00344-020-10241-x

Hind S, Strickler S, Boyle P et al (2016) Tomato receptor FLAGELLIN-SENSING 3 binds flgII-28 and activates the plant immune system. Nature Plants 2:16128. https://doi.org/10.1038/nplants.2016.128

Ilyas N, Mumtaz K, Akhtar N et al (2020) Exopolysaccharides producing bacteria for the amelioration of drought stress in wheat. Sustainability. https://doi.org/10.3390/su12218876

Ivanova MF, Kostina EE, Filip'echeva YA et al (2024) Impact of lipopolysaccharides from rhizospheric bacteria on the growth of potato microplants under vitro conditions. Ann Agrar Sci 3:23–28. https://doi.org/10.28983/asj.y2024i3pp23-28

Jofré E, Fischer S, Rivarola V (1998) Saline stress affects the attachment of *Azospirillum brasilense* Cd to maize and wheat roots. Can J Microbiol 44:416–422. https://doi.org/10.1139/w98-024

Jofré E, Lagares A, Mori G (2004) Disruption of dTDP-rhamnose biosynthesis modifies lipopolysaccharide core, exopolysaccharide production, and root colonization in *Azospirillum brasilense*. FEMS Microbiol Lett 231:267–275. https://doi.org/10.1016/S0378-1097(04)00003-5

Kawaharada Y, Kelly S, Nielsen MW et al (2015) Receptor-mediated exopolysaccharide perception controls bacterial infection. Nature 523:308–312. https://doi.org/10.1038/nature14611

Knirel YA, Valvano MA (eds) (2011) Bacterial lipopolysaccharides: structure, chemical synthesis, biogenesis and interaction with host cell. Springer, Vienna

Konnova SA, Makarov OE, Skvortsov IM et al (1994) Isolation, fractionation and some properties of polysaccharides produced in a bound form by *Azospirillum brasilense* and their possible involvement in *Azospirillum*-wheat root interactions. FEMS Microbiol Lett 118:93–100. https://doi.org/10.1111/j.1574-6968.1994.tb06809.x

Konnova ON, Boiko AS, Burygin GL (2008) Chemical and serological studies of liposaccharides of bacteria of the genus *Azospirillum*. Mikrobiologiia 77:350–357. https://doi.org/10.1134/S0026261708030090

Kotecha H, Poduval PB (2019) Microbial lectins: roles and applications. In: Nandan Meena S, Mohan Naik M (eds) Advances in biological science research: a practical approach, 1st edn. Academic, London, pp 135–147. https://doi.org/10.1016/B978-0-12-817497-5.00009-4

Lerner A, Castro-Sowinski S, Valverde A et al (2009) The *Azospirillum brasilense* Sp7 noeJ and noeL genes are involved in extracellular polysaccharide biosynthesis. Microbiology 155:4058–4068. https://doi.org/10.1099/mic.0.031807-0

Mazgaeen L, Gurung P (2020) Recent advances in lipopolysaccharide recognition systems. Int J M Sci. https://doi.org/10.3390/ijms21020379

Méndez-Gómez M, Castro-Mercado E, Peña-Uribe CA et al (2020) *Azospirillum brasilense* Sp245 lipopolysaccharides induce target of rapamycin signaling and growth in *Arabidopsis thaliana*. J Plant Physiol 253:153270. https://doi.org/10.1016/j.jplph.2020.153270

Merino S, Tomás JM (2014) Gram-negative flagella glycosylation. Int J Mol Sci 15:2840–2857. https://doi.org/10.3390/ijms15022840

Michiels KW, Vanderleyden J, Van Gool AP et al (1988) Isolation and characterization of *Azospirillum brasilense* loci that correct *Rhizobium meliloti* exoB and exoC mutations. J Bacteriol 170:5401–5404. https://doi.org/10.1128/jb.170.11.5401-5404.1988

Michiels KW, Croes CL, Vanderleyden J (1991) Two different modes of attachment of *Azospirillum brasilense* Sp7 to wheat roots. J Gen Microbiol 137:2241–2247. https://doi.org/10.1099/00221287-137-9-2241

Moens S, Michiels K, Vanderleyden J (1995) Glycosylation of the flagellin of the polar flagellum of *Azospirillum brasilense*, a Gram-negative nitrogen-fixing bacterium. Microbiology 141:2651–2657. https://doi.org/10.1099/13500872-141-10-2651

Mora P, Rosconi F, Franco Fraguas L et al (2008) *Azospirillum brasilense* Sp7 produces an outer-membrane lectin that specifically binds to surface-exposed extracellular polysaccharide produced by the bacterium. Arch Microbiol 189:519–524. https://doi.org/10.1007/s00203-007-0343-5

Morales-Ascencio C, Castro-Mercado E, García-Pineda E (2019) Effect of lipopolysaccharides from *Azospirillum brasilense* on the growth of cucumber plants. Ciencia Nicolaita 78:75–86. https://doi.org/10.35830/cn.vi78.469

Netrusov AI, Liyaskina EV, Kurgaeva IV et al (2023) Exopolysaccharides producing bacteria: a review. Microorganisms 11:1541. https://doi.org/10.3390/microorganisms11061541

Newman MA, Sundelin T, Nielsen JT (2013) MAMP (microbe-associated molecular pattern) triggered immunity in plants. Front Plant Sci 4:139. https://doi.org/10.3389/fpls.2013.00139

Nguyen MP, Tran LVH, Namgoong H et al (2019) Applications of different solvents and conditions for differential extraction of lipopolysaccharide in Gram-negative bacteria. J Microbiol 57:644–654. https://doi.org/10.1007/s12275-019-9116-5

Oliveira ALM, Santos OJAP, Marcelino PRF et al (2017) Maize inoculation with *Azospirillum brasilense* Ab-V5 cells enriched with exopolysaccharides and polyhydroxybutyrate results in high productivity under low N fertilizer input. Front Microbiol 8:1873. https://doi.org/10.3389/fmicb.2017.01873

Paradis G, Chevance FFV, Liou W et al (2017) Variability in bacterial flagella re-growth patterns after breakage. Sci Rep 7:1282. https://doi.org/10.1038/s41598-017-01302-5

Park K, Kloepper JW, Ryu CM (2008) Rhizobacterial exopolysaccharides elicit induced resistance on cucumber. J Microbiol Biotechnol 18:1095–1100

Reinhold B, Hurek T, Fendrik I (1985) Strain-specific chemotaxis ofA zospirillum spp. J Bacteriol 162:1. https://doi.org/10.1128/jb.162.1.190-195.1985

Renukadevi KP, Angayarkanni J, Karunakaran G (2012) Extraction and characterization of lipopolysaccharide from *Serratia rubidaea* and its cytotoxicity on lung cancer cell line-nci-h69. Acta Tech Corviniensis 2:97–101

Rossez Y, Wolfson EB, Holmes A (2015) Bacterial flagella: twist and stick, or dodge across the kingdoms. PLoS Pathog 11:e1004483. https://doi.org/10.1371/journal.ppat.1004483

Sanguankiattichai N, Buscaill P, Preston GM (2022) How bacteria overcome flagellin pattern recognition in plants. Curr Opin Plant Biol 67:102224. https://doi.org/10.1016/j.pbi.2022.102224

Sankari JU, Dinakar S, Sekar C (2011) Dual effect of *Azospirillum* exopolysaccharides (EPS) on the enhancement of plant growth and biocontrol of blast (*Pyricularia oryzae*) disease in upland rice (var. ASD-19). J Phytol 3:16–19

Shang-Guan K, Wang M, Sin Htwe NMP et al (2018) Lipopolysaccharides trigger two successive bursts of reactive oxygen species at distinct cellular locations. Plant Physiol 176:2543–2556. https://doi.org/10.1104/pp.17.01637

Shirokov A, Budanova A, Burygin G et al (2020) Flagellin of polar flagellum from *Azospirillum brasilense* Sp245: isolation, structure, and biological activity. Int J Biol Macromol 147:1221–1227. https://doi.org/10.1016/j.ijbiomac.2019.10.092

Shiu S-H, Bleecker AB (2001) Receptor-like kinases from *Arabidopsis* form a monophyletic gene family related to animal receptor kinases. Proc Natl Acad Sci USA 98:10763–10768. https://doi.org/10.1073/pnas.181141598

Sigida EN, Fedonenko YP, Zdorovenko EL et al (2013a) Structure of repeating units of a polysaccharide(s) from the lipopolysaccharide of *Azospirillum brasilense* SR80. Carbohydr Res 371:40–44. https://doi.org/10.1016/j.carres.2013.01.005

Sigida EN, Fedonenko YP, Shashkov AS et al (2013b) Structural studies of the O-specific polysaccharide(s) from the lipopolysaccharide of *Azospirillum brasilense* type strain Sp7. Carbohydr Res 380:76–80. https://doi.org/10.1016/j.carres.2013.07.013

Sigida EN, Fedonenko YP, Zdorovenko EL et al (2014a) Characterization of the lipopolysaccharides of serogroup II *Azospirillum* strains. Microbiology 83:326–334. https://doi.org/10.1134/S0026261714040158

Sigida EN, Fedonenko YP, Shashkov AS et al (2014b) Structural studies of the polysaccharides from the lipopolysaccharides of *Azospirillum brasilense* Sp246 and SpBr14. Carbohydr Res. https://doi.org/10.1016/j.carres.2014.05.008

Sigida EN, Fedonenko YP, Shashkov AS et al (2015) Structure of the polysaccharides from the lipopolysaccharide of *Azospirillum brasilense* Jm125A2. Carbohydr Res 416:37–40. https://doi.org/10.1016/j.carres.2015.08.011

Sigida EN, Fedonenko YP, Shashkov AS (2018) Structure of the O-specific polysaccharide from *Azospirillum fermentarium* CC-LY743. Carbohydr Res 465:40–43. https://doi.org/10.1016/j.carres.2018.06.003

Sigida EN, Fedonenko YP, Shashkov AS et al (2019a) Structure of the O-specific polysaccharide of *Azospirillum doebereinerae* type strain GSF71^{T}. Carbohydr Res 478:54–57. https://doi.org/10.1016/j.carres.2019.04.009

Sigida EN, Fedonenko YP, Shashkov AS et al (2019b) Structural studies of O-specific polysaccharide(s) and biological activity toward plants of the lipopolysaccharide from *Azospirillum brasilense* SR8. Int J Biol Macromol 126:246–253. https://doi.org/10.1016/j.ijbiomac.2018.12.229

Sigida EN, Shashkov AS, Zdorovenko EL et al (2020) Structure of the O-specific polysaccharide from *Azospirillum formosense* CC-Nfb-7(T). Carbohydr Res 494:108060. https://doi.org/10.1016/j.carres.2020.108060

Sigida EN, Grinev VS, Zdorovenko EL et al (2022) O-Antigens of *Azospirillum zeae* N7(T), *Azospirillum melinis* TMCY 0552(T), and *Azospirillum palustre* B2(T): structure elucidation and analysis of biosynthesis genes. Russ J Bioorg Chem 48:519–528. https://doi.org/10.1134/S1068162022030177

Sigida EN, Grinev VS, Kokoulin MS et al (2024) Structural and genetic relationship of the O-antigens of the type strains *Azospirillum agricola* CC-HIH038 and *Azospirillum doebereinerae* GSF71. Microbiology 93:855–863. https://doi.org/10.1134/S0026261724606948

Skvortsov IM, Ignatov VV (1998) Extracellular polysaccharides and polysaccharide-containing biopolymers from *Azospirillum* species: properties and the possible role in interaction with plant roots. FEMS Microbiol Lett 165:223–229. https://doi.org/10.1111/j.1574-6968.1998.tb13150.x

Smol'kina ON, Kachala VV, Fedonenko YP et al (2010) Capsular polysaccharide of the bacterium *Azospirillum lipoferum* Sp59b: structure and antigenic specificity. Biochem Mosc 75:606–613. https://doi.org/10.1134/s000629791005010x

Song WY, Wang GL, Chen LL et al (1995) A receptor kinase-like protein encoded by the rice disease resistance gene, *Xa21*. Science 270:1804–1806. https://doi.org/10.1126/science.270.5243.1804

Steenhoudt O, Vanderleyden J (2000) *Azospirillum*, a free-living nitrogen-fixing bacterium closely associated with grasses: genetic, biochemical and ecological aspects. FEMS Microbiol Rev 24:487–506. https://doi.org/10.1111/j.1574-6976.2000.tb00552.x

Sumayo M, Hahm MS, Ghim SY (2013) Determinants of plant growth-promoting *Ochrobactrum lupini* KUDC1013 involved in induction of systemic resistance against *Pectobacterium carotovorum* subsp. *carotovorum* in tobacco leaves. Plant Pathol J 29:174–181. https://doi.org/10.5423/PPJ.SI.09.2012.0143

Tarrand JJ, Krieg NR, Dobereiner J (1978) A taxonomic study of the *Spirillum lipoferum* group, with descriptions of a new genus, *Azospirillum* gen. nov. and two species. *Azospirillum lipoferum* (Beijerinck) comb. nov. and *Azospirillum brasilense* sp. nov. Can J Microbiol 24:967–980

Tkachenko OV, Burygin GL, Evseeva NV et al (2021) Morphogenesis of wheat calluses treated with *Azospirillum* lipopolysaccharides. Plant Cell Tissue Organ Cult 147:147–155. https://doi.org/10.1007/s11240-021-02114-2

Vallejo-Ochoa J, López-Marmolejo M, Hernández-Esquivel AA et al (2017) Early plant growth and biochemical responses induced by *Azospirillum brasilense* Sp245 lipopolysaccharides in wheat (*Triticum aestivum* L.) seedlings are attenuated by procyanidin B2. Protoplasma 255:685–694. https://doi.org/10.1007/s00709-017-1180-2

Van Bastelaere E, De Mot R, Michiels K et al (1993) Differential gene expression in *Azospirillum* spp. by plant root exudates: analysis of protein profiles by two-dimensional polyacrylamide gel electrophoresis. FEMS Microbiol Lett 112:335–342. https://doi.org/10.1111/j.1574-6968.1993.tb06472.x

Vlot AC, Sales JH, Lenk M et al (2021) Systemic propagation of immunity in plants. New Phytol 229:1234–1250. https://doi.org/10.1111/nph.16953

Wisniewski-Dyé F, Borziak K, Khalsa-moyers G, Alexandre G et al (2011) *Azospirillum* genomes reveal transition of bacteria from aquatic to terrestrial environments. PLoS Genet 7:e1002430. https://doi.org/10.1371/journal.pgen.1002430

Wisniewski-Dyé F, Lozano L, Acosta-Cruz E et al (2012) Genome sequence of *Azospirillum brasilense* CBG497 and comparative analyses of *Azospirillum* core and accessory genomes provide insight into niche adaptation. Gene 3:576–602. https://doi.org/10.3390/genes3040576

Wisniewski-Dyé F, Drogue B, Prigent-Combaret C (2013) Azospirillum-plant interaction: from root colonization to plant growth promotion. In: Rodelas González MB, Gonzalez-Lopez J (eds) Beneficial plant-microbial interactions, 1st edn. CRC Press, Boca Raton

Yegorenkova I, Konnova S, Sachuk V et al (2001) *Azospirillum brasilense* colonization of wheat roots and the role of lectin–carbohydrate interactions in bacterial adsorption and root-hair deformation. Plant Soil 231:275–282. https://doi.org/10.1023/A:1010340700694

Zhu GY, Dobbelaere S, Vanderleyden J (2002) Use of green fluorescent protein to visualize rice root colonization by *Azospirillum irakense* and *A. brasilense*. Funct Plant Biol 29:1279–1285. https://doi.org/10.1071/FP02120

Zhulin IB, Armitage JP (1993) Motility, chemokinesis, and methylation-independent chemotaxis in *Azospirillum brasilense*. J Bacteriol 175:952–958. https://doi.org/10.1128/jb.175.4.952-958.1993

Chapter 12
Azospirillum as a Biocontrol Agent

Raúl O. Pedraza and María P. Filippone

Abstract Over the past 30 years, the use of plant growth-promoting bacteria (PGPB) has gained significant relevance in agricultural microbiology worldwide due to their proven ability to enhance crop yields, preserve agricultural ecosystems, and reduce the impact of agrochemicals. Among PGPB, the genus *Azospirillum* has been identified in various regions across the globe, establishing itself as a true plant colonizer and a model PGPB. Until not many years ago, *Azospirillum* was not considered a true biocontrol agent due to the lack of evidence supporting mechanisms that could explain this function. However, growing scientific evidence now supports several mechanisms by which *Azospirillum* exerts biocontrol activity against phytopathogens and insects, while others remain to be fully understood. This chapter explores the background of *Azospirillum* as a biocontrol agent, the mechanisms it employs to suppress plant pathogens, the role of its polar flagellum flagellin as an emerging subject of study in the control of phytopathogens, methods of applying *Azospirillum*, and future projections for its use in plant disease management.

Keywords Biocontrol · Pathogens · Insects · Disease management · Antimicrobial activity

---

R. O. Pedraza (✉)
Facultad de Agronomía, Zootecnia y Veterinaria, Universidad Nacional de Tucumán, San Miguel de Tucumán, Tucumán, Argentina

M. P. Filippone
Facultad de Agronomía, Zootecnia y Veterinaria, Universidad Nacional de Tucumán, San Miguel de Tucumán, Tucumán, Argentina

Consejo Nacional de Investigaciones Científicas y Tecnológicas (CONICET), San Miguel de Tucumán, Tucumán, Argentina

F. D. Cassán et al. (eds.), *The Definitive Handbook of Azospirillum*,
https://doi.org/10.1007/978-3-032-09222-9_12

12.1 Introduction

Plant pathogens and pests can significantly impact agricultural productivity. Several economically important plant diseases have become increasingly widespread, and one of the most urgent challenges to global environmental sustainability and public health is the heavy reliance on chemical pesticides (Fones et al. 2020). Raising awareness about the environmental and health risks linked to synthetic chemical pesticides is strongly recommended for promoting sustainable crop management and reducing chemical use (Donley 2019).

Synthetic agrochemicals are often considered unsustainable, prompting the search for more environmentally friendly alternatives (El-Saadony et al. 2022). Plant growth-promoting bacteria (PGPB) are naturally occurring microorganisms that are effective, environmentally safe, and non-toxic, offering a promising alternative to chemical pesticides (Bashan and de-Bashan 2010). Biological control is one of the most cost-effective and long-lasting strategies for managing plant diseases and minimizing crop losses (El-Saadony et al. 2022).

Microbial biocontrol agents combat plant pathogens through a variety of mechanisms, including inducing plant resistance, priming plant defenses against pathogens, competing for space and nutrients, and disrupting environmental conditions needed for pathogen growth (Boro et al. 2022). The most commonly used biocontrol agents include biofungicides (*Bacillus* sp., *Pseudomonas* sp., and *Trichoderma* sp.) and bioinsecticides (*Bacillus thuringiensis*) (Gupta and Dikshit 2010). Many microorganisms not only inhibit pathogens but also promote plant growth, as they colonize different plant surfaces and form beneficial plant-microbe interactions.

Over the last 30 years, the use of PGPB has gained significant relevance in agricultural microbiology worldwide due to its proven ability to enhance crop yields, preserve agricultural ecosystems, and reduce the impact of synthetic agrochemicals (Kaschuk and Hungria 2017). Among PGPB, the genus *Azospirillum* has been identified in agricultural soils from different parts of the world, making it a true plant colonizer and a model PGPB. Its beneficial effects have been reported in more than 100 plant species (Pereg et al. 2016; Cassán et al. 2015; Mehnaz 2015).

In addition to its well-documented growth-promoting qualities, *Azospirillum* has also been reported to provide protection against abiotic stress factors, such as salinity and water stress (Barassi et al. 2006; Rodríguez-Salazar et al. 2009; Carrozzi et al. 2012; Saeed et al. 2016), as well as biotic stress, enhancing plant responses to harmful organisms of various natures (Yasuda et al. 2009; Fukami et al. 2017; Elías et al. 2018). However, in the latter case, research on the use of *Azospirillum* as a biocontrol agent remains less extensive than its use as a growth promoter, although it continues to increase. *Azospirillum* has demonstrated the ability to control various organisms that affect a wide range of agricultural crops.

This chapter explores the background of *Azospirillum* as a biocontrol agent, the mechanisms it employs to suppress plant pathogens, the role of its polar flagellum flagellin as an emerging area of study in the control of phytopathogens, methods of applying *Azospirillum*, and future projections for its use in plant disease management.

12.2 Background on the Use of *Azospirillum* as a Biocontrol Agent

The main mechanisms of biocontrol in *Azospirillum* include competition for nutrients, the production of antimicrobial compounds, and the stimulation of natural plant defenses. Among the pathogens most frequently reported to be controlled by *Azospirillum* are various species of *Fusarium* spp., which cause wilt and rot in many plants. *Azospirillum* can inhibit the growth of these pathogens through the production of siderophores and secondary metabolites such as phenylacetic acid (Sultana et al. 2023).

Azospirillum can suppress the growth of the fungus *Rhizoctonia solani*, which causes root and stem rot in various crops, by competing for resources in the soil and producing antimicrobial compounds (Fendrihan et al. 2018). *Alternaria* spp., a fungal genus responsible for leaf spots and fruit and vegetable rot, can also be partially controlled by *Azospirillum* through the production of antifungal substances (Tahir Mehmood et al. 2021). Additionally, while less studied, some strains of *Azospirillum* have shown the ability to limit the growth of *Sclerotium rolfsii*, although their effectiveness depends on the strain and environmental conditions (Sultana et al. 2023). Recently, the effectiveness of *A. argentinense* REC3, applied by seed inoculation, in improving soybean nodulation with native rhizobia, as well as grain yield, and protecting against *Fusarium* fungi was reported (Martínez et al. 2024). Following soil infestation, *F. tucumaniae* was confirmed as a phytopathogen, with REC3 inhibiting in vitro its mycelial growth. Greenhouse assays showed significant reductions in disease severity (78.2%), root rot (32.3%), and disease incidence (50.9%) in soybean plants susceptible to sudden death syndrome (SDS) when REC3 was applied. Field experiments, under drought stress conditions, showed increased nodulation (37.2%) and grain yield (2077 ± 272.8 kg ha^{-1}) in REC3-treated plants compared to controls (1516 ± 272.8 kg ha^{-1}). Furthermore, a low disease incidence (0.90%) was observed only in the SDS susceptible soybean variety without REC3 treatment (Martínez et al. 2024). These results showed the significant potential of REC3 inoculation as a biotechnological strategy for enhancing agricultural productivity, managing diseases, and improving crop resilience under challenging environmental conditions, thereby ensuring more sustainable and resilient agricultural practices. Another current example is the fungus *Neopestalotiopsis clavispora*, the etiological agent of the root and crown rot disease in strawberry plants. It is known as a pathogen that causes severe damage to strawberry crops and is considered worldwide as an emerging disease (Baggio et al. 2021). Ramallo et al. (2024) reported that applications of *A. argentinense* REC3 significantly reduced disease severity and also stimulated leaf biomass, increasing the fresh and dry weight of aerial tissues and the leaf area index.

Among bacterial pathogens, different species of the genus *Xanthomonas*, known to cause diseases such as leaf spots and cankers in various crops, can have their impact reduced by *Azospirillum* through resource competition and antimicrobial compound production (Sultana et al. 2023). Against *Pseudomonas syringae*, a pathogen responsible for leaf and fruit diseases, *Azospirillum* helps protect plants by stimulating systemic defenses (Drogue et al. 2014).

It has also been reported that *Azospirillum* may contribute to soil pathogen control through potential synergistic roles. In some microbial consortia, *Azospirillum* may help create an unfavorable environment for pests by altering soil microbial communities or producing antifungal or antibacterial compounds that could indirectly affect pest populations (Vishwakarma et al. 2020).

The effect of *Azospirillum* against other crop enemies, such as nematodes, insect pests, and viruses, has been less explored. There is evidence that *Azospirillum* can reduce the incidence of plant-parasitic nematodes by improving root system health and reducing conditions that favor their proliferation (Dias-Arieira et al. 2012; Elkelany et al. 2020).

Reports also suggest that *Azospirillum* may play an indirect role in mitigating viral diseases by enhancing overall plant vigor and stimulating systemic defenses. So far, no direct effect of *Azospirillum* on plant viruses has been demonstrated. For example, Lade et al. (2019) found that *A. brasilense* Sp7 did not reduce viral load in tomato/PVX or maize/MDMV systems; however, a positive growth-promoting effect was observed in both plant species. Additionally, it has been reported that *Azospirillum* inoculation in mixed systems with other biological agents has enhanced the action of beneficial bacteria that may have more direct antiviral effects (Cassán et al. 2009).

Reports on the effect of *Azospirillum* in weed control are even scarcer. Given that *Azospirillum* can interact with many plant species, it is assumed that it could also interact with weeds, benefiting them in the same way as the crops (Bashan and Holguin 1995). However, Miché et al. (2000) published a study in which *Azospirillum* improved the performance of sorghum crops to the detriment of the weed *Striga hermonthica*. Although the exact reason for this phenomenon was not determined, the authors hypothesized that it was due to competition between *Azospirillum* strains and the weed for the same host plants (Miché et al. 2000). In another study, although no direct effect on weeds was determined, it was reported that a strain of *A. lipoferum* showed tolerance to herbicides (butachlor alone or in combination with bensulfuron methyl) used for weed control in rice cultivation. This strain contributed to an almost 20% increase in crop yields compared to plots treated only with herbicides (Seddigui Kiasari et al. 2018). Some examples of the biocontrol exerted by some *Azospirillum* species on different targets and crops are shown in Table 12.1.

12.3 Biocontrol Mechanisms Exerted by *Azospirillum*

Until not many years ago, *Azospirillum* was not considered a true biological control agent due to the lack of evidence regarding mechanisms that could explain this characteristic. Although many reports attributed biocontrol properties to this genus, they did so without explaining the mechanisms of action (Bashan and de Bashan and de-Bashan 2010). Currently, scientific evidence supports some mechanisms developed by *Azospirillum* to exert biocontrol activity against phytopathogens and insects, while others remain undetermined. Some aspects of this topic were

Table 12.1 Some examples of the biocontrol action exerted by *Azospirillum* species on different types of crop enemies

Species	Target	Crop	References
A. brasilense	Fungus: *Colletotrichum acutatum*	*Fragaria ananassa*	Tortora et al. (2011)
A. lipoferum	Fungus: *Alternaria solani*	*Solanum tuberosum*	Tahir Mehmood et al. (2021)
A. argentinense	Fungus: *Fusarium tucumaniae*	*Glycine max*	Martínez et al. (2024)
A. argentinense	Fungus: *Neopestalotiopsis clavispora*	*Fragaria ananassa*	Ramallo et al. (2024)
A. brasilense	Bacterium: *Xanthomonas oryzae pv oryzae*	*Oryza sativa*	Islam and Bora (1998)
A. brasilense	Bacterium: *Pseudomonas syringae* pv. *Tomato*	*Solanum lycopersicum*	Bashan and de-Bashan (2002)
A. brasilense	Virus: potato virus X (PVX) and maize dwarf mosaic virus (MDMV)	*Solanum lycopersicum* and *Zea mays*	Lade et al. (2019)
A. brasilense	Virus: yellow dwarf virus (BYDV)	*Triticum aestivum*	Santos and Simões Bento (2019)
A. brasilense	Nematode: *Pratylenchus brachyurus*	*Glycine max* and *Zea mays*	Dias-Arieira Belyakov et al. (2012)
A. brasilense	Nematodo: *Meloidogyne incognita*	*Solanum melongena*	Elkelany et al. (2020)
A. brasilense	Insect: *Mayetiola destructor* (Say)	*Triticum aestivum*	Prischmann-Voldseth et al. (2020)
A. brasilense	Insect: *Diabrotica speciosa*	*Zea mays*	Santos et al. (2014)
A. lipoferum	Weed: *Echinochloa* sp.	*Oriza sativa*	Seddigui Kiasari et al. (2018)
A. brasilense	Weed: *Striga hermonthica*	*Sorghum bicolor*	Miché et al. (2000)

previously reviewed by Bashan and de-Bashan (2010), Fendrihan et al. (2018), and Fukami et al. (2018). Below, some key mechanisms supporting the biocontrol activity of *Azospirillum* are mentioned.

12.3.1 Competition for Nutrients and Space

Azospirillum, as a rhizosphere bacterium, occupies spaces in the rhizosphere and consumes essential nutrients, limiting the availability of these resources for pathogens. This creates a form of competitive exclusion in which pathogens, especially soil-borne ones, face greater difficulties in establishing and thriving. For instance, when *A. brasilense* Cd was included in a culture with the pathogenic mangrove rhizosphere bacterium *Staphylococcus* spp., the population of the latter was significantly reduced (Holguin and Bashan 1996). Also, it was observed that tomato seed inoculation with *A. brasilense* Cd combined with a single streptomycin foliar treatment and two foliar bactericide applications at 5-day intervals reduced disease severity produced by *Pseudomonas syringae* pv. *tomato*, the causal agent of bacterial speck of tomato, in tomato seedlings by over 90% after 4 weeks, and significantly slowed disease development under misty conditions (Bashan and de-Bashan 2002). Plant growth-promoting activity of two *Azospirillum* strains and their effects on foliar and vascular bacterial diseases (bacterial canker caused by *Clavibacter michiganensis* subsp. *michiganensis*, or bacterial spot caused by *Xanthomonas campestris* pv. *vesicatoria*) were evaluated on fresh market and cherry tomatoes. Plant growth promotion was observed on both types of tomatoes, and *Azospirillum* treatments altered tomato susceptibility to bacterial diseases. For example, the cherry tomato was more resistant to bacterial canker but more susceptible to bacterial spot than the fresh-market tomato. Canker severity was not affected by *Azospirillum* seed treatments (Romero et al. 2003).

12.3.2 Synthesis of Siderophores

Siderophores are low-molecular mass compounds with high iron affinity (Neilands 1995) that allow soil microorganisms to sequester and solubilize ferric iron in iron-poor environments. Tortora et al. (2011) have demonstrated that under iron-limited conditions, strains REC2 and REC3 of *A. brasilense* produce siderophores. Chemical assays revealed that they secreted catechol-type siderophores, including salicylic acid, detected by thin-layer chromatography coupled with fluorescence spectroscopy and gas chromatography-mass spectrometry analysis. Siderophores produced by these bacterial strains showed in vitro antifungal activity against the fungus *Colletotricum acutatum* M11, an etiological agent of the disease anthracnose in strawberry plants. Additionally, this latter coincided with results obtained from phytopathological tests performed *in planta*, where a reduction of anthracnose symptoms on strawberry plants previously inoculated with *A. brasilense* was observed (Tortora et al. 2011, 2012).

12.3.3 Induced Systemic Resistance

When plant roots are inoculated with PGPB, a systemic response known as induced systemic resistance (ISR) is activated (Pieterse and Van Loon 1999; Pieterse et al. 2009). *Azospirillum* can activate ISR in plants, a process that enhances the plant's ability to defend itself against a wide range of pathogens and pests. ISR is a plant-wide response that stimulates the production of defensive enzymes such as peroxidases, phenylalanine ammonia-lyases, and phenolic compounds. This mechanism strengthens natural defenses and primes the plant to respond more quickly to upcoming infections or attacks. Nevertheless, it was also reported that some PGPB may also induce the development of an acquired systemic response (SAR) (Chen et al. 1999; Tortora et al. 2012). It has been informed that *A. brasilense* REC3 induced the accumulation of salicylic acid (SA) and the production of ethylene (ET) in strawberry plants, both phytohormones related to SAR and ISR, respectively (Tortora et al. 2012; Elías et al. 2018).

Tortora et al. (2012) reported that endophytic root colonization of strawberry plants with *A. brasilense* REC3 confers systemic protection against the fungus *Colletotrichum acutatum* M11 by the direct activation of some plant defense reactions and also primes the plant for a stronger defense reaction when exposed to further infection. Defense mechanisms induced by *A. brasilense* REC3 included the reinforcement of the plant cell wall by increasing the content of total soluble phenolic compounds and callose depositions, and the transient accumulation of SA. The latter brings about the upregulation of defense-related genes, such as those encoding pathogenesis-related proteins like PR1, chitinases, and glucanase. Therefore, the activation of a systemic defense response, together with the plant growth-promoting effect exerted by *A. brasilense* REC3 strain, can explain the increase in strawberry plants' tolerance to anthracnose disease caused by *C. acutatum* M11. Furthermore, Elías et al. (2021) have shown that roots of strawberry plants treated with *A. brasilense* REC3 caused the accumulation of ROS, SA, callose, lignin, and leaf stomatal closure. Besides, the expression of genes associated with defense response (e.g., FaPR1, FaCAT, FaRBOH-D, FaRBOH-F, FaCHI23, FaCHI2-2, and FaGSL5) indicated the activation of the innate immunity of these plants. After a phytopathogenic assay against the necrotrophic fungus *Macrophomina phaseolina*, the etiological agent of the charcoal rot disease, root inoculations with REC3 reduced plant mortality, showing the effectiveness of REC3 to control the fungus. Similar results on strawberry plants, REC3 treatments, and infection with *M. phaseolina* were also observed by Viejobueno et al. (2021). Foliar treatment with *A. lipoferum* AL-3 reduced by 68.1% the early blight disease severity of potato, caused by *Alternaria solani*, through induced systemic resistance (Tahir Mehmood et al. 2021). Total phenolics, defense-related enzyme (peroxidase, polyphenol oxidase, and phenylalanine ammonia lyase) activities, and PR-gene transcriptomic levels were significantly increased (Tahir Mehmood et al. 2021).

12.3.4 Type VI Secretion System

Genome analyses revealed that the *A. brasilense* Az39 genome contains two complete sets of genes encoding type VI secretion systems (T6SS), including the T6SS1 that is induced by the indole-3-acetic acid (IAA) phytohormone (Rivera et al. 2014). The T6SS is a multiprotein machine, widespread in gram-negative bacteria, that delivers protein effectors in both prokaryotic and eukaryotic cells. Cassán et al. (2021) have reported that the *A. brasilense* type VI secretion system promotes cell aggregation, biocontrol protection against phytopathogens, and attachment to the microalgae *Chlorella sorokiniana*. They showed that the *Azospirillum* T6SS gene provides antibacterial activity against *Escherichia coli* and several plant pathogens, including *Agrobacterium tumefaciens*, *Pectobacterium carotovorum,* and *Dickeya dadantii*, thus conferring protection of the microalgae against pathogens.

12.3.5 Production of Phytohormones and Bioactive Compounds

Azospirillum produces phytohormones such as auxins, cytokinins, and gibberellins, which stimulate root growth and enhance nutrient absorption, increasing plant vigor (Bashan and de-Bashan 2010). Additionally, the production of certain bioactive compounds has been identified, including bacteriocins, phenylacetic acid, and siderophores. The latter sequester iron in the rhizosphere, limiting access to this nutrient for iron-dependent pathogens (Tortora et al. 2011). Tapia-Hernandez et al. (1990) have reported that some *Azospirillum* strains produced bacteriocins that inhibited the growth of several indicator bacteria in solid media. Partially purified bacteriocins were inactivated at pH 4 but were very stable between pH 5 and 10. An antimicrobial compound from *A. brasilense* supernatant extract was isolated and identified as the auxin-like molecule, phenylacetic acid (Somers et al. 2005); its synthesis is mediated by the indole-3-pyruvate decarboxylase, a key enzyme in the IAA production in *A. brasilense*. Using an *Agrobacterium tumefaciens* bioreporter assay, it was demonstrated that the antimicrobial activity of this compound against *A. tumefaciens*. Also, the effect of antifungal compounds produced by *A. brasilense* Sp245 could play a role in the reduction of fungal rhizospheric population of *Prunus cerasifera* L. clone Mr.S 2/5 plantlets against *Rhizoctonia* damping-off during the acclimatization phase of micropropagation (Russo et al. 2008). Figure 12.1 shows the inhibition of mycelial growth of some phytopathogenic fungi caused by *Azospirillum* under in vitro conditions.

12.3.6 Modulation of the Rhizosphere Microbiome

The presence of *Azospirillum* in the rhizosphere can influence the composition of root-associated microbiota. This shift may promote the presence of other beneficial or less virulent microorganisms, displacing pathogenic ones and establishing a more balanced and resilient microbiome against external invasions. Coniglio et al. (2022) have reported that *A. argentinense* Az39 is able to colonize the maize rhizosphere after inoculation and that this colonization process induces an increase in the relative

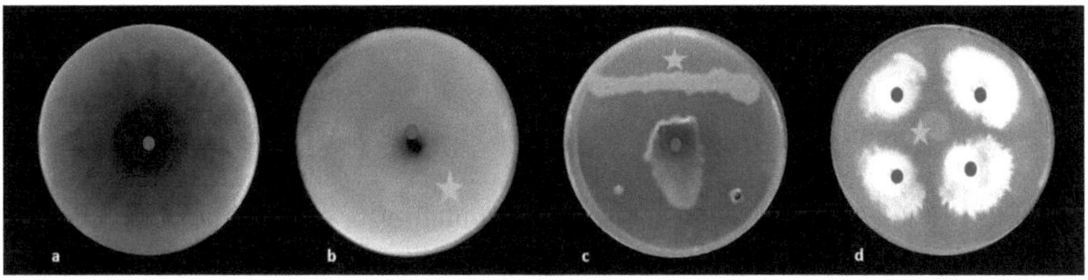

Fig. 12.1 Fungal development and in vitro inhibition of mycelial growth of three phytopathogenic fungi caused by *Azospirillum argentinense* REC3 on potato glucose agar medium. Growth of *Macrophomina phaseolina* without the presence of REC3 (**a**); inhibition of *M. phaseolina* growth in the presence of REC3 (**b**); inhibition of *Botrytis cinerea* growth in the presence of a streak with REC3 (**c**); inhibition of *Fusarium tucumaniae* growth at four locations of the fungus, with respect to the central application of REC3 (**d**). The red circle indicates the presence of the fungus, while the yellow star indicates the presence of *A. argentinense* REC3

abundance of beneficial bacterial genera, commonly used as biofertilizers in agriculture. After a high-throughput 16S rRNA gene sequence analysis under experimentally controlled conditions, they observed that *Azospirillum* was the most abundant genus occurring in the rhizosphere of inoculated plants. According to the LEfSe index, *Burkholderia*, *Massilia*, and *Sphingobium* were the three most abundant genera associated with Az39 inoculation. An increase in the relative abundance of some members of the hyphomicrobiales order (e.g., Rhizobiales) was also detected in inoculated plants. Besides, co-occurrence networks showed a positive interaction between *Azospirillum* and *Pseudomonas* and a negative correlation with *Hyphomicrobium*. By DNA sequencing and bioinformatic analysis, Coniglio et al. (2024) have recently explored the prokaryotic community composition and predicted metabolic functions to understand the impact of exogenously supplemented IAA or IAA produced by *A. argentinense* Az39 on the maize rhizosphere microbiome. They observed that prokaryotic groups specifically increase in the maize rhizosphere, according to the treatments applied, and that these groups could be considered specific markers of the IAA activity in the rhizosphere.

12.3.7 Indirect Effects on Pests and Weeds

By strengthening the root system and overall plant health, *Azospirillum* makes crops less susceptible to insect attacks and reduces pest attraction. However, it was also observed that colonization by *A. brasilense* affects the host selection and performance of the insect *Diabrotica speciosa* (Santos et al. 2014). Root larvae preferentially orient toward non-inoculated plants against inoculated roots and gain less weight when feeding on inoculated plants. As the inoculation with *A. brasilense* induces higher emissions of (E)-β-caryophyllene compared with non-inoculated plants, probably the non-preference of *D. speciosa* for inoculated plants is linked to this sesquiterpene, which is well known to mediate belowground insect-plant interactions.

Although *Azospirillum* does not directly affect weeds, its presence can limit their access to soil resources, creating a less favorable environment for their growth. It was reported that *A. brasilense* cells contain a low-molecular-weight compound that inhibits germination and growth of the radicle of Egyptian broomrape seeds (*Orobanche aegyptiaca*), a specific weed parasite of sunflower (Dadon et al. 2004). Also, *Azospirillum* spp. inhibited germination of the parasitic striga weed seeds (*Striga hermonthica*) that infest fields of tropical sorghum, thus promoting growth of sorghum (Bouillant et al. 1997). However, according to Bashan (1998), *Azospirillum* has not been reported to induce any negative effect on healthy plants. If this is the case, these toxic compounds can be in vitro artifacts, or they were induced only in the presence of pathogens (Bashan and de-Bashan 2010).

12.3.8 Yet Unknown Mechanisms

There are many reports of possible "biological control" exerted by *Azospirillum* without specifying the mechanisms (reviewed by Bashan and de-Bashan 2010). For instance, inoculation of rice plants with *Azospirillum* sp. B510 induced resistance against diseases caused by the fungus *Magnaporthe oryzae* and the bacteria *Xanthomonas oryzae*, probably by activating a novel type of resistance mechanism independent of salicylic acid-mediated defense without signal accumulation or expression of pathogenesis-related genes (Yasuda et al. 2009). It was also reported that *A. brasilense* mitigates water stress imposed by a vascular disease caused by *Clavibacter michiganensis* subsp. *michiganensis* by increasing xylem vessel area and stem hydraulic conductivity in tomato and delays wilting caused by this pathogen by yet unknown mechanisms (Romero et al. 2014).

As we can see, these biocontrol mechanisms give *Azospirillum* a complementary role in crop protection, alongside plant growth promotion. It integrates well into integrated management systems, although its focus is more on promoting plant health rather than directly controlling specific pathogens or pests.

12.4 The Polar Flagellum Flagellin of *Azospirillum* as an Emerging Subject of Study in the Control of Phytopathogens

The bacterial flagellum is a key component of prokaryotic cell motility. For many bacteria, it determines the possibility of moving in space and adapting to the conditions of the life environment (Belyakov et al. 2012). The building block of the filament of the bacterial flagellum is flagellin, one of the most widespread proteins in bacteria (Macnab 2003).

It was reported that *A. brasilense* has a polar flagellum and several lateral flagella. Whereas the latter are synthesized only when the bacteria grow on solid or semi-solid surfaces (Moens et al. 1995a; Steenhoudt and Vanderleyden 2000) and are involved in a special movement known as swarming (Hall and Krieg 1983), the polar flagellum is responsible for bacterial movement in a liquid medium, known as swimming. It is also known that the flagellin of the polar flagellum of *A. brasilense* Sp7 is a glycoprotein of approximately 100 kDa (Moens et al. 1995b). In 1995, glycosylation of the flagellin of the polar flagellum was demonstrated for the first time for *A. brasilense* type strain Sp7 (Moens et al. 1995b). Subsequently, the presence of an O-linked polysaccharide in Sp7′s glycosylated flagellin was ascertained, and the structure of its tetrasaccharide repeating unit was reported (Belyakov et al. 2012). Also, it was noted that glycosylation of flagellin is not a rare phenomenon, but is rather widespread among gram-negative bacteria and plays an important role in the processes of filament assembly and molecular recognition during bacterial interactions with various partners (Merino and Tomás 2014).

Some beneficial bacteria and/or microbe-associated molecular patterns (MAMPs) are known to induce priming and allow it to react more efficiently to different stresses (Koornneef and Pieterse 2008). Bacterial flagellin was found among MAMPs, and it is recognized by specific receptors (Zipfel 2009).

The polar flagellin of *Azospirillum* has been mainly studied for its ability to induce plant defenses, and its use in the control of phytopathogens is being explored since it can activate immune responses in plants without requiring the direct application of live bacteria. Elías et al. (2021) have evaluated the capacity of *A. brasilense* REC3 polar flagellin AzFlap as a MAMP, eliciting biochemical, histological, and molecular defense responses that can provide strawberry plants with protection against the pathogenic fungus *Macrophomina phaseolina*. They showed that plants treated with AzFlap brought about the accumulation of salicylic acid, mainly conjugated, to induce stomatal closure, to produce lignin accumulation at the xylem of strawberry leaves, as well as to induce callose depositions in the periphery of the cell wall in the conducting elements, vessels, or tracheids of strawberry leaves. Also, the relative expression of defense-related genes confirmed the activation of the strawberry innate immune system. The level of gene expression was strongly time and treatment dependent, suggesting a complex regulation of defense signaling. Phytopathogenic assays in plants treated with AzFlap or REC3 and infected with *M. phaseolina* were also performed. Root inoculations with REC3 or foliar treatment with AzFlap were able to reduce plant mortality, showing the effectiveness of both treatments to control *M. phaseolina*. These results indicate that flagellin AzFlap from *A. brasilense* REC3 behaves as an MAMP that activates a defense response against *M. phaseolina* in strawberry plants. The ability of the flagellin AzFlap to limit the negative impact of the charcoal root rot disease in strawberry plants sets a basis to study more profoundly the mechanisms induced by this flagellar protein. Correa Deza et al. (2023) also observed that both REC3 and AzFlap reduced lesions caused by the fungus *Botrytis cinerea*, the etiological agent of gray mold disease, on detached strawberry leaves.

Although the use of *Azospirillum* flagellin for the control of phytopathogens is still in the experimental phase, a lower incidence of foliar and root diseases has been observed in plants treated with this protein (Elías et al. 2021; Correa Deza et al. 2023). Since flagellin is a protein rather than a live bacterium, its application is particularly promising in situations where introducing live microorganisms may be limited or undesirable (e.g., in controlled cultivation systems or postharvest applications). Moreover, the direct application of flagellin to crops could stimulate plant defenses without the risk of external organism proliferation, and its use may be of interest in combination with other bioelicitors.

Additionally, the use of flagellin could reduce the need for fungicides and bactericides in phytopathogen control, offering a more sustainable alternative for disease management. Overall, although the use of *Azospirillum* polar flagellin in phytopathogen control is an emerging and actively researched area, initial studies indicate its potential to activate innate plant defenses and contribute indirectly to phytopathogen control.

It is also worth noting that, beyond its role in pathogen control, *Azospirillum* flagellin is also involved in promoting plant growth, as evidenced by changes in the root architecture of *Arabidopsis thaliana* (Mora et al. 2023) and wheat (Correa Deza et al. 2024).

12.5 Methods of Applying *Azospirillum* spp.

The ultimate goal and economic rationale for long-term studies on the effects of PGPB, such as *Azospirillum* spp., on plant growth and productivity is the practical application of inoculants by growers. To this end, inoculants consist of viable cultures with high concentrations of *Azospirillum* spp., embedded, suspended, or mixed in a carrier of choice (Bashan and de-Bashan 2015).

As *Azospirillum* contributes to plant growth in part through its biocontrol activity, there is no specific application method designed solely for biocontrol purposes. In any case, inoculation with *Azospirillum* serves as a preventive, not curative, measure against pathogen activity. Supporting this, Guerrero-Molina et al. (2014) reported that early responses of strawberry plants to *Azospirillum* inoculation included increased levels of phenolic compounds, reduced lipid peroxidation, and activation of genes associated with defense (FaPR1), reactive oxygen species (ROS) detoxification (FaCAT and FaAPXc), and bacterial perception (FaFLS2), indicating that strawberry plants can sense the presence of nearby bacteria. Structural changes, such as callose deposition and cell wall reinforcement, occurred later and are considered delayed responses. These findings suggest that *Azospirillum* inoculation can trigger physiological, structural, and molecular modifications that promote beneficial traits in plants. Because these induced signals play essential roles in plant defense, they may provide strawberry plants with broad-spectrum and long-lasting resistance to pathogens, ultimately improving plant health and productivity.

In field crops, inoculation with *Azospirillum* spp. is commonly carried out by treating seeds with liquid or solid carriers containing the bacteria or by applying them directly into the furrow at sowing (Cassán and Díaz-Zorita 2016). This method increases bacterial cell numbers in the rhizosphere, ensuring early root colonization and enhancing the likelihood that the selected strains will exert their beneficial effects (Balandreau 2002). Currently, seed treatment at sowing is the preferred inoculation method for *Azospirillum* spp., typically applied without mixing with other products such as micronutrients, fungicides, insecticides, or polymers (Cassán and Díaz-Zorita 2016). Excessive seed handling is discouraged, as desiccation, high temperatures, UV exposure, and toxic substances can negatively impact microbial viability (Díaz-Zorita et al. 2015). Moreover, sowing and furrow treatments should be performed in moist seedbeds to prevent seed desiccation or direct sunlight exposure.

Additional inoculation methods include soaking seedlings in *Azospirillum*-containing solutions before transplanting, foliar and soil spray applications, seed coating treatments combined with chemical additives that enhance long-term bacterial survival, drip irrigation applications, and soaking potato tubers in bacterial inoculum. Some of these examples are shown in Table 12.2.

In the context of biocontrol of phytopathogens, evaluating disease symptoms in plants is essential to determine the level of protection conferred by *Azospirillum* in crop systems. Beyond the specific symptoms associated with each disease, standardized scales are available for assessing the incidence and severity of plant diseases. For example, the **gray mold** is a widespread disease in strawberry plants caused by *Botrytis cinerea*, a necrotrophic fungus known for its ubiquity and prevalence

Table 12.2 Some methods of applying *Azospirillum* in different crops

Method	Crop	Crop condition	References
Seed inoculation	*Triticum aestivum* L. *Glycine max* (L.) Merr.	Field Field	Díaz-Zorita and Fernández-Canigia (2009) and Martínez et al. (2024)
Soil spray in furrow	*Zea mays* L., *Triticum aestivum* L.	Greenhouse and Field	Fukami et al. (2016)
Seedlings soaked in bacterial inoculum	Petunia	Greenhouse	Toffoli et al. (2021)
Foliar spray	*Lactuca sativa*, *Eruca sativa* L. *Zea mays* L., *Triticum aestivum* L.	Hydroponic system Greenhouse and Field	Oliveira et al. (2023) and Fukami et al. (2016)
Drip irrigation system	*Saccharum* spp.	Field	Simões et al. (2019)
Tubers soaked in bacterial inoculum	*Solanum tuberosum* L.	Field	Naqqash et al. (2022)

worldwide (Breen et al. 2022). This pathogen affects all aerial parts of the plant, although it most commonly manifests as fruit rot. **Anthracnose** is a fungal disease caused by *Colletotrichum* spp., which produces various symptoms in strawberry plants, including root, fruit, and crown rot; lesions on petioles and runners; and irregular black spots on leaves (Denoyes-Rothan et al. 2003). **Charcoal rot** of roots and crowns in strawberry is caused by *Macrophomina phaseolina*, a persistent soil-borne pathogen that infects more than 500 plant species, both cultivated and wild (Ijaz et al. 2012). Typical symptoms in strawberry include foliage wilting, drying, and defoliation of older leaves, crown rot, and plant death (Koike 2008). **Sudden death syndrome (SDS)** is a significant disease in soybeans caused by *Fusarium* spp. (Martínez et al. 2024). Characteristic symptoms include foliar chlorosis and necrosis, root rot, vascular discoloration in stems and roots, defoliation, and plant death. Table 12.3 summarizes the diseases, pathogens, symptoms, and disease severity rating (DSR) scales for the examples mentioned above.

12.6 Projections for the Use of *Azospirillum* in the Biocontrol of Phytopathogens

The use of *Azospirillum* as a biocontrol agent in agricultural crops has a promising future, especially with advancements in biotechnology and a growing trend toward sustainable agriculture. Some projections and future directions for its application include:

12.6.1 Development of More Effective Formulations

The creation of advanced microbial formulations that combine *Azospirillum* spp. with other beneficial microorganisms, or compounds that improve their survival, may enhance their efficacy in controlling pathogens and pests. These formulations, optimized for field conditions and different soil types, could help strengthen their biocontrol and growth-promoting effects. Alternative methods of inoculation, as effective as the standard seed inoculation procedure, may represent a significant approach to avoiding potential incompatibility between inoculants and pesticides used during seed treatment (Cassán et al. 2020).

12.6.2 Optimization of Specific Strains

With advances in genetics and bioengineering, it is possible to select or enhance *Azospirillum* strains with specific resistance properties against certain pathogens.

Table 12.3 Some examples of diseases, pathogens, main symptoms, and the corresponding disease severity rating (DSR) scale for evaluation

Disease	Pathogen	Main symptoms	DSR scale
Gray mold in strawberry (*Fragaria ananassa* Duch.)	*Botrytis cinerea*	Affects aerial parts, primarily fruit rot 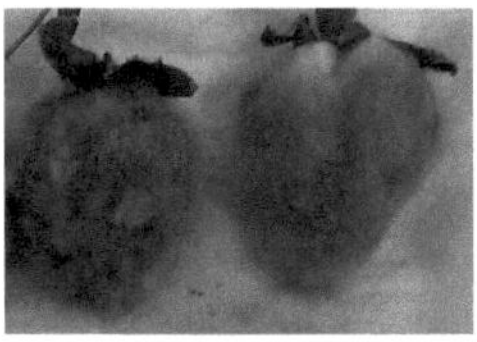	(Vellicce et al. 2006): 1—no lesion 2—lesion 1 mm 3—lesions 1–5 mm 4—lesions 5–10 mm 5—>10 mm or dead leaflet
Anthracnose in strawberry (*Fragaria ananassa* Duch.)	*Colletotrichum* spp.	Root, fruit, and crown rot; lesions on petioles and runners; black spots on leaves 	(Delp and Milholland 1980): 1—healthy petiole 2—lesions <3 mm 3—lesions 3–10 mm 4—lesions 10–20 mm, girdling 5—necrotic petiole, dead plant
Charcoal rot in strawberry (*Fragaria ananassa* Duch.)	*Macrophomina phaseolina*	Foliage wilting, drying, defoliation, crown rot, plant death 	(Fang et al. 2011): 0—healthy 1—slightly stunted 2—stunted, yellowing 3—severely stunted/wilting 4—most leaves wilted 5—plant dead
Sudden death syndrome (SDS) in soybean (*Glycine max* (L.) Merr.)	*Fusarium* spp.	Foliar chlorosis and necrosis, root rot, vascular discoloration, defoliation, plant death 	**Foliar** (Huang and Hartman 1998): 1—0% foliage affected 2—1–20% affected 3—21–50% 4—51–80% 5—81–100% **Root** (Gray

(continued)

Table 12.3 (continued)

Disease	Pathogen	Main symptoms	DSR scale
			and Achenbach 1996): 1—healthy roots 2—<25% necrosis 3—25–50% necrosis 4—51–90% necrosis 5—>90% necrosis (dead)

These specialized strains could be better adapted to particular crop conditions or have a greater ability to activate specific plant defenses (El-Saadony et al. 2022). By integrating omics-based and metagenomic studies, knowledge of interactions between beneficial bacteria and phytopathogens could be expanded, revealing key factors and mechanisms involved in disease suppression (Ayaz et al. 2023).

12.6.3 Use in Combination with Other Bioproducts

Current research highlights the potential of combining *Azospirillum* with other biofertilizers and biocontrol agents, such as mycorrhizal fungi and PGPB. This approach could enhance the effects of *Azospirillum*, providing a broader and more effective strategy for pest and disease management.

12.6.4 Better Understanding of ISR Mechanisms

The induction of systemic resistance by *Azospirillum* is a key area of interest for understanding how to modulate this response in different crop species. Further studies in this field could help optimize its application at critical growth stages, ensuring that plants develop stronger defenses against diseases and stress.

12.6.5 Application in High-Value Crops

The trend toward organic farming and the use of bioproducts opens the possibility of using *Azospirillum* in high-value crops, such as fruits, vegetables, and ornamental plants, where pest and disease control is crucial for product quality and profitability.

12.6.6 Reduction in Chemical Pesticide Use

With improved biocontrol efficacy, *Azospirillum* could contribute to reducing the use of conventional pesticides and fungicides. This would not only benefit the environment but also lower costs for farmers and promote more sustainable agricultural practices. Educating producers about using microorganisms for biocontrol instead of chemical pesticides is also an important aspect to consider (Boro et al. 2022).

12.6.7 Climate-Resilient Agriculture

Climate change imposes additional stress on crops, making them more vulnerable to pathogens and pests. Future research could focus on using *Azospirillum* spp. to enhance crop resilience to adverse conditions such as droughts and extreme temperatures, strengthening plants' natural defenses in these scenarios (Martínez et al. 2024).

12.6.8 Integration into Precision Agriculture Systems

With the development of monitoring technologies and precision agriculture, *Azospirillum* spp. applications could be adjusted more accurately and timely. Sensors and predictive models could help determine the best application times, maximizing their effectiveness and reducing bioinoculant waste.

These projections highlight part of the potential of *Azospirillum* not only as a biocontrol agent but also as a key tool in developing more sustainable and efficient agricultural practices, adapted to the current challenges of global agriculture.

Acknowledgments We thank Agr. Eng. Victoria del V. González, from the EEAOC Phytopathology Laboratory, Tucumán, Argentina, for providing the images shown in Table 12.3.

References

Ayaz M, Li CH, Ali Q et al (2023) Bacterial and fungal biocontrol agents for plant disease protection: Journey from lab to field, current status, challenges, and global perspectives. Molecules 28:6735. https://doi.org/10.3390/molecules28186735

Baggio JS, Forcelini BB, Wang NY et al (2021) Outbreak of leaf spot and fruit rot in Florida strawberry caused by *Neopestalotiopsis* spp. Plant Dis. https://doi.org/10.1094/PDIS-06-20-1290-RE

Balandreau J (2002) The spermosphere model to select for plant growth promoting bacteria. In: Kennedy IR, Choudhury ATMA (eds) Biofertilisers in action. Rural Industries Research and Development Corporation, Barton, p 55e63

Barassi CA, Ayrault G, Creus CM et al (2006) Seed inoculation with *Azospirillum* mitigates NaCl effects on lettuce. Sci Hortic 109(1):8–14. https://doi.org/10.1016/j.scienta.2006.02.025

Bashan Y (1998) Inoculants of plant growth-promoting bacteria for use in agriculture. Biotechnol Adv 16:729–770

Bashan Y, de-Bashan LE (2002) Protection of tomato seedlings against infection by *Pseudomonas syringae* pv. Tomato by using the plant growth-promoting bacterium *Azospirillum brasilense*. Appl Environ Microbiol 68(6). https://doi.org/10.1128/AEM.68.6.2637-2643.2002

Bashan Y, de-Bashan LE (2010) Chapter two—how the plant growth-promoting bacterium *Azospirillum* promotes plant growth—a critical assessment. Adv Agron 108:77–136. https://doi.org/10.1016/S0065-2113(10)08002-8

Bashan Y, de-Bashan LE (2015) Inoculant preparation and formulations for *Azospirillum* spp. In: Cassán FD, Okon Y, Creus CM (eds) Handbook for Azospirillum, technical issues and protocols. Springer, Cham, pp 469–485

Bashan Y, Holguin G (1995) Inter-root movement of *Azospirillum brasilense* and subsequent root colonization of crop and weed seedlings growing in soil. Microb Ecol 29:269–281. https://doi.org/10.1007/BF00164890

Belyakov AY, Burygin GL, Arbatsky NP et al (2012) Identification of an O-linked repetitive glycan chain of the polar flagellum flagellin of *Azospirillum brasilense* Sp7. Carbohydr Res 361:127–132. https://doi.org/10.1016/j.carres.2012.08.019

Boro M, Sannyasi S, Chettri D et al (2022) Microorganisms in biological control strategies to manage microbial plant pathogens: a review. Arch Microbiol 204:666. https://doi.org/10.1007/s00203-022-03279-w

Bouillant ML, Miche L, Ouedraogo O et al (1997) Inhibition of Striga seed germination associated with sorghum growth promotion by soil bacteria. CR Acad Sci Paris-Sciences de la vie 320: 159–162

Breen J, Mur LAJ, Sivakumaran A et al (2022) *Botrytis cinerea* loss and restoration of virulence during in vitro culture follows flux in global DNA methylation. Int J Mol Sci 23:3034. https://doi.org/10.3390/ijms23063034

Carrozzi LE, Creus CM, Barassi CA et al (2012) Reparation of aged lettuce (*Lactuca sativa*) seeds by osmotic priming and *Azospirillum brasilense* inoculation. Botany 90:1093–1102. https://doi.org/10.1139/b2012-087

Cassán F, Díaz-Zorita M (2016) *Azospirillum* sp. in current agriculture: from the laboratory to the field. Soil Biol Biochem 103(2016):117–130. https://doi.org/10.1016/j.soilbio.2016.08.020

Cassán F, Perrig D, Sgroy V et al (2009) *Azospirillum brasilense* Az39 and *Bradyrhizobium japonicum* E109, inoculated singly or in combination, promote seed germination and early seedling growth in corn (*Zea mays* L.) and soybean (*Glycine max* L). Eur J Soil Biol 45:28–35. https://doi.org/10.1016/j.ejsobi.2008.08.005

Cassán FD, Okon Y, Creus CM (2015) Handbook for Azospirillum, technical issues and protocols. Springer, Cham

Cassán F, Coniglio A, López G et al (2020) Everything you must know about *Azospirillum* and its impact on agriculture and beyond. Biol Fertil Soils 56:461–479. https://doi.org/10.1007/s00374-020-01463-y

Cassán FD, Coniglio A, Amavizca E et al (2021) The *Azospirillum brasilense* type VI secretion system promotes cell aggregation, biocontrol protection against phytopathogens and attachment to the microalgae *Chlorella sorokiniana*. Environ Microbiol 23(10):6257–6274. https://doi.org/10.1111/1462-2920.15749

Chen C, Bélanger RR, Benhamou N et al (1999) Role of salicylic acid in systemic resistance induced by *Pseudomonas* spp. against *Pythium aphanidermatum* in cucumber roots. Eur J Plant Pathol 105:477–486. https://doi.org/10.1023/A:1008743502784

Coniglio A, Larama G, Molina R et al (2022) Modulation of maize rhizosphere microbiota composition by inoculation with *Azospirillum argentinense* Az39 (formerly *A. brasilense* Az39). J Soil Sci Plant Nutr 22:3553–3567. https://doi.org/10.1007/s42729-022-00909-6

Coniglio A, Larama G, Nievas S et al (2024) Auxin-mediated modulation of maize rhizosphere microbiome: insights from *Azospirillum* inoculation and indole-3-acetic acid treatment. J Soil Sci Plant Nutr 24:6906–6921. https://doi.org/10.1007/s42729-024-02013-3

Correa Deza MA, Venegas Tarancón SG, Amigo JA et al (2023) Efecto de flagelina polar AzFlap de *Azospirillum argentinense* REC3 en el control de *Botrytis cinerea* en frutilla. In: Libro de Resúmenes XL Jornadas Científicas de la Asociación de Biología de Tucumán. Tucumán, Argentina, 25–26 October 2023

Correa Deza MA, Venegas Tarancón SG, Pedraza RO et al (2024) Root growth-promoting effect in wheat plants exerted by *Azospirillum argentinense* REC3, its flagellar protein AzFlap and a strawberry hydroalcoholic extract in a simple rhizotron system. J Microbiol Biotechnol Food Sci. https://doi.org/10.55251/jmbfs.10825

Dadon T, Bar Nun N, Mayer AM (2004) A factor from *Azospirillum brasilense* inhibits germination and radicle growth of *Orobanche aegyptiaca*. Isr J Plant Sci 52:83–86

Delp BR, Milholland RD (1980) Evaluating strawberry plants for resistance to *Colletotrichum fragariae*. Plant Dis 64:1071–1073

Denoyes-Rothan B, Guérin G, Délye C et al (2003) Genetic diversity, and pathogenic variability among isolates of *Colletotrichum* species from strawberry. Phytopathology 93:219–228

Dias-Arieira CR, Marini PM, Fontana LF et al (2012) Efeito de *Azospirillum brasilense*, Stimulate® e fosfito de potássio no controle de *Pratylenchus brachyurus* em soja e milho. Nematropica 42: 170–175

Díaz-Zorita M, Fernández-Canigia MV (2009) Field performance of a liquid formulation of *Azospirillum brasilense* on dryland wheat productivity. Eur J Soil Biol 45:3–11. https://doi.org/10.1016/j.ejsobi.2008.07.001

Díaz-Zorita M, Fernández-Canigia MV, Bravo OA et al (2015) Field evaluation of extensive crops inoculated with *Azospirillum* sp. In: Cassán FD, Okon Y, Creus CM (eds) Handbook for Azospirillum, technical issues and protocols. Springer, Cham, pp 435–445

Donley N (2019) The USA lags behind other agricultural nations in banning harmful pesticides. Environ Health 18:44. https://doi.org/10.1186/s12940-019-0488-0

Drogue B, Sanguin H, Chamam A et al (2014) Plant root transcriptome profiling reveals a strain-dependent response during *Azospirillum*-rice cooperation. Front Plant Sci 5:607. https://doi.org/10.3389/fpls.2014.00607

Elías JM, Guerrero-Molina MF, Martínez-Zamora MG et al (2018) Role of ethylene and related gene expression in the interaction between strawberry plants and the plant growth-promoting bacterium *Azospirillum brasilense*. Plant Biol 20:490–496. https://doi.org/10.1111/plb.12697

Elías JM, Ramírez-Mata A, Albornóz PL et al (2021) The polar flagellin of *Azospirillum brasilense* REC3 induces a defense response in strawberry plants against the fungus *Macrophomina phaseolina*. J Plant Growth Regul. https://doi.org/10.1007/s00344-021-10490-4

Elkelany US, El-Mougy NS, Abdel-Kader MM (2020) Management of root-knot nematode *Meloidogyne incognita* of eggplant using some growth-promoting rhizobacteria and chitosan under greenhouse conditions. Egypt J Biol Pest Control 30:134. https://doi.org/10.1186/s41938-020-00334-w

El-Saadony MT, Saad AM, Soliman SM et al (2022) Plant growth promoting microorganisms as biocontrol agents of plant diseases: mechanisms, challenges and future perspectives. Front Plant Sci 13:923880. https://doi.org/10.3389/fpls.2022.923880

Fang X, Phillips D, Li H et al (2011) Comparisons of virulence of pathogens associated with crown and root diseases of strawberry in Western Australia with special reference to the effect of temperature. Sci Hortic 131:39–48

Fendrihan S, Constantinescu F, Sicuia O et al (2018) *Azospirillum* strains as biofertilizers and biocontrol agents—a practical review type (method/approach). J Adv Agric. https://doi.org/10.24297/jaa.v7i3.6324

Fones HN, Bebber DP, Chaloner et al (2020) Threats to global food security from emerging fungal and oomycete crop pathogens. Nat Food 1:332–342. https://doi.org/10.1038/s43016-020-0075-0

Fukami J, Nogueira MA, Araujo RS et al (2016) Accessing inoculation methods of maize and wheat with *Azospirillum brasilense*. AMB Express 6(3):13. https://doi.org/10.1186/s13568-015-0171-y

Fukami J, Ollero FJ, Megías M et al (2017) Phytohormones and induction of plant stress tolerance and defense genes by seed and foliar inoculation with *Azospirillum brasilense* cells and metabolites promote maize growth. AMB Express 7:153. https://doi.org/10.1186/s13568-017-0453-7

Fukami J, Cerezini P, Hungria M (2018) *Azospirillum*: benefits that go far beyond biological nitrogen fixation. AMB Express 8:73. https://doi.org/10.1186/s13568-018-0608-1

Gray LE, Achenbach LA (1996) Severity of foliar symptoms and root and crown rot of soybean inoculated with various isolates and inoculum rates of *Fusarium solani*. Plant Dis 80(10): 1197–1199. https://doi.org/10.1094/PD-80-1197

Guerrero-Molina MF, Lovaisa NC, Salazar SM et al (2014) Physiological, structural and molecular traits activated in strawberry plants after inoculation with the plant growth-promoting bacterium *Azospirillum brasilense* REC3. Plant Biol. https://doi.org/10.1111/plb.12270

Gupta S, Dikshit AK (2010) Biopesticides: an ecofriendly approach for pest control. J Biopest 1: 186–188

Hall PG, Krieg NR (1983) Swarming of *Azospirillum brasilense* on solid media. Can J Microbiol 29:1592–1594. https://doi.org/10.1139/m83-243

Holguin G, Bashan Y (1996) Nitrogen fixation by *Azospirillum brasilense* Cd is promoted when co-cultured with by the mangrove rhizosphere bacterium *Staphylococcus* sp. Soil Biol Biochem 28:1651–1660

Huang YH, Hartman GL (1998) Reaction of selected soybean genotypes to isolates of *Fusarium solani* f. sp. glycines and their culture filtrates. Plant Dis 82:999–1002. https://doi.org/10.1094/PDIS.1998.82.9.999

Ijaz S, Sadaqat HA, Khan MN (2012) A review of the impact of charcoal rot (*Macrophomina phaseolina*) on sunflower. J Agric Sci 151:222–227

Islam N, Bora LC (1998) Biological management of bacterial leaf blight of rice (*Oryza sativa*) with plant growth promoting rhizobacteria. Indian J Agric Sci 68:798–800

Kaschuk G, Hungria M (2017) Diversity and importance of diazotrophic bacteria to agricultural sustainability in the tropics. In: de Azevedo J, Quecine M (eds) Diversity and benefits of microorganisms from the tropics. Springer, Cham. https://doi.org/10.1007/978-3-319-55804-2_12

Koike ST (2008) Crown rot of strawberry caused by *Macrophomina phaseolina* in California. Plant Dis 92:1253

Koornneef A, Pieterse CMJ (2008) Cross talk in defense signaling. Plant Physiol 146:839–844. https://doi.org/10.1104/pp.107.112029

Lade SB, Román C, del Cueto-Ginzo AI et al (2019) Differential proteomics analysis reveals that *Azospirillum brasilense* (Sp7) promotes virus tolerance in maize and tomato seedlings. Eur J Plant Pathol 155:1241–1263. https://doi.org/10.1007/s10658-019-01852-6

Macnab RM (2003) How bacteria assemble flagella. Ann Rev Microbiol 57:77–100. https://doi.org/10.1146/annurev.micro.57.030502.090832

Martínez CV, González V, Ploper DL et al (2024) Soybean inoculation with *Azospirillum argentinense* improves nodulation, grain yield, and alleviates the incidence of *Fusarium* at field with drought stress. J Crop Improv 38(6):591–617. https://doi.org/10.1080/15427528.2024.2378358

Mehnaz S (2015) *Azospirillum*: A biofertilizer for every crop. In: Arora N (ed) Plant-microbes symbiosis: applied facets. Springer, New Delhi. https://doi.org/10.1007/978-81-322-2068-8_15

Merino S, Tomás JM (2014) Gram-negative flagella glycosylation. Int J Mol Sci 15:2840–2857. https://doi.org/10.3390/ijms15022840

Miché L, Bouillant ML, Rohr R et al (2000) Physiological and cytological studies on the inhibition of Striga seed germination by the plant growth-promoting bacterium *Azospirillum brasilense*. Eur J Plant Pathol 106:347–351. https://doi.org/10.1023/A:1008734609069

Moens S, Michiels K, Keijers V et al (1995a) Cloning, sequencing, and phenotypic analysis of laf1, encoding the flagellin of the lateral flagella of *Azospirillum brasilense* Sp7. J Bacteriol 177: 5419–5426. https://doi.org/10.1128/jb.177.19.5419-5426.1995

Moens S, Michiels K, Vanderleyden J (1995b) Glycosylation of the flagellin of the polar flagellum of *Azospirillum brasilense*, a gram-negative nitrogen-fixing bacterium. Microbiology 141: 2651–2657. https://doi.org/10.1099/13500872-141-10-2651

Mora V, López G, Molina R et al (2023) *Azospirillum argentinense* modifes *Arabidopsis* root architecture through auxin dependent pathway and flagellin. J Soil Sci Plant Nutr 23:4543–4557. https://doi.org/10.1007/s42729-023-01371-8

Naqqash T, Malik KA, Imran A et al (2022) Inoculation with *Azospirillum* spp. acts as the liming source for improving growth and nitrogen use efficiency of potato. Front Plant Sci 13:929114. https://doi.org/10.3389/fpls.2022.929114

Neilands JB (1995) Siderophores: structure and function of microbial iron transport compounds. J Biol Chem 270:26723–26726. https://doi.org/10.1074/jbc.270.45

Oliveira CES, Gato IMB, Moreira VA et al (2023) Inoculation methods of *Azospirillum brasilense* in lettuce and arugula in the hydroponic system. Rev Bras Eng Agríc Ambient 27(9):653–662. https://doi.org/10.1590/1807-1929/agriambi.v27n9p653-662

Pereg L, de-Bashan LE, Bashan Y (2016) Assessment of affinity and specificity of *Azospirillum* for plants. Plant Soil 399:389–414. https://doi.org/10.1007/s11104-015-2778-9

Pieterse CMJ, Van Loon LC (1999) Salicylic acid-independent plant defence pathways. Trends Plant Sci 4:52–58. https://doi.org/10.1016/S1360-1385(98)01364-8

Pieterse CMJ, Leon-Reyes A, Van Der Ent S et al (2009) Networking by small-molecule hormones in plant immunity. Nat Chem Biol 5:308–316. https://doi.org/10.1038/nchembio.164

Prischmann-Voldseth DA, Özsisli T, Aldrich-Wolfe L et al (2020) Microbial inoculants differentially influence plant growth and biomass allocation in wheat attacked by gall-inducing Hessian fly (Diptera: Cecidomyiidae). Environ Entomol 49(5):1214–1225. https://doi.org/10.1093/ee/nvaa102

Ramallo AC, Namen S, Salazar SM et al (2024) New approaches for the sustainable management of *Neopestalotiopsis clavispora*, the causal agent of strawberry crown and root rot in Tucumán, Argentina. Rev Agron Noroeste Argent 44(2):10–14. https://doi.org/10.61914/ranar.4402.02

Rivera D, Revale S, Molina R et al (2014) Complete genome sequence of the model rhizosphere strain *Azospirillum brasilense* Az39, successfully applied in agriculture. Genome Announc 2(6): e00614–e00683

Rodríguez-Salazar J, Suárez R, Caballero-Mellado J et al (2009) Trehalose accumulation in *Azospirillum brasilense* improves drought tolerance and biomass in maize plants. FEMS Microbiol Lett 296:52–59. https://doi.org/10.1111/j.1574-6968.2009.01614.x

Romero AM, Correa OS, Moccia S (2003) Effect of *Azospirillum* mediated plant growth promotion on the development of bacterial diseases on fresh-market and cherry tomato. J Appl Microbiol 95:832–838. https://doi.org/10.1046/j.1365-2672.2003.02053.x

Romero AM, Vega D, Correa OS (2014) *Azospirillum brasilense* mitigates water stress imposed by a vascular disease by increasing xylem vessel area and stem hydraulic conductivity in tomato. Appl Soil Ecol 82:38–43. https://doi.org/10.1016/j.apsoil.2014.05.010

Russo A, Vettori L, Felici C et al (2008) Enhanced micropropagation response and biocontrol effect of *Azospirillum brasilense* Sp245 on *Prunus cerasifera* L. clone Mr.S 2/5 plants. J Biotechnol 134:312–319. https://doi.org/10.1016/j.jbiotec.2008.01.020

Saeed M, Ilyas N, Mazhar R et al (2016) Drought mitigation potential of *Azospirillum* inoculation in canola (*Brassica napus*). J Appl Bot Food Qual 278:270–278. https://doi.org/10.5073/JABFQ.2016.089.035

Santos F, Simões Bento JM (2019) Use of the plant growth promotion rhizobacteria as biocontrol agents: perspectives for the use of *Azospirillum brasilense* to control Barley Yellow Dwarf Virus. Schizophr Bull. https://doi.org/10.21276/sb.2019.5.6.1

Santos F, Penãflor MFGV, Pare PW et al (2014) A novel interaction between plant-beneficial rhizobacteria and roots: colonization induces corn resistance against the root herbivore *Diabrotica speciosa*. PLoS ONE 9(11):e113280. https://doi.org/10.1371/journal.pone.0113280

Seddigui Kiasari A, Aminpanah H, Sharifi P (2018) Chemical weed control in paddy fields inoculated with *Azospirillum lipoferum*. Planta Dan. https://doi.org/10.1590/S0100-83582018360100085

Simões WL, Oliveira ARD, Souza MAD et al (2019) Efficient inoculation of diazotrophic bacteria into sugarcane coating treatments also in combination with other chemical compounds providing a long term survival of azospirila. Pesq Agrop Brasileira. https://doi.org/10.1590/S1678-3921.pab2019.v54.01110

Somers E, Ptacek D, Gysegom P et al (2005) *Azospirillum brasilense* produces the auxin-like phenylacetic acid by using the key enzyme for indole-3-acetic acid biosynthesis. Appl Environ Microbiol 71:1803–1810

Steenhoudt O, Vanderleyden J (2000) *Azospirillum*, a free-living nitrogen-fixing bacterium closely associated with grasses: genetic, biochemical and ecological aspects. FEMS Microbiol Rev 24: 487–506. https://doi.org/10.1016/S0168-6445(00)00036-X

Sultana U, Desai S, Pinisetty S (2023) *In vitro* screening for abiotic stress tolerance and biocontrol ability of plant growth promoting strains of *Azotobacter* and *Azospirillum* spp. Curr Agric Res. https://doi.org/10.12944/CARJ.11.3.14

Tahir Mehmood GL, Tehmina A, Waheed A (2021) *Azospirillum lipoferum* strain AL-3 reduces early blight disease of potato and enhances yield. Crop Prot. https://doi.org/10.1016/j.cropro.2020.105349

Tapia-Hernandez A, Mascarua-Esparza M, Caballero-Mellado J (1990) Production of bacteriocins and siderophore-like activity by *Azospirillum brasilense*. Microbios 64:73–83

Toffoli LM, Martínez-Zamora MG, Medrano NN et al (2021) Natural occurrence of *Azospirillum brasilense* in petunia with capacity to improve plant growth and flowering. J Basic Microbiol. https://doi.org/10.1002/jobm.202100064

Tortora ML, Díaz-Ricci JC, Pedraza RO (2011) *Azospirillum brasilense* siderophores with antifungal activity against *Colletotrichum acutatum*. Arch Microbiol 193(4):275–286. https://doi.org/10.1007/s00203-010-0672-7

Tortora ML, Díaz-Ricci JC, Pedraza RO (2012) Protection of strawberry plants (*Fragaria ananassa* Duch) against anthracnose disease induced by *Azospirillum brasilense*. Plant Soil 356:279–290. https://doi.org/10.1007/s11104-011-0916-6

Vellicce GR, Díaz Ricci JC, Hernández L et al (2006) Enhanced resistance to *Botrytis cinerea* mediated by the transgenic expression of the chitinase gene *ch5*B in strawberry. Transgenic Res 15:57–68. https://doi.org/10.1007/s11248-005-2543-6

Viejobueno J, Albornoz PL, Camacho M et al (2021) Protection of strawberry plants against charcoal rot disease (*Macrophomina phaseolina*) induced by *Azospirillum brasilense*. Agronomy 11(2):195. https://doi.org/10.3390/agronomy11020195

Vishwakarma K, Kumar N, Shandilya C et al (2020) Revisiting plant–microbe interactions and microbial consortia application for enhancing sustainable agriculture: a review. Front Microbiol 11:560406. https://doi.org/10.3389/fmicb.2020.560406

Yasuda M, Isawa T, Shinozaki S et al (2009) Effects of colonization of a bacterial endophyte, *Azospirillum* sp. B510, on disease resistance in rice. Biosci Biotechnol Biochem 73(12): 2595–2599. https://doi.org/10.1271/bbb.90402

Zipfel C (2009) Early molecular events in PAMP-triggered immunity. Curr Opin Plant Biol 12(4): 414–420

Part IV
Interaction of *Azospirillum* spp. with Other Organisms

Chapter 13
Interaction of *Azospirillum* spp. with Plants: Molecular Plant Growth Responses

Hexon Ángel Contreras-Cornejo, Juan Manuel Sanchez-Yañez, Elizabeth Carrillo-Flores, Ma. Elena Mellado-Rojas, and Elda María Beltrán-Peña

Abstract As has been widely reported, the auxin indole-3-acetic acid (IAA) is the plant hormone responsible for the changes in root architecture induced by the plant growth-promoting rhizobacteria (PGPR) of the *Azospirillum* genus. Such changes increase the area of root exploration, with greater uptake of minerals and water, resulting in increased leaf area development and grain productivity. Therefore, biofertilizers containing only *Azospirillum* or combined with other PGPR genera are used in vegetable seeds to reduce and optimize nitrogen fertilizer dosages and to prevent its excessive use. For this reason, some molecular studies on the effects of *Azospirillum* during its interaction with plants have focused on the auxin response pathway. This pathway is a complex process that depends on the homeostasis, transport, perception, and signaling of auxin. In addition, it has been reported that the Target of Rapamycin (TOR) signaling pathway, activated by auxins, also participates in changes of root architecture in the model plant *Arabidopsis thaliana*. Due to the importance of this PGPR in various areas, from agriculture to the bioenhancement of contaminated soils, in this chapter, we will review different aspects, mechanisms, and processes involved in the remarkable effects of *Azospirillum* during its interaction with plants.

Keywords Auxin signaling · Root architecture · Perception · TOR pathway · Biofertilizer

H. Á. Contreras-Cornejo
Tecnológico Nacional de México, Instituto Tecnológico de Morelia, Morelia, Michoacán, México

J. M. Sanchez-Yañez
Environmental Microbiology Laboratory, Chemical Biological Research Institute, Universidad Michoacana de San Nicolas de Hidalgo, Morelia, Michoacan, México

E. Carrillo-Flores · M. E. Mellado-Rojas · E. M. Beltrán-Peña (✉)
Signal Transduction Laboratory, Chemical Biological Research Institute, Universidad Michoacana de San Nicolas de Hidalgo, Morelia, Michoacan, México
e-mail: eldabelt@umich.mx

F. D. Cassán et al. (eds.), *The Definitive Handbook of Azospirillum*,
https://doi.org/10.1007/978-3-032-09222-9_13

13.1 Diversity, Distribution and Root Colonization of *Azospirillum*

Species within the genus *Azospirillum* have been isolated and identified in soils from all geographical regions worldwide, including freshwater and extreme environments such as the Arctic, deserts, and mountains, and have also been used for agricultural production on both temperate and tropical soil. *Azospirillum* is particularly reported to exhibit positive ecological interactions with more than 33 plant families and over 120 species (Aeron et al. 2020; Sun et al. 2021; Rabara et al. 2023; Xu et al. 2023). Enhanced root system growth not only reduces the need for high fertilizer doses but also helps achieve maximum nutrient use efficiency (Pedraza et al. 2020; Housh et al. 2023). This improvement contributes to maintaining soil productivity while minimizing air and water pollution problems caused by excessive fertilization (Havlin et al. 2017; Cruz-Hernandez et al. 2022; Utkin et al. 2023).

Azospirillum colonization is a complex mechanism that initially involves chemotaxis (Ganusova et al. 2021, 2024). In this process, host plants release important organic molecules through root exudates (i.e., carbohydrates and acids). These compounds are detected by bacterial cells, stimulating bacterial migration toward the point of emission. A gradient of chemoeffectors is thereby established near the root surface. In *Azospirillum brasilense*, Tlp, a chemoreceptor of the 4-helix bundle family, has been identified and its activity is modulated by the *NtrC* and *RpoN* regulons (Ganusova et al. 2024).

Once the bacteria have located the host root, colonization occurs. The colonization pattern of *Azospirillum* is on the epidermis of the root tissue in the case of ectophytic strains (Bashan et al. 1991), while endophytes are found within the cells (Naqqah et al. 2022). Colonization particularly focuses on the root elongation zone and at the base of root hairs, and they have also been observed within xylem vessels, and as new root structures such as young hairs and lateral roots emerge, they are also colonized (Okon and Kapulnik 1986). Colonization by *Azospirillum* has also been observed in the emergence zone of lateral roots (Vande Broek et al. 1993).

13.2 *Azospirillum* Plant Benefits

Azospirillum species establish an associative symbiosis with the root system of higher plants, which begins with the induction of seed germination, followed by invasion of the rhizosphere and rhizoplane by a chemotactic attraction of organic compounds that allow *Azospirillum* to grow in these sites and facilitate the exchange of minerals and organic compounds (Shahrajabian et al. 2021; Sun et al. 2021). Consequently, the excretion of tryptophan from the plant causes this PGPR to convert into indole-3-acetic acid (IAA), which induces a significant growth of the root system to increase mineral absorption, which favors healthy plant growth

(Cassan and Diaz-Zorita 2016; Sun et al. 2021; Sivasakthivelan et al. 2023). Although there is evidence that *Azospirillum* has the genetic ability to transform other plant compounds into other phytohormones such as gibberellins and cytokinins, which positively adjust the dose of nitrogen- and phosphorus-based fertilizers for sustainable production of a wide range of foods of plant origin (Aeron et al. 2020; Housh et al. 2023; Sivasakthivelan et al. 2023). A fundamental aspect of the genetics of *Azospirillum* species is their nitrogenase system, which is induced in the absence of organic and inorganic nitrogen sources (Housh et al. 2023; Wang et al. 2024). This trait is beneficial in hydroponic systems, where the concentration of essential nutrients such as nitrogen—a limiting factor for healthy plant growth and yield—is controlled (Rozier et al. 2019; Pedraza et al. 2020; Rabara et al. 2023). There is also evidence of the positive effect of *Azospirillum* species on the growth and activity of the rhizoplane and rhizosphere, as it promotes a balance between the dynamics of essential plant elements and other soil microbial groups, that are fundamental to biogeochemical cycles (Housh et al. 2023; Utkin et al. 2023; Wang et al. 2024). For the above reasons, useful biofertilizers have existed since the end of the late twentieth century for the sustainable production of plant-based foods that are safe for human and animal consumption and for the environment (Cassan and Diaz-Zorita 2016; Thomas and Singh 2019; Aeron et al. 2020; Shahrajabian et al. 2021). *Azospirillum* also has other useful properties for preventing plant diseases caused by actinomycetes, bacteria, fungi, and nematodes through a mechanism known as induced systemic resistance (Pagnussat et al. 2023; Pedraza et al. 2020).

13.2.1 Enhanced Biomass Production

Azospirillum species are a biological tool to reduce and optimize nitrogen and phosphorus fertilization doses for plant health associated with healthy growth that generates profitable yields in harmony with the environment (Thomas and Singh 2019; Sivaskthivelan et al. 2023; Utkin et al. 2023). Since *Azospirillum* has the ability to synthesize the nitrogenase system that reduces atmospheric nitrogen (N_2) to ammonium (NH_4^+), in an environment that compromises *Azospirillum* and allows the survival of the plant (Thomas and Singh 2019; Shahrajabian et al. 2021; Sun et al. 2021). While *Azospirillum* uses the products of plant root metabolism for the synthesis of IAA, which induces to increase the density of lateral and hairs root that uptake the main minerals that give health to the plant and a significant increase in plant development, resulting in greater productivity in plants of economic interest (Cassan and Diaz-Zorita 2016; Amavizca et al. 2017; Aeron et al. 2020). With this positive interaction between *Azospirillum* and the plant, it contributes to increased tolerance against biotic stress caused by root pathogens, as well as abiotic stress such as soil salinity (Pagnussat et al. 2023; Sivasakthivelan et al. 2023; Utkin et al. 2023).

13.2.2 Root Growth and Branching

Azospirillum can transform tryptophan released by the plant root into IAA, an auxin that promotes cell elongation and division in the root system. Among the PGPRs studied to date, including *Azospirillum*, the most commonly observed effect on root architecture is a reduction in primary root growth and an increase in the number and length of lateral roots, as well as stimulation of the elongation of root hairs (Aeron et al. 2020; Housh et al. 2023; Wang et al. 2024). The effect of *Azospirillum* on plant development has been attributed to the microbial IAA it produces (Sun et al. 2021; Shahrajabian et al. 2021; Sivasakthivelan et al. 2023). Through this mechanism, *Azospirillum* modifies the morphology and physiology of these roots to increase the surface area of contact and exploration of root tissue to maximize the uptake of essential minerals and water (Thomas and Singh 2019; Palacios et al. 2021; Pagnussat et al. 2023; Utkin et al. 2023). With this change in plant growth stage, shortening the vegetative cycle has a positive effect on yield and even the nutritional quality of grain for human consumption (Rozier et al. 2019; Aeron et al. 2020; Pedraza et al. 2020; Sun et al. 2021). This does not imply loss of soil fertility or release of greenhouse gases, which allows sustainable land use and an ecological action that mitigates global warming (Havlin et al. 2017; Pedraza et al. 2020; Shahrajabian et al. 2021; Sun et al. 2021).

13.2.3 Abiotic Stress Tolerance

This topic has been extensively discussed in Chap. 14 of the *Handbook*. However, a brief introduction is provided here. *Azospirillum* spp. supports plant tolerance to various abiotic stresses through multiple physiological and molecular mechanisms. Under salinity stress, it helps maintain a favorable K^+/Na^+ ratio, increases antioxidant enzyme activity, and reduces oxidative damage (Degon et al. 2023). During waterlogging, it improves photosynthetic efficiency and ethanol metabolism, thereby mitigating toxicity. In drought conditions, *Azospirillum* stimulates root proliferation, modulates hormone profiles (increasing abscisic acid (ABA) and reducing ethylene), and enhances water retention and water-use efficiency (Curá et al. 2017). Trehalose-overproducing mutants further strengthen drought resilience (Rodriguez-Salazar et al. 2009). At the molecular level, *Azospirillum* spp. possesses a sophisticated reactive oxygen species (ROS) defense system regulated by sigma factors and antioxidant-related genes (Gupta et al. 2018). Moreover, through interaction with other organisms, it reduces oxidative stress and activates plant defense responses.

13.3 Mechanisms of Plant Growth Promotion by *Azospirillum*

Plant growth promotion by member of the genus *Azospirillum* can occur mainly through two conditions: (i) without physical contact between *Azospirillum* and the root, and (ii) during root colonization, which leads to a significant improvement in plant nutrition and growth (Bashan et al. 1991). *Azospirillum* has been observed to induce root elongation and branching, increase leaf surface area and stem diameter, and alter plant metabolism including nitrogen fixation and biochemical changes, as evidenced by flavonoid signaling (Fibach-Paldi et al. 2011). Plant growth stimulation by *Azospirillum* involves one or several growth regulators that are produced by the bacteria, and these in turn activate molecular mechanisms (Fukami et al. 2018; Sun et al. 2025). Plant growth promotion by rhizosphere microorganisms includes changes in [Ca^{2+}] ion concentrations and second messenger level, activation of protein kinases and transcription factors, modulation of gene expression, and protein synthesis (Spaepen et al. 2014). The physical interaction of bacteria with root cells (colonization process) can also stimulate endogenous signaling mechanisms in the plant.

13.3.1 Nutrient Acquisition

Root colonization can lead to a variety of biochemical, molecular, physiological, and nutritional (i.e., NO_3^-), potassium (K), and phosphate ($H_2PO_4^-$) changes in the host plant. Several *Azospirillum* strains induced morphological changes in the roots of sorghum, wheat, and maize plants after germination. In particular, *Azospirillum* can significantly increase the concentrations of N, P, and K contents of maize, sorghum, and wheat, in addition to dry matter (Okon and Kapulnik 1986). *Azospirillum* also can promote the uptake of other plant nutrients such as Fe, Ca, Mg, B, S, and Mn (Pii et al. 2016; Oliveira et al. 2023).

The genus *Azospirillum* is widely known for its extraordinary nitrogen-fixing abilities with its host (Von Bulow and Döbereiner 1975). Oxygen and carbon sources are crucial factors influencing nitrogen fixation during the *Azospirillum*-plant interaction (Vande-Broek et al. 1993). It has been proposed that nitric oxide (NO_2) released by *Azospirillum* during nitrate respiration plays a key role in growth promotion (Zimmer et al. 1988). In particular, diazotrophic strains such as TN03 and TN09 possess extraordinary nitrogen-fixing potential because potato plants inoculated with these bacteria increased their nitrogen content compared to non-inoculated plants (Naqqash et al. 2022). There is evidence that *Azospirillum* has the ability to fix nitrogen in host plants even in the presence of additional nitrogen sources (i.e., nitrate or nitrate). For instance, nitrogen fixation evaluations were carried out using an atmosphere enriched with $^{15}N_2$, which was subsequently detected in the roots and shoots of wheat plants, finding that the highest amounts of nitrogen were detected in the foliage. It is important to highlight that in conjunction with nitrogen fixation by *Azospirillum* spp., the accumulation of total biomass in the host plant also takes place and that the mechanism of absorption and biological fixation of nitrogen can be modified by the form of application of the element and the bacteria that will assist in this process (Ruppel and Merbach 1997). Phosphorus is a key element in plant growth and development because it is required for several energy-demanding functions. Therefore, it is part of the molecular structure of key compounds such as adenosine triphosphate (ATP), genetic material, and protein kinases. At the developmental level, phosphorus can modulate processes such as the development of the root system, which involves the growth of the main root and the formation of lateral roots in *A. thaliana*. Interestingly, the diazotrophic *Azospirillum* strains TN03 and TN09 have extraordinary potential for phosphorus solubilization and the production of organic acids (Naqqash et al. 2022), which may favor the availability of the nutrient niche for plants.

There is evidence that *Azospirillum* can simultaneously enhance the uptake of nitrogen and phosphorus, as demonstrated in corn plants inoculated with this bacterium. In the case of phosphorus concentrations, it was detected that corn plants absorb approximately 40% more of this element in their P_2O_5 form as compared to non-inoculated plants (Sangoquiza-Caiza et al. 2019). Iron (Fe) is an essential micronutrient for plant growth. Two mechanisms have been proposed for Fe uptake: strategy I, which involves the absorption of this metal after its reduction, and strategy II, which involves the chelation of the element. This element is mainly used in chlorophyll biosynthesis and participates as a cofactor in enzymatic activity. In terms of plant development, Fe modulates the formation of roots and root hairs. Iron is also involved in nitrogen fixation in leguminous plants. It has been proposed that unavailable Fe in the soil may be sequestered by microbial metabolites with siderophore activity. This way, they chelate it (i.e., Fe^{3+}) and aid in its uptake, presenting iron in a more assimilable form for plants (Leong 1986). Specifically, the cultivation of cucumber plants (*Cucumis sativus* L.) under Fe deficiency in the presence of *A. brasilense* showed an increase in acidification and overexpression of ferric-chelate reductase gene (*CsFRO1*) and plasma membrane H^+-ATPase gene (*CsHA1*), an effect that is not observed under conditions of efficiency of this metal. Importantly, it was observed that plants not inoculated with the bacteria grown under ^{57}Fe deficiency grew about 30% more than plants with Fe without the bacteria. In addition, it was observed that in the presence of ^{57}Fe. *A. brasilense* promoted the growth of the root system 30% more than plants without the bacteria but with the presence of ^{57}Fe (Pii et al. 2016), showing that the modulation of the root system by this bacterial genus largely involves Fe as a conditioning factor for root growth and development.

13.4 Hormonal Regulation of Plant Growth

This topic has been extensively addressed in Chap. 10 of the *Handbook*. However, an introductory excerpt is described here. Plants possess several hormones involved in the development process, such as auxins, gibberellins, and cytokinin, while defense responses are regulated by salicylic acid, jasmonic acid, and ethylene, which, together with ABA, also modulate developmental processes and senescence. Depending on the growth conditions, plants can activate the participation of one hormone over another or enhance the activity of multiple hormones simultaneously (Bashan and de-Bashan 2010). Interestingly, soil microorganisms such as *Azospirillum* can manipulate the molecular mechanisms of these hormones, inducing hormonal crosstalk that results in an adjustment in growth and defense responses (Fukami et al. 2018). Indole-3-acetic acid (IAA) induces the formation of lateral roots and the accumulation of leaf biomass. IAA is also involved in phototropic and gravitropic responses. *A. brasilense* is a prolific producer of IAA (Ona et al. 2005; Spaepen et al. 2007). In the bacterium itself, IAA appears to be involved in the protection of bacterial cells under chemical stress that can be generated by the production of other indole metabolites (Gasunova et al. 2025). During the bacteria-plant interaction, the IAA produced modulates the growth of the host plant's root system and the accumulation of leaf biomass (Contreras-Cornejo et al. 2009). Carrillo-Flores et al. (2022a) showed evidence for the participation of auxins produced by *Azospirillum* in the modulation of the growth of the model plant *Arabidopsis thaliana*. This was deduced from in vitro experiments where transgenic *A. thaliana* plants expressing the auxin-responsive gene DR5::GUS within 2–6 days showed an increase in the expression of said gene in the lateral roots and in the tip of the primary root compared to control plants that did not receive the stimulus of the bacteria at a density of 2.5×10^5 CFU/mL. Furthermore, it is evident that *Azospirillum* did not induce the formation of lateral roots in the double mutant of *A. thaliana arf7 arf19* affected in transcriptional factors of response to auxins compared to the wild line Columbia-0. However, many biochemical and molecular components remain to be elucidated during the promotion of plant growth by this bacterial genus. Cytokinins (CKs) are a group of molecules that act as phytohormones in plants. Increased concentrations of these phytohormones initiate the signaling mechanism in which CKs are perceived by histidine kinase receptors located on the cell membrane; the signal subsequently induces the activation of transcription factors in the nucleus (Kieber and Schaller 2018). Specifically, the *A. brasilense* RA-17 strain produces zeatin that is synthesized from isopentenyl adenine; in the interaction of these strains, it will be observed that the phenotype induced by the bacteria on wheat plants (*Triticum aestivum*) depends largely on bacterial CK (Zaheer et al. 2022). ABA is an isoprenoid molecule that moves in the xylem from the root to the foliage, where it modulates transcription (Hartung et al. 2022). During *Azospirillum*-plant interactions, ABA concentrations can increase, contributing to plant stress tolerance through activation of stress-related genes (Sun et al. 2025). In radish (*Raphanus sativus*), *Azospirillum* can mitigate the negative effects on stomatal conductance, chlorophyll content, photosystem II (PSII) efficiency, and growth rate in terms of tuberous root diameter, while enhancing the number of leaves, leaf area, and fresh and dry weight, compared to the non-inoculated treatment under waterlogging (Salazar-García et al. 2022).

13.5 Molecular Plant Responses to *Azospirillum*

13.5.1 Genomic and Proteomic Signatures

One of the key molecules involved in the *Azospirillum*-plant interaction is indole-3-acetic acid (IAA), a phytohormone that regulates virtually all aspects of plant development. Initial reports of *Azospirillum*-induced changes in root architecture focused on various grasses. However, the first attempt to unravel the molecular mechanisms behind this interaction was carried out by Spaepen et al. (2014) using the model plant *Arabidopsis thaliana*. Although their study, and others using different plant species, observed similar root system alterations driven by microbial IAA, no specific molecular mechanism was established at the time. Plant auxin response is highly complex and involves three main processes: homeostasis, transport, and perception. Homeostasis refers to the regulation of auxin biosynthesis, degradation, and inactivation. IAA is synthesized in the foliage and transported to the demanding tissues through the phloem and polar auxin transport (PAT), which requires the influx carriers AUX1/LIKE AUX1 (AUX1/LAX) and the efflux carrier PIN-FORMED/P-GLYCOPROTEIN/ATP-Binding Cassette (PIN/PGP/ABCB). This cell-to-cell transport creates localized auxin maxima at specific sites, which is essential for triggering the development processes of certain tissues as occur with the system root (Tanaka et al. 2006; Brumos et al. 2018; Omary et al. 2022; Favre et al. 2024).

More recently, it has been shown that auxins can also move through plasmodesmata, cytoplasmic channels connecting adjacent cells (Mellor et al. 2020). In *Arabidopsis,* auxin perception occurs in the nucleus via the TIR1/AFB receptor complex (TRANSPORT INHIBITOR RESPONSE1/AUXIN SIGNALING F-BOX). Upon auxin binding, this receptor targets the AUX/IAA transcriptional repressors for degradation, thereby releasing auxin response factors (ARFs) that activate downstream gene expression (Schaller et al. 2015). It has been reported that *A. brasilense* rapidly alters *Arabidopsis* root architecture by reducing primary root length and enhancing lateral root formation within a few days of exposure. Since these phenotypes have been attributed to bacterial IAA, Carrillo-Flores et al. (2022a) analyzed auxin accumulation in *Arabidopsis DR5::GUS* seedlings exposed to the bacterium. They observed increased auxin levels in the vascular tissue of the primary root, particularly in the lateral root differentiation zone. This effect became more pronounced by the fourth day of exposure. Interestingly, by the fifth day, auxins were no longer detectable in the primary root meristem. Instead, the auxin observed in vascular tissues was redistributed toward lateral root meristems (Fig. 13.1). This

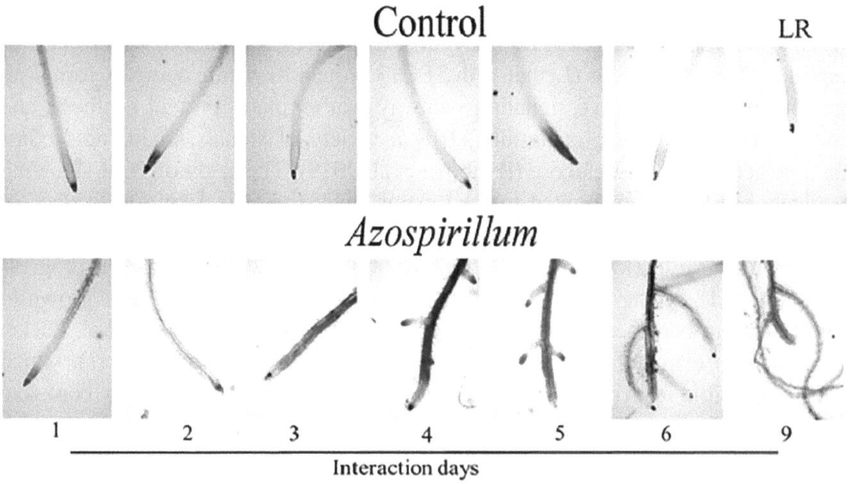

Fig. 13.1 Effect of *Azospirillum brasilense* Sp245 on auxin levels in roots of *A. thaliana*. *DR5::uidA* seedlings were transferred to control or supplemented with 2.5×10^5 CFU/mL of *Azospirillum*. These media were incubated for 1, 2, 3, 4, 5, 6, and 9 days. At the end of the incubation time, the seedlings were treated with X-Gluc, followed by a clearing process. The seedlings were mounted on slides, and microscopic observations were made at 4×. Representative photographs of the roots from the seedlings in the control medium and that inoculated with *Azospirillum*. LR, lateral root. (Carrillo-Flores et al. 2022a)

shift in auxin localization suggests that PAT might play a role in the redistribution of auxins during the *Arabidopsis*-*Azospirillum* interaction. To test this, the authors used N-1-naphthylphthalamic acid (NPA), a known PAT inhibitor, and found that it blocked the bacterium's stimulatory effect on lateral root formation. These results strongly suggest that PAT is indeed involved in mediating the plant's response to *Azospirillum*.

13.5.1.1 Target of Rapamicyn Signaling

The Target of Rapamycin (TOR) is an evolutionarily conserved kinase found in organisms ranging from yeast to mammals and in plants. TOR is a highly conserved 250-kDa atypical serine/threonine protein kinase that belongs to the PI3K family and is made up of the following domains: HEAT, Huntington repeats, Elongation factor 3 regulatory, subunit A of PP2A, TOR1; FAT, FRAM-ATM-TRAP; FKBP-Rapamycin-Binding (FRB), the catalytic kinase domain, and the C-terminal FATC domain. In various plant species, TOR is highly similar, especially in FRB and kinase domains. TOR senses energy and nutrient levels and is a regulator of cell growth, in addition to modulating protein synthesis and metabolism. TOR is also involved in restoring cell wall integrity, proliferation, and cell size (Rexin et al. 2015).

In plants, TOR functions as part of a complex known as TORC1, which includes mTOR (mammalian Target of Rapamycin), RAPTOR (Regulatory-Associated Protein of mTOR), and LST8 (Lethal with SEC13 protein 8). These complex forms are part of a nutrient-sensitive signaling pathway that regulates protein synthesis and growth. TORC1 activity is stimulated by a variety of signals, including auxins, abscisic acid, light, and glucose (Ryabova et al. 2019). The study of TOR signaling in plants has long been limited by two major challenges: the natural resistance of plants to rapamycin and the embryo lethality caused by null mutations in the *TOR* gene in *Arabidopsis* (Menand et al. 2002; Ryabova et al. 2019). However, progress has been made. Xiong and Sheen (2012) demonstrated that *Arabidopsis* grown in liquid medium becomes sensitive to rapamycin. Subsequently, Xiong et al. (2013) developed an estradiol-inducible *tor* transgenic *Arabidopsis* line (*tor-es1*), which made it possible to bypass the embryo-lethal phenotype associated with complete loss of TOR function. In addition, second-generation TOR inhibitors that target the ATP-binding site, such as AZD-8055, Torin1, and Torin2 have been shown to significantly alter plant growth (Xiong et al. 2013; Deng et al. 2016; Li et al. 2016; Montané and Menand 2018). The role of TOR during the *Arabidopsis*-*Azospirillum* interaction was investigated using two complementary approaches: a pharmacological strategy and other genetics. In the pharmacological approach, co-treatment of seedlings with AZD-8055 and *A. brasilense* resulted in a reduced number and length of lateral roots. Similarly, in the genetic approach, *tor-es1* seedlings showed a clear reduction in both lateral root formation and elongation after 3 and 6 days of exposure to the bacterium (Carrillo-Flores et al. 2022b). Together, these findings support the conclusion that TOR is involved in the stimulation of lateral root development in *A. thaliana* seedlings in response to *A. brasilense*. This role of TOR was also reported by Méndez-Gómez et al. (2020), who observed a similar inhibitory effect on root development in seedlings exposed to *Azospirillum* in the presence of AZD-8055.

13.6 Analysis of *Arabidopsis*-*Azospirillum* Interaction and TOR Signaling Pathway

Three experimental systems have been developed to study the *A. thaliana*-*Azospirillum* interaction under various growth conditions. These systems allow for phenotypic and molecular analysis of *Arabidopsis* seedlings and include (1) direct root-bacteria contact, (2) chemical signaling via diffusible bacteria compounds, and (3) signaling via volatile organic compounds (VOCs).

13.6.1 Preliminary Procedure

Experimental procedure

1. Surface-sterilize *Arabidopsis* seeds by immersing them in 95% (v/v) ethanol for 5 min.
2. Transfer the seeds to 20% (v/v) bleach for an additional 5 min.
3. Rinse the seeds five times with sterile distilled water.
4. Stratify the seeds at 4 °C for 48 h to synchronize germination.
5. Sow the seeds on agar plates containing 0.2× Murashige and Skoog (MS) medium (Murashige and Skoog 1962), adjusted to pH 7, with 0.9 g/L MS salt, 6 g/L sucrose, and 10 g/L Agar Plant TC (PhytoTechnology Laboratories, St, Lenexa, KS, USA) (Méndez-Gómez et al. 2021) or MS adjusted to pH 5.7 and supplemented with 0.8% (w/v) agar (Mora et al. 2023).
6. Place the plates vertically in a growth chamber at 22 °C with a 16 h light/8 h dark photoperiod and a light intensity of 100 μmol m^{-2} s^{-1}.
7. Use the bacterial strain under study (wild-type or mutant), previously cultured under optimal conditions. Alternatively, *Azospirillum*'s flagellin may be used according to Mora et al. (2023).

The different interactions can be performed as follows:

Direct Root-Bacteria Contact Assay
Experimental procedure

1. Mix the *Azospirillum* bacterial inoculum (10^9 CFU/mL) with warm MS medium (37–39 °C) to achieve final concentrations (10^3–10^8 CFU·mL^{-1}).
2. Place the mixture into sterile Petri dishes and allow it to solidify. Use control plates using the same medium with an equal volume of 0.01 M $MgSO_4$ instead of bacterial inoculum.
3. Aseptically transplant 4-day-old *Arabidopsis* seedlings onto the medium containing the bacterial inoculum.
4. Seal the Petri plates with Parafilm and incubate them vertically in a growth chamber at 22 °C with a 16 h light/8 h dark photoperiod.
5. Evaluate the growth parameters after 5–7 days of inoculation.

Diffusible Compound Assay
Experimental procedure

1. Prepare sterile 0.5 cm diameter filter paper discs.
2. Autoclave the *Azospirillum* bacterial suspension (121 °C, 15 psi, 20 min).
3. Pipette 10 μL of drops of one of the following onto separate filter paper discs: autoclaved *Azospirillum* suspension, *Escherichia coli* XL1 culture ($OD_{600} = 1.0$), or IAA concentrations (0.5 and 1 μM)
4. Place each disc 5 cm from the root tip.
5. Evaluate the growth parameters after 5–7 days of inoculation.

Volatile Organic Compound Assay

Experimental procedure

1. Use divided Petri dishes to physically separate *Arabidopsis* seedlings and bacterial cultures while allowing them to share the same airspace.
2. Place seedlings on one side and the *Azospirillum* culture on the other.
3. Seal the dishes with Parafilm and incubate them under the same growth conditions as above.
4. Evaluate the growth parameters after 5–7 days of inoculation.

13.6.2 Analysis of Root Colonization and Plant Growth Parameters

Experimental procedure

1. Use *Azospirillum* strains expressing the *GUS* reporter gene to visualize root colonization.
2. Perform colonization analyses after treatments involving both direct contact and exposure to diffusible compounds.
3. Quantify bacterial colonization by counting colony-forming units (CFU) per root:

   3.1. Aseptically remove the roots from the seedlings.
   3.2. Shake the roots in 500 μL of sterile liquid LB medium for 10 min to detach the bacteria.
   3.3. Spot aliquots (100 μL) on LB agar and incubate at 28 °C for 48 h.
   3.4. Count the number of CFU per root.

4. Analyze plant growth parameters, such as primary root length, number and density of lateral roots, shoot fresh weight, and number and length of root hairs. Wash the root system and photograph it for digital analysis. Use software such as WinRHIZO with a scanner or ImagenJ for quantification of the root system.

13.7 Analysis of Participation TOR During *Arabidopsis thaliana*-*Azospirillum* Interaction

The estradiol-conditioned mutant line *tor-es1* of *A. thaliana* (Xiong and Sheen 2012) was used to determine the role of TOR signaling during root development induced by *Azospirillum* (Carrillo-Flores et al. 2022b).

Experimental procedure

1. *Arabidopsis tor-es1* seeds were disinfected by incubating with a 5% chlorine and 1% SDS solution for 5 min.
2. Transfer the seeds to 95% (v/v) ethanol for 1 min.
3. Rinse the seeds seven times with sterile distilled water.
4. Stratify the seeds at 4 °C for 48 h to synchronize germination.
5. Sow seeds on agar plates containing 0.2× MS culture medium, which was prepared by dissolving in distilled water the amount of salts corresponding to a 0.2× concentration and 0.5% of sucrose. After adjusting the pH to 5.7, 0.8% (w/v) agar was added. Thereafter, 200 μL/L of a vitamin solution was added to the warm, previously sterilized medium.
6. Place plates vertically in a growth chamber at 22 °C with a 16 h light/8 h dark photoperiod and a light intensity of 100 μmol m^{-2} s^{-1}, for 5 days.
7. Prepare 0.2× MS culture medium and supplemented with: (i) DMSO, (ii) DMSO plus 2.5 × 10^5 CFU/mL *Azospirillum*, (iii) 10 μM estradiol dissolved in DMSO, and (iv) estradiol together with 2.5 × 10^5 CFU/mL *Azospirillum*.
8. Aseptically transplant 5-day-old *Arabidopsis* seedlings into the above-mentioned growth medium.
9. Seal Petri plates with Parafilm and incubate them vertically in a growth chamber at 22 °C with a 16 h light/8 h dark photoperiod.
10. Evaluate growth parameters after 1, 3, and 6 days of post-inoculation.

Analyze plant growth parameters, such as primary root length (PRL), number and density of lateral roots (LRN and LRD, respectively). The PRL is measured with a ruler (cm), the LRN is obtained by counting the roots emerging from the primary root under a stereomicroscope and the LRD was obtained by calculating the ratio between LRN and PRL.

13.7.1 TOR Signaling Pathway Analysis

TOR Expression and Cell Division

This protocol includes a standardized methodology to assess the role of *Azospirillum* in modulating the TOR pathway, a key regulator of plant cell growth, division, and metabolism. TOR activity between other molecules is tightly linked to the auxin signal and is influenced by direct bacterial colonization and perception of bacterial molecules such as lipopolysaccharides (LPS).

Experimental procedure

1. Use *A. thaliana* transgenic lines *TOR/tor-1* (Menand et al. 2002) and *pCycB1:1::uidA* (Colon-Carmona et al. 1999).
2. Perform assays 1, 3, or 6 days after transplanting seedlings onto 0.2× MS medium containing the *Azospirillum* inoculum.

3. Stain transgenic lines expressing the *uidA* reporter gene with 0.1% X-Gluc (5-bromo-4-chloro-3-indolyl, β-D-glucuronide) dissolved in phosphate buffer (0.1 M, pH 7; NaH_2PO_4 and Na_2HPO_4), supplemented with 2 mM potassium ferrocyanide and 2 mM potassium ferricyanide.
4. Incubate seedlings at 37 °C overnight, except for *TOR/tor-1* (incubated for 10 h).
5. Rinse and fix plants as described by Malamy and Benfey (1997).
6. Mount samples on glass slides and seal them with nail polish and observe samples under a light microscope.
7. Quantify GUS signal intensity using ImageJ. For *TOR::GUS*, define a 50 × 400 μm region from the quiescent center through the elongation zone. For *CYCB1::GUS*, analyze a 100 × 100 μm region in the center of the meristem.
8. To assess auxin involvement, include *A. brasilense* ipdC FAJ009 mutant strain (impaired in IAA production) and add exogenous IAA at 1 μM as a positive control for TOR activation.

TOR Activity Analysis via S6K Phosphorylation

Experimental procedure

1. TOR activity is analyzed by examining the phosphorylation of the S6 kinase (S6K), a downstream target.
2. Transplant 3-day-old *Arabidopsis* seedlings onto Petri dishes containing 0.2× MS medium.
3. After 6 days of growth, immerse the roots in 2 mL of 0.2× MS liquid medium containing 0, 100, or 150 μg/mL LPS for 24 h.
4. Collect 100 mg of plant tissue, freeze in liquid nitrogen, and grind to a fine powder.
5. Resuspended the powder in 100 μL of extraction buffer (50 mM HEPES, pH 7.6, 1 mM sodium vanadate, 1 mM sodium molybdate, 1 mM benzamidine, 20 mM NaF, 0.2 mM PMSF in methanol, and 2 mM DTT).
6. Vortex briefly and centrifuge at 9000 rpm for 30 min at 4 °C.
7. Transfer the supernatant to a sterile Eppendorf tube.
8. Quantify total protein content using the Bradford assay (Bradford 1976).
9. Separate 20 μg of protein by SDS-PAGE (12%) and transfer onto a polyvinylidene fluoride (PVDF) membrane.
10. Wash the membrane for 10 min in 0.15 M TBS-T buffer (50 mM Tris-HCl, 0.15 M NaCl, 0.1% Tween 20, pH 7.4).
11. Block the membrane overnight with 5% skim milk in TBS-T in constant agitation.
12. Incubate the membrane with mouse monoclonal antibodies against p70 S6 kinase α and phospho-p70 S6 kinase α (Cat. sc-8418 and sc-8416, Santa Cruz Biotechnology), diluted 1:5000.
13. Wash the membrane sequentially with agitation at 4 °C: first with 0.15 M TBS-T (5 min), then with 1 M TBS-T (5 min), and finally with 0.15 M TBS-T (5 min).
14. Incubate with anti-mouse IgG (LI-COR Biosciences, Lincoln, NE, USA) secondary antibody (1:10,000 dilution) in 0.15 M TBS-T for 2 h at 4 °C under constant agitation.

15. Wash three times for 5 min each with 0.15 M TBS-T.
16. Develop the signal using the WesternSure PREMIUM chemiluminescent substrate (LI-COR Biosciences, Lincoln, NE, USA) and visualize with a ChemiDoc™ MP imaging system.

13.8 Conclusions

In this chapter, we review the beneficial effects of inoculating plants with *Azospirillum* and focus on the molecular mechanisms that have been analyzed in the model plant *A. thaliana*. Early reports on wheat and *Arabidopsis* demonstrated that *Azospirillum* decreased the primary root length while increasing the root branching after 6–7 days of interactions Spaepen and Vanderleyden 2009; Spaepen et al. 2014; Méndez-Gómez et al. 2020; Méndez-Gómez et al. 2021). Carrillo-Flores et al. (2022a, b) evaluated what happened at shorter interaction times. Surprisingly, they found dramatic changes on the first day of exposure to the bacteria, such as the arrest of the growth of the primary root, whose length is maintained throughout the experiment (Fig. 13.1), and stimulation of maturation of the lateral root primordia. After 3 days of exposure of *Arabidopsis* to *Azospirillum*, the observed changes were a very noticeable increase of lateral root number, arrest of cell cycle in the meristem of the primary root, and disappearance of the elongation zone of the primary root. The aforementioned data clearly indicate that the effect of bacteria on *Arabidopsis* root architecture development begins from early times of exposure. Upon detailed revision of the literature on the effect of PGPR in different plants, none of these interactions has reported that the arrest of primary root growth is affected by the two events that determine development of this organ (elongation and cellular division) (Fibatch-Paldi et al. 2011; Spaepen et al. 2014), as happens during interaction with *Azospirillum*. In addition, the bacteria caused a depletion of the root meristem of the primary root after 3 days of exposure. It has been widely reported that *A. brasilense* alters the root architecture of many plants, and this effect has been attributed to the auxins it produces (Spaepen and Vanderleyden 2009; Mendez-Gomez et al. 2021). In the *Azospirillum-Arabidopsis* interaction reported by Spaepen et al. (2014), they evaluated the effect of the auxins produced by the bacteria when inoculating *A. thaliana* with two strains of *A. brasilense* Sp245: one wild and the other mutant in auxin biosynthesis (FAJ009). Their results allowed them to suggest that the auxins produced by the bacteria could be involved in the morphological changes of the root architecture. These authors also analyzed gene expression by microarrays of *Arabidopsis* roots exposed for 3 and 7 days to *Azospirillum*. and their analysis did not allow them to propose a mechanism by which auxins could be involved in the changes of the root system. The auxin levels in *Arabidopsis* seedlings exposed to *Azospirillum* indicate that auxins begin to increase after 3 days of exposure in the primary root, specifically in the differentiation zone of the lateral roots, reaching a maximum in the same zone one day later. Subsequently, auxins were shown to be distributed to meristems of the lateral roots, and from the fifth day on, auxins disappeared from the root meristem of the primary root (Fig. 13.1) (Carrillo-Flores et al. 2022a, b). Since auxins appear to be involved in the effect of *Azospirillum* on root architecture development, Carrillo-Flores et al. (2022a, b) decided to investigate what elements of the auxin response pathway might be participating in the outcome. For this, they analyzed the effect of the inoculation of the bacteria on the root architecture of several mutant lines of this pathway; however, we did not observe differences in the parameters of the root architecture of the mutants with respect to the control. This was unexpected, especially in the auxin carrier mutants, since other results with the *DR5::uidA* line indicated that auxins increased in the vascular bundle at short exposure times and were eventually mobilized to the meristems of lateral roots (Fig. 13.1). To directly analyze the participation of PAT during the *Azospirillum-Arabidopsis* interaction, they used the PAT inhibitor, NPA. The experiments with NPA strongly suggested that PAT was one of the factors necessary to prompt *Azospirillum*-induced changes in root architecture. So, they decided to test the *pgp1, pgp4,* and *pgp9* mutant lines of other auxin efflux carriers, and in this way, to determine that the PGP1 carrier seemed to be involved in this effect. As we can see, we have only just begun to study the biochemical and molecular mechanisms involved during the promotion of plant growth by this bacterial genus, so there is still much to discover to understand this phenomenon.

Acknowledgments The work of EMB-P was financially supported by the Secretaría de Ciencias, Humanidades, Tecnología e Innovación (CBF 2023-2024-2234).

References

Aeron A, Kharel E, Jha CK, Meena VS, Aziz SMA, Islam MT, Kim MSK, Pattanayak A, Rajashekara H (2020) Revisiting the plant growth-promoting rhizobacteria: Lessons from the past and objectives for the future. Arch Microbiol 202:665–676

Amavizca E, Bashan Y, Ryu C, Farag MA, Bebout BM (2017) Enhanced performance of the microalga *Chlorella sorokiniana* remotely induced by the plant growth-promoting bacteria *Azospirillum brasilense* and *Bacillus pumilus*. Sci Rep 7:1–11

Balaguera-Lopez HE, Hernandez JP (2022) Effect of plant growth-promoting bacteria *Azospirillum brasilense* on the physiology of radish (*Raphanus sativus* L.) under waterlogging stress. Agronomy 12:726

Bashan Y, de-Bashan LE (2010) How the plant growth-promoting bacterium *Azospirillum* promotes plant growth-a critical assessment. Adv Agron 108:77–135

Bashan Y, Levanony H, Whitmoyer RE (1991) Root surface colonization of non-cereal crop plants by pleomorphic *Azospirillum brasilense* Cd. J Gen Microbiol 137:187–196

Bradford MM (1976) A rapid and sensitive method for the quantitation of microgram quantities of protein utilizing the principle of protein-dye binding. Anal Biochem 72:248–254

Brumos J, Robles LM, Yun J, Vu TC, Jackson S, Alonso JM, Stepanova AN (2018) Local auxin biosynthesis is a key regulator of plant development. Dev Cell 47:306–318

Carrillo-Flores E, Arreola-Rivera J, Pazos-Solís DM, Bocanegra-Mondragón F, Fierros-Romero G, Mellado-Rojas ME, Beltrán-Peña E (2022a) Participation of auxin transport in the early response of the *Arabidopsis* root system to inoculation with *Azospirillum brasilense*. Phyton 91: 2384–2401

Carrillo-Flores E, Arreola-Rivera J, Pazos-Solís DM, Bocanegra-Mondragón M, Fierros-Romero G, Mellado-Rojas ME, Beltrán-Peña E (2022b) TOR participation on the root system changes of *Arabidopsis* during its interaction with *Azospirillum*. J App Biotechnol Bioeng 9:18–23

Cassan F, Diaz-Zorita M (2016) *Azospirillum* sp. in current agriculture: from the laboratory to the field. Soil Biol Biochem 103:117–130

Cohen AC, Travaglia CN, Bottini R, Piccoli PN (2009) Participation of abscisic acid and gibberellins produced by endophytic *Azospirillum* in the alleviation of drought effects in maize. Botany 87:455–462

Colón-Carmona A, You R, Haimovitch-Gal T, Doerner P (1999) Spatio-temporal analysis of mitotic activity with a labile cyclin-GUS fusion protein. Plant J 20:503–508. Colon-Carmona

Contreras-Cornejo HA, Macías-Rodríguez L, Cortés-Penagos C, López-Bucio J (2009) *Trichoderma virens*, a plant beneficial fungus, enhances biomass production and promotes lateral root growth through auxin-dependent mechanism in *Arabidopsis*. Plant Physiol 149: 1579–1592

Cruz-Hernández MA, Mendoza-Herrera A, Bocanegra-Garcia V, Rivera G (2022) *Azospirillum* spp. from plant growth-promoting bacteria to their use in bioremediation. Microorganisms 10: 1057

Curá JA, Franz DR, Filosofía JE, Balestrasse KB, Burgueño LE (2017) Inoculation with *Azospirillum* sp. and *Herbaspirillum* sp. bacteria increases the tolerance of maize to drought stress. Microorganisms 5(3):41. https://doi.org/10.3390/microorganisms5030041

Degon Z, Dixon S, Rahmatallah Y, Galloway M, Gulutzo S, Price H, Cook J, Glazko G, Mukherjee A (2023) *Azospirillum brasilense* improves rice growth under salt stress by regulating the expression of key genes involved in salt stress response, abscisic acid signaling, and nutrient transport, among others. Front Agron 5:1216503. https://doi.org/10.3389/fagro.2023.1216503

Deng K, Yu L, Zheng X, Zhang K, Wang W, Dong P, Zhang J, Ren M (2016) Target of rapamycin is a key player for auxin signaling transduction in Arabidopsis. Front Plant Sci 7:291

Favre P, van Schaik E, Schorderet M, Yerly F, Reinhardt D (2024) Regulation of tissue growth in plants-a mathematical modeling study on shade avoidance response in Arabidopsis hypocotyls. Front Plant Sci 15:1285655

Fibach-Paldi S, Burdman S, Okon Y (2011) Key physiological properties contributing to rhizosphere adaptation and plant growth promotion abilities of *Azospirillum brasilense*. FEMS Microbiol Lett 326:99–108

Fukami J, Cerezini P, Hungria M (2018) Azospirillum: benefits that go far beyond biological nitrogen fixation. AMB Express 8:73

Ganusova EE, Vo LT, Mukherjee T, Alexandre G (2021) Multiple CheY proteins control surface-associated lifestyles of *Azospirillum brasilense*. Front Microbiol 12:664826

Ganusova EE, Russell MH, Patel S, Seats T, Alexandre G (2024) An *Azospirillum brasilense* chemoreceptor that mediates nitrate chemotaxis has conditional roles in the colonization of plant roots. Appl Environ Microbiol 90:1–18

Ganusova EE, Banerjee I, Seats T, Alexandre G (2025) Indole-3-acetic acid (IAA) protects *Azospirillum brasilense* from indole-induced stress. Appl Environ Microbiol 91:e02384–24

Gupta DK, Palma JM, Corpas FJ (eds) (2018) Antioxidants and antioxidant enzymes in higher plants. Springer, Berlin, pp 1–300. https://doi.org/10.1007/978-3-319-75088-0

Hartung W, Sauter A, Hose E (2022) Abscisic acid in the xylem: where does it come from, where does it go to? J Exp Bot 53:27–32

Havlin JL, Beaton JD, Tisdale SL, Nelson WL (2017) Soil fertility and fertilizers: an introduction to nutrient management, 8th edn. Pearson, Chennai

Housh AB, Noel R, Powell A, Waller S, Wilder SL, Sopko S, Benoit M, Powell G, Schueller MJ, Ferrieri RA (2023) Studies using mutant strains of *Azospirillum brasilense* reveal that atmospheric nitrogen fixation and auxin production are light dependent processes. Microorganisms 11:1727

Kieber JJ, Schaller GE (2018) Cytokinin signaling in plant development. Development 145: dev149344

Leong J (1986) Siderophores: their biochemistry and possible role in the biocontrol of plant pathogens. Annu Rev Phytopathol 24:187–209

Malamy JE, Benfey PN (1997) Organization and cell differentiation in lateral roots of *Arabidopsis thaliana*. Development 124:33–44

Mellor NL, Voß U, Janes G, Bennett MJ, Wells DM, Band LR (2020) Auxin fluxes through plasmodesmata modify root.tip auxin distribution. Development 147:dev181669

Menand B, Desnos T, Nussaume L, Berber F, Bouchez D, Meyer C, Robaglia C (2002) Expression and disruption of the *Arabidopsis TOR (target of rapamycin)* gene. Proc Natl Acad Sci USA 99: 6422–6427

Méndez-Gómez M, Castro-Mercado E, Peña-Uribe A, Reyes-de-la Cruz H, López-Bucio J, García-Pineda E (2020) Target of rapamycin signaling plays a role in *Arabidopsis* growth promotion by *Azospirillum brasilense* Sp245. Plant Sci 293:110416

Méndez-Gómez M, Barrera-Ortiz S, Castro-Mercado E, López-Bucio J, García-Pineda E (2021) The nature of the interaction *Azospirillum-Arabidopsis* determine the molecular and morphological change in root and plant growth promotion. Protoplasma 258:179–189

Montané MH, Menand B (2018) TOR inhibitors: from mammalian outcomes to pharmacogenetics in plants and algae. J Exp Bot 70:2297–2312

Mora MV, Lopez-Gaston A, Molina-Romina M, Coniglio-Nayla A, Nievas-Sofia M, de-Diego N, Zeljković-Sanja Ć, Sarmiento-Sara S, Spíchal L, Robertson S, Wilkins O, Elías-Juliana M, Pedraza-Raúl O, Estévez-José M, Belmonte-Marl F, Cassan FD (2023) *Azospirillum argentinense* modifies Arabidopsis root architecture through auxin-dependent pathway and flagellin. J Soil Sci Plant Nutr 23:4543–4557

Murashige T, Skoog F (1962) A revised medium for rapid growth and bioassays with tobacco tissue cultures. Physiol Plant 15(3):473–497

Naqqash T, Malik KA, Imran A, Hameed S, Shahid M, Hanif MK, Majeed A, Iqbal MJ, Qaisrani MM, van Elsas JD (2022) Inoculation with *Azospirillum* spp. acts as the liming source for improving growth and nitrogen efficiency of potato. Front Plant Sci 13:929114

Okon Y, Kapulnik Y (1986) Development and function of *Azospirillum*-inoculated roots. Plant Soil 90:3–16

Oliveira CEDS, Jalal A, Vitoria LS, Giolo VM, Oliveira TJSS, Aguilar JV, Camargos LSD, Brambilla MR, Fernandes GC, Vargas PF, Zoz T, Teixeira-Filho MCM (2023) Inoculation with *Azospirillum brasilense* strains AbV5 and AbV6 increases nutrition, chlorophyll, and leaf yield of hydroponic lettuce. Plants 12:3107

Omary M, Gil-Yarom N, Yahav C, Steiner E, Efron I (2022) A conserved super locus regulates above-and below-ground root initiation. Science 375

Ona O, van Impe J, Prinsen E, Vanderleyden J (2005) Growth and indole-3-acetic acid biosynthesis of *Azospirillum brasilense* Sp245 is environmentally controlled. FEMS Microbiol Lett 246: 125–132

Pagnussat LA, Nascimento MD, Maroniche G, Gonorazky G, Rizza LS, Creus C, Curatti L (2023) *Azospirillum baldaniorum* improves acclimation, lipid productivity and oxidative response of a microalga under salt stress. Algal Res 74:103192

Palacios OA, Lopez BR, Palacios-Espinosa A, Hernandez-Sandoval FE, de-Bashan LE (2021) The immediate effect of riboflavin and lumichrome on the mitigation of saline stress in the microalga *Chlorella sorokiniana* by the plant-growth-promoting bacterium *Azospirillum brasilense*. Algal Res 58:102424

Pedraza RO, Filippone MP, Fontana C, Salazar SM, Ramírez-Mata A, Sierra-Cacho D, Baca BE (2020) *Azospirillum*. in beneficial microbes in agro-ecology: bacteria and fungi. In: Amaresan N, Kumar MS, Annapurna K, Kumar K, Sankaranarayanan A (eds). Academic/ Elsevier, New York/London, pp 73–105

Pham T-M, Bui XD, Trang LVK, Le T-M, Nguyen ML, Tringh D-M, Phuong NTD, Khoo KS, Chew KW, Show PL (2022) Isolation of indole-3-acetic acid-producing *Azospirillum brasilense* from Vietnamese wet rice: Co-immobilization of isolate and microalgae as a sustainable biorefinery. J Biotechnol 349:12–20

Pii Y, Marastoni L, Springeth C, Fontanella MC, Beone GM, Cesco S, Mimmo T (2016) Modulation of Fe acquisition process by *Azospirillum brasilense* in cucumber plants. Environ Exp Bot 130:216–225

Rabara S, Vishwakarma NP, Patel S (2023) Isolation and biochemical identification of N_2 fixing bacteria (*Azospirillium sp.*) from Saurashtra region. Curr Agri Res J 11:277–286

Rexin D, Meyer C, Robaglia C, Veit B (2015) TOR signalling in plants. Biochemist J7:1–14

Rodríguez-Salazar J, Suarez R, Caballero-Mellado J, Iturriaga G (2009) Trehalose accumulation in *Azospirillum brasilense* improves drought tolerance and biomass in maize plants. FEMS Microbiol Lett 296:52–59. https://doi.org/10.1111/j.1574-6968.2009.01614.x

Rozier C, Gerin F, Czarnes S, Legendre L (2019) Biopriming of maize germination by the plant growth-promotion rhizobacterium *Azospirillum lipoferum* CRT1. J Plant Physiol 237:111–119

Ruppel S, Merbach W (1997) Effect of ammonium and nitrate on 15N2-fixation of *Azospirillum* spp., and *Pantoea agglomerans* in association with wheat plants. Microbiol Res 152:377–383

Ryabova LA, Robaglia C, Meyer C (2019) Target of Rapamycin kinasel: central regulatory hub for plant growth and metabolism. J Exp Bot 70:2211–2216

Salazar-García G, Balaguera-López HE, Hernández JP (2022) Effect of plant growth-promoting bacteria *Azospirillum brasilense* on the physiology of Radish (*Raphanus sativus* L.) under waterlogging stress. Agronomy 12:726

Sangoquiza-Caiza CA, Yánez-Guzman CF, Borges M (2019) Response of a maize variety in the absorption of nitrogen and phosphorus when inoculating *Azospirillum* sp. and *Pseudomonas fluorescens*. Avances en Ciencias e Ingenierias 11:84–95

Schaller GE, Bishopp A, Kieber JJ (2015) The yin-yang of hormones: cytokinin and auxin interactions in plant development. Plant Cell 27(1):44–63

Shahrajabian MH, Chaski C, Polyzos N, Petropoulos SA (2021) Biostimulants application: a love input cropping management tool for sustainable farming vegetables. Biomolecules 11:698

Shi L, Wu Y, Sheen J (2018) TOR signaling in plants: conservation and innovation. Development 145:1–13

Sivasakthivelan P, Saranraj P, Sayyed RZ, Arivukkarasu K, Kokila M, Manigandan M, Seifi S (2023) Inoculant production and formulation of *Azospirillum* species. In: Plant growth promoting microorganisms of arid region, pp 423–455

Spaepen S, Vanderleyden J (2009) Plant Growth-promoting actions of rhizobacteria. Adv Bot Res 51:283–320

Spaepen S, Vanderleyden J, Remans R (2007) Indole-3-acetic acid in microbial and microorganism–plant signaling. FEMS Microbiol Rev 31:425–448

Spaepen S, Bossuyt S, Engelen K, Marchal K, Vanderleyden J (2014) Phenotypical and molecular responses of *Arabidopsis thaliana* roots as a result of inoculation with the auxin-producing bacterium *Azospirillum brasilense*. New Phytol 201:850–861

Sun G, Elowsky C, Li G, Wilson RA (2018) TOR–autophagy branch signaling via Imp1 dictates plant–microbe biotrophic interface longevity. PLoS Genet 14:e1007814

Sun W, Shahrajabinan MH, Cheng Q (2021) Nitrogen fixation and diazotrophs—a review. Rom Biotechnol Lett 26:2834–2845

Sun W, Hesam-Shahrajabian M, Wang N (2025) A study of the different strains of the genus *Azospirillum* spp. on increasing productivity and stress resilience in plants. Plants 14:267

Tanaka H, Dhonukshe P, Brewer B, Friml J (2006) Spatiotemporal asymmetric auxin distribution: a means to coordinate plant development. Cell Mol Life Sci 63:2738–2754

Thomas L, Singh I (2019) Microbial biofertilizers: types and applications. In: Giri B, Prasad R, Wu QS, Varma A (eds) Biofertilizers for sustainable agriculture and environment. Soil biology, vol 55, pp 1–19

Utkin YN, Ma Q-G, Gureeva MV, Gureev AP (2023) Molecular mechanisms determining the role of bacteria from the genus *Azospirillum* in plant adaptation to damaging environmental factors. Int J Mol Sci 24:9122

Vande-Broek A, Michiels J, Van Gool A, Vanderleyden J (1993) Spatial-temporal colonization patterns of *Azospirillum brasilense* on the wheat root surface and expression of the bacterial nifH gene during association. Mol Plant-Microbe Interact 6:592–600

Von Bulow JGD, Döbereiner J (1975) Potential for nitrogen fixation in maize genotypes in Brazil. Proc Natl Acad Sci USA 72:2389–2393

Wang H, Jin H, Chen Z, Li W, Ma J, Hu T, Liu Q, Zhang Y, Lin X, Xie Z (2024) Azospirillum isscasi sp.nov., a bacterium isolated from rhizosphere soil of rice. Int J Syst Evol Microbiol 74:006218

Xiaojuan L, Wenguo C, Yanlin L, Hui L, Liwen F, Zengyu L, Lin X, Hongtao L, Tongda X, Yan X (2016) Differential TOR activation and cell proliferation in *Arabidopsis* root and shoot apexes. Proc Natl Acad Sci USA 10:2765–2770

Xiong Y, Sheen J (2012) Rapamycin and glucose–Target of Rapamycin (TOR) protein signaling in plants. J Biol Chem 287:2836–2842

Xiong Y, McCormack M, Li L, Hall Q, Xiang C, Sheen J (2013) Glucose–TOR signaling reprograms the transcriptome and activates meristem. Nature 496:81–186

Xu CH, Chen G, Liu Y, Xiao W, Cui XL, Ding ZG, Liu WH, Wang YX (2023) *Azospirillum aestuarii* sp. nov., a novel nitrogen-fixting and aerobic denitrifying bacteria isolated from an stuary of a freshwater river. Curr Microbiol 80:1–9

Zaheer MS, Ali HH, Iqbal MA, Erinle KO, Javed T, Iqbal J, Hashmi MIU, Mumtaz MZ, Salama EAA, Kalaji HM, Wróbel J, Dessoky ES (2022) Cytokinin production by *Azospirillum brasilense* contributes to increase in growth, yield, antioxidant, and physiological systems of wheat (*Triticum aestivum* L.). Front Microbiol 13:886041

Zimmer W, Roeben K, Bothe H (1988) An alternative explanation for plant growth promotion by bacteria of the genus Azospirillum. Planta 176(3):333–342

Chapter 14
Interaction of *Azospirillum* spp. with Plants: Alleviation of Stress

Maria V. Gureeva, Artem P. Gureev, Stella S. Evstigneeva, Yulia P. Fedonenko, and Margarita Y. Grabovich

Abstract Bacteria of the genus *Azospirillum* belong to plant growth-promoting rhizobacteria. They are actively used as biofertilizers in agriculture and have the ability to mitigate stress caused by various abiotic environmental factors. This chapter examines the role of bacteria of the genus *Azospirillum* in reducing stress in crop plants caused by heavy metals, pesticides, perchlorate, drought, salinity, hydrocarbons, and radiation. Mechanisms for reducing stress caused by various pollutants during inoculation of plants with these bacteria are described. The methodological part of the chapter systematizes data on methods of inoculation of various crop plants with bacteria of the genus *Azospirillum* separately and in combination with other microorganisms or fertilizers. Methods for assessing the level of stress in plants are also described: what indicators can be used and what methods can be used to determine them.

Keywords Heavy metals · Drought · Salinity · Pollutants · Bioremediation

---

M. V. Gureeva (✉) · M. Y. Grabovich
Department of Biochemistry and Cell Physiology, Voronezh State University, Voronezh, Russia

A. P. Gureev
Department of Genetics, Cytology and Bioengineering, Voronezh State University, Voronezh, Russia

S. S. Evstigneeva
Laboratory of Biochemistry, Institute of Biochemistry and Physiology of Plants and Microorganisms, Saratov Scientific Centre of the Russian Academy of Sciences, Saratov, Russia

Department of Biochemistry and Clinical Laboratory Diagnostics, Saratov State Medical University of V. I. Razumovsky, Ministry of Health of the Russian Federation, Saratov, Russia

Y. P. Fedonenko
Laboratory of Biochemistry, Institute of Biochemistry and Physiology of Plants and Microorganisms, Saratov Scientific Centre of the Russian Academy of Sciences, Saratov, Russia

F. D. Cassán et al. (eds.), *The Definitive Handbook of Azospirillum*,
https://doi.org/10.1007/978-3-032-09222-9_14

14.1 Introduction

Azospirillum is a genus of bacteria belonging to the plant growth-promoting rhizobacteria (PGPR) group. Many representatives of this genus have been isolated from agricultural soil and the rhizosphere of cereal plants (dos Santos Ferreira et al. 2022). Even strains isolated from other habitats can be used as plant inoculants (Gureeva et al. 2023; Gureev et al. 2024). Moreover, strains initially isolated from the rhizosphere of cereals are also used for inoculation of many noncereal plants (Mariotti et al. 2021; Tiepo et al. 2020; Evseeva et al. 2019; Khalid et al. 2017; Rabiei et al. 2020; Caamal-Chan et al. 2023; Alzate Zuluaga et al. 2022; Farhangi-Abriz et al. 2020; Omer et al. 2022; Bahgat et al. 2023; Xu et al. 2018; Pan et al. 2019; Marastoni et al. 2019). In addition to stimulating plant growth, these microorganisms are able to mitigate the stress effects of various abiotic factors on plants (Gureeva and Gureev 2023). *Azospirillum* mainly modulates the synthesis of secondary metabolites and phytohormones in plants, which can reduce stress-induced inhibition of photosynthesis and oxidative damage (Alzate Zuluaga et al. 2022). These bacteria are capable of both producing phytohormones themselves (Gomez et al. 1998; Cohen et al. 2015) and influencing their biosynthesis in plants by synthesizing signaling molecules, such as NO (Creus et al. 2005; Molina-Favero et al. 2008), or by regulating the expression of corresponding genes (Cura et al. 2017).

Some *Azospirillum* strains can participate in the biodegradation of pollutants, such as perchlorate (Waller et al. 2004; Coates et al. 1999; Nozawa-Inoue et al. 2005), hydrocarbons (Muratova et al. 2005; Wu et al. 2021), pesticides (Foster et al. 2004), which can also indirectly reduce the stress caused by these compounds in plants. In addition, *Azospirillum* can activate xenobiotic detoxification mechanisms in plants (Gureeva et al. 2025), affect the content of osmolytes, and change the elasticity of membranes in root cells (Dimkpa et al. 2009).

The first part of this chapter summarizes the data on the alleviation of various types of abiotic stress in plants by *Azospirillum*. The second part describes the methods of inoculating plants with bacteria of the genus *Azospirillum*. The third part focuses on the parameters used to determine the level of stress in plants.

14.2 *Azospirillum* Participation in Protecting Plants from Stress Factors

Bacteria of the genus *Azospirillum*, as representatives of a large group of plant growth promoting rhizobacteria (PGPR), help plants survive unfavorable environmental conditions (Gureeva and Gureev 2023). In doing so, they activate a number of protective mechanisms that reduce stress in plants. This section will examine various stress factors for plants, against which *Azospirillum* is used.

14.2.1 Heavy Metals

Heavy metals (HMs) pose a serious threat to agriculture and plant health. These elements can accumulate in soil as a result of industrial activities, mining, irrational use of chemical fertilizers, and the use of motor vehicles (Wu et al. 2022). However, HM stress in plants can be reduced by rhizosphere bacteria from the genus *Azospirillum* (Fig. 14.1).

Cadmium, which accumulates in soil due to the combustion of solid and liquid fuels, as well as near metallurgical plants, is one of the most toxic HMs. It negatively affects plants at the morphological, physiological, biochemical, and molecular levels (Danilova et al. 2023). *Azospirillum brasilense* CGMCC1.10379 promotes the production of abscisic acid (ABA), which inhibits cadmium uptake in wild-type

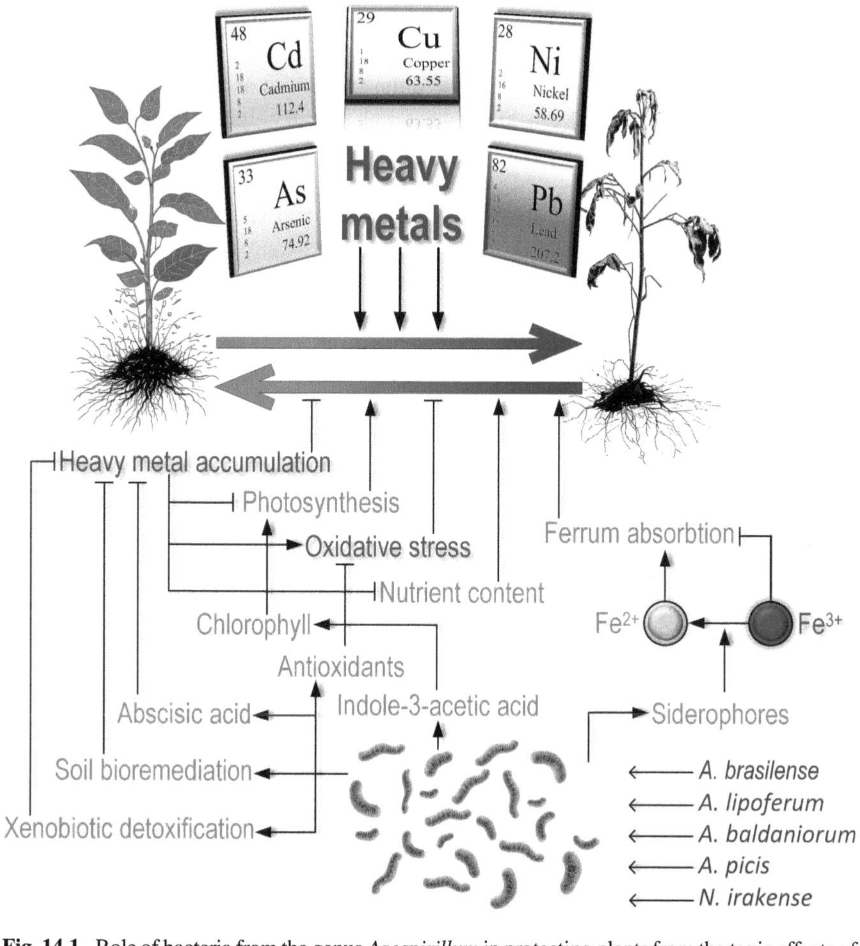

Fig. 14.1 Role of bacteria from the genus *Azospirillum* in protecting plants from the toxic effects of heavy metals

Arabidopsis thaliana, whereas no such effect was observed in abscisic acid (ABA)-insensitive mutants. Moreover, inoculation of *A. brasilense* CGMCC1.10379 downregulated the expression of iron-regulated transporter 1 (IRT1), which can also function as a cadmium transporter (Xu et al. 2018). Similar results were obtained for pakchoi (*Brassica chinensis* L.) grown in cadmium-contaminated soil. *A. brasilense* Yu-62 ACCC10109 in an ABA-dependent manner alleviated the Cd-induced photosynthesis inhibition and oxidative damage (Pan et al. 2019). At the same time, co-application of *A. brasilense* CGMCC1.10379 and H_2S enhanced pakchoi resistance to cadmium stress (Cui et al. 2023). Although *Azospirillum lipoferum* 137 increased cadmium concentrations in barley seedling roots, it stimulated root growth, biomass growth, and nutrient content (Belimov and Dietz 2000). *Azospirillum* biofertilizer packs (species not specified) improved switchgrass (*Panicum virgatum*) growth and normalized soil pH in the presence of cadmium and lead contamination (Arora et al. 2016).

Arsenic is also a highly toxic metal that causes damage to root cells and a decrease in the size of chlorophyll parenchyma in leaves (Burachevskaya et al. 2021). *A. brasilense* Cd was shown to be quite tolerant to arsenic at concentrations found in soils. Arsenic did not affect bacterial motility, biofilm and exopolymer formation, auxin and siderophore production, and N_2 fixation (Vezza et al. 2020). *A. brasilense* (strain not specified) increased growth rate, leaf area, photosynthesis, and yield of wheat grown in arsenic-supplemented soil (Zaheer et al. 2022). The protective mechanisms of two *A. brasilense* strains against arsenic toxicity were investigated in maize. Plants inoculated with strain CD showed an increase in the activities of superoxide dismutase, glutathione (GSH) S-transferase, and GSH reductase. Arsenic decreased the GSH content in roots inoculated with strain Az39, and plants inoculated with CD showed a decrease in oxidized GSH (glutathione disulfide [GSSG]) levels (Armendariz et al. 2015).

Nickel and lead can significantly accumulate in wheat seedlings, which is accompanied by dysfunction of the photosynthetic apparatus and a significant decrease in wheat yield in field experiments. *Azospirillum picis* B-2897^T and *Niveispirillum irakense* B-2893^T contributed to a decrease in the concentration of nickel and lead in wheat seedlings and an increase in the expression of genes involved in xenobiotic detoxification in roots. *A. picis* B-2897^T, *N. irakense* B-2893^T, and *A. brasilense* B-1547^T restored wheat yields to varying degrees when grown on soil contaminated with nickel and lead (Gureeva et al. 2025).

Copper is an essential microelement. However, even a slight excess of permissible copper concentrations in the soil leads to a significant toxic effect on plants (Kulikova et al. 2015). It was shown that Cu was significantly less toxic than nickel and lead in relation to seven studied *Azospirillum* strains. At concentrations from 1.5 to 15 mg/L, copper even stimulated bacterial growth (with the exception of *A. brasilense* B-1547^T). The studied strains to varying degrees contributed to the accumulation of copper in the roots (Gureeva et al. 2025). Inoculation of plants with *A. brasilense* Sp7 in the presence of copper promoted copper uptake and activation of root peroxidases and oxidases to the greatest extent compared to *Azospirillum baldaniorum* B-3036^T (=Sp245^T), and also suppressed the photosynthetic apparatus of plants (Muratova et al. 2022). *A. picis* B-2897^T prevented copper-induced decrease in chlorophyll *b* and increased beta-carotene content in wheat seedlings, and also promoted increased expression of the GSH S-transferase gene, which may help reduce copper toxicity in roots (Gureeva et al. 2025). *A. brasilense* Cd (DSM-1843) caused significant inhibition of the expression of genes encoding ferric-chelate reductase oxidase (CsFRO2 and CsFRO3) in cucumber, which was accompanied by a decrease in iron and copper accumulation and an increase in root growth against the background of copper toxicity (Marastoni et al. 2019). It is important to consider that copper deficiency can also cause impaired plant growth. Copper deficiency caused oxidative stress and chlorophyll deficiency, while *A. brasilense* (strain not specified) reduced oxidative stress and restored chlorophyll levels due to the secretion of indoleacetic acid (IAA) in *Chlorella sorokiniana* (Peng et al. 2021).

14.2.2 Pesticides

Pesticides pose a significant threat not only to pests but also to crops due to incomplete specificity. Pesticides can accumulate in plants, causing disturbances in growth and functioning of the photosynthetic apparatus (Pandey et al. 2022), and mitochondrial dysfunctions (Alimova et al. 2024) alter polyphenol metabolism (Jakl et al. 2021), and can also negatively affect rhizosphere bacteria (Yu et al. 2023). For example, disturbances in growth, hormone production, and nitrogen fixation were shown in *A. brasilense* ATCC 29145 when treated with methadone (Gomez et al. 1998). At the same time, *Azospirilla* are capable of decomposing some types of pesticides. *Azospirillum* isolates capable of completely biodegrading ethion were found (Foster et al. 2004).

In addition, it was shown that *Azospirillum* can partially neutralize the toxic effect of pesticides on plants. Thiamethoxam, lambda-cyhalothrin, propiconazole, difenoconazole, clopyralid, and dicamba inhibited wheat growth and increased oxidative stress. However, the strains *Azospirillum thiophilum* BV-S^T, *A. brasilense* JCM1224^T, *A. picis* DSM 19922^T, *Azospirillum halopraeferens* DSM 3675^T, *A. lipoferum* JCM 1247^T, *A. baldaniorum* Sp245^T, *Niveispirillum irakense* DSM 11586^T neutralized the toxic effect of these pesticides (Gureev et al. 2024). *A. thiophilum* BV-S^T improved wheat growth, reduced oxidative stress and chlorophyll deficiency in cypermethrin-induced toxicity, and increased the expression of genes responsible for xenobiotic detoxification (Gureeva et al. 2023).

14.2.3 Perchlorate

Perchlorate (ClO_4^-) is a persistent pollutant formed by natural and anthropogenic processes (Acevedo-Barrios and Olivero-Verbel 2021). It has been shown to readily accumulate in plants (Andraski et al. 2014), affecting chlorophyll content and root systems (He et al. 2013), and inhibiting plant growth (Xie et al. 2014) and survival (Acevedo-Barrios et al. 2018). One method of removing perchlorate from ecosystems is microbial degradation. Perchlorate-reducing bacteria (PRB) reduce ClO_4^- or chlorate (ClO_3^-) to chlorite (ClO_2^-) via perchlorate reductase PcrABCD and then disproportionate ClO_2^- to Cl^- and O_2 via chlorite dismutase Cld (Butler et al. 2010). Electron donors for ClO_4^- reduction can be organic compounds such as methanol and acetate (Bardiya and Bae 2011), as well as methane (Luo et al. 2015; Lv et al. 2019; Xie et al. 2018), or inorganic compounds such as H_2 and S (Ju et al. 2007; Nerenherg et al. 2002). ClO_4^- reduction is usually inhibited by the presence of nitrate (Coates and Achenbach 2004; Wan et al. 2016) or oxygen (Coates and Jackson 2009; Xu et al. 2015), as these electron acceptors are preferred by many microorganisms (Coates and Jackson 2009).

ClO_4^--degrading *Azospirillum* strains have been repeatedly isolated from perchlorate-contaminated samples (Waller et al. 2004; Coates et al. 1999; Nozawa-Inoue et al. 2005), and *Azospirillum* has also been found to be among the main genera involved in perchlorate and nitrate reduction in a batch membrane biofilm reactor (Lv et al. 2019).

Currently, there are not many bacterial genomes known that contain all the genes required for ClO_4^- reduction. However, these genes have been identified in a representative of the genus *Azospirillum*, *Azospirillum* sp. TTI. They are encoded jointly with genes encoding other proteins required for bacterial growth in the presence of ClO_4^-: those responsible for electron transport, oxidative stress (necessary due to the unintended accumulation of the by-product hypochlorite from chlorite dismutation), the synthesis of the molybdenum cofactors Moco (since perchlorate reductase belongs to the dimethyl sulfoxide (DMSO) reductase family, which requires a molybdenum cofactor for its functioning), and the *crp*-type transcriptional regulator (Melnyk and Coates 2015).

Azospirillum ability to biodegrade ClO_4^- allows them to be used for the remediation of contaminated soils, and therefore, the negative impact of perchlorate on plants can be reduced (Gureev and Gureeva 2023).

14.2.4 Osmotic Stress

Osmotic stress in a plant cell occurs when the concentration of the solvent (water) in the environment is lower than in the cell. This is possible in two cases: salinization and drought. A physical way to reduce osmotic stress is the synthesis of osmolytes—low-molecular organic substances that are soluble in the intracellular environment and change the properties of biological fluids (Zhu et al. 1997). *Azospirillum* strains can accumulate glycine, proline, trehalose, and betaine, thereby supporting the plant's ability to tolerate osmotic stress (Zahra et al. 2024).

14.2.4.1 Drought

Droughtstress is a major constraint on global agricultural production. Approximately one-third of the Earth's land area is in arid and semi-arid regions, while most other areas are often subject to periodic and unexpected climatic droughts that can be fatal to plants (Farooq et al. 2012).

Drought stress results in decreased nutrient diffusion, induces free radical formation, results in decreased chlorophyll content, and affects nitrate reductase activity due to lower nitrate uptake from soil (Caravaca et al. 2005). Drought also increases ET biosynthesis, which inhibits plant growth (Ali et al. 2014).

Azospirillum species are able to increase plant tolerance to drought stress by producing auxins (Arzanesh et al. 2011; Dimkpa et al. 2009) or by synthesizing nitric oxide, which acts as a signaling molecule in the IAA-inducing pathway (Creus et al. 2005; Molina-Favero et al. 2008). Auxins at low concentrations enhance root growth and stimulate the formation of lateral roots, which leads to an increase in the root area, and therefore increases water uptake and prevents the occurrence of drought stress (Dimkpa et al. 2009).

ABA is considered one of the most important plant growth regulators involved in osmotic stress and tolerance signaling (Tiwari et al. 2017). Data on the effect of *Azospirillum* on ABA levels in plants are contradictory. On the one hand, it was shown that *A. lipoferum* reduces drought stress due to the production of ABA and gibberellins (Cohen et al. 2009). The ABA level also increased in *Arabidopsis* plants inoculated with *A. brasilense* Sp245 (Cohen et al. 2008). The production of this hormone by *A. lipoferum* increased the ABA concentration in inoculated maize (*Z. mays*) seedlings, which led to stomatal closure (Cohen et al. 2015). However, stomatal closure suppresses photosynthesis, which leads to inhibition of plant growth (Arkhipova et al. 2020). On the other hand, inoculation of maize with *A. brasilense* SP-7 in combination with *Herbaspirillum seropedicae* Z-152 under drought conditions resulted in a decreased expression of the *ZmVP14* gene, which is involved in ABA biosynthesis, and decreased ABA levels in the plant. In addition, in this work, inoculation caused a decrease in the ET level in maize (Cura et al. 2017).

One of the ways in which bacteria stimulate drought tolerance in plants is by altering the elasticity of root cell membranes (Dimkpa et al. 2009). *A. brasilense* has been shown to reduce the membrane potential of wheat seedlings and the phospholipid content of cowpea cell membranes due to altered proton efflux activity (Bashan et al. 1992). Inoculation with *Azospirillum* can prevent the increase in phosphatidylcholine and the decrease in phosphatidylethanolamine levels under water deficit conditions in wheat seedlings (Pereyra et al. 2006).

Many PGPB have been recognized as containing trehalose biosynthetic pathway genes; trehalose plays an important role as an osmoprotectant under salt stress (Zahra et al. 2024). Inoculation of maize with *Azospirillum argentinense* Az19, which overexpresses the trehalose biosynthetic gene, conferred drought tolerance to maize and significantly increased plant biomass (Garcia et al. 2017, 2023, 2024). Very little trehalose is thought to be translocated to maize roots and signaling pathways for plant stress tolerance (Rodríguez-Salazar et al. 2009).

Another indicator of osmotic stress reduction during drought, namely, a decrease in proline levels, was observed when plants were inoculated with bacteria from the genus *Azospirillum* (Cohen et al. 2015; Curá et al. 2017; Tiepo et al. 2018). Inoculation of maize plants with *A. lipoferum* improved plant growth through the accumulation of free amino acids and soluble sugars compared to untreated plants under drought stress (Bano et al. 2013).

It has also been shown that under drought conditions, inoculation with bacteria from the genus *Azospirillum* leads to the activation of enzymatic (Akhtar et al. 2021; Muhammad et al. 2022; Tiepo et al. 2020; Heidari and Golpayegani 2012) and nonenzymatic (Tiepo et al. 2020; Kasim et al. 2013) antioxidant pathways in plants. Bacterial inoculation also resulted in a decrease in hydrogen peroxide and lipid peroxidation levels in plants (Tiepo et al. 2018).

14.2.4.2 Salinization

Salinity affects more than 6% of the world's total land area (Munns and Tester 2008). Soil salinity has increased due to inefficient irrigation, improper fertilizer application, and industrial pollution (Ouhibi et al. 2014). Salinity causes Na^+ toxicity and ion imbalance and disrupts vital metabolic processes in plant cells such as protein synthesis, enzymatic reactions, and ribosome functions (Angon et al. 2022).

Representatives of the genus *Azospirillum* have repeatedly demonstrated their effectiveness in reducing salt stress in plants. The possibility of their use as inoculants under saline conditions is due to the halotolerance of some strains (Rivarola et al. 1998; Saleena et al. 2002).

The mitigating effect of inoculation may be manifested in the modulation of osmolyte concentrations in plants. For example, one of the responses of maize to salinity is the accumulation of a powerful osmolyte, raffinose, in leaves. Inoculation of plants with *A. brasilense* HM053 resulted in a decrease in the raffinose content in leaves and an increase in the sucrose content (Powell et al. 2022). Inoculation with *Azospirillum* also improves the content of soluble sugars and proline in plants (Omer et al. 2022; Alzate Zuluaga et al. 2022) and increases the content of glycine betaine (Checchio et al. 2021) under salt stress.

In addition, *Azospirillum* can increase the K^+/Na^+ ratio in plants (Rabiei et al. 2020; Omer et al. 2022; Khalid et al. 2017; Rabie et al. 2005; Del Amor and Cuadra-Crespo 2011; Alamri and Mostafa 2009); increase the nitrogen, phosphorus, calcium (Rabie et al. 2005; Alamri and Mostafa 2009), and magnesium (Rabie et al. 2005) content in the yield; increase the nitrate content; and decrease the chloride content (Del Amor and Cuadra-Crespo 2011), as well as increase the activity of nitrogenase and phosphatase (Rabie et al. 2005) under salinity.

Also, inoculation with *Azospirillum* leads to improvement of plant morphological characteristics (Rabiei et al. 2020; Omer et al. 2022; Khalid et al. 2017; Alzate Zuluaga et al. 2022) and an increase in yield (Rabiei et al. 2020; Khalifa et al. 2021) and protein content (Omer et al. 2022; Rabie et al. 2005) under saline conditions.

Azospirillum also affects the level of oxidative stress in plants under saline conditions. This leads to a decrease in the content of MDA (Omer et al. 2022; Khalid et al. 2017) and hydrogen peroxide (Omer et al. 2022). Data on the effect of *Azospirillum* on antioxidant enzymes under salt stress are contradictory. They can cause both an increase (Rabiei et al. 2020; Alzate Zuluaga et al. 2022; Khalifa et al. 2021; Checchio et al. 2021) and a decrease (Rabiei et al. 2020; Omer et al. 2022) in the activity of antioxidant enzymes.

In recent years, the modulation of gene expression involved in salt stress responses has been increasingly assessed by transcriptome analysis. *A. brasilense* Sp245 has been shown to improve rice growth under salt stress by regulating the expression of key genes involved in stress defense and response, abscisic acid and jasmonic acid signaling, and ion and nutrient transport (Degon et al. 2023). PGPB *A. brasilense* Cd increased the expression of stress-responsive genes *CaMYB*, *CaJAR1*, *CaWRKY*, and *CaLOX2*. Although this strain did not act as a PGPB under salt stress, it upregulated genes associated with the ISR-type response (*CaMYB72*) and also induced the expression of *CaDnaJ02* and *CaDnaJ04* genes (heat shock proteins associated with jasmonic acid synthesis) (Caamal-Chan et al. 2023).

Most studies have shown an increase in chlorophyll and carotenoid contents when plants were inoculated with *Azospirillum* under salt stress (Rabiei et al. 2020; Omer et al. 2022; Khalifa et al. 2021). However, Del Amor and Cuadra-Crespo showed that co-inoculation of sweet pepper with *A. brasilense* and *Pantoea diversa* did not affect the photochemical efficiency of photosystem II and the relative chlorophyll content, but did maintain higher stomatal conductance; therefore, they concluded that the effect of inoculants on salinity response was mainly due to stomatal regulation of photosynthesis rather than the effect on biochemical limitations of photosynthesis (Del Amor and Cuadra-Crespo 2011).

It was also shown that the most important secondary metabolism compounds (phenylpropanoids, alkaloids, and other N-containing metabolites, as well as membrane lipids) and phytohormones (brassinosteroids, cytokinins, and methyl salicylate) showed the most pronounced modulation in response to *Azospirillum* treatment of tomatoes under salt stress conditions (Alzate Zuluaga et al. 2022). Thus, the effect that *Azospirillum* inoculation has on plants may vary, but in most cases, it results in the alleviation of salt stress.

14.2.5 Hydrocarbons

Petroleum hydrocarbons are pervasive environmental toxins. Due to oil spills from anthropogenic accidents, these compounds can cause pollution of water, air, and soil, including agricultural land. Hydrocarbons can impair soil aeration, which can both harm the soil microbiome and interfere with plant development (Mohanta et al. 2023). At the same time, many microorganisms are involved in hydrocarbon degradation and soil bioremediation. Studies have shown that in the early stages of biodegradation, the *α-Proteobacteria* group (particularly the genera *Sphingomonas* and *Azospirillum*) was the dominant group in soil contamination with polycyclic aromatic hydrocarbon, and the total petroleum hydrocarbon (Viñas et al. 2005). Soil samples artificially contaminated with oil showed an increase in the growth of nitrogen-fixing bacteria belonging to the genera *Gordonia*, *Rhizobium*, *Kocuria*, and *Azospirillum*. It was suggested that such diazotrophs could enrich soils with fixed nitrogen and stimulate hydrocarbon biodegradation (Al-Mailem et al. 2019).

Initially, the ability to degrade oil was demonstrated for the strain *A. brasilense* SR80. Under hydroponic conditions, crude oil stimulated the growth of *A. brasilense* SR80. This promoted the development of the root system of wheat in the presence of oil and increased the level of oil degradation by the plant-microbe association (Muratova et al. 2005). Somewhat later, species living in oil-polluted conditions were discovered, for example, the species *Amauroderma rugosum* IMMIB AFH-6T (Young et al. 2008). The isolate *Acremonium zeae* S13D1 was isolated from an oil-polluted region. And it can use diesel fuel and gasoline as the sole carbon source (Nanekar and Kokitkar 2024). *Azospirillum oleiclasticum* (two strains RWY-5-1-1T and ROY-1-1-2) not only can grow under oil-polluted conditions but also can biodegrade crude oil (Wu et al. 2021). *Aspergillus oryzae* MS6, as part of the microbiome in the rhizosphere of maize, contributed to the bioremediation of crude oil-polluted soil (Saeed et al. 2021).

14.2.6 *Radiation*

An attempt was made to inoculate *Azospirillum* sp. strain TS13 plants in contaminated soil in Fukushima for bioremediation by transferring radioactive cesium to the above-ground parts of plants. Positive effects of inoculation were obtained, but the concentrations of radioactive cesium during their transfer to the tested plants were not very high (Djedidi et al. 2015).

14.3 Methodological Part on *Azospirillum* Inoculation

Various researchers have used various methods of inoculating plants with *Azospirillum* in the presence of stress factors in their work. This chapter discusses the plants used and the methods of their inoculation.

The protective effect of *Azospirillum* bacteria under water stress has been demonstrated in cereals (Table 14.1), including maize (Curá et al. 2017; García et al. 2017, 2024; Souza et al. 2023; Tiepo et al. 2024), wheat (Arzanesh et al. 2011; Akhtar et al. 2021; Alhammad et al. 2023; Valente et al. 2023), soybean (Silva et al. 2019), sorghum (Kamali and Mehraban 2020), rye (Rahnama et al. 2023), ryegrass (Cortés-Patiño et al. 2021), and millet (Yaadesh et al. 2023). Also, *A. baldaniorum* and *A. lipoferum* species stimulated the growth and development of *Arabidopsis*

Table 14.1 Bacteria of the genus *Azospirillum* and conditions of plant inoculation under stress conditions

	Inoculation conditions				
Azospirillum	Plant	Bacterial cell concentration	Cocultivation	Period inoculation	References
Drought and water stress					
A. argentinense Az19	Commercial maize (*Zea mays* L.) hybrid (DOW 510 PW), seeds	5.0×10^7 CFU/mL	–	–	García et al. (2024)
A. baldaniorum (brasilense) Sp245	*Arabidopsis thaliana* wild-type Columbia (Col-0 ecotype) and the aba2-1 mutant, plants with two fully expanded leaves	1.0×10^6 CFU/mL	–	–	Cohen et al. (2015)
	Purple basil (*Ocimum basilicum* L.) cv. Red Rubin, seedlings	1.0×10^7 CFU/mL	–	–	Mariotti et al. (2021)
	Common wheat (*Triticum aestivum* L.), seeds	2.0×10^7 cells	–	–	Valente et al. (2023)
A. brasilense	Wheat (*Triticum aestivum* L.) variety Galaxy-2013, seedlings	–	*Rhizobium pisi*	–	Alhammad et al. (2023)
A. brasilense Ab-V_5 and Ab-V_6	Soybean *Glycine max* L. Merrill.) 5D 615 RR, seeds	2.0×10^8 viable cells per mL	*Bradyrhizobium japonicum* SEMIA 5079 and SEMIA 5080 strains	–	Silva et al. (2019)
	Maize (*Zea mays* L.), seeds	–	–	–	Souza et al. (2023)
A. brasilense Ab-V5	Neotropical trees *Cecropia pachystachya* Trécul and *Cariniana estrellensis* (Raddi) Kuntze, seeds	1.0×10^6 cells per mL	–	–	Tiepo et al. (2020)
	Maize (*Zea mays* L.), seedlings	1.0×10^6 cells per mL	–	–	Tiepo et al. (2024)
A. brasilense Az39, *Azospirillum* isolate Az19	Commercial maize (*Zea mays* L.) hybrid (DOW 510 PW), seedlings	5.0×10^7 cells per seedlings	–	2 h	García et al. (2017)

(continued)

Table 14.1 (continued)

Azospirillum	Inoculation conditions: Plant	Bacterial cell concentration	Cocultivation	Period inoculation	References
A. brasilense SP7	Maize (*Zea mays* L.), seeds	1.6×10^9 CFU/mL	–	24 h	Curá et al. (2017)
A. brasilense D7	Perennial ryegrass (*Lolium perenne* L.) var. One50, seeds	10^8 CFU/mL	*Herbaspirillum* sp. AP02, *Herbaspirillum* sp. AP21, *Pseudomonas fluorescens* N7	–	Cortés-Patiño et al. (2021)
A. brasilense (MT742666) and *A. lipoferum* (MT742664)	Wheat (*Triticum aestivum* L.) variety Pak 13, seeds	10^8 CFU/mL	*Bacillus sp.* MT482404	8 h	Akhtar et al. (2021)
Azospirillum formosense AIM57	Pearl millet *Cenchrus americanus* (L.) Morrone (Pusa composite 443), seeds	10^8 CFU/mL	*Bacillus* sp. IMSB1, AB4, RP24, IMBJ3, IMJ4, IMJ7, IMJ12	60 min	Yaadesh et al. (2023)
A. lipoferum AZ1, AZ9 and AZ45	Chamran wheat (*Triticum aestivum* L.), seedlings	3.2×10^9/mL	–	3 h	Arzanesh et al. (2011)
A. lipoferum PTCC12D	*Secale montanum* Guss., seeds	5.0×10^8 CFU/mL	–	1 h	Rahnama et al. (2023)
Azospirillum spp.	Sorghum (*Sorghum bicolor* L.), seeds	1.0×10^7 living cells	*Azotobacter* spp. (nitroxin)	–	Kamali and Mehraban (2020)
Salinization and osmotic stress					
A. baldaniorum (*brasilense*) Sp245	Potato (*Solanum tuberosum* L. cv. Nevsky) microplants, microcuttings	10^6 cells per mL	–	5 days	Evseeva et al. (2019)
	Potato (*Solanum tuberosum* L. cv. Nevsky) microplants, microcuttings	10^6 cells per mL	–	5 days	Arkhipova et al. (2020)
	Rice (*Oryza sativa* cv. Nipponbare), seedlings	10^8 cells per mL	–	–	Degon et al. (2023)

A. brasilense	White clover (*Trifolium repens* L.), seedlings	1.0×10^9 CFU/mL	–	–	Khalid et al. (2017)
A. brasilense	Coriander (*Coriandrum sativum* L.), seeds	10^7 CFU/mL	*Azotobacter chroococcum*	4 h	Rabiei et al. (2020)
A. brasilense Ab-V5 and Ab-V6	Maize (*Zea mays* L.) hybrid DKB330 VT PRO2, seeds	3.0×10^5 cells per seed	*Rhizobium tropici* CIAT899	–	Fukami et al. (2018a)
	Maize (*Zea mays* L.) hybrid DKB330 VT PRO2, seeds	3.0×10^5 cells per seed	–	–	Fukami et al. (2018b)
	Maize (*Zea mays* L.), seeds	5.0×10^8 cells per mL	–	–	Checchio et al. (2021)
A. brasilense Cd	Chili pepper (*Capsicum annuum* L. var. Caballero), roots	10^6 CFU/mL	–	–	Caamal-Chan et al. (2023)
A. brasilense CW903	Red pepper (*Capsicum annuum* L.), seeds	1.0×10^9 CFU/mL	–	4 h	Joe et al. (2022)
A. brasilense DSM-1843	Tomato (*Solanum lycopersicum* L. cv MicroTom), roots	10^6 cells per mL	–	–	Alzate Zuluaga et al. (2022)
A. brasilense RS-SP7	Rapeseed (*Brassica napus* L. cv. okapi), seeds	10^8 CFU/mL	–	–	Farhangi-Abriz et al. (2020)
A. brasilense strain 7000	Flax (*Linum usitatissimum* L. cv. Giza 9), seeds	10^6 CFU/mL	*Pseudomonas geniculata* strain 60786	–	Omer et al. (2022)
A. lipoferum (ATCC 29707)	Roselle *Hibiscus sabdariffa* L. (balady, dark red), roots	10^9 CFU/mL	–	–	Bahgat et al. (2023)
Heavy metal stress					
A. brasilense	Wheat (*Triticum aestivum* L.) seeds (arsenic stress)	–	–	–	Zaheer et al. (2024)

(continued)

Table 14.1 (continued)

	Inoculation conditions				
Azospirillum	Plant	Bacterial cell concentration	Cocultivation	Period inoculation	References
A. argentinense Az39	Soybean (*Glycine max* L.) cv. DM 4670, seeds (arsenic stress)	$OD_{620nm} = 0.5$	*Bradyrhizobium japonicum* E109	2 h	Armendariz et al. (2019)
A. argentinense Az39 and *A. brasilense* CD	Hybrid maize (*Zea mays* L.) DK72-10 VT3pro, seedlings (arsenic stress)	–	–	–	Peralta et al. (2021)
A. brasilense 91	*Arabidopsis* Col-0 ecotype and mutants, seedlings (cadmium stress)	10^8 cells per mL	–	–	Xu et al. (2018)
A. brasilense Yu-62	pakchoi (*Brassica chinensis* L.), seedlings (cadmium stress)	10^8 CFU/mL	–	–	Pan et al. (2019)
A. brasilense (CGMCC1.10379)	pakchoi (*Brassica chinensis* L.), seedlings (cadmium stress)	10^8 CFU/mL	–	–	Cui et al. (2023)
A. lipoferum 137	Spring barley (*Hordeum vulgare* L.), seeds (cadmium stress)	3.0×10^8 cells per mL	–	–	Belimov and Dietz (2000)
A. brasilense EMCC1454	Chickpea (*Cicer arietinum* L., cv. Giza 3), seedlings (chromium stress)	10^8 CFU/mL	–	25 min	El-Ballat et al. (2023)
A. brasilense Cd	Cucumber (*Cucumis sativus* L. cv Chinese Long), seedlings (copper stress)	10^6 CFU/mL	–	–	Marastoni et al. (2019)
A. brasilense Sp7 and Cd, and *A. baldaniorum* Sp245	Wheat (*Triticum aestivum* L. cv. Saratovskaya 29), seedling (copper stress)	10^7 cells per mL	–	–	Muratova et al. (2022)

(Cohen et al. 2015), basil (Mariotti et al. 2021), and various tropical trees (Tiepo et al. 2020) when grown under dry conditions. In experiments with bacterial inoculation, standard suspensions with cell concentrations in the range of 10^6 to 10^9 CFU/mL^{-1} were used. In most cases, the inoculation material was introduced by treating seeds or seedlings, and the duration of exposure of plants to bacteria varied from several minutes to a day (Table 14.1).

Mixed inoculants that combine bacteria of the genus *Azospirillum* and other microorganisms of the PGPR group can be used to protect plants from drought. The protective properties of mixed cultures of *azospirilla* with *rhizobia* or soil *bacilli* were assessed for wheat (*Triticum aestivum* L.). In the work (Akhtar et al. 2021), a combination of *A. brasilense* and *A. lipoferum* bacteria with representatives of *Bacillus* spp. inoculation of wheat seeds with combined cultures for 8 h, added in equivalent concentrations of 10^8 CFU/mL, under drought conditions significantly improved germination characteristics. It should be noted that the addition of SiO_2 nanoparticles to mixed bacterial cultures and inoculation of wheat seeds with such a preparation stimulated plant growth and yield. In turn, a mixed culture of *A. brasilense* with *Rhizobium pisi* (Alhammad et al. 2023) was used to increase wheat yield under drought conditions in field experiments.

Combinations of *Azospirillum* and *Bacillus* spp. have also been used in other cereals. For example, the effect of *A. formosense* and a number of *Bacillus* isolates with growth-promoting properties on the growth of pearl millet (*Cenchrus americanus* L.) at early stages of development under water deficit conditions was assessed (Yaadesh et al. 2023). Millet seeds were primed with a bacterial suspension of *Azospirillum* and *Bacillus* cultures with a total cell concentration of 10^8 CFU/mL for an hour. The seeds were then placed in a medium with a lack of moisture, which was simulated by adding PEG6000.

Combinations of *A. brasilense* bacteria with bacteria of the genera *Herbaspirillum* spp., *Pseudomonas* spp., and *Bradyrhizobium* spp. are known in inoculation material for soybean (*Glycine max* L. Merrill.) and ryegrass (*Lolium perenne* L.) under drought conditions (Silva et al. 2019). For the combined bacterial suspension, the stock cultures were diluted to a concentration of 2.0×10^8 viable cells per mL. The plants were cultivated under water deficit conditions for 10 days.

Combinations of nitroxin (a biofertilizer composed of *Azospirillum* spp. and *Azotobacter* spp. bacteria with a viable cell count of about 1.0×10^7) and arbuscular mycorrhizal fungus-based biofertilizers were studied for the growth and development of sorghum (*Sorghum bicolor* L.) under drought conditions (Kamali and Mehraban 2020).

Soil salinity is one of the key factors affecting plant growth and productivity in agricultural cultivation. The protective effect of bacteria of the genus *Azospirillum* on the growth and development of plants belonging to the families *Poaceae* (rice, corn, wheat), *Fabaceae* (clover), *Solanaceae* (potato, chili pepper), *Apiaceae* (coriander), *Cruciferae* (rapeseed), *Linaceae* (flax), *Malvaceae* (hibiscus), and others was studied under salinization of substrates and/or soils (Table 14.1). In most inoculation experiments, representatives of the species *A. brasilense* were used with cell concentrations from 10^5 CFU/mL (Fukami et al. 2018a, 2018b) to 10^9 CFU/mL (Khalid et al. 2017; Joe et al. 2022) and different periods of treatment of seeds, seedlings, or plant roots. *A. baldaniorum* strain Sp245 (previously *A. brasilense* Sp245) bacteria were successfully used to treat potato microplants (*Solanum tuberosum* L. cv. Nevsky) under osmotic stress. The inoculation bacterial suspension contained 10^6 cells per mL, while the incubation time of microcuttings did not exceed 5 days.

The use of *A. lipoferum* cell suspension at a concentration of 10^9 CFU/mL together with aloe vera leaf extract contributed to the protection of hibiscus (*Hibiscus sabdariffa* L.) growing in saline soil (Bahgat et al. 2023).

Bacteria of the species *A. brasilense* and *A. formosense* can be used for co-inoculation with other rhizosphere growth-promoting bacteria when treating plants under saline conditions. Thus, pre-activation of coriander (*Coriandrum sativum* L.) seeds with a joint culture of *A. brasilense* and *Azotobacter chroococcum* cells when sown in saline media was performed (Rabiei et al. 2020). In the work (Fukami et al. 2018a), maize (*Zea mays* L.) seeds were primed with joint cultures of *A. brasilense* Ab-V5 or Ab-V6 bacteria with the *Rhizobium tropici* strain CIAT 899 and placed in media with a salinity level of about 170 mM NaCl. A joint bacterial culture of endophytes *A. brasilense* and *Pseudomonas geniculata* was used to treat flax (*Linum usitatissimum* L.) seeds and their subsequent cultivation under saline conditions (Omer et al. 2022).

Azospirillum bacteria are also able to increase the resistance of plants with which they enter into associative mutually beneficial relationships to toxic stress caused by HMs (Belimov et al. 2004). Table 14.1 presents examples of the use of various species of *Azospirillum* bacteria to protect plants from high concentrations of arsenic, cadmium, chromium, and copper in the environment.

It was shown that treatment of hybrid maize DK72-10 VT3pro seedlings with *A. brasilense* strains Az39 and CD neutralized the negative effects of adding arsenate (As^V) to the medium (Peralta et al. 2021). To solve the problem of reduced wheat yield in the presence of toxic arsenic, a strategy was developed for the combined use of a bacterial suspension of *azospirilla* and seaweed extracts for seed priming (Zaheer et al. 2024). In the study (Armendariz et al. 2019), the effectiveness of a combination of PGPR bacteria *Bradyrhizobium japonicum* E109 and *A. brasilense* Az39 was assessed in the inoculation of soybean seeds and subsequent cultivation in the presence of 25 μM As^V and As^{III}.

To optimize the development of barley (*Hordeum vulgare* L.) under conditions of elevated cadmium concentrations in the medium, a number of rhizobacteria, including *A. lipoferum*, were tested. Plants were grown in hydroponic and quartz sand cultures in the presence of 50 μM $CdCl_2$. Before sowing, barley seeds were primed with a bacterial cell suspension at a concentration of 3.0×10^8 cells per mL (Belimov and Dietz 2000). Bacteria *A. brasilense* were used to protect Chinese cabbage pakchoi (*Brassica chinensis* L.) from elevated concentrations of cadmium salts in the soil. In the previous work (Pan et al. 2019), cabbage seedlings were treated with a bacterial suspension of *azospirilla* at a concentration of 10^8 CFU/mL. Subsequently, a strategy combining bacterial inoculation with *A. brasilense* and exogenous signaling stimulation with hydrogen sulfide (H_2S) was developed for pakchoi (Cui et al. 2023).

Phytoprotective properties of bacteria of the genus *Azospirillum* were tested when cucumber (*Cucumis sativus* L. cv Chinese Long) and wheat (*Triticum aestivum* L. cv. Saratovskaya 29) plants were exposed to toxic concentrations of copper salts. Thus, in a study (Marastoni et al. 2019) on a hydroponic cucumber culture model, inoculation with a bacterial suspension of *A. brasilense* Cd at a concentration of 10^6 CFU/mL was used. In relation to wheat, the inoculation material of *A. brasilense* Sp7 and Cd, and *A. baldaniorum* Sp245 bacteria was used and the growth rates of 14-day-old seedlings were assessed under conditions of increased concentrations of copper salts (Muratova et al. 2022). The work (El-Ballat et al. 2023) is devoted to the investigation of the potential of *A. brasilense* bacteria as a useful bioinoculant to promote growth, yield, and resistance to chromium stress (0, 130, and 260 μM $K_2Cr_2O_7$) in chickpea (*Cicer arietinum* L.).

The simultaneous use of pesticides and biological growth promoters, including bacteria of the genus *Azospirillum*, is a promising approach to increase the productivity of many crops. However, to ensure successful inoculation, special attention is recommended to the time between seed inoculation and sowing. Studies conducted by Takahashi et al. (2022) using *A. brasilense* FP2 (Sp7 ATCC 29145 Nif+ Smr Nalr), a spontaneous mutant of the Sp7 strain and insecticide (imidacloprid and thiodicarb) and fungicide (triadimenol), showed that the time between inoculation and subsequent sowing should be no more than 4 h to maintain the viability of microorganisms. PCR showed that only 20 inoculated cells remained viable 48 h after inoculation (Takahashi et al. 2022). Similarly, *A. thiophilum* BV-S^T successfully survived treatment of wheat seeds with the insecticide cypermethrin. The seeds were first soaked in the insecticide solution for 20 min and then incubated in a bacterial suspension of *A. thiophilum* BV-S^T for 20 min. After drying, the seeds were immediately planted in petri dishes. A week later, PCR confirmed successful inoculation of the bacteria into wheat roots (Gureeva et al. 2023).

14.4 Methodological Part on Plants: What Indicators Are Studied in Plants to Determine the Level of Stress

Exposure to biotic and abiotic stresses results in changes in plant transcriptomics and metabolomics and significantly reduces crop growth and productivity. Stress-induced reactions in plants are manifested at all levels (organs, tissues, and cells) and affect physiological and biochemical processes, and therefore developmental processes (Mittler 2006). These changes affect the composition of symbiotic microbiota through root and leaf exudates, but commensal microbes can exert a protective effect by significantly reducing the impact of stress factors on plants (Calvo et al. 2014; Liu et al. 2020; Cellini et al. 2021). *Azospirillum* spp. have demonstrated great potential in mitigating the effects of various stresses, which is enhanced when co-inoculated with other PGPB (Vacheron et al. 2015; Cura et al. 2017; Fukami et al. 2018a; Filipini et al. 2021; Cortés-Patiño et al. 2021).

To determine the positive effect of bacterial inoculation, morphological, physiological, and molecular markers of plants (e.g., yield, morphological parameters, surged ROS production, antioxidant protection enzymes, photosynthesis efficiency, pigment content, micro- and macroelement content, VOC content) are usually detected.

14.4.1 Morphometric Parameters of Plants

Plant cell growth is very sensitive to stress due to the decrease in turgor pressure, which affects all physiological processes. Several parameters are used as biomarkers: the number of leaves per plant, the size of individual leaves, the change in leaf surface area, including the length-to-width ratio of leaves, since these parameters are related to the intensity of photosynthesis, production of fresh and dry biomass and separately shoots and roots, stem diameter, and root surface area. The range of determined parameters depends on the phase of plant development. For mature plants, the most important is the determination of yield (dry biomass and its composition, wheat spike morphogenesis and grain threshability, grain rows/spike ratio, grains/row, 100-grain weight, grains/spike, grain yield/plant, yield index, etc.)

Evaluation of plant growth parameters after inoculation with *Azospirillum* spp. usually includes morphometry (plant height, number of fully developed true leaves, dry (65 °C oven for 72 h, or 80 °C oven for 48 h) weight of turgor and wilted leaves, stems, and roots), for which the roots must be carefully and thoroughly freed from the soil (Romero et al. 2014; Cura et al. 2017). Parameters are determined for all plants or for randomly selected plants to obtain statistically reliable values.

Leaf area is a reliable index of growth factors determining dry matter accumulation, carbohydrate metabolism, crop yield, and crop quality (Lal and Subba Rao 1951). Therefore, different techniques have been developed for measuring the total leaf surface area of a growing plant (Edward et al. 1956; Schurer 1971; Obiefuna and Ndubizu 1979; Pandey and Singh 2011; Singh et al. 2018). In recent years, various programs for analyzing scanned images and an electronic planimeter have been widely used (Dorofeeva and Bonetskaya 2020).

14.4.2 Content of the Main Pigments of Photosynthesis

A reliable indicator of stress-induced changes in plants is the content of photosynthetic pigments (total chlorophyll, chlorophyll *a* and *b*) and carotenoids. Various modifications of the destructive method for chlorophyll determination using organic solvent extraction and subsequent spectrophotometry are common (Lichtenthaler and Wellburn 1983).

1. Add 2 ml of dimethyl sulfoxide to crushed fresh leaf fragments (200 mg) in Eppendorf tubes.
2. Incubate the tubes in a water bath at 65 °C for 2 h.
3. After cooling to room temperature, centrifuge the extracts (13,000 *g*, 5 min).
4. The absorbance of supernatant was read at 480, 649.1, and 665.1 nm using a spectrophotometer.
5. The concentrations (μg ml^{-1}) of chlorophyll *a* (C_a), chlorophyll *b* (C_b), and total carotenoids (C_{x+c}) were calculated from the following formulas and were then determined in leaf biomass (Wellburn 1994):

$$C_a = 12.47A_{665.1} - 3.62A_{649.1}$$

$$C_b = 25.06A_{649.1} - 6.5 \cdot A_{665.1}$$

$$c_{x+C} = (1000A_{480} - 1.29C_a - 53.78C_b)/220$$

The pigment content was calculated using formulas and expressed as mg total chlorophyll per g fresh weight of leaves.

Another powerful tool for determining photosynthetic activity at the cellular, leaf, and plant level is chlorophyll fluorescence imaging (Chl-F) in both laboratory and field settings. Chl-F can be analyzed or inferred by measurements using a variety of devices, which can be classified by the type of light source used, LED (PAM measurements), laser (LiDAR, LIFT, and LIFS), and natural light (SiF and hyperspectral reflectance at the canopy surface), and by controlled or natural conditions (Pérez-Bueno et al. 2019).

In addition, various devices (e.g., CL-01, SPAD-502, Dualex, and CCM-200) were used to estimate chlorophyll content in plants, which showed reliable results under optimal plant nutrition conditions, but not under nutrient deficiency. The results confirmed the need for additional testing of these devices by parallel spectrophotometric analysis (Kalaji et al. 2017). A good association between the chlorophyll meter Soil Plant Analysis Development (SPAD) index and the chlorophyll content (obtained by the destructive method) in plants has been reported in some crops (Shibaeva et al. 2020; Xavier et al. 2021). When using chlorophyll meters, follow the manufacturer's instructions, but to obtain reliable data, measurements must be taken on at least 10 plants and at least five points on each leaf.

Often, changes in the total carbon and total nitrogen contents are estimated for stressed plants using a modified Walkley-Black method (Walkley and Black 1934; Bahadori and Tofighi 2016) and the Kjeldahl method (Goyal et al. 2022). It was shown the existence of a relationship between the index of chlorophyll and the N status in the leaves of several species, including wheat, rice, and oats and maize (Kalaji et al. 2017).

14.4.3 Malondialdehyde

Malondialdehyde (MDA) as a biomarker is widely used for examining oxidative stress in various organisms, including plants. Lipid peroxidation is a chain phenomenon resulting in the formation of various active compounds that result in cellular damage. However, as rightly noted by Morales and Munné-Bosch (2019), lipid peroxidation is not only caused by ROS, but can also be the result of increased lipoxygenase activity. In addition to MDA formation, lipid peroxidation can induce the formation of jasmonates, which are an important component of stress resistance (Weber et al. 2004; Farmer and Mueller 2013).

The accumulation of MDA as a product of the breakdown of endoperoxides of unsaturated fatty acids caused by radicals is determined by photocolorimetry (Buege and Aust 1978). The method is based on the formation in an acidic and hot environment between MDA and thiobarbituric acid (TBA):

1. Frozen roots (100 mg) were homogenized in 2 ml trichloroacetic acid (TCA; 0.1%, w/v) in a pre-chilled mortar and then centrifuged (5000 *g*, 10 min).
2. A mixture of 1 ml aliquots of the supernatant with 2 ml TBA (0.5%, w/v) in 20% TCA (w/v) was heated in a water bath (100 °C, 20 min), quickly cooled in an ice bath, and then centrifuged (5000 *g*, 10 min).
3. The absorbance of the supernatant was recorded at 532 nm and corrected for nonspecific absorbance at 600 nm.
4. The absorbance of the sample is determined at 535 nm against a blank that contains all the reagents minus the biological sample. The malondialdehyde concentration of the sample can be calculated using an extinction coefficient of $1.56 \times 10^5\ M^{-1}\ cm^{-1}$.

The main limitation of this method was the presence of interfering compounds in plants that also absorb at 532 nm, which led to an overestimation of MDA values. Hodges et al. (1999) proposed a spectrophotometric method in which the absorbance at 532 nm of a solution containing plant extract free of TBA was subtracted from the absorbance of an identical solution incubated with TBA, although this method also has some limitations. Several sensitive methods for the determination of MDA content have been developed using derivatization in combination with different separation and/or detection methods (gas chromatography, liquid chromatography with UV/fluorescence detection, or mass spectrometry) (Davey et al. 2005; Mendonça et al. 2017).

1. Plant tissues (500 mg) were frozen in liquid nitrogen and homogenized in a pre-cooled mortar using 2 ml metaphosphoric acid/EDTA/PVPP extraction buffer and then centrifuged (5000 *g*, 10 min). A 100 μl aliquot of the extract was stabilized by the addition of 10 μl of 2% butylated hydroxytoluene in ethanol, and the extracts were stored at −20 °C until analysis.
2. Sample separation is performed on columns with C18 stationary phase for excellent peak symmetry, performance, and pH stability.

3. The column eluted at 2.5 ml/min with a 1-min linear gradient of 38–70% methanol in 50 mM potassium phosphate (pH 6.8), and after 30 s, it was returned to the starting conditions. Absorption spectra were collected over the range of 400–600 nm with a spectral resolution of 2.4 nm and a frequency of 5 Hz. MDA $(TBA)_2$ was quantified at its absorption maximum of 532 nm.
4. Standard curves of 0–2 μM MDA were prepared on the day of the analyses from a stock solution of 1,1,3,3-tetraethoxypropane in 40% ethanol according to Korchazhkina et al. (2003).

14.4.4 Total Soluble Carbohydrates

An increase in free proline and soluble sugars in plants can be caused by biotic or abiotic stress conditions (drought, salinity, cold, HMs, or certain pathogens) (Chen and Murata 2002; Mattioli et al. 2009; Vacheron et al. 2015). Total soluble carbohydrates are determined according to McCready et al. (1950) using glucose solution (0.1 g/L) as standard. HPLC methods are increasingly used to determine the total amount of soluble carbohydrates, which are characterized by high reproducibility of results (Da Ros et al. 2020):

1. Pre-washed leaves with water are dried on filter paper and ground in liquid nitrogen in a pre-cooled mortar. Weigh 100 mg of the ground sample in a 15-ml polypropylene tube, and add 50 μl of galactitol as an internal standard. For each sample, at least three replicates are analyzed.
2. Add 4 ml of the extracting solution of methanol/chloroform/water (12:5:3) to each tube, mix, and keep for 16 h at 4 °C. Then, the extracts are thoroughly mixed on a vortex and centrifuged (5000 *g*, 10 min, at 4 °C). Carefully remove the supernatant and transfer it to a 50-ml polypropylene tube. Add 4 ml of methanol/chloroform/water extraction solution to the sediment, mix, and centrifuge. Remove the supernatant and combine it with the sample in a 50-ml tube. Repeat the procedure one more time.
3. Add 5 ml of distilled water to the combined supernatants and mix thoroughly, then centrifuge the samples (5000 *g*, 5 min at 4 °C) to separate the phases.
4. Remove 2 ml of the upper polar phase into a microcentrifuge tube and remove all solvent under vacuum and resuspend the dried sample in 1 ml of deionized water. Pass the sample through a 0.45-μm filter into a vial before HPLC for quantification.
5. The analysis is carried out using a Dionex PA1 column (4 × 250 mm) and pulse amperometry detection. Mobile phase: 16 mM NaOH and 2 mM NaOAc for 30 min; 200 mM NaOH from 30 to 40 min; 16 mM NaOH and 2 mM NaOAc from 40 to 60 min; flow rate 0.8 ml min^{-1}. Sugar standards, including sugar alcohols, are used depending on the material being analyzed or the specific carbohydrates of interest (glucose, sucrose, fructose, etc.) in concentrations from 0.1 to 2 mg ml^{-1}.

14.4.5 Free Proline

Three methods are suitable for the determination of free proline (Ábrahám et al. 2010). When screening a large number of samples, the simple and convenient isatin paper assay can be used. The isatin method is based on the interaction of a color reagent (1H-indole-2,3-dione), in which the chromatographic paper is pre-soaked, with proline contained in the ethanol extract of plants. Before analysis, it is necessary to prepare the Whatman 3MM chromatographic paper. Prepare a fresh solution of 1 g isatin in 100 ml methanol, 2.5 ml glacial acetic acid, soak the chromatographic paper in it, and then dry in air for 1 h. The paper can be stored in the dark for several weeks.

1. Plant tissue samples (100 mg) are frozen in liquid nitrogen in a pre-cooled mortar, then 20% ethanol (1 ml) is added, and the plant material is ground. The extract is centrifuged at room temperature (13,000 *g*, 5 min).
2. 10 μl of the solution is applied to isatin paper, dried in air for 30 min, and then oven-heated at 90 °C until a blue color appears. The intensity of the blue color correlates with the proline content. A series of L-proline concentrations in 20% ethanol (0–5 mg/ml) are used as reference samples, and a similar reaction is performed.
3. For semi-quantitative evaluation, isatin paper is washed in 50% ethanol for 1–2 min, dried, and scanned. Densitometry and freely available software such as ImageJ or NIH Image software are used to obtain data.

The colorimetric measurement according to Bates et al. (1973) using L-proline as standard is quantitative and provides reliable data on the proline content. This method does not require complex instruments and expensive reagents and is also suitable for analyzing multiple samples.

1. Freeze fresh plant tissue samples (100 mg) in liquid nitrogen, add 3% sulfosalicylic acid (1 ml), and grind in a mortar. Centrifuge the extracts at room temperature (13,000 *g*, 5 min).
2. Prepare the reaction mixture in a separate test tube: 100 μl 3% sulfosalicylic acid, 200 μl glacial acetic acid, 200 μl acid ninhydrin.
3. Add 100 μl of the plant extract supernatant, mix well but carefully to avoid high pressure and accidental opening of the tubes in the subsequent reaction. Pierce the lids of the tubes or use special clamps for microtubes.
4. Incubate the tubes in a boiling water bath for 60 min. To stop the reaction, place the tubes on ice. Add 1 ml of toluene to the reaction mixture and shake the samples for 20 s. After separation of the organic and aqueous phases (5 min), collect the organic phase in a clean tube.
5. Measure the absorbance at 520 nm in a spectrophotometer using toluene as a standard. Determine the concentration of proline using a standard curve with different concentrations of L-proline. Proline concentration is usually expressed in μg or μM per gram of plant wet weight.

For comparative analysis of the concentration of all amino acids, the HPLC method is optimal.

1. Freeze plant material (100 mg) in liquid nitrogen and grind in a pre-chilled mortar in 2 ml of cold ethanol. Transfer the solution to a 2-ml microcentrifuge tube and mix well on a vortex and centrifuge (13,000 *g*, 5 min).
2. Transfer 1 ml of the supernatant to a new tube, add 1 ml of 120 μM α-aminobutyrate (dissolved in deionized distilled water), and mix well.
3. Incubate the mixture at 70 °C for 5 min, then at 4 °C for 1 h, then centrifuge (14,000 *g*, 30 min at 4 °C).
4. Take aliquots of supernatants (200 μl), dry in vacuum, add 1.2 ml of deionized water, keep at room temperature for 1 h, then dilute the solution 10-fold, and analyze by HPLC.

Carillo and Gibon (2011) proposed a simple, rapid, and harmless method for the determination of proline based on ninhydrin using a microplate reader or a cuvette spectrophotometer.

14.4.6 Glycine Betaine

Another biomarker of plant stress is the accumulation of glycine betaine, which causes osmotic adjustment and maintains normal cellular metabolism during stress (Giri 2011; Annunziata et al. 2019). Quantitative determination of glycine betaine in plant leaf tissues using cation exchange HPLC under isocratic conditions was described by Di Martino et al. (2003).

Leaf disks (100 mg) were ground in liquid nitrogen to a powder, mixed with 2 ml of deionized water, incubated for 24 h, and centrifuged (10,000 *g*, 15 min). The extraction procedure was repeated. The supernatants were pooled and used for analysis. Glycine betaine was determined by HPLC using a cation exchange column (Dionex Hypersil SCX, 5 μm, 250 × 4.6 mm) at 30 °C under isocratic conditions at a flow rate of 1 ml min^{-1}. Eluent phase: sodium phosphate buffer 0.05 M, pH 3.7 (95%), methanol (5%). Glycine betaine elution was detected using a diode array detector at 195 nm and quantified by elution of a standard solution of the pure compound under the same conditions.

14.4.7 Antioxidant Enzymes

Antioxidant enzymes also play an important role in scavenging and detoxifying ROS in plants, and therefore, their activity correlates with stress levels. Superoxide dismutase (SOD), ascorbate peroxidase (APX) and catalase (CAT), monodehydroascorbate reductase (MDHAR), dehydroascorbate reductase

(DHAR), and GSH reductase (GR) have the ability to regulate ROS accumulation in plant cells (Gupta et al. 2018).

The first step of all methods used to evaluate various antioxidant enzymes in higher plants is their extraction from plant material. Fresh plant samples are ground into a fine powder with liquid nitrogen, followed by homogenization in buffer solutions and centrifugation. The collected supernatants are stored at 4 °C for various antioxidant enzyme assays (Pisoschi et al. 2016).

14.4.7.1 Catalase Assay

To prepare the extraction mixture, add dithiothreitol (DTT) to a final concentration of 5 mM into Tris-HCl buffer (pH 7.5) containing 0.1 mM EDTA, 0.1% Triton X-100 (v/v), and 10% glycerol (v/v). This extraction buffer can be stored for no more than 24–48 h. Fresh plant samples (200 mg) homogenize in 5 ml extraction buffer. Incubate the mixture under gentle stirring for 30 min at 4 °C. Transfer crude extracts into 2-mL microcentrifuge tube and centrifuge (15,000 *g*, 30 min at 4 °C).

In a quartz cuvette, add 100 μl of the plant extract supernatant to 900 μl of 50 mM potassium phosphate buffer (pH 7.0) containing 10.6 mM hydrogen peroxide (54.1 μl of 30% hydrogen peroxide solution was added to 50 ml of phosphate buffer) and mix quickly and thoroughly.

The absorbance was measured continuously with a spectrophotometer at 240 nm for 2 min using ultrapure water as a blank. Calculate the enzyme activity based on the decrease in absorbance using the molar extinction coefficient of hydrogen peroxide at 240 nm (43.6 M^{-1} cm^{-1}).

14.4.7.2 Superoxide Dismutase Assay

Fresh homogenized plant samples (200 mg) were extracted in 5 ml of 100 mM potassium phosphate buffer (pH 7.8) containing 0.1% (v/v) Triton X-100, 2% (w/v) polyvinylpyrrolidone (PVP), and 0.1 mM EDTA. The extract was then centrifuged (18,000 *g*, 10 min at 4 °C); the supernatant was collected and dialyzed using a membrane tube (cut-off 14 kDa) for 4 h against cold extraction buffer. The incubation mixture for enzyme activity determination (3 ml) contains 50 mM potassium phosphate buffer (pH 7.8), 45 μM methionine, 20 μM potassium cyanide, 84 μM nitroblue tetrazolium (NBT), and 5.3 mM riboflavin. After the addition of an aliquot of the plant extract, the mixture is placed in a box covered with aluminum foil and incubated under 15 W fluorescent lamps (10 min, at 25 °C). Then, using a spectrophotometer, the absorbance at 600 nm is measured. One unit of enzyme activity is defined as the amount of enzyme that causes 50% inhibition of NBT reduction (Misra and Fridovich 1972).

14.4.7.3 Ascorbate Peroxidase Assay

Fresh plant samples (200 mg) were homogenized in 5 ml of 50 mM potassium phosphate buffer (pH 7.8) containing 1 mM ascorbic acid, 1 mM phenylmethanesulfonyl fluoride (PMSF), and 1% PVP. The reaction mixture was centrifuged (18,000 *g*, 10 min at 4 °C); the supernatant was collected and dialyzed. 900 μl of 100 mM Tris-acetate buffer (pH 7.0) containing 2 mM ascorbic acid and 100 μl of the extract was added to a quartz cuvette; 2 mM hydrogen peroxide was added to initiate the reaction. The decrease in absorbance was measured using a spectrophotometer at 290 nm for 100 s. An extinction coefficient of 2.8 mM^{-1} cm^{-1} is used to calculate the reaction. The activity of the enzyme is expressed in units per milligram of protein (Ali et al. 2005).

14.4.7.4 Ascorbate Oxidase Assay

A plant tissue sample (200 mg) was homogenized in 5 ml of 20 mM potassium phosphate (pH 7.4) containing 1.5% PVP and 0.5 mM PMSF using a Polytron (10 min) in an ice bath. The mixture was then centrifuged (15,000× *g*, 15 min at 4 ° C). The collected supernatant was dialyzed.

1 ml of 20 mM potassium phosphate buffer (pH 7.0) containing 2.5 mM ascorbic acid was added to a quartz cuvette, followed by the addition of 10 μl of enzyme extract to initiate the reaction. The decrease in absorbance was measured for 3 min at 265 nm, and the activity was calculated using an extinction coefficient of 14 mM^{-1} cm^{-1} (Diallinas et al. 1997).

14.4.7.5 Glutathione Reductase Assay

Fresh plant samples (200 mg) were homogenized in a cool mortar and pestle in 5 ml of 50 mM Tris-HCl buffer (pH 7.6), then centrifuged (18,000 *g*, 5 min at 4 °C). The supernatant was collected and dialyzed. 200 μl of the enzyme extract and 800 μl of the reaction mixture of 50 mM Tris-HCl buffer (pH 7.6) containing 1 mM glutathione disulfide (GSSG), 0.15 mM NADPH, and 3 mM $MgCl_2$ were added to a quartz cuvette. The decrease in NADPH absorption is determined at 340 nm using the reaction mixture as a control. The specific enzyme activity is expressed as units per milligram of protein.

It is impossible to ignore the rapidly developing multi-omics approaches, including genomics, transcriptomics, proteomics, metabolomics, and phenomics, which allow integrating studies of plant interactions with microbes and their external environment and monitoring changes in cells in real time. The progress of these technologies allows us to hope for large-scale results and their implementation in the field. Meena et al. (2017) summarized plant responses to abiotic stresses in terms of biochemical and molecular mechanisms, followed by the phenomenon of microbe-mediated stress mitigation. Combining these high-throughput approaches will significantly expand our knowledge of the mechanisms involved in stress mitigation in plant-bacteria systems.

References

Ábrahám E, Hourton-Cabassa C, Erdei L et al (2010) Methods for determination of proline in plants. In: Sunkar R (eds) Plant stress tolerance. Methods in molecular biology, vol 639, pp 317–331. https://doi.org/10.1007/978-1-60761-702-0_20

Acevedo-Barrios R, Olivero-Verbel J (2021) Perchlorate contamination: sources, effects, and technologies for remediation. Rev Environ Contam Toxicol 256:103–120

Acevedo-Barrios R, Sabater-Marco C, Olivero-Verbel J (2018) Ecotoxicological assessment of perchlorate using *in vitro* and *in vivo* assays. Environ Sci Pollut Res 25:13697–13708. https://doi.org/10.1007/s11356-018-1565-6

Akhtar N, Ilyas N, Mashwani ZU et al (2021) Synergistic effects of plant growth promoting rhizobacteria and silicon dioxide nano-particles for amelioration of drought stress in wheat. Plant Physiol Biochem 166:160–176. https://doi.org/10.1016/j.plaphy.2021.05.039

Alamri SA, Mostafa YS (2009) Effect of nitrogen supply and *Azospirillum brasilense* Sp-248 on the response of wheat to seawater irrigation. Saudi J Biol Sci 16:101–107. https://doi.org/10.1016/j.sjbs.2009.10.009

Alhammad BA, Zaheer MS, Ali HH et al (2023) Effect of co-application of *Azospirillum brasilense* and *Rhizobium pisi* on wheat performance and soil nutrient status under deficit and partial root drying stress. Plants 12(17):3141. https://doi.org/10.3390/plants12173141

Ali MB, Hahn E-J, Paek K-Y (2005) Effects of light intensities on antioxidant enzymes and malondialdehyde content during short-term acclimatization on micropropagated *Phalaenopsis plantlet*. Environ Exp Bot 54:109–120. https://doi.org/10.1016/j.envexpbot.2004.06.005

Ali SZ, Sandhya V, Venkateswar Rao L (2014) Isolation and characterization of drought-tolerant ACC deaminase and exopolysaccharide-producing fluorescent *Pseudomonas* sp. Ann Microbiol 64:493–502. https://doi.org/10.1007/s13213-013-0680-3

Alimova AA, Gureeva MV, Gladkikh MI et al (2024) Alternative oxidase of plants mitochondria is related with increased resistance of tomato mtDNA to the difenoconazole exposure. Plant Gene 40:100475. https://doi.org/10.1016/j.plgene.2024.100475

Al-Mailem DM, Kansour MK, Radwan SS (2019) Cross-bioaugmentation among four remote soil samples contaminated with oil exerted just inconsistent effects on oil-bioremediation. Front Microbiol 10:2827. https://doi.org/10.3389/fmicb.2019.02827

Alzate Zuluaga MY, Miras-Moreno B, Monterisi S et al (2022) Integrated metabolomics and morpho-biochemical analyses reveal a better performance of *Azospirillum brasilense* over plant-derived biostimulants in counteracting salt stress in tomato. Int J Mol Sci 23(22):14216. https://doi.org/10.3390/ijms232214216

Andraski BJ, Jackson WA, Welborn TL et al (2014) Soil, plant, and terrain effects on natural perchlorate distribution in a desert landscape. J Environ Qual 43:980–994. https://doi.org/10.2134/jeq2013.11.0453

Angon PB, Tahjib-Ul-Arif M, Samin SI et al (2022) How do plants respond to combined drought and salinity stress? A systematic review. Plants 11:2884. https://doi.org/10.3390/plants11212884

Annunziata MG, Ciarmiello LF, Woodrow P et al (2019) Spatial and temporal profile of glycine betaine accumulation in plants under abiotic stresses. Front Plant Sci 10:230. https://doi.org/10.3389/fpls.2019.00230

Arkhipova TN, Evseeva NV, Tkachenko OV et al (2020) Rhizobacteria inoculation effects on phytohormone status of potato microclones cultivated *in vitro* under osmotic stress. Biomolecules 10(9):1231. https://doi.org/10.3390/biom10091231

Armendariz AL, Talano MA, Oller ALW et al (2015) Effect of arsenic on tolerance mechanisms of two plant growth-promoting bacteria used as biological inoculants. J Environ Sci 33:203–210. https://doi.org/10.1016/j.jes.2014.12.024

Armendariz AL, Talano MA, Olmos Nicotra MF et al (2019) Impact of double inoculation with *Bradyrhizobium japonicum* E109 and *Azospirillum brasilense* Az39 on soybean plants grown under arsenic stress. Plant Physiol Biochem 138:26–35. https://doi.org/10.1016/j.plaphy.2019.02.018

Arora K, Sharma S, Monti A (2016) Bio-remediation of Pb and Cd polluted soils by switchgrass: a case study in India. Int J Phytoremed 18:704–709. https://doi.org/10.1080/15226514.2015.1131232

Arzanesh MH, Alikhani HA, Khavazi K et al (2011) Wheat (*Triticum aestivum* L.) growth enhancement by *Azospirillum* sp. under drought stress. World J Microbiol Biotechnol 27:197–205. https://doi.org/10.1007/s11274-010-0444-1

Bahadori M, Tofighi H (2016) A modified Walkley-Black method based on spectrophotometric procedure. Commun Soil Sci Plant Anal 47(2):213–220. https://doi.org/10.1080/00103624.2015.1118118

Bahgat AR, Dahab AA, Elhakem A et al (2023) Integrated action of rhizobacteria with *Aloe vera* and moringa leaf extracts improves defense mechanisms in *Hibiscus sabdariffa* L. cultivated in saline soil. Plants 12(21):3684. https://doi.org/10.3390/plants12213684

Bano QUDSIA, Ilyas N, Bano A et al (2013) Effect of *Azospirillum* inoculation on maize (*Zea mays* L.) under drought stress. Pak J Bot 45(S1):13–20

Bardiya N, Bae JH (2011) Dissimilatory perchlorate reduction: a review. Microbiol Res 166:237–254. https://doi.org/10.1016/j.micres.2010.11.005

Bashan Y, Alcaraz-Melendez L, Toledo G (1992) Responses of soybean and cowpea root membranes to inoculation with *Azospirillum brasilense*. Symbiosis 13:217–228

Bates LS, Waldren RP, Teare ID (1973) Rapid determination of free proline for water stress studies. Plant Soil 39:205–207. https://doi.org/10.1007/BF00018060

Belimov AA, Dietz KJ (2000) Effect of associative bacteria on element composition of barley seedlings grown in solution culture at toxic cadmium concentrations. Microbiol Res 155:113–121. https://doi.org/10.1016/S0944-5013(00)80046-4

Belimov AA, Kunakova AM, Safronova VI et al (2004) Employment of rhizobacteria for the inoculation of barley plants cultivated in soil contaminated with lead and cadmium. Microbiology 73:99–106. https://doi.org/10.1023/B:MICI.0000016377.62060.d3

Buege JA, Aust SD (1978) Microsomal lipid peroxidation. Methods Enzymol 52:302–310. https://doi.org/10.1016/s0076-6879(78)52032-6

Burachevskaya M, Minkina T, Fedorenko A et al (2021) Accumulation, translocation, and toxicity of arsenic in barley grown in contaminated soil. Plant Soil 467:91–106. https://doi.org/10.1007/s11104-021-05067-9

Butler CS, Clauwaert P, Green SJ et al (2010) Bioelectrochemical perchlorate reduction in a microbial fuel cell. Environ Sci Technol 44:4685–4691. https://doi.org/10.1021/es901758z

Caamal-Chan MG, Loera-Muro A, Romero-Geraldo RJ et al (2023) Bacterial strains from saline environment modulate the expression of saline stress-responsive genes in pepper (*Capsicum annuum*). Plants 12(20):3576. https://doi.org/10.3390/plants12203576

Calvo P, Nelson L, Kloepper JW (2014) Agricultural uses of plant biostimulants. Plant Soil 383:3–41. https://doi.org/10.1007/s11104-014-2131-8

Caravaca F, Alguacil MM, Hernández JA et al (2005) Involvement of antioxidant enzyme and nitrate reductase activities during water stress and recovery of mycorrhizal Myrtus communis and Phillyrea angustifolia plants. Plant Sci 169:191–197. https://doi.org/10.1016/j.plantsci.2005.03.013

Carillo P, Gibon Y (2011) Protocol: extraction and determination of proline. Prometheus Wiki 2011:1–5

Cellini A, Spinelli F, Donati I et al (2021) Bacterial volatile compound-based tools for crop management and quality. Trends Plant Sci 26(9):968–983. https://doi.org/10.1016/j.tplants.2021.05.006

Checchio MV, de Cássia AR, de Oliveira KR et al (2021) Enhancement of salt tolerance in corn using *Azospirillum brasilense*: an approach on antioxidant systems. J Plant Res 134(6):1279–1289. https://doi.org/10.1007/s10265-021-01332-1

Chen THH, Murata N (2002) Enhancement of tolerance of abiotic stress by metabolic engineering of betaines and other compatible solutes. Curr Opin Plant Biol 5:250–257. https://doi.org/10.1016/s1369-5266(02)00255-8

Coates JD, Achenbach LA (2004) Microbial perchlorate reduction: rocket-fuelled metabolism. Nat Rev Microbiol 2:569–580. https://doi.org/10.1038/nrmicro926

Coates J, Jackson WA (2009) Principles of perchlorate treatment. In: Stroo HF, Ward CH (eds) In situ bioremediation of perchlorate in groundwater. Springer, New York, pp 29–53

Coates JD, Michaelidou U, Bruce RA et al (1999) Ubiquity and diversity of dissimilatory (per) chlorate-reducing bacteria. Appl Environ Microbiol 65:5234–5241. https://doi.org/10.1128/AEM.65.12.5234-5241.1999

Cohen AC, Bottini R, Piccoli PN (2008) *Azospirillum brasilense* Sp 245 produces ABA in chemically-defined culture medium and increases ABA content in arabidopsis plants. Plant Growth Regul 54:97–103. https://doi.org/10.1007/s10725-007-9232-9

Cohen AC, Travaglia CN, Bottini R et al (2009) Participation of abscisic acid and gibberellins produced by endophytic *Azospirillum* in the alleviation of drought effects in maize. Botany 87:455–462. https://doi.org/10.1139/B09-023

Cohen AC, Bottini R, Pontin M et al (2015) *Azospirillum brasilense* ameliorates the response of *Arabidopsis thaliana* to drought mainly via enhancement of ABA levels. Physiol Plant 153(1):79–90. https://doi.org/10.1111/ppl.12221

Cortés-Patiño S, Vargas C, Álvarez-Flórez F et al (2021) Potential of *Herbaspirillum* and *Azospirillum* consortium to promote growth of perennial ryegrass under water deficit. Microorganisms 9(1):91. https://doi.org/10.3390/microorganisms9010091

Creus CM, Graziano M, Casanovas EM et al (2005) Nitric oxide is involved in the *Azospirillum brasilense*-induced lateral root formation in tomato. Planta 221:297–303. https://doi.org/10.1007/s00425-005-1523-7

Cui Q, Liu D, Chen H et al (2023) Synergistic interplay between *Azospirillum brasilense* and exogenous signaling molecule H2S promotes Cd stress resistance and growth in pak choi (*Brassica chinensis* L.). J Hazard Mater 444(Pt B):130425. https://doi.org/10.1016/j.jhazmat.2022.130425

Curá JA, Franz DR, Filosofía JE et al (2017) Inoculation with *Azospirillum* sp. and *Herbaspirillum* sp. bacteria increases the tolerance of maize to drought stress. Microorganisms 5(3):41. https://doi.org/10.3390/microorganisms5030041

Da Ros L, Unda F, Mansfield SD (2020) Determination of soluble carbohydrates. In: Bärlocher F, Gessner M, Graça M (eds) Methods to study litter decomposition. Springer, Cham. https://doi.org/10.1007/978-3-030-30515-4_15

Danilova ED, Zlobin IE, Efimova MV (2023) Short-term pretreatment with melatonin increases the resistance of barley plants to the subsequent action of cadmium. Russ J Plant Physiol 70:95. https://doi.org/10.1134/S1021443723600800

Davey MW, Stals E, Panis B et al (2005) High-throughput determination of malondialdehyde in plant tissues. Anal Biochem 347:201–207. https://doi.org/10.1016/j.ab.2005.09.041

Degon Z, Dixon S, Rahmatallah Y et al (2023) *Azospirillum brasilense* improves rice growth under salt stress by regulating the expression of key genes involved in salt stress response, abscisic acid signaling, and nutrient transport, among others. Front Agron 5:1216503. https://doi.org/10.3389/fagro.2023.1216503

Del Amor FM, Cuadra-Crespo P (2011) Plant growth-promoting bacteria as a tool to improve salinity tolerance in sweet pepper. Funct Plant Biol 39:82–90. https://doi.org/10.1071/FP11173

Di Martino C, Delfine S, Pizzuto R et al (2003) Free amino acids and glycine betaine in leaf osmoregulation of spinach responding to increasing salt stress. New Phytol 158(3):455–463. https://doi.org/10.1046/j.1469-8137.2003.00770.x

Diallinas G, Pateraki I, Sanmartin M et al (1997) Melon ascorbate oxidase: cloning of a multigene family, induction during fruit development and repression by wounding. Plant Mol Biol 34: 759–770. https://doi.org/10.1023/A:1005851527227

Dimkpa C, Weinand T, Asch F (2009) Plant–rhizobacteria interactions alleviate abiotic stress conditions. Plant Cell Environ 32:1682–1694. https://doi.org/10.1111/j.1365-3040.2009.02028.x

Djedidi S, Terasaki A, Aung HP et al (2015) Evaluation of the possibility to use the plant–microbe interaction to stimulate radioactive 137 Cs accumulation by plants in a contaminated farm field in Fukushima, Japan. J Plant Res 128:147–159. https://doi.org/10.1007/s10265-014-0678-3

Dorofeeva MM, Bonetskaya SA (2020) Comparative analysis of some classical and contemporary techniques for determining leaf blade area. Rastitelnye resursy 56(2):182–192. https://doi.org/10.31857/S0033994620020041

dos Santos Ferreira N, Coniglio A, Puente M et al (2022) Genome-based reclassification of *Azospirillum brasilense* Az39 as the type strain of *Azospirillum argentinense* sp. nov. Int J Syst Evol Microbiol 72(8). https://doi.org/10.1099/ijsem.0.005475

Edward ME, Shadbolt CA, Holm L (1956) Use of an optical planimeter for measuring leaf area. Plant Physiol 31(6):484–486. https://doi.org/10.1104/pp.31.6.484

El-Ballat EM, Elsilk SE, Ali HM et al (2023) Metal-resistant PGPR strain *Azospirillum brasilense* EMCC1454 enhances growth and chromium stress tolerance of chickpea (*Cicer arietinum* L.) by modulating redox potential, osmolytes, antioxidants, and stress-related gene expression. Plants 12(11):2110. https://doi.org/10.3390/plants12112110

Evseeva NV, Tkachenko OV, Denisova AY et al (2019) Functioning of plant-bacterial associations under osmotic stress *in vitro*. World J Microbiol Biotechnol 35(12):195. https://doi.org/10.1007/s11274-019-2778-7

Farhangi-Abriz S, Tavasolee A, Ghassemi-Golezani K et al (2020) Growth-promoting bacteria and natural regulators mitigate salt toxicity and improve rapeseed plant performance. Protoplasma 257(4):1035–1047. https://doi.org/10.1007/s00709-020-01493-1

Farmer EE, Mueller MJ (2013) ROS-mediated lipid peroxidation and RES-activated signaling. Annu Rev Plant Biol 64:429–450. https://doi.org/10.1146/annurev-arplant-050312-120132

Farooq M, Hussain M, Wahid A et al (2012) Drought stress in plants: an overview. In: Plant responses to drought stress: from morphological to molecular features. Springer, Berlin/Heidelberg, pp 1–33

Filipini LD, Pilatti FK, Meyer E et al (2021) Application of *Azospirillum* on seeds and leaves, associated with *Rhizobium* inoculation, increases growth and yield of common bean. Arch Microbiol 203:1033–1038. https://doi.org/10.1007/s00203-020-02092-7

Foster LJR, Kwan BH, Vancov T (2004) Microbial degradation of the organophosphate pesticide, Ethion. FEMS Microbiol Lett 240:49–53. https://doi.org/10.1016/j.femsle.2004.09.010

Fukami J, de la Osa C, Ollero FJ, Megías M, Hungria M (2018a) Co-inoculation of maize with *Azospirillum brasilense* and *Rhizobium tropici* as a strategy to mitigate salinity stress. Funct Plant Biol 45(3):328–339. https://doi.org/10.1071/FP17167

Fukami J, Ollero FJ, de la Osa C et al (2018b) Antioxidant activity and induction of mechanisms of resistance to stresses related to the inoculation with *Azospirillum brasilense*. Arch Microbiol 200(8):1191–1203. https://doi.org/10.1007/s00203-018-1535-x

García JE, Maroniche G, Creus C et al (2017) *In vitro* PGPR properties and osmotic tolerance of different *Azospirillum* native strains and their effects on growth of maize under drought stress. Microbiol Res 202:21–29. https://doi.org/10.1016/j.micres.2017.04.007

García JE, Ruiz M, Maroniche GA et al (2023) Inoculation with *Azospirillum argentinense* Az19 improves the yield of maize subjected to water deficit at key stages of plant development. Rev Argent Microbiol 55(3):255–261. https://doi.org/10.1016/j.ram.2023.01.002

García JE, Pagnussat LA, Amenta MB et al (2024) Maize drought protection by *Azospirillum argentinense* Az19 requires bacterial trehalose accumulation. Appl Microbiol Biotechnol 108(1):543. https://doi.org/10.1007/s00253-024-13391-0

Giri J (2011) Glycinebetaine and abiotic stress tolerance in plants. Plant Signal Behav *6*(11): 1746–1751. https://doi.org/10.4161/psb.6.11.17801

Gomez F, Salmeron V, Rodelas B et al (1998) Response of *Azospirillum brasilense* to the pesticides bromopropylate and methidathion on chemically defined media and dialysed-soil media. Ecotoxicology 7:43–47. https://doi.org/10.1023/A:1008807701523

Goyal K, Singh N, Jindal S et al (2022) Kjeldahl method. Adv Tech. Anal Chem 1:105–112

Gupta DK, Palma JM, Corpas FJ (eds) (2018) Antioxidants and antioxidant enzymes in higher plants. Springer, Berlin, pp 1–300. https://doi.org/10.1007/978-3-319-75088-0

Gureev AP, Kryukova VA, Eremina AA et al (2024) Plant-growth promoting rhizobacteria *Azospirillum* partially alleviate pesticide-induced growth retardation and oxidative stress in wheat (*Triticum aestivum* L.). Plant Growth Regul 104:503–521. https://doi.org/10.1007/s10725-024-01186-2

Gureeva MV, Gureev AP (2023) Molecular mechanisms determining the role of bacteria from the genus *Azospirillum* in plant adaptation to damaging environmental factors. Int J Mol Sci 24(11): 9122. https://doi.org/10.3390/ijms24119122

Gureeva MV, Alimova AA, Eremina AA et al (2023) Enhancing wheat seedling tolerance to cypermethrin through *Azospirillum thiophilum* pretreatment. Russ J Plant Physiol 70:184. https://doi.org/10.1134/S102144372360215X

Gureeva MV, Kirillova MS, Trandina VA et al (2025) Effect of bacteria from the genus *Azospirillum* on oxidative stress levels in wheat *Triticum aestivum* L. in the presence of copper, nickel, and lead. Microorganisms 13:334. https://doi.org/10.3390/microorganisms13020334

He H, Gao H, Chen G et al (2013) Effects of perchlorate on growth of four wetland plants and its accumulation in plant tissues. Environ Sci Pollut Res 20:7301–7308. https://doi.org/10.1007/s11356-013-1744-4

Heidari M, Golpayegani A (2012) Effects of water stress and inoculation with plant growth promoting rhizobacteria (PGPR) on antioxidant status and photosynthetic pigments in basil (*Ocimum basilicum* L.). J Saudi Soc Agric Sci 11:57–61. https://doi.org/10.1016/j.jssas.2011.09.001

Hodges DM, DeLong JM, Forney CF et al (1999) Improving the thiobarbituric acid-reactive-substances assay for estimating lipid peroxidation in plant tissues containing anthocyanin and other interfering compounds. Planta 207:604–611. https://doi.org/10.1007/s004250050524

Jakl M, Kovač I, Ćavar Zeljković S et al (2021) Triazole fungicides in soil affect the yield of fruit, green biomass, and phenolics production of *Solanum lycopersicum* L. Food Chem 351:129328. https://doi.org/10.1016/j.foodchem.2021.129328

Joe MM, Benson A, Walitang DI et al (2022) Development of ACCd producer *A. brasilense* mutant and the effect of inoculation on red pepper plants. 3 Biotech 12(10):252. https://doi.org/10.1007/s13205-022-03300-5

Ju X, Field JA, Sierra-Alvarez R et al (2007) Chemolithotrophic perchlorate reduction linked to the oxidation of elemental sulfur. Biotechnol Bioeng 96:1073–1082. https://doi.org/10.1002/bit.21197

Kalaji MH, Dabrowski P, Cetner MD et al (2017) A comparison between different chlorophyll content meters under nutrient deficiency conditions. J Plant Nutr 40:1024–1034. https://doi.org/10.1080/01904167.2016.1263323

Kamali S, Mehraban A (2020) Effects of Nitroxin and arbuscular mycorrhizal fungi on the agro-physiological traits and grain yield of sorghum (*Sorghum bicolor* L.) under drought stress conditions. PLoS One 15(12):e0243824. https://doi.org/10.1371/journal.pone.0243824

Kasim WA, Osman ME, Omar MN et al (2013) Control of drought stress in wheat using plant-growth-promoting bacteria. J Plant Growth Regul 32:122–130. https://doi.org/10.1007/s00344-012-9283-7

Khalid M, Bilal M, Hassani D et al (2017) Mitigation of salt stress in white clover (*Trifolium repens*) by *Azospirillum brasilense* and its inoculation effect. Bot Stud 58(1):5. https://doi.org/10.1186/s40529-016-0160-8

Khalifa T, Elbagory M, Omara AED (2021) Salt stress amelioration in maize plants through phosphogypsum application and bacterial inoculation. Plants 10:2024. https://doi.org/10.3390/plants10102024

Korchazhkina O, Exley C, Spencer SA (2003) Measurement by reversed-phase high-performance liquid chromatography of malondialdehyde in normal human urine following derivatisation with 2, 4-dinitrophenylhydrazine. J Chromatogr B 794(2):353–362. https://doi.org/10.1016/S1570-0232(03)00495-1

Kulikova AL, Kuznetsova NA, Burmistrova NA (2015) Change in growth and physiological parameters in soybean seedlings in response to toxic action of copper. Russ J Plant Physiol 62:455–464. https://doi.org/10.1134/S1021443715040111

Lal K, Subba Rao M (1951) A rapid method of leaf area determination. Nature 167:72. https://doi.org/10.1038/167072a0

Lichtenthaler HK, Wellburn AR (1983) Determinations of total carotenoids and chlorophylls a and b of leaf extracts in different solvents. Biochem Soc Trans 11(5):591–592. https://doi.org/10.1042/bst0110591

Liu H, Brettell LE, Qiu Z et al (2020) Microbiome-mediated stress resistance in plants. Trends Plant Sci 25(8):733–743. https://doi.org/10.1016/j.tplants.2020.03.014

Luo YH, Chen R, Wen LL et al (2015) Complete perchlorate reduction using methane as the sole electron donor and carbon source. Environ Sci Technol 49:2341–2349. https://doi.org/10.1021/es504990m

Lv PL, Shi LD, Wang Z et al (2019) Methane oxidation coupled to perchlorate reduction in a membrane biofilm batch reactor. Sci Total Environ 667:9–15. https://doi.org/10.1016/j.scitotenv.2019.02.330

Marastoni L, Pii Y, Maver M et al (2019) Role of *Azospirillum brasilense* in triggering different Fe chelate reductase enzymes in cucumber plants subjected to both nutrient deficiency and toxicity. Plant Physiol Biochem 136:118–126. https://doi.org/10.1016/j.plaphy.2019.01.013

Mariotti L, Scartazza A, Curadi M et al (2021) *Azospirillum baldaniorum* Sp245 induces physiological responses to alleviate the adverse effects of drought stress in purple basil. Plants 10(6):1141. https://doi.org/10.3390/plants10061141

Mattioli R, Costantino P, Trovato M (2009) Proline accumulation in plants: not only stress. Plant Signal Behav 4(11):1016–1018. https://doi.org/10.4161/psb.4.11.9797

McCready RM, Guggolz J, Silviera V et al (1950) Determination of starch and amylose in vegetables. Anal Chem 22(9):1156–1158. https://doi.org/10.1021/ac60045a016

Meena KK, Sorty AM, Bitla UM et al (2017) Abiotic stress responses and microbe-mediated mitigation in plants: the omics strategies. Front Plant Sci 8:172. https://doi.org/10.3389/fpls.2017.00172

Melnyk RA, Coates JD (2015) The Perchlorate reduction genomic island: mechanisms and pathways of evolution by horizontal gene transfer. BMC Genomics 16:862. https://doi.org/10.1186/s12864-015-2011-5

Mendonça R, Gning O, Di Cesaré C et al (2017) Sensitive and selective quantification of free and total malondialdehyde in plasma using UHPLC-HRMS. J Lipid Res 58:1924–1931. https://doi.org/10.1194/jlr.D076661

Misra HP, Fridovich I (1972) The role of superoxide anion in the autoxidation of epinephrine and a simple assay for superoxide dismutase. J Biol Chem 247:3170–3175

Mittler R (2006) Abiotic stress, the field environment and stress combination. Trends Plant Sci 11(1):15–19. https://doi.org/10.1016/j.tplants.2005.11.002

Mohanta S, Pradhan B, Behera ID (2023) Impact and remediation of petroleum hydrocarbon pollutants on agricultural land: a review. Geomicrobiol J 41:345–359. https://doi.org/10.1080/01490451.2023.2243925

Molina-Favero C, Creus CM, Simontacchi M et al (2008) Aerobic nitric oxide production by *Azospirillum brasilense* Sp245 and its influence on root architecture in tomato. Mol Plant-Microbe Interact 21:1001–1009. https://doi.org/10.1094/MPMI-21-7-1001

Morales M, Munné-Bosch S (2019) Malondialdehyde: facts and artifacts. Plant Physiol 180(3): 1246–1250. https://doi.org/10.1104/pp.19.00405

Muhammad F, Raza MAS, Iqbal R et al (2022) Ameliorating drought effects in wheat using an exclusive or co-applied rhizobacteria and ZnO nanoparticles. Biology 11:1564. https://doi.org/10.3390/biology11111564

Munns R, Tester M (2008) Mechanisms of salinity tolerance. Annu Rev Plant Biol 59:651. https://doi.org/10.1146/annurev.arplant.59.032607.092911

Muratova AY, Turkovskaya OV, Antonyuk LP et al (2005) Oil-oxidizing potential of associative rhizobacteria of the genus *Azospirillum*. Microbiology 74:210–215. https://doi.org/10.1007/s11021-005-0053-4

Muratova AY, Lyubun EV, Golubev SN et al (2022) Effect of copper ions on the associations of *Azospirillum* bacteria with wheat seedlings (*Triticum aestivum* L.). Vavilovskii Zhurnal Genet Selektsii 26(5):477–485. https://doi.org/10.18699/VJGB-22-58

Nanekar RD, Kokitkar SS (2024) Enzyme profiling of petroleum hydrocarbon-degrading bacteria isolated from oil contaminated soils. Indian J Sci Technol 17:2412–2420. https://doi.org/10.17485/IJST/v17i23.642

Nerenberg R, Rittmann BE, Najm I (2002) Perchlorate reduction in a hydrogen-based membrane–biofilm reactor. J Am Water Works Assoc 94:103–114. https://doi.org/10.1002/j.1551-8833.2002.tb10234.x

Nozawa-Inoue M, Scow KM, Rolston DE (2005) Reduction of perchlorate and nitrate by microbial communities in vadose soil. Appl Environ Microbiol 71:3928–3934. https://doi.org/10.1128/AEM.71.7.3928-3934.2005

Obiefuna JC, Ndubizu TOC (1979) Estimating leaf area of plantain. Sci Hortic 11(1):31–36. https://doi.org/10.1016/0304-4238(79)90051-7

Omer AM, Osman MS, Badawy AA (2022) Inoculation with *Azospirillum brasilense* and/or *Pseudomonas geniculata* reinforces flax (*Linum usitatissimum*) growth by improving physiological activities under saline soil conditions. Bot Stud 63(1):15. https://doi.org/10.1186/s40529-022-00345-w

Ouhibi C, Attia H, Rebah F et al (2014) Salt stress mitigation by seed priming with UV-C in lettuce plants: growth, antioxidant activity and phenolic compounds. Plant Physiol Biochem 83:126–133. https://doi.org/10.1016/j.plaphy.2014.07.019

Pan W, Lu Q, Xu QR et al (2019) Abscisic acid-generating bacteria can reduce Cd concentration in pakchoi grown in Cd-contaminated soil. Ecotoxicol Environ Saf 177:100–107. https://doi.org/10.1016/j.ecoenv.2019.04.010

Pandey SK, Singh H (2011) A simple, cost-effective method for leaf area estimation. J Bot 2011: 658240. https://doi.org/10.1155/2011/658240

Pandey JK, Dubey G, Gopal R (2022) Prolonged use of Insecticide dimethoate inhibits growth and photosynthetic activity of wheat seedlings: a study by laser-induced chlorophyll fluorescence spectroscopy. J Fluoresc 32:2159–2172. https://doi.org/10.1007/s10895-022-03010-4

Peng H, De-Bashan LE, Higgins BT (2021) *Azospirillum brasilense* reduces oxidative stress in the green microalgae *Chlorella sorokiniana* under different stressors. J Biotechnol 325:179–185. https://doi.org/10.1016/j.jbiotec.2020.10.029

Peralta JM, Bianucci E, Romero-Puertas MC et al (2021) Targeting redox metabolism of the maize-*Azospirillum brasilense* interaction exposed to arsenic-affected groundwater. Physiol Plant 173(3):1189–1206. https://doi.org/10.1111/ppl.13514

Pereyra MA, Zalazar CA, Barassi CA (2006) Root phospholipids in *Azospirillum*-inoculated wheat seedlings exposed to water stress. Plant Physiol Biochem 44:873–879. https://doi.org/10.1016/j.plaphy.2006.10.020

Pérez-Bueno ML, Pineda M, Barón M (2019) Phenotyping plant responses to biotic stress by chlorophyll fluorescence imaging. Front Plant Sci 10:1135. https://doi.org/10.3389/fpls.2019.01135

Pisoschi AM, Pop A, Cimpeanu C et al (2016) Antioxidant capacity determination in plants and plant-derived products: a review. Oxidative Med Cell Longev 2016(1):9130976. https://doi.org/10.1155/2016/9130976

Powell A, Wilder SL, Housh AB et al (2022) Examining effects of rhizobacteria in relieving abiotic crop stresses using carbon-11 radiotracing. Physiol Plant 174:e13675. https://doi.org/10.1111/ppl.13675

Rabie GH, Aboul-Nasr MB, Al-Humiany A (2005) Increased salinity tolerance of cowpea plants by dual inoculation of an arbuscular mycorrhizal fungus *Glomus clarum* and a nitrogen-fixer *Azospirillum brasilense*. Mycobiology 33:51–60. https://doi.org/10.4489/MYCO.2005.33.1.051

Rabiei Z, Hosseini SJ, Pirdashti H et al (2020) Physiological and biochemical traits in coriander affected by plant growth-promoting rhizobacteria under salt stress. Heliyon 6(10):e05321. https://doi.org/10.1016/j.heliyon.2020.e05321

Rahnama S, Ghehsareh Ardestani E, Ebrahimi A et al (2023) Seed priming with plant growth-promoting bacteria (PGPB) improves growth and water stress tolerance of *Secale montanum*. Heliyon 9(4):e15498. https://doi.org/10.1016/j.heliyon.2023.e15498

Rivarola V, Castro S, Mori G et al (1998) Response *of Azospirillum brasilense* Cd to sodium chloride stress. Antonie Leeuwenhoek 73:255–261. https://doi.org/10.1023/A:1001053414777

Rodríguez-Salazar J, Suarez R, Caballero-Mellado J et al (2009) Trehalose accumulation in *Azospirillum brasilense* improves drought tolerance and biomass in maize plants. FEMS Microbiol Lett 296:52–59. https://doi.org/10.1111/j.1574-6968.2009.01614.x

Romero AM, Vega D, Correa OS (2014) *Azospirillum brasilense* mitigates water stress imposed by a vascular disease by increasing xylem vessel area and stem hydraulic conductivity in tomato. Appl Soil Ecol 82:38–43. https://doi.org/10.1016/j.apsoil.2014.05.010

Saeed M, Ilyas N, Arshad M et al (2021) Development of a plant microbiome bioremediation system for crude oil contamination. J Environ Chem Eng 9:105401. https://doi.org/10.1016/j.jece.2021.105401

Saleena LM, Rangarajan S, Nair S (2002) Diversity of *Azospirillum* strains isolated from rice plants grown in saline and nonsaline sites of coastal agricultural ecosystem. Microb Ecol 44:271–277. https://doi.org/10.1007/s00248-002-2019-7

Schurer K (1971) Direct reading optical leaf area planimeter. Acta Bot Neerl 20(1):132–140. https://doi.org/10.1111/j.1438-8677.1971.tb00693.x

Shibaeva TG, Mamaev AV, Sherudilo EG (2020) Evaluation of a SPAD-502 plus chlorophyll meter to estimate chlorophyll content in leaves with interveinal chlorosis. Russ J Plant Physiol 67:690–696. https://doi.org/10.1134/S1021443720040160

Silva ER, Zoz J, Oliveira CES et al (2019) Can co-inoculation of *Bradyrhizobium* and *Azospirillum* alleviate adverse effects of drought stress on soybean (*Glycine max* L. Merrill.)? Arch Microbiol 201(3):325–335. https://doi.org/10.1007/s00203-018-01617-5

Singh MC, Singh KG, Singh JP (2018) Indirect method for measurement of leaf area and leaf area index of soilless cucumber crop. Adv Plants Agric Res 8(2):188–191. https://doi.org/10.15406/apar.2018.08.00311

Souza LC, Monteiro GGTN, Marinho RKM et al (2023) Nitrogen metabolism in maize plants submitted to drought, brassinosteroids and *Azospirillum*. Braz J Biol 83:e276264. https://doi.org/10.1590/1519-6984.276264

Takahashi WY, Galvão CW, Urrea-Valencia S et al (2022) Impact of seed-applied fungicide and insecticide on *Azospirillum brasilense* survival and wheat growth-promoting ability. Lett Appl Microbiol 74:604–612. https://doi.org/10.1111/lam.13645

Tiepo AN, Hertel MF, Rocha SS et al (2018) Enhanced drought tolerance in seedlings of Neotropical tree species inoculated with plant growth-promoting bacteria. Plant Physiol Biochem 130: 277–288. https://doi.org/10.1016/j.plaphy.2018.07.021

Tiepo AN, Constantino LV, Madeira TB et al (2020) Plant growth-promoting bacteria improve leaf antioxidant metabolism of drought-stressed Neotropical trees. Planta 251(4):83. https://doi.org/10.1007/s00425-020-03373-7

Tiepo AN, Fávaro MH, Amador TS et al (2024) Associative bacteria and arbuscular mycorrhizal fungus increase drought tolerance in maize (*Zea mays* L.) through morphoanatomical, physiological, and biochemical changes. Plants 13(12):1667. https://doi.org/10.3390/plants13121667

Tiwari S, Lata C, Singh Chauhan P et al (2017) A functional genomic perspective on drought signalling and its crosstalk with phytohormone-mediated signalling pathways in plants. Curr Genomics 18:469–482. https://doi.org/10.2174/1389202918666170605083319

Vacheron J, Renoud S, Muller D et al (2015) Alleviation of abiotic and biotic stresses in plants by *Azospirillum*. In: Cassán F, Okon Y, Creus C (eds) Handbook for *Azospirillum*. Springer, Cham. https://doi.org/10.1007/978-3-319-06542-7_19

Valente J, Gerin F, Mini A et al (2023) Symbiotic variations among wheat genotypes and detection of quantitative trait loci for molecular interaction with auxin-producing *Azospirillum* PGPR. Microorganisms 11(6):1615. https://doi.org/10.3390/microorganisms11061615

Vezza ME, Olmos Nicotra MF, Agostini E et al (2020) Biochemical and molecular characterization of arsenic response from *Azospirillum brasilense* Cd, a bacterial strain used as plant inoculant. Environ Sci Pollut Res 27:2287–2300. https://doi.org/10.1007/s11356-019-06959-1

Viñas M, Sabaté J, Espuny MJ et al (2005) Bacterial community dynamics and polycyclic aromatic hydrocarbon degradation during bioremediation of heavily creosote-contaminated soil. Appl Environ Microbiol 71:7008–7018. https://doi.org/10.1128/AEM.71.11.7008-7018.2005

Walkley A, Black I (1934) An examination of the Degtjareff method and a proposed modification of the chromic matter and a proposed modification of the chromic acid titration method. Soil Sci 34:29–38

Waller AS, Cox EE, Edwards EA (2004) Perchlorate-reducing microorganisms isolated from contaminated sites. Environ Microbiol 6:517–527. https://doi.org/10.1111/j.1462-2920.2004.00598.x

Wan D, Liu Y, Niu Z et al (2016) Perchlorate reduction by hydrogen autotrophic bacteria and microbial community analysis using high-throughput sequencing. Biodegradation 27:47–57. https://doi.org/10.1007/s10532-015-9754-1

Weber H, Chételat A, Reymond P et al (2004) Selective and powerful stress gene expression in *Arabidopsis* in response to malondialdehyde. Plant J 37:877–888. https://doi.org/10.1111/j.1365-313X.2003.02013.x

Wellburn AR (1994) The Spectral Determination of chlorophylls a and b, as well as total carotenoids, using various Solvents with spectrophotometers of different resolution. J Plant Physiol 144:307–313. https://doi.org/10.1016/S0176-1617(11)81192-2

Wu D, Zhang XJ, Liu HC et al (2021) *Azospirillum oleiclasticum* sp. nov, a nitrogen-fixing and heavy oil degrading bacterium isolated from an oil production mixture of Yumen Oilfield. Syst Appl Microbiol 44:126171. https://doi.org/10.1016/j.syapm.2020.126171

Wu Y, Li X, Yu L et al (2022) Review of soil heavy metal pollution in China: spatial distribution, primary sources, and remediation alternatives. Resour Conserv Recycl 181:106261. https://doi.org/10.1016/j.resconrec.2022.106261

Xavier WD, Castoldi G, Cavalcante TJ et al (2021) Portable chlorophyll meter for indirect evaluation of photosynthetic pigments and nitrogen content in sweet sorghum. Sugar Tech 23:560–570. https://doi.org/10.1007/s12355-020-00922-y

Xie Y, Tao G, Chen Q et al (2014) Effects of perchlorate stress on growth and physiological characteristics of rice (*Oryza sativa* L.) seedlings. Water Air Soil Pollut 225:1–8. https://doi.org/10.1007/s11270-014-2077-8

Xie T, Yang Q, Winkler MK et al (2018) Perchlorate bioreduction linked to methane oxidation in a membrane biofilm reactor: performance and microbial community structure. J Hazard Mater 357:244–252. https://doi.org/10.1016/j.jhazmat.2018.06.011

Xu X, Gao B, Jin B et al (2015) Study of microbial perchlorate reduction: considering of multiple pH, electron acceptors and donors. J Hazard Mater 285:228–235. https://doi.org/10.1016/j.jhazmat.2014.10.061

Xu Q, Pan W, Zhang R et al (2018) Inoculation with *Bacillus subtilis* and *Azospirillum brasilense* produces abscisic acid that reduces Irt1-mediated cadmium uptake of roots. J Agric Food Chem 66(20):5229–5236. https://doi.org/10.1021/acs.jafc.8b00598

Yaadesh S, Tomar GS, Kaushik R et al (2023) *Azospirillum-Bacillus* associations: synergistic effects on *in vitro* PGP traits and growth of pearl millet at early seedling stage under limited moisture conditions. 3 Biotech 13(3):90. https://doi.org/10.1007/s13205-023-03503-4

Young CC, Hupfer H, Siering C et al (2008) *Azospirillum rugosum* sp. nov., isolated from oil-contaminated soil. Int J Syst Evol Microbiol 58:959–963. https://doi.org/10.1099/ijs.0.65065-0

Yu Z, Lu T, Qian H (2023) Pesticide interference and additional effects on plant microbiomes. Sci Total Environ 888:164149. https://doi.org/10.1016/j.scitotenv.2023.164149

Zaheer MS, Ali HH, Erinle KO et al (2022) Inoculation of *Azospirillum brasilense* and exogenous application of trans-zeatin riboside alleviates arsenic induced physiological damages in wheat (*Triticum aestivum*). Environ Sci Pollut Res 29:33909–33919

Zaheer MS, Aijaz N, Hameed A et al (2024) Cultivating resilience in wheat: mitigating arsenic toxicity with seaweed extract and *Azospirillum brasilense*. Front Microbiol 15:1441719. https://doi.org/10.3389/fmicb.2024.1441719

Zahra ST, Tariq M, Abdullah M et al (2024) Salt-tolerant plant growth-promoting bacteria (ST-PGPB): an effective strategy for sustainable food production. Curr Microbiol 81(10):304. https://doi.org/10.1007/s00284-024-03830-6

Zhu JK, Hasegawa PM, Bressan RA et al (1997) Molecular aspects of osmotic stress in plants. Crit Rev Plant Sci 16:253–277. https://doi.org/10.1080/07352689709701950

Chapter 15
Interaction of *Azospirillum* with Microalgae

Luz de-Bashan, Oskar Palacios, Juan Pablo Hernandez, and Lina María González-González

Abstract Beyond its established role in agriculture, *Azospirillum* can enhance the growth and metabolism of photosynthetic organisms such as microalgae. The underlying hypothesis is that *Azospirillum,* as a non-specific plant growth-promoting bacterium (PGPB), influences green microalgae in a manner similar to higher plants. These effects occur at multiple physiological and metabolic levels, suggesting new applications for *A. brasilense*. Physiological studies indicate that indole-3-acetic acid (IAA) produced by A. *brasilense* promotes population growth in several Chlorophyceae species. Metabolic analyses reveal that co-cultivation alters the activity of nitrogen cycle enzymes, including glutamine synthetase and glutamate dehydrogenase, in *Chlorella* spp, leading to increased nitrogen uptake and intracellular nitrogen accumulation. IAA also enhances ADP-glucose pyrophosphorylase activity, leading to increased starch accumulation, and stimulates acetyl-CoA carboxylase, which in turn increases fatty acid synthesis and total lipid content in the microalgae. A direct exchange of nitrogen and carbon between *Azospirillum brasilense* Cd and *Chlorella sorokiniana* was demonstrated using nanoscale secondary ion mass spectrometry (nanoSIMS). In addition, studies have shown that volatile compounds produced by *A. brasilense* positively affect the growth and

Dedicated to the memory of Prof. Yoav Bashan.

L. de-Bashan (✉)
The Bashan Institute of Science, Auburn, AL, USA

Department of Entomology and Plant Pathology, Auburn University, Auburn, AL, USA
e-mail: luz@bashanfoundation.org

O. Palacios
The Bashan Institute of Science, Auburn, AL, USA

Facultad de Ciencias Químicas, Universidad Autónoma de Chihuahua, Chihuahua, México

J. P. Hernandez
Universidad Rosario Castellanos, Ciudad de México, México

L. M. González-González
The Bashan Institute of Science, Auburn, AL, USA

F. D. Cassán et al. (eds.), *The Definitive Handbook of Azospirillum*,
https://doi.org/10.1007/978-3-032-09222-9_15

metabolism of *Chlorella* and that riboflavin and lumichrome produced by *A. brasilense* significantly influence the photosynthetic and auxiliary pigments in microalgae. The combined use of *Azospirillum* and microalgae has proven effective for wastewater treatment and for restoring desert-degraded soils after the application of wastewater-derived debris. The following section provides a summary of these findings.

Keywords Microalgae · Nitrogen uptake · Lipid synthesis · Pigments · Wastewater treatment

15.1 Introduction

In 2000, Yoav Bashan and his team began an intriguing exploration to expand the use of *Azospirillum* from agricultural plants to aquatic green microalgae (González and Bashan 2000). They specifically created a synthetic association between the chlorophyte *Chlorella* spp. and *Azospirillum brasilense*, proposing it as a straightforward, quantitative experimental model for studying beneficial interactions between plants and bacteria (de-Bashan and Bashan 2008).

The logic behind this model was to simplify the initial screening of growth promotion traits in *Azospirillum*, creating a simpler biological system. Microalgae have a small genome (approx. 40 Mb compared to 125 Mb of *Arabidopsis,* the primary model in plant biology) and can serve as the plant partner in the interaction. The most basic definition of a green plant is that it contains chlorophylls *a* and *b*, starch as a storage material inside the chloroplast, and a cell wall made of cellulose, all characteristics shared with green algae. Moreover, higher plants and green algae are systematically part of the same group (Chlorobionta), and there is 70–98% genetic similarity between land plants and green algae (Devereux et al. 1990). *Chlorella* spp. are simple, nonmotile, unicellular, aquatic microalgae that have been used in photosynthesis and respiration studies. Much of the knowledge on synthesizing carbohydrates in microalgae has been derived from studies on this genus. Furthermore, since the late 1940s, attention has been drawn to the potential of mass cultivation of this microalga to produce high-value, low-volume compounds, such as pigments for food industries, including the health-food market, and application in wastewater treatment (de-Bashan and Bashan 2008). Thus, choosing this genus to test the underlying hypothesis that *A. brasilense*, a non-specific plant growth-promoting bacterium, would influence green microalgae similarly to higher plants seems straightforward. As it turns out, the effects were significant at all levels, paving the way for new applications of *A. brasilense* (Cassan et al. 2020).

Physiological studies have shown that, similar to higher plants, mechanisms such as the production of indole-3-acetic acid and vitamins play crucial roles in influencing microalgae. *Azospirillum* has been found to enhance the growth of various microalgae species and influence the activity of enzymes involved in metabolic pathways for nitrogen and phosphorus assimilation, as well as starch and fatty acid synthesis. In most cases, co-culturing with different strains of *Azospirillum* leads to increased production of pigments, lipids, and carbohydrates in microalgae (González-González and de-Bashan 2021; Cassan et al. 2020). Also, the mitigating effect of *Azospirillum* on environmental stresses such as salinity in microalgae cultures has been explored. Recent studies have revealed that *Azospirillum* also benefits from its relationship with microalgae. Exudates from microalgae can promote the growth and metabolism of *Azospirillum* and even modulate its gene expression. A direct exchange of nitrogen and carbon between *Azospirillum brasilense* and *Chlorella sorokiniana* has been demonstrated, along with the exchange of metabolites exuded into the medium by both microorganisms (de-Bashan et al. 2016; Palacios et al. 2023a). This combination has been successfully applied in wastewater treatment, the recovery of degraded desert soils, and as a biofertilizer. This chapter summarizes 25 years of contributions to this field.

15.2 Establishment of the Association Between *Azospirillum* and Microalgae: Attachment and Remote Interaction

A key aspect of studying the effects of *Azospirillum* is its common structural and phenotypic interaction with the roots of higher plants through fibrils and sheath material. However, the mechanisms underlying such physical interactions with organisms like microalgae were previously unknown. Therefore, the initial question that arose during the development of this synthetic association was whether a similar physical attachment could be formed and maintained between the bacterium and the surface of rootless, single-celled microalgae.

The initial visualization of this physical interaction was documented by de-Bashan et al. (2011), where analysis using fluorescent in situ hybridization (FISH) and scanning electron microscopy (SEM) revealed an increase in the populations of both partners and the formation of strong structures that kept the bacteria and microalgal cells firmly attached. Further isotopic imaging with nanoscale secondary ion mass spectrometry (NanoSIMS) demonstrated the direct transfer of metabolites between the two organisms, including organic carbon from the microalgae to the bacteria and nitrogen compounds from the bacteria to the microalgae (de-Bashan et al. 2016).

To confirm the effectiveness of the attachment of *A. brasilense* and its impact on the growth, aggregation capacity, and metabolite production of the microalga *Chlorella sorokiniana*, a mutant of *A. brasilense* Az39, defective in the type VI secretion system (T6SS), was constructed and characterized. This work specifically focused on aggregation and attachment processes during the microalgae-bacteria interaction (Cassan et al. 2021). The mutant exhibited a reduced capacity for cell aggregation and biofilm formation. The mutant and its parental wild-type strain were cultured independently or with *C. sorokiniana*. Interestingly, the results showed that the microalgae populations were significantly enhanced through interaction with both strains of *A. brasilense*. Moreover, both strains significantly improved the total lipid and pigment production in the microalgae; however, the effect was greater with the wild-type strain, indicating that a reduction in attachment did not negate the positive effects of *Azospirillum,* although these effects were significantly lower in the absence of physical contact.

Based on these findings, Amavizca et al. (2017) examined the remote effects—occurring without physical contact—of *Azospirillum brasilense* Cd on the growth of *C. sorokiniana*. In a system where a septum physically separated the cultures, *Azospirillum* enhanced the growth of the microalga up to sixfold and increased its cell volume by about threefold. Additionally, *A. brasilense* remotely induced increases in the levels of total lipids, carbohydrates, and chlorophyll in the microalgae, indicating significant alterations in its physiology.

All these results confirmed that the bacteria can positively influence green microalgae's growth and metabolic parameters, even without physical attachment.

15.3 Effect of *Azospirillum* on Microalgae Growth Under Different Nutritional and Stress Conditions

15.3.1 Nutrients

The positive effects of *Azospirillum* strains on the growth of microalgae have been observed under various growth conditions, including stress situations. Research on the interactions between *Azospirillum* and microalgae has primarily been conducted in growth media designed to replicate the nutrient conditions found in domestic or aquaculture wastewater. This is significant because one of the main applications of microalgae is to treat wastewater by removing inorganic nutrients such as nitrogen (in the form of ammonium or nitrate) and phosphorus (as phosphate) (González and Bashan 2000; Ruiz-Güereca and Sánchez-Saavedra 2016). The concentration of nutrients, particularly carbon and nitrogen, plays a crucial role in establishing the interaction between *Azospirillum* and microalgae. When carbon or nitrogen concentrations are low, a rapid physical interaction occurs between *Azospirillum brasilense* and *Chlorella* cells. This results in a higher microalgal cell density during the early stages of incubation compared to when both microorganisms are grown in a nutrient-rich medium, such as nutrient broth (Palacios et al. 2019).

15.3.1.1 Nitrogen

Nitrogen is a significant nutritional stressor for microalgae cultures, and *Azospirillum* helps to mitigate this stress. Research has shown that when nitrogen is present in the culture medium, *A. brasilense* can enhance growth and biomass productivity. However, under conditions of nitrogen absence or low nitrogen concentrations, the growth-promoting effect of *Azospirillum* becomes even more pronounced. In one study, during the first 72 h of incubation, *Chlorella sorokiniana* showed a 36% increase in biomass when interacting with *A. brasilense* in nitrogen-sufficient conditions, compared to when the microalgae were grown independently. Interestingly, this increase in biomass productivity was not observed at later incubation times. Conversely, under nitrogen-depleted conditions, the biomass of *Chlorella* continued to increase even after 120 h of incubation with *A. brasilense* (Peng et al. 2021). Ramos-Ibarra et al. (2019) reported similar findings, observing enhanced growth in *Chlorella vulgaris* and *Tetradesmus obliquus* when co-cultured with *A. brasilense* under nitrogen-limited conditions. Microalgae interacting with *A. brasilense* exhibited increased cell density, growth rate, and biomass productivity. Furthermore, the interaction between *A. brasilense* and *Scenedesmus* sp. under nitrogen-limiting conditions led to greater biomass accumulation and larger microalgal cell sizes (Contreras-Angulo et al. 2019).

15.3.1.2 Carbon

Carbon is a key nutrient necessary for the growth of microalgae. Under autotrophic conditions, microalgae use inorganic carbon (CO_2) for their development. However, CO_2 occurs in very low concentrations in nature, at about 0.004% in the atmosphere. To address this limitation, microalgae have evolved a carbon concentration mechanism (CCM) that helps deliver concentrated CO_2 to the CO_2-fixing enzyme RUBISCO (Contreras et al. 2024).

While using microalgal cultures to capture CO_2 during biotechnological processes, such as biogas upgrading (González-González and de-Bashan 2021), high CO_2 concentrations can create stressful conditions for microalgae. These conditions lead to a persistent acidic pH due to the formation of carbonic acid when CO_2 dissolves in the microalgal culture (Contreras et al. 2024). Research has shown that *Azospirillum* helps alleviate this stress in *Chlorella*, *Scenedesmus*, and *Chlamydomonas* under elevated CO_2 levels (15%, 25%, and 35%). In all instances, *A. brasilense* enhanced microalgal growth, resulting in higher biomass productivity (Choix et al. 2018). Furthermore, when exposed to synthetic biogas (comprising 75% methane and 25% CO_2), the interaction between *Chlorella, Scenedesmus*, and *Arthrospira* with *A. brasilense* promotes increased growth and biomass productivity in microalgae (Barbosa-Nuñez et al. 2022; Choix et al. 2024).

In addition to autotrophic growth, some microalgae can utilize organic carbon as a carbon and energy source under heterotrophic conditions (Perez-Garcia et al. 2011). While the positive impact of *Azospirillum* on the cell density of *C. vulgaris* was not evident under heterotrophic conditions (Choix et al. 2012b), a beneficial effect has been observed in biomass productivity due to increased metabolite accumulation in *Chlorella* sp. and *Scenedesmus* sp. (Choix et al. 2023).

15.3.2 pH Stress

pH is a critical factor for growth because it influences enzymatic activities. Studies have shown that a neutral pH of 7.0 enhances the interaction between *Azospirillum* and microalgae, resulting in a moderate improvement in microalgae growth. However, under acidic (pH 6.0) or alkaline conditions (pH 8.0), the growth of *Chlorella* spp. is significantly enhanced by *A. brasilense* (de-Bashan et al. 2005; Palacios et al. 2016a).

15.3.3 Salinity Stress

Salinity is a stress condition that is commonly associated with increased lipid content in microalgae cultures. As a result, it has been proposed as a viable alternative for producing biofuels from microalgae biomass. However, salinity stress can reduce microalgae growth, which negatively impacts biomass productivity (Church et al. 2017). Research has shown that the presence of *A. brasilense* affects *Chlorella* and *Scenedesmus*, grown under high salinity, by mitigating saline stress through osmoregulation. This process is accompanied by an increase in carbohydrate concentration and antioxidant activity (Pagnussat et al. 2020; Palacios et al. 2021). Furthermore, the recently identified species *Azospirillum baldanorium* Sp245 (formerly known as *Azospirillum brasilense* Sp245) has demonstrated a positive effect on the growth and biomass production of *Scenedesmus obliquus* under saline stress. It helps mitigate salt-induced oxidative stress and cell damage by reducing the levels of reactive oxygen species (Pagnussat et al. 2023).

15.3.4 Stress for Emergent Contaminants

Emerging contaminants, such as nanoparticles, have been studied in relation to the interaction between *Azospirillum* and microalgae. These nanoparticles can enter the natural environment through various human activities, potentially impacting the growth of microorganisms like microalgae in natural water bodies (Markus et al. 2016). Given *Azospirillum*'s ability to mitigate diverse stresses, it is reasonable to hypothesize that the interaction between microalgae and bacteria could have a positive effect on these microorganisms, even in the presence of nanoparticles. For instance, cultures of *Chlorella* and *Scenedesmus* grown alongside *Azospirillum* in the presence of metallic nanoparticles such as ZnO and CuO demonstrated increased microalgal cell density and biomass productivity (Alonso et al. 2024; Palacios et al. 2023b).

While the positive impact of *Azospirillum* on microalgae growth—such as cell density, cell size, and biomass productivity—has been established under various growth conditions, this effect is particularly pronounced in stressful environments, including those characterized by nutrient deficiency, pH variations, salinity, and the presence of emerging contaminants.

15.4 Effect of *Azospirillum* on the Physiology and Metabolism of Microalgae

The significant increase in microalgae growth observed during co-cultivation with *Azospirillum* spp. has led not only to notable enhancements in cell size, biomass, growth rate, and productivity but also to increased metabolic activity and metabolite production (de-Bashan et al. 2002a; González and Bashan 2000) (Fig. 15.1).

Initial metabolic studies indicated that *Azospirillum* influences the activities of key enzymes involved in ammonium assimilation in microalgae. Specifically, under co-culture conditions, the activities of glutamine synthetase (GS) and glutamate dehydrogenase (GDH) in *Chlorella* spp. were significantly higher than in those grown alone, leading to enhanced ammonium uptake and higher intracellular nitrogen accumulation (de-Bashan et al. 2008a; Meza et al. 2015a). Additionally, *C. vulgaris* showed increased intracellular phosphate accumulation as polyphosphate when interacting with *A. brasilense* (Meza et al. 2015b).

The impact of *Azospirillum* on metabolic pathways also results in higher intracellular levels of lipids, carbohydrates, pigments, and proteins. For instance,

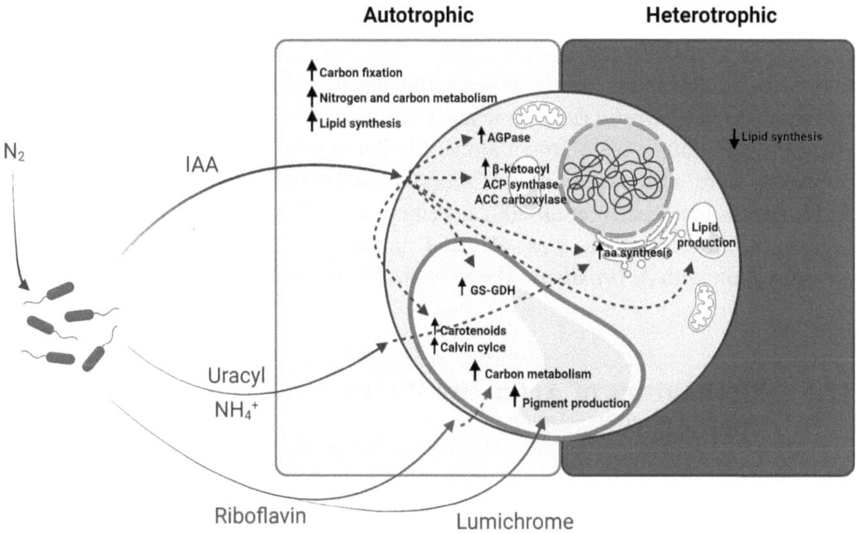

Fig. 15.1 Model of the effect of *A. brasilense* on metabolism of *Chlorella* spp, under autotrophic and heterotrophic conditions. Indole-3-acetic acid (IAA), glutamine synthetase (GS), glutamine dehydrogenase (GDH), amino acid (aa), ADP-glucose pyrophosphorilase (AGPase)

de-Bashan et al. (2002a) found a substantial increase in lipid content in three *Chlorella* strains co-cultivated with *A. brasilense*. Similarly, Peng et al. (2020) reported increased crude lipid content in *C. sorokiniana* and *Auxenochlorella protothecoides* when growing with *A. brasilense* for 72 and 120 h. Furthermore, both the content and variety of fatty acids increased in *Chlorella* strains growing alongside *A. brasilense* (Leyva et al. 2014). In subsequent studies, Leyva et al. (2015) found that the activity of acetyl-CoA carboxylase (ACCase), a pivotal enzyme in de novo fatty acid biosynthesis, is enhanced by the co-cultivation of *Chlorella vulgaris* with *A. brasilense* under autotrophic and heterotrophic conditions. However, despite the higher lipid levels observed in co-cultures (up to a fivefold increase), no direct correlation was established between this increase and the ACCase activity. Lastly, *A. baldaniorum* Sp245 improves biomass and lipid productivity in *S. obliquus* under salt stress (Pagnussat et al. 2023).

A. brasilense also positively affected carbohydrate production in *C. vulgaris* and *C. sorokiniana* (Choix et al. 2012a, 2012b). The authors reported increases of up to 72% and 90% in total carbohydrates for *C. vulgaris* under autotrophic and heterotrophic conditions, respectively, while *C. sorokiniana* exhibited a carbohydrate production increase of around 55% under autotrophic conditions and 21% under heterotrophic conditions. Later, Choix et al. (2014) measured the impact of *A. brasilense* on ADP-glucose pyrophosphorylase (AGPase, a key enzyme in starch synthesis activity and starch synthesis) and starch production in *C. vulgaris* under heterotrophic conditions using D-glucose or Na-acetate as carbon sources. The AGPase activity during the first 72 h of incubation was higher in *C. vulgaris* co-cultured with *A. brasilense*, coinciding with increased starch accumulation and carbon uptake by the microalgae.

Co-cultivating microalgae with *Azospirillum* can significantly boost pigment production. González and Bashan (2000) observed that chlorophyll b, ß-carotene, lutein, and violaxanthin levels increased by up to 82%, 76%, 103%, and 118%, respectively, in *Chlorella vulgaris* when co-cultured with *A. brasilense*, compared to when the microalgae were grown alone. Similar significant increases in these four pigments were also reported in *Chlorella sorokiniana* when interacting with *A. brasilense* (de-Bashan et al. 2002a; Peng et al. 2020). Additionally, the concentration of chlorophyll a was elevated in *A. protothecoides* during a 72-h co-cultivation with *A. brasilense* (Peng et al. 2020).

15.5 Mechanisms of Action by Which *Azospirillum* Affects Microalgae

As already presented, the positive effect of *Azospirillum* on the growth and metabolism of microalgae has been demonstrated under different cultivation conditions. Now, the quest is to define which mechanisms of growth promotion are involved in the interaction (Fig. 15.1).

15.5.1 Indole-3-Acetic Acid (IAA)

As one of the main phytohormones produced by *Azospirillum*, it is reasonable to hypothesize that IAA plays a key role in the interaction between the bacteria and microalgae. Indeed, a positive effect of *Azospirillum* on the growth of *C. vulgaris* was demonstrated using the wild-type strains *A. brasilense* sp245 and *A. brasilense* Sp6, along with their IAA-deficient mutants, *A. brasilense* FAJ0009 and *A. brasilense* SpM7819. The findings indicated that the bacteria positively influenced *C. vulgaris* growth only when the wild-type strains were utilized (de-Bashan et al. 2008b).

During the interaction, *Azospirillum* actively produces IAA due to an exchange of molecules, in which tryptophan and thiamine are secreted by the microalgae cells. *Azospirillum* utilizes these compounds to synthesize IAA, which, in turn, promotes higher growth rates in the microalgae (Palacios et al. 2016b). Although the specific mechanism by which IAA enhances microalgal growth remains unclear, its positive influence on enzymatic activities and gene expression in microalgae has been documented. For instance, IAA stimulates nitrogen and carbon metabolism as well as pigment production in microalgae. Studies involving the wild-type and IAA-deficient mutants of *A. brasilense* previously described have shown that IAA production stimulates the activities of key enzymes such as glutamine synthetase (GS), glutamate dehydrogenase (GDH), and adenosine diphosphate-glucose pyrophosphorylase (AGPase) (Meza et al. 2015a, 2015b; Palacios et al. 2016c) (Fig. 15.1).

In the case of AGPase, *Azospirillum* promotes increased glucose uptake and starch accumulation under heterotrophic conditions. However, in autotrophic environments, the IAA produced by *A. brasilense* positively affects the growth of *C. sorokiniana* while leading to a consistent reduction in energy storage compounds such as neutral lipids and starch (Peng et al. 2020).

Barbosa-Nuñez et al. (2022) reported the positive effect of IAA produced by *Azospirillum* on the growth of *Chlorella* and its carbon dioxide fixation during biogas upgrading through the microalga-bacterium interaction.

The expression of genes related to lipid synthesis, such as β-ketoacyl ACP synthase I, is stimulated in *C. vulgaris* when IAA is introduced (Jusoh et al. 2015). Additionally, genes associated with carotenoid production are overexpressed in *Chlorella* in the presence of IAA, establishing a connection between the phytohormone and the mitigation of oxidative stress in microalgae (Alsenani et al. 2019). Recent research has also shown that IAA produced by bacteria can induce biofilm formation between microalgal and bacterial cells by stimulating the expression of genes related to the Calvin cycle, thereby promoting carbon assimilation, amino acid biosynthesis, and protein release (Xie et al. 2024; Chen et al. 2025).

15.5.2 Nitrogen Fixation

Although nitrogen is one of the most abundant elements in the environment, it primarily exists in forms that are unavailable to most organisms. Nitrogen fixation is a key mechanism by which *Azospirillum* promotes plant growth (Fukami et al. 2018), and it is also considered one of the primary mechanisms for promoting the growth of microalgae (Palacios et al. 2022). However, *A. brasilense* can only fix a small fraction of nitrogen, which means it is not one of its main traits for promoting growth (Bashan and de-Bashan 2010).

Despite this limitation, the growth of *Scenedesmus* under nitrogen-depleted conditions has been attributed to the nitrogen fixation activity of *Azospirillum*, which releases ammonium into the medium (Contreras-Angulo et al. 2019). Additionally, it has been suggested that *Azospirillum* may enable microalgae to acquire nitrogen through a mechanism other than atmospheric nitrogen fixation. Research using nuclear magnetic resonance (NMR) has demonstrated that *A. brasilense* produces and excretes nitrogen sources such as uracil. It was found that *C. sorokiniana* can utilize this compound as its sole nitrogen source (Palacios et al. 2023a).

15.5.3 Vitamins

Vitamins are essential micronutrients for all organisms. Some vitamins, such as cobalamin, are produced exclusively by prokaryotes, meaning that eukaryotic organisms rely directly on them to obtain this vitamin (Palacios et al. 2014). The production and exchange of vitamins during interactions among microorganisms, even those that are not auxotrophic for a particular vitamin, provide beneficial effects. This exchange facilitates the availability of essential nutrients, thereby reducing the energy expended by microorganisms on de novo synthesis.

While research on the role of vitamins in microalga-bacteria interactions is limited, studies have shown that interactions between *Azospirillum* and microalgae result in the production and extrusion of vitamins like thiamine and riboflavin (Palacios et al. 2016a; López et al. 2019). Thiamine serves an important function as a cofactor for IAA production in *A. brasilense*, and its production during the interaction between *Azospirillum* and *Chlorella* is crucial for maintaining their symbiotic relationship (Palacios et al. 2016b).

Riboflavin, another vital cofactor, plays a significant role in carbohydrate metabolism and has been shown to promote the growth of microalgae under saline stress conditions (López et al. 2019). Additionally, lumichrome, a degradation product of riboflavin, is produced during the interaction between *Azospirillum* and microalgae. Lumichrome has been found to mitigate stress in microalgae growing under saline conditions and leads to increased pigment production by the microalgae (Palacios et al. 2021; Peng et al. 2021).

15.5.4 Volatiles

As discussed earlier, the positive effects of *Azospirillum* on the growth and metabolism of microalgae have been observed even without direct physical contact between the two microorganisms. In this scenario, *Azospirillum* produces volatile compounds such as CO_2, 2,3-butanediol, and acetoin, which positively influence microalgae growth and metabolism (Amavizca et al. 2017). The importance of these molecules, which can diffuse through the media and generate positive effects at a distance on microalgae, must be studied to elucidate the mechanism and their importance in *Azospirillum*-microalga interactions.

15.5.5 Other Mechanisms

Research on the growth-promoting traits of microorganisms that benefit microalgae has identified several areas needing further investigation (Palacios et al. 2022). Specifically, the roles of siderophore production, microbial competition, and the breakdown of complex carbon sources to facilitate the assimilation of simple carbon sources by microalgae require more study, particularly concerning the interactions with *Azospirillum.* Gaining a comprehensive understanding of the complexities in *Azospirillum*-microalga interactions could lead to various biotechnological applications.

15.6 Changes in Gene Expression in *Azospirillum* Interacting with Microalgae

Azospirillum-microalga interactions were studied mainly by analyzing both microorganisms' growth behavior and metabolite production. Gene expression analysis has been used to analyze the interaction between *Azospirillum* and plants, elucidating the genetic determinants of this bacterium during its interaction and giving a real view of the importance of the bacterium's different growth promotion traits (Drogue et al. 2014). In the *Azospirillum*-microalga interaction, specific genes such as *ipdC*, which encodes the key enzyme for IAA production, have been analyzed under different growth conditions (Palacios et al. 2016b; Barbosa-Núñez et al. 2022). Nonetheless, a global analysis of gene expression of *Azospirillum* interacting with microalgae has not been done yet. A recent microarray analysis of *A. brasilense* demonstrated its ability to adapt to the growth conditions used in microalgae-based biogas upgrading processes, supporting its potential use to enhance microalgal growth and metabolism during methane purification (Garciglia-Mercado et al. 2024). As more mechanisms by which *Azospirillum* promotes microalgal growth are elucidated, analyzing the expression of genes associated with each mechanism will be key to understanding the specific growth conditions that activate them and whether all mechanisms are involved in interactions with different microalgal genera.

15.7 Application of the *Azospirillum*-Microalgae Association

15.7.1 Use of the Azospirillum*-Microalgae Association for Wastewater Treatment*

The growing demand for efficient wastewater treatment is driven by population growth, urbanization, and the need to minimize environmental pollution. Biological systems have emerged as sustainable biotechnological alternatives for nutrient removal due to their ability to assimilate pollutants while simultaneously producing biomass. A promising approach involves using *A. brasilense* in combination with microalgae from the genus *Chlorella*. This combination leverages the metabolic capabilities of both microorganisms: the microalgae absorb nutrients, while *Azospirillum* promotes plant growth. Together, they enable effective nutrient removal from wastewater and produce biomass that can be utilized for various purposes.

Many wastewaters often have low nitrogen levels, which limits the growth of microalgae. However, *Azospirillum* can effectively address this limitation by providing a nitrogen source through biological fixation, thereby promoting microalgae growth and aiding nutrient uptake (Hernandez et al. 2009). Additionally, co-cultivating microalgae with *Azospirillum* has been shown to enhance the removal of nutrients, such as nitrogen and phosphorus, from wastewater (de-Bashan et al. 2002b). Research indicates that microalgae can assimilate these nutrients during their growth, thus reducing the overall nutrient load in treated effluents (de-Bashan et al. 2004; Geremia et al. 2021; Gómez-Serrano et al. 2015). The combination of *Azospirillum*'s nitrogen fixation capabilities and the nutrient uptake efficiency of microalgae creates a synergistic effect that improves the overall efficiency of wastewater treatment systems. This is particularly important considering regulations requiring the reduction of nutrient discharges from wastewater treatment plants (Gómez-Serrano et al. 2015).

As a technological way to enhance the efficiency in wastewater treatment using *Azospirillum* and microalgae, a combined immobilization in alginate beads was developed (de-Bashan et al. 2004) (Fig. 15.2). The co-immobilization of these organisms provides a barrier that protects them from native microbial competition in wastewater. This stability is important to maintain long-term functionality and efficiency in large-scale systems (Covarrubias et al. 2012). A detailed description of the co-immobilized system is provided in de-Bashan et al. (2015).

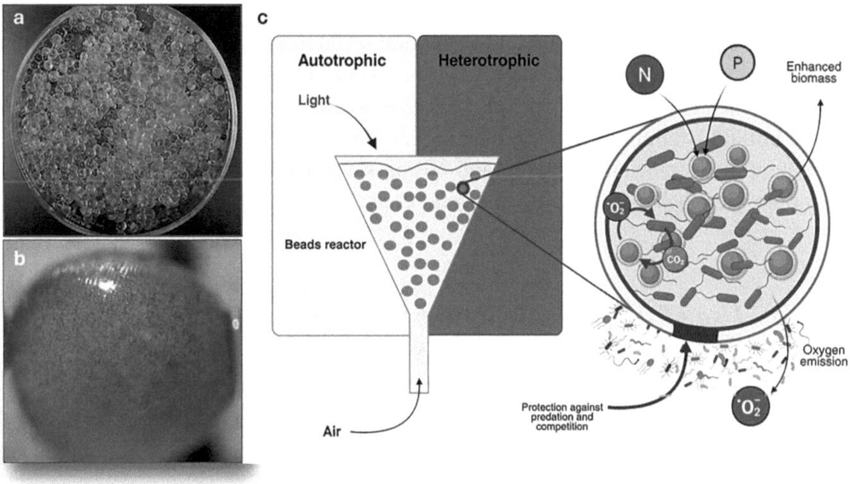

Fig. 15.2 (**a, b**). Alginate beads containing *C. sorokiniana* co-immobilized with *A. brasilense*. (**c**). A conceptual model of the co-immobilized system used for wastewater treatment

The efficiency of nutrient removal by the *Azospirillum-Chlorella* immobilized system is affected by the cultivation mode. As discussed earlier in this chapter, one of the main characteristics of microalgae is their ability to grow under different conditions—autotrophic, heterotrophic, and mixotrophic. The immobilization of *A. brasilense* together with the microalga *C. vulgaris* leads to significant variations in both growth and the removal of ammonium and phosphorus, depending on the growth regime. Ammonium removal under autotrophic and mixotrophic conditions is significantly more efficient in co-immobilized microalgae than in microalgae alone, while co-immobilized microalgae showed higher efficiency in removing phosphorus under all growth conditions (Pérez-García et al. 2010). These results highlight the ability of *A. brasilense* to stimulate microalgal metabolic activity and support its use in settings where large ponds cannot be constructed or artificial light sources installed, thereby making wastewater treatment more economical. Another way to improve nitrogen and phosphorus removal is the application of starvation periods, which induce physiological adaptations in microalgae that prioritize nutrient uptake. When the microalgae *C. vulgaris* and *C. sorokiniana* were each co-immobilized with *A. brasilense* and subjected to starvation periods, they exhibited greater phosphorus removal from both synthetic and domestic wastewater (Hernández et al. 2006).

Despite its broad capabilities, the *Azospirillum*-microalgae system faces challenges related to the durability of the alginate matrix and the potential biological deterioration over time. Decomposition of alginate beads during tertiary wastewater treatment can compromise system efficiency and require further optimization to ensure structural integrity (Cruz et al. 2013). In any case, the *Azospirillum*-microalgae partnership offers a sustainable, efficient, and versatile approach to wastewater treatment. By combining the metabolic capabilities of both organisms, this system addresses environmental pollution and contributes to the generation of valuable bioproducts, making it a fundamental technology for the circular economy. Another important consideration is the economic viability of integrating *Azospirillum*-microalgae systems into existing wastewater treatment infrastructures. While initial installation costs may be higher than conventional treatment methods, the potential for nutrient recovery and biomass valorization may offset these costs over time (de-Bashan et al. 2015; Ferreira et al. 2019; Mahdy et al. 2015).

15.7.2 Use of the **Azospirillum**-*Microalgae Association for Soil Restoration and Sustainable Agriculture*

Soil degradation is an escalating global issue, leading to reduced agricultural productivity, loss of biodiversity, and increased dependence on chemical inputs. Sustainable agriculture and soil restoration strategies are essential for maintaining ecosystem stability and ensuring long-term food security. One promising approach involves the microbial consortium of *A. brasilense* and microalgae, which has demonstrated significant potential in enhancing plant growth and soil health. Given their ability to promote soil fertility and plant development, the synergistic interaction between these microorganisms may further amplify their effectiveness as agents of soil restoration. This association not only improves soil quality but also enhances plant growth by supplying bioavailable nutrients and biostimulatory compounds (González-González and Bashan 2023).

Microalgae contribute to soil health by improving structure and aggregate stability through exopolysaccharide production, facilitating oxygen supply to deeper layers, and aiding in nutrient solubilization and mineralization. They are a valuable source of organic matter and nutrients (Mutum et al. 2022; Ramakrishnan et al. 2023). Their fertilizer potential is well established, as they produce plant growth-promoting compounds such as phytohormones, amino acids, vitamins, and polysaccharides, which act as biostimulants (Ronga et al. 2019) (Fig. 15.3). Additionally, microalgal extracts function as slow-release organic fertilizers, improving soil properties while optimizing nutrient availability and resource efficiency (Carrillo et al. 2020).

Beyond agriculture, *Azospirillum* has been applied to environmental restoration, particularly in rehabilitating degraded desert soils (Bashan et al. 2012; Moreno et al. 2017; González et al. 2018, see Chap. 23). Due to its ability to enhance plant resilience under adverse conditions, *Azospirillum*—along with other plant growth-promoting bacteria—is especially valuable in resource-limited environments, making it an effective tool for revegetating eroded areas.

The combined use of microalgae and bacteria in synthetic consortia for soil restoration remains largely unexplored despite growing evidence that their interactions—such as substrate exchange, signaling molecule communication, and horizontal gene transfer—enhance their adaptability to environmental challenges

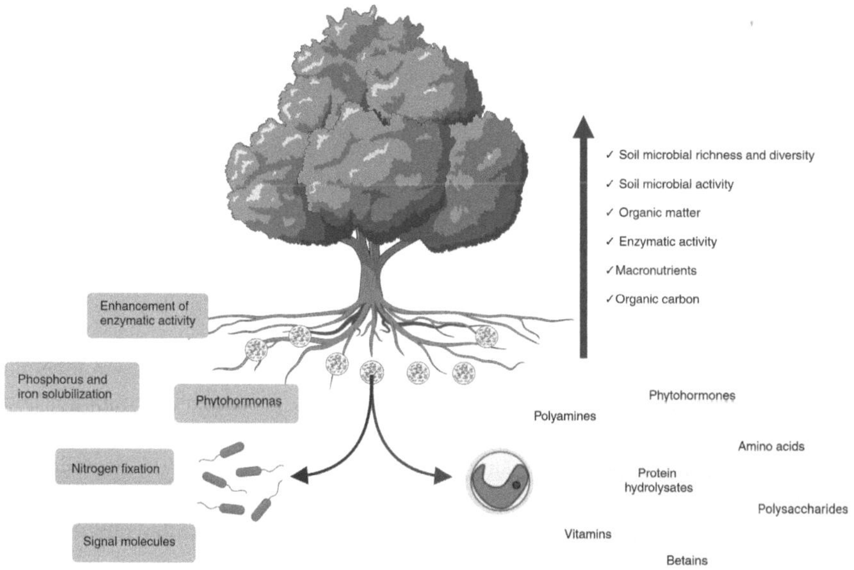

Fig. 15.3 Schematic diagram of the microalgae-bacteria interaction mechanisms to enhance plant growth and restore degraded soils

and improve soil conditions (González-González and de-Bashan 2021; González-González and de-Bashan 2023). In a few published studies on this topic, Trejo et al. (2012) demonstrated that debris from *C. sorokiniana* and *A. brasilense* can effectively improve infertile soils with low organic matter content. Following wastewater treatment, applying dried alginate beads containing *C. sorokiniana* and *A. brasilense* significantly increased organic matter, organic carbon, and microbial carbon levels in degraded desert soils. Later, López et al. (2013) reported that inoculating soil with the same microbial consortium led to a notable increase in microbial richness and diversity (Fig. 15.3). Additionally, sorghum growth improved in the treated soil, with *A. brasilense* extensively colonizing the plant's root surface. An improvement in soil fertility has also been reported with the association between *Azospirillum lipoferum* and the filamentous cyanobacteria *Nostoc linckia*, due to increased soil nitrogen content and optimized nutrient utilization (Solomon et al. 2024a).

Other studies on the interaction between *Azospirillum* and microalgae have primarily focused on their effects on plant growth and yield for agricultural purposes. For instance, co-immobilized inoculants of *A. brasilense* and *C. vulgaris* have enhanced salt tolerance in hydroponically grown tomato seedlings, resulting in improved seedling growth. de Oliveira et al. (2025) evaluated the effect of co-inoculation with *A. brasilense* and *Rhizobium tropici* in combination with two cyanobacteria and six green microalgae strains on common bean grain yield and profitability. The study found that the highest profitability was achieved with the triple inoculation of *Chlorella vulgaris* + *Rhizobium* + *Azospirillum*, due to a significant increase in grain yield. Similarly, *A. brasilense* and *Chlorella* sp. inoculation in *Hymenaea courbaril* L. seedlings, where *Chlorella* was applied at 14- and 21-day intervals, demonstrated a synergistic effect, promoting seedling quality and development (Biazatti et al. 2024). Additionally, numerous studies have explored the role of filamentous cyanobacteria in plant growth, reporting increased growth yield and quality in staple crops such as maize, corn, and beans (Gavilanes et al. 2020; Horácio et al. 2020; Matsuo et al. 2022; Solomon et al. 2024b).

Despite these promising findings, the potential of synthetic *Azospirillum*-microalgae consortia for soil restoration and agricultural enhancement remains underexplored. Future research should focus on understanding the mechanisms governing these interactions, optimizing inoculation strategies, and assessing their long-term effects on soil health and crop productivity under diverse environmental conditions. Integrating these microbial consortia into sustainable restoration and agricultural practices could provide an innovative approach to improving soil fertility, enhancing plant resilience, and mitigating the impacts of soil degradation. Expanding field studies and scaling up these biotechnological applications will be crucial to unlocking their full potential for regenerative agriculture and environmental sustainability.

15.8 Methods to Study Metabolites Involved in the Interaction of *Azospirillum*-Microalgae

Many of the methods utilized to study this interaction, including the visualization of the physical interaction through fluorescent in situ hybridization (FISH) and scanning electron microscopy (SEM), the production of the alginate beads and their use for wastewater treatment, and the application of the interaction to increase fertility in eroded soils, were thoroughly detailed in the **Handbook for *Azospirillum*** (de-Bashan et al. 2015). Thus, in this chapter, we will focus specifically on the methods used to study the metabolites involved in the interaction.

The metabolite exchange forms the foundation for interactions between microalgae and bacteria. Consequently, studies on *Azospirillum*-microalga interactions have focused on analyzing the production and secretion of metabolites from these cells into the culture media. To gain insight into this interaction, it is essential to obtain cell-free media (exudates) from the microalgae-bacteria association, select appropriate methods for analysis, and devise strategies to determine the role of these metabolites.

15.8.1 Obtention of Free-Cell Medium (Exudates)

The metabolites involved in the interaction between microalgae and bacteria are produced and released into the culture medium. To analyze these metabolites, the cells must first be separated from the medium. This can be achieved by centrifuging

30 mL of the culture at 6000× *g* and then filtering the supernatant through a 0.22 μm membrane filter (GSWP02500 EMD Millipore). The filtered medium can then be used to analyze the presence and concentration of metabolites released by the microorganisms (Palacios et al. 2023a).

15.8.2 Analysis of Exudates by Spectrophotometry

Once obtained, exudate metabolites, carbohydrates, and proteins can be quantified by spectrophotometry using the phenol-sulfuric acid method (carbohydrates) and Bradford (proteins).

(a) Phenol-sulfuric method

1. At 5 mL of exudates, add 5 mL of H_2SO_4 1 M and heat at 100 °C for 1 h.
2. After this time, transfer 50 μL of the sample to a microplate.
3. Add 150 μL of concentrated H_2SO_4 (be careful in this step since this is an exothermic reaction) and mix the sample with the micropipette.
4. Add 30 μL of a solution of 5% phenol and mix the sample with the micropipette.
5. Incubate the microplate at 90 °C for 5 min; after this, let the sample cool at room temperature.
6. Read the microplate at 490 nm.

The quantification of carbohydrates in samples must be determined using a standard glucose curve (Choix et al. 2012b).

(b) Bradford

1. Transfer 50 μL of the exudates to a microplate.
2. Add 200 μL of Bradford reagent and mix the sample with the micropipette.
3. Incubate in the dark at room temperature for 30 min.
4. Read the microplate at 540 nm.

The quantification of proteins in samples must be determined using a standard curve of albumin (Bradford 1976).

15.8.3 Analysis of Exudates by Nuclear Magnetic Resonance (NMR)

Nuclear magnetic resonance (NMR) has been reported to elucidate the profile of metabolites presented in the media during the interaction between *Azospirillum* and microalgae such as *Chlorella*, finding the presence of glycerol produced by microalgae and uracil by *Azospirillum* (Palacios et al. 2023a). This method is not destructive or specific and can help show a complete analysis of the structural information of organic molecules in the culture medium. However, the detection limit of NMR is low, which could mask molecules at too low concentrations (Ben-Tal et al. 2022). For this reason, a concentration step for the exudates must be done; lyophilization is recommended to avoid denaturalizing metabolites by high temperatures through concentration by evaporation. Once concentrated, 20 mg of the lyophilized sample is dissolved in 0.6 mL D_2O (pH 7.0 phosphate buffer containing 1 mM TMSP-d_4 and 2 mM NaN_3) and analyzed using NMR equipment.

15.8.4 Analysis of Exudates by High-Pressure Liquid Chromatography (HPLC)

Molecules, such as vitamins, amino acids, or phytohormones, which are produced at low concentrations during the interactions of *Azospirillum*-microalgae, are key for the interaction and are even considered signal molecules (Palacios et al. 2016a). Strategies like liquid chromatography have been proposed to detect and quantify molecules at low concentrations during microorganism interactions. For *Azospirillum*-microalgae interactions, high-pressure liquid chromatography (HPLC) has been used to detect the presence of phytohormones (IAA), vitamins, amino acids, and organic acids during the interaction (Palacios et al. 2016b; Barbosa-Nuñez et al. 2022; Greipel et al. 2024). In general, the injection of 100 μL of exudates in an HPLC equipped with reverse-phase columns and operated under isocratic methods is reported.

A reverse-phase column (5 μm particle size, 150 × 46 mm) is required for the determination of tryptophan and indole-3-acetic acid in exudates by HPLC. An isocratic method is proposed, where the mobile phase is composed of methanol, water, and acetic acid (36:63:1 v/v), with a flow rate of 0.5 mL·min^{-1} for 20 min. IAA and tryptophan are detected at 290 nm (Palacios et al. 2016b).

For the determination of B-group vitamins (thiamine and riboflavin), a reverse-phase column (10 μm particle size, 250 × 4 mm) is recommended. An isocratic method is used with a mobile phase consisting of 0.1 M phosphate buffer (pH 7.0) and acetonitrile (72:28, v/v), at a flow rate of 0.65 mL·min^{-1} for 25 min. Riboflavin, flavin mononucleotide (FMN), flavin dinucleotide (FDN), and lumichrome (degradation compound of riboflavin) could be detected by fluorescence at 370 nm for excitation and 520 nm for emission. For vitamins such as thiamine and its derivatives (thiamine monophosphate, thiamine diphosphate, and thiamine triphosphate), an oxidation step prior to column separation is necessary due to the natural lack of fluorescence in these compounds. To achieve this, 500 μL of 1% potassium ferrocyanide (dissolved in 15% aqueous NaOH) is added to the sample and mixed for 10 s. The oxidized compounds obtained (thiochrome, thiochrome monophosphate, and thiochrome diphosphate) are detected at 370 nm for excitation and 435 nm for emission (Palacios et al. 2016a).

For carboxylic acid determination, a phase reverse column (5 μm particle size, 250 × 4.6 mm) is used. The isocratic method uses 9 volumes of 50 mM K_3PO_4 with 10 % H_3PO_4 and a volume of methanol with a flow rate of 1 mL·min^{-1} for 15 min. Carboxylic acids are detected at 208 nm (Greipel et al. 2024).

15.9 Conclusions

The interaction between *Azospirillum* spp. and photosynthetic single-cell microalgae offers valuable insights into how this plant growth-promoting bacterium (PGPB) interacts with plants more broadly. This relationship is crucial for exploring its physiological, biochemical, and molecular characteristics. As a distinct subfield within *Azospirillum* research, this interaction has become a significant area of research within *Azospirillum* studies. The biomass generated from the association between *Azospirillum* and microalgae can be utilized in various applications, presenting a promising opportunity for producing valuable metabolites such as lipids, pigments, carbohydrates, and proteins. However, this biotechnological application must be tested under scaled-up conditions to fully assess its potential.

Also, the *Azospirillum*-microalgae partnership offers a sustainable, efficient, and versatile approach to wastewater treatment, soil restoration, and sustainable agriculture. Combining the metabolic capabilities of both organisms contributes to the generation of valuable bioproducts, making it a fundamental technology for the circular economy.

References

Alonso KA, Choix FJ, Nevarez-Moorillón GV et al (2024) Microalgae-bacteria interaction in alginate beads prevents the negative effect of copper oxide nanoparticles on the growth and metabolism of *Scenedesmus* sp. J Appl Phycol 36(1):155–166. https://doi.org/10.1007/s10811-023-03150-5

Alsenani F, Wass TJ, Ma R et al (2019) Transcriptome-wide analysis of *Chlorella* reveals auxin-induced carotenogenesis pathway in green microalgae. Algal Res 37:320–335. https://doi.org/10.1016/j.algal.2018.12.002

Amavizca E, Bashan Y, Ryu CM et al (2017) Enhanced performance of the microalga *Chlorella sorokiniana* remotely induced by the plant growth-promoting bacteria *Azospirillum brasilense* and *Bacillus pumilus*. Sci Rep 7:41310. https://doi.org/10.1038/srep41310

Barbosa-Nuñez JA, Palacios OA, de-Bashan LE et al (2022) Active indole-3-acetic acid biosynthesis by the bacterium *Azospirillum brasilense* cultured under a biogas atmosphere enables its beneficial association with microalgae. J Appl Microbiol 132:3650–3663. https://doi.org/10.1111/jam.15509

Bashan Y, de-Bashan LE (2010) How the plant growth-promoting bacterium *Azospirillum* promotes plant growth—a critical assessment. Adv Agron 108:77–136. https://doi.org/10.1016/S0065-2113(10)08002-8

Bashan Y, Salazar BG, Moreno M et al (2012) Restoration of eroded soil in the Sonoran Desert with native leguminous trees using plant growth-promoting microorganisms and limited amounts of compost and water. J Environ Manag 102:26–36. https://doi.org/10.1016/j.jenvman.2011.12.032

Ben-Tal Y, Boaler PJ, Dale HJ et al (2022) Mechanistic analysis by NMR spectroscopy: a users guide. Prog Nucl Magn Reson Spectrosc 129:28–106. https://doi.org/10.1016/j.pnmrs.2022.01.001

Biazatti RM, Santos SC, Santos CC et al (2024) Do *Azospirillum brasiliense* and *Chlorella* sp. microalgae benefit the production of Hymenaea courbaril L. seedlings? Can J For Res 55: 1–11. https://doi.org/10.1139/cjfr-2024-0069

Bradford MM (1976) A rapid sensitive method for the quantization of microgram quantities of protein utilizing the principle of protein-dye binding. Ann Biochem 72:248–257. https://doi.org/10.1016/0003-2697(76)90527-3

Carillo P, Ciarmiello LF, Woodrow P et al (2020) Enhancing sustainability by improving plant salt tolerance through macro- and micro-algal biostimulants. Biology 9(9):253. https://doi.org/10.3390/biology9090253

Cassan F, Coniglio A, López G et al (2020) Everything you must know about *Azospirillum* and its impact on science, research, agriculture and beyond. Biol Fertil Soils 56:461–479. https://doi.org/10.1007/s00374-020-01463-y

Cassan FD, Coniglio A, Amavizca E et al (2021) The *Azospirillum brasilense* Type VI secretion system promotes cell aggregation, biocontrol protection against phytopathogens and attachment to the microalgae *Chlorella sorokiniana*. Environ Microbiol 23(10):6257–6274. https://doi.org/10.1111/1462-2920.15749

Chen YT, Zhao DS, Liu XL et al (2025) Physiological, transcriptomic and metabolomic responses of the marine diatom *Phaeodactylum tricornutum* to the auxin IAA. Algal Res 102:103993. https://doi.org/10.1016/j.algal.2025.103993

Choix FJ, de-Bashan LE, Bashan Y (2012a) Enhanced accumulation of starch and total carbohydrates in alginate-immobilized *Chlorella* spp. induced by *Azospirillum brasilense.* I Autotrophic conditions. Enzyme Microb Technol 51:294–299. https://doi.org/10.1016/j.enzmictec.2012.07.013

Choix FJ, de-Bashan LE, Bashan Y (2012b) Enhanced accumulation of starch and total carbohydrates in alginate-immobilized Chlorella spp. induced by *Azospirillum brasilense*: II. Heterotrophic conditions. Enzyme Microb Technol 51(5):300–309. https://doi.org/10.1016/j.enzmictec.2012.07.012

Choix FJ, Bashan Y, Mendoza A et al (2014) Enhanced activity of ADP glucose pyrophosphorylase and formation of starch induced by *Azospirillum brasilense* in Chlorella vulgaris. J Biotechnol 177:22–34. https://doi.org/10.1016/j.jbiotec.2014.02.014

Choix FJ, López-Cisneros CG, Méndez-Acosta HO (2018) *Azospirillum brasilense* increases CO_2 fixation on microalgae *Scenedesmus obliquus*, *Chlorella vulgaris*, and *Chlamydomonas reinhardtii* cultured on high CO_2 concentrations. Microb Ecol 76:430–442. https://doi.org/10.1007/s00248-017-1139-z

Choix FJ, Palacios OA, Contreras CA et al (2023) Growth and metabolism enhancement in microalgae co-cultured in suspension with the bacterium *Azospirillum brasilense* under heterotrophic conditions. J Appl Phycol 35(1):57–71. https://doi.org/10.1007/s10811-022-02891-z

Choix FJ, Palacios OA, Mondragón-Cortez P et al (2024) Synergic association of the consortium *Arthrospira* maxima with the microalga growth-promoting bacterium *Azospirillum* cultured under the stressful biogas composition. Bioprocess Biosyst Eng 47(2):181–193. https://doi.org/10.1007/s00449-023-02947-5

Church J, Hwang JH, Kim KT et al (2017) Effect of salt type and concentration on the growth and lipid content of *Chlorella vulgaris* in synthetic saline wastewater for biofuel production. Bioresour Technol 243:147–153. https://doi.org/10.1016/j.biortech.2017.06.081

Contreras CA, Palacios OA, de-Bashan LE et al (2024) Microalga growth-promoting bacteria as strategy to improve CO_2 removal from biogas. Bioenerg Res 17(4):2082–2099. https://doi.org/10.1007/s12155-024-10800-0

Contreras-Angulo JR, Mata TM, Cuellar-Bermudez SP et al (2019) Symbiotic co-culture of Scenedesmus sp. and *Azospirillum brasilense* on N-deficient media with biomass production for biofuels. Sustainability 11(3):707. https://doi.org/10.3390/su11030707

Covarrubias SA, de-Bashan LE, Moreno M et al (2012) Alginate beads provide a beneficial physical barrier against native microorganisms in wastewater treated with immobilized bacteria and microalgae. Appl Microbiol Biotechnol 93(6):2669–2680. https://doi.org/10.1007/s00253-011-3585-8

Cruz I, Bashan Y, Hernández-Carmona G et al (2013) Biological deterioration of alginate beads containing immobilized microalgae and bacteria during tertiary wastewater treatment. Appl Microbiol Biotechnol 97(22):9847–9858. https://doi.org/10.1007/s00253-013-4703-6

de Oliveira KS, Volsi B, Telles TS et al (2025) Co-inoculation with *Rhizobium*, *Azospirillum*, and microalgae increases common bean yield and profitability. Agron J 117(1):e21719. https://doi.org/10.1002/agj2.21719

de-Bashan LE, Bashan Y (2008) Joint immobilization of plant growth-promoting bacteria and green microalgae in alginate beads as an experimental model for studying plant-bacterium interactions. Appl Environ Microbiol 74:6797–6802. https://doi.org/10.1128/AEM.00518-08

de-Bashan LE, Bashan Y, Moreno M et al (2002a) Increased pigment and lipid content, lipid variety, and cell and population size of the microalgae *Chlorella* spp. when co-immobilized in alginate beads with the microalgae-growth-promoting bacterium *Azospirillum brasilense*. Can J Microbiol 48:514–521. https://doi.org/10.1139/w02-051

de-Bashan LE, Moreno M, Hernandez J-P et al (2002b) Removal of ammonium and phosphorus ions from synthetic wastewater by the microalgae *Chlorella vulgaris* coimmobilized in alginate beads with the microalgae growth-promoting bacterium *Azospirillum brasilense*. Water Res 36: 2941–2948. https://doi.org/10.1016/S0043-1354(01)00522-X

de-Bashan LE, Hernandez JP, Morey T et al (2004) Microalgae growth-promoting bacteria as "helpers" for microalgae: a novel approach for removing ammonium and phosphorus from municipal wastewater. Water Res 38(2):466–474. https://doi.org/10.1016/j.watres.2003.09.022

de-Bashan LE, Antoun H, Bashan Y (2005) Cultivation factors and population size control the uptake of nitrogen by the microalgae *Chlorella vulgaris* when interacting with the microalgae growth-promoting bacterium *Azospirillum brasilense*. FEMS Microbiol Ecol 54(2):197–203. https://doi.org/10.1016/j.femsec.2005.03.014

de-Bashan LE, Magallon P, Antoun H et al (2008a) Role of glutamate dehydrogenase and glutamine synthetase in *Chlorella vulgaris* during assimilation of ammonium when jointly immobilized with the microalgae-growth-promoting bacterium *Azospirillum brasilense*. J Phycol 44:1188–1196. https://doi.org/10.1111/j.1529-8817.2008.00572.x

de-Bashan LE, Antoun H, Bashan Y (2008b) Involvement of indole-3-acetic-acid produced by the growth-promoting bacterium *Azospirillum* spp. in promoting growth of *Chlorella vulgaris*. J Phycol 44:938–947. https://doi.org/10.1111/j.1529-8817.2008.00533.x

de-Bashan LE, Schmid M, Rothballer M et al (2011) Cell-cell interaction in the eukaryote-prokaryote model using the microalgae *Chlorella vulgaris* and the bacterium *Azospirillum brasilense* immobilized in polymer beads. J Phycol 47:1350–1359. https://doi.org/10.1111/j.1529-8817.2011.01062.x

de-Bashan LE, Hernandez JP, Bashan Y (2015) Interaction of *Azospirillum* spp. with microalgae: a basic eukaryotic–prokaryotic model and its biotechnological applications. In: Handbook for *Azospirillum*: technical issues and protocols. Springer, Cham, pp 367–388. https://doi.org/10.1007/978-3-319-06542-7_20

de-Bashan LE, Mayali X, Bebout BM et al (2016) Establishment of stable synthetic mutualism without co-evolution between microalgae and bacteria demonstrated by mutual transfer of metabolites (NanoSIMS isotopic imaging) and persistent physical association (Fluorescent in situ hybridization). Algal Res 15:179–186. https://doi.org/10.1016/j.algal.2016.02.019

Devereux R, Loeblich AR, Fox GE (1990) Higher plant origins and the phylogeny of green algae. J Mol Evol 31:18–24. https://doi.org/10.1007/BF02101788

Drogue B, Sanguin H, Borland S et al (2014) Genome-wide profiling of *Azospirillum lipoferum* 4B gene expression during interaction with rice roots. FEMS Microbiol Ecol 87(2):543–555. https://doi.org/10.1111/1574-6941.12244

Ferreira A, Ribeiro B, Ferreira AF et al (2019) *Scenedesmus obliquus* microalga-based biorefinery—from brewery effluent to bioactive compounds, biofuels and biofertilizers—aiming at a circular bioeconomy. Biofuels Bioprod Biorefin 13(5):1169–1186. https://doi.org/10.1002/bbb.2032

Fukami J, Cerezini P, Hungria M (2018) *Azospirillum*: benefits that go far beyond biological nitrogen fixation. AMB Express 8(1):73. https://doi.org/10.1186/s13568-018-0608-1

Garciglia-Mercado C, Contreras CA, Choix FJ et al (2024) Metabolic and physiological adaptations of microalgal growth-promoting bacterium *Azospirillum brasilense* growing under biogas atmosphere: a microarray-based transcriptome analysis. Arch Microbiol 206(4):173. https://doi.org/10.1007/s00203-024-03890-z

Gavilanes FZ, Andrade DS, Zucareli C et al (2020) Co-inoculation of *Anabaena cylindrica* with *Azospirillum brasilense* increases grain yield of maize hybrids. Rhizosphere 15:100224. https://doi.org/10.1016/j.rhisph.2020.100224

Geremia E, Ripa M, Catone CM et al (2021) A review about microalgae wastewater treatment for bioremediation and biomass production—a new challenge for Europe. Environment 8(12):136. https://doi.org/10.3390/environments8120136

Gómez-Serrano C, Morales-Amaral MM, Acién FG et al (2015) Utilization of secondary-treated wastewater for the production of freshwater microalgae. Appl Microbiol Biotechnol 99(16): 6931–6944. https://doi.org/10.1007/s00253-015-6694-y

Gonzalez LE, Bashan Y (2000) Increased growth of the microalga *Chlorella vulgaris* when coimmobilized and cocultured in alginate beads with the plant growth-promoting bacterium *Azospirillum brasilense*. Appl Environ Microbiol 66:1527–1531. https://doi.org/10.1128/AEM.66.4.1527-1531.2000

Gonzalez EJ, Hernandez JP, de-Bashan LE et al (2018) Dry micro-polymeric inoculant of *Azospirillum brasilense* is useful for producing mesquite transplants for reforestation of degraded arid zones. Appl Soil Ecol 129:84–93. https://doi.org/10.1016/j.apsoil.2018.04.011

González-González LM, de-Bashan LE (2021) Toward the enhancement of microalgal metabolite production through microalgae–bacteria consortia. Biology 10(4):282. https://doi.org/10.3390/biology10040282

González-González LM, de-Bashan LE (2023) The potential of microalgae–bacteria consortia to restore degraded soils. Biology 12(5):693. https://doi.org/10.3390/biology12050693

Greipel E, Nagy K, Csákvári E et al (2024) Chemotactic interactions of Scenedesmus sp. and *Azospirillum brasilense* investigated by microfluidic methods. Microb Ecol 87(1):52. https://doi.org/10.1007/s00248-024-02366-3

Hernandez JP, de-Bashan LE, Bashan Y (2006) Starvation enhances phosphorus removal from wastewater by the microalga *Chlorella* spp. co-immobilized with *Azospirillum brasilense*. Enzym Microb Technol 38(1–2):190–198. https://doi.org/10.1016/j.enzmictec.2005.06.005

Hernandez J-P, de-Bashan LE, Rodriguez DJ et al (2009) Growth promotion of the freshwater microalga *Chlorella vulgaris* by the nitrogen-fixing, plant growth-promoting bacterium *Bacillus pumilus* from arid zone soils. Eur J Soil Biol 45:88–93. https://doi.org/10.1016/j.ejsobi.2008.08.004

Horácio EH, Zucareli C, Gavilanes FZ et al (2020) Co-inoculation of rhizobia, azospirilla and cyanobacteria for increasing common bean production. Semina: Ciênc Agrár 41(5 Suppl 1): 2015–2028. https://doi.org/10.5433/1679-0359.2020v41n5Supl1p2015

Jusoh M, Loh SH, Chuah TS et al (2015) Indole-3-acetic acid (IAA) induced changes in oil content, fatty acid profiles and expression of four fatty acid biosynthetic genes in Chlorella vulgaris at early stationary growth phase. Phytochemistry 111:65–71. https://doi.org/10.1016/j.phytochem.2014.12.022

Leyva LA, Bashan Y, Mendoza A et al (2014) Accumulation of fatty acids in Chlorella vulgaris under heterotrophic conditions in relation to activity of acetyl-CoA carboxylase, temperature, and co-immobilization with *Azospirillum brasilense*. Naturwissenschaften 101(10):819–830. https://doi.org/10.1007/s00114-014-1223-x

Leyva LA, Bashan Y, de-Bashan LE (2015) Activity of acetyl-CoA carboxylase is not directly linked to accumulation of lipids when *Chlorella vulgaris* is co-immobilised with *Azospirillum brasilense* in alginate under autotrophic and heterotrophic conditions. Ann Microbiol 65:339–349. https://doi.org/10.1007/s13213-014-0866-3

Lopez BR, Bashan Y, Trejo A et al (2013) Amendment of degraded desert soil with wastewater debris containing immobilized *Chlorella sorokiniana* and *Azospirillum brasilense* significantly modifies soil bacterial community structure, diversity, and richness. Biol Fertil Soils 49:1053–1063. https://doi.org/10.1007/s00374-013-0799-1

Lopez BR, Palacios OA, Bashan Y et al (2019) Riboflavin and lumichrome exuded by the bacterium *Azospirillum brasilense* promote growth and changes in metabolites in *Chlorella sorokiniana* under autotrophic conditions. Algal Res 44:101396. https://doi.org/10.1016/j.algal.2019.101696

Mahdy A, Mendez L, Ballesteros M et al (2015) Algaculture integration in conventional wastewater treatment plants: anaerobic digestion comparison of primary and secondary sludge with microalgae biomass. Bioresour Technol 184:236–244. https://doi.org/10.1016/j.biortech.2014.09.145

Markus AA, Parsons JR, Roex EWM et al (2016) Modelling the transport of engineered metallic nanoparticles in the river Rhine. Water Res 91:214–224. https://doi.org/10.1016/j.watres.2016.01.003

Matsuo O, Zucareli C, Horácio EH et al (2022) Co-inoculation of Anabaena cylindrica and Azospirillum brasilense during initial growth and chloroplast pigments of corn. Rev Bras Eng Agríc Ambient 26:97–102. https://doi.org/10.1590/1807-1929/agriambi.v26n2p97-102

Meza B, de-Bashan LE, Bashan Y (2015a) Involvement of indole-3-acetic acid produced by *Azospirillum brasilense* in accumulating intracellular ammonium in *Chlorella vulgaris*. Res Microbiol 166:72–83. https://doi.org/10.1016/j.resmic.2014.12.010

Meza B, de-Bashan LE, Hernandez J-P et al (2015b) Accumulation of intracellular polyphosphate in *Chlorella vulgaris* cells is related to indole-3-acetic acid produced by *Azospirillum brasilense*. Res Microbiol 166: 399–407. https://doi.org/10.1016/j.resmic.2015.03.001

Moreno M, De-Bashan LE, Hernandez JP et al (2017) Success of long-term restoration of degraded arid land using native trees planted 11 years earlier. Plant Soil 421:83–92. https://doi.org/10.1007/s11104-017-3438-z

Mutum L, Janda T, Ördög V et al (2022) Biologia Futura: potential of different forms of microalgae for soil improvement. Biol Futura:1–8. https://doi.org/10.1007/s42977-021-00103-2

Pagnussat LA, Maroniche G, Curatti L et al (2020) Auxin-dependent alleviation of oxidative stress and growth promotion of Scenedesmus obliquus C1S by *Azospirillum brasilense*. Algal Res 47: 101839. https://doi.org/10.1016/j.algal.2020.101839

Pagnussat LA, Do Nascimento M, Maroniche G et al (2023) *Azospirillum baldaniorum* improves acclimation, lipid productivity and oxidative response of a microalga under salt stress. Algal Res 74:103192. https://doi.org/10.1016/j.algal.2023.103192

Palacios OA, Bashan Y, de-Bashan LE (2014) Proven and potential involvement of vitamins in interactions of plants with plant growth-promoting bacteria—an overview. Biol Fertil Soils 50: 415–432. https://doi.org/10.1007/s00374-013-0894-3

Palacios OA, Bashan Y, Schmid M et al (2016a) Enhancement of thiamine release during synthetic mutualism between *Chlorella sorokiniana* and *Azospirillum brasilense* growing under stress conditions. J Appl Phycol 28:1521–1531. https://doi.org/10.1007/s10811-015-0697-z

Palacios OA, Gomez-Anduro G, Bashan Y et al (2016b) Tryptophan, thiamine and indole-3-acetic acid exchange between *Chlorella sorokiniana* and the plant growth-promoting bacterium *Azospirillum brasilense*. FEMS Microbiol Ecol 92(6):fiw077. https://doi.org/10.1093/femsec/fiw077

Palacios OA, Choix FJ, Bashan Y et al (2016c) Influence of tryptophan and indole-3-acetic acid on starch accumulation in the synthetic mutualistic *Chlorella sorokiniana-Azospirillum brasilense* system under heterotrophic conditions. Res Microbiol 167:367–379. https://doi.org/10.1016/j.resmic.2016.02.005

Palacios OA, Lopez BR, Bashan Y et al (2019) Early changes in nutritional conditions affect formation of synthetic mutualism between *Chlorella sorokiniana* and the bacterium *Azospirillum brasilense*. Microb Ecol 77(4):980–992. https://doi.org/10.1007/s00248-018-1282-1

Palacios OA, Lopez BR, Palacios-Espinosa A et al (2021) The immediate effect of riboflavin and lumichrome on the mitigation of saline stress in the microalga *Chlorella sorokiniana* by the plant-growth-promoting bacterium *Azospirillum brasilense*. Algal Res 58:102424. https://doi.org/10.1016/j.algal.2021.102424

Palacios OA, López BR, de-Bashan LE (2022) Microalga Growth-Promoting Bacteria (MGPB): a formal term proposed for beneficial bacteria involved in microalgal–bacterial interactions. Algal Res 61:102585. https://doi.org/10.1016/j.algal.2021.102585

Palacios OA, Espinoza-Hicks JC, Camacho-Dávila AA et al (2023a) Differences in exudates between strains of *Chlorella sorokiniana* affect the interaction with the microalga growth-promoting bacteria *Azospirillum brasilense*. Microb Ecol 85(4):1412–1422. https://doi.org/10.1007/s00248-022-02026-4

Palacios OA, León-Vega RA, López BR et al (2023b) Microalga-bacteria interaction mitigates adverse effects on microalga produced by ZnO nanoparticles. Algal Res 74:103198. https://doi.org/10.1016/j.algal.2023.103198

Peng H, de-Bashan LE, Bashan Y et al (2020) Indole-3-acetic acid from *Azospirillum brasilense* promotes growth in green algae at the expense of energy storage products. Algal Res 47:101845. https://doi.org/10.1016/j.algal.2020.101845

Peng H, de-Bashan LE, Higgins BT (2021) *Azospirillum brasilense* reduces oxidative stress in the green microalgae *Chlorella sorokiniana* under different stressors. J Biotechnol 325:179–185. https://doi.org/10.1016/j.jbiotec.2020.10.029

Perez-Garcia O, de-Bashan LE, Hernandez JP et al (2010) Efficiency of growth and nutrient uptake from wastewater by heterotrophic, autotrophic, and mixotrophic cultivation of Chlorella vulgaris immobilized with *Azospirillum brasilense*. J Phycol 46(4):800–812. https://doi.org/10.1111/j.1529-8817.2010.00862.x

Perez-Garcia O, Bashan Y, Puente ME (2011) Organic carbon supplementation of sterilized municipal wastewater is essential for heterotrophic growth and removing ammonium by the microalga *Chlorella vulgaris*. J Phycol 47(1):190–199. https://doi.org/10.1111/j.1529-8817.2010.00934.x

Ramakrishnan B, Maddela NR, Venkateswarlu K et al (2023) Potential of microalgae and cyanobacteria to improve soil health and agricultural productivity: a critical view. Environ Sci Adv 2(4):586–611. https://doi.org/10.1039/d2va00158f

Ramos-Ibarra JR, Rubio-Ramírez TE, Mondragón-Cortez P et al (2019) *Azospirillum brasilense*-microalga interaction increases growth and accumulation of cell compounds in *Chlorella vulgaris* and *Tetradesmus obliquus* cultured under nitrogen stress. J Appl Phycol 31:3465–3477. https://doi.org/10.1007/s10811-019-01862-1

Ronga D, Biazzi E, Parati K et al (2019) Microalgal biostimulants and biofertilisers in crop productions. Agronomy 9(4):192. https://doi.org/10.3390/agronomy9040192

Ruiz-Güereca DA, Sánchez-Saavedra MDP (2016) Growth and phosphorus removal by *Synechococcus elongatus* co-immobilized in alginate beads with *Azospirillum brasilense*. J Appl Phycol 28:1501–1507. https://doi.org/10.1007/s10811-015-0728-9

Solomon W, Mutum L, Janda T et al (2024a) Microalgae–bacteria interaction: a catalyst to improve maize (*Zea mays* L.) growth and soil fertility. Cereal Res Commun:1–13. https://doi.org/10.1007/s42976-024-00558-8

Solomon W, Janda T, Molnar Z (2024b) Impact of the microalgae-bacteria interaction on maize (*Zea mays* L.) health and yield. In: BIO Web Conference, vol 125. EDP Sciences, p 03004. https://doi.org/10.1051/bioconf/202412503004

Trejo A, de-Bashan LE, Hartmann A et al (2012) Recycling waste debris of immobilized microalgae and plant growth-promoting bacteria from wastewater treatment as a resource to improve fertility of eroded desert soil. Environ Exp Bot 75:65–73. https://doi.org/10.1016/j.envexpbot.2011.08.007

Xie Z, Ou Z, Zhang M et al (2024) Indole-3-acetic acid regulating the initial adhesion of microalgae in biofilm formation. Environ Res 252:119093. https://doi.org/10.1016/j.envres.2024.119093

Chapter 16
Interaction of *Azospirillum* spp. with Other Bacteria and Fungi

Elisabeth Derollez, Pablo R. Díaz, Florencia Donadio, Fabricio Dario Cassán, Guillermo A. Maroniche, and Claire Prigent-Combaret

Abstract *Azospirillum* has attracted considerable attention as a microbial inoculant due to its diverse plant-beneficial effects, such as nitrogen-fixing, enhanced nutrient availability, production of phytohormones such as indole-3-acetic acid (IAA), cytokinins, and gibberellins, and improvement of plant tolerance to abiotic and biotic stresses. Furthermore, their co-inoculation with other plant-beneficial microorganisms can lead to more consistent and significant yield improvements, and it has emerged as a promising strategy for sustaining plant growth. This review chapter first explores the in vitro interactions between *Azospirillum* and other microbial partners, highlighting antagonism interaction on *Azospirillum* or of *Azospirillum* on other rhizosphere microbial members, synergistic interactions, and even exploitation by *Azospirillum* of public goods produced by other beneficial rhizobacteria. It then discusses how these interactions can be exploited in planta to improve *Azospirillum* root colonization, enhance its antifungal activity, or improve symbiotic relationships through co-inoculation with mycorrhizal fungi or nitrogen-fixing bacteria such as *Bradyrhizobium*. Finally, optimized protocols are proposed for evaluating the in vitro compatibility of *Azospirillum* strains with other microorganisms and assessing the efficiency of co-inoculation on plant performance. Altogether, the

E. Derollez · C. Prigent-Combaret (✉)
Laboratoire d'Ecologie Microbienne, Université Claude Bernard Lyon 1, UMR CNRS 5557, Villeurbanne, France
e-mail: claire.prigent-combaret@univ-lyon1.fr

P. R. Díaz · G. A. Maroniche
Facultad de Ciencias Agrarias, Universidad Nacional de Mar del Plata (UNMdP), Buenos Aires, Argentina

Consejo Nacional de Investigaciones Científicas y Técnicas (CONICET), Buenos Aires, Argentina

F. Donadio · F. D. Cassán
Laboratorio de Fisiología Vegetal y de la Interacción Planta-Microorganismo, Instituto de Investigaciones Agrobiotecnológicas (INIAB-CONICET), Universidad Nacional de Río Cuarto (UNRC), Río Cuarto, Córdoba, Argentina

F. D. Cassán et al. (eds.), *The Definitive Handbook of Azospirillum*,
https://doi.org/10.1007/978-3-032-09222-9_16

data presented in this review underscore the value of integrating *Azospirillum* with complementary plant growth-promoting rhizobacteria (PGPR) partners to sustainably enhance crop productivity.

Keywords Co-inoculation · Synergy · Antagonism · Mycorrhizae · Bradyrhizobium

16.1 Introduction

Azospirillum strains are free N_2-fixing bacteria well known for their ability to positively influence the growth of plants by consistently stimulating the root branching of the host plant, the root biomass, and the soil area explored by roots (Bashan et al. 2004; Cassán et al. 2020; Fukami et al. 2018). Over the past 50 years, the ability of a single *Azospirillum* inoculation to alleviate both abiotic and biotic stresses in plants has been demonstrated, showing efficacy against drought (Bano et al. 2013; Cohen et al. 2015; Creus et al. 2004), high salinity (Alamri and Mostafa 2009; Bacilio et al. 2004; Fasciglione et al. 2015; Omar et al. 2009), low nitrogen levels (Esquivel-Cote et al. 2010), and several plant diseases by producing siderophores, thereby limiting the iron available for the growth of plant pathogens (Tortora et al. 2011), or by triggering ISR defense mechanisms (for reviews, see Fukami et al. 2018; Vacheron et al. 2015) (Fig. 16.1). Based on these positive

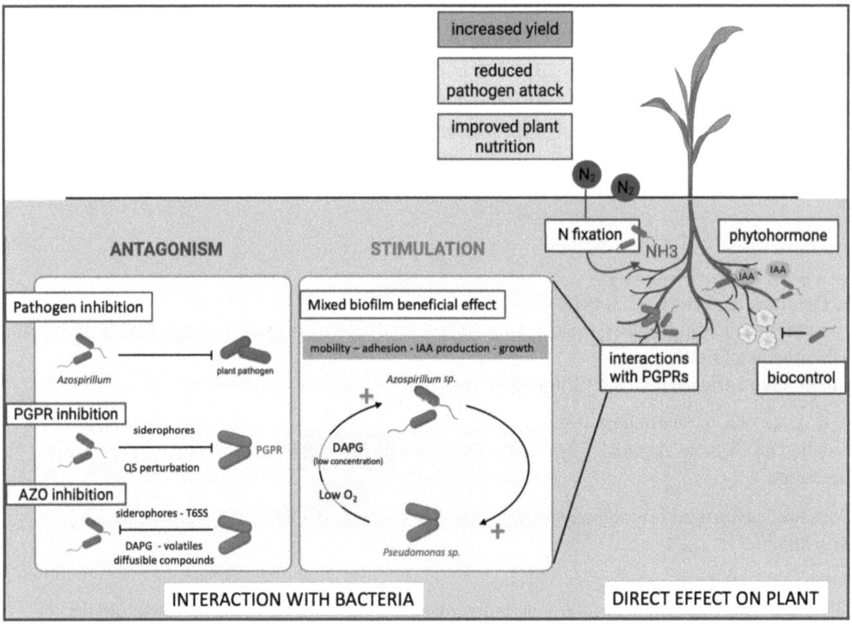

Fig. 16.1 Main mechanisms used by *Azospirillum sp.* to improve plant growth directly or indirectly through interactions with other PGPR present in the rhizosphere.

effects, the first commercial inoculant containing *A. brasilense* was commercialized in 2009 in Brazil. They are now being commercialized more widely, not only in South America but also in Europe. However, what is very remarkable about *Azospirillum*-based bioinoculants is that, despite their positive effects on crop yield, their population size in the rhizosphere usually declines quite rapidly after inoculation onto plants. This suggests that their interactions with other members of the rhizomicrobiota may not only determine their ecological behavior but also their plant-beneficial activities. In addition, studying their relationships with other plant-beneficial microorganisms may help to improve their efficacy on plants as bioinoculants.

Indeed, a lot of the current research focuses on the effects of combinations between *Azospirillum* and other bacteria to enhance these beneficial outcomes. This approach aims to optimize plant growth and resilience through synergistic microbial interactions in agriculture. Microbial groups of agricultural importance that have been combined with *Azospirillum* include rhizobia, pseudomonads and mycorrhizal fungi, mainly to explore their potential additive effects on plant growth. For instance, co-inoculation of *Azospirillum* with *Rhizobium* has been shown to increase nitrogen fixation and crop yield in common bean (*Phaseolus vulgaris* L.), though genotype-dependent responses were observed (Remans et al. 2008). Similarly, the combined inoculation of *Azospirillum brasilense* and *Trichoderma harzianum* improved yield, nutrient uptake, and biofortification in hydroponic systems (Moreira et al. 2022). In salt-stressed flax plants, *A. brasilense* and *Pseudomonas geniculata* co-inoculation enhanced plant growth and antioxidant enzyme activity, highlighting its potential for abiotic stress mitigation (Omer et al. 2022). The synergistic effects of co-inoculation were further evidenced in cowpea, where *Bradyrhizobium sp.* combined with *A. brasilense* increased nitrogen uptake and grain yield (average increase of 8% compared to *Bradyrhizobium* alone), reducing the need for mineral nitrogen fertilization that should remain below 20 kg $N.ha^{-1}$ (Galindo et al. 2022). Similar benefits were reported in maize, where dual inoculation with *Azospirillum* strains improved biomass production and nutrient content (Ribeiro et al. 2022).

Even when these co-inoculation experiments were not specifically designed for studying microbe-microbe interactions, in the end they may reflect the dynamics of interaction between *Azospirillum* and the other partner during colonization. This is based on the assumption that the compatibility of *Azospirillum* with other microbial species plays a critical role in determining co-inoculation outcomes. For instance, *A. argentinense* Az39 and *P. fluorescens* A506, which show a compatible interaction in vitro, exhibited strong synergy, enhancing lettuce growth and root system architecture. However, when Az39 was co-inoculated with *P. fluorescens* Pf-5, no beneficial effects were observed, emphasizing the importance of selecting compatible microbial partners (Díaz et al. 2023). Moreover, the optimal inoculum dose also requires further investigation (Da Silva Oliveira et al. 2023).

In contrast to the vast literature describing how *Azospirillum* co-inoculation with other PGPM impacts plant growth, basic research on the biology of the interaction of *Azospirillum* and other microorganisms is not abundant. So far, most of the in vitro work done in this regard has focused on the interaction of *Azospirillum* with fluorescent *Pseudomonas*, a bacterial group that is also of high importance as members of the plant microbiome (Shalev et al. 2022). Comparatively, the knowledge of the mechanisms that govern the interaction of azospirilla with other important PGPM, such as rhizobia, *Bacillus*, *Trichoderma,* or mycorrhizal fungi, is poor. The first works on the study of the molecular mechanisms of interaction between bacteria of the genus *Azospirillum* and those of the genus *Pseudomonas* began in the 2010s in France (Combes-Meynet et al. 2011; Couillerot et al. 2011). They evidence a key role of the *Pseudomonas* secondary metabolite 2,4-diacetylphloroglucinol (DAPG) on the physiology of *Azospirillum* with pleiotropic effects depending on its concentration. At high concentrations, this compound perturbates the cell membrane, inducing the formation of reactive oxygen species (Biessy and Filion 2021), while at low concentration, it acts as a signaling compound, enhancing the expression of plant-beneficial encoding genes in *Azospirillum.* Thus, DAPG induces different responses in prokaryotic and eukaryotic cells depending on its concentrations, known as the hormesis concept (Bondy 2023). Later, G. Maroniche and L. Pagnussat in Argentina showed that, depending on the strain compatibility between *Azospirillum and Pseudomonas* strains, *Pseudomonas* can antagonize *Azospirillum* or vice versa, and *Azospirillum* can exploit biofilms and public goods of certain *Pseudomonas* strains (Díaz et al. 2023; Maroniche et al. 2018; Pagnussat et al. 2016). Evidence accumulated over the years indicates that the ratio of interacting populations and strains genotype, along with specific environmental factors, such as physical support, nutrient availability, temperature, and atmospheric composition, are key determinants in shaping the outcome of *Azospirillum* interaction with *Pseudomonas* (Combes-Meynet et al. 2011; Couillerot et al. 2011; Díaz et al. 2022, 2023; Maroniche et al. 2018; Pagnussat et al. 2016). The conjunction of these genetic and environmental variables leads to negative and/or positive associations, in turn resulting in different social dynamics such as exclusion/antagonism, competition, cooperation/synergism, or exploitation, some of which will be described in the next sections (Fig. 16.1).

16.2 In Vitro Interactions of *Azospirillum* with Other Microbial Partners: Promoting Microbial Compatibility with Knowledge of the Different Types of Interactions Described Between *Azospirillum* and Other Bacteria Under In Vitro Conditions

16.2.1 Antagonism Interactions on Azospirillum

Antagonism consists of the growth inhibition of one microorganism on another one. Numerous microbial species have an antagonistic action on *Azospirillum* that occurs through different processes (Fig. 16.1). It is well known that some PGPR, like

Pseudomonas from the fluorescent group, produce antibiotic molecules acting against soil-disease microbial agents, such as 2,4-diacetylphloroglucinol (DAPG) (Almario et al. 2013; Keel et al. 1990). DAPG can act by altering multiple basic cell processes (Kwak et al. 2011), including the disruption of the proton gradient across the mitochondrial membrane of eukaryotes (Troppens et al. 2013). It was also shown that it could have negative effects on bacteria. From a collection of around twenty *Azospirillum* strains belonging to 6 different species, isolated from host plants and of varied geographical origins, we observed that sensitivity to DAPG is strain-specific (Couillerot et al. 2011). Using electron microscopy, it has been reported that *A. brasilense* Cd strain (a strain with intermediate sensitivity to DAPG) was sensitive to the addition of the synthetic antimicrobial metabolite DAPG, which induced carotenoid accumulation, formation of poly-β-hydroxybutyrate-like granules, cytoplasmic membrane damage, and growth inhibition. Most *Azospirillum* strains are moderately sensitive to DAPG, but anyway they are more sensitive to DAPG than many other saprophytic rhizobacteria (Keel et al. 1992; Moënne-Loccoz et al. 2001; Walsh et al. 2003), or phytopathogenic root-associated microorganisms (Cronin et al. 1997; Keel et al. 1992; Schouten et al. 2004). Various DAPG-producing strains of *Pseudomonas* were observed as having, in vitro, an antagonistic effect on *Azospirillum*, while mutants of *Pseudomonas* unable to synthesize phloroglucinols had reduced antagonistic effects. Fluorescent *Pseudomonas* producing DAPG are therefore capable of inhibiting the growth of *Azospirillum* in vitro, and this antagonistic effect involves the production of DAPG. Not all *Azospirillum* strains show the same sensitivity to this molecule, and it would be interesting to investigate the genetic determinants involved in this resistance. In addition to DAPG, another antibiotic regulated by the Gac/Rsm system that inhibits the growth of *A. brasilense* has been detected, although its identity is still unknown (Maroniche et al. 2018).

Siderophores can also confer competitive advantages to bacteria for colonizing plant tissues by excluding other microorganisms. Analyses of siderophore-producing bacteria (SPB) on rice plants show that *Pseudomonas* isolate G20 was able to inhibit *A. brasilense* in vitro through the production of siderophores (Loaces et al. 2011). It has been proposed that native endophytic bacteria in rice seed, like *Bacillus pumilus* and *Corynebacterium flavescens*, competed with *A. brasilense,* which was excluded from the rhizoplane and affected in its early colonization of roots (Bacilio-Jiménez et al. 2001). It is sometimes observed that discrete colonies of *A. baldaniorum* Sp245 that grow in direct contact with a lawn of fluorescent *Pseudomonas* spp. show a reduced size (Maroniche et al. 2018; Pagnussat et al. 2016). It was shown that inhibition of Sp245 growth by the presence of pseudomonads was overall stronger on King's B media that was developed to enhance the production of *Pseudomonas* siderophores (King et al. 1954). However, when *P. protegens* Pf-5 isogenic mutants impaired in the production of pyoverdine (Δ*pvdL*) and/or enantio-pyochelin (Δ*pchA*), the two major siderophores produced by this strain, were used, the exclusion halo was significantly reduced (Maroniche et al. 2018), confirming the prominent role of *Pseudomonas* siderophores in their antagonistic action against *Azospirillum*.

In some studies, a contact-dependent inhibition of *Azospirillum* was observed. *A. baldaniorum* Sp245 growth was affected in the presence of wild-type *P. putida* KT2440, but not with a mutant lacking functional type-6 secretion systems (T6SS). The lethal effect of the antibacterial K1-T6SS belonging to the subclass 4B against *Azospirillum* was demonstrated (Maroniche et al. 2018). Interestingly, A. *baldaniorum* is predicted to encode a subtype 4A T6SS that could improve the aggressivity of the response from *Pseudomonas* through the "T6SS dueling" mechanism (Basler et al. 2013).

16.2.2 Antagonism Interaction of Azospirillum *on Other Rhizosphere Microbial Members*

Azospirillum can also act as an antagonist for other species like plant pathogens, and even if it remains much less well-known, it could play a role as a biocontrol agent (Fig. 16.1). Its antagonistic action against microbial or fungal plant pathogens can be evaluated in vitro by confronting *Azospirillum* and plant pathogen strains, on agar plate, and quantifying the growth inhibition of the pathogen (Loaces et al. 2011; Russo et al. 2008; Sultana et al. 2024; Tortora et al. 2011). In another study, six strains of *Azospirillum* (Asp2, 10, 22, 30, 32, and 39) were shown to inhibit mycelial development of five major soil-borne phytopathogens in dual culture assays (*Macrophomina phaseolina, Sclerotium rolfsii, Rhizoctonia solani, Fusarium oxysporum f.sp. ricini, and Alternaria tenuissima)* (Sultana et al. 2024). *A. brasilense* has also been shown to inhibit several SPBs (siderophore-producing bacteria) isolated from rice during dual culture assays, like *Pseudomonas oryzihabitans, Pantoea ananatis* JA04, *Pantoea sp.* CWB600, and *Sphingomonas sp*. EM0335 (Loaces et al. 2011). The biocontrol activity of *A. baldaniorum* Sp245 was demonstrated both in vitro and in vivo against the pathogenic fungus *Rhizoctonia spp*. On the agar plate, the hyphal growth was 7.8 ± 0.3 cm in the control and 3.5 ± 0.1 cm in the presence of *A. baldaniorum* Sp245. In planta, all inoculated treatments showed a 100% rate of plant survival, compared with a 0% rate of plant survival as seen in the control (Russo et al. 2008). Similarly, the tomato bacterized with *Azospirillum* Sp7 or BNM-65 strains shows a higher survival rate (all fresh-market tomato control plants were dead 30 days after inoculation with *C. michiganensis pv. michiganensis*, while 40–30% of the plants pre-treated with Sp7 or BNM-65 were still alive) (Romero et al. 2003). Also, inoculation of *Azospirillum* sp. B510 exhibited enhanced resistance against the rice blast fungus *Magnaporthe oryzae* and the bacterial pathogen *Xanthomonas oryzae* (Yasuda et al. 2009). Sankari *et al.* demonstrated that the exopolysaccharides (EPS) of *Azospirillum* sp., applied with a concentration of 200 ppm, enhance the biocontrol against *Pyricularia oryzae* (Sankari et al. 2011).

Notably, evidence indicates that *Azospirillum* can also antagonize other PGPR. Experiments suggest that diffusible metabolites *from A. baldaniorum* Sp245 can reduce *P. fluorescens* A506 biofilm development (Díaz et al. 2022). *Azospirillum* is able to secrete factors, such as type 6 secretion system effectors (Cassan et al. 2021), bacteriocins or siderophores (Combes-Meynet et al. 2011; Tapia-Hernández et al. 1990), or to interfere with the quorum sensing (QS) process of pseudomonads by quenching the signals (Gualpa et al. 2019). All these *Azospirillum* functional traits are strong candidates to explain A506 growth restriction. The complex interaction between *A. baldaniorum* Sp245 and *P. fluorescens* A506 will be more thoroughly explored later.

Given the complexity of these interactions, it is clear that the selection of strains to mix with *Azospirillum* in multi-strain inoculants needs special attention, since in some cases, antagonistic interactions may occur and could lead to the absence of phytobeneficial effects (Felici et al. 2008; García de Salamone et al. 2012; Trabelsi et al. 2011).

16.2.3 Synergism Between Azospirillum *and Other Rhizobacteria*

Some synergistic effects were observed between *Azospirillum* and *Pseudomonas* strains when using lower population levels of *Pseudomonas* than *Azospirillum* populations (Fig. 16.1). To address this issue at an experimental level, we asked whether DAPG at low concentrations could promote the mechanisms by which *Azospirillum* adapts to the rhizosphere of its host plant. To confirm this hypothesis, we developed a differential fluorescence induction approach coupled with cytometric sorting to identify *Azospirillum* genes induced by low concentrations of DAPG (0.1 and 10 μM) (Combes-Meynet et al. 2011). We showed that DAPG is able to induce in vitro the expression of genes involved in the processes of root colonization and phytostimulation of *A. baldaniorum* Sp245, especially the *ppdC* and *nirK* genes. An analysis was also carried out on the various phenotypes associated with these processes, such as mobility, adhesion, and auxin production, for which stimulation by low concentrations of DAPG was observed.

Later, Pagnussat et al. (2016) reported the synergistic interaction of *A. baldaniorum* Sp245 and *Pseudomonas protegens* CHA0, a DAPG-producing strain. When they are co-cultured as biofilms under static conditions, higher total populations of both species are achieved. In turn, the higher growth of *Azospirillum* was found to be due to an increase in biofilm formation. Again, this cooperation was shown to be dependent on a lower initial population of *Pseudomonas* and required a particular growth medium to favor *Azospirillum* biofilm formation and prevent *Pseudomonas* siderophore production. Notably, when these same strains were put to interact on an agarized medium, the cooperation turned into competition, leading to the inhibition of *Azospirillum* growth (Pagnussat et al. 2016). This exemplifies how high the plasticity of the interaction between these two bacterial groups can be.

16.2.4 Exploitation by Azospirillum *of Other Beneficial Rhizobacteria*

The high complexity in bacterial interactions that occurs in single- or multi-species biofilms has been a focus of research in recent years (Sadiq et al. 2021). It is well-established that in microbial communities, it is common for certain genotypic variants (in the case of single-species biofilms) or species (in the case of multi-species biofilms) to produce public goods, defined as compounds or functions that provide collective benefits, while others (i.e., cheaters) use them to enhance their growth, thus improving their fitness at the expense of the producers but without contributing to the community. If, in addition, the cheaters affect the producer's fitness, the relationship turns into an exploitation, which might generate instability within the microbial community. This intricate social dynamic between bacteria is likewise manifested when *Azospirillum* and *Pseudomonas* interact on agarized media. While it is common to observe *Pseudomonas* antagonism against *Azospirillum*, superimposed effects arise in certain strain combinations, unveiling a strikingly more complex interaction (Couillerot et al. 2011; Maroniche et al. 2018; Pagnussat et al. 2016). As aforementioned, *Azospirillum* growth could be inhibited or enhanced depending on the concentration of DAPG in the medium (Couillerot et al. 2011). Later, Maroniche et al. (2018) reported a similar observation: when *A. baldaniorum* Sp245 colonies interact with *P. fluorescens* A506 biofilms on RC medium, diffusible compounds produced by A506 biofilms stimulate the growth of nearby *Azospirillum* colonies but inhibit the more distant ones. This *Azospirillum*-stimulatory effect was most evidently produced by A506 but is not exclusive to this strain; other ones such as *Pseudomonas* sp. LSR1 can also induce it (Maroniche et al. 2018). More comprehensive research on the interaction of *A. baldaniorum* Sp245 with *P. fluorescens* A506 or *Pseudomonas* sp. LSR1 revealed that not only these strains can coexist in mixed macrocolony-type biofilms but also that *Azospirillum* is highly favored by the interaction while, unexpectedly, *Pseudomonas* is negatively affected. For example, strain A506 or LSR1 can stimulate the growth of *A. baldaniorum* Sp245 by approximately 250% or 600%, respectively, when seeded in a 1:1 ratio while building the mixed biofilm, resulting at the same time in lower *Pseudomonas* final populations in both cases (Díaz et al. 2022, Díaz et al., unpublished results). Moreover, in addition to increasing its growth, *A. baldaniorum* Sp245 benefits from the interaction with A506 or LSR1 in dual-species macrocolonies by acquiring an increased tolerance to UV irradiation (Díaz et al., unpublished results).

Given the well-known biocontrol capacity of *Pseudomonas*, it is really surprising that *A. baldaniorum* Sp245 is not inhibited at all by *P. fluorescens* A506 on NNi medium but rather experiences a growth boost directly proportional to the initial population of pseudomonads building the mixed biofilm. On RC medium, the exploitation dynamic is less stable and highly dependent on the initial populations of both bacteria forming the biofilm, turning into mutual competition when A506 is in equal or higher initial numbers than Sp245. Notably, competition on RC can be overridden if nitrate is present, uncovering the vital role of this nutrient, in addition to O_2 levels (see Chap. 6), in regulating bacterial physiology to allow the manifestation of *Azospirillum* exploitation capacity against *Pseudomonas* (Díaz et al. 2022).

All attempts to detect diffusible compounds responsible for *Azospirillum* growth stimulation failed, evidencing that direct physical contact is a requisite to observe the effect. Instead, confocal microscopy of *A. baldaniorum* Sp245 and *P. fluorescens* A506 cells within the biofilm, along with further experimentation with different culture conditions, *P. fluorescens* A506 biofilm mutants and a microaerophilia reporter strain of Sp245, demonstrated that the increased growth of *Azospirillum* in the mixed biofilm is driven by at least two mechanisms: (1) its ability to make use of the matrix produced by *Pseudomonas*, which forms a massive scaffold that *Azospirillum* can adhere to and proliferate within and (2) the reduction in oxygen concentration inside the biofilm, presumably due to the accelerated aerobic metabolism of A506, which creates conditions in which *Azospirillum* can grow more favorably (Díaz et al. 2022). This last conclusion is supported by preliminary transcriptomic studies that show a downregulation of *A. baldaniorum* Sp245 genes coding for oxidative phosphorylation components when A506 is present (Amenta et al. unpublished results). In addition, the ultrastructure of Sp245-A506 biofilms unveiled a more complex relationship between both strains, being segregated (suggesting competence) or mixed (suggesting cooperation) depending on the sector of the biofilm (Díaz et al. 2022). In contrast to *Azospirillum* response to the interaction, the mechanisms behind the reduced fitness of the interacting *Pseudomonas* is not well understood yet and requires more research.

16.3 Improvement of the Interactions of *Azospirillum* with Plants by Combining *Azospirillum* with Other Microbial Partners

16.3.1 Allowing Increased Root Colonization with Biofilm Established by Pseudomonas*: Exploitation by* Azospirillum *of* Pseudomonas *Public Goods*

As mentioned earlier, it is of great interest to investigate the potential of combining *Azospirillum* with other plant-beneficial microbes to produce synergistic effects on plant development. However, predicting how the interaction with a partner might impact *Azospirillum* establishment in the rhizosphere and the successive root colonization process is not trivial. Minimally, it might be desirable to test strain compatibility in vitro before formulating mixed inoculants in order to discard potentially harmful partners. In this regard, Diaz et al. (2023) contrasted in vitro results obtained from mixed biofilms produced by different combinations of *Azospirillum* and *Pseudomonas* strains and their performance for lettuce root colonization and growth promotion when co-inoculated. It was shown that *A. argentinense*

Az39 + *P. fluorescens* A506 combination, which has a high in vitro compatibility when forming mixed biofilms, is able to stimulate lettuce growth synergically, with its impact being most pronounced in the root system. Similar results have been observed when co-inoculating *A. baldaniorum* Sp245 and *P. fluorescens* A506 on lettuce (Diaz et al., unpublished results). Although this synergism for plant growth stimulation might be attributed to the capacity of *Azospirillum* to exploit *P. fluorescens* A506 in mixed biofilms (Díaz et al. 2022), further experiments carried out in our lab under rhizospheric environments indicate that *A. baldaniorum* Sp245 and *P. fluorescens* A506 interaction dynamic in vivo is different. *A. baldaniorum* Sp245 is not able to boost its growth in mixed biofilms when agarized root exudates are used as the sole nutrient source, nor is it negatively affected. In contrast, *P. fluorescens* A506 growth in the biofilm decreases in the presence of *Azospirillum*, similar to what is observed in vitro. In line with these results, if Sp245 + A506 mixed biofilms are seeded in situ over lettuce roots, both strains tend to segregate and colonize different niches. Even when this suggests competition between them, images show that they not only co-exist in this segregated manner but are also able to colonize the roots more abundantly when co-inoculated than when inoculated individually, a result that is corroborated by viable cell counting experiments (Diaz et al., unpublished results). Thus, the fact that both strains benefit from each other's presence in terms of root colonization might partially account for the synergistic effect on plant growth.

16.3.2 Azospirillum *Root Colonization Could Be Impaired by Other PGPR Partners*

In 2017 Abo-baker et al. observed that mixing of *P. fluorescens* with *Azospirillum lipoferum* in the wheat rhizosphere resulted in a highly significant 97.87% reduction in the population of *A. lipoferum* compared to *A. lipoferum* inoculated individually and consequently suggested that a compatibility test has to be done before recommending any multibiofertilizer inoculants (Abo–baker Basha and Khalaphallah 2017). In addition, depending on the sensitivity of the *Azospirillum* strains to DAPG produced by certain *Pseudomonas* strains, differences in the coexistence of *Pseudomonas* and *Azospirillum* on the root system of the same plant were observed. For example, the distribution of the *A. baldaniorum* Sp245 population (intermediate sensitivity to DAPG) on wheat roots is affected by the *P. ogarae* F113 strain: these two populations colonize different root zones, whereas they develop close to each other when DAPG biosynthesis is inactivated. Populations of DAPG-producing fluorescent *Pseudomonas* therefore have a major impact on the spatial distribution of *Azospirillum* populations interacting with the plant, with consequences on the positive impact of *Azospirillum* strains on the plant growth (Couillerot et al. 2011; personal communication).

The T6SS of *Pseudomonas spp.* is also found to be involved in the control of root colonization of *Azospirillum*. The number of *A. baldaniorum* Sp245Gm cells recovered from the roots of wheat plantlets was significantly lower by 1 log unit when co-inoculated with the T6SS-expressing wild-type strain *P. putida* KT2440R, but when all the T6SS systems were nonfunctional the reduction of CFU was not significant in comparison to single *A. baldaniorum*-inoculated plants (Maroniche et al. 2018).

It is interesting to know that the opposite effect, i.e., a reduction in the populations of the *Azospirillum* partner, can also exist. Indeed, when mixing *A. lipoferum* and *P. fluorescens* in the rhizosphere of wheat, the results showed a significant decrease in the number of *P. fluorescens*, reaching 28.6% under sterilized soil and 82.94% in unsterilized soil compared to *P. fluorescens* inoculated alone (Abo–baker Basha and Khalaphallah 2017). Likewise, *A. argentinense* Az39 can negatively affect the early establishment of *P. fluorescens* A506 in lettuce seedlings, leading to lower counts of the pseudomonads in the roots (Díaz et al. 2023).

16.3.3 Improving the Antagonistic Activities of Azospirillum *Strains Against Fungal Pathogens with PGPR Partners*

Various studies demonstrate the efficiency of *Azospirillum* in preventing bacterial or fungal plant diseases. *Azospirillum* can reduce the incidence and severity of *Rhizoctonia solani* damping-off on tomato (Gupta 1995), prevent bacterial speck disease development of tomato caused by *Pseudomonas syringae pv.* tomato (Bashan and de-Bashan 2002), improve resistance of cherry tomato to bacterial canker induced by *Clavibacter michiganensis* (Romero et al. 2003), crown gall disease, *Cucumis sativus* disease (Hassouna et al. 1998), bacterial leaf blight of mulberry (Sudhakar et al. 2000), *Prunus cerasifera* disease (Russo et al. 2008), and anthracnose symptoms on strawberry plants (Tortora et al. 2012), and can enhance disease resistance in rice (Yasuda et al. 2009). *Azospirillum brasilense* REC3 is also able to reduce symptoms of anthracnose caused by *Colletotrichum acutatum* M11, a strawberry pathogen, by participating actively in the induction of plant systemic protection, including the reinforcement of plant cell walls, by increasing the content of total soluble phenolic compounds and callose depositions and the transient accumulation of salicylic acid (Tortora et al. 2011).

16.3.4 Mycorrhization/Bacterial Colonization Synergism

Consortia comprising one *Pseudomonas* strain, a mycorrhizal inoculant mix composed of three *Glomus* strains, and one strain of *Azospirillum* (i.e., three different consortia containing three different *Azospirillum* strains were constituted) were inoculated on maize seeds. The effects on maize were measured notably using secondary metabolic profiling and biomass measurements. All the three-component consortia have similar phytostimulation effects, despite different secondary metabolic responses (Couillerot et al. 2013; Walker et al. 2012). Maize secondary metabolite profiles varied between treatments, and especially when plants were inoculated by single PGPR and by consortia, pointing out that consortium effects could not be predicted based on knowledge of single inoculation effects. In addition, maize elaborated specific metabolic patterns according to the *Azospirillum* strains present within the three-component consortia (Couillerot et al. 2013).

16.3.5 Co-inoculation of Azospirillum *with Nitrogen Fixers for Improving Nitrogen-Fixing Symbiosis*

Bioinoculant products have been composed of only one microorganism for a long time. Several studies have stressed the positive impact of multi-species consortia on plant growth, plant health, and plant physiology. By combining microorganisms with different plant-beneficial functions, one can expect better effects on the plant due to synergistic effects of the microorganisms rather than when one plant-beneficial microorganism is inoculated. The impact of multi-species bioinoculants on plant growth was studied as early as 1995, with *Azospirillum* and *Arthrobacter* strains used as nitrogen fixers and *Agrobacterium* as a phosphate-solubilizing bacterium (Belimov et al. 1995). Indeed, inoculation of barley with the consortium allowed increased nitrogen and phosphate plant uptakes, compared to single inoculations. The presence of *Azospirillum* could promote the establishment of symbiosis between rhizobia and different legume species, generally leading to increased nodulation and an earlier onset of biological nitrogen fixation due to the presence of the free-living bacterium in the system (Burdman et al. 1997; González and Lluch 1991; Groppa et al. 1998; Okon and Vanderleyden 1997). Numerous studies have reported that *Azospirillum* not only enhances the establishment of the rhizobia-legume symbiosis under agronomic conditions but also increases crop yields of these species by promoting plant growth and development and improving resistance to unfavorable conditions (Andreeva and Okorokov 1993; Benintende et al. 2010; Cassán et al. 2020; Fabbri and Del Gallo 1995; Itzigsohn et al. 1993; Marks et al. 2013). Galal (1997) demonstrated that co-inoculating soybean seeds with *A. brasilense* and *B. japonicum* significantly increased dry matter content, nitrogen percentage, and assimilated nitrogen compared to plants inoculated only with *B. japonicum* (Galal 1997). This effect was attributed to the bacterium's ability to produce growth-regulating compounds that modify the plant's morphophysiology, particularly its roots. Similarly, Benintende et al. (2010) reported that soybean seed co-inoculation with *B. japonicum* and *A. brasilense* resulted in significant increases in nodule weight, shoot biomass, and nitrogen accumulation in the aerial biomass compared to rhizobia-only inoculation. Molla et al. (2001) found that inoculation with *Azospirillum* or its co-inoculation with *B. japonicum* in soybean crops increased root volume, dry root weight, and aerial biomass. Later, Cassán et al. (2009) observed that co-inoculation with *B. japonicum* and *A. argentinense* altered soybean seed germination capacity and plant growth compared to the uninoculated control. Additionally, under water- and nitrogen-limited conditions, Burdman et al. (1997) demonstrated that co-inoculation with *B. japonicum* and *Azospirillum* could enhance the yield of certain legumes. Hungria et al. (2013) and Nogueira et al. (2018) reported that co-inoculation with *Bradyrhizobium* and *Azospirillum* increased soybean (Glycine max (L.) Merrill) grain yield by over 15% and 200 kg·ha^{-1}, respectively. Studies on soybean co-inoculation in Argentina and Brazil also found respective increases in nodulation of over 5% (Fipke et al. 2016; Galindo et al. 2018; Hungria et al. 2015) and 10% (Benintende et al. 2010; Ferraris and Couretot 2013; Ferraris and Couretot 2011) over the *Bradyrhizobium* inoculation.

16.3.6 *Proposed Interaction Models Between* Azospirillum *and Rhizobial Symbiont and Combination of PGP Properties from the Different Members for Increased Effects on Plants*

Members of the genus *Azospirillum* can enhance rhizobia-legume symbiosis through various potential mechanisms (Ferri et al. 2017; Remans et al. 2008; Rodelas et al. 1999). The production of phytohormones, particularly auxins and cytokinins, by *Azospirillum* has been the most commonly proposed mechanism to explain the improvement in symbiosis establishment (Ferri et al. 2017; Molla et al. 2001; Srinivasan et al. 1996; Vessey and Buss 2002). Okon and Vanderleyden (1997) suggested that the plant growth promotion induced by *Azospirillum* is primarily due to its ability to produce phytohormones, especially auxins such as indole-3-acetic acid (IAA). Since Thimann (1936) first proposed that auxins play a crucial role in nodule ontogeny in rhizobia-legume symbiosis, numerous studies have indicated that changes in auxin concentrations—whether through inoculation with auxin-producing bacteria or exogenous application of these compounds—can significantly alter nodule organogenesis patterns (Schmidt et al. 1988). For instance, the co-inoculation of an *Ensifer meliloti* strain inefficient in IAA production with *A. brasilense* (IAA producer) on alfalfa seeds led to a significant increase in the number of nodules on the primary root. Remans et al. (2008) provided direct evidence of the growth-promoting effects of IAA through co-inoculation studies involving *A. brasilense* and *R. etli* in beans, using an *A. brasilense* *ipd*C gene knockout mutant. Other studies have confirmed the beneficial effects of co-inoculation but have proposed alternative mechanisms to explain them (Yahalom et al. 1990). For example, co-inoculation of bean seedlings with *A. brasilense* resulted in increased flavonoid production at the root level, leading to higher expression of *nod* genes in the inoculated *Rhizobium* strains (Burdman et al. 1997). Dardanelli et al. (2008) demonstrated the positive effects of *A. brasilense* co-inoculation on bean development and nodule organogenesis under hydroponic conditions, associating these effects with the presence of such molecules.

Based on the results presented so far, we can assume that the current biological model explaining the beneficial effects of legume co-inoculation with *Azospirillum* remains limited and uncertain. It has been analyzed and understood from the perspective of a hypothetical interaction between *Azospirillum* and the inoculated plant, primarily through the production of specific phytohormones and other biologically active growth-regulating compounds, particularly affecting root development. However, the current model considered the interaction between *Azospirillum* and rhizobia as an effector in improving rhizobia-legume symbiosis establishment.

The recent publication of Torres et al. (2022) confirmed that previous incubation of *B. japonicum* and *A. argentinense* improves the *Bradyrhizobium*-soybean symbiosis and increases both plant growth under greenhouse and crop productivity under field conditions in comparison with the single inoculation with the bradyrhizobia or the combination of both bacterial strains during the seeds sowing (co-inoculation). The higher values of exopolysaccharides and cell biomass obtained for the bradyrhizobia by the previous incubation with *A. argentinense* correlated with the fact that *B. japonicum* increased its survival on the soybean seeds after inoculation and improved the symbiosis process with the legume. The physiological changes observed in bradyrhizobia behavior and their impact on the *Bradyrhizobium*-soybean symbiosis depend, at least partly, on the *A. argentinense* capacity to produce and release indole-3-acetic acid into the culture medium. These changes are also affected by the time of *B. japonicum* exposition to the *Azospirillum* strain or the IAA molecule.

16.3.7 Synergism Allows Better PGP Gene Expression or Activity: Metabolic Interaction, Signaling Effect

In planta, to study the impact of the signaling effects produced by *Pseudomonas* on *Azospirillum* during the interaction with the plant partner, we carried out successive inoculations of the two PGPR populations: first *Azospirillum*, then 48 h later *Pseudomonas* at a lower population level. Under these conditions, we observed a positive effect of inoculation with a DAPG-producing *Pseudomonas* strain on the expression in the rhizoplane of two *Azospirillum* phytobeneficial genes, *ppdC* and *nirK*, involved in the production of phytohormones, whereas inoculation with a *Pseudomonas* strain incapable of producing DAPG had no effect. A positive effect of successive inoculations of strains Sp245 and F113 was then observed on wheat growth. These results corroborate the hypothesis that DAPG can act as a signal molecule by promoting the establishment of functional beneficial interactions between PGPRs and their host plant (personal communication). Hence, synergism interactions between *Pseudomonas* and *Azospirillum* could improve their phytobeneficial properties *in planta*.

16.4 "Optimized Protocol" for Combining *Azospirillum* with Other Microbial Members to Establish Mixed Inoculant

The protocols will cover first the evaluation of the compatibility between *Azospirillum* strains with other microorganisms under in vitro conditions and second the in planta co-inoculation of *Azospirillum* with other microorganisms even if its compatibility with other microorganisms is not fully met.

16.4.1 Evaluating on Petri Plates the Compatibility Between Azospirillum *and Other Microorganisms*

The antagonistic relation between *Azospirillum* and other bacteria or fungi can be addressed in different ways.

16.4.1.1 Bacteria Antagonism at a Distance

This capacity can be observed by plating *Azospirillum* (10^3 CFU) on a plate, letting it dry, and putting a drop of 10 μL of the other bacteria on the plate. After 48–72 h at 28 °C, the size of the inhibition radius and the size and morphology of *Azospirillum* colonies are analyzed with 7× magnification or a transparency scanning unit (Maroniche et al. 2018; Pagnussat et al. 2016). *Azospirillum* cells (2×10^7 cells.mL^{-1}) can also be cultured in a molten agar overlay (1.5% w/v) and drops of 15 μL from bacterial cultures of another strain (at 10^9 cells.mL^{-1}), spotted on the *Azospirillum* overlay (Couillerot et al. 2011). After 72 h incubation at 28 °C, the width (radius) of the inhibition zone surrounding the bacterial colony is measured. The compatibility between *Azospirillum* and the other strain can then be assessed by quantifying the inhibition zone in the *Azospirillum* overlay. The potential inhibitory activity of *Azospirillum* strains against other strains can also be tested using a similar protocol but reversing partners.

16.4.1.2 Direct Contact Interaction

This capacity can be analyzed by plating 10^9 CFU of a bacterium on a plate, letting it dry, and putting drops of 10 μL from a dilution culture of *Azospirillum* spp. (5×10^8 CFU/mL). After 3 days at 28 °C, the number and morphology of the colonies are analyzed with an optical magnifier. The *Azospirillum-Pseudomonas* in-contact interaction can also be analyzed in mixed macrocolonies (Maroniche et al. 2018), as described below.

16.4.1.3 Production of Diffusible Compounds

The implication of diffusible compounds can also be determined by the "medium slice assay" as described by Díaz et al. (2022). A drop of 10 μL of a bacterial culture of the strain producing the metabolite (tagged with GFP) is seeded on a plate. After drying, a rectangular portion of the medium with the dried drop in its center is excised, rotated 90° and then placed upside-down, forming a "bridge" across the hole that is left in the medium. Then, a drop of *Azospirillum* tagged with mCherry is deposited on top of the slice, aligned to the previous macrocolony located below on the opposite side of the medium. Plates are then incubated 2–3 days at 28 °C and fluorescence of both strains is measured (Díaz et al. 2022).

16.4.1.4 Fungal Antagonism

For in vitro antagonism tests against fungal pathogens, potato dextrose agar (PDA) is currently used as media for fungal growth. *Azospirillum* is inoculated on PDA and incubated 24 h at 28 °C before putting fungal plugs in the middle of the plate. Seven days of incubation are required before analyzing the inhibition of hyphal growth compared to a control without bacteria (Russo et al. 2008). Similar assays can be performed on iron-limited PGA media supplemented with CAS solution for the detection of *Azospirillum* siderophores (Tortora et al. 2011).

16.4.2 *Evaluating the Behavior of* Azospirillum *Cells with Other Bacteria in Mixed Macrocolony*

Interaction in dual-species macrocolony biofilms (Díaz et al. 2022) is performed by using a GFP-tagged *Pseudomonas* (or other interacting species of interest) and mCherry-tagged *Azospirillum*. Bacterial suspensions of both strains are used individually or combined at different ratios to seed macrocolonies on NNi or any other suitable medium (King's B, LB, and NA should not be used for analyzing *Azospirillum* interaction with *Pseudomonas* due to siderophore production by the latter). After 48–96 h of incubation at 28 °C, biofilms can be analyzed macroscopically. Macroscopic photographs of the macrocolonies can be obtained with a magnifying glass or digital microscope. To analyze bacterial growth and distribution within the biofilm, plates are photographed on a UV/blue-transilluminator with appropriate emission filters. Growth estimation by fluorescence images is done as described by Díaz et al. (2022) and should be confirmed by CFU quantification. To that end, macrocolonies are excised, collected in 1.5 ml microcentrifuge tubes containing 900 μl of saline solution, and disassembled with a pestle for subsequent CFU count.

These different types of in vitro experiments, and especially the mixed macrocolony study, are useful to determine the potential compatibility between *Azospirillum* and other microorganisms in mixed inoculants prior to plant co-inoculation.

16.4.3 Co-inoculation of Azospirillum *with Other PGPR on Plant*

Azospirillum is grown in liquid media that could be Oxoid Nutrient medium (Russo et al. 2008), NFb semi-solid medium supplemented with 1 g l^{-1} ammonium chloride (Romero et al. 2003), Nutrient Broth (Maroniche et al. 2018), or Luria Bertani (Couillerot et al. 2011). The other PGPR strain is also grown in any suitable medium. After overnight or 2 days of growth, depending on the medium used, the cells of both *Azospirillum* and the other strain are washed twice in sterile water, 0.9% NaCl solution, or $MgSO_4$ 10 mM solution and adjusted to a specific CFU/mL (often 10^7 CFU mL^{-1}) before being used as inocula (Couillerot et al. 2011). Cells from the *Azospirillum* strain and from the other genus could be inoculated simultaneously on seeds at 10^6 c.f.u. each. However, if the two bacterial strains are incompatible, sequentially inoculating the plants first with *Azospirillum* cells and two days later with the other bacterial strain could result in a greater stimulation of plant growth compared to when simultaneous inoculation is used (Vacheron et al. 2012). Shoot and root biomasses, such as root development parameters, could be assessed after several days of growth under specific greenhouse conditions depending on the plant species by measuring root and shoot biomasses and by characterizing the root system architecture using WinRHIZO (Regent Instruments, Canada) or an equivalent system. Real-time PCR with specific markers for each strain could be used to quantify the population levels of each member of the inoculant as described in (Couillerot et al. 2013).

16.4.4 *Co-inoculation of* Azospirillum *with Nitrogen-Fixing Bacteria on Plants: The* Azospirillum-Bradyrhizobium-*Soybean Model*

In the following experimental section, we will describe a procedure to evaluate the prior interaction between *Azospirillum* and *Bradyrhizobium* and its effect on Bradyrhizobium-soybean symbiosis establishment.

16.4.4.1 Bacterial Cultures

B. japonicum (or *B. diazoefficiens*) are inoculated in 250 ml flasks containing 100 ml of yeast extract manitol (YEM) culture medium (Vincent 1970). Cultures are incubated at 30 °C with orbital shaking (180 rpm) until the exponential growth phase is reached. The purity of the cultures is checked routinely on YEM agar (Vincent 1970) and TSA plating (Trypticase Soybean Agar) at 10% (v/v). *A. argentinense, A. brasilense, or A. baldaniorum* are inoculated in 250 ml flasks containing 100 ml in sterile Luria Bertani broth medium (Bertani 1951) in darkness until OD_{595} reaches 1.5 at 36 °C overnight. The purity of the cultures was checked routinely on Luria Bertani agar medium (Bertani 1951).

16.4.4.2 Experimental Design

Naked soybean seeds (*Glycine max* (L.) Merrill) without previous treatment with fungicides, insecticides, micronutrients, or other products are used for these studies. Seeds are inoculated with individual or combined cultures of *Bradyrhizobium* and *Azospirillum*, according to the treatments described below. The *Bradyrhizobium* titer is adjusted to 5.0×10^9 cfu.ml^{-1} obtained at the late exponential growth phase in YEM, as described by Vincent (1970). The *Azospirillum* titer is adjusted to 5.0×10^8 cfu ml^{-1} obtained at late exponential growth phase in Luria broth, as described by (Molina et al. 2018). Inoculation doses are adjusted to obtain a final volume of 12 ml kg^{-1} of soybean seeds. The following treatments are performed: (i) uninoculated seeds treated with phosphate-buffered solution (12 ml kg^{-1}) (control); (ii) seeds inoculated with equal volume (1:1) of *Bradyrhizobium* and phosphate-buffered solution; (iii) seeds inoculated with equal volume (1:1) of *Azospirillum* and phosphate-buffered solution; (iv) seeds co-inoculated with *Bradyrhizobium* and *Azospirillum* in a 1:1 ratio; (v) seeds inoculated with a mix, in a 1:1 ratio, of *Bradyrhizobium* and *Azospirillum*. In the treatment (iv), each microorganism is applied separately during the inoculation process, while for the treatment (v) equal volumes of both microorganisms are mixed and maintained under static conditions at room temperature (25 °C) during 24 h before the inoculation of the seeds. After the application of the treatments, the seeds are maintained under sterile laminar air flow conditions at room temperature (25 °C) until analysis. Treatments are performed in triplicate from three independent experiments.

16.4.4.3 Bacterial Physiological State

To determine the number of viable cells (cfu ml^{-1}) in each culture or in the mix of both, the microdrop quantification method is performed (Miles et al. 1938). Plates containing YEM medium (Vincent 1970) or Congo Red medium (Cáceres 1982) are used for *Bradyrhizobium* and *Azospirillum*, respectively. The plates are analyzed after incubation at 30 °C for 7 days in the case of *Bradyrhizobium* and 37 °C for 4 days in the case of *Azospirillum*, as described by Cassán et al. (2009) modifications. In the case of the combined cultures, plates containing both YEM and Congo Red medium are used at the temperatures and incubation time mentioned before. Quantification of IAA is performed by spectrophotometry (Glickmann and Dessaux 1995) and confirmed by HPLC following the protocols described in Chap. 10. Briefly, aliquots of 1000 μl of bacterial culture are centrifuged at 11.3 g for 10 min. Next, the samples are filtered (0.2 μm) and 500 μl of the supernatant are mixed with 500 μl of Salkowski's reagent and gently shaken in an inverted position at least 10 times. The samples are incubated in the dark for 30 min, and absorbance is measured at 530 nm. An aliquot of filtered supernatants is injected into the HPLC equipment. Quantification of EPS is performed following Torres et al. (2018). Briefly, the samples are centrifuged, and then the supernatants are filtered and treated with DNase I and proteinase K as previously described (Bergmaier et al. 2001; Read and Costerton 1987). The EPS precipitation is obtained after application of a threefold volume of high-purity ethanol (95% v/v) at −20 °C and overnight incubation. A second ethanol precipitation is performed to ensure proper recovery of EPS (Chang et al. 2007). EPS is dried at room temperature before resuspending in deionized water. Total carbohydrate content (EPS mg^{-1} biomass) is measured with the phenol–sulfuric acid method, following DuBois et al. (1956) with glucose as a standard.

16.4.4.4 Seed Recovery Assays

Soybean seeds that had been either inoculated or co-inoculated are maintained in aseptic environmental conditions at room temperature (≅25 °C) for 4 h before initial analysis and then 6 days before a second analysis, as described by Torres et al. (2018). Bacterial cell count and survival factor percentage (SFP) are determined for *Bradyrhizobium* and *Azospirillum* following the criteria by Penna et al. (2011). The plates were incubated at 30 °C for 7 days in the case of *Bradyrhizobium* and at 37 °C for 4 days in that of *Azospirillum*. Additional counts are performed on uninoculated seeds (control). Results are expressed as the number of viable cells recovered from the seeds (cfu $seed^{-1}$) and the SFP as the cfu $seed^{-1}$ 4 h after inoculation, according to Penna et al. (2011). Treatments are performed in triplicate from three independent experiments.

16.4.4.5 Plant-Microbe Interaction Assays

The treated soybean seeds from each treatment are sown in plastic pots (300-ml volume capacity) containing vermiculite as a solid substrate. Three seeds per pot are planted in six pots and irrigated with sterile N-free Hoagland's solution (25% v/v) (Hoagland and Broyer 1936), keeping them at field capacity during the study. The seedlings are maintained inside a growth chamber for 21 days with a photoperiod of 16/8-h light regime, 30/20 °C temperature, and a relative humidity of 80%. At the end of the experiment, the following parameters are measured: (1) number of nodules located on main root (MRN) per plant, (2) number of nodules located on secondary roots (SRN) per plant, (3) total number of nodules on roots (RN) per plant, (4) percentage of plants with three or more nodules on the main root (BT), following Burton (1978), and (5) shoot and root dry weight (SDW and RDW). Treatments are performed with six replicates per treatment.

Acknowledgments We thank the ECOS-SUD program for their support on the project A23B03 untitled 'A metabolomic approach for the identification of chemical signals from *Pseudomonas* that stimulate the growth and efficacy of *Azospirillum* as a crop biostimulant'. The drafting of this chapter was initiated during the mission of C. Prigent-Combaret and E. Derollez, in Argentina, in the team of G. Maroniche (FCA-UNMdP, Balcarce) in November 2024. This article is based upon work from COST Action MiCropBiomes, CA22158, supported by COST (European Cooperation in Science and Technology). This research was also supported by funding from the Joint Programming Initiative on Antimicrobial Resistance (JPIAMR), under the JPIAMR-ACTION Joint Transnational Call 2021 (Grant No. JPIAMR2021-194 to C.PC. and E.D.).

References

Abo-baker Basha AA-E, Khalaphallah R (2017) The biological control bacteria pseudomonas fluorescens inhibits free nitrogen fixing bacteria in the rhizosphere. Int J Curr Microbiol App Sci 6:223–231. https://doi.org/10.20546/ijcmas.2017.601.027

Alamri SA, Mostafa YS (2009) Effect of nitrogen supply and Azospirillum brasilense Sp-248 on the response of wheat to seawater irrigation. Saudi J Biol Sci 16:101–107. https://doi.org/10.1016/j.sjbs.2009.10.009

Almario J, Prigent-Combaret C, Muller D et al (2013) Effect of clay mineralogy on iron bioavailability and rhizosphere transcription of 2,4-diacetylphloroglucinol biosynthetic genes in biocontrol *Pseudomonas protegens*. Mol Plant-Microbe Interactions® 26:566–574. https://doi.org/10.1094/MPMI-11-12-0274-R

Andreeva NA, Okorokov LA (1993) Purification and characterization of highly active and stable polyphosphatase from Saccharomyces cerevisiae cell envelope. Yeast 9:127–139. https://doi.org/10.1002/yea.320090204

Bacilio M, Rodriguez H, Moreno M et al (2004) Mitigation of salt stress in wheat seedlings by a gfp-tagged Azospirillum lipoferum. Biol Fertil Soils 40. https://doi.org/10.1007/s00374-004-0757-z

Bacilio-Jiménez M, Aguilar-Flores S, del Valle MV et al (2001) Endophytic bacteria in rice seeds inhibit early colonization of roots by *Azospirillum brasilense*. Soil Biol Biochem 33:167–172. https://doi.org/10.1016/S0038-0717(00)00126-7

Bano Q, Ilyas N, Bano A et al (2013) Effect of azospirillum inoculation on maize (*Zea mays* l.) under drought stress. Pak J Bot:13–20

Bashan Y, de-Bashan LE (2002) Protection of tomato seedlings against infection by *Pseudomonas syringae* pv. tomato by using the plant growth-promoting bacterium *Azospirillum brasilense*. Appl Environ Microbiol 68:2637–2643. https://doi.org/10.1128/AEM.68.6.2637-2643.2002

Bashan Y, Holguin G, de-Bashan LE (2004) *Azospirillum* -plant relationships: physiological, molecular, agricultural, and environmental advances (1997–2003). Can J Microbiol 50:521–577. https://doi.org/10.1139/w04-035

Basler M, Ho BT, Mekalanos JJ (2013) Tit-for-tat: type VI secretion system counterattack during bacterial cell-cell interactions. Cell 152:884–894. https://doi.org/10.1016/j.cell.2013.01.042

Belimov AA, Kojemiakov AP, Chuvarliyeva CV (1995) Interaction between barley and mixed cultures of nitrogen fixing and phosphate-solubilizing bacteria. Plant Soil 173:29–37. https://doi.org/10.1007/BF00155515

Benintende S, Uhrich W, Herrera M et al (2010) Comparison of coinoculation with *Bradyrhizobium japonicum* and *Azospirillum brasilense* with inoculation with *Bradyrhizobium japonicum* on nodulation, growth, and N accumulation in the soybean crop. Agriscientia 27:71–77

Bergmaier D, Lacroix C, Guadalupe Macedo M et al (2001) New method for exopolysaccharide determination in culture broth using stirred ultrafiltration cells. Appl Microbiol Biotechnol 57: 401–406. https://doi.org/10.1007/s002530100764

Bertani G (1951) Studies on lysogenesis i. J Bacteriol 62:293–300. https://doi.org/10.1128/jb.62.3.293-300.1951

Biessy A, Filion M (2021) Phloroglucinol derivatives in plant-beneficial Pseudomonas spp.: biosynthesis, regulation, and functions. Metabolites 11:182. https://doi.org/10.3390/metabo11030182

Bondy SC (2023) The hormesis concept: strengths and shortcomings. Biomolecules 13:1512. https://doi.org/10.3390/biom13101512

Burdman S, Kigel J, Okon Y (1997) Effects of *Azospirillum brasilense* on nodulation and growth of common bean (*Phaseolus vulgaris* L.). Soil Biol. Biochem. Int Symp Sustain Agric Tropics Role Biol Nitrogen Fix 29:923–929. https://doi.org/10.1016/S0038-0717(96)00222-2

Burton JC (1978) Monitoring quality in legume inoculants and preinoculated seed

Cáceres EAR (1982) Improved medium for isolation of *Azospirillum* spp. Appl Environ Microbiol 44:990–991. https://doi.org/10.1128/aem.44.4.990-991.1982

Cassán F, Perrig D, Sgroy V et al (2009) *Azospirillum brasilense* Az39 and *Bradyrhizobium japonicum* E109, inoculated singly or in combination, promote seed germination and early seedling growth in corn (*Zea mays* L.) and soybean (*Glycine max* L.). Eur J Soil Biol Ecol Appl Azospirillum Plant Growth Promoting Bact (PGPB) 45:28–35. https://doi.org/10.1016/j.ejsobi.2008.08.005

Cassán F, Coniglio A, López G et al (2020) Everything you must know about Azospirillum and its impact on agriculture and beyond. Biol Fertil Soils 56:461–479. https://doi.org/10.1007/s00374-020-01463-y

Cassan FD, Coniglio A, Amavizca E et al (2021) The *Azospirillum brasilense* type VI secretion system promotes cell aggregation, biocontrol protection against phytopathogens and attachment to the microalgae *Chlorella sorokiniana*. Environ Microbiol 23:6257–6274. https://doi.org/10.1111/1462-2920.15749

Chang W-S, van de Mortel M, Nielsen L et al (2007) Alginate production by *Pseudomonas putida* creates a hydrated microenvironment and contributes to biofilm architecture and stress tolerance under water-limiting conditions. J Bacteriol 189:8290–8299. https://doi.org/10.1128/jb.00727-07

Cohen AC, Bottini R, Pontin M et al (2015) *Azospirillum brasilense* ameliorates the response of *Arabidopsis thaliana* to droughtmainly via enhancement of ABA levels. Physiol Plant 153:79–90. https://doi.org/10.1111/ppl.12221

Combes-Meynet E, Pothier JF, Moënne-Loccoz (2011) The *Pseudomonas* secondary metabolite 2,4-diacetylphloroglucinol is a signal inducing rhizoplane expression of *Azospirillum* genes involved in plant-growth promotion. Mol Plant-Microbe Interactions® 24:271–284. https://doi.org/10.1094/MPMI-07-10-0148

Couillerot O, Combes-Meynet E, Pothier JF et al (2011) The role of the antimicrobial compound 2,4-diacetylphloroglucinol in the impact of biocontrol *Pseudomonas fluorescens* F113 on *Azospirillum brasilense* phytostimulators. Microbiology 157:1694–1705. https://doi.org/10.1099/mic.0.043943-0

Couillerot O, Ramírez-Trujillo A, Walker V et al (2013) Comparison of prominent *Azospirillum* strains in *Azospirillum–Pseudomonas–Glomus* consortia for promotion of maize growth. Appl Microbiol Biotechnol 97:4639–4649. https://doi.org/10.1007/s00253-012-4249-z

Creus CM, Sueldo RJ, Barassi CA (2004) Water relations and yield in *Azospirillum*- inoculated wheat exposed to drought in the field. Can J Bot 82:273–281. https://doi.org/10.1139/b03-119

Cronin D, Moëenne-Loccoz Y, Fenton A (1997) Ecological interaction of a biocontrol Pseudomonas. £uorescens strain producing 2,4-diacetylphloroglucinol with the soft rot potato pathogen Erwinia carotovora subsp. atroseptica. FEMS Microbiol Ecol 23:95–106. https://doi.org/10.1111/j.1574-6941.1997.tb00394.x

Da Silva Oliveira CE, Jalal A, Vitória LS (2023) Inoculation with *Azospirillum brasilense* strains AbV5 and AbV6 increases nutrition, chlorophyll, and leaf yield of hydroponic lettuce. Plants 12:3107. https://doi.org/10.3390/plants12173107

Dardanelli MS, Fernández de Córdoba FJ, Espuny MR (2008) Effect of *Azospirillum brasilense* coinoculated with *Rhizobium* on *Phaseolus vulgaris* flavonoids and Nod factor production under salt stress. Soil Biol Biochem 40:2713–2721. https://doi.org/10.1016/j.soilbio.2008.06.016

Díaz PR, Romero M, Pagnussatt L et al (2022) *Azospirillum baldaniorum* Sp245 exploits *Pseudomonas fluorescens* A506 biofilm to overgrow in dual-species macrocolonies. Environ Microbiol 24:5707–5720. https://doi.org/10.1111/1462-2920.16195

Díaz PR, Merlo F, Carrozzi L (2023) Lettuce growth improvement by *Azospirillum argentinense* and fluorescent *Pseudomonas* co-inoculation depends on strain compatibility. Appl Soil Ecol 189:104969. https://doi.org/10.1016/j.apsoil.2023.104969

DuBois M, Gilles KA, Hamilton JK et al (1956) Colorimetric method for determination of sugars and related substances. Anal Chem 28:350–356. https://doi.org/10.1021/ac60111a017

Esquivel-Cote R, Ramírez-Gama RM, Tsuzuki-Reyes G (2010) *Azospirillum lipoferum* strain AZm5 containing 1-aminocyclopropane-1-carboxylic acid deaminase improves early growth of tomato seedlings under nitrogen deficiency. Plant Soil 337:65–75. https://doi.org/10.1007/s11104-010-0499-7

Fabbri P, Del Gallo M (1995) Specific interaction between chickpea (Cicer arietinum L.) and three chickpea-Rhizobium strains inoculated singularly and in combination with *Azospirillum brasilense* Cd. In: Fendrik I, del Gallo M, Vanderleyden J, de Zamaroczy M (eds) Azospirillum VI and related microorganisms. Springer, Berlin/Heidelberg, pp 257–267. https://doi.org/10.1007/978-3-642-79906-8_28

Fasciglione G, Casanovas EM, Quillehauquy V (2015) Azospirillum inoculation effects on growth, product quality and storage life of lettuce plants grown under salt stress. Sci Hortic 195:154–162. https://doi.org/10.1016/j.scienta.2015.09.015

Felici C, Vettori L, Giraldi E (2008) Single and co-inoculation of *Bacillus subtilis* and *Azospirillum brasilense* on *Lycopersicon esculentum*: Effects on plant growth and rhizosphere microbial community. Appl Soil Ecol 40:260–270. https://doi.org/10.1016/j.apsoil.2008.05.002

Ferraris G, Couretot L (2011) Interacción entre microorganismos fijadores de nitrógeno y promotores de crecimiento (PGPM) en soja. I. Bradyrhizobium japonicum y Azospirillum brasilense: efectos sobre la nodulación el rendimiento y su interacción con prácticas de manejo

Ferraris G, Couretot L (2013) Evaluación de tratamientos biológicos con Bradyrhizobium japonicum y Azospirillum brasilense en soja: efectos sobre la fijación de nitrógeno y el rendimiento

Ferri GC, Braccini AL, Anghinoni FBG (2017) Effects of associated co-inoculation of *Bradyrhizobium japonicum* with *Azospirillum brasilense* on soybean yield and growth. Afr J Agric Res 12:6–11. https://doi.org/10.5897/AJAR2016.11711

Fipke GM, Conceição GM, Grando LFT (2016) Co-inoculation with diazotrophic bacteria in soybeans associated to urea topdressing. Ciênc E Agrotecnolog 40:522–533. https://doi.org/10.1590/1413-70542016405001316

Fukami J, Cerezini P, Hungria M (2018) Azospirillum: benefits that go far beyond biological nitrogen fixation. AMB Express 8:73. https://doi.org/10.1186/s13568-018-0608-1

Galal YGM (1997) Dual inoculation with strains of Bradyrhizobium japonicum and Azospirillum brasilense to improve growth and biological nitrogen fixation of soybean (Glycine max L.). Biol Fertil Soils 24:317–322. https://doi.org/10.1007/s003740050250

Galindo FS, Teixeira Filho MCM, Buzetti S (2018) Technical and economic viability of co-inoculation with *Azospirillum brasilense* in soybean cultivars in the Cerrado. Rev Bras Eng Agríc E Ambient 22:51–56. https://doi.org/10.1590/1807-1929/agriambi.v22n1p51-56

Galindo FS, Pagliari PH, Da Silva EC (2022) Co-inoculation with Azospirillum brasilense and Bradyrhizobium sp. enhances nitrogen uptake and yield in field-grown cowpea and did not change N-fertilizer recovery. Plants 11:1847. https://doi.org/10.3390/plants11141847

García de Salamone IE, Funes JM, Di Salvo LP (2012) Inoculation of paddy rice with *Azospirillum brasilense* and *Pseudomonas fluorescens*: Impact of plant genotypes on rhizosphere microbial communities and field crop production. Appl Soil Ecol Microorg Sustain Manag Soil 61:196–204. https://doi.org/10.1016/j.apsoil.2011.12.012

Glickmann E, Dessaux Y (1995) A critical examination of the specificity of the salkowski reagent for indolic compounds produced by phytopathogenic bacteria. Appl Environ Microbiol 61:793–796. https://doi.org/10.1128/aem.61.2.793-796.1995

González J, Lluch C (1991) Biología del nitrógeno [www Document]. Editor. RUEDA. URL https://www.editorialrueda.es/bibliografia/biologia-del-nitrogeno-p572823972. Accessed 4 Apr 2025

Groppa MD, Zawoznik MS, Tomaro ML (1998) Effect of co-inoculation with *Bradyrhizobium japonicum* and *Azospirillum brasilense* on soybean plants. Eur J Soil Biol 34:75–80. https://doi.org/10.1016/S1164-5563(99)90004-3

Gualpa J, Lopez G, Nievas S (2019) *Azospirillum brasilense* Az39, a model rhizobacterium with AHL quorum-quenching capacity. J Appl Microbiol 126:1850–1860. https://doi.org/10.1111/jam.14269

Gupta S (1995) Growth promotion of tomato plants by rhizobacteria and imposition of energy stress on *Rhizoctonia solani*. Soil Biol Biochem 27:1051–1058. https://doi.org/10.1016/0038-0717(95)00011-3

Hassouna MG, El-Saedy MAM, Saleh HMA (1998) Biocontrol of soil-borne plant pathogens attacking cucumber *(Cucumis sativus)* by Rhizobacteria in a semiarid environment. Arid Soil Res Rehabil 12:345–357. https://doi.org/10.1080/15324989809381523

Hoagland DR, Broyer TC (1936) General nature of the process of salt accumulation by roots with description of experimental methods. Plant Physiol 11:471–507. https://doi.org/10.1104/pp.11.3.471

Hungria M, Nogueira MA, Araujo RS (2013) Co-inoculation of soybeans and common beans with rhizobia and azospirilla: strategies to improve sustainability. Biol Fertil Soils 49:791–801. https://doi.org/10.1007/s00374-012-0771-5

Hungria M, Nogueira MA, Araujo RS (2015) Soybean seed co-inoculation with Bradyrhizobium spp. and Azospirillum brasilense: a new biotechnological tool to improve yield and sustainability:811–817

Itzigsohn R, Kapulnik Y, Okon Y (1993) Physiological and morphological aspects of interactions between Rhizobium meliloti and alfalfa (Medicago sativa) in association with Azospirillum brasilense. Can J Microbiol 39:610–615. https://doi.org/10.1139/m93-088

Keel C, Oberhansli T, Haas D (1990) Pseudomonads as antagonists of plant pathogens in the rhizosphere: role of the antibiotic 2,4-diacetylphloroglucinol in the suppression of black root rot of tobacco. Symbiosis (Rehovot) 9(1–3):327–341. ref. 14327-341

Keel C, Schnider U, Maurhofer M et al (1992) Suppression of root diseases by *Pseudomonas fluorescens* CHA0: importance of the bacterial secondary metabolite

2,4-diacetylphloroglucinol. Mol Plant-Microbe Interact 5:4. https://doi.org/10.1094/MPMI-5-004

King EO, Ward MK, Raney DE (1954) Two simple media for the demonstration of pyocyanin and fluorescin. J Lab Clin Med 44:301–307

Kwak Y-S, Han S, Thomashow LS et al (2011) *Saccharomyces cerevisiae* genome-wide mutant screen for sensitivity to 2,4-diacetylphloroglucinol, an antibiotic produced by *Pseudomonas fluorescens*. Appl Environ Microbiol 77:1770–1776. https://doi.org/10.1128/AEM.02151-10

Loaces I, Ferrando L, Fernández Scavino A (2011) Dynamics, diversity and function of endophytic siderophore-producing bacteria in rice. Microb Ecol 61:606–618. https://doi.org/10.1007/s00248-010-9780-9

Marks BB, Megías M, Nogueira MA et al (2013) Biotechnological potential of rhizobial metabolites to enhance the performance of *Bradyrhizobium* spp. and *Azospirillum brasilense* inoculants with soybean and maize. AMB Express 3:21. https://doi.org/10.1186/2191-0855-3-21

Maroniche GA, Diaz PR, Borrajo MP (2018) Friends or foes in the rhizosphere: traits of fluorescent *Pseudomonas* that hinder *Azospirillum brasilense* growth and root colonization. FEMS Microbiol Ecol. https://doi.org/10.1093/femsec/fiy202

Miles AA, Misra SS, Irwin JO (1938) The estimation of the bactericidal power of the blood. Epidemiol Infect 38:732–749. https://doi.org/10.1017/S002217240001158X

Moënne-Loccoz Y, Tichy H-V, O'Donnell A et al (2001) Impact of 2,4-diacetylphloroglucinol-producing biocontrol strain *Pseudomonas fluorescens* F113 on intraspecific diversity of resident culturable fluorescent pseudomonads associated with the roots of field-grown sugar beet seedlings. Appl Environ Microbiol 67:3418–3425. https://doi.org/10.1128/AEM.67.8.3418-3425.2001

Molina R, Rivera D, Mora V (2018) Regulation of IAA biosynthesis in Azospirillum brasilense under environmental stress conditions. Curr Microbiol 75:1408–1418. https://doi.org/10.1007/s00284-018-1537-6

Molla AH, Shamsuddin ZH, Halimi MS (2001) Potential for enhancement of root growth and nodulation of soybean co-inoculated with *Azospirillum* and *Bradyrhizobium* in laboratory systems. Soil Biol Biochem 33:457–463. https://doi.org/10.1016/S0038-0717(00)00186-3

Moreira VDA, Oliveira CEDS, Jalal A (2022) Inoculation with Trichoderma harzianum and Azospirillum brasilense increases nutrition and yield of hydroponic lettuce. Arch Microbiol 204:440. https://doi.org/10.1007/s00203-022-03047-w

Nogueira MA, Prando AM, de Oliveira AB et al (2018) Ações de transferência de tecnologia em inoculação/coinoculação com Bradyrhizobium e Azospirillum na cultura da soja na safra 2017/18 no estado do Paraná. Portal Embrapa [www Document]. URL https://www.embrapa.br/busca-de-publicacoes/-/publicacao/1095314/acoes-de-transferencia-de-tecnologia-em-inoculacaocoinoculacao-com-bradyrhizobium-e-azospirillum-na-cultura-da-soja-na-safra-201718-no-estado-do-parana. Accessed 4 Apr 2025

Okon Y, Vanderleyden J (1997) Root-associated Azospirillum species can stimulate plants. Asm News

Omar MNA, Osman MEH, Kasim WA (2009) Improvement of salt tolerance mechanisms of barley cultivated under salt stress using Azospirillum brasilense. In: Ashraf M, Ozturk M, Athar HR (eds) Salinity and water stress, tasks for vegetation science. Springer, Dordrecht, pp 133–147. https://doi.org/10.1007/978-1-4020-9065-3_15

Omer AM, Osman MS, Badawy AA (2022) Inoculation with Azospirillum brasilense and/or Pseudomonas geniculata reinforces flax (Linum usitatissimum) growth by improving physiological activities under saline soil conditions. Bot Stud 63:15. https://doi.org/10.1186/s40529-022-00345-w

Pagnussat LA, Salcedo F, Maroniche G (2016) Interspecific cooperation: enhanced growth, attachment and strain-specific distribution in biofilms through *Azospirillum brasilense-Pseudomonas protegens* co-cultivation. FEMS Microbiol Lett 363:fnw238. https://doi.org/10.1093/femsle/fnw238

Penna C, Massa R, Olivieri F (2011) Erratum to: A simple method to evaluate the number of bradyrhizobia on soybean seeds and its implication on inoculant quality control. AMB Express 1:25. https://doi.org/10.1186/2191-0855-1-25

Read RR, Costerton JW (1987) Purification and characterization of adhesive exopolysaccharides from *Pseudomonas putida* and *Pseudomonas fluorescens*. Can J Microbiol 33:1080–1090. https://doi.org/10.1139/m87-189

Remans R, Ramaekers L, Schelkens S (2008) Effect of Rhizobium–Azospirillum coinoculation on nitrogen fixation and yield of two contrasting Phaseolus vulgaris L. genotypes cultivated across different environments in Cuba. Plant Soil 312:25–37. https://doi.org/10.1007/s11104-008-9606-4

Ribeiro VP, Gomes EA, de Sousa SM et al (2022) Co-inoculation with tropical strains of *Azospirillum* and *Bacillus* is more efficient than single inoculation for improving plant growth and nutrient uptake in maize. Arch Microbiol 204:143. https://doi.org/10.1007/s00203-022-02759-3

Rodelas B, González-López J, Martínez-Toledo MV (1999) Influence of Rhizobium/Azotobacter and Rhizobium/Azospirillum combined inoculation on mineral composition of faba bean (Vicia faba L.). Biol Fertil Soils 29:165–169. https://doi.org/10.1007/s003740050540

Romero AM, Correa OS, Moccia S et al (2003) Effect of Azospirillum-mediated plant growth promotion on the development of bacterial diseases on fresh-market and cherry tomato. J Appl Microbiol 95:832–838. https://doi.org/10.1046/j.1365-2672.2003.02053.x

Russo A, Basaglia M, Casella S et al (2008) *Pseudomonas fluorescens* 134 as a biological control agent (BCA) model in cell immobilization technology. Biotechnol Prog 21:309–314. https://doi.org/10.1021/bp040030w

Sadiq FA, Burmølle M, Heyndrickx M (2021) Community-wide changes reflecting bacterial interspecific interactions in multispecies biofilms. Crit Rev Microbiol 47:338–358. https://doi.org/10.1080/1040841X.2021.1887079

Sankari JU, Dinakar S, Sekar C (2011) Dual Effect of *Azospirillum* Exopolysaccharides (EPS) on the enhancement of plant growth and biocontrol of blast (*Pyricularia oryzae*) disease in upland rice (var. ASD-19)

Schmidt W, Martin P, Omay SH et al (1988) Influence of Azospirillum brasilense on nodulation of legumes. In: Klingmüller W (ed) Azospirillum IV. Springer, Berlin/Heidelberg, pp 92–100. https://doi.org/10.1007/978-3-642-73072-6_12

Schouten A, Van Den Berg G, Edel-Hermann V (2004) Defense responses of *Fusarium oxysporum* to 2,4-diacetylphloroglucinol, a broad-spectrum antibiotic produced by *Pseudomonas fluorescens*. Mol Plant-Microbe Interactions® 17:1201–1211. https://doi.org/10.1094/MPMI.2004.17.11.1201

Shalev O, Karasov TL, Lundberg DS (2022) Commensal Pseudomonas strains facilitate protective response against pathogens in the host plant. Nat Ecol Evol 6:383–396. https://doi.org/10.1038/s41559-022-01673-7

Srinivasan M, Holl FB, Petersen DJ (1996) Influence of indoleacetic-acid-producing Bacillus isolates on the nodulation of Phaseolus vulgaris by Rhizobium etli under gnotobiotic conditions. Can J Microbiol 42:1006–1014. https://doi.org/10.1139/m96-129

Sudhakar, P., Chattopadhyay, G.N., Gangwar, S.K. et al (2000) Effect of foliar application of *Azotobacter* , *Azospirillum* and *Beijerinckia* on leaf yield and quality of mulberry (*Morus alba*). J Agric Sci 134, 227–234. https://doi.org/10.1017/S0021859699007376

Sultana U, Pinisetty S, Desai S (2024) In vitro screening for antagonistic ability of plant growth promoting strains of Pseudomonas, Azotobacter and Azospirillum Spp. Adv Agric Technol Plant Sci 7:180150

Tapia-Hernández A, Mascarúa-Esparza MA, Caballero-Mellado J (1990) Production of bacteriocins and siderophore-like activity by Azospirillum brasilense. Microbios 64:73–83

Thimann KV (1936) Auxins and the growth of roots. Am J Bot 23:561–569. https://doi.org/10.1002/j.1537-2197.1936.tb09026.x

Torres D, Benavidez I, Donadio F (2018) New insights into auxin metabolism in *Bradyrhizobium japonicum*. Res Microbiol 169:313–323. https://doi.org/10.1016/j.resmic.2018.04.002

Torres D, Donadio F, López G et al (2022) Previous Incubation of *Bradyrhizobium japonicum* E109 and *Azospirillum argentinense* Az39 (formerly A. brasilense Az39) Improves the *Bradyrhizobium*-Soybean Symbiosis. J Soil Sci Plant Nutr 22:4669–4682. https://doi.org/10.1007/s42729-022-00948-z

Tortora ML, Díaz-Ricci JC, Pedraza RO (2011) *Azospirillum brasilense* siderophores with antifungal activity against *Colletotrichum acutatum*. Arch Microbiol 193:275–286. https://doi.org/10.1007/s00203-010-0672-7

Tortora ML, Díaz-Ricci JC, Pedraza RO (2012) Protection of strawberry plants (Fragaria ananassa Duch.) against anthracnose disease induced by *Azospirillum brasilense*. Plant Soil 356:279–290. https://doi.org/10.1007/s11104-011-0916-6

Trabelsi D, Mengoni A, Ben Ammar H (2011) Effect of on-field inoculation of Phaseolus vulgaris with rhizobia on soil bacterial communities. FEMS Microbiol Ecol 77:211–222. https://doi.org/10.1111/j.1574-6941.2011.01102.x

Troppens DM, Chu M, Holcombe LJ et al (2013) The bacterial secondary metabolite 2,4-diacetylphloroglucinol impairs mitochondrial function and affects calcium homeostasis in Neurospora crassa. Fungal Genet Biol 56:135–146. https://doi.org/10.1016/j.fgb.2013.04.006

Vacheron J, Combes-Meynet E, Bellvert F et al (2012) Plant growth-promoting properties of *Pseudomonas* biocontrol agent producing 2,4-diacetylphloroglucinol

Vacheron J, Renoud S, Muller D et al (2015) Alleviation of abiotic and biotic stresses in plants by *Azospirillum*. In: Cassán FD, Okon Y, Creus CM (eds) Handbook for Azospirillum: technical issues and protocols. Springer, Cham, pp 333–365. https://doi.org/10.1007/978-3-319-06542-7_19

Vessey JK, Buss TJ (2002) *Bacillus cereus* UW85 inoculation effects on growth, nodulation, and N accumulation in grain legumes: Controlled-environment studies. Can J Plant Sci 82:282–290. https://doi.org/10.4141/P01-047

Vincent JA (1970) Manual for the practical study of root-nodule bacteria, 2nd edn

Walker V, Couillerot O, Von Felten A (2012) Variation of secondary metabolite levels in maize seedling roots induced by inoculation with *Azospirillum*, *Pseudomonas* and *Glomus* consortium under field conditions. Plant Soil 356:151–163. https://doi.org/10.1007/s11104-011-0960-2

Walsh UF, Moënne-Loccoz Y, Tichy H-V et al (2003) Residual impact of the biocontrol inoculant *Pseudomonas fluorescens* F113 on the resident population of rhizobia nodulating a red clover rotation crop. Microb Ecol 45:145–155. https://doi.org/10.1007/s00248-002-2026-8

Yahalom E, Okon Y, Dovrat A (1990) Possible mode of action of *Azospirillum brasilense* strain Cd on the root morphology and nodule formation in burr medic (Medicago polymorpha). Can J Microbiol 36:10–14. https://doi.org/10.1139/m90-003

Yasuda M, Isawa T, Shinozaki S (2009) Effects of colonization of a bacterial endophyte, *Azospirillum* sp. B510, on disease resistance in rice. Biosci Biotechnol Biochem 73:2595–2599. https://doi.org/10.1271/bbb.90402

Part V
Azospirillum spp. Inoculant Formulation, Quality Control and Agronomic Evaluation

Chapter 17
Agronomic Evaluation of Extensive Crops Inoculated with *Azospirillum* sp.

Martín Díaz-Zorita, María Virginia Fernández Canigia, Oscar Bravo, Ariadna Berger, and Emilio Horacio Satorre

Abstract The inoculation of extensive crops with *Azospirillum* spp. has been extensively evaluated worldwide, with abundant information describing its role in enhancing early growth, nitrogen uptake, and final grain yield under diverse field conditions. This chapter presents the design and outcomes of inoculation with *Azospirillum brasilense* and *Azospirillum argentinense* field trials performed across different continents, under a wide range of environmental and agronomic management. The chapter covers inoculation studies performed on seeds, in-furrow during sowing, or as foliar on top of the crop canopies using diverse formulations. On average, the introduction of *Azospirillum* spp. to large-scale crops resulted in a 10% increase in grain yield over non-inoculated controls. The responses were variable, reflecting interactions among crop types, environmental conditions, and inoculation strategies. In rainfed wheat (*Triticum aestivum* L.) trials conducted in semiarid regions, 60–200 observations per season were required to detect statistically significant yield effects depending on rainfall patterns, highlighting the importance of robust experimental designs. The analysis of response distributions offers a realistic approach to evaluate productivity under abiotic stress while enabling probabilistic assessments of inoculation outcomes. Studies on soybean (*Glycine max* L.) showed co-inoculation with *Azospirillum* and *Bradyrhizobium* spp. enhances nodulation and

M. Díaz-Zorita (✉)
Facultad de Agronomía, Universidad Nacional de La Pampa, Santa Rosa, La Pampa, Argentina

CONICET, Universidad Nacional de La Pampa, Santa Rosa, La Pampa, Argentina
e-mail: mdzorita@agro.unlpam.edu.ar

M. V. F. Canigia
Independent Adviser, Santa Rosa, La Pampa, Argentina

O. Bravo
Departamento de Agronomía, Universidad Nacional del Sur, Bahía Blanca, Buenos Aires, Argentina

A. Berger
Universidad del CEMA, Buenos Aires, Argentina

E. H. Satorre
Facultad de Agronomía, Universidad de Buenos Aires, Buenos Aires, Argentina

F. D. Cassán et al. (eds.), *The Definitive Handbook of Azospirillum*,
https://doi.org/10.1007/978-3-032-09222-9_17

yield stability, particularly under stressed environments. The use of *Azospirillum* spp. inoculants is a complementary practice to enhance crop performance mostly under temperate and semi-arid cropping systems worldwide.

Keywords Field trials · Yield · Nitrogen uptake · Crops · Agronomy

17.1 Introduction

The various mechanisms by which *Azospirillum* sp. enhance plant growth and crop productivity have been extensively studied across different experimental systems. These mechanisms, which partially explain the positive effects of inoculation on crop growth and yield, include both direct and indirect interactions. The direct effects involve improvements in plant nutritional status, primarily through biological nitrogen fixation, while indirect effects include the mitigation of abiotic stresses. Additionally, *Azospirillum* sp. produces phytohormones and enzymes that stimulate root development and promote the growth of other beneficial microorganisms in the rhizosphere, contributing to overall plant health and vigor (Bayan and Levanony 1990; Okon and Labandera-Gonzalez 1994; Dobbelare et al. 2003). Despite the abundance of studies on these mechanisms, most have been conducted under regulated conditions in greenhouses, using pot experiments that provide a highly regulated environment (Van Dommelen et al. 2009; Rodrigues et al. 2008; Puente et al. 2017). While these studies are essential for understanding the fundamental processes involved, they often have limitations when extrapolating results to large-scale agricultural systems. The controlled nature of greenhouse experiments does not fully replicate the complexity of field conditions, where environmental variability and multiple stress factors can influence the interaction between *Azospirillum* sp. and crops.

Field trials Field trials evaluating the effects of *Azospirillum* sp. inoculation on extensive crops under actual production conditions are relatively scarce. Moreover, their findings are often inconsistent and sometimes inconclusive, likely due to the inherent complexity of field studies and because yield is a multifactorial-dependent outcome highly affected by eventual abiotic stresses. The interactions between *Azospirillum* sp., plant species, soil characteristics, and environmental factors create a dynamic system where multiple abiotic stresses may be simultaneously mitigated by the microorganism. This complexity makes it challenging to design field experiments that accurately capture the full range of potential benefits. Given these challenges, the objective of this study is to provide an overview of field trial methodologies and data analysis approaches for evaluating crop production responses to *Azospirillum* sp. inoculation. By addressing key aspects of experimental design, variability in environmental conditions, and data interpretation, this chapter aims to improve our understanding of how *Azospirillum* sp. performs under real-world agricultural conditions and how its benefits can be maximized for sustainable crop production.

17.2 Methodological Approach and General Outcomes

Systematic and regular surveys were carried out using Google Scholar and other platforms from May 2014 to June 2025, combining the following keywords "*Azospirillum*," "inoculation," and "field evaluation." Only studies in which field trials under standard production conditions in annual crops were performed were included. Initially, we reviewed 47 articles from 12 countries worldwide published since early in the twenty-first century, documenting grain crop production responses to *Azospirillum* sp. inoculants, encompassing 347 cases. Most studies were focused on seed-applied treatments on cereal crops (86.7 %), particularly rainfed maize (*Zea mays* L.) and were conducted mainly in Latin America (especially Brazil) and Asia. On average, across all reported crops, grain yield increased by 10% following inoculation with *Azospirillum* sp. Winter cereals, such as wheat (*Triticum aestivum* L.) and barley (*Hordeum vulgare* L.), showed a higher response (14.0%) than summer cereals (9.5%), including maize, sorghum (*Sorghum bicolor* L.), and rice (*Oryza sativa* L.), or legumes (6.6%), such as beans (*Phaseolus vulgaris*), and soybean (*Glycine max* (L.) Merrill). A particular case study is the practice of co-inoculation of *Azospirillum* sp. with rhizobial strains in legume crops. The preliminary reviewed studies showed that under regulated and field conditions, soybean crops co-inoculated *Azospirillum* sp. and *Bradyrhizobium* sp. show better nodulation (Benintende et al. 2010), plant growth performance, and grain yield responses greater than 15% (Hungria et al. 2013). The contribution of the co-inoculation practice on soybean production was recently validated under growers' management conditions in large Brazilian soybean production areas that, averaged over a 5-year period, provided US$ 111.5 ha^{-1} of mean profit per cropping season, among other environmental benefits (Prando et al. 2024). The combination of *Azospirillum argentinense* with *Ensifer meliloti* in pelletized alfalfa (*Medicago sativa*) seeds promoted, in 21 experimental sites in the Pampas, to improved early pasture establishment and greater dry matter yield in the first growing season (Díaz-Zorita et al. 2012). Few studies have examined in-furrow applications of *Azospirillum* sp. during sowing (Hungria et al. 2013), although this method is suggested as an alternative to mitigate compatibility issues with seed-applied treatments that could reduce microbial survival. In-furrow inoculants are typically diluted in chlorine-free water with a pH between 6.0 and 8.0 and applied at a rate that ensures large droplets are delivered in a near-continuous stream beneath the seedbed (e.g., 30–40 L ha^{-1} at 1.0–2.0 bar using perforated disc nozzles). In-furrow application at sowing time using various inoculant formulations is a less frequently used method to introduce selected *Azospirillum* sp. strains in field crops' rhizosphere. Studies performed in the Brazilian Cerrado showed that the in-furrow inoculation using Ab-V5 and Ab-V6 strains at a dose of 200 mL ha^{-1} when planting maize increased the nitrogen content of the leaves. Even on reducing the doses of nitrogen fertilization, this treatment kept adequate crop growth and grain production greater than the control without inoculation (Galindo et al. 2022). Similarly, Fukami et al. (2016) found that furrow inoculation when planting maize or wheat improved N uptake and biomass accumulation but only described positive responses on maize grain production. Additionally, Hungria et al. (2010) reported that the inoculation with the strains Ab-V5 and Ab-V6 increased maize and wheat yields and is mainly applied to the seeds, particularly under reduced fertilizer input. In annual crops, the in-furrow application of *Azospirillum* sp. inoculants is recommended to be a valid alternative to seed inoculation, especially when the latter is limited by incompatible seed treatments, keeping its agronomic benefits on early plant growth and nitrogen crop economy. A recent field study conducted by de Souza et al. (2025) evaluated the potential of *A. brasilense* applied in the planting furrow in sugarcane (*Saccharum officinarum*) cultivation under a range of nitrogen fertilization rates. Their results showed that combining the application of the inoculant with half of the recommended N fertilization rate resulted in similar biomass and yield levels to the full N fertilization rate without inoculation. The inoculated plants show enhanced root growth and more nitrogen use efficiency, supporting the expected biological contribution of *A. brasilense* to early plant establishment and nutrient uptake.

Some studies have reported transplanting seedlings soaked in *Azospirillum* sp. solutions, particularly in rice production (Govindarajan et al. 2008; Islam et al. 2012; Khalid et al. 2011). However, due to its impracticality and limited applicability in extensive farming, this method was not considered in our deep review. Other cases of study were foliar applications of *Azospirillum* sp., although with dispersed results and limited recommendations regarding the optimal timing, the adequate combination with other agronomic practices (like herbicide spraying) or the key environmental conditions for effective use (El Habbasha et al. 2013; Cardozo et al. 2021).

17.2.1 *Soybean Combined Inoculation with* **Azospirillum** *sp. and Rhizobia*

The combined inoculation in soybeans involves the application of *Bradyrhizobium japonicum* or *Bradyrhizobium diazoefficiens* strains with the plant growth-promoting bacteria *Azospirillum*, mostly *A. brasilense* or *A. argentinense*. The most widespread method is the treatment of the seeds at planting, using liquid or peat-based formulations compatible with diverse chemical seed treatments (Hungria et al. 2015; Fukami et al. 2016). The frequent concentrations of these microorganisms in the inoculants are $\geq 10^9$ CFU mL^{-1} for *Bradyrhizobium* sp. and $\geq 10^8$ CFU mL^{-1} for *Azospirillum* sp. The goal is to ensure at least 1.2×10^6 rhizobial cells and 1.2×10^5 *Azospirillum* sp. cells per seed at the application. In Brazil, for example, standard strains of *B. diazoefficiens* (SEMIA 5079, SEMIA 5080) are applied together with *A. brasilense* (Ab-V5 and Ab-V6), adjusting doses to prevent mutual interference. In Argentina, although the available information about this practice is still limited, the combination is using registered *B. japonicum*, mostly the E109 strain, and *A. argentinense Az39T* (former *A. brasilense* AZ39 INTA, Maroniche et al. 2024). After applying the treatment, the inoculated seeds are shade-dried before planting within less than 24 h.

Field trials are usually arranged in randomized complete blocks with at least six replicates. Plot size is typically 24 m^2, and when is it feasible to be separated by buffer zones to avoid cross-contamination? Soybean crop management includes plants stand greater than 300 plants m^{-2} at emergence and when it is needed a base fertilization with P or K without N fertilizers. In several locations, foliar application of micronutrients (Mo and Co) at the V4 stage is recommended to avoid early limitations of these elements strongly needed for efficient biological nitrogen fixation (Galindo et al. 2018). In all studies, crops were kept free of weed and pest competition using diverse combinations of registered chemical and biological control products. In general, the evaluation of the treatments includes nodule number and biomass, shoot biomass, nitrogen content and total N accumulation at early vegetative growth stages (approximately 35–50 days after sowing), and grain yield at harvest.

Prando et al. (2024) analyzed the contribution of seed inoculation to the growth and grain production of soybeans under farmer's scale. The study was performed during five consecutive growing seasons in 273 smallholder farms in Paraná State (Brazil), establishing technical reference units of approximately 1 ha under homogeneous soil conditions and regular regional soybean agronomic practices. In each of the farms, two treatments were evaluated: a non-inoculated control and a co-inoculation treatment where soybean seeds were inoculated with *Bradyrhizobium* spp. (strains SEMIA 5079 and 5080) and *A. brasilense* (strains Ab-V5 and Ab-V6). The inoculants were applied on the day of sowing using either peat-based or liquid formulations, with doses of inoculants adjusted to deliver 2.4×10^6 cells per seed of *Bradyrhizobium* and 1.2×10^5 cells per seed of *A. brasilense*. Statistical analyses between treatments were performed using the Mann–Whitney *U* test due to the non-normal data distribution. Results were analyzed both globally and by categories of grain yield as well as by macro-region.

In soils with native rhizobial populations greater than 10^4 CFU g^{-1} , applying co-inoculation matched or exceeded single inoculation in terms of nodulation, biomass, and nitrogen accumulation. Hungria et al. (2015) showed from studies performed in Brazil that soybean grain yield increases from 7% to 17% in co-inoculated crops reaching levels comparable to the control with 200 kg N ha^{-1}. In a five-site analysis, García et al. (2021) found an average increase of +14.7% in grain yield and +16% in grain nitrogen with co-inoculation compared to rhizobia strains alone. Studies performed in Argentina also showed benefits due to the combined application of *B. japonicum* strains and *A. argentinense*. For example, in a multilocation trial conducted in the central Pampas region, Morla et al. (2019) reported an additional gain yield attributable to *Azospirillum* above the single application of rhizobia was approximately 105 kg ha^{-1}. In Tocantins (Brazil), under severe water stress, Naoe et al. (2020) recorded a +77% yield gain with co-inoculation compared to single inoculation. Benintende et al. (2010), in the Pampas region, also concluded about the benefit of co-inoculation increasing in nodulation and biological nitrogen fixation especially under drought conditions.

These authors measured that the nitrogen derived from biological fixation due to the co-inoculation increases from 26% to 30% in a dry year, while this value improves from 41% to 47% in wet season.

In the United States, de Borja Reis et al. (2022) conducted a multistate study across 25 trials finding positive responses to co-inoculation in only 10% of the sites, with no significant improvements in biological nitrogen fixation (BNF) BNF or in the concentration of protein in the harvested grains. The low response was attributed to soils with adequate N availability and high native rhizobial populations. In Mozambique, Chibeba et al. (2020) observed approximately 22% yield increase with the inoculation using *B. diazoefficiens* strain USDA 110, but the addition of *A. brasilense* strains Ab-V5 and Ab-V6 provided no further benefit. These authors considered that *Azospirillum* strains used were not well adapted to the tropical environment and further studies were needed before the widespread application of this practice.

The agronomic effectiveness of co-inoculation with *Bradyrhizobium* sp. and *Azospirillum* sp. in soybean seed treatments depends on multiple soil, climate, and management factors. Current field studies showed that the more frequent and consistent responses are observed under environmental limitations that restrict biological nitrogen fixation during the early crop growth. Under high yielding environments for soybean growth, because of still not clearly described negative effects, increasing the inoculum of *A. brasilense* co-inoculated with rhizobia in soybean compromised both N nutrition and grain yield should not be encouraged as an agronomic practice (Caires et al. 2024). In these environments attaining high yields and adequate early crop growth without water limitations, the demand of nitrogen could be satisfied exclusively through the optimization of the biological fixation of N without increasing the early presence of *Azospirillum* sp.

Recent findings by Torres et al. (2022) demonstrated that the co-inoculation of *B. japonicum* strain E109 and *A. argentinense* strain Az39, particularly through a short-term pre-incubation, enhances microbial interaction and improves symbiotic efficiency in soybean. This approach increased the survival of *B. japonicum* on the seed surface and stimulated nodulation, resulting in increased nodule number, size, and biomass. Field experiments with soybean seeds treated with pre-incubated formulations with these microorganisms showed a grain yield increase of approximately 17% compared to the single inoculation with *B. japonicum*. The authors attribute these improvements to synergistic effects between both strains, including co-production of indole 3 acetic acid and enhanced exopolysaccharide production promoting root colonization.

17.2.2 Foliar Application of Azospirillum *sp.*

Several studies have recently evaluated the foliar application of *Azospirillum* spp. strains as a complementary strategy to seed inoculation, aiming to enhance the growth and productivity of major field crops such as maize, wheat, and soybean.

These applications have used strains like *A. argentinense* T39 in Argentina and Ab-V5 or Ab-V6 in Brazil in liquid formulations with concentrations ranging from 10^8 to 10^9 CFU mL^{-1}. The foliar spraying of these products mostly occurs between growth stages V4 and V6 in maize, and prior to canopy closure in wheat or soybeans. The volumes of spray solution, also named as tank mix, ranged from 100 to 200 L ha^{-1} containing between 200 and 300 mL ha^{-1} of the inoculant with *Azospirillum* sp. The applications, to favor microbial survival and leaf penetration, were mostly performed early in the morning under low solar radiation and with high relative humidity.

Experimental conditions have included conventional crop management systems with intensive herbicide use (e.g., glyphosate, atrazine), variable water availability, and contrasting fertilization levels. Most field trials were conducted under randomized block designs, with plot sizes ranging from 30 to 60 m^2 and multiple replications, in locations across Argentina and Brazil.

In maize crops, Cardozo et al. (2021), over three growing seasons in Argentina, reported that foliar application of *A. brasilense* strain Az39 (currently known as *A. argentinense*) restored root biomass and chlorophyll content reduced by herbicide use, and improved grain yield. In Brazil, Galindo et al. (2022) and Fukami et al. (2016) showed that foliar spray of *A. brasilense* on top of maize crops during early vegetative growth stages allowed nitrogen fertilizer rates to be reduced by up to 25% without reducing the attained grain production. These authors also documented improved colonization of leaf surfaces, the stimulation of stress-tolerance-related genes and the increment in phytohormone levels. A meta-analysis conducted by Barbosa et al. (2022), based on 103 field trials performed across 54 Brazilian sites, showed that the foliar application with *A. brasilense* strains Ab-V5 and Ab-V6 increased root biomass by 12%, leaf nitrogen concentration by 4%, and grain yield also around 5%. In wheat crops, the spraying of *A. brasilense* strains during tillering provides more benefits on crop growth, dry matter accumulation and N use than when the inoculant was applied on the seeds or in-furrow at sowing (Fukami et al. 2016). The results from this study showed that combining foliar applications with 75% of the recommended N fertilization doses the crop attained the highest grain yield.

The currently recommended Az39, Ab-V5, and Ab-V6 strains, with validated beneficial effects on plant growth when applied to the rhizosphere, showed adequate phyllosphere adaptation and effective colonization capacity under field conditions (Hungria et al. 2015; Fukami et al. 2016). But the advantage of the foliar applications avoiding the negative interactions between seed-applied pesticides and microbial inoculants still need to identify key environmental and crop management conditions related to the effectiveness of this treatment. For example, humid and subtropical climates tend to enhance bacterial survival and colonization when applied to the leaf surface (Galindo et al. 2022). This condition, combined with limited nitrogen availability or other abiotic stress the treated crops could lead to beneficial growth responses. But when the abiotic stress limits the growth of the crops that occur in dry and warm environments the efficacy of applying foliar microorganisms could be reduced and without beneficial effects on the growth of the treated plants. Also, under high-input conditions with adequate nutrient supply and no significant environmental stress, crop responses to foliar inoculation tend to be modest or inconsistent

17.2.3 *Seed Applied Inoculation with* **Azospirillum** *sp.*

Seed treatment with *Azospirillum* spp. formulations is generally recommended without mixing in slurry with other products (e.g., micronutrients, fungicides, insecticides, and polymers) to prevent potential interactions with active ingredients or formulation components (e.g., colorants, dispersants, and stabilizers).

Several formulations have been developed to allow seed inoculation well before sowing (Díaz-Zorita et al. 2012), but most studies reported that inoculated seeds were planted within 6 h of treatment. Regardless of application timing, excessive seed handling should be avoided, as *Azospirillum* sp. cells are sensitive to desiccation, high temperatures, ultraviolet light, and toxic compounds. Additionally, sowing should be conducted in moist soil to prevent seed desiccation and direct sunlight exposure.

As in other inoculation studies, field experiments typically use large plots (>10 m in length), with measurements taken from the central portion while excluding border rows to minimize cross-contamination. This setup also reduces seed manipulation by allowing the use of larger seed quantities. To further minimize contamination between treatments, machinery should be cleaned between sowing operations, and control (non-inoculated) treatments should be planted first.

Crop management practices that influence growth under abiotic stress conditions (e.g., nutrient deficiencies, water stress) also affect *Azospirillum* sp. responses. Therefore, defining appropriate agronomic practices, such as irrigation, fertilization, and planting dates, according to study objectives is essential. For a comprehensive assessment of *Azospirillum* sp. inoculation effects, crop management should mimic standard production practices, avoiding excessive nutrient availability or irrigation that could mask the microorganism's potential to alleviate moderate abiotic stress. Under extreme stress conditions (e.g., severe drought or major nutrient deficiencies) or in the absence of growth limitations, inoculation responses are minimal (Lana et al. 2012; Mehnaz et al. 2010; Naiman et al. 2009; Naseri et al. 2013). Most reviewed studies ensured sufficient fertilization, mainly NPK, to meet crop requirements, with only a few conducted under nutrient-limited conditions.

Field experiments were predominantly conducted at single locations over a few production seasons or at a limited number of locations within a single season, using randomized complete block designs. However, this design has limitations for analyzing *Azospirillum* sp. contributions to crop production due to random spatial and temporal interactions. Only a few studies evaluated biological treatment across multiple locations and growing seasons, capturing its effects under varying environmental conditions and standard production practices. These studies revealed variable responses depending on fixed factors (e.g., soil type) and random factors (e.g., production seasons) (Díaz-Zorita and Fernández-Canigia 2009). In the central part of the Semiarid Pampas region, field studies conducted over three consecutive growing seasons in seven sites established under similar soil and wheat management conditions provided widely different production responses to inoculation (Gallace and Díaz-Zorita 2025). These authors concluded that the greater responses were attained when water limitations for growth were observed only during the vegetative growth stages of the crops and under adequate water conditions during seed filling. The minimal increase in grain yield occurred under high productive seasons without water limitations all along the growing season and intermediate responses were described in the season with water limitations from sowing until grain maturity. The variability in weather conditions during the growing season increases the uncertainty in the responses of the inoculations requiring complex probabilistic approaches to describe the expected yearly contribution of this practice.

To address these complexities, the following sections of this chapter will present two approaches for interpreting grain yield responses to *Azospirillum* sp. inoculation, considering both spatial (e.g., soil type and crop management) and temporal (e.g., production season) variability.

17.2.4 Soil Type and Temporal Variability Effects on Crop Responses to Inoculation

To quantify the effect of wheat seed inoculation with a formulation containing *A. argentinense* (formerly *A. brasilense* strain INTA Az39, Maroniche et al. 2024), a field trial was conducted over three consecutive growing seasons in a 50-ha area within the "Napostá" experimental field (−38° 25′ 32″, −62° 17′ 07″). This site is located close to Bahía Blanca (Buenos Aires province, Argentina) within the semiarid Pampas region. The plot under evaluation is situated in a rolling landscape in the upper valley of the Saladillo Dulce creek, characterized by complex slight to moderate slopes and soils classified as Entic Haplustolls, saline to sodic saline Entic Haplustolls, moderately deep Petrocalcic Paleustolls, and shallow to deep Petrocalcic Calciustolls (Soil Survey Staff 1999). These soils consistently exhibit loam to sandy loam textures in their upper profiles.

Each season, wheat crops were managed following best practices for high-yield production in the semiarid region, including zero tillage with chemical weed control, late June sowing, and fertilization with 40 kg ha^{-1} of diammonium phosphate (18: 46:0). Two treatments, control (non-inoculated) and inoculated seeds, were applied annually in paired 8 ha strips across the different soil types. Shoot dry matter and grain yield were measured at 15 geopositioned sites each year, then classified into the four predominant soil types for comparison between inoculated and control treatments. Descriptive statistics (mean and dispersion parameters) were calculated for each treatment, considering spatial (soil type) and temporal (growing season) variability. A factorial split-plot ANOVA was performed, with growing seasons and seed treatments as main factors and soil type as the split factor. Shoot dry matter data were transformed into homogenized variances (Di Rienzo et al. 2013).

Table 17.1 Shoot dry matter and grain production of wheat crops grown in a field during three seasons from the semiarid pampas inoculated with *Azospirillum argentinense*

		Mean	SD	Min.	Max.	CV
Variable	**Treatment**	Mg ha^{-1}				%
Shoot dry matter	Control without inoculation	5.8 a	2.7	2.5	14.2	46.6
	Inoculated	6.7 b	3.1	2.5	13.6	46.9
Grain yield	Control without inoculation	1.2 a	0.5	0.5	2.8	39.0
	Inoculated	1.4 b	0.5	0.6	2.6	36.0

Abbreviations: *SD* standard deviation, *CV* coefficient of variation
In each variable, mean values with different letters show least significant differences ($P < 0.05$)
Reproduced from Díaz-Zorita et al. (2016)

On average, across soil types and growing seasons, wheat crops inoculated with *A. argentinense* exhibited higher biomass and yield in 74% of the sites, with 15% greater shoot dry matter and 17% higher grain yield compared to the control (Table 17.1). The ANOVA revealed significant effects of growing season ($P < 0.01$) and biological seed treatment ($P < 0.05$) on grain yield, with no significant interaction between them ($P = 0.21$). Differences among soil types were not statistically significant ($P = 0.28$), likely due to the limited number of samples.

Sixty-five percent ($P < 0.01$) of the variability in grain yield was attributed to yearly weather conditions, primarily the amount and distribution of rainfall during the growing period. Inoculation with *A. argentinense* accounted for 4% ($P < 0.05$) of the variability, while soil type contributed only 1% ($P > 0.50$), with the remaining 30% explained by interactions among factors and the experimental error. The greatest response to inoculation was observed in sites with the highest soil productivity indices, primarily deep Entic Haplustolls. However, the lack of significant differences between biological treatments within each soil type was attributed to the limited sample size.

The results of this study revealed that deviations in mean yields across seasons impact the number of samples required to detect treatment differences with 95% confidence. In seasons with normal growing conditions, 60–75 samples were needed to distinguish differences between means, whereas in the driest and lowest-yielding season, over 200 samples were required. This high number of observations could be reduced by restricting sample collection to spatially homogeneous sites.

17.3 Applying Experimental Data to Evaluate Crop Management Decisions

Experimental data is particularly valuable when used to support management decisions. Previous sections of this chapter have shown that *Azospirillum* spp. inoculation generally increases crop yields. However, the yield response to *Azospirillum* sp. is unpredictable, as it is influenced by multiple factors, including its diverse modes of action and ability to mitigate various abiotic stresses. In fact, studies have reported negative, neutral, and positive responses to inoculation across different crop species (Araujo et al. 2013; Hungria et al. 2013; Piccinin et al. 2011; Turan and Fikrettin 2013). This raises an important question: how can experimental data be used to inform management decisions?

One approach is to select experiments based on predefined criteria, such as soil type or growing season, as discussed in a previous section. However, this method can be misleading, especially if applied by inexperienced individuals. Instead, this section explores an alternative approach to analyzing experimental data for evaluating the benefits of crop inoculation. For this purpose, we examined results from 316 field trials conducted in Argentina, where maize grain yields were measured under control (non-inoculated) and inoculated (*A. argentinense*) treatments. These trials were carried out across various locations in the Pampas region under standard production practices (e.g., NP fertilization, crop rotation, and predominantly zero tillage) over 12 consecutive growing seasons (M. Díaz-Zorita, unpublished data). Since ANOVA analysis showed no significant differences ($P = 0.05$) among individual trial responses, the entire dataset (316 cases) was considered for analysis.

Crop management decisions typically rely on the balance between incremental revenue and cost. If the additional revenue from technology exceeds its cost, its adoption is economically justified. In this case, incremental revenue is calculated as the yield increase due to inoculation multiplied by the net price (gross price minus marketing expenses). Since both yield response and price fluctuate, incremental revenue is also a random variable and should be analyzed using probability distribution rather than as a single, fixed value.

Using empirical data, the probability distributions of maize yields for both control and inoculated treatments can be described across percentiles (Table 17.2). Moreover, statistical comparisons of maize yields across different percentiles, provide deeper insights (Table 17.3). In the analyzed dataset, yields ranged from as low as 2,020 to as high as 18,654 kg ha^{-1} (Table 17.2). Although yield differences varied throughout the distribution, significant differences were observed only between the 5th and 95th percentiles (Table 17.3). The probability of finding differences at the

Table 17.2 Maize distribution parameters of control and *Azospirillum argentinense* seed-inoculated maize crops in a database from 316 field experiments performed in Argentina

	0	**5**	**25**	**50**	**75**	**95**	**100**	**Mean**	**SD**
Treatment	Grain yield (kg ha^{-1})								
Control without inoculation	2020	4563	7169	8792	10,545	12,855	15,924	8794	2512
With on seed inoculation	2394	5092	7583	9196	11,015	13,095	18,654	9237	2562
	Relative grain yield increases to inoculation (%)								
	−43	−8	1	4	9	24	66	6	11

SD standard deviation

Reproduced from Díaz-Zorita et al. (2016)

Table 17.3 Mean values of percentile ranges in the empirical distribution of maize grain yields from control and *Azospirillum argentinense* on seed-inoculated crops in a database from 316 field experiments performed in Argentina

	Range of the empirical distribution				
	0–5	**5–25**	**25–75**	**75–95**	**95–100**
Treatment	Grain yield (kg ha^{-1})				
Control without inoculation	3691	6118	8755	11,541	13,838
With on seed inoculation	3912	6566	9216	11,948	14,448
P	ns	*<0.001*	*<0.001*	*<0.001*	ns
n	16	63	157	64	16
SE	209	92	78	82	316

P statistical significance, *n* number of sites considered in each range, *SE* standard errors of the means, *ns* non-significant differences between means
Reproduced from Díaz-Zorita et al. (2016)

extremes of the distribution was low. However, within the central 90% of cases, where crop yields ranged between 6,000 and 12,000 kg ha^{-1}, *A. argentinense*-inoculated crops consistently showed a higher probability of achieving greater yields than non-inoculated crops (Table 17.3).

Monte Carlo simulation is a powerful tool for extending and analyzing data. This technique estimates the probability of distribution of an outcome influenced by random variables. Although initially developed for operations research, it has been widely applied to agricultural decision-making under uncertainty (Clow and Flaskerud 2001; Ferreyra et al. 2001; Berger and Peña 2013).

To implement Monte Carlo simulation, probability distributions must be defined for each random variable involved. Using experimental data on maize yield responses to *Azospirillum argentinese* inoculation in Argentina, a general probability distribution was constructed, showing a response range from −43% to 66%, with less than a 20% chance of negative responses (Table 17.2). However, since crop response data is not normally distributed, alternative probability distributions, such as betapert and cumulative, were also tested. The betapert distribution is useful in various scenarios, including those based on observed data, limited information, or expert opinion. Cumulative distribution, on the other hand, is preferable when large datasets are available, as it requires no data manipulation. These distributions require subjective estimation of minimum and maximum values, which cannot be directly sampled in a Monte Carlo simulation. For this study, these values were set at −45% and 75%. Among the tested distributions, the cumulative distribution best replicated the observed maize yield responses to inoculation across 316 sites in Argentina (Table 17.2). While the normal and general distributions performed better than the betapert distribution, they tended to overestimate the probability of no response.

For decision-making, maize prices must also be accounted for as a random variable. In this analysis, its probability distribution was modeled using a betapert function, with parameters estimated based on inflation-adjusted historical prices and expert opinions. Due to structural changes in commodity markets, long-term price trends may not be relevant for Argentine producers. Therefore, maize price distribution was based on the last 7 years of harvest prices, resulting in a betapert distribution with a minimum of $118.7, a most probable value of $170.0, and a maximum of $200.0 per ton. Marketing expenses were set at current values, including fixed costs of $38 per ton (freight and drying) and a 3% deduction for commissions and taxes. The net price was then calculated as 97% of the gross price minus fixed expenses per ton produced.

A betapert distribution was also defined for the base maize yield (without *A. argentinese* inoculation), incorporating data from agricultural simulation models and expert opinion. The estimated parameters were 5,009 kg ha^{-1} (minimum), 10,630 kg ha^{-1} (most probable), and 12,623 kg ha^{-1} (maximum), with an expected grain yield of 10,025 kg ha^{-1}. Revenue was calculated as the product of base yield, percentage yield response, and net price. Using the resulting distribution (Fig. 17.1), the probability of incremental revenue exceeding the cost of inoculation was determined. In over 81% of cases, the additional revenue from inoculation outweighed its cost, supporting its adoption as a crop management practice under current production conditions.

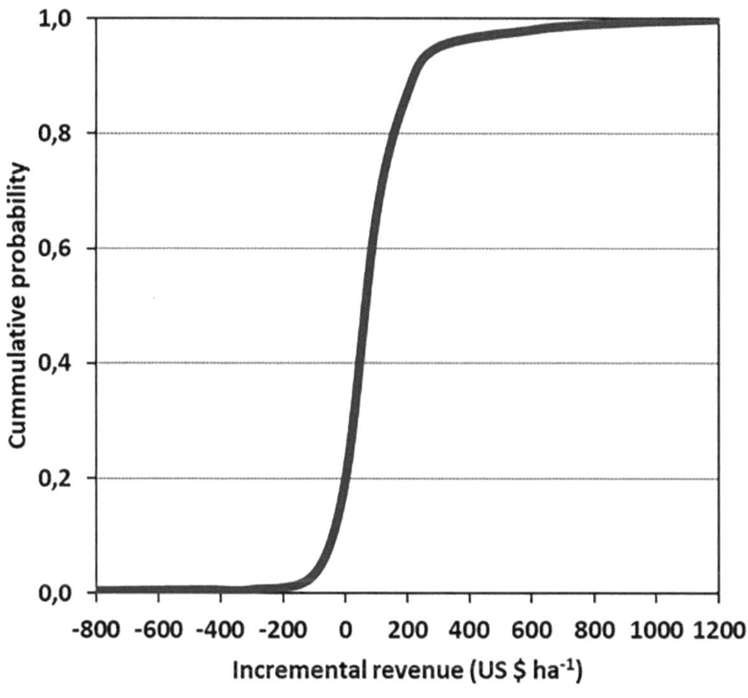

Fig. 17.1 Incremental revenue distribution for *A. argentinense* inoculation in maize crops from the north of Buenos Aires province, Argentina. (Reproduced from Díaz-Zorita et al. 2016)

This approach differs significantly from traditional methods of using experimental results for decision-making. Instead of relying on a single deterministic outcome, typically the mean response, it provides a distribution of possible outcomes and a probabilistic framework for stakeholders. This technique offers a more realistic assessment of maize yield response to *A. argentinense* inoculation in Argentina.

17.4 Conclusions

Numerous field studies, particularly in Latin America and Asia, have shown that *Azospirillum* spp. inoculation increases grain yields, mainly in cereals. While this microbe enhances the nutritional status of crops and mitigates minor water and other abiotic stresses, experimental designs should follow standard production practices, avoiding major nutrient deficiencies and extreme water stress. Inoculation is typically applied to seeds, requiring careful handling and storage to ensure high microbial survival. Compatibility with other seed treatments and the long-term viability of *Azospirillum* sp. depend on the specific formulation used. A proper interpretation of results should consider the frequency of responses and distinguish between hierarchical productivity factors. Given the high variability in crop responses to *Azospirillum* spp. inoculation, many observations are needed to detect significant differences between treatments. Analyzing response distributions offers a more realistic assessment of crop performance under complex, interacting factors, providing stakeholders with a probabilistic framework for decision-making.

References

Araujo A, Baldani V, Galisa P et al (2013) Response of traditional upland rice varieties to inoculation with selected diazotrophic bacteria isolated from rice cropped at the Northeast region of Brazil. Appl Soil Ecol 64:49–55

Barbosa JZ, Roberto LDA, Hungria M et al (2022) Meta-analysis of maize responses to *Azospirillum brasilense* inoculation in Brazil: benefits and lessons to improve inoculation efficiency. Appl Soil Ecol 170:104276. https://doi.org/10.1016/j.apsoil.2021.104276

Bayan Y, Levanony H (1990) Current status of *Azospirillum* inoculation technology: *Azospirillum* as a challenge for agriculture. Can J Microbiol 36:591–608

Benintende S, Uhrich W, Herrera M et al (2010) Comparison between co-inoculation with *Bradyrhizobium japonicum* and *Azospirillum brasilense* and single inoculation with *Bradyrhizobium japonicum* on nodulation, growth, and nitrogen accumulation in soybean crops). Agriscientia 27(2):71–77. https://doi.org/10.31047/1668.298x.v27.n2.2768

Berger A, Peña B (2013) Decisiones riesgosas en empresas agropecuarias (Risky decisions in agricultural enterprises). Editorial Facultad de Agronomía, Universidad de Buenos Aires, Buenos Aires

Caires EF, Bini AR, Duart VM et al (2024) Co inoculation of *Azospirillum brasilense* and late nitrogen fertilization for no till soybean production. Agron J 116(4):2034–2047. https://doi.org/10.1002/agj2.21602

Cardozo P, Di Palma A, Martin S et al (2021) Improvement of maize yield by foliar application of *Azospirillum brasilense* Az39. J Plant Growth Regul. https://doi.org/10.1007/s00344-021-10356-9

Chibeba AM, Kyei-Boahen S, de Fátima Guimarães M et al (2020) Towards sustainable yield improvement: field inoculation of soybean with *Bradyrhizobium* and co-inoculation with *Azospirillum* in Mozambique. Arch Microbiol 202(9):2579–2590. https://doi.org/10.1007/s00203-020-01976-y

Clow A, Flaskerud G (2001) Marketing and crop insurance combined to manage risk on a Cass County representative farm. Agribusiness and Applied Economics Report 455, Ag Exp St North Dakota State University

de Borja Reis AF, Moro Rosso LH, Adee E et al (2022) Seed inoculation with *Azospirillum brasilense* in the U.S. soybean systems. Field Crop Res 283:108537. https://doi.org/10.1016/j.fcr.2022.108537

de Souza JAL, Teixeira LS, da Silva Freitas G et al (2025) *Azospirillum brasilense* in the planting furrow of sugarcane to minimize the use of N fertilizer. Plants 14(11):1599. https://doi.org/10.3390/plants14111599

Di Rienzo J, Casanoves F, Balzarini M et al (2013) InfoStat versión 2013. Grupo InfoStat, FCA, Universidad Nacional de Córdoba, Córdoba

Díaz-Zorita M, Fernández Canigia M (2009) Field performance of a liquid formulation of *Azospirillum brasilense* on dryland wheat productivity. Eur J Soil Biol 45:3–11

Díaz-Zorita M, Baliña R, Fernández Canigia M (2012) *Azospirillum brasilense* enhances alfalfa productivity: field evaluation. In: Pérez JC, Soler Arango J, Posada Uribe LF (eds) PGPR. 9th International and 1st Latin-American PGPR Workshop. 'Returning to our roots', Quirama, Medellín, 3–8 June 2012

Díaz-Zorita M, Fernández Canigia MV, Bravo O et al (2016) Field evaluation of extensive crops inoculated with *Azospirillum* sp. In: Cassán D, Okon Y (eds) Handbook for *Azospirillum*: technical issues and protocols. Springer, Cham, pp 435–445. https://doi.org/10.1007/978-3-319-32004-2_28

Dobbelaere S, Vanderleyden J, Okon Y (2003) Plant growth-promoting effects of diazotrophs in the rhizosphere. Crit Rev Plant Sci 22:107–149

El Habbasha S, Tawfik M, El Kramany M (2013) Comparative efficacy of different bio-chemical foliar applications on growth, yield and yield attributes of some wheat cultivars. World J Agric Sci 9:345–353

Ferreyra R, Podestá G, Messina C et al (2001) A linked-modeling framework to estimate maize production risk associated with ENSO-related climate variability in Argentina. Agric For Meteorol 107:177–192

Fukami J, Nogueira MA, Araujo RS et al (2016) Accessing inoculation methods of maize and wheat with *Azospirillum brasilense*. AMB Express 6:3. https://doi.org/10.1186/s13568-015-0171-y

Galindo FS, Teixeira Filho MCM, Buzetti S et al (2018) Technical and economic viability of co inoculation with *Azospirillum brasilense* in soybean cultivars in the Cerrado. Rev Bras Eng Agric Ambient 22(1):51–56. https://doi.org/10.1590/1807-1929/agriambi.v22n1p51-56

Galindo FS, Rodrigues WL, Fernandes GC et al (2022) Enhancing agronomic efficiency and maize grain yield with *Azospirillum brasilense* inoculation under Brazilian savannah conditions. Eur J Agron 134:126471. https://doi.org/10.1016/j.eja.2022.126471

Gallace ME, Díaz-Zorita M (2025) Mejoras del rendimiento de trigo con *Azospirillum argentinense* en la región semiárida pampeana (Enhancing wheat yield with *Azospirillum argentinense* in the semi-arid Pampas). Semiarida 35(2):33–40

Garcia MVC, Nogueira MA, Hungria M (2021) Combining microorganisms in inoculants is agronomically important but industrially challenging: case study of a composite inoculant containing *Bradyrhizobium* and *Azospirillum* for the soybean crop. AMB Express 11(1):71. https://doi.org/10.1186/s13568-021-01230-8

Govindarajan M, Balandreau J, Kwon S et al (2008) Effects of the inoculation of *Burkholderia vietnamensis* and related endophytic diazotrophic bacteria on grain yield of rice. Microb Ecol 55:21–37

Hungria M, Campo RJ, Souza EM et al (2010) Inoculation with selected strains of *Azospirillum brasilense* and *A. lipoferum* improves yields of maize and wheat in Brazil. Plant Soil 331:413–425. https://doi.org/10.1007/s11104-009-0262-0

Hungria M, Nogueira M, Araujo R (2013) Co-inoculation of soybeans and common beans with rhizobia and azospirilla: strategies to improve sustainability. Biol Fertil Soils 49:791–801

Hungria M, Nogueira MA, Araujo RS (2015) Soybean seed co-inoculation with *Bradyrhizobium spp.* and *Azospirillum brasilense*: a new biotechnological tool to improve yield and sustainability. Am J Plant Sci 6:811–817. https://doi.org/10.4236/ajps.2015.66087

Islam M, Sattar M, Ashrafuzzaman M et al (2012) Improvement of yield potential of rice through combined application of biofertilizer and chemical nitrogen. Afr J Microbiol Res 6:745–750

Khalid A, Sultana S, Arshad M et al (2011) Performance of auxin producing rhizobacteria for improving growth and yield of wheat and rice grown in rotation under field conditions. Int J Agric Appl Sci 3:44–50

Lana M, Dartora J, Marini D et al (2012) Inoculation with *Azospirillum*, associated with nitrogen fertilization in maize. Rev Ceres Vicosa 59:10–17

Maroniche GA, Puente ML, García JE et al (2024) Phenogenetic profile and agronomic contribution of *Azospirillum argentinense* Az39T, a reference strain for the South American inoculant industry. Microbiol Res 283:127650. https://doi.org/10.1016/j.micres.2024.127650

Mehnaz S, Kowalik T, Reynolds B et al (2010) Growth promoting effects of corn (*Zea mays*) bacterial isolates under greenhouse and field conditions. Soil Biol Biochem 42:1848–1856

Morla FD, Cerioni GA, Giayetto O et al (2019) Evaluación de la co-inoculación en soja con *Bradyrhizobium japonicum* y *Azospirillum brasilense* (Evaluation of soybean co-inoculation with *Bradyrhizobium japonicum* and *Azospirillum brasilense*). Mercosoja 2019

Naiman A, Latrónico A, García de Salamone I (2009) Inoculation of wheat with *Azospirillum brasilense* and *Pseudomonas fluorescens*: impact on the production and culturable rhizosphere microflora. Eur J Soil Biol 45:44–51

Naoe AML, Peluzio JM, Campos LJM et al (2020) Coinoculação com *Azospirillum brasilense* em cultivares de soja submetidas a déficit hídrico (Co-inoculation with *Azospirillum brasilense* in soybean cultivars under water déficit). Rev Bras Eng Agric Ambient 24(2):89–94. https://doi.org/10.1590/1807-1929/agriambi.v24n2p89-94

Naseri R, Moghadam A, Darabi F et al (2013) The effect of deficit irrigation and *Azotobacter chroococcum* and *Azospirillum brasilense* on grain yield, yield components of maize (S.C. 704) as a second cropping in western Iran. Bull Environ Pharmacol Life Sci 2:104–112

Okon Y, Labandera-González C (1994) Agronomic applications of *Azospirillum*: an evaluation of 20 years worldwide field inoculation. Soil Biol Biochem 26:1591–1601

Piccinin G, Dan L, Braccini A et al (2011) Agronomic efficiency of *Azospirillum brasilense* in physiological parameters and yield components in wheat crop. J Agron 10:132–135

Prando AM, Barbosa JZ, de Oliveira AB et al (2024) Benefits of soybean co-inoculation with *Bradyrhizobium* spp. and *Azospirillum brasilense*: large-scale validation with farmers in Brazil. Eur J Agron 127:Article 127112

Puente ML, Gualpa JL, Lopez GA et al (2017) The benefits of foliar inoculation with *Azospirillum brasilense* in soybean are explained by an auxin signaling model. Symbiosis. https://doi.org/10.1007/s13199-017-0536-x

Rodrigues E, Rodrigues L, Martínez de Oliveira A et al (2008) *Azospirillum amazonense* inoculation: effects on growth, yield and N2 fixation of rice (Oryza sativa L.). Plant Soil 302:249–261

Soil Survey Staff (1999) Soil taxonomy: a basic system of soil classification for making and interpreting soil surveys, 2nd edn. USDA Agric Handbook 436, Washington, DC

Torres D, Donadio F, López GA et al (2022) Previous incubation of *Bradyrhizobium japonicum* E109 and *Azospirillum argentinense* Az39 (formerly *A. brasilense* Az39) improves the *Bradyrhizobium*–soybean symbiosis. J Soil Sci Plant Nutr 22(4):4669–4682. https://doi.org/10.1007/s42729-022-00948-z

Turan M, Fikrettin S (2013) Role of plant growth promoting rhizobacter strain reduce application rates mineral fertilizer in barley. ProEnvironment 6:324–331

Van Dommelen A, Croonenborghs A, Spaepen S et al (2009) Wheat growth promotion through inoculation with an ammonium-excreting mutant of *Azospirillum brasilense*. Biol Fertil Soils 45:549–553

Chapter 18
Field Evaluation on Intensive Crops. What to Expect from the Use of *Azospirillum* Genus in Intensive Crops

Cecilia M. Creus, Maria Flavia Luna, Ana Carmen Cohen, and Santiago Adolfo Vio

Abstract Vegetable production is on the rise worldwide. In addition, consumers are demanding improvements in the nutritional and health benefits of these vegetables. At the same time, there is a growing trend toward sustainable agriculture that does not harm the environment and reduces pollution from agrochemicals. Compared to extensive crops, there is less information available on the use of *Azospirillum* inoculation in vegetable crops and even less under productive conditions. Here, we summarize the information on the effects of *Azospirillum* spp. application on lettuce and tomato as main model species and advances in situations tested in greenhouses and in the productive field under different conditions. Inoculation with *Azospirillum* alone or in combination with other microorganisms shows significant benefits during the seedling and productive stages. Improved plant growth and development, higher yields and improved leaf and fruit quality, and extended storage life are achieved. Crop performance can be enhanced even during periods of stress

---

C. M. Creus (✉)
Laboratorio de Bioquímica Vegetal y Microbiana, Facultad de Ciencias Agrarias, Universidad Nacional de Mar del Plata, Balcarce, Buenos Aires, Argentina
e-mail: creuspi@hotmail.com

M. F. Luna
Centro de Investigación y Desarrollo en Fermentaciones Industriales (CINDEFI, CONICET/UNLP), La Plata, Buenos Aires, Argentina

Comisión de Investigaciones Científicas de la Provincia de Buenos Aires (CIC-PBA), La Plata, Buenos Aires, Argentina

A. C. Cohen
Grupo de Fisiología Vegetal y Microbiología, Instituto de Biología Agrícola de Mendoza, IBAM (Conicet-UNCuyo), Chacras de Coria, Mendoza, Argentina

Facultad de Ciencias Agrarias, Universidad Nacional de Cuyo, Chacras de Coria, Mendoza, Argentina

S. A. Vio
Centro de Investigación y Desarrollo en Fermentaciones Industriales (CINDEFI, CONICET/UNLP), La Plata, Buenos Aires, Argentina

F. D. Cassán et al. (eds.), *The Definitive Handbook of Azospirillum*,
https://doi.org/10.1007/978-3-032-09222-9_18

and moderate soil salinity. *Azospirillum* has a high potential to be used as a low-cost and environmentally friendly tool to increase the productivity and quality of vegetable crops.

Keywords Vegetables · Greenhouse · Productivity · Quality · Sustainability

18.1 Introduction: A Brief History on the Use of *Azospirillum* spp. in Vegetable Intensive Crops

In the mid-1970s, Johanna Döbereiner and her colleagues isolated a microorganism associated with the roots of several grasses, including *Digitaria*, *Panicum maximum,* and *Cynodon dactylon*, as well as grain crops such as wheat, maize, sorghum, and rice (Day et al. 1975). The initial interest in this bacterium was, of course, due to its capacity for biological nitrogen fixation (BNF) with non-leguminous plants. The genus was later reclassified as *Azospirillum,* with two model species described (Tarrand et al. 1978). This initial focus on describing the BNF with gramineous plants obscured for a time the idea that *Azospirillum* colonizes a wide range of plants and could therefore be used in practical approaches for horticultural crops and other non-grass vegetable plants. It was not until the 1980s that its ubiquity in different soils and environments was investigated, and it was seen as a general colonizer of plants, including intensively cultivated vegetable species (Bashan and Holguin 1997).

The term plant growth promoting rhizobacteria (PGPR) was coined by Joe Kloepper while studying soil bacteria at Auburn University, USA. These bacteria were first isolated from potato soils. These consortia of PGPR, probably including the genus *Azospirillum*, were then applied for the first time to vegetable crops, with PGPR being added to irrigation water on radish, sugar beet, and potato crops in Florida and California (Kloepper and Schroth 1978). Later, the name PGPR was expanded to PGPB to include all the plant beneficial bacteria, other than rhizospheric, according to their specific role and colonization capacity (Bashan and Holguin 1998). However, commercial PGPB-based plant inoculants (bio-inputs) have been developed first for extensive agriculture and then for intensive agriculture, and there is still a growing demand to extend their use to horticulture.

18.2 General Overview of *Azospirillum* spp. Actions Focused on Vegetable Crops

Vegetables play a fundamental role in human diet and nutrition, as the consumption of their leaves, fruits, and roots helps to meet the body's needs due to their important contribution of vitamins, minerals, and proteins. They are therefore considered to be an essential part of a healthy diet, which has led to an increase in their consumption in recent years (FAO 2020; Queral et al. 2024). Vegetables account for 12% of global agricultural production after cereals and sugar crops (Novello et al. 2024). In general, most vegetable production is divided into two stages: (1) a seedling stage from sowing to transplanting, carried out in greenhouse nurseries, mainly in trays and (2) a production stage, from transplanting to harvesting, performed in a producer's farm, either in the greenhouse or in the field. Transplanting, as a transitional practice between these two stages of horticultural production, represents a critical step with a great impact on the establishment, development, and yield of the crop, due to the stress caused by manipulating the plants and subjecting them to a drastic change in the environment (Finch-Savage and Bassel 2016). Therefore, all efforts are focused on obtaining a quality seedling (nutritional and sanitary) capable of facing stressful situations without major inconveniences. This has led to production practices in seedling nurseries where the irrational use of fertilizers and fungicides is commonly overexpressed to achieve this objective. The application of bio-inputs based on PGPB at the seedling stage in horticultural species (both at sowing and transplanting) is a good alternative to increase the emergence rate, improve the growth of shoots and roots, obtain more robust and vigorous seedlings, and accelerate production times, among other beneficial effects (Abd El-Samad et al. 2019; Berger et al. 2017; Garbi et al. 2016), which gives added value to the seedling. Clearly, it is also possible to apply bio-inputs based on PGPB in the vegetable production stage, not only as biofertilizers, biostimulants, or biocontrol agents but also for stress amelioration. Some authors also demonstrated that PGPB could be able to improve the nutritional value of fruits and vegetables by increasing the antioxidant capacity and phenolics content (Nautiyal et al. 2008; Koheler et al. 2010; Cappellari et al. 2013; Fasciglione et al. 2015). Therefore, the use of bio-inputs in horticultural crops should be prioritized in a sustainable agricultural context, with the added benefit of improving the nutritional quality of vegetables produced with these agronomic practices.

Currently, there are several species in the genus *Azospirillum* with the ability to enhance plant growth through a variety of mechanisms, many of which occur in a synergistic fashion. These microorganisms improve plant growth, development, health, and fitness through various direct and indirect mechanisms summarized in Fig. 18.1 (Sun et al. 2025).

18.3 Inoculation Responses on Different Intensive Crops in Soil Systems Under Cover and in Field Conditions

Although less information is found about application trials in horticultural crops compared with extensive crops, it has been reported that inoculation with *Azospirillum* shows remarkable benefits at seedling and productive stages in different vegetables such as tomato, lettuce, pepper, onion, cucumber, among others.

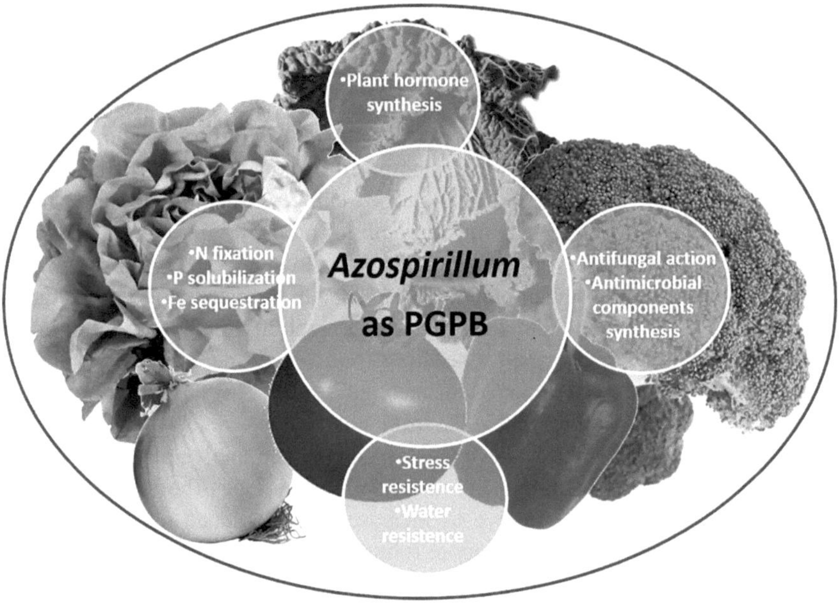

Fig. 18.1 Modes of action of *Azospirillum* spp. as PGPB

Several *Azospirillum* strains improve plant vigor, germination rates, leaf area and total biomass production, root development, and nutrient uptake. They contribute to higher yields and quality crops (Vendruscolo and de Lima 2021). Moreover, other characteristics of the management practices, including abiotic and biotic factors such as substrate, fertilization plan, irrigation practices, botanical variety, formulation and mode of application of the bio-input, as well as the presence of other PGPB, produced different results regarding the response of *Azospirillum* spp. (Cassán and Diaz-Zorita 2016). Because of this, before evaluating the in vivo action of PGPB under field conditions on the target crop, trials are first conducted under laboratory conditions with controlled light, temperature, and humidity to minimize the mentioned effects that may mask their action. Doses and application methods can then be adjusted and validated in field trials. Below is a compilation of various investigations conducted around the world on this topic.

18.3.1 Tomato

Tomato (*Solanum lycopersicum* L.) is one of the most widely consumed vegetables worldwide, due to the wide versatility of its use (fresh and processed) and due to its high nutraceutical value (Martínez-Damian et al. 2019). Among vegetables, tomatoes represent the largest production of horticultural crops (FAO 2022) and remained stable at 16% over the period 2000–2021. Given the global significance of the

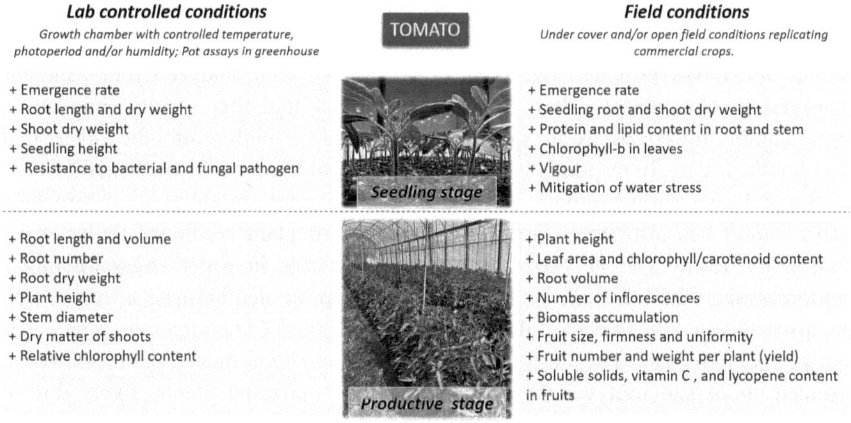

Fig. 18.2 Positive effect of *Azospirillum* spp. applied on tomato plants at different production stages

tomato crop, research has focused on the potential of *Azospirillum* inoculation to enhance plant growth, improve resilience to biotic and abiotic stresses, and optimize crop productivity in a sustainable manner (Fig. 18.2).

Early research by Bashan et al. (2002) demonstrated that the inoculation of tomato plants with *A. brasilense* encapsulated in alginate microbeads significantly increased plant height and biomass accumulation. This was further supported by Díaz and Iglesias (2002), who observed that field-grown tomato plants treated with *A. brasilense* exhibited greater leaf area, height, and root volume compared to uninoculated plants. These initial findings laid the groundwork for understanding the growth-promoting effects of *Azospirillum* on tomato plants. The influence of *Azospirillum* on physiological parameters has also been extensively examined. de Lima et al. (2018) explored the effects of different inoculant doses containing *A. brasilense* Ab-V5 and Ab-V6 on two tomato cultivars, finding that optimal doses (4–6 mL kg^{-1} of seed) maximized plant height, stem diameter, root development, and chlorophyll content. These findings were echoed in a study by Andrade-Sifuentes et al. (2020), which reported that inoculation with *A. brasilense* DSM 1843 not only improved seedling emergence and growth but also enhanced the nutraceutical quality of tomato fruits by increasing soluble solids, vitamin C, and lycopene content. Yield improvements associated with *Azospirillum* inoculation have also been well documented. Pérez-Rodríguez et al. (2020) assessed the effect of root inoculation with *A. argentinense* Az39 and *Pseudomonas fluorescens* Rt6M10 in industrial tomato varieties under a reduced fertilization regime. Their results showed significant increases in root and shoot biomass, improved transplant stress tolerance, and enhanced fruit yield and quality, comparable to or even surpassing chemically fertilized plants. Similarly, Guerrieri et al. (2021) confirmed the positive effects of *A. brasilense* Sp7 in greenhouse-grown tomato seedlings under both organic and conventional management. They observed that *Azospirillum* application at different levels (seed, root, or combined) promoted biomass accumulation, likely due to its nitrogen-fixing capabilities, with the most pronounced benefits under conventional conditions. A deeper understanding of the interaction between *Azospirillum* and tomato roots was provided by Nisha et al. (2024), who analyzed root exudates involved in chemotaxis. Their findings confirmed that this specific interaction significantly improved root system development, reinforcing the idea that *Azospirillum* actively responds to plant signals to enhance symbiotic benefits.

Beyond the improvement in plant growth and production parameters, *Azospirillum* has also been shown to contribute to plant resilience under stress conditions. Romero et al. (2014) investigated its role in water stress tolerance, demonstrating that *A. brasilense* BNM65 not only promoted biomass accumulation but also mitigated the impact of the pathogenic bacterium *Clavibacter michiganensis* subsp. *michiganensis*. The study revealed that *Azospirillum*-treated plants exhibited reduced shoot and root weight loss compared to untreated plants, likely due to structural modifications such as increased xylem vessel area, a more extensive root system, and reduced specific leaf area, all of which enhance water uptake and transport efficiency.

Disease resistance is another crucial aspect of *Azospirillum*-mediated plant benefits. Fujita et al. (2017) demonstrated that endophytic colonization of tomato roots by *Azospirillum* sp. B510, originally isolated from rice, significantly reduced symptoms of bacterial leaf spot (*Pseudomonas syringae* pv. *tomato* DC3000) and gray mold (*Botrytis cinerea*). This highlights the role of *Azospirillum* not only in direct plant growth promotion but also in activating plant defense mechanisms against both bacterial and fungal pathogens.

Taken together, these studies underscore the multifaceted role of *Azospirillum*-based bio-input in tomato crops. Beyond promoting vegetative growth, this bacterial genus contributes to stress tolerance, disease resistance, and improved fruit quality, offering a sustainable biotechnological tool as an alternative and/or complement to conventional fertilization and plant protection strategies. As research continues, *Azospirillum* stands as a promising biofertilizer and biostimulant, with the potential to enhance productivity in modern tomato agriculture while reducing reliance on synthetic inputs.

18.3.2 Lettuce

Lettuce (*Lactuca sativa* L.) is often used as a model for studying PGPB-plant interactions because of its many advantages, including its small size, short time to harvest, and responsiveness to microbial effects. This vegetable has a rapid production cycle, making the initial establishment important in defining its yield capacity. In this sense, several research and development works have been carried out in the laboratory and in the field to evaluate the action of *Azospirillum* on this crop, summarized in Fig. 18.3. Barassi et al. (2006) reported greater germination of lettuce seeds when inoculated with *Azospirillum* sp. It was also shown that inoculation of lettuce seeds with *A. baldaniorum* Sp245 increased seed vigor and seedling

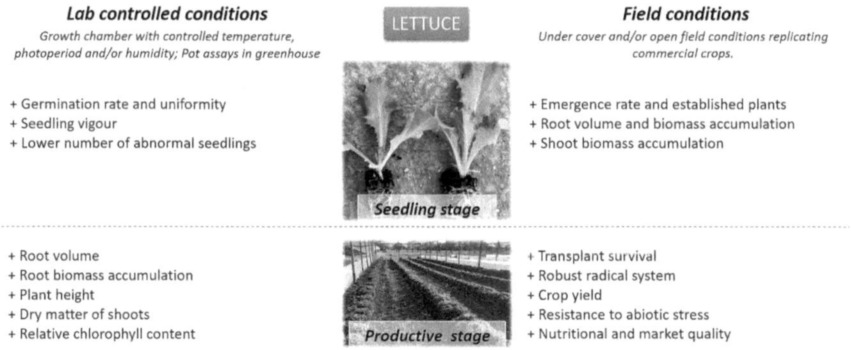

Fig. 18.3 Positive effect of *Azospirillum* spp. applied on lettuce at different stages of the production process

emergence percentages (Carrozzi et al. 2012). Furthermore, in field trials, de Lima et al. (2017) showed that the combined application of nitrogen and *A. brasilense* resulted in an increase in the percentage of established plants.

Many works reported greater root and shoot production on lettuce after single inoculation with different strains of *A. brasilense*. evaluated the effect of inoculation with *Azospirillum* on lettuce plants and obtained an increase in the root surface, which favorably influenced the posttransplant behavior of the seedling and impacted the yield. In turn, Chamangasht et al. (2012) observed a significant increase in crop yield by inoculating seeds with *Azospirillum* sp. Lettuce plants cv. Mantecosa inoculated with *A. brasilense* applied to the seed and grown in a greenhouse showed better transplant survival than uninoculated controls under moderate soil salinity conditions (Fasciglione et al. 2012). Several studies have shown the palliative effect that the use of *Azospirillum* has on abiotic stresses, as better and more robust radical systems induced by the bacterium inoculation enable the plant to cope with posttransplantation and salt stress (Barassi et al. 2006; Fasciglione et al. 2012).

Fasciglione et al. (2015) studied the influence of salt stress produced by 40 mol m^{-3} NaCl on the relative growth and net assimilation rates of lettuce from transplanting to harvest. Even though salt stress decreased the relative growth rate in control plants, this parameter did not change in *A. baldaniorum* Sp 245 inoculated plants. They also showed positive effects on inoculated plants at harvest and postharvest, improving nutritional and market quality. A decrease in chlorophyll content due to salt stress was observed at harvest and after 10 days of storage in non-inoculated plants. However, no changes in total chlorophyll content and its components *a* and *b* were observed in inoculated plants at the end of the storage period. Another important antioxidant, ascorbic acid, decreased after 20 days of storage due to the effect of salinity, while its level remained higher than the control in *Azospirillum*-inoculated plants. According to sensory data on overall visual quality, salt stress deteriorated the appearance of lettuce at harvest and during 10 days of storage. *Azospirillum* improved this parameter at harvest and postharvest. Objective data on color characteristics such as hue, chroma, and L values provided by a chromatometer were superior in inoculated plants (Fasciglione et al. 2015).

In addition, since horticultural seeds are expensive in the marketplace, the effects of *Azospirillum* on aged seeds were studied because of their potential impact (Carrozzi et al. 2012). Seed quality, defined as the ability to germinate, can decrease with prolonged storage (Bradford 2006). Changes in metabolic activities associated with seed aging are complex and include lipid peroxidation and free radical damage to key cellular molecules and cell structures (Walters et al. 2005). The technique of osmotic conditioning or osmopriming is a process in which seeds are allowed to imbibe enough water to allow early events in the germination process to occur, but not enough to allow radicle emergence through the seed coat. Typically, seeds are pre-imbibed in polyethylene glycol or inorganic salts, and this technique has been proposed to stimulate and improve the rate and uniformity of germination (Ghiyasi et al. 2008; Salehzade et al. 2009). *Azospirillum* inoculation on one-year aged lettuce seeds of the cultivar 'Crimor INTA' without previous osmopriming enhanced seed vigor and seedling emergence percentages and decreased the proportion of abnormal seedlings. Furthermore, there was an additional effect of osmopriming as a previous treatment of *Azospirillum* inoculation on the germination percentage of aged lettuce seeds, producing a lower number of abnormal seedlings (Carrozzi et al. 2012). This work demonstrated that both treatments can overcome the negative effects of aging on lettuce seeds, helping farmers to extend the period seeds can be used.

18.3.3 Pepper

In a variety of pepper (*Capsicum annum* L.), it has been seen that plants inoculated 7 days after sowing with *A. brasilense* reached greater height and dry weight (Di-Barbaro et al. 2005).

Samyukta et al. (2024) carried out a pot culture experiment with bell peppers treated with biochar and biofertilizers, including *Azospirillum*, Vesicular Arbuscular Mycorrhiza, and *Serendipita indica,* applied directly to the soil during transplanting. Results showed that plants treated with biochar + *Azospirillum* had higher flavonoid, total soluble protein, and ascorbic acid content. Lobato Ureche et al. (2021) evaluated *Azospirillum argentinense* Az39 inoculated on pepper (*Capsicum annum* L.) seeds. Inoculation significantly improved germination, seedling growth, root length, and diameter of pepper seedlings.

18.3.4 Fruit Production

A. brasilense colonizes strawberry plants, particularly the internal tissues of stolons and the surface and internal tissues of roots (Pedraza et al. 2007). Three strains of *A. brasilense* (REC3, RLC1, PEC5) promoted strawberry plant growth, expressed as root length, root area, and root and shoot dry weight. These effects were dependent on the genotypes (cultivars 'Milsei', 'Selva', and 'Camarosa') used in the plant-bacterial strain association. The total sugars contained in the root exudates of strawberry plants play an important role in the chemotactic effect of *A. brasilense*, which could partially explain the observed plant growth-promoting effect (Pedraza et al. 2010). The ability to colonize the internal tissues of roots and stolons of *A. brasilense* would allow growers to have multiple generations of plants in the nursery already inoculated and in better conditions for planting in the field, which would be a contribution to sustainable agricultural practices (Guerrero-Molina et al. 2012). The inoculated plants with *A. brasilense* RLC1 and REC3 under field conditions exhibited more vigorous vegetative growth and increased the total fruit yield and marketable yield as compared with non-inoculated plants (Salazar et al. 2012).

Bartolini et al. (2017) evaluated the influence of *A. brasilense* Sp245 on the propagation of young grapevine plants cv. Colorini grafted on two rootstocks (420A and 157/11) during the vegetative development of young grapevine (*Vitis vinifera* L.). The bacterization of the rootstocks showed a positive influence on the improvement of the final quality of the root apparatus. However, the effects are not uniform, suggesting an influence of genetic diversity and nutritional and hormonal status of the cutting. In both rootstocks, a significant positive effect of *A. brasilense* Sp245 on total phenolics and iron (Fe) concentrations was observed compared to the control. Tofanini et al. (2014) continued these studies and evaluated the effects of *A. brasilense* Sp245 on grapevine rootstocks grafted with 'Sangiovese'. They evaluated the ability of the inoculation to improve the rooting parameters of some rootstocks that are not easily rooted by conventional techniques. Nine rootstocks were tested in a conventional nursery, while two rootstocks were tested in an organic nursery: *V. berlandieri* × *V. riparia* 420A, 161-49, 157-11, SO4; *V. berlandieri* × *V. rupestris* 140Ru, 775P, 1103P; *V. riparia* × *V. rupestris* 101-14, 3309C. The quality of the root system was improved in terms of the number of roots, root architecture, and total biomass of the vines produced in the organic nursery, while in the conventional nursery the results depended on the rootstock. However, further studies are needed to understand the mode of action and to determine how Sp245 could be used for the sustainable production of grapevine plants.

In olive (*Olea europaea* L., cultivar 'Leccino'), preliminary results indicated that *A. baldaniorum* Sp245 induced anatomical changes in cuttings of this cultivar. Sp245 induced significant cellular activities (stimulation of induction processes related to adventitious root formation) and could be considered as a valid alternative rooting agent to reduce the use of indole-3-butyric acid (IBA), the most used synthetic auxin in nursery production of olive plants (Bartolini et al. 2023).

18.3.5 Ornamental Crops

The use of bio-inputs in ornamental crops is an emerging trend that offers sustainable and environmentally friendly alternatives to synthetic fertilizers and pesticides in this important sector. In Petunia plants, inoculation with *A. brasilense* strain REC3 increased root growth, shoot weight, and number of flowers. The time to flowering was also shortened in inoculated plants. These results indicate that inoculation increases vegetative growth and early flowering, thus reducing fertilization costs and environmental pollution (Toffoli et al. 2018). Similar results were obtained by Santos et al. (2022), who studied the inoculation of *Lisianthus (Eustoma grandiflorum)* seeds with *A. brasilense* Az39. The results showed an increase in the in vitro germination rate, which produced precocity and promoted aerial and root development, evidenced by increases in biometric parameters and anatomical modifications. Also, improved root development, producing homogeneous seedlings, larger, healthy, and vigorous. However, Avinash et al. (2019) studied five strains, and only the strain *Azospirillum* ATR-39 had significantly higher plant height, number of tillers, and other growth parameters when it was inoculated on tuberose (*Polianthes tuberosa* L.) cv. Mexican Single.

18.3.6 Other Important Intensive Crops

As a general colonizer, *Azospirillum* strains have been tested in many important vegetable crops. Yield and quality of various crops such as cucumber (*Cucumis sativus* L.) and sweet potato (*Ipomoea batatas* L.) were increased by inoculation combined with different types and doses of inorganic fertilizer application. Fruit quality and yield could be increased in cucumber by inoculating with *Azospirillum* sp. along with N, P, and K-based fertilizers (Kumar and Kumar 2019). In addition, the inoculation of sweet potato with *Azospirillum* sp. combined with the use of inorganic fertilizers (N, P, and K) allowed an increase in protein content and yield (Pérez-Pazos and Sanchez-Lopez 2018).

The use of *Azospirillum* has also shown beneficial effects in the *Brassicaceae* family, such as cauliflower (*Brassica oleracea* var. *Botrytis*) and broccoli (*Brassica oleracea* var. *Italica*). In cauliflower, the application of *A. brasilense* together with inorganic fertilizers based on N and P resulted in increases in plant weight and yield (Salim et al. 2021). While applying *Azospirillum* sp. during sowing favored increased broccoli stem diameter (Romero et al. 2016) and higher yield (Vendruscolo and de Lima 2021).

Bahgat et al. (2023) reported that inoculation with *A. lipoferum* increased the yield of roselle (*Hibiscus subdariffa* L.), an herbaceous plant widely used worldwide in the medicinal, beverage, and food industries. *A. lipoferum* inoculation revealed significantly higher root length and fresh and dry root weights compared to uninoculated control plants. Also, the highest productivity and yield of calyx sepals (the most economically important part) were obtained by *Azospirillum* treatment compared to *Bacillus* and *Azotobacter* strains.

Despite many positive results in various horticultural crops, there are few reports of species that do not respond positively to inoculation with *Azospirillum*. Such is the case of onion (*Allium cepa* L.) seedlings inoculated with two strains of *Azospirillum*, which showed no relevant effects on bulb diameter, leaf area, and

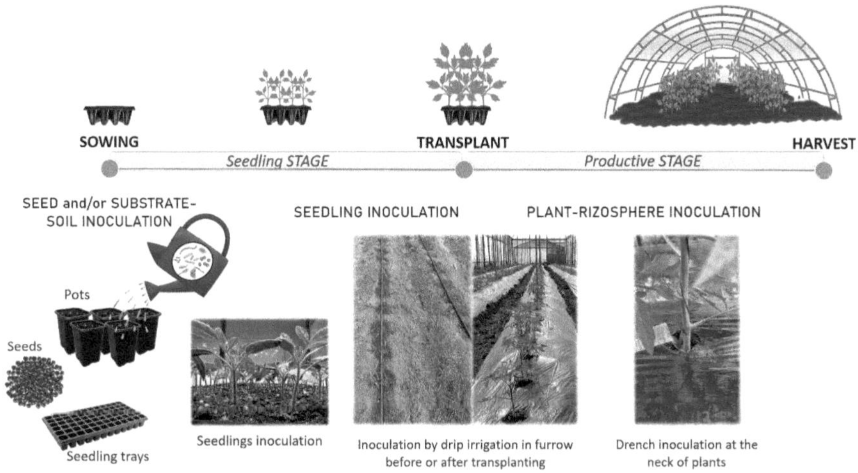

Fig. 18.4 Inoculation times and methods commonly used to apply bio-inputs in intensive crops

plant dry weight (Romero et al. 2016). In the same way, chili pepper (*Capsicum chinense* L.) inoculated with *A. brasilense* at the time of transplant showed no significant effect on plant growth, fruit size, and crop yield (Reyes-Ramírez et al. 2014), while in cassava (*Manihot esculenta*) the application of *A. brasilense* 21 days after transplant did not affect development or yield (Sharara and Abd El-Aal 2016).

The diversity of responses that exist in vegetable species to inoculation with *Azospirillum* could be explained by many and varied factors that also affect any type of agronomic practices. These include not only the inoculated strain but also the edaphic environment and the occurrence of adverse climatic events such as short water supply or periods of extremely high temperature. Certainly, inoculation can be performed in different agricultural production systems, which has implications for the inoculation methods, the dose, and the dilution to apply (Fig. 18.4) (Vio et al. 2025). However, in the case of commercial bio-inputs, this is handled and indicated by the bio-input provider.

18.4 General Guidelines for the Use of *Azospirillum* spp. in Horticultural Crops

In general terms, depending on the moment and way of inoculation (Fig. 18.4), several protocols might be possible for horticultural microbial application, leaving aside foliar inoculation. Considering that a liquid concentration of 10^9 UFC/ml is a common title for bio-inputs based on *Azospirrillum* spp., it is possible to calculate the required dose of bio-input and the necessary amount of water to be used, particularly for this microorganism, taking into account the number of microorganisms that are intended to be reached per seed or per plant, as well as other considerations mentioned by Vio et al. (2025).

Some examples of protocols for *Azospirillum* inoculation are mentioned below:

18.4.1 On Seed

To reach ~10^6 CFU per seed, 1 ml of concentrated bio-input is commonly used for every 1000 seeds (Fasciglione et al. 2012).

18.4.2 In Substrate (Tray or Pot)

Inoculation is usually done with the first irrigation of the substrate, either before or after sowing, at a ratio of 1 ml of 1:5 diluted bio-input for every 10 ml of substrate, to achieve a population of ~10^6–10^7 CFU per g of substrate (Vio et al. 2023).

18.4.3 In Plant

The protocol varies depending on the crop stage and the inoculation methodology used.

18.4.3.1 Seedlings Before Transplanting

Inoculation can be performed as previously described for the substrate (tray or pot) with the aim of achieving the same bacterial population. This can be done by either watering or immersing the trays or pots in the diluted bio-input. The usual ratio is 1 ml of 1:5 diluted bio-input for every 10 ml of substrate, with the bio-input and water requirements adjusted according to the amount of substrate present in the total number of trays or pots to be inoculated (Pérez-Rodriguez et al. 2020; Vio et al. 2023).

18.4.3.2 Plants After Transplanting

At the productive stage, inoculation can be carried out via the irrigation system (mostly drip irrigation) or by drench application to the neck of the plants. As a reference for drench inoculation of tomato plants 30 days posttransplant, a dose of 30 ml of concentrated bio-input per 100 plants could be used, using an appropriate amount of water (e.g., 25–40 ml per plant). According to Vio et al. (2025), this dose would allow a population of 10^8 CFU per g of rhizosphere soil. For drip irrigation, these same authors recommend a dose of 15 ml of concentrated bio-input per 60 m long furrow, using ~27 L of water. This would result in a population of ~10^4 CFU g^{-1} of irrigated soil.

In most of these experiments, where positive responses of increased growth and plant production were found, *Azospirillum*-based inoculants were used together with the application of inorganic fertilizers, which allowed increasing the effectiveness of this PGPB with fertilizer doses lower than those recommended (Cassán et al. 2020). In these cases, *Azospirillum* could have promoted root development, optimizing nutrient uptake, plant growth, and crop yield.

18.5 *Azospirillum* in Soilless Systems

Cultivation of vegetables in soilless systems is becoming especially important because of the significant decrease in land suitable for agriculture. PGPB also offers a solution to several of the largest problems in hydroponic production by preventing pathogen outbreaks, improving the plants' response to environmental stress, and increasing crop yield per m^2 all of which reduces the payback period of initial capital investment. In this sense, *A. brasilense* has been evaluated on the production of pepper (*Capsicum annuum* L.), tomato (*Solanum lycopersicum* L.) and basil (*Ocimum bacilicum* L.) in a semihydroponic culture system (Di-Barbaro et al., 2023). High levels of microbial survival were obtained, and better-quality seedlings were observed in the inoculated treatments (Di-Barbaro et al. 2023). Chau and Hung (2023) examined the beneficial effects of inoculation methods of *A. brasilense* on plant growth and nutrition of two lettuce cultivars grown in hydroponics. Four treatments were evaluated: non-inoculated, foliar application of *Azospirillum*, application of *Azospirillum* in nutrient solution, and *Azospirillum* in nutrient solution + foliar application. All inoculation methods improved plant height, shoot fresh and dry mass, fresh leaf yield, and accumulation of potassium, calcium, phosphorus, iron, manganese, and copper. Inoculation via nutrient solution + foliar application of *A. brasilense* was the most suitable for hydroponic lettuce GR-607 cultivation because it provides greater fresh leaf yield. All inoculation methods increased leaf yield in hydroponic lettuce under these assay conditions. In other lettuce hydropic study using multiple nitrogen fixers, including *A. brasilense*, *P. fluorescens*, *Azotobacter chroococcum,* and *Bacillus subtilis*, the authors observed that the amount of nitrogen uptake almost doubled, especially when (non-nitrogen-fixing) arbuscular mycorrhizal fungi were also added with the bacteria (Stegelmeier et al. 2022).

18.6 Impact of the Use of *Azospirillum* spp. in Synthetic Mixtures

The use of microbial multistrain mixtures as bio-inputs for intensive crops is a growing practice in agriculture. The combination of two or more beneficial microorganisms within a multistrain formulation might produce additive and/or synergistic effects on crops (Thomloudi et al. 2019). However, the selection of strains for dual-species formulations is usually based on the individual characteristics of each strain, without considering their compatibility. In general, compatibility tests are carried out in vitro by studying mutual growth inhibition or antagonistic effects. Also, the effect of bacteria applied in combination may be different from the effect of each bacterium alone. Furthermore, when used in combination with beneficial microorganisms other than bacteria, such as mycorrhizal fungi or microalgae, these positive outcomes can be enhanced and contribute to integrated crop management practices (Vio et al. 2023). Considering the complex and diverse scenario that makes up the rhizosphere, this becomes logical.

An experimental formulation with three different *Azospirillum* strains inoculated at the time of sowing produced a greater number of leaves and increased the fresh weight of the plants (Carletti et al. 2011). Similar results were obtained by Garbi et al. (2016) after inoculation of lettuce with three other strains of *Azospirillum* (Az39, Pl64 y Pl3); an increase in root absorption area and plant fresh weight was observed in lettuce plants regarding uninoculated plants, with no significant changes in primary root length or dry weight.

The genera *Azospirillum* and *Pseudomonas* have been used for many years in the development of commercial inoculants that combine strains belonging to both genera (Valverde et al. 2015). Inoculation with *A. argentinense* Az39 or *Pseudomonas* Rt6M10 increased root and shoot dry weight of tomato seedlings when used alone, but the combination did not differ from the uninoculated in shoot dry weight (Pérez-Rodriguez et al. 2020). Vio et al. (2023) showed that multispecies bacterial bio-input based on *Azospirillum argentinense* Az39, *Gluconacetobacter diazotrophicus* PAL-5, *Pseudomonas protegens* Pf-5, and *Bacillus* sp. Dm-B10 improved lettuce plant growth more effectively than when these strains were inoculated as single species. Bacteria persisted together, showing they are compatible and colonized seedling roots of lettuce plants grown in controlled conditions. Interestingly, colonization was highly related to an early and enhanced growth of seedlings grown in the nursery. This synthetic bacterial mixture showed their synergistic beneficial effect both in the nursery greenhouse as well as the commercial production farm.

Díaz et al. (2023) add complexity to this picture, demonstrating how combining two PGPBs into a single inoculant can result in synergistic effects or a loss of the benefits produced by each strain separately, depending on the chosen partner strains. *A. argentinense* Az39 and *P. fluorescens* A506 showed high compatibility both in vitro and in vivo when co-inoculated in lettuce, while *A. argentinense* Az39 and *P. protegens* Pf-5 have a lower in vitro compatibility, and their combination undermine the plant growth stimulatory capacity that each strain has individually. This evidence confirms that bacterial strain compatibility is crucial for mixed inoculant effectiveness.

Novello et al. (2024) evaluated the potential of a Synthetic Community (SynCom) characterized and composed of free-living nitrogen-fixing bacteria (*Azospirillum*, *Azotobacter*, and *Rhizobium*) and a consortium of three species of mycorrhizal fungi and two strains belonging to the genus *Bacillus* and *Paenibacillus* (Micomix) in supporting tomato field production by reducing the input of chemical fertilizer. The results demonstrated that the use of SynComs could support marketable berry production under reduced fertilization conditions. SynComs not only support tomato production under reduced fertilizer inputs but also contribute to the enhancement of soil microbial biodiversity. This dual benefit underscores the potential of SynComs as a soil management strategy, offering a pathway toward sustainable agriculture and environmental stewardship.

Di-Barbaro et al. (2021) evaluated the effect of *A. brasilense* and mycorrhizal fungi on the nutrition of the yacon crop (*Smallanthus sonchifolius* [(Poeppig and Endlicher) H. Robinson]). The trial was carried out under field conditions, and agronomic parameters and health status of the plants at the time of implantation of the culture were determined. The propagules were inoculated with *A. brasilense* and with native mycorrhizal fungi, generating four treatments, including the control and the co-inoculation of the mixture of microorganisms under study. The results indicated that the co-inoculation with *A. brasilense* and with native mycorrhizal fungi significantly increased the growth of plants in height, leaf area, biomass, dry matter, and yields in field production.

However, negative results have also been found in the co-inoculation of *Azospirillum* strains with other bacterial genera: Lobo et al. (2022) tested the effect of indole acetic acid (IAA) on the inoculation of *B. subtilis* and *A. brasilense* in three tomato genotypes in lab-controlled experiments. They reported that the inoculation of tomato seedlings with a mixture of *B. subtilis* and *A. brasilense* can reduce some plant parameters, such as plant height, stem diameter, and the number of roots. They also mentioned that the negative effect is not related to the excess of IAA produced by the bacteria but is probably an interference of each bacterium in the mode of interaction and in the growth promotion effect with the plant.

18.7 Concluding Remarks

Although plant inoculants based on PGPB were initially developed for extensive agriculture, there is an increasing demand to extend their use to intensive farming systems such as horticulture. Organic, hydroponic, and greenhouse growing are some of the common methods used to grow vegetables. All of these are conducted in a controlled environment, a condition that in turn could help to achieve more stable PGPB effects than in the open field (Malusá et al. 2012). More variable results and less positive effects are expected for field crops, which are also used for horticultural species such as lettuce, swiss chard, cauliflower, broccoli and potatoes, among others. Despite its benefits, the adoption of *Azospirillum* in commercial horticulture faces challenges, including variability in farmer environmental awareness, management of application methods, and interaction with other soil microorganisms (Pedrino et al. 2024). Even though there are some companies that market products that have been duly evaluated for the horticultural crops they are intended to be used for and are registered with regulatory agencies, there are also many formulations used in extensive crops that have been extrapolated for their use in horticultural crops without much prior experimentation or monitoring, and are even recommended for sequential or calendered as we are used to doing with synthetic inputs. One of the current challenges for *Azospirillum*-based bio-inputs is the need for inoculant companies to develop effective formulations that can be used under different ways of application, environmental conditions, storage, and handling.

What can be expected from the use of *Azospirillum* in horticultural crops? Certainly, more developed root systems, which in turn produce more surface area to explore in search of water and mineral resources. This could alleviate transplant stress, which would be beneficial when growing in moderately saline soil, and also help overcome short periods of water stress. On the other hand, in the field, N, P, and K fertilization could be reduced when applied in combination with *Azospirillum* bio-input. To what extent could chemical fertilization be reduced? This is a matter of debate; it depends on the edaphic characteristics, the vegetable species grown, and other factors related to the management of agronomic practices. In addition to all these factors, environmental and, of course, economic benefits can also be expected from increased yields and reduced costs of chemical inputs.

Nevertheless, more research is needed on horticultural crops and *Azospirillum* interactions, especially in open field studies. But there is also a need for greater involvement and oversight by government agencies and intermediaries between bio-input producers and horticultural producers, as well as greater consumer demand for healthy food without the addition of chemicals that are harmful to health and the environment (Brunelle et al. 2024). Addressing these issues is critical to maximizing the benefits of *Azospirillum* and ensuring its successful integration into modern agricultural practices, particularly in the context of increasing environmental concerns and demands for sustainable food production.

References

Abd El-Samad EH, Glala AA, Abd El Baset A et al (2019) Improving the establishment, growth and yield of tomato seedlings transplanted during summer season by using natural plant growth bio-stimulants. Middle East J Agric Res 8(1):311–329

Andrade-Sifuentes A, Fortis-Hernández M, Preciado-Rangel P et al (2020) *Azospirillum brasilense* and solarized manure on the production and phytochemical quality of tomato fruits (*Solanum lycopersicum* L.). Agronomy 10(12):1956. https://doi.org/10.3390/agronomy10121956

Avinash M, Mahadeva S, Vendan KT et al (2019) Influence of *Azosprillum* isolates on growth parameters of tuberose (Polianthes tuberosa L.) cv. Mexican single. Int J Curr Microbiol App Sci 8(3):664–670

Bahgat AR, Dahab AA, Elhakem A et al (2023) Integrated action of rhizobacteria with aloe vera and moringa leaf extracts improves defense mechanisms in *Hibiscus sabdariffa* L. cultivated in saline soil. Plants 12(21):3684

Barassi CA, Ayrault G, Creus CM et al (2006) Seed inoculation with *Azospirillum* mitigates NaCl effects on lettuce. Sci Hortic 109:8–14

Bartolini S, Carrozza GP, Scalabrelli G et al (2017) Effectiveness of *Azospirillum brasilense* Sp245 on young plants of *Vitis vinifera* L. Open Life Sci 12(1):365–372

Bartolini S, Pappalettere L, Toffanin A (2023) *Azospirillum baldaniorum* Sp245 induces anatomical changes in cuttings of Olive (*Olea europaea* L., cultivar Leccino): preliminary results. Agronomy 13(2):301

Bashan Y, Holguin G (1997) *Azospirillum*–plant relationships: environmental and physiological advances (1990–1996). Can J Microbiol 43(2):103–121

Bashan Y, Holguin G (1998) Proposal for the division of plant growth-promoting rhizobacteria into two classifications: biocontrol-PGPB (plant growth-promoting bacteria) and PGPB. Soil Biol Biochem 30(8–9):1225–1228

Bashan Y, Hernandez JP, Leyva LA et al (2002) Alginate microbeads as inoculant carriers for plant growth-promoting bacteria. Biol Fertil Soils 35:359–368

Berger B, Baldermann S, Ruppel S (2017) The plant growth-promoting bacterium *Kosakonia radicincitans* improves fruit yield and quali of *Solanum lycopersicum*. J Sci Food Agric 97: 4865–4871

Bradford KJ (2006) Germination: internal factors affecting (physical, physiological, molecular and biochemical). In: Bewley JD, Black M, Halmer P (eds) The encyclopedia of seeds: science, technology and uses. CABI Publishing, Wallingford, pp 127–228

Brunelle T, Chakir R, Carpentier A (2024) Reducing chemical inputs in agriculture requires a system change. Commun Earth Environ 5:369

Cappellari LR, Santoro MV, Nievas F (2013) Increase of secondary metabolite content in marigold by inoculation with plant growth-promoting rhizobacteria. Appl Soil Ecol 70:16–22

Carletti S, Garbi M, Vita F (2011) Efecto del momento de aplicación de un inoculante formulado con *Azospirillum brasilense* sobre un cultivo de lechuga bajo invernadero. Horticult Argentina 30(73):88

Carrozzi L, Creus CM, Barassi CA (2012) Reparation of aged lettuce (*Lactuca sativa*) seeds by osmotic priming and *Azospirillum brasilense* inoculation. Botany 90:1093–1102

Cassán F, Diaz-Zorita M (2016) *Azospirillum* sp. in current agriculture: from the laboratory to the field. Soil Biol Biochem 103:117–130

Cassán F, Coniglio A, López G (2020) Everything you must know about *Azospirillum* and its impact on agriculture and beyond. Biol Fertil Soils 56:461–479

Chamangasht S, Ardakani MR, Khavazi K (2012) Improving lettuce (*Lactuca sativa* L.) growth and yield by the application of biofertilizers. Ann Biol Res 3(4):1876–1879

Chau MH, Hung VV (2023) Effect of inoculation methods of *Azospirillum brasilense* on growth and yield of lettuce in the hydroponic system. J For Sci Technol 8(2):3–9

Day JM, Neves MCP, Dobereiner J (1975) Nitrogenase activity on the roots of tropical forage grasses. Soil Biol Biochem 7:107–112

de Lima AA, dos Reis Venturoso L, Silva BAA et al (2017) Eficiência da inoculação de Azospirillum brasilense associado com enraizador no crescimento e na produção de alface. Revista Verde de Agroecologia e Desenvolvimento Sustentável 12(2):233–240

de Lima NDSA, Vogel GF, Fey R (2018) Rates of application of *Azospirillum* brasilense in tomato crop. Rev Agric Neotropical 5(4):81–87

Di-Barbaro G, Pernasetti S, Stegmayer A (2005) Evaluación del efecto de Azospirillum brasilense en la germinación y emergencia del pimiento pimentonero (Capsicum annum L. var. Trompa de elefante). Revista del CIZAS 6:74–85

Di-Barbaro MG, Andrada H, del Valle E et al (2021) Biofertilization of yacon with *Azospirillum brasilense* and native mycorrhizal fungi, cultivated in the Central Valley of Catamarca, Argentina. Int J Food Sci Agric 5(4):737–747

Di-Barbaro MG, Andrada HE, Centeno Toledo F, Espeche Acosta ER, Plaza J, González Basso MV & Guzmán PS (2023) Effect of native strain of Azospirillum brasilense in the production of horticultural seedlings in semihydroponic system of culture in bag. J Appl Biotechnol Bioeng 10(3):90-92

Díaz I, Iglesias MC (2002) Inoculación con *Azospirillum* sp. en cultivo de tomate (*Lycopersicum esculentum* vr. Coloso), bajo invernáculo. In: XIII Reunión de Comunicaciones Científicas y Técnicas. Fac. de Cs. Agrarias–UNNE, Corrientes, Argentina

Díaz PR, Merlo F, Carrozzi L et al (2023) Lettuce growth improvement by *Azospirillum argentinense* and fluorescent *Pseudomonas* co-inoculation depends on strain compatibility. Appl Soil Ecol 189:104969

FAO (2020) Fruit and vegetables—your dietary essentials. The International Year of Fruits and Vegetables, 2021, background paper. Rome. https://doi.org/10.4060/cb2395en

FAO (2022) Agricultural Production Statistics 2000–2021; FAOSTAT Analytical Brief Series No. 41. FAO, Rome

Fasciglione G, Casanovas EM, Yommi A et al (2012) *Azospirillum* improves lettuce growth and transplant under saline conditions. J Sci Food Agric 92:2518–2523

Fasciglione G, Casanovas EM, Quillehauquy V et al (2015) *Azospirillum* inoculation effects on growth, product quality and storage life of lettuce plants grown under salt stress. Sci Hortic 195: 154–162

Finch-Savage WE, Bassel GW (2016) Seed vigour and crop establishment: extending performance beyond adaptation. J Exp Bot 67(3):567–591

Fujita M, Kusajima M, Okumura Y et al (2017) Effects of colonization of a bacterial endophyte, Azospirillum sp. B510, on disease resistance in tomato. Biosci Biotechnol Biochem 81:1657–1662

Garbi M, Carletti S, Sillon C et al (2016) Respuesta de plántulas de lechuga mantecosa (*Lactuca sativa* L.) a la inoculación con una formulación compuesta por tres cepas de *Azospirillum brasilense*. Horticult Argentina 35(86):19–28

Ghiyasi M, Abbasi AS, Tajbakhsh M et al (2008) Effect of osmopriming with polyethylene glycol 8000 (PEG8000) on germination and seedling growth of wheat (*Triticum aestivum* L.) seed under salt stress. Res J Biol Sci 3:1249–1251

Guerrero-Molina MF, Winik BC, Pedraza RO (2012) More than rhizosphere colonization of strawberry plants by *Azospirillum brasilense*. Appl Soil Ecol 61:205–212

Guerrieri MC, Fiorini A, Fanfoni E et al (2021) Integrated genomic and greenhouse assessment of a novel plant growth-promoting rhizobacterium for tomato plant. Front Plant Sci 12:660620

Kloepper JW, Schroth MN (1978) Plant growth-promoting rhizobacteria in radish. In: Proceedings of the 4th Internet Conference on Plant Pathogenic Bacter, Station de Pathologie Vegetale et Phytobacteriologie, INRA, Angers, France, 1978, vol 2, pp 879–882

Kumar M, Kumar K (2019) Role of Bio-fertilizers in vegetables production: A review. JPharmacognosy Phytother 8(1):328–334

Kohler J, Caravaca F, Roldán A (2010) An AM fungus and a PGPR intensify the adverse effects of salinity on the stability of rhizosphere soil aggregates of Lactuca sativa. Soil Biol Biochem 42: 429–434

Lobato Ureche MA, Pérez-Rodriguez MM, Ortiz R (2021) Rhizobacteria improve the germination and modify the phenolic compound profile of pepper (*Capsicum annum* L.). Rhizosphere 18: 100334

Lobo LLB, de Andrade da Silva MSR, Castellane TCL et al (2022) Effect of indole-3-acetic acid on tomato plant growth. Microorganisms 2022(10):2212

Malusá E, Sas-Paszt L, Ciesielska J (2012) Technologies for beneficial microorganisms inocula used as biofertilizers. Sci World J 2012:1–12. https://doi.org/10.1100/2012/491206. Article ID 491206

Martínez-Damián MT, Cano-Hernández R, Moreno-Pérez EDC et al (2019) Effect of preharvest growth bioregulators on physicochemical quality of saladette tomato. Revista Chapingo Serie Horticultura 25(1):29–43

Nautiyal CS, Govindarajan R, Lavania M et al (2008) Novel mechanism of modulating natural antioxidants in functional foods: involvement of plant growth promoting rhizobacteria NRRL B-30488. J Agric Food Chem 56(12):4474–4481

Nisha FA, Tagoe JNA, Pease AB et al (2024) Plant seedlings of peas, tomatoes, and cucumbers exude compounds that are needed for growth and chemoattraction of *Rhizobium leguminosarum* bv. *viciae* 3841 and *Azospirillum brasilense* Sp7. Can J Microbiol 70(5):150–162

Novello G, Bona E, Nasuelli M et al (2024) The impact of nitrogen-fixing bacteria-based biostimulant alone or in combination with commercial inoculum on tomato native rhizosphere microbiota and production: an open-field trial. Biology 13(6):400. https://doi.org/10.3390/biology13060400

Pedraza RO, Motok J, Tortora ML et al (2007) Natural occurrence of *Azospirillum brasilense* in strawberry plants. Plant Soil 295:169–178

Pedraza RO, Motok J, Salazar SM et al (2010) Growth-promotion of strawberry plants inoculated with *Azospirillum brasilense*. World J Microbiol Biotechnol 26:265–272

Pedrinho A, Mendes LW, do Rêgo Barros, F. M. et al (2024) The interplay between *Azospirillum brasilense* and the native bacterial communities in the soil and rhizosphere of maize (*Zea mays* L.). Soil Biol Biochem 189:109292

Pérez-Pazos JV, Sánchez-López DB (2018) Influence of plant growth promoting bacteria in seed yields of super-elite sweet potato (*Ipomoea batata*s Lam) in the field. Biotecnol Apl 35(2): 2211–2216

Pérez-Rodriguez MM, Pontin M, Lipinski V et al (2020) *Pseudomonas fluorescens* and *Azospirillum brasilense* increase yield and fruit quality of tomato under field conditions. J Soil Sci Plant Nutr 20:1614–1624

Queral J, Wargers A, Tarro L et al (2024) Changes over time of fruit, vegetable and sweets consumption among European adolescents and associations with sociodemographic characteristics: an HBSC study. J Public Health 46(3):392–402

Reyes-Ramírez A, López-Arcos M, Ruiz-Sánchez E et al (2014) Effectiveness of microbial inoculants on growth and productivity of habanero pepper (*Capsicum chinense* Jacq.). Agrociencia 48(3):285–294

Romero AM, Vega D, Correa OS (2014) *Azospirillum brasilense* mitigates water stress imposed by a vascular disease by increasing xylem vessel area and stem hydraulic conductivity in tomato. Appl Soil Ecol 82:38–34

Romero WC, Barrios Díaz JM, Rodríguez Mendoza MDN et al (2016) Producción de plántulas de hortalizas con *Azospirillum* sp. y aspersión foliar de miel de abeja. Revista mexicana de ciencias agrícolas 7(1):59–70

Salazar SM, Lovaisa NC, Guerrero Molina MF et al (2012) Fruit yield of strawberry plants inoculated with *Azospirillum brasilense* RLC1 and REC3 under field conditions. Revista Agronómica del Noroeste Argentino 32(1–2):63–66

Salehzade H, Shishvan MI, Ghiyasi M et al (2009) Effect of seed priming on germination and seedling growth of wheat (*Triticum aestivun* L.). J Res Biol Sci 4:629–631

Salim H, Ali A, Alsaady M et al (2021) Effect of plant growth promoting rhizobacteria (PGPR) on growth of cauliflower (*Brassica oleracea* L. var. botrytis). Plant Arch 20(1):782–786

Samyukta SMS, Viji MM, Manju RV et al (2024) Synergy of biochar and biofertilizers to improve bell pepper fruit biochemical quality with increased soil carbon, *Azospirillum* population and mycorrhization. Plant Sci Today 11(4):1605–1614

Santos MP, Martínez SJ, Yarte ME et al (2022) Effect of Azospirillum brasilense on the *in vitro* germination of *Eustoma grandiflorum* (Raf.) Schinn.(Gentianaceae). Sci Hortic 299:111041

Sharara MS, Abd El-Aal HA (2016) Effect of some bio-fertilizers on growth, productivity, chemical composition and processing of cassava tubers. J Agric Sci 61(6):529–539

Stegelmeier AA, Rose DM, Joris BR et al (2022) The use of PGPB to promote plant hydroponic growth. Plants 11(20):2783

Sun W, Shahrajabian MH, Wang N (2025) A study of the different strains of the genus *Azospirillum* spp. on increasing productivity and stress resilience in plants. Plants 14:267

Tarrand J, Krieg N, Döbereiner J (1978) A taxonomic study of the *Spirillum lipoferum* group, with descriptions of a new genus, Azospirillum gen. nov., and two species, *Azospirillum lipoferum* (Beijerinck) comb. nov. and *Azospirillum brasilense* sp. nov. Can J Microbiol 24:967–980

Thomloudi EE, Tsalgatidou PC, Douka D et al (2019) Multistrain versus single-strain plant growth promoting microbial inoculants-the compatibility issue. Hellenic Plant Protect J 12(2):61–77

Toffanin A, D'Onofrio C, Carrozza GP et al (2014) Use of beneficial bacteria *Azospirillum brasilense* Sp245 on grapevine rootstocks grafted with 'Sangiovese'. In: International symposium on grapevine roots, vol 1136, pp 177–184

Toffoli LM, Medrano NN, Martinez Zamora MG (2018) Performance of petunia crop in response to inoculation with *Azospirillum brasilense*. Revista Agronomía Noroeste Argentino [online] 38(1):45–49

Valverde C, Gonzalez Anta G, Ferraris G (2015) Chapter 21: *Pseudomonas* and *Azospirillum*. In: Cassán, Okon, Creus (eds) Handbook for Azospirillum: technical issues and protocols, pp 389–409

Vendruscolo EP, de Lima SF (2021) The *Azospirillum* genus and the cultivation of vegetables. A review. BASE

Vio SA, Galar ML, Gortari MC et al (2023) Multispecies bacterial bio-input: tracking and plant-growth-promoting effect on *Lettuce* var. sagess. Plants 12(4):736

Vio SA, Martínez Sbrancia C, Gortari C et al (2025) Inoculation of microorganisms into the soil and rhizosphere. Part IV, chapter 50. In: Dharumadurai D, Sankara Narayanan A (eds) Plant microbiome engineering: methods and protocols in food science. Springer Nature, pp 457–467

Walters C, Landré P, Hill L et al (2005) Organization of lipid reserves in cotyledons of primed and aged sunflower seeds. Planta 222:397–407

Chapter 19
Synthetic Consortia with *Azospirillum*

Maddalena Del Gallo

Abstract Since its rediscovery in the 1970s, *Azospirillum* has created countless expectations. At the beginning, it was thought that an *Azospirillum* inoculation at sowing was the definitive solution to provide nitrogen to grasses, avoiding the use of nitrogen fertilizers. However, the results were not always visible. It was still unclear why about half of the field trials, unlike the experiments conducted in vitro under controlled conditions, did not produce positive effects. The first experiments conducted in vitro on *Azospirillum* and *Arthrobacter* co-cultures showed how *Azospirillum* was stimulated in some of its interactive functions with the plant by other bacteria. Subsequent field experiments conducted on various plants inoculated with *Azospirillum* in consortiums with different selected bacteria have instead shown consistently positive results.

Azospirillum, being the most intensively studied plant growth-promoting rhizobacteria (PGPR), remains one of the best and most reliable PGP bacteria, and it is the best documented case of a plant-hormone producer. It associates with many crop plants, works very well in consortium with other bacteria, and penetrates inside the host plant, where it can interact very closely with the host. In conclusion, especially in consortiums with other selected bacteria, *Azospirillum* remains one of the most interesting plant probiotic to be used in field inoculations.

Keywords Consortia · Co-inoculation · Synergy · Plant probiotics · Soil fertility

19.1 Introduction

Since its rediscovery in the 1970s (Döbereiner and Day 1975; Döbereiner et al. 1976; Döbereiner 1983) and its use in agronomic applications, *Azospirillum* has created countless expectations. After its first description by Beijerinck in 1925 (1925), it was

M. Del Gallo (✉)
Environmental Microbiology, Environmental Sciences, University of L'Aquila, L'Aquila, Italy
e-mail: mariamaddalena.delgallo@univaq.it

F. D. Cassán et al. (eds.), *The Definitive Handbook of Azospirillum*,
https://doi.org/10.1007/978-3-032-09222-9_19

never obtained in pure culture, and when Schröder (1932) recovered a pure culture, she was unable to demonstrate its N_2 fixation. However, Becking (1963) reisolated this microorganism and demonstrated its nitrogen-fixation ability. But again, this organism was largely ignored until Döbereiner and Day et al. (1975) made new observations on *Digitaria decumbens* grown in N-free medium. They described a *Spirillum lipoferum* N_2-fixer at low O_2 concentration and high colonizer of the roots of this plant.

After this final rediscovery and description, it was thought that an *Azospirillum* inoculation at sowing was the definitive solution to provide nitrogen to grasses, thus avoiding the use of nitrogen fertilizers. Numerous laboratory studies and field experiments were conducted all over the world. In addition to Brazil (Day and Döbereiner 1976), *Azospirillum* was tested with numerous grasses such as wheat, corn, sorghum, millet, various forage grasses, rice, and other grasses in the USA, in various parts of Europe, in Egypt, and in India (Okon et al. 1976; Smith et al. 1976; Albrecht et al. 1983; Hegazi et al. 1983; Watanabe and Roger 1984; Alexander and Zuberer 1988; Lakshmi et al. 1977). However, the results were not always visible (Döbereiner et al. 1978). While the first experiments conducted in Florida on pearl millet gave extremely positive results (Smith et al. 1976), as did those obtained in Belgium (Reyders and Vlassak 1982) and in Egypt (Hegazi et al. 1983), no effects of inoculation were seen on corn in Wisconsin (Albrecht et al. 1981) and in England (O'Hara et al. 1981). In Italy, various inoculation experiments were initially conducted. The late prof. Franco Favilli at the University of Florence had already isolated a *Spirillum lipoferum* Sp6 in 1974 and had conducted some physiological studies and inoculation tests in greenhouses with positive results (personal communication). However, it was still unclear why about half of the field trials, unlike the experiments conducted in vitro and in greenhouses under controlled conditions, did not produce positive effects.

Two factors seemed to favor a positive inoculation effect: experiments conducted in soils poor in organic matter and, to a less clear extent, the genotype of the inoculated plant.

The first factor could be explained by the low competitiveness of *Azospirillum* against various bacteria such as *Pseudomonads*, *Bacillus,* and others. In poor soil, *Azospirillum*, being not competitive in rich soil, encountered fewer competitors present in the autochthonous microflora.

In the second hypothesis, in the case that it was the genotype of the plant that influenced the colonization by this bacterium, the explanation was more complex. The plant genotype certainly had an influence, especially in the case of corn, but other factors also came into play, including genetic improvement: plants that were little genetically manipulated responded better to inoculation. An experiment conducted by García de Salomone and Döbereiner (1996) on the inoculation of different varieties of maize conducted in Argentina with four strains of Azospirillum belonging to both the *A. brasilense* and *A. lipoferum* species showed large differences in the response to the inoculation between the different maize varieties.

19.2 *Azospirillum* Associated with Other Bacteria

Of particular interest, but in some respects in conflict with previous theories, were the studies conducted with *Azospirillum* in association with *Arthrobacter*, grown in mixed culture in a chemostat: both bacteria coexisted in a wide range of growth rates (Cacciari et al. 1986). Furthermore, the presence of *Arthrobacter giacomelloi* determined not only an increase in the nitrogenase activity of *Azospirillum brasilense* at all oxygen pressures studied, including the higher ones generally not tolerated (Fig. 19.1) (Cacciari et al. 1989) but also stimulated its functionality, such as, for example, the production of phytohormones (Lippi et al. 1992).

In the context of some studies on the interactions between *Azospirillum* and autochthonous bacteria, Quatrini et al. (1992) set up a model that could simulate the characteristic microenvironment of the plant-soil system. They isolated some bacteria directly from the rhizosphere of sorghum, and the sequencing of the 16S rDNA gene indicated that the isolates belonged to the genera *Pseudomonas*, *Klebsiella,* and *Stenotrophomonas*, often present in the sorghum rhizosphere. In the presence of these three strains, *Azospirillum* declined rapidly, both when inoculated onto sorghum seedling in a gnotobiotic system and when grown in a chemostat, where it became competitive only when the nutrient flow was abundant. Since the carbon sources, especially carbohydrates, used by *A. brasilense* are much lower in number than those used by the other three bacteria, the authors hypothesized that

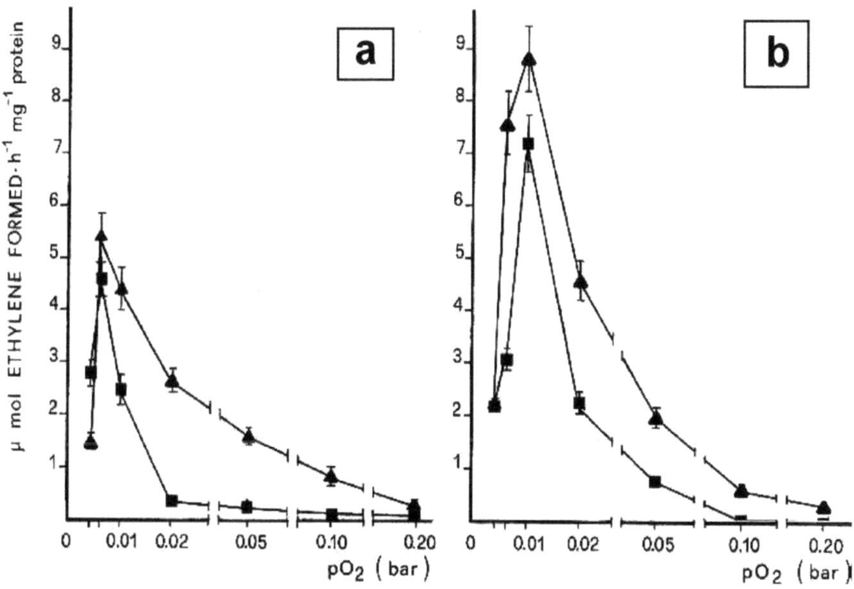

Fig. 19.1 Nitrogenase activity of *Azospirillum brasilense* in pure culture (squares) and in the presence of *Arthrobacter giacomelloi* (triangles) after 24 (**a**) and 48 (**b**) hours of growth in batch cultures at different partial pressures of O_2. (From: Cacciari et al. 1989 Arch. Microbiol. 152, 111–114)

both the quality and quantity of the substrate are among the most important factors for the proliferation of this bacterium on the plant. However, the first mixed inoculations with *Azospirillum* and other nitrogen-fixing and non-nitrogen-fixing microorganisms but always associated with the plant as endophytes (and not rhizospheric) plant growth-promoting bacteria (PGPBs) seemed to have more reproducible positive results.

In the search for consortium bacterial formulations, however, it is always necessary to keep in mind that the plant has its own microbiota. The more biodiversified it is, the more in balance. The introduction of more selected PGPBs on a plant leads to an enrichment of its microbiome and to greater well-being and greater resistance to environmental stress and attacks by pathogens.

We are associated with a greater number of bacteria and other microorganisms than cells in our body. We are an ecosystem. Why not consider plants similar to us? All surfaces of our body are covered in microorganisms, most of which are found inside, especially in our intestines. The same happens with plants. All surfaces are covered by microorganisms, and also their interior, the xylem vessels in particular, which can be vaguely assimilated to a plant intestine. The plant microbiome is proving to be increasingly important for the well-being of the plant. Furthermore, microorganisms can stimulate and activate various biosynthetic pathways of the plant, promoting its well-being (Fig. 19.2).

Since the 1990s, especially after discussing with Dr. Johanna Döbereiner and following her suggestions, we had concluded that *Azospirillum* was found inside plants, in particular sorghum and sugar cane, always in association with *Herbaspirillum*, with various Pseudomonadaceae, and, especially in the case of sugar cane, with *Gluconacetobacter*.

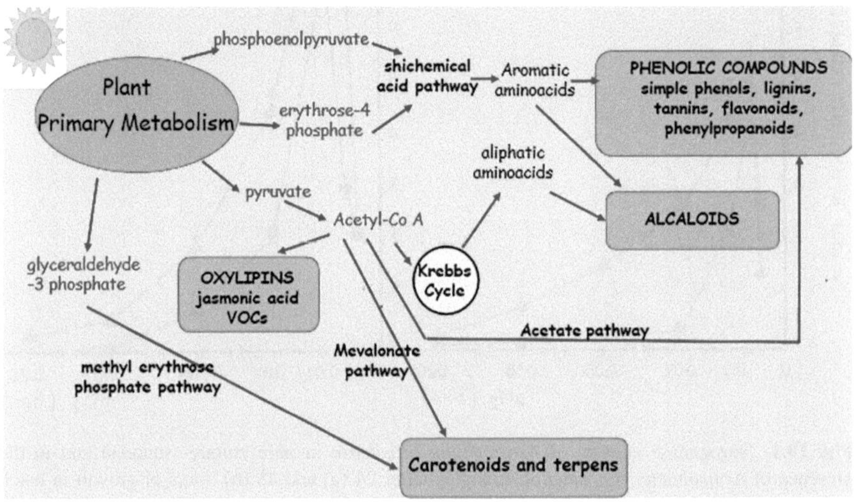

Fig. 19.2 Plant biosynthetic pathways that can be activated by PGPBs, endophytes in particular

After several tests conducted both in vitro and in association with various plants, we have developed in our laboratory a consortium formed by four bacteria, *Azospirillum brasilense*, *Herbaspirillum seropedicae*, *Gluconacetobacter diazotrophicus,* and *Burkholderia ambifaria*, which has proven to be particularly active and reproducible in numerous plants. We have been using this consortium since the 1990s with always reproducible results (Di Fiore et al. 1994; Fabbri et al. 1998; Canini et al. 2006; Botta et al. 2013; Stagnari et al. 2018; Pace et al. 2020; Pagnani et al. 2020; Pellegrini et al. 2020; Pellegrini et al. 2021a, b). At Dr. Johanna Döbereiner's suggestion, we also tried mixed inoculations with arbuscular mycorrhizae, always with extremely positive results (Pennelli et al. 1994). In this consortium, *Azospirillum* seems to have mainly a role of hormonal stimulation on the plant, *Gluconacetobacter* that of fixing nitrogen, and *Burkholderia* in addition to producing auxins—but in smaller quantities than *Azospirillum*—produces siderophores, ACC-deaminase, and some volatile substances with inhibitory activity against some pathogenic fungi. It is interesting to note that *Herbaspirillum*, an excellent nitrogen fixer and producer of auxins, is also a producer of large quantities of AI2, the autoinducing molecules of quorum sensing that allow communication between different species (unpublished results). This bacterium could therefore have a connecting function between the different bacterial species in the colonization of the plant. Furthermore, these molecules stimulate different functions in the microbial community, such as the formation of biofilm, very important in the colonization of the plant.

19.3 Methodological Approach to the Development of Consortia

Today, the idea of inoculating more than one bacterium has become common practice. Mixed consortia of bacteria of different species are increasingly inoculated. Interestingly, *Azospirillum* always penetrates inside the plant, especially in the xylem vessels (Pagnani et al. 2020), and, even if it does not directly supply nitrogen to the plant, it produces a large amount of auxins through different metabolic pathways (Spaepen et al. 2014). However, *Azospirillum*, despite being one of the bacteria most frequently associated with the plant on which it has extremely positive effects, has been isolated inside the most varied plant species and in the most disparate environments, it is little studied in consortium with other microorganisms.

One of the negative aspects of agriculture is the reduction of biodiversity: plant biodiversity if crop rotations are not introduced, and microbial biodiversity, both fungal and bacterial, depends directly and indirectly on plant biodiversity. Increasing microbial biodiversity with mycorrhizae and bacteria is essential for the health of the agroecosystem. However, developing a microbial consortium is not that simple. Theoretically, a PGP-bacterial consortium should be able to grow all the bacteria together without negative interactions; each of them should produce a different substance that can act as a stimulus for the plant or contribute to the efficiency of the consortium (N_2 fixation, auxin production, phosphate solubilization, antibiotic, and siderophore production). It is essential, however, if you select a good consortium, to test it on different plants and on different soils. In the literature you often find consortia tested only on a single crop. If you want to move to a higher level of industrial-scale production, it is necessary that the inoculum is valid for more crops to increase the profit margin. To move to an industrial production level, it is obviously necessary to fine-tune the entire scale-up process, including production methods, growth media, and, above all, valid carriers that allow the inoculum to survive for valid times and in valid ways.

One of the most difficult problems to face is finding a common growth medium. Especially in the case of *Azospirillum brasilense* that uses few sugars and prefers some organic acids such as malic acid. It is not really essential that the bacteria grow together, especially if they have different duplication times. Mixed inoculation experiments can be conducted directly on potted plants, but coculture interaction studies are extremely useful to identify aspects of the consortium such as stimulation or inhibition of some functions. Furthermore, the consortium must be composed of the greatest possible number of bacteria and must be capable of being industrially produced. It is necessary to select the right bacteria that can coexist within the plant and perform their functions. In general, the selection of bacterial species isolated inside the same plant can lead to the development of efficient microbial consortia, but their selection is not always easy. Initially, Cacciari et al. (1983, 1986), for example, had selected an *Azospirillum* and an *Arthrobacter* that were able to grow together and fix nitrogen, but in vitro. When inoculated on corn in the field, the consortium did not prove to be as effective as in vitro. Indeed, in the plots inoculated with the single bacteria, better results were obtained than in those inoculated with the consortium (Del Gallo et al. 1989).

A problem arises when it is necessary to produce a mixed bacterial inoculant on an industrial scale. It is obviously necessary to fine-tune the entire scaling-up process, including production methods, soils, and, above all, valid carriers that allow the inoculum to survive for valid times and in valid ways. Biofertilizers manufactured frequently are carrier based and, frequently, suffer from short shelf life, poor quality, high contamination, and low and unpredictable field performances. Death of the organisms in the inoculated seeds is one of the important factors contributing to the failure of the inoculation response in field conditions (see Chap. 21).

Thanks to molecular biology techniques that are now readily available on the market and easy to use, it is theoretically possible to construct interesting bacterial consortia. By extracting DNA from a given plant, for example, or from a given environment, it is possible to identify the different bacterial species present. Then, through bioinformatic analyses and accessible and open databases, for example, Kyoto Encyclopedia of Genes and Genomes (KEGG), it is possible to identify genes corresponding to interesting agronomic traits in the individual species, select the species, and build a possible microbial consortium at the table. But further laboratory and field tests are obviously necessary. It is not enough to do a pot test to develop a microbial consortium. It must then be tested in the field for at least three growing seasons to be able to define it as effective. It must then be tested in different environments and on different plant species, unless you want to produce it for a single species.

19.4 Recent Literature Describing *Azospirillum* Inoculations in Consortium

Nowadays, the use of mixed bacterial inoculants is increasingly present. Many works are published on the isolation of new PGPB species, and in international journals, more and more works are published on field inoculation trials with mixed consortia (Santoyo et al. 2021).

The consortia that include *Azospirillum* are applied on various plants of agronomic interest and for different purposes. For example, Marimuthu et al. (2013) reported the combined application of *Azospirillum* sp. and *Pseudomonas fluorescens* in some field experiments, with the final goal to study the influences of intercropping system in cotton with inorganic fertilizer and the two bioinoculants on the incidence of root rot and yield. The results revealed that the inoculated cotton *Sesbania aculeata* had the lowest incidence of root rot. Furthermore, this intercropping recorded the maximum number of sympodial branches and the highest cotton seed yield.

Jha et al. (2013) reported the use of spent waste of mushroom (mycostraw) enriched with a microbial consortium to reduce the dependence of rice on chemical fertilizers. The components of the microbial consortium were *Pseudomonas*, *Azospirillum,* and cyanobacteria. The application of the bioinoculant for four consecutive years on rice resulted in an increase in fertilizer use efficiency, acetylene reduction activity (as an index of nitrogen fixation), and the number of *Azospirillum* and soil nutrients such as total nitrogen, total phosphorus, and organic matter by 14–16% compared to the initial value. Furthermore, grain yield increased by 17% and straw yield by 14.1%. Varietal difference had no influence. Finally, the authors proposed a novel approach for the conversion of rice straw into mycostraw (organic manure) and its enrichment with plant growth-promoting rhizobacteria (PGPR).

Another interesting experiment was conducted on sunflowers by Amalraj et al. (2015). The aim of the study was to develop a microbial consortium composed of *Azospirillum brasilense*, *Azotobacter chroococcum*, *Bacillus megaterium* var. *phosphaticum,* and *Trichoderma viride*. The formulation was developed based on their ability to solubilize nutrients and promote plant growth in sunflowers. Various carriers were also tested in the study to find one that would provide good shelf life and good ability to colonize the plant. *B. megaterium* var. *phosphaticum* showed good in vitro solubilization in agar medium amended with tricalcium phosphate, potassium aluminum silicate, zinc oxide, and zinc carbonate. Similar solubilization potential was also observed in the formulation compared to the other formulations.

However, microencapsulated bioinoculants showed significant growth promotion in sunflowers after 60 days from sowing.

In Mexico, Molina et al. (2017) developed a mixed inoculum on maize. In the work, a bacterial consortium containing four PGPB strains compatible with each other and tolerant to dehydration was formulated. The formulation included a moderately dehydration-tolerant strain (*Pseudomonas putida*) and three highly dehydration-tolerant bacteria (*Sphingomonas* sp., *Azospirillum brasilense,* and *Acinetobacter* sp.). The four bacterial strains were able to adhere to the seeds and colonize the rhizosphere of the plants when applied in both mono-inoculation and multi-inoculation treatments, demonstrating that these bacteria can also coexist without antagonistic effects in association with the plants. The effects of the bacterial consortium on the growth of blue maize were evaluated. Seeds inoculated with single bacterial strains or with the bacterial consortium were subjected to two experimental conditions before sowing: normal hydration or dehydration. Overall, inoculation with the bacterial consortium increased shoot and root dry weight, plant height, and diameter compared to uninoculated or mono-inoculated control treatments. The bacterial consortium formulated in the work had greater benefits for blue corn plants even when the inoculated seeds were subjected to dehydration stress before germination, making this formulation certainly attractive for future field applications.

Jayashree and Jagadeesh (2017) tested a bacterial consortium for nursery use consisting of an *Azospirillum*, a phosphate-solubilizing bacterium (*Pseudomonas striata*), a potassium-solubilizing PGP (*Pseudomonas* sp.) and—interestingly—a lactic acid bacterium (*Leuconostoc mesenteroides*) with antifungal activity. Its effects on the growth and biomass of different vegetable seedlings raised in trays were evaluated in a commercial farmer's nursery. Microbial inoculation resulted in improved plant height and biomass in eggplant and chili. From this study, the authors concluded that the microbial consortium consisting of strains with multiple functions can be used to treat coco peat to produce quality vegetable seedlings grown in trays.

Inoculation with *Azospirillum* and three other bacteria also has a positive effect on soil (Pellegrini et al. 2021b). In this work, the effect of inoculation of *A. brasilense* together with *Gluconacetobacter diazotrophicus, Herbaspirillum seropedicae*, and *Burkholderia ambifaria* on *Allium cepa* L. and on soil health was analyzed. Two different potato varieties were inoculated in the field, and the plants were followed until harvest. Then, soil physicochemical and molecular profiles (DNA extraction and 16S community sequencing) were analyzed, comparing them to the initial conditions before sowing. Soil fertility and microbial community structure and diversity were positively affected by the bacterial inoculum. At harvest, the soil with the presence of the bacterial consortium showed an increase in total organic carbon, organic matter, and available phosphorus, as well as higher concentrations of nutrients than the control. The ecological indexes calculated from the molecular profiles showed that community diversity was positively affected by the bacterial treatment.

There are many articles in the literature on mixed inocula of *Azospirillum* and various Rhizobia. The first experiments conducted in the 90s by Del Gallo and Fabbri (1990, 1991) showed an extremely positive interaction between *Azospirillum* and *Rhizobium*. More recently, paper published by Moretti et al. (2020) is particularly interesting because they evaluated the effect of different combinations of beneficial bacteria with and without microbial secondary metabolites produced by rhizobia on two soybean cultivars over three growing seasons under tropical field conditions. *Bradyrhizobium*, *japonicum*, and *Bradyrhizobium diazoefficiens* were inoculated with different combinations of *Bacillus subtilis*, *Azospirillum brasilense,* and metabolites enriched in lipo-chitooligosaccharides extracted from *B. diazoefficiens* and *Rhizobium tropici.* The standard inoculation of Bradyrhizobium combined with *Azospirillum brasilense* and microbial secondary metabolites increased total N in leaves, total P and N-urcide, number of nodules, and dry weight of the plant. The 100-seed weight, grain yield, and crude protein concentration in grain also increased.

PGPB consortia can also mitigate the effect of abiotic stresses on plant growth and development. However, Cortés-Patiño et al. (2021) found that the degree of plant response is host-specific. They conducted experiments by inoculating *Herbaspirillum, Azospirillum*, and *Pseudomonas* (single and co-inoculated) into perennial ryegrass plants subjected to drought. The plants were grown under controlled conditions and subjected to water deficit for 10 days. A statistically significant increase of approximately 30% in dry biomass production was observed using three co-inoculation combinations. Genomic analysis identified representative genes associated with plant colonization and growth promotion. Co-inoculations mostly reduced growth-promoting traits tested in vitro. However, co-inoculation of *Herbaspirillum* and *Azospirillum brasilense* resulted in the highest indolic compound production. Although the *Azospirillum* strain showed the highest potential in in vitro and in silico tests, plants responded better when PGPBs were co-inoculated, demonstrating the importance of integrating in silico, in vitro, and in vivo results when selecting PGPR to mitigate drought stress.

In a recent article, Sandini et al. (2024) report the results of four inoculation experiments of *Azospirillum brasilense* and *Pseudomonas fluorescens* on maize in different locations characterized by different edaphoclimatic conditions. Four treatments were carried out in the different locations: (1) without inoculation and without applying N in coverage; (2) without inoculation and with application of 100% of the recommended dose of N (200 kg ha^{-1} of N); (3) *Azospirillum brasilense* + application of 75% of the recommended dose of N; (4) bacterial consortium of *Azospirillum brasilense* + *Pseudomonas fluorescens* + application of 75% of the recommended dose of N. In all locations, with the use of the bacterial consortium and 75% of the recommended dose of N, it was possible to reduce the N dose by up to 25% without compromising crop productivity. In this way, the authors concluded that the use of the consortium of the two bacteria favors a savings of up to 25% of N fertilization in top dressing and the possibility of using two microorganisms with the capacity to stimulate plant growth with a single application.

A study has just been published by the working group of the Russian Academy of Sciences in Saratov on the production of biofilm by two species belonging to the genus *Azospirillum* in association with a strain of *Enterobacter cloacae*. The authors focused on this aspect of *Azospirillum* because the formation of biofilm is of great importance for both epiphytic and endophytic rhizobacteria, due to their affinity with the surface of plant roots and, therefore, for their ability to colonize the host (Shelud'ko et al. 2025).

Biofilm formation is the basis of the early stages of plant colonization. This work is interesting because the two *Azospirillum* species used in the experiments, *A. brasilense* and *A. baldaniorum*, formed biofilms on wheat roots, especially on the root tip and root hairs, as well as at the sites of lateral root formation in association with Enterobacter. Biofilm production on wheat roots was not characteristic of the specific strain of *Enterobacter cloacae* isolated from Jerusalem artichoke (*Helianthus tuberosus* L.) roots. However, biofilms formed on the root system of wheat seedlings inoculated with mixed cultures of all three species included areas with multicellular aggregates. The widespread distribution of *Azospirillum* cells among Enterobacteria or *Enterobacter* cells among *Azospirillum* species in the experiment is interesting because there is not much literature on biofilm formation by mixed microbial communities of *Azospirillum* and other bacteria.

In the literature, there is also a patent by Fukami et al. (2023), which describes an agricultural composition comprising a consortium of *Azospirillum* sp. and *Pseudomonas* sp. for promoting plant growth. The patent also details a production process and emphasizes increased stability of the composition. This consortium is designed to enhance plant growth through mechanisms such as nitrogen fixation, auxin biosynthesis, siderophore production, and phosphorus solubilization.

The patented process involves both fermentation (cultivation of *Azospirillum* and *Pseudomonas*) and a stabilization process (developing a technical solution to stabilize the mixture of microorganisms, likely in a single package). The patent emphasizes that the production process leads to increased stability and shelf-life of the microbial consortium.

The consortium is designed to promote plant growth through N_2 fixation (*Azospirillum* is a well-known nitrogen-fixing bacterium), auxin biosynthesis (both *Azospirillum* and *Pseudomonas* produce plant hormones), siderophore production, and P solubilization (*Pseudomonas* can solubilize phosphorus, making it available to plants. The patent also ensures a synergistic effect (the interaction between *Azospirillum* and *Pseudomonas* is highlighted, with *Azospirillum*'s exopolysaccharide production being utilized by *Pseudomonas* and *Pseudomonas*'s organic acids contributing to *Azospirillum*'s viability).

19.5 Conclusions

Microorganisms associated or symbiotic with plants can and must be used in agriculture to replace agrochemicals. Especially in the field, better and more reliable results are obtained when inoculants composed of multiple selected bacterial species are used.

Further studies are still needed to manage them successfully:

- We have yet to develop reliable inoculation methods and reliable ways to keep inocula active and long-lasting.
- There are many products on the market, but many have problems, also contain different species, and do not work in environmental situations differently.
- There are many publications circulating on the subject; many bacteria with interesting PGP activities are continuously being isolated, but they often refer to particular situations (soil, plants) and cannot function in different situations.

Azospirillum, being the most intensively studied PGPR, remains one of the best and most reliable PGP bacteria (Okon 1985). It is the best documented case of plant-hormone producers. It associates with many crop plants, works very well in consortium with other bacteria, and penetrates inside the host plant, where it can interact very closely with the host, which can thus control it. Furthermore, by also colonizing niches inside the host, it thus prevents possible attacks by pathogens.

Azospirillum certainly has some limitations. For example, when inoculated singly, it often fails to colonize the plant. This fact can easily be overcome if inoculated in combination with other bacteria. The use of inocula composed of multiple selected microorganisms is therefore the future path to follow.

References

Albrecht SL, Okon Y, Lonnquist J et al (1981) Nitrogen fixation by corn-*Azospirillum* associations in a temperate climate. Crop Sci 21:301–306. https://doi.org/10.2135/cropsci1981.0011183X002100020024x

Albrecht SL, Gaskins MH, Milam JR et al (1983) Ecological factors affecting survival and activity of *Azospirillum* in the rhizosphere. Experientia Suppl 48:138–154

Alexander DB, Zuberer D (1988) Impact of soil environmental factors on rates of N_2-fixation associated with roots of intact maize and sorghum plants. Plant Soil 110:303–315. https://doi.org/10.1007/BF02226811

Amalraj ELD, Mohanty D, Praveen Kumar G et al (2015) Potential microbial consortium for plant growth promotion of sunflower (*Helianthus annuus* L.). Proc Natl Acad Sci, India, Sect B Biol Sci 85(2):635–642. https://doi.org/10.1007/s40011-014-0361-4

Becking JH (1963) Fixation of molecular nitrogen by an aerobic Vibrio or *Spirillum*. Antonie van Leeuvenhoek 29:326. https://doi.org/10.1007/BF02046082

Beijerinck MW (1925) Uber ein *Spirillum*, welches freien Stickstoffbinden kann? Zentbl Bakt Parasitkde 11(63):353–359

Botta AL, Santacecilia A, Ercole C et al (2013) In vitro and in vivo inoculation of four endophytic bacteria on *Lycopersicon esculentum*. New Biotechnol 18:829–839. https://doi.org/10.1016/j.nbt.2013.01.001

Cacciari I, Lippi D, Ippoliti S et al (1983) Improvement of nitrogen fixation by mixed cultures of *Azospirillum brasilense* and *Arthrobacter giacomelloi*. In: Biological processes and soil fertility. Department of Soil Science, University of London, Reading

Cacciari I, Del Gallo M, Ippoliti S et al (1986) Growth and survival of *Azospirillum brasilense* and *Arthrobacter giacomelloi* in binary continuous culture. Plant Soil 90:107–116. https://doi.org/10.1007/BF02277390

Cacciari I, Lippi D, Ippoliti S et al (1989) Response to oxygen of diazotrophic *Azospirillum brasilense—Arthrobacter giacomelloi* mixed batch culture. Arch Microbiol 152:111–114. https://doi.org/10.1007/BF00456086

Canini A, Canuti L, Grilli Caiola M, Del Gallo M (2006) Immunocytochemical characterisation of endophytic bacteria *Azospirillum brasilense*, *Herbaspirillum seropedicae*, *Burkholderia ambifaria* and *Gluconacetobacter diazotrophicus*. Ann Microbiol 56:393–399. https://doi.org/10.1007/BF03175039

Cortés-Patiño S, Vargas C, Álvarez-Flórez F et al (2021) Potential of *Herbaspirillum* and *Azospirillum* consortium to promote growth of perennial ryegrass underwater deficit. Microorganisms 9:91. https://doi.org/10.3390/microorganisms9010091

Day JM, Döbereiner J (1976) Physiological aspects of nitrogen fixation by a *Spirillum* from Digitaria roots. Soil Biol Biochem 24:1061–1064

Day JM, Neves MCP, Döbereiner J (1975) Nitrogenase activity on the roots of tropical forage grasses. Soil Biol Biochem 7:107–112

Del Gallo M, Fabbri P (1990) Inoculation of *Azospirillum brasilense* Cd on Chick-pea (*Cicer arietinum*). Symbiosis 9:283–287

Del Gallo M, Fabbri P (1991) Effect of soil organic matter on chickpea inoculated with *Azospirillum brasilense* Cd and *Rhizobium leguminosarum* bv Cicer. Plant Soil 137:171–175. https://www.jstor.org/stable/42937186

Del Gallo M, Cacciari I, Lippi D et al (1989) Primi risultati della batterizzazione di *Zea mays* L. in campo con colture singole e miste di azotofissatori. Riv Agron 23:116–120

Di Fiore S, Fabbri P, Marini F, Del Gallo M, Bozzini A (1994) Tripartite interaction between *Zea mays*, *Glomus* spp. and nitrogen-fixing bacteria introduced into plant rhizosphere via mycorrhiza. NATO Advanced Research Workshop on Azospirillum and Related Microorganisms, Sárvár, Hungary, September 4–7

Döbereiner J (1983) Ten years *Azospirillum*. Exp Suppl 48:9–23

Döbereiner J, Day JM (1975) Nitrogen fixation in the rhizosphere of tropical grasses. In: Stewart WDP (ed) Nitrogen fixation by free-living micro-organisms. Cambridge University Press, Cambridge, pp 39–56

Döbereiner J, Marriel IE, Nery M (1976) Ecological distribution of *Spirillum lipoferum* Beijerinck. Can J Microbiol 22:1464–1473

Döbereiner J, Burris RH, Hollaender A et al (eds) (1978) Limitation and potential for biological nitrogen fixation in the tropics. Plenum Press, p 398

Fabbri P, Di Fiore S, del Gallo M (1998) Establishment of an association between N_2-fixing endophytic bacteria and *Zea mays*. In: Elmerich C, Kondorosi A, Newton WE (eds) Biological nitrogen fixation for the 21st century. Kluver Academic Publication, New York, p 410

Fukami J, Gomes DF, Gomes JM et al (2023) Agricultural composition composed by consortium between *Azospirillum* sp and *Pseudomonas* sp, production process and increased stability thereof, and use as promotor of plant growth for agricultural application. International Patent WO2023039646 https://patentscope.wipo.int/search/ja/detail.jsf;jsessionid=7613B1FD1EF093599AEDBA6CB193BE8C.wapp1nB?docId=WO2023039646&_gid=202312#:~:text=The%20present%20invention%20refers%20to,cellular%20viability%20of%20the%20Azospirillum

García de Salomone I, Döbereiner J (1996) Maize genotype effects on the response to *Azospirillum* inoculation. Biol Fertil Soils 21:193–196. https://doi.org/10.1007/BF00335934

Hegazi NA, Monib M, Amer HA et al (1983) Response of maize plants to inoculation with azospirilla and (or) straw amendment in eastern Egypt. Can J Microbiol 29:888–894

Jayashree C, Jagadeesh KS (2017) Testing the effect of the microbial consortium on growth of vegetable seedlings in a farmer's nursery. Int J Curr Microbiol App Sci 6:1636–1639. https://doi.org/10.20546/ijcmas.2017.602.182

Jha M, Chourasia S, Sinha S (2013) Microbial consortium for sustainable rice production. Agroecol Sust Food Syst 37:340–362. https://doi.org/10.1080/10440046.2012.672376

Lakshmi V, Satyanarayana Rao A, Vijayalakshmi K et al (1977) Establishment and survival of *Spirillum lipoferum*. Proc Indian Acad Sci 86 B:397–404

Lippi D, Cacciari I, Pietrosanti T et al (1992) Interactions between *Azospirillum* and *Arthrobacter* in diazotrophic mixed cultures. Symbiosis 13:107–114

Marimuthu S, Ramamoorthy V, Samiyappan R et al (2013) Intercropping system with combined application of *Azospirillum* and *Pseudomonas fluorescens* reduces root rot incidence caused by *Rhizoctonia bataticola* and increases seed cotton yield. J Phytopathol 161:405–411. https://doi.org/10.1111/jph.12084

Molina-Romero D, Baez A, Quintero Hernandez V et al (2017) Compatible bacterial mixture, tolerant to desiccation, improves maize plant growth. PLoS ONE 12(11):e0187913. https://doi.org/10.1371/journal.pone.0187913

Moretti LG, Crusciol CAC, Bossolani JW et al (2020) Bacterial consortium and microbial metabolites increase grain quality and soybean yield. J Soil Sci Plant Nutr 20:1923–1934. https://doi.org/10.1007/s42729-020-00263-5

O'Hara GW, Davey MR, Lucas JA (1981) Effect of inoculation of *Zea mays* with *A. brasilense* under temperate conditions. Can J Microbiol 27:871–877. https://doi.org/10.1139/m81-138

Okon Y (1985) Azospirillum as a potential inoculant for agriculture. Trends Biotechnol 3:223–228. https://doi.org/10.1016/0167-7799(85)90012-5

Okon Y, Albrecht SL, Burris RH (1976) Factors affecting growth and nitrogen fixation of *Spirillum lipoferum*. J Bacteriol 127:1248–1254

Pace P, Pellegrini M, Palmieri S et al (2020) Plant growth-promoting rhizobacteria for in vitro and ex vitro performance enhancement of Apennines' Genepì (*Artemisia umbelliformis* subsp. *eriantha*), an endangered phytotherapeutic plant. In vitro Cell Develop Biol Plant Plant 56: 134–142. https://doi.org/10.1007/s11627-019-10035-1

Pagnani GC, Galieni A, Stagnari F et al (2020) Open field inoculation with PGPR as a strategy to manage fertilization of ancient *Triticum* genotypes. Biol Fertil Soils 56:111–124. https://doi.org/10.1007/s00374-019-01407-1

Pellegrini M, Ercole C, Di Zio C et al (2020) In vitro and in planta antagonistic effects of plant growth promoting rhizobacteria consortium against soilborne plant pathogens of *Solanum tuberosum* and *S. lycopersicum*. FEMS Microbiol Lett 367(13):fnaa099. https://doi.org/10.1093/femsle/fnaa099

Pellegrini M, Pagnani G, Rossi M et al (2021a) *Daucus carota* L. seed inoculation with a consortium of bacteria improves plant growth, soil fertility status and microbial community. Appl Sci 11:3274. https://doi.org/10.3390/app11073274

Pellegrini M, Spera DM, Ercole C et al (2021b) *Allium cepa* L. inoculation with a consortium of plant growth-promoting bacteria: effects on plant, soil and on the autochthonous microbial community. Microorganisms 9:639. https://doi.org/10.3390/microorganisms9030639

Pennelli B, Fabbri P, Isopi R et al (1994) Coinoculation of the mycorrhizal fungus *Glomus constrictum* Trappe (*Glomales*) and plant-growth promoting rhizobacteria on sorghum. G Bot Ital 128(1):264. https://doi.org/10.1080/11263509409437105

Quatrini P, Vettraino AM, Carcupino M et al (1992) Colonization of Sorghum roots by *Azospirillum* and three bacteria isolated from sorghum rhizosphere. Symbiosis 13:255–269

Reynders L, Vlassak K (1982) Use of *Azospirillum brasilense* as biofertilizer in intensive wheat cropping. Plant Soil 66:217–223. http://www.jstor.org/stable/42933985

Sandini IE, Pacentchuk F, Andrade de Siqueira FD et al (2024) Bacterial consortium of *Azospirillum brasilense* and *Pseudomonas fluorescens* on the stimulation of growth of corn culture. Ciência Rural 54(9):e20230123. https://doi.org/10.1590/0103-8478cr20230123

Santoyo G, Guzmán-Guzmán P, Parra-Cota FI et al (2021) Plant growth stimulation by microbial consortia. Agronomy 11(2):219. https://doi.org/10.3390/agronomy11020219

Schroder M (1932) Zentbl Bakt. Parasitkde 85:177–212

Shelud'ko AV, Mokeev DI, Petrova LP et al (2025) Bacteria of the genus *Azospirillum* form biofilms with *Enterobacter cloacae* K7 in the root system of wheat seedlings. Microbiology 94: 404–414. https://doi.org/10.1134/S0026261724610273

Smith RL, Bouton JH, Schank SC et al (1976) Nitrogen fixation in grasses inoculated with *Spirillum lipoferum*. Science 193(4257):1003–1005. https://doi.org/10.1126/science.193.4257.1003

Spaepen S, Bossuyt S, Engelen K et al (2014) Phenotypical and molecular responses of *Arabidopsis thaliana* roots as a result of inoculation with the auxin-producing bacterium *Azospirillum brasilense*. New Phytol 201:850–861. https://doi.org/10.1111/nph.12590

Stagnari F, Pisante M, Pagnani G et al (2018) Plant growth-promoting rhizobacteria (PGPR) in *Cannabis sativa* var. Finola greenhouse cultivation: an alternative fertilization strategy to improve plant growth and quality characteristics. Ind Crop Prod 123:75–83. https://doi.org/10.1016/j.indcrop.2018.06.033

Watanabe I, Roger PA (1984) Nitrogen fixation in wetland rice field. In: Subba Rao NS (ed) Corrent development in biological nitrogen fixation. Edward Arnold, Baltimore

Chapter 20
From Petri Dishes to Deep Learning Networks: New Roads in *Azospirillum* Research with Artificial Intelligence

Victor Hugo Buttrós and Joyce Dória

Abstract Artificial intelligence (AI) is reshaping the landscape of microbiological research, offering unprecedented tools for studying complex microbial interactions and optimizing the application of plant growth-promoting bacteria like *Azospirillum*. This chapter presents an integrative framework for applying AI to *Azospirillum* research, highlighting both conceptual advances and methodological pipelines. The first half of the chapter addresses theoretical challenges inherent to microbiome data analysis, including high dimensionality, compositionality, and multi-omics integration. It explores the application of machine learning and deep learning techniques for predictive modeling, phenotyping, and simulation-based optimization of bacterial inoculants. The second half offers a detailed, step-by-step experimental protocol for researchers aiming to implement AI-driven workflows, including batch effect correction, transfer learning, and federated learning across multiple field trials. Emphasis is placed on data standardization, variable harmonization, and reproducibility—key factors for ensuring that AI models are robust, interpretable, and biologically meaningful. This chapter serves as both a conceptual guide and a practical resource for microbiologists and agronomists seeking to integrate AI into *Azospirillum*-based sustainable agriculture.

Keywords Artificial intelligence · Machine learning · Deep learning · Microbiome · Predictive modeling

---

V. H. Buttrós (✉)
Department of Biology, Agricultural Microbiology Graduate Program, Federal University of Lavras, Lavras, Brazil

J. Dória
Department of Agriculture, Agrotechnology Sector, Federal University of Lavras, Lavras, Brazil

F. D. Cassán et al. (eds.), *The Definitive Handbook of Azospirillum*,
https://doi.org/10.1007/978-3-032-09222-9_20

20.1 Introduction

Artificial intelligence (AI) transforms microbiology by enabling researchers to analyze complex microbial interactions with unprecedented precision. In the context of *Azospirillum* research, AI provides a robust framework for identifying microbial patterns, predicting plant growth responses, and optimizing bioinoculant applications (Zhao et al. 2023). Unlike conventional statistical methods, which struggle with high-dimensional, sparse, and nonlinear microbiome data, AI techniques can integrate diverse datasets, extract meaningful biological signals, and generate predictive models for plant–microbe interactions (Pan 2021).

Traditional approaches to studying *Azospirillum* rely on labor-intensive experimental setups and controlled field trials to assess inoculation success. While these methods are indispensable, they lack scalability and often fail to capture the full complexity of microbial networks in real agricultural environments (Deng et al. 2021; Walsh et al. 2024). AI addresses this limitation by leveraging machine learning (ML) and deep learning (DL) algorithms that can process multi-omics data—including metagenomics, transcriptomics, metabolomics, and soil parameters—to uncover hidden relationships between *Azospirillum*, soil microbiota, and plant physiology. This data-driven approach allows researchers to predict inoculation outcomes before field implementation, significantly reducing the need for costly and time-consuming wet lab experiments specifically for model training (Huang et al. 2024).

Several key AI advancements are particularly relevant to *Azospirillum* studies. Federated Learning (FL) enables collaborative model training across multiple datasets while preserving data privacy—allowing researchers worldwide to improve predictive models without sharing proprietary data (Qi et al. 2024). Edge AI and the Internet of Things (IoT) provide real-time monitoring of soil conditions, offering immediate insights into microbial activity and plant responses. Additionally, reinforcement learning and simulation-based AI can model the long-term effects of *Azospirillum* inoculation under various environmental conditions, optimizing application strategies (Rupanetti and Kaabouch 2024).

Despite its promise, AI adoption in *Azospirillum* research faces significant challenges, particularly regarding data transparency, standardization, and ethical AI deployment. AI models are only as good as the data they are trained on; biases in experimental design, data scarcity, and lack of reproducibility can lead to unreliable predictions. Global efforts in data sharing, standardized databases, and interdisciplinary collaboration are essential to harness AI's potential fully. Without open, high-quality datasets, AI-driven models will require extensive field validation, slowing progress and limiting their practical application (Högberg 2024; Stark et al. 2025).

This chapter explores how AI can revolutionize *Azospirillum* research by addressing computational challenges, integrating multi-omics data, and enabling predictive modeling of microbial interactions. The first half of the chapter provides a theoretical foundation, covering the limitations of conventional statistical approaches and the key AI techniques suited for microbial studies. The second half presents practical methodologies for applying AI to *Azospirillum* studies, including experimental workflows, AI-powered predictive modeling, and real-world case studies. By bridging computational advancements with microbiological research, AI offers a transformative path toward precision microbial applications in sustainable agriculture.

20.2 Key Computational Challenges in Conventional Statistics

Understanding *Azospirillum* interactions with plants and soil microbiomes poses several computational challenges that limit accurate prediction and modeling. Microbiome datasets are often high-dimensional, sparse, and context-dependent, making traditional statistical methods inadequate for capturing the complexity of plant–microbe systems (Busato et al. 2023). In *Azospirillum* studies, where outcomes depend on soil type, native microbiota, environmental variables, and plant genotype, conventional models cannot easily integrate these interacting factors (Deng et al. 2021).

A key obstacle is the curse of dimensionality: the number of variables—such as microbial taxa, gene expression, or metabolite levels—typically exceeds the number of samples. This leads to overfitting and weak generalizability. Classical tools like linear regression, ANOVA, and correlation assume linearity and independence, which are rarely valid in microbial networks. For example, *Azospirillum* inoculation may shift root exudate profiles, modifying the microbial community and nutrient cycling in nonlinear ways that traditional methods cannot capture (Anuragi et al. 2024; Pan 2021).

Microbiome data is also compositional, meaning taxon abundances are relative rather than absolute, which complicates analysis with standard statistics (Busato et al. 2023). AI techniques, including machine and deep learning, offer alternatives that do not rely on strict assumptions and can infer latent structures within the data (Walsh et al. 2024; Zhao et al. 2023).

Another challenge lies in integrating multi-omics data. Modern *Azospirillum* studies incorporate metagenomic, transcriptomic, metabolomic, and soil chemical datasets—each differing in scale, format, and noise. Deep learning architectures can synthesize these diverse inputs to reveal functional links between microbial genes and plant traits (Huang et al. 2024; Xu et al. 2024).

Lastly, the scarcity of large, well-annotated datasets hinders AI's full potential in this field. Many microbiome studies are limited by small sample sizes, batch effects, and experimental variability. Robust AI models require broader data coverage and consistency. Initiatives in data sharing, transparency, and federated learning—where models are trained across institutions without moving raw data—are essential for building predictive systems that generalize across crops, soils, and climates (Högberg 2024; Qi et al. 2024; Stark et al. 2025).

Overcoming these barriers with AI will enable a shift from descriptive analyses to predictive, actionable insights, enhancing the use of Azospirillum as a microbial inoculant in sustainable agriculture.

Overall, the AI-driven approaches address these challenges by:

Handling high dimensionality: Algorithms can incorporate thousands of features (taxa, genes, metabolites, and edaphic factors) without manual reduction, capturing the full complexity of the data.

Robustness to noise: Many machine learning models (e.g., ensemble methods) are resilient to outliers, environmental variability, and sequencing errors.

Pattern recognition: AI can detect subtle multivariate patterns (e.g., a combination of microbial taxa indicative of a healthy rhizosphere) that conventional analysts might miss.

20.3 AI-Powered Computational Tools in *Azospirillum* Research

20.3.1 Machine Learning for Microbial Ecology

Machine learning (ML) has become a fundamental tool in microbial ecology, providing methods to classify microbial communities, infer ecological interactions, and identify key drivers of microbial dynamics (Walsh et al. 2024). In *Azospirillum* research, ML techniques help uncover the environmental and microbial factors influencing inoculation success and plant growth promotion.

Tree-based algorithms, such as random forest (RF) and gradient boosting machines (GBM), are widely used for microbiome classification and feature selection. These models excel at identifying which microbial taxa or soil properties are most predictive of *Azospirillum* abundance or function. For example, an RF model trained on soil microbiome data can determine whether a specific microbial consortium enhances or inhibits *Azospirillum* colonization. Similarly, extreme gradient boosting (XGBoost) has demonstrated superior predictive accuracy in microbial ecology studies by capturing complex nonlinear interactions between environmental variables and microbial communities (Breiman 2001; Noble 2006; Virmani et al. 2024).

Beyond supervised learning, unsupervised approaches such as clustering and network analysis play a critical role in understanding *Azospirillum* ecology. Community detection algorithms, such as graph-based neural networks or co-occurrence networks, help identify potential microbial partners or competitors of *Azospirillum* in different soils (Pan et al. 2024). These insights inform precision agriculture strategies by predicting the most favorable conditions for inoculation success.

20.3.2 AI in Multi-omics Data Integration

A major computational challenge in *Azospirillum* studies is integrating diverse biological datasets, including metagenomics, transcriptomics, metabolomics, and phenotypic traits. AI-driven multi-omics analysis enables researchers to decipher the functional roles of *Azospirillum* beyond taxonomic identification (Huang et al. 2024).

Deep learning, particularly autoencoders and neural network-based feature fusion models, has emerged as a powerful approach for handling heterogeneous, high-dimensional omics datasets. These models can reduce noise, extract meaningful latent representations, and link microbial gene expression profiles to plant growth responses. For example, deep neural networks trained on 16S rRNA sequencing and soil metabolomics data can predict which chemical signatures are associated with successful *Azospirillum* colonization (Li et al. 2024; Schumann et al. 2025).

Additionally, graph neural networks (GNNs) are being explored for modeling microbial interactions, offering a way to represent multi-omics data as structured networks rather than tabular datasets (Corso et al. 2024). This is particularly valuable for understanding how *Azospirillum* integrates into native microbial communities and how its activity is influenced by soil chemistry or plant root exudates.

20.3.3 Predictive Modeling of Azospirillum*–Plant Interactions*

AI enables predictive modeling of how *Azospirillum* inoculation will impact plant performance under different soil and climatic conditions. Supervised learning models, trained on field trial datasets, can forecast variables such as plant biomass, nitrogen uptake efficiency, and resistance to abiotic stress following inoculation.

For instance, random forests or gradient-boosted models can integrate data on soil composition, native microbiota, and environmental factors to predict whether *Azospirillum* will enhance plant growth in a given agricultural system. These models optimize inoculant application strategies and help farmers anticipate the best conditions for microbial biofertilizer use (Mahmud et al. 2024).

Beyond static predictions, reinforcement learning (RL) and simulation-based AI are being explored to model dynamic plant–microbiome interactions. In a novel approach, an RL framework could simulate *Azospirillum* performance in response to changing soil nutrient levels, helping researchers identify optimal inoculation timing, dosage, and co-inoculant combinations (García et al. 2025). These AI-driven simulations reduce the need for extensive field trials by testing thousands of potential scenarios *in silico* before real-world validation (Ekundayo 2024).

20.3.4 AI-Driven Phenotyping and Trait Prediction

Advancements in computer vision and deep learning are transforming plant phenotyping, allowing researchers to quantify the effects of *Azospirillum* inoculation with high precision. Convolutional Neural Networks (CNNs) are handy for analyzing plant images and detecting changes in root architecture, shoot biomass, and leaf health. For example, CNN-based image recognition models can distinguish subtle root morphology changes in inoculated versus non-inoculated plants, providing quantitative metrics on root branching, elongation, and biomass accumulation (Meraj et al. 2024; Peyal et al. 2025). This non-invasive, high-throughput phenotyping approach accelerates trait selection in breeding programs to enhance plant–microbe interactions.

Additionally, AI-based genomic selection models integrate microbiome data with plant genetic profiles to predict which crop varieties will benefit most from *Azospirillum* inoculation. This predictive capability enables precision agriculture approaches where microbial biofertilizers are tailored to specific plant genotypes, enhancing nitrogen fixation efficiency while reducing fertilizer dependency (Wu et al. 2025).

20.3.5 Considerations for AI Implementation in Azospirillum Research

Despite AI's transformative potential, its application in *Azospirillum* research is hindered by limited high-quality datasets, variability in experimental conditions, and a lack of standardization across studies. Unlike traditional statistical methods, which can sometimes generate insights from smaller datasets, AI models require large, diverse, and well-annotated training datasets to ensure generalizability across different soil types, crops, and environmental conditions. The absence of standardized methodologies often leads to model biases, making it difficult to extrapolate results from controlled experiments to field applications (Zha et al. 2025).

To address these challenges, multi-experiment comparability techniques such as batch effect correction, transfer learning, and federated learning enable AI models to integrate data from multiple studies while preserving biological signals. These methods allow AI-driven *Azospirillum* research to scale effectively, ensuring that models remain robust and applicable across diverse agricultural contexts.

20.3.5.1 Batch Effect Correction for Harmonizing Multi-source Data

Batch effects are systematic biases from differences in sequencing platforms, sample preparation, or environmental conditions that can obscure real *Azospirillum* signals and mislead AI predictions (Yu et al. 2024). Correction methods include ComBat (R:

sva, Python: pyComBat), which uses an empirical Bayesian framework to remove non-biological variation while preserving biological signals, and SVA (R: sva), which uses latent variable modeling to eliminate hidden batch effects. Harmony (R: harmony, Python: harmonypy) aligns biological features across datasets while maintaining within-group variation and is especially useful for transcriptomic, metagenomic, and metabolomic integration. Quantile normalization (R: preprocessCore, Python: QuantileTransformer) standardizes signal intensities and is commonly applied to metabolomics and transcriptomics. Best practices include correcting feature-level matrices (e.g., OTU tables or gene expression), using PCA or UMAP pre- and post-correction to visualize variance retention, and validating model accuracy on independent datasets to confirm improvements.

20.3.5.2 Transfer Learning for Cross-Experiment Adaptation

Transfer learning improves model generalization by adapting models trained on one dataset to perform accurately on new data with minimal retraining (Wang et al. 2024). This is particularly valuable for *Azospirillum* studies where soil, climate, or plant genotype varies. The approach involves pretraining a model on broad datasets capturing *Azospirillum*–plant interactions, fine-tuning it with smaller, site-specific datasets, and applying domain adaptation techniques to align differing feature distributions. Applications include localizing microbiome models to specific agroecosystems, adapting plant growth predictions to new crop varieties, and generalizing multi-omics models to distinct soil types. Recommended tools include pretrained neural networks (R: torch, Python: TensorFlow), domain adaptation libraries (Python: DomainBed, R: daexp), and feature normalization (R: mixOmics, Python: MinMaxScaler). Best practices involve ensuring biological similarity between datasets, freezing early model layers during fine-tuning, and validating on independent test data to ensure predictive reliability.

20.3.5.3 Federated Learning for Cross-Study AI Model Training Without Data Sharing

Federated learning (FL) allows multiple research groups to collaboratively train AI models on local data without sharing raw files, preserving data privacy and intellectual property—a major advantage for *Azospirillum* research conducted across institutions or private companies (Qi et al. 2024). In FL, each site trains a local model, shares only updates with a central aggregator, and contributes to the improvement of a global model that learns from distributed data. Applications include multi-institutional AI development, integration of field trial data across environments, and collective modeling for microbial inoculant design. Core tools include TensorFlow Federated (Python: tff) for decentralized learning, PySyft (Python: syft) for secure aggregation with differential privacy, and H2O.ai (R and Python) for distributed biological modeling. Best practices include adjusting learning rates to ensure convergence across sites, using weighted updates based on dataset sizes, applying privacy-preserving techniques, and validating final models on external datasets to confirm generalizability.

20.3.5.4 Advancing AI-Driven *Azospirillum* Research Through Standardization and Open Science

Beyond computational methods, standardizing experimental protocols and data repositories is critical for advancing AI applications in *Azospirillum* research. Metadata standardization ensures that datasets collected by different research groups can be meaningfully compared. The genomic standards consortium (GSC) MIxS metadata schema is a practical framework for ensuring consistency across microbiome studies.

Open-access repositories such as Qiita (for microbiome data) and DataDryad (for omics datasets) enable broader accessibility and reproducibility. Additionally, initiatives promoting findable, accessible, interoperable, reusable (FAIR) data principles should be prioritized to ensure AI models are trained on high-quality, standardized datasets that reflect real-world agricultural conditions.

By adopting batch effect correction, transfer learning, and federated learning, *Azospirillum* research can fully harness AI's potential, leading to scalable, precise, and data-driven biofertilizer applications in sustainable agriculture.

20.4 Successful Applications of AI in *Azospirillum* Research

The integration of artificial intelligence techniques has yielded meaningful advances in the study of *Azospirillum*, enabling novel insights into plant–microbe interactions, genomic functions, ecological distributions, and applied biotechnological innovations. This section summarizes representative cases of AI use across four application domains.

20.4.1 Plant Trait Prediction and Breeding

One line of research has focused on applying AI to understand how different maize genotypes respond to *Azospirillum brasilense* inoculation under nitrogen stress (Vidotti et al. 2019). Using genomic prediction models such as G-BLUP and RKHS, machine learning was applied to breeding data to estimate shoot and root development traits. The findings revealed that inoculated plants exhibited distinct genetic controls compared to uninoculated ones, with genomic prediction achieving

R^2 values between 0.42 and 0.78. These models facilitated the identification of hybrid genotypes with superior responses to inoculation, demonstrating the potential of machine learning in accelerating breeding programs tailored to beneficial plant–microbe associations.

Another breeding-related study by Yassue et al. (2023) combined hyperspectral imaging and genomic SNP data to classify inoculation status and predict phenotypic traits in tropical maize. Machine learning algorithms, including partial least squares, Bayesian regression models, and AutoML ensembles, enabled high-resolution phenotyping. Spectral data captured physiological responses to *Azospirillum*, enhancing predictions of traits like shoot dry mass and stalk diameter. Genomic features were more effective for plant height prediction. Together, the multimodal AI approach enabled improved detection of inoculation effects and accurate forecasting of growth outcomes.

20.4.2 Genomic Function Prediction

AI has also contributed to gene function annotation in *Azospirillum*. One study by Rubel et al. (2016) implemented a multilayer perceptron classifier trained on conserved protein sequence features to reclassify the draB gene, previously annotated as hypothetical. The classifier predicted this gene to be a member of the NifO protein family, suggesting a role in nitrogenase regulation. The approach, implemented using the Weka platform, successfully identified over 80 homologous proteins in diverse bacterial genomes. This application highlights how AI can guide the reannotation of microbial genes and support hypothesis generation for experimental validation.

20.4.3 Environmental Modeling and Microbial Ecology

Large-scale ecological modeling has employed machine learning to predict the abundance of nitrogen-fixing microbes, including *Azospirillum*, across diverse soil and climate contexts by Hao et al. (2025). A random forest model, optimized through particle swarm techniques, was trained on global microbiome datasets. Its predictions identified environmental tipping points such as a soil pH near 7.84, beyond which the diazotroph community structure undergoes a marked shift. SHAP-based interpretability methods revealed that organic carbon, pH, and precipitation were the most influential features. The model showed that *Azospirillum* thrives in neutral-to-alkaline soils, distinguishing it ecologically from acidophilic diazotrophs. These findings support the application of AI tools in developing systems-level understanding of microbial distributions in response to environmental gradients.

20.5 AI-Integrated Experimental Workflow for *Azospirillum* Research

To maximize the potential of AI in *Azospirillum* research, experimental workflows must be carefully structured to facilitate data integration, preprocessing, model training, and validation (Walsh et al. 2024; Zhao et al. 2023). This section provides a structured step-by-step approach detailing essential AI methods, preprocessing techniques, and helpful R and Python packages for each analysis stage.

Step 1: Data collection involves integrating multi-omics profiles, plant traits, and environmental metadata to model Azospirillum–plant interactions. Microbial data includes 16S rRNA sequencing for community composition, shotgun metagenomics for functional annotation, and metatranscriptomics for gene expression. Plant data includes root architecture (via imaging), shoot biomass, yield, and nutrient uptake (measured from tissue). Environmental variables such as soil pH, organic matter, nitrogen content, climate (temperature, precipitation), and management practices (e.g., fertilization, irrigation) must be recorded. Experimental design should include uninoculated controls, standardized metadata formats, and barcode-based sample tracking to improve reproducibility and harmonization. Useful packages include phyloseq, microbiome, and vegan in R and biopython, pandas, and scikit-bio in Python.

Step 2: Preprocessing aims to reduce technical noise and improve dataset compatibility. For microbiome data, remove low-quality reads using DADA2 (R) or Deblur (Python), denoise and collapse OTUs to ASVs, and filter rare taxa occurring in fewer than 5% of samples. For transcriptomics and metabolomics, missing values can be imputed using missForest (R) or KNNImputer (Python) and normalized via quantile transformation using normalize.quantiles() (R) or QuantileTransformer() (Python). For plant phenotypes, outliers can be removed using interquartile range filtering, and root traits can be standardized with scaler() from caret (R) or StandardScaler() from scikit-learn (Python). Feature selection can be performed using random forest importance (randomForest in R or sklearn. ensemble in Python), while batch effects can be corrected using ComBat (R) or pyComBat (Python). Common issues include data sparsity (addressed via ZINB-WaVE in R or scvi-tools in Python), multi-batch variation, and high dimensionality (mitigated using Boruta in R or feature selection in Python).

Step 3: AI model implementation involves preparing training and test sets, selecting models, and tuning. Partition datasets using createDataPartition() in caret (R) or train_test_split() in scikit-learn (Python), and balance class distributions with SMOTE using smotefamily (R) or imbalanced-learn (Python). For microbiome or soil data, random forest models are effective; for multi-omics integration, neural networks using nnet (R), Keras, or TensorFlow (Python) are recommended. Root image traits can be modeled using convolutional neural networks (EBImage in R, OpenCV in Python), and inoculation success predicted using gradient boosting methods like XGBoost (R) or lightgbm (Python). Hyperparameter tuning can be done with tune() in tidymodels (R) or with Optuna and GridSearchCV in Python. To prevent overfitting, dropout regularization should be applied (torch in R, Keras, or TensorFlow in Python).

Step 4: Model validation ensures generalizability and interpretability. For classification tasks, use accuracy() and roc_auc() from yardstick (R), or classification_report() in sklearn.metrics (Python). For regression, use R2() and mae() in R, or r2_score() and mean_absolute_error() in Python. Perform k-fold cross-validation with trainControl(method = "cv", number = 5) in caret (R) or cross_val_score() in Python. For explainability, apply SHAP analysis using SHAPforxgboost (R) or shap (Python), and local interpretability with lime (R or Python). These tools allow tracing model predictions back to biological features, improving trust and insight in *Azospirillum*-related AI models.

To further enhance AI-driven research in *Azospirillum* applications, several scalability strategies can be adopted. Federated learning allows model training across multiple institutions without sharing raw data, maintaining privacy while improving model generalizability (e.g., using the federatedtools package in Python). Automated AI pipelines enable real-time data processing and deployment through APIs, using tools like MLflow in Python or plumber in R. Additionally, standardized and open-access microbiome repositories—such as Qiita for microbiome profiles and DataDryad for broader omics datasets—promote reproducibility and enhance training data availability. By incorporating these approaches into a structured AI-integrated workflow, *Azospirillum* research can evolve from descriptive studies to predictive, scalable, and data-driven applications, advancing microbial inoculant strategies in sustainable agriculture.

20.6 Example Script

A simple script (Table 20.1) can be used to integrate and analyze multiple *Azospirillum* experiments using deep learning. Before running the script, ensure all metadata tables use consistent column names and units across experiments. Variables should be prefixed as follows: P_ for predictors (e.g., soil traits, nutrient uptake), Y_ for response variables (e.g., yield, NUE), and T_ for additional traits used for annotation or plotting. ASV tables for both 16S and ITS regions must be formatted with samples as rows and features as columns, and each experiment must be saved in a clearly named file structure within a single working directory.

Table 20.1 # Example Python script: Deep learning pipeline for combining multiple experiments # with metadata + 16S + ITS + taxonomy + soil + plant traits and predicting desired responses

```
import os
import pandas as pd
import numpy as np
from sklearn.preprocessing import StandardScaler, QuantileTransformer
from sklearn.impute import KNNImputer
from sklearn.model_selection import train_test_split
from sklearn.decomposition import PCA
from pycombat import pycombat
import tensorflow as tf
from tensorflow.keras.models import Sequential
from tensorflow.keras.layers import Dense, Dropout
from tensorflow.keras.callbacks import EarlyStopping
import shap

# ----------------------------
# Set your working directory where all input files are located
# Replace the path below with the actual path on your system
# ----------------------------
working_dir = 'data'  # Change this to your actual working directory

# ----------------------------
# Adjustable Parameter: Number of experiments
# Set this to the number of separate experiments or batches you want to merge
# ----------------------------
num_experiments = 5  # Change this number based on your actual experiment
count

# ----------------------------
# IMPORTANT: Harmonize variable naming conventions before running
# Add 'P_' prefix to predictor columns (used as inputs for modeling)
# Add 'Y_' prefix to response variables (e.g., Y_Yield, Y_NUE)
# Add 'T_' prefix to non-predictive traits used for annotation or plotting
only
```

(continued)

Table 20.1 (continued)

```
# Ensure that all variables with the same biological meaning are in the same
units
# For variables measured using different methods (e.g., total N by Kjeldahl
vs. combustion), transform them using z-score or unit conversion beforehand
# ----------------------------

# Define paths to the data files based on working directory
metadata_files = [os.path.join(working_dir, f"experiment_{i}_metadata.csv")
for i in range(1, num_experiments + 1)]
asv16s_files = [os.path.join(working_dir, f"experiment_{i}_16s_asv.csv") for i
in range(1, num_experiments + 1)]
asvits_files = [os.path.join(working_dir, f"experiment_{i}_its_asv.csv") for i
in range(1, num_experiments + 1)]
taxonomy_files = [os.path.join(working_dir, f"experiment_{i}_taxonomy.csv")
for i in range(1, num_experiments + 1)]

# Load all metadata, 16S ASV, ITS ASV, and taxonomy tables
meta_list = [pd.read_csv(f) for f in metadata_files]
asv16s_list = [pd.read_csv(f, index_col=0) for f in asv16s_files]
asvits_list = [pd.read_csv(f, index_col=0) for f in asvits_files]
taxa_list = [pd.read_csv(f) for f in taxonomy_files]  # Optional, not used
here but may be useful later

# Annotate each dataset with batch information
for i in range(num_experiments):
    meta_list[i]['batch'] = f'batch_{i+1}'
    asv16s_list[i]['batch'] = f'batch_{i+1}'
    asvits_list[i]['batch'] = f'batch_{i+1}'

# Combine datasets into unified tables
meta_df = pd.concat(meta_list, ignore_index=True)
asv16s_df = pd.concat(asv16s_list, ignore_index=False).fillna(0)
asvits_df = pd.concat(asvits_list, ignore_index=False).fillna(0)

# Step 2: Preprocessing and quality filtering
# Filter low prevalence taxa from 16S and ITS tables
asv16s_df = asv16s_df.loc[:, (asv16s_df > 0).sum(axis=0) > (0.05 *
asv16s_df.shape[0])]
```

(continued)

Table 20.1 (continued)

```
asvits_df = asvits_df.loc[:, (asvits_df > 0).sum(axis=0) > (0.05 *
asvits_df.shape[0])]

# Quantile normalization of ASV tables
qt_16s = QuantileTransformer(output_distribution='normal')
asv16s_normalized = qt_16s.fit_transform(asv16s_df.drop(columns=['batch']))
qt_its = QuantileTransformer(output_distribution='normal')
asvits_normalized = qt_its.fit_transform(asvits_df.drop(columns=['batch']))

# Ensure only P_ variables (predictors) and Y_ variables (responses) are used
for modeling
predictors = [col for col in meta_df.columns if col.startswith('P_')]
responses = [col for col in meta_df.columns if col.startswith('Y_')]
meta_predictors = meta_df[predictors]
meta_responses = meta_df[responses]

# Impute missing predictor values
meta_predictors_imputed =
pd.DataFrame(KNNImputer(n_neighbors=5).fit_transform(meta_predictors),
                                       columns=meta_predictors.columns)

# Merge imputed metadata predictors and normalized ASV features into one
dataframe
full_data = pd.concat([
    meta_predictors_imputed.reset_index(drop=True),
    pd.DataFrame(asv16s_normalized, columns=asv16s_df.columns[:-1]),
    pd.DataFrame(asvits_normalized, columns=asvits_df.columns[:-1])
], axis=1)

# Apply batch effect correction across all features (excluding response
variables)
batch_labels = meta_df['batch'].astype(str)
combat_corrected = pycombat(full_data, batch_labels)

# Define input (X) and output (y) for model training
X = combat_corrected
y = meta_responses.reset_index(drop=True)
```

(continued)

Table 20.1 (continued)

```
# Standardize input features
scaler = StandardScaler()
X_scaled = scaler.fit_transform(X)

# Split data into training and test subsets
X_train, X_test, y_train, y_test = train_test_split(X_scaled, y,
test_size=0.3, random_state=42)

# Deep learning model
model = Sequential([
    Dense(128, activation='relu', input_shape=(X_train.shape[1],)),
    Dropout(0.3),
    Dense(64, activation='relu'),
    Dropout(0.2),
    Dense(y.shape[1])  # Output size matches number of Y_ variables
])

model.compile(optimizer='adam', loss='mean_squared_error', metrics=['mae'])
es = EarlyStopping(monitor='val_loss', patience=10, restore_best_weights=True)
model.fit(X_train, y_train, validation_split=0.2, epochs=200, batch_size=32,
callbacks=[es])

# Evaluate model
loss, mae = model.evaluate(X_test, y_test)
print(f'MAE on test set: {mae}')

# Interpret model using SHAP
explainer = shap.Explainer(model, X_train)
shap_values = explainer(X_test[:100])
shap.plots.beeswarm(shap_values)

# This script now supports both 16S and ITS data integration, standardized
metadata structure,
# flexible response variable selection, and consistent variable naming for
future annotations.
```

References

Anuragi A, Sisodia DS, Pachori RB (2024) Mitigating the curse of dimensionality using feature projection techniques on electroencephalography datasets: an empirical review. Artif Intell Rev 57(3):75

Breiman L (2001) Random forests. Mach Learn 45(1):5–32. https://doi.org/10.1023/A:1010933404324

Busato S, Gordon M, Chaudhari M et al (2023) Compositionality, sparsity, spurious heterogeneity, and other data-driven challenges for machine learning algorithms within plant microbiome studies. Curr Opin Plant Biol 71:102326. https://doi.org/10.1016/j.pbi.2022.102326

Corso G, Stark H, Jegelka S et al (2024) Graph neural networks. Nature Rev Methods Primers 4(1): 17. https://doi.org/10.1038/s43586-024-00294-7

Deng Z, Zhang J, Li J et al (2021) Application of deep learning in plant–microbiota association analysis [Review]. Front Genet 12. https://doi.org/10.3389/fgene.2021.697090

Ekundayo F (2024) Leveraging AI-driven decision intelligence for complex systems engineering. Int J Res Publ Rev 5(11):1–10

García G, Carlin M, Cano RDJ (2025) Holobiome harmony: linking environmental sustainability, agriculture, and human health for a thriving planet and one health. Microorganisms 13(3):514

Hao Y, Liu H, Li J et al (2025) Environmental tipping points for global soil nitrogen-fixing microorganisms. Science 28(1):111634. https://doi.org/10.1016/j.isci.2024.111634

Högberg C (2024) Stabilizing translucencies: governing AI transparency by standardization. Big Data Soc 11(1):20539517241234298

Huang HH, Shu J, Liang Y (2024) MUMA: a multi-omics meta-learning algorithm for data interpretation and classification. IEEE J Biomed Health Inf 28(4):2428–2436. https://doi.org/10.1109/JBHI.2024.3363081

Li Y, Ren X, Yu H et al (2024) A penalized integrative deep neural network for variable selection among multiple omics datasets. Quant Biol 12(3):313–323. https://doi.org/10.1002/qub2.51

Mahmud T, Datta N, Chakma R et al (2024) An approach for crop prediction in agriculture: integrating genetic algorithms and machine learning. IEEE Access 12:173583–173598. https://doi.org/10.1109/ACCESS.2024.3478739

Meraj T, Sharif MI, Raza M (2024) Computer vision-based plants phenotyping: a comprehensive survey. iScience 27(1). https://doi.org/10.1016/j.isci.2023.108709

Noble WS (2006) What is a support vector machine? Nat Biotechnol 24(12):1565–1567. https://doi.org/10.1038/nbt1206-1565

Pan AY (2021) Statistical analysis of microbiome data: the challenge of sparsity. Curr Opin Endocrine Metab Res 19:35–40. https://doi.org/10.1016/j.coemr.2021.05.005

Pan S, Jiang X, Zhang K (2024) WSGMB: weight signed graph neural network for microbial biomarker identification. Brief Bioinform 25(1):bbad448

Peyal MMK, Shihab AUI, Ocean SI et al (2025) Monitoring leafy plant root morphology with pattern recognition imaging techniques. In: 2025 IEEE 15th annual computing and communication workshop and conference (CCWC)

Qi P, Chiaro D, Guzzo A et al (2024) Model aggregation techniques in federated learning: a comprehensive survey. Futur Gener Comput Syst 150:272–293

Rubel ET, Raittz RT, Coimbra NADR et al (2016) ProClaT, a new bioinformatics tool for in silico protein reclassification: case study of DraB, a protein coded from the draTGB operon in Azospirillum brasilense. BMC Bioinf 17(18):455. https://doi.org/10.1186/s12859-016-1338-5

Rupanetti D, Kaabouch N (2024) Combining edge computing-assisted internet of things security with artificial intelligence: applications, challenges, and opportunities. Appl Sci 14(16):7104

Schumann Y, Gocke A, Neumann JE (2025) Computational methods for data integration and imputation of missing values in omics datasets. Proteomics 25(1–2):e202400100. https://doi.org/10.1002/pmic.202400100

Stark Z, Glazer D, Hofmann O et al (2025) A call to action to scale up research and clinical genomic data sharing. Nat Rev Genet 26(2):141–147

Vidotti MS, Matias FI, Alves FC et al (2019) Maize responsiveness to Azospirillum brasilense: Insights into genetic control, heterosis and genomic prediction. PLoS One 14(6):e0217571. https://doi.org/10.1371/journal.pone.0217571

Virmani D, Ghori MAS, Tyagi N (2024) Machine learning: the driving force behind intelligent systems and predictive analytics. In: 2024 international conference on trends in quantum computing and emerging business technologies

Walsh C, Stallard-Olivera E, Fierer N (2024) Nine (not so simple) steps: a practical guide to using machine learning in microbial ecology. MBio 15(2):e0205023. https://doi.org/10.1128/mbio.02050-23

Wang N, Wang T, Ning K (2024) Refining biome labeling for large-scale microbial community samples: leveraging neural networks and transfer learning. Environ Sci Ecotechnol 17:100304. https://doi.org/10.1016/j.ese.2023.100304

Wu H, Han R, Zhao L et al (2025) AutoGP: an intelligent breeding platform for enhancing maize genomic selection. Plant Commun

Xu C, Cao S, Alam MA (2024) Application of novel statistical and machine-learning methods to high-dimensional clinical cancer and (multi-) omics data volume II. Frontiers Media SA 15: 1487893

Yassue RM, Galli G, Fritsche-Neto R et al (2023) Classification of plant growth-promoting bacteria inoculation status and prediction of growth-related traits in tropical maize using hyperspectral image and genomic data. Crop Sci 63(1):88–100. https://doi.org/10.1002/csc2.20836

Yu Y, Mai Y, Zheng Y et al (2024) Assessing and mitigating batch effects in large-scale omics studies. Genome Biol 25(1):254

Zha D, Bhat ZP, Lai K-H et al (2025) Data-centric artificial intelligence: a survey. ACM Comput Surv 57(5):1–42

Zhao L, Walkowiak S, Fernando WGD (2023) Artificial intelligence: a promising tool in exploring the phytomicrobiome in managing disease and promoting plant health. Plants 12(9):1852. https://www.mdpi.com/2223-7747/12/9/1852

Chapter 21
Inoculant Preparation and Formulations for *Azospirillum* spp.

Fabricio Dario Cassán, Mauricio Schoebitz, Juan A. Moreno-Cid, Andrea Figueiredo Giroldo, Daniela Scarabel, Mariana Puente, Julia García, and Solon Araujo

Abstract *Azospirillum*-based inoculants are widely used around the world, particularly across South America, due to their well-documented agronomic benefits in crops such as maize, wheat, and soybean. When co-inoculated with rhizobia, they enhance nitrogen fixation, root development, and tolerance to abiotic stresses. Liquid formulations dominate the market because of their ease of use and extended shelf life, which is supported by the inclusion of sugars, polyalcohols, or other osmolytes. Industrial-scale production has been made possible through optimized growth media and the use of bioreactors. Solid formulations and encapsulation techniques (such as alginate-starch beads) offer added protection and controlled release of bacteria into the soil. Application methods, including seed coating, in-furrow, and foliar spraying, significantly influence efficacy, with in-furrow and foliar applications offering better compatibility with agrochemicals. Nanotechnology is beginning to play a role in

F. D. Cassán (✉)
Instituto de Investigaciones Agrobiotecnológicas (INIAB-CONICET), Universidad Nacional de Río Cuarto, Río Cuarto, Córdoba, Argentina
e-mail: fcassan@exa.unrc.edu.ar

M. Schoebitz
Departamento de Suelos y Recursos Naturales, Facultad de Agronomía, Universidad de Concepción, Concepción, Chile

J. A. Moreno-Cid
Bioprocess Research and Development Pilot Plant, Bionet, Spain

A. F. Giroldo
Research and Development, Agrocete, Brazil

D. Scarabel
Consultant, BioScale Consulting, Brasília, Brazil

M. Puente · J. García
Instituto Nacional de Tecnología Agropecuaria (INTA), Instituto de Microbiología y Zoología Agrícola, Buenos Aires, Argentina

S. Araujo
Consultant Associação Nacional de Promoção e Inovação da Indústria de Biológicos, ANPII Bio, Brasília, Brazil

F. D. Cassán et al. (eds.), *The Definitive Handbook of Azospirillum*,
https://doi.org/10.1007/978-3-032-09222-9_21

improving delivery and nutrient uptake, though careful control of dosage is essential to avoid phytotoxic effects. Precision agriculture tools, such as drones and geospatial modeling, are being used to better assess inoculant performance in the field. Regulatory frameworks in various countries ensure quality through microbial counts, purity tests, and strain identification. Agronomic effectiveness is evaluated through germination rates, biomass and yield measurements, and nitrogen use efficiency. Stress resilience is further supported by poly-β-hydroxybutyrate (PHB) accumulation and biofilm formation. Together, these innovations are promoting the successful integration of *Azospirillum* into sustainable agriculture.

Keywords Inoculants · Formulations · Application · Shelf life · Bioreactors

21.1 Introduction

The global market for *Azospirillum*-based inoculants has seen significant growth, particularly in South America, where their use has expanded beyond cereals to include legumes such as soybeans. Co-inoculation with rhizobia has gained traction, especially in Brazil, due to its positive effects on plant growth and resilience to abiotic stressors (Barbosa et al. 2021). A comprehensive meta-analysis of 347 trials across 12 countries, including Brazil and Argentina, demonstrated that *Azospirillum* spp. inoculation significantly increased grain yield, with the highest impact observed in winter cereals, followed by summer cereals (Díaz-Zorita et al. 2015). In Argentina, 316 field experiments in the Pampas region reported a 66–80% increase in maize yield due to *A. brasilense* inoculation, particularly under early-season water stress conditions (Díaz-Zorita et al. 2015). In wheat, inoculation benefits were more pronounced during early growth stages but diminished under favorable conditions during seed filling (Kazi et al. 2016). Inoculation success rates across multiple studies ranged from 70% to 80% (Díaz-Zorita and Fernández-Canigia 2009), highlighting the influence of environmental and management factors on inoculant efficacy. Co-inoculation of legumes with rhizobia and *Azospirillum* spp. has been shown to enhance nitrogen fixation and plant growth. In soybean, co-inoculation accelerated nodule development, increased shoot nitrogen content, and improved plant performance, particularly under drought conditions (Chibeba et al. 2020; Cerezini et al. 2016). In Argentina, field trials with alfalfa demonstrated that co-inoculation with *Ensifer meliloti* and *A. brasilense* nearly doubled productivity compared to single rhizobia inoculation (Díaz-Zorita 2012). Similarly, soybean and common bean yields in Brazil increased by 16.1% and 19.6%, respectively, with co-inoculation compared to single rhizobia inoculation (Hungria et al. 2013). The adoption of co-inoculation in Brazil has grown significantly, with a 30% increase in use between 2016 and 2018 (Prando et al. 2016, 2018). New application methods, such as foliar inoculation and in-furrow application, are being explored to overcome potential compatibility issues with seed-applied pesticides. Research has confirmed the benefits of foliar inoculation in soybean, showing improved growth and nutrient uptake mediated by auxin signaling (Puente et al. 2018, 2019). These alternative strategies may further enhance the effectiveness of *Azospirillum*-based inoculants. The success of *Azospirillum*-based inoculants in agriculture is closely linked to their formulation, which impacts their stability, viability, and field performance. The first registered Azospirillum inoculant in South America using A. brasilense dates back to 1996 in Argentina (SENASA), and the industry has since expanded, with more than one hundred products currently available from around hundred companies across Argentina, Brazil, and Uruguay (Cassán et al. 2020). Most of these inoculants (94%) are formulated as liquid carriers, reflecting industry preferences for ease of use and longer shelf life. Ensuring high-quality formulations involves a multistage process: laboratory development and validation, industrial-scale production, and rigorous quality control to maintain bacterial viability and its effectiveness until application. In Argentina and Brazil, regulatory standards require a minimum bacterial concentration of 1×10^7 CFU·mL^{-1} to 2×10^8 CFU·mL^{-1}, respectively, with a typical shelf life of 6 months. Market data from 2022 indicate that 23 million doses of *Azospirillum* spp. inoculants were sold in South America, covering nearly 20 million hectares. Ultimately, the widespread adoption of *Azospirillum* spp. inoculants underscores their agronomic value. However, continued investment in formulation research, field validation, and farmer education is essential to maximize their benefits. In Brazil, for example, the use of this inoculant has been growing continuously: in 2020, companies affiliated with ANPII Bio used 14,8 million doses, and in 2024, they have already used 43.5 million. Distributed in corn, wheat, and soybean co-inoculation. The integration of *Azospirillum* spp. into sustainable agricultural practices highlights its potential to enhance crop productivity while reducing reliance on chemical fertilizers.

The following section outlines the procedures for *Azospirillum* spp. multiplication and inoculant formulation at various scales, from laboratory cultivation to industrial bioreactors, employing different formulation strategies. It covers liquid and solid media for bacterial propagation, scale-up processes, and traditional inoculation methods such as seed, soil, and foliar applications. Additionally, innovative technologies, including nanotechnology-based formulations, inoculant shelf-life optimization, compatibility assessments, and adaptation strategies for extreme conditions, are explored within the context of precision agriculture.

21.2 Liquid Media for Propagation of *Azospirillum*

21.2.1 Chemically Defined Media

The discovery of several species belonging to the genus *Azospirillum* occurred with the use of a chemically defined culture media formulated without mineral nitrogen to induce bacterial nitrogen fixation, such as the Nitrogen-Free bromothymol blue (NFb) medium (Döbereiner et al. 1996). For metabolic and nutritional studies, the use of chemically defined culture media is essential, as with a more precise composition of its components, there will be greater reproducibility of the results. The nutritional needs of *Azospirillum* spp. have been described by several authors (for further details, see Chap. 1), and its species exhibit a range of metabolic capabilities, including the ability to grow on various organic acids and carbohydrates as carbon sources. They can utilize malate, succinate, lactate, and pyruvate, while some species can also metabolize d-fructose and d-glucose; biotin is also a vital nutrient for these bacteria (Döbereiner and Day 1976). Some examples of other chemically defined culture media, in addition to NFb, have been proposed. This is the case of the AAM-defined medium ($g{\cdot}L^{-1}$): K_2HPO_4, 2.0; KH_2PO_4, 6.0; $MgSO_4.7H_2O$, 0.2; NaCl, 0.1; $CaCl_2.2H_2O$, 0.026; $Na_2MoO_4.2H_2O$, 0.002; $FeCl_3$, 0.01; NH_4Cl, 1.0; distilled water, 1000 mL. The pH is adjusted to 6.0. The carbon sources are sterilized separately (D,L-sodium malate) or filter sterilized (sucrose) and added aseptically to the minimal medium to give a final concentration of 0.03 M; or the **LGI medium** ($g{\cdot}L^{-1}$): sucrose, 5.0; K_2HPO_4, 0.2; KH_2PO_4, 0.6; $MgSO_4.7H_2O$, 0.2; $CaCl_2.2H_2O$, 0.02; $Na_2MoO_4.2H_2O$, 0.002; bromothymol blue 0.5% in 0.2 N KOH, 5ml; FeEDTA (1.64% solution), 4 mL; vitamin solution (biotin, 0.01 g; pyridoxal-HCl, 0.02g; distilled water, 100ml), 1 mL; pH adjusted to 6.0–6.2 with H_2SO_4 and the final volume is brought to 1000 mL with distilled water.

21.2.2 Complex Media

Complex culture media allow enrichment of growth media with a variety of nutrients and growth factors at a lower cost, with the use of extracts and hydrolysates, which makes them convenient for use in industries. Preliminary in vitro studies indicated that, in complex media, the bacterium prefers compounds of gluconate and glycerol (unpublished data) as supplements over the organic acids, such as malate and succinate, which are commonly used for cultivation in specific media for this genus (Bashan and de-Bashan 2015). Moreno-Galvan et al. (2012) tested several nutritional sources for the development of a culture medium for mass production of *A. brasilense* and through the application of statistical designs, found the optimal composition of the medium ($g\ L^{-1}$): 28.33 glutamate, 2.92 yeast extract, 1.34 $K_2HPO_4\ 3H_2O$, 0.5 $MgSO_4\ 7H_2O$, and 0.02 $FeCl_3$. This alternative medium produced 1.71 times more proteins and 1.68 times more biomass when compared to *DYGS medium* ($g\ L^{-1}$): D,L-malate 1.0; yeast extract 2.0; glucose 1.0; glutamate 1.5; peptone 1.5 and $MgSO_4\ 7H_2O$ 0.5. The composition of the *OAB medium* (also known as *NFb modified*) is described by Bashan et al. (1993): solution A ($g.L^{-1}$): DL-malic acid, 5.0; NaOH, 3.0; $MgSO_47H_2O$, 0.2; $CaCl_2$, 0.02; NaCl, 0.1; NH_4Cl, 1.0; yeast extract, 0.1; $FeCl_3$, 0.01 and ($mg{\cdot}L^{-1}$): $NaMoO_42H_2O$, 2.0; $MnSO_4$, 2,1; H_3BO_3, 2,8; Cu $(NO_3)_2\ 3H_2O$, 0,04; $ZnSO_4{\cdot}7H_2O$, 0,24; plus 900 mL of distilled water; solution B ($g.L^{-1}$): K_2HPO_4, 6.0; KH_2PO_4, 4.0 and 100 mL of distilled water. After autoclaving and cooling, the two solutions are mixed. The pH of the medium is 6.8 (Bashan et al. 1993). The *BTB-1* and *BTB-2* media are the result of modifications of the TYG and modified TYG media, aimed at a massive production of cells in a shorter period, already considering carbon sources such as ($g.L^{-1}$) Na-gluconate, 5.0, and ($mL.L^{-1}$) glycerol, 8.0, preferred by the majority of *Azospirillum* species (Bashan et al. 2016). The choice of carbon sources in the quantities described above differentiates version 1 and 2 of the BTB medium, which is completed with the addition of ($g.L^{-1}$): tryptone, 5.0; yeast extract, 5.0; NaCl, 1.2; $MgSO_4{\cdot}7H_2O$, 0.25; K_2HPO_4, 0.13; $CaCl_2$, 0.22; K_2SO_4, 0.17; Na_2SO_4, 2.4; $NaHCO_3$, 0.5; Na_2CO_3, 0.09; and Fe(III)EDTA, 0.07. The pH is adjusted to 7.0 after sterilization (Bashan et al. 2002).

21.2.3 Inoculant Formulations

An inoculant is a complex biological formulation that combines two elements: cultured microorganisms, and compounds secreted into their growth medium. A commercially available inoculant should not be considered as a carrier of microorganisms but rather as a complex mixture resulting from bacterial biotransformation of components added to the growth medium into different metabolites. Liquid inoculants are special liquid formulations containing not only the desired microorganisms and their nutrients but also special cell protectants or substances that encourage the formation of resting spores or cysts for longer shelf life and tolerance to adverse conditions (Sivasakthivelan and Saranraj 2013). Most media used to produce commercial inoculants are considered proprietary and are therefore not revealed. Based on the knowledge of carbon metabolism by *Azospirillum* spp., Bashan et al. (1993) replaced or supplemented the media already described with sources preferred by the bacteria. Later, these same authors, based on the replacement of glucose in tryptone–yeast extract–glucose medium by Na-gluconate or glycerol, tested two new culture media for mass cultivation of *Azospirillum* spp. After 18 hours of incubation, these modifications increased the populations of different strains (to $\sim 10^{11}$ cells and $\sim 5 \times 10^9$ $CFU{\cdot}mL^{-1}$) in addition to significantly reducing the generation time, showing that they are suitable options to produce common synthetic inoculants (Bashan and de-Bashan 2015). These sources should consider the optimal supply mainly of carbon, nitrogen, mineral salts, and vitamins for the growth and maintenance of the viability of these microorganisms during shelf life. It is also necessary to consider the addition of compounds that serve as cell protectors to improve the quality of the inoculant, such as the following characteristics also mentioned by Zayed (2016): (a) prevention of osmolysis; (b) better adhesion to seed; (c) stabilizing the product; (d) binding or inactivating of soluble seed coat toxins; (e) enhancing of microorganism survival during storage; and (f) protection of inoculum after exposure to extreme environmental conditions upon inoculation to seed and planting. The liquid formulations of *Azospirillum* spp. with the addition of trehalose, polyvinylpyrollidine, and glycerol enhanced and maintained the population level at 10^8 up to 10 months of storage. The liquid formulation shows better adherence and survival on seeds, roots of seedlings, and in the rhizosphere soil than the solid carrier-based *Azospirillum* spp. inoculants (Sivasakthivelan and Saranraj 2013). The nature and concentration of additives affect the performance of inoculum. Dayamani (2010) has tried different osmolytes in different concentrations to optimize it for liquid inoculants with various microorganisms, including *A. brasilense*. The results of the study showed that each organism responds variably to different osmolytes and their concentration, with the population of *Azospirillum* sp. remaining more stabilized with polyvinylpyrrolidone (PVP) and polyethylene glycol (PEG) at both 1% and 2%. In a study to identify polymers that maintain the cell viability of the inoculant during storage, Cortés-Patiño and Bonilla (2015) compared the protective effect of five polymers: carrageenan, sodium alginate (SA), trehalose, polyvinylpyrrolidone (PVP), and glycerol, as well as a phosphate-buffered saline control. It was shown that 0.75% carrageenan and 1% sodium alginate had the best effect on cell viability and were suitable for formulating a liquid *A. brasilense* inoculant (Zayed 2016). Kumaresan and Reetha (2011) also tested the effect of different polymers on the concentration of *Azospirillum* spp. over shelf life. The results showed that maximum population was recorded in trehalose (10 mM) 4.00×10^9 followed by glycerol (10 mM) 3.33×10^9; gum arabica (0.3%) 2.67×10^9; PVP (2%) 2.33×10^9; PEG (1.0) 10.33×10^8, and polyvinyl alcohol (PVA, 0.5%) 5.00×10^8 cells.mL^{-1} of *Azospirillum* spp. liquid inoculants during 6 months of storage at room temperature (28 ± 2 °C) whereas a minimum population 6.33×10^2 was recorded in the control (without chemical additives) N_2 free malate broth alone during 5 months. For further details, see the full report of Kumaresan and Reetha (2011). Döbereiner and Day (1976) had previously reported that the accumulation of poly-β-hydroxybutyrate (PHB) in *Azospirillum* spp. cells increases under various stress conditions and that this accumulation would be related to the formation of cysts and flocs, also suggesting that inoculants generated from flocculated cells seemed more efficient than those originating from non-flocculated cells. Survival of cyst cells was observed for up to 1 year and 2 months at a population level of 10^8 by Vendan and Thangaraju (2007), who studied the induction of cyst formation in *Azospirillum* spp. In this study, the cyst cells accumulated poly-b-hydroxybutyrate (PHB) granules and exhibited desiccation tolerance of up to 20 days and temperature tolerance of up to 40 °C. Thus, the cyst-based liquid formulation would have a dual advantage: longer shelf life and tolerance to harmful environmental conditions. They also reported that media containing fructose and KNO_3 favored flocculation and the formation of cyst cells, providing nutritional stress to *Azospirillum* spp. cells.

21.2.3.1 Liquid Inoculant Formulations

In South America—Brazil, Argentina, Uruguay, and Paraguay—liquid carriers are most often used to formulate the biological products (94%); 6% of the products are formulated in solid carriers, such as peat or bentonite. This clearly shows the formulation preferences of the companies that are manufacturing these products (Cassán et al. 2020). Liquid inoculant formulations may use various broth cultures amended with agents that promote cell survival in the package and after application to seed or soil. Liquid inoculant is easily adapted to advanced seeding equipment, since it can be sprayed onto the seed as it passes through the seed auger and dries before it travels into the seed bin on the planter (Tittaburt et al. 2007). Since the liquid biofertilizer has high cell counts, each seed receives more than thousands of cells. Additives in liquid biofertilizer protect the cells on the inoculated seeds against toxicity, desiccation, and osmotic shock (Vithal Navi 2004). Liquid formulations can be produced by a simple fermentation process, packed directly from the fermenter aseptically, and stored for a long time without loss of viability. It is cost-effective as it avoids processing and sterilization of solid carrier material. No contamination during storage can be detected, as complete sterilization could be achieved with liquid formulations. The quantity of inoculum required is also lower compared to carrier-based formulations, hence easier for farmers to handle (Sahu and Brahmaprakash 2016). Other advantages of liquid inoculants, according to the authors, are (1) less amount of inoculant needed; (2) no need of any sticker material unlike carrier-based inoculants; (3) supports higher number of cells for longer time; (4) easier to produce; (5) easier to sterilize completely, thus prevents contamination; (6) compatible with modern agriculture machinery for its application; (7) easier transport of large number of inoculum in small bottles; and (8) easy to apply also as fertigation.

21.2.3.2 Liquid Inoculants

21.2.3.2.1 Inoculant Scale-Up in Submerged Fermenters

This protocol describes the industrial-scale cultivation of *A. brasilense* for the production of microbial inoculants intended for plant biostimulants. The scale-up process follows a fermentation train from a working cell bank (WCB) to a 4000 L production fermenter while maintaining microaerophilic conditions. This strategy is implemented to direct the metabolism towards bacterial growth with enhanced nitrogen-fixing capacity. The cultivation process progresses through the following stages: (a) revival from Working Cell Bank (WCB); (b) shake flask pre-culture; (c) Seed Fermenter 1 (10 L); (d) Seed Fermenter 2 (200 L); and (e) Production Fermenter (4000 L). The inoculum volume is maintained at 5% (v/v) between scales, with aeration and agitation parameters optimized for biomass growth.

21.2.3.2.2 Culture Media

This section includes Tables 21.1, 21.2, and 21.3 outlining the composition of both the chemically defined NFb culture and the industrial-scale production medium. These tables provide detailed information on the ingredients and their concentrations essential for optimizing the growth of *Azospirillum* sp. throughout the scale-up process. They serve as a valuable reference for maintaining consistent microbial performance at a large scale.

Table 21.1 Nitrogen-Free bromothymol blue (NFb) medium composition for the selective cultivation of *Azospirillum* sp.

Component	Concentration	Units
Malic acid	5	$g{\cdot}L^{-1}$
K_2HPO_4	0.5	$g{\cdot}L^{-1}$
$MgSO_4{\cdot}7H_2O$	0.2	$g{\cdot}L^{-1}$
NaCl	0.1	$g{\cdot}L^{-1}$
$CaCl_2{\cdot}2H_2O$	0.02	$g{\cdot}L^{-1}$
Micronutrients solution	2	$mL{\cdot}L^{-1}$
H_2O (eqf)	1000	$g{\cdot}L^{-1}$

Table 21.2 Stock solution of micronutrients to be used for the NFb and PM

Component	Concentration	Units
$CuSO_4{\cdot}5H_2O$	0.04	$g{\cdot}L^{-1}$
$ZnSO_4{\cdot}7H_2O$	0.12	$g{\cdot}L^{-1}$
H_3BO_3	1.40	$g{\cdot}L^{-1}$
$Na_2MoO_4{\cdot}2H_2O$	1.0	$g{\cdot}L^{-1}$
$MnSO_4$. H_2O	1.175	$g{\cdot}L^{-1}$
$FeSO_4{\cdot}7H_2O$	0.5	$g{\cdot}L^{-1}$
H_2SO_4	5	$mL{\cdot}L^{-1}$
H_2Oeqf	1000	$g{\cdot}L^{-1}$

Abbreviations: Nitrogen-Free bromothymol blue medium (NFb); Production medium (PM)

Note: Store in an opaque container to prevent direct exposure to light and keep at 4 °C

Table 21.3 Production media (PM) to cultivate *Azospirillum* sp. at a commercial scale

Component	Concentration	Units
Malic acid	4	$g{\cdot}L^{-1}$
Glycerol	1	$g{\cdot}L^{-1}$
Yeast extract	2	$g{\cdot}L^{-1}$
$(NH_4)_2SO_4$	0.5	$g{\cdot}L^{-1}$
K_2HPO_4	3	$g{\cdot}L^{-1}$
KH_2PO_4	2	$g{\cdot}L^{-1}$
$MgSO_4{\cdot}7H_2O$	0.5	$g{\cdot}L^{-1}$
NaCl	0.1	$g{\cdot}L^{-1}$
$CaCl_2{\cdot}2H_2O$	0.02	$g{\cdot}L^{-1}$
Micronutrients solution	2	$mL{\cdot}L^{-1}$
H_2O(eqf)	1000	$g{\cdot}L^{-1}$

Note: The PM must be sterilized by autoclaving at 121 °C for 30 min in the bioreactor/fermenter before being used. The pH must be adjusted to 7 ± 0.2 with KOH before autoclaving

Table 21.4 Fermentation conditions for *Azospirillum* sp. propagations and scale-up of the process

Parameter	Shake flask	Seed fermenter 1	Seed fermenter 2	Production fermenter
Volume (L)	0.5	10	200	4000
Media	NFb	Production media	Production media	Production media
Temperature (°C)	30 ± 1	30 ± 1	30 ± 1	30 ± 1
pH control	Uncontrolled	Uncontrolled	Uncontrolled	Uncontrolled
DO_2 cascade (%)	Uncontrolled	5	5	5
Back pressure (barg)	NA	0.2	0.2	0.2
Gas constant (vvm)	NA	0.2	0.2	0.2
Airflow (NL/min)	NA	2.4	48	960
Agitation Cascade (rpm)	110	135–340	60–145	20 50
kLa (h^{-1})	NA	20–80	20–80	20–80
Time (hours)	24	18–24	18–24	36–48

21.2.3.2.3 Cultivation Conditions

The fermentation is performed under microaerophilic conditions with a dissolved oxygen (DO_2) level of 5%, controlled via constant air sparging and cascade agitation. The scale-up criterion is to maintain a kLa between 20 and 80 h^{-1} to ensure optimal metabolism, directing bacterial growth through the pathway that enhances nitrogen fixation. Table 21.4 summarizes the fermentation conditions of propagations and scale-up of the process

21.2.3.2.4 Fermenter Preparation

Step 1: Equipment inspection and component verification

Inspect all ports, inoculation lines, and propagation lines to ensure that all components, including addition septums and sterile addition ports and valves, are installed, clean, and functional. Verify that all gaskets, seals, and valves are properly fitted to prevent contamination and leakage.

Step 2: Probe calibration

Perform the calibration test on the pH and DO_2 probes. (a) pH probe: Perform a two-point calibration using pH 4.0 and 7.0 standard buffers; ensure proper response and drift correction before installation in the fermenter, and rinse with distilled water before insertion. (b) Dissolved oxygen (DO_2) probe: perform the air calibration before sterilization. For that, expose the optical DO_2 probe to ambient air and perform a 100% air-saturated calibration following the manufacturer's instructions. The probe will undergo a second in-process sterilization cycle after the fermenter medium is sterilized and before inoculation to ensure sterility.

Step 3: Leak test

Fill the fermenter with 50% of the final medium volume using decalcified water. Secure all ports and fittings, pressurize the system, and inspect all connections for leaks. Note: If any leaks are detected, adjust fittings or replace faulty components.

Step 4: Media preparation and addition

Prepare the fermentation medium at 95% of the final working volume to accommodate the 5% inoculum addition. Ensure that this 95% volume contains the full 100% concentration of the Process Medium (PM) composition to maintain proper nutrient balance post-inoculation. Transfer the prepared medium or prepare the medium directly into the fermenter under clean conditions, confirming that all input and output ports are closed before sterilization. Note: Use decalcified water for media preparation.

Step 5: Fermenter sterilization (sterilization-in-place [SIP] cycle)

In the case of an automatic sterilization sequence, the sterilization process is controlled by the fermenter's software (SW) in a fully automated cycle. In that case, configure the SIP parameters, such as sterilization temperature—121 °C, duration—30 min, and cooling temperature—30 °C (set as the process operating temperature for *Azospirillum* sp.). The system will automatically execute pre-heating, heating, holding, and cooling phases. After completion, verify temperature uniformity and ensure proper depressurization before opening any sterile connections.

Step 6: Post-sterilization cooling and ready-to-use conditions

Allow the fermenter to cool down to 30 °C, ensuring it reaches the target process temperature and perform the final DO_2 probe calibration. For that, perform a post-SIP 100% process calibration before inoculation and sterility check. Inspect for any visible contamination and collect a sample of sterile medium for microscopic examination. Once sterility is validated, the fermenter is ready for inoculation.

21.2.3.2.5 Fermenters Preparation and Fermentation Seed Train (Fig. 21.1)

Step 1: Revival from working cell bank

Inoculate per duplicate an *Azospirillum* sp. Strain from a −80 °C glycerol stock into 50 mL of NFb medium into two 250 mL flasks. Incubate overnight at 30 °C and 100 rpm until the optical density at 600 nm (OD600) reaches approximately 1.0.

Step 2: Shake flask pre-culture (500 mL)

Transfer the reactivated *Azospirillum* sp. cells into two 1000 mL shake flasks, each containing 250 mL NFb medium and incubate at 30 °C, 100 rpm for 24 h.

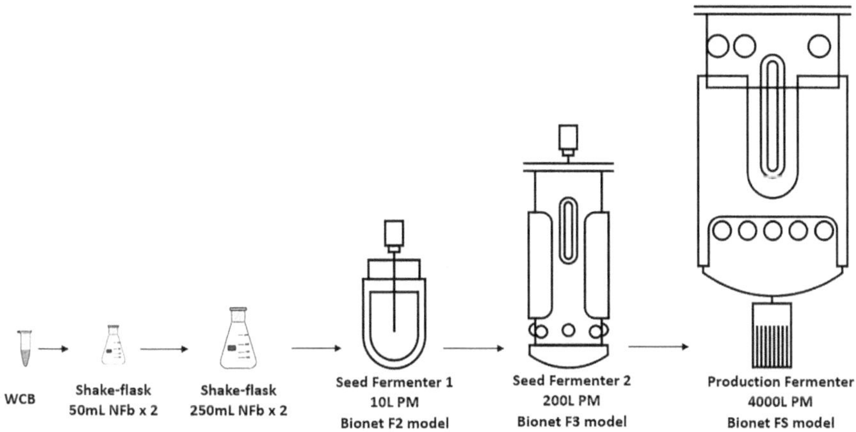

Fig. 21.1 Example of a fermentation scale-up train for *A. brasilense*, starting from a Working Cell Bank (WCB) and progressing through shake flasks (NFb medium) to seed and production fermenters (PM medium). The fermenter models shown are an example from the European manufacturer Bionet, including the F2 (10 L), F3 (200 L), and FS (4000 L) models

Step 3: Seed fermenter 1 (10 L)

Inoculate 10 L fermenter containing production medium PM with 5% (v/v) shake flask culture. Operate at 30 °C, with DO_2 at 5%, 0.2 vvm air, and agitation adjusted in a cascade mode to maintain a kLa between 20 and 80 h^{-1}. Cultivate for 18–24 h.

Step 4: Seed fermenter 2 (200 L)

Inoculate 200 L fermenter containing production medium PM with 5% (v/v) seed fermenter 1 culture. Operate under the same DO_2, and equivalent aeration, and agitation conditions. Cultivate for 18–24 h.

Step 5: Production fermenter (4000 L)

Inoculate 4000 L fermenter containing production medium PM with 5% (v/v) Seed Fermenter 2 culture. Operate at 30 °C, maintain DO at 5%, apply 0.2 vvm airflow, and adjust agitation in cascade mode to maintain a kLa between 20 and 80 h^{-1}. Cultivate for 36–48 h and harvest at OD_{600} 2–4 and CFU around 1×10^9 and 5×10^9 CFU·mL^{-1}

21.2.3.2.6 Storage

Store the inoculant at 4–10 °C in clean, sterile containers. The typical shelf-life ranges from 6 to 12 months.

21.2.3.2.7 Quality Control

Perform viable cell count using serial dilution and plating as described in Chap. 22. The cell counting is higher than 1×10^9 CFU·mL^{-1} at the time of the production. Ensure contamination-free culture through microscopic examination.

21.2.3.3 Solid Inoculant Formulations

The introduction of *Azospirillum* sp. into the soil relies on various soil characteristics, including the amount of organic matter, soil structure, water retention capacity, pH, and nutrient levels. Enhancing the success of rhizobacteria in soil can be achieved by examining the genetic and metabolic diversity of the strains, as well as their interactions with plant genotypes, stages of plant development, root exudates, plant hormone signaling, and native microbial communities. A major factor to consider is the development of correct formulations for bacterial delivery (Schoebitz and López Belchí 2016). The use of *Azospirillum* sp. in agriculture is mainly carried out in liquid form, as it is an efficient and cost-effective method for creating microbial inoculants. This process involves mixing the inoculant with water and applying it to seeds, planting furrows, or spraying it on leaves. Despite the high-water content in liquid applications, challenges arise in terms of storage, transportation, and shelf life. Additionally, liquid inoculants do not ensure the survival of rhizobacteria in the soil for successful colonization of the plant's rhizosphere. Solid inoculants using organic carriers offer an alternative to liquid formulations (de-Bashan et al. 2024). The bio-encapsulation technique improves the protection and controlled release of bacteria, enabling successful root and rhizosphere colonization. The principle of bio-encapsulation involves safeguarding the microorganisms introduced into the soil and ensuring their gradual and sustained release. The rate at which the encapsulation matrix degrades is directly linked to the biological activity of the microorganisms, as faster degradation results in less protection and reduced activity. These dried capsules can be stored at room temperature for extended periods, providing a conducive environment for bacteria and minimizing the risk of decreased survival rates. Enhancements to these inoculants can be made by adding essential nutrients for bacterial growth, effectively turning the capsules into bioreactors that can increase the number of encapsulated bacteria introduced into soil. Bacterial inoculants have addressed many issues associated with traditional peat inoculants, which often exhibit significant variability in quality. There are numerous benefits to the bio-encapsulation of *Azospirillum*, such as providing physical protection from stressful environmental conditions and the gradual release of bacteria for successful root colonization (Cesari et al. 2020). Dried beads containing *A. brasilense* presented 76% viable cells after one year of storage (Schoebitz et al. 2013).

For effective bio-encapsulation, the chosen materials should possess qualities such as handling for formulation, compatibility with PGPB, protection from environmental factors, assured release of PGPB, facilitation of inoculation, and high cell viability during storage (Schoebitz et al. 2012). Biopolymers, especially polysaccharides such as alginate, carrageenan, Arabic gum, xanthan gum, starch, cellulose, and chitosan, are commonly used to meet these criteria. Alginate and chitosan salts are particularly favored because of their cost-effectiveness, water solubility, non-toxicity, biodegradability, and ability to form physically gelled systems at room temperature (Giraldo et al. 2023). Advances in solid formulations featuring bio-encapsulation materials and techniques for rhizobacteria are presented. Conventional microbial inoculants cannot ensure high cell viability during formulation, storage, and soil inoculation processes. They must be stored at room temperature, avoiding extreme temperature oscillations. The shelf life of liquid inoculants is very short, with viability decreasing by one or two log units. Liquid inoculants cannot protect rhizobacteria from biotic and abiotic soil stress. Bio-encapsulation provides a niche in which rhizobacteria are protected from soil stress. Furthermore, liquid inoculum has an instantaneous release with bacteria delivered only initially (de-Bashan et al. 2024). Encapsulated inoculants enable a slow release of rhizobacteria, resulting in prolonged fertilization benefits. In addition to comparing capsules with liquid inoculants, fermentation is one of the costliest processes, requiring large quantities of water, broth, and sterile medium. Nevertheless, each capsule in the soil can function as a bioreactor, boosting the concentration of initial viable cells, which can help reduce the expenses related to bacterial multiplication. Moreover, several capsules distributed around the rhizosphere soil can operate as separate units, increasing cell concentration, enhancing root colonization, and boosting the effectiveness of microbial inoculants in the field. The selection of the appropriate bio-encapsulation technique depends on the bacterial characteristics, required stability during storage and processing, and production costs. Spray-drying is a flexible technique for formulating bacteria, although temperature can be critical. The fluid-bed process is a promising technique for large-scale production of microbial inoculants in agriculture. However, ionic gelation is currently the most suitable method for encapsulating rhizobacteria. Alginate-starch beads can load high cell concentrations of 10^9 CFU·g^{-1} with average diameters of over 4 mm. The size of the beads is influenced by the encapsulation method, matrix composition, and internal diameter of the diffuser. From the perspective of farmers and the agricultural sector, the ideal bead diameter should correspond to the size of the seeds. This alignment allows for the simultaneous planting of seeds and introduction of microbial inoculants using the same seed drill, ensuring that beads and seeds are placed closely together in the soil to boost the effectiveness of the inoculants. Many devices are designed to produce beads in laboratory conditions and on a small scale, allowing testing of encapsulated formulations only in growth chambers and greenhouses. To produce large amounts of inoculants, trials on large fields require innovative bio-encapsulation devices. GeniaLab developed jet cutter technology, allowing the production of high-viscosity beads by transferring fluid through a needle and cutting it into segments using a rotating cutting wheel. The bead diameter is conditioned by the wheel rotation speed and fluid mass passing through. Jet cutter technology enables the scaling up of bead production to broaden the application of formulated solid inoculants in farming. Future advancements in formulation techniques will be guided by industry needs, transitioning research from the laboratory to large-scale production.

21.2.3.3.1 Bio-encapsulation

Bacterial biomass production by *Azospirillum* sp. has been described by several authors (for further details,, see Sect. 21.2.1). All materials used for the bio-encapsulation process must be sterilized, and the process should be carried out under aseptic conditions. The matrix solution is prepared following the protocol described by (Benítez et al. 2024). A solution of sodium alginate (Gelymar, Puerto Montt, Chile) with a medium content of L-guluronic acid at 2.5% w/w and 1.0% w/w potato starch is prepared in distilled water. Sodium alginate is dissolved in distilled water at 60 °C with continuous stirring for 15 min, and starch was added, which is stirred until completely dissolved. To homogenize the mixture, an Ultra Turrax IKA T-18 disperser is used at 10,000 rpm for 15 min until it completely dissolved. The resulting solution was allowed to cool to room temperature.

The specific parameters used in the JetCutter setup are (Martin-Díaz et al. 2025):

1. *Device type:* JetCutter Type S 1 (GeniaLab GmbH, Braunschweig, Germany)
2. *Nozzle diameter:* 0.9 mm
3. *Cutting tool:* 16 wires with a diameter of 0.59 mm (0.00059 m)
4. *Pump:* An eccentric screw pump from GeniaLab used for forming microparticles
5. *Fluid velocity:* Targeted for a fluid velocity of 10 m s^{-1}
6. *Pump velocities tested:* 300, 600, 900, and 1200 rpm
7. *Jet inclination:* Adjusted to 2.45° to reduce spray losses and achieve a homogeneous jet
8. *Polymer solution:* A solution of alginate with a medium content of L-guluronic acid at 2.5% w/w and 1.0% w/w potato starch was prepared in distilled water
9. *Cross-linking solution:* The cut segments were placed in a 2.5% w/w aqueous $CaCl_2$ solution at a mass ratio of 1:5 (polymer solution/$CaCl_2$ solution)

Microbial Inoculant Immobilization

The polymer solution is prepared as described above and separated into two samples: one with L-tryptophan (0.2 g L^{-1}) because L-tryptophan is a precursor of auxins in bacteria. The separated solutions are injected into a JetCutter at a bomb velocity of 445 rpm. The velocity of the cutting tool (GeniaLab, Braunschweig, Germany) was set so that the ratio of the fluid velocity to the cutting tool velocity is 1. The resulting microparticles are left in a $CaCl_2$ solution for the previously determined optimal cross-linking time. The formed microparticles are then removed and subjected to agitated washing in 0.85% w/w NaCl solution for 20 min at 140 rpm. After washing, microparticles are spread on the trays and exposed to UV radiation for 30 min to reduce contamination (Fig. 21.2). They were then suspended in a flask containing bacterial culture broth at a ratio of 1:4 (g of microparticles·mL^{-1} of culture broth) with agitation for 30 min at 28 °C and 140 rpm. Finally, the resulting microbial inoculant is removed from the culture broth, placed in trays on Kraft paper, and dried in an oven at 30 °C for 24 h.

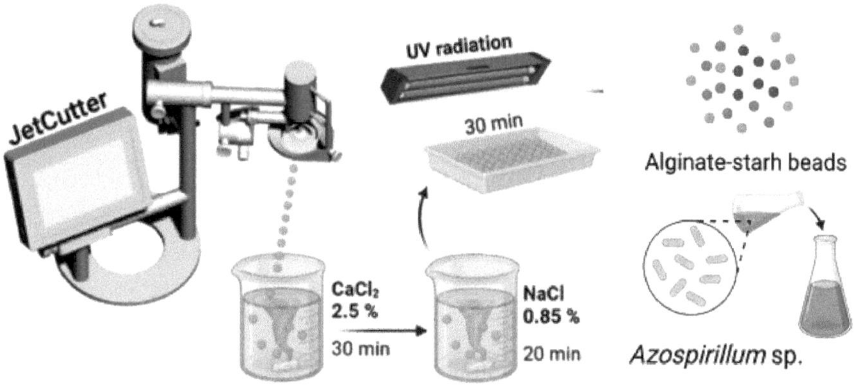

Fig. 21.2 Innovative JetCutter technology to scale the production of a solid *Azospirillum* microbial inoculant for sustainable agriculture

Bead Bacterial Counts

Ten beads are dissolved in 10 mL sterilized sodium citrate solution (pH 8.5; 60 g L^{-1}) for 30 min at 20 °C in a rotary shaker until complete dissolution. The number of released bacteria is determined using the standard plate count method on plate count agar. The results are expressed as total CFU (obtained from 30 mL of cell growth culture or matrix solution).

21.2.4 Inoculation Procedures

The colonization of *Azospirillum* sp. occurs on root surfaces, in the rhizosphere, and in the phyllosphere (Hungria 2016). When developing a formulation, it is essential to select the most efficient strains for plant association, considering the application method and research-based evidence regarding doses and results. A concentration of viable *Azospirillum* spp. cells below the minimum required for efficacy will cause the inoculant to underperform, while uncontrolled dose increases may lead to excessive phytohormone production, potentially harming plant growth. Therefore, it is essential to follow the manufacturer's instructions for bacteria application, including the recommended dose and method. *Azospirillum*-based inoculants in solid peat medium can be applied via seed treatment, while those in liquid medium can be applied not only through seed treatment but also in the planting furrow or via foliar spraying. Producers have shown a preference for liquid inoculants due to their ease of application. A study conducted in 2025 by ANPII Bio (Brazilian Association for the Promotion and Innovation of the Biological Industry), in collaboration with associated companies, estimated the distribution of inoculant application methods in Brazil. Considering Azospirillum revenues in 2024, 51% was applied through on-farm seed treatment, 30% in the planting furrow, 16% through industrial seed treatment, and 3% via foliar spraying. More than just a technology that enhances crop productivity and quality, the use of *Azospirillum*-based inoculants is a sustainable alternative for agriculture. Inoculation of maize with *A. brasilense* reduces nitrogen topdressing fertilization by 25% without compromising productivity (Hungria et al. 2022).

21.2.4.1 Traditional Applications

21.2.4.1.1 Seed Inoculation

Seed treatment is the most widely used inoculant application technique and can be performed manually or using electric machines, rotary drums, or concrete mixers. Direct treatment in the seeder box is not recommended, as it is difficult to achieve uniform seed coverage. For peat-based inoculants, an adhesive such as a 10% sugar solution is recommended. The inoculant should always be the last product applied to the seeds, following chemical pesticides and micronutrients, and no mixing of products should be performed. After the inoculation, seeds should be left to dry and sown within a maximum of 24 h, avoiding exposure to sunlight and temperatures above 32 °C during treatment and sowing. *Azospirillum* spp. is highly sensitive to heat and desiccation. A key factor is the seed's water content, which, if excessive, can negatively affect seed quality. The Brazilian Agricultural Research Corporation (Embrapa) recommends not exceeding 300 mL of liquid per 50 kg of maize seeds. The regular dose for seed treatment of *A. brasilense* in Brazil is 200 g or 200 mL per 50 kg of maize or wheat seeds, but in Argentina the agronomical doses of inoculants formulated with *A. argentinense* for seed treatment have been established between 300 and 500 mL per 50 kg of maize or wheat seeds. However, it's important to note that these doses can vary based on environmental conditions, such as soil moisture and temperature, and chemical seed treatments, which may affect inoculant viability. Adjusting doses accordingly could improve inoculant performance in different crops and climates (Santini et al. 2018).

21.2.4.1.2 On-Furrow Inoculation

A practice that has been growing in the field is the application of inoculants in the planting furrow. Producers report advantages such as the reduction of labor required to perform seed treatment on the farm and the possibility of purchasing seeds that have already been industrially treated with fungicides, insecticides, nematicides, and micronutrients. The greatest benefit is undoubtedly the possibility of reducing *Azospirillum* spp. and other biological contact with agrochemicals. In this sense, Fukami et al. (2016) demonstrated that in-furrow inoculation with *A. brasilense* is a promising alternative to traditional seed inoculation, particularly in mitigating the incompatibility between bacterial inoculants and pesticide-treated seeds. The findings indicated that this method effectively increased the abundance of diazotrophic bacteria in plant tissues, particularly in the roots and rhizosphere, leading to enhanced nitrogen assimilation and overall plant growth. Furthermore, field experiments showed that in-furrow inoculation enabled a 25% reduction in nitrogen fertilizer requirements. The equipment for applying a water jet directly into the furrow is attached to the planter, and the spray tank must have thermal insulation to reduce the exposure of biological agents to sunlight and high temperatures, ensuring greater levels of bacteria survival. The products used in the spray with *Azospirillum* spp. must be demonstrably compatible. The inoculant is sprayed onto the seed as it is sown, and the furrow is then closed. Therefore, it is essential that the soil be moist for the bacteria to be able to multiply. The regular dose for on-furrow application of *A. brasilense* in Brazil is 200 mL per hectare. In Argentina, the agronomical doses of inoculants formulated with *A. argentinense* for on-furrow treatment have been established between 2000 and 4000 mL per hectare of maize or wheat crops.

21.2.4.1.3 Foliar Inoculation

The *Azospirillum* spp. foliar application strategy is relatively recent. Some products registered for this modality, with proven agronomic efficiency studies, already exist in Brazil for maize and pastures such as *Brachiaria*. Research has investigated this form of application of *Azospirillum* as another strategy to avoid contact with chemical products and incompatibility problems, and there is still the need for further research to understand its *modus operandi*. Fukami et al. (2017) observed positive responses with foliar applications of metabolites produced by two strains of *A. brasilense* to maize at growth stage 2.5. The foliar application of bacterial metabolites in general improved not only plant growth but also upregulated plant genes related to defense mechanisms. The Brazilian Agricultural Research Corporation (Embrapa) recommends 200 mL per hectare as a regular dose for foliar treatment of *A. brasilense*; however, in Argentina, the agronomic doses of inoculants formulated with *A. argentinense* for foliar applications vary according to the specific crop. In this context, Cardozo et al. (2022) evaluated the impact of foliar inoculation with *A. argentinense* on maize, in combination with various herbicide treatments, to enhance crop productivity. The inoculum was prepared as a bacterial suspension with a concentration of 1.10^8 CFU·mL^{-1} and applied at a rate of 1 L ha^{-1} during the V6 growth stage. Over the years 2015/2016 and 2016/2017 growing seasons, the results showed that foliar inoculation with *A. argentinense* at the V6 stage, along with herbicide treatments involving atrazine and glyphosate, significantly improved chlorophyll content, cell membrane stability, and key yield components compared to other treatments. The overall average yield across both seasons was 16,992.6 kg ha^{-1}, with the combined foliar inoculation and herbicide treatment achieving the highest values for ear length, grain weight, and number of grains per ear.

21.2.4.1.4 Co-inoculation

The joint inoculation of two or more beneficial microorganisms is called co-inoculation and is widely used for legumes such as soybeans and beans. *Azospirillum* spp., with its action on root growth, act symbiotically and help *Bradyrhizobium* spp. with enhancing nodulation and biological nitrogen fixation (see Chap. 16, section "Interaction of *Azospirillum* with Rhizobia"). Co-inoculation can be done through seed treatment, delivering approximately 120,000 cells of *Azospirillum* spp. per seed (1 dose, 100 g or 100 mL per 50 kg of soybean seeds) or doubled by applying in the planting furrow for two doses per hectare. In soybean, the average productivity gains with the use of co-inoculation are around 16%, indicating an increase of 8% when compared to the inoculation of *Bradyrhizobium* spp. alone (Hungria and Nogueira 2014). In line with these findings, Puente et al. (2019) evaluated the effectiveness of different inoculation techniques in soybean crops at two sites in the Buenos Aires province (Argentina), aiming to investigate the potential benefits of co-inoculating *A. argentinense* either at sowing or via foliar application, in combination with *B. japonicum*. Foliar inoculation was carried out by spraying soybean leaves at the V3 growth stage with a bacterial suspension of *A. argentinense* at a concentration of 3×10^8 CFU·mL^{-1}, applying 6 mL per plant. For seed inoculation, seeds were immersed for 5 min in cultures containing *B. japonicum* at 1×10^6 CFU·mL^{-1} and *A. argentinense* strain at 1×10^7 CFU·mL^{-1}, either individually or in combination, depending on the specific treatment (single or co-inoculation). The results revealed that both inoculation methods involving *Azospirillum* led to higher yields compared to the single inoculation with *B. japonicum*. Furthermore, plants treated with *Azospirillum* exhibited increased 1000-grain weight and higher nitrogen, phosphorus, and protein content in their grains, in contrast to those subjected to either single inoculation or co-inoculation with *B. japonicum* alone. Similarly, during the 2018/2019 growing season, Torres et al. (2022) conducted field trials to assess the impact of various inoculation treatments on soybean, including both single and co-inoculations with *A. argentinense* (5×10^8 CFU·mL^{-1}) and *B. japonicum* (5×10^9 CFU·mL^{-1}), applied either at sowing or 24 h prior to sowing, with an inoculation rate adjusted to achieve a final volume of 12 mL per kg of seeds. The results indicated that both co-inoculation treatments, at T0 and T24, significantly enhanced soybean plant growth. Specifically, the co-inoculation applied 24 h before sowing increased nodule biomass by 18.2%, enhanced shoot biomass by 17.6%, and resulted in a 9.1% increase in yield compared to inoculation with *B. japonicum* alone. For further details, see Chap. 16.

21.2.4.2 Innovative Applications and Technologies

21.2.4.2.1 Nanotechnology-Based Formulations

Agriculture is undergoing continuous development, increasing the demand for innovative technological solutions aimed at reducing the use of synthetic chemicals that negatively impact the environment (Shang et al. 2019). In this context, biofertilizers have emerged as a promising alternative; however, they face challenges such as limited shelf life, vulnerability to extreme temperatures, and competition with the soil's native microbiota (Chakdar and Pabbi 2019). In response to these challenges, nanotechnology, which manipulates materials at scales below 100 nm, is revolutionizing the biofertilizer field by enhancing the synergistic interaction between nanoparticles (NPs) and plant growth-promoting rhizobacteria (PGPR). Nanoformulations offer significant advantages, including increased nutrient absorption rates, enhanced nutrient use efficiency, and a marked reduction in losses, making them a key tool in achieving more sustainable agriculture (Sharma et al. 2023; Usman et al. 2020). Nanoparticles are typically categorized into carbon-based, metallic, organic, and semiconductor types (Buzea and Pacheco 2017). This includes nanoparticles of silver, titanium, zinc oxide, silica, carbon, boron, gold, and zeolite, which are synthesized through a variety of methods (Table 21.5).

Nanobiofertilizers are advanced nanoformulations that either contain essential nutrients (micro- and macronutrients) or function as carriers and additives when combined with minerals and microorganisms (Kah et al. 2018). Additionally, these nanomaterials can be engineered to encapsulate nutrients, thereby optimizing the release and efficacy of fertilizers (DeRosa 2010). The integration of nanoparticles with biofertilizers is referred to as nanobiofertilizers. The use of nanofertilizers provides several benefits, including enhanced nutrient uptake, improved nutrient use efficiency, and a substantial reduction in nutrient losses, making them a pivotal tool for promoting sustainable agriculture (Sharma et al. 2023). Manivannan et al.

Table 21.5 Classification and synthesis of nanomaterials

Nanomaterial	Abbreviation	Synthesis methods	References
Silver	Ag	CRS, LS, SGM, and PS	Sadeghi et al. (2010) and Ghaffari-Moghaddam et al. (2014)
Titanium	Ti	SGM, CVD, TRM, and HSM	Chen and Mao (2007) and Sharma et al. (2014)
Zinc oxide	ZnO	SGM, HSM, PS, and CVD	Noman et al. (2022) and Awasthi (2021)
Silica	SiO_2	SGM, HSM, PS, and AM	Brinker and Scherer (2013) and Rahman and Padavettan (2012)
Carbon	C	CVD, SuM, ADM, and PS	Maduraiveeran and Jin (2021) and Ando and Zhao (2006)
Boron	B	CRS, SGM HSM, and LS	Tatiya et al. (2020) and Demirci et al. (2016)
Gold	Au	CRS, SGM, PS, and LS	Carabineiro (2014) and Verma et al. (2014)
Zeolite	–	HSM, SGM, SiGS, and PS	Cejka et al. (2010) and Larsen (2007)

Abbreviations: *ADM* arc discharge method, *AM* aerosol method, *CRS* chemical reduction synthesis (wet chemistry), *CVD* chemical vapor deposition, *HSM* hydrothermal synthesis method, *LS* laser synthesis, *PS* precipitation synthesis, *SGM* sol-gel method, *SiGS* silica gel synthesis, *SuM* sublimation method, *TRM* thermal reduction method

(2021) explored the effects of zinc oxide nanoparticles (ZnO NPs) synthesized by *Azospirillum* on mung bean (*Vigna radiata*) growth. Their results revealed that the application of *Azospirillum*-synthesized ZnO NPs significantly improved several growth parameters, including seed germination rate, leaf area index, root and shoot development, biomass accumulation, and chlorophyll content. However, concentrations above 30 mg.L^{-1} of ZnO NPs were found to have detrimental effects, negatively impacting plant growth and development. Similarly, Zaheer de-Bashan et al. (2024) demonstrated that the inoculation of wheat seeds with titanium dioxide nanoparticles (nTiO_2) at a concentration of 30 mg·mL^{-1}, combined with *A. brasilense*, resulted in significant improvements in key growth parameters, such as plant height, spike length, number of spikelets per spike, photosynthetic rate, leaf area index, and stomatal conductance. However, they also observed that concentrations exceeding 30 mg·L^{-1} of nTiO_2 adversely affected these growth characteristics. Abbas et al. (2023) demonstrated that the application of ZnO-NPs to wheat seeds effectively alleviated water stress, resulting in enhanced growth parameters, increased chlorophyll content, and elevated antioxidant activity and proline accumulation. In a subsequent study, El-Beltagi et al. (2025) explored the synergistic effects of ZnO-NPs combined with *Azospirillum* sp. on cotton (*Gossypium barbadense* L.) under water deficit conditions, reporting similar positive outcomes. Their findings revealed that this combined treatment significantly mitigated water stress, leading to substantial improvements in plant height, dry weight, leaf area, chlorophyll content, seed index, seed yield, and fiber percentage. Notably, this combined approach outperformed the individual application of ZnO-NPs or *Azospirillum* sp. Under reduced irrigation conditions, antioxidant enzyme activity and proline content increased, while cotton fiber quality remained largely unaffected by the water deficit. Nanobiofertilizers offer a distinct advantage over traditional chemical fertilizers, as they do not contribute to nutrient leaching or loss following application. Furthermore, their minimal required quantities significantly reduce the risk of soil and water contamination, promoting more sustainable agricultural practices (Nayana et al. 2020).

21.2.4.2.2 Precision Agriculture

The Fourth Industrial Revolution is transforming agriculture through the integration of advanced technologies such as artificial intelligence (AI), the internet of things (IoT), robotics, and unmanned vehicles, significantly enhancing farming practices and productivity. At the core of this transformation is a farm management strategy focused on observing, measuring, and responding to spatial and temporal variations in agricultural inputs. By adhering to the "5 R's" (right amount, right rate, right place, right time, and right manner), this approach enables more efficient resource utilization and sustainability (Bhardwaj 2023). Key technologies used in this strategy include global positioning system (GPS) for accurate field mapping and data collection, geographic information systems (GIS) for spatial data analysis to optimize crop management, unmanned aerial vehicles (UAVs) for enhanced monitoring and reduced costs, and variable rate technology (VRT) for tailored input application. Additionally, yield monitoring systems and remote sensing help track crop performance and detect field conditions, allowing for early disease detection. The integration of IoT and AI technologies further enhances soil management, pest control, and market demand forecasting, driving productivity and sustainability in modern agriculture (Choudhury et al. 2022; Bhardwaj 2023; Karunathilake et al. 2023). Despite the well-documented advantages of PA in optimizing farming practices and resource management, there is a notable lack of comprehensive studies exploring the synergistic effects of combining PA techniques with *Azospirillum* inoculation. Often, the effect of biological products on crop yield is subtle, with increases that are not easily detectable. Traditional models struggle to identify differences in productivity between treatments because they overlook key physiological factors and the spatial variability of yield within a plant population in a given plot (Thai et al. 2020). Given the importance of *Azospirillum*-based inoculants, Martins et al. (2023) evaluated the effectiveness of geospatial techniques, including kriging and spectral models, to estimate maize yield enhancement in field conditions following the application of *A. brasilense*. Management practices play a significant role in yield variability, with Kriging models offering more accurate predictions than image-based models, which are limited by smaller sample sizes and spectral resolution. These differences arise from the distinct approaches of each method: kriging utilizes spatial inference to model yield variations, while image-based models rely on spectral responses. The study's findings suggest that spatial analysis can identify yield increases even when statistical differences are not evident in control areas. Moreover, kriging is the preferred method for generating accurate yield increase surfaces in mechanized harvesting systems with extensive georeferenced data. In contrast, high-resolution image-based models can produce reliable yield surfaces under specific conditions, though their accuracy may vary. Brazil is the top global producer of soybean (*Glycine max*), thanks to advancements in genetic improvement. However, high-throughput phenotyping faces challenges due to genetic diversity and environmental variability, which affect accuracy and increase costs. Alves de-Bashan et al. (2024) examined the use of vegetation indices (VIs) for phenotyping soybean varieties inoculated with *Bradyrhizobium* spp. + *Azospirillum* spp. at different growth stages using drone-captured RGB images (digital cameras that capture images using the primary color channels: red, green, and blue). The study found that UAVs provide high-precision, large-scale data that aids in evaluating plant health, growth, and stress responses. VIs derived from RGB images are useful for non-invasive phenotyping, helping identify superior lines and select traits with precision. The optimal times for phenotyping are during early and late growth stages, when VIs strongly correlate with agronomic traits. These findings highlight the importance of choosing the right indices and stages for effective soybean management, demonstrating the value of drone-based RGB imagery for crop improvement.

21.2.4.2.3 Inoculants Shelf Life and Storage

The development of efficient biofertilizer carriers is crucial for maintaining microbial viability and ensuring long-term commercial success. The stability of biofertilizers plays a vital role as microbial inoculants. Several approaches have been investigated to enhance the stability and shelf life of biofertilizers during storage. A common technique involves encapsulating microbial suspensions or pellets in protective coatings, such as fats, proteins, or polysaccharides, to improve their stability. Furthermore, the use of coated and expanded clay particles as carriers in biological concrete has been explored, with the goal of preserving bacterial viability within the concrete structure (Yadav and Yadav 2024). *Azospirillum* sp. bioinoculants are widely recognized for their ability to fix nitrogen and promote plant growth. However, conventional carrier-based bioinoculants often suffer from limited shelf life, contamination, and suboptimal field performance. To address these issues, Vendan and Thangaraju (2007) developed a liquid cyst-based formulation by introducing *Azospirillum* sp. into a minimal salt medium (MSM) that induces cyst formation. Complete conversion of vegetative cells into cysts occurred within 96 h. The cysts remained viable in MSM for over a year, maintaining a stable population of 10^8 throughout the study. These cyst cells accumulated poly-β-hydroxybutyrate (PHB) granules and demonstrated tolerance to desiccation for up to 20 days and to temperatures up to 40 °C. This cyst-based formulation offers the advantages of an extended shelf life and increased resistance to environmental stress. Additionally, cyst regeneration into vegetative cells was successfully observed in several media, including tap water, sterile water, rice gruel, and nutrient broth. Several authors have investigated and addressed the challenges associated with *Azospirillum*-based formulations. In this context, the studies by Kumaresan and Reetha (2011) and Santhosh (2015) provide complementary approaches to the formulation of *Azospirillum*-based inoculants, specifically focusing on the incorporation of cell protectants to enhance the viability of microbial populations during storage. Kumaresan and Reetha (2011) evaluated chemical additives, including polyvinylpyrrolidone (PVP), glycerol, gum Arabic, trehalose, polyethylene glycol (PEG), and polyvinyl alcohol (PVA), to enhance the growth and survival of *A. brasilense* in nitrogen-free malic acid broth. The results demonstrated that formulations containing trehalose (10 mM), glycerol (10 mM), gum arabic (0.3%), and PVP (2%) were the most effective, enhancing the survival of *Azospirillum* for up to 11 months while maintaining populations of 10^8 cells·mL^{-1} at room temperature (28–32 °C). Similarly, Santhosh (2015) focused on the formulation and evaluation of liquid biofertilizers based on *Azospirillum*, *Rhizobium*, *Azotobacter*, and *Bacillus megaterium*, testing the efficacy of cell protectants such as glycerol (0.5%), PVP (0.5%), PEG (0.5%), glutaraldehyde (GA, 0.5%), and sodium alginate (SA, 0.1%). In this study, formulations with 0.5% PVP and 0.5% glycerol were observed to be the most effective, maintaining the highest number of colonies across all strains for over 180 days. Therefore, both studies complement each other by highlighting that liquid bioinoculants, especially those formulated with cell protectants like glycerol and PVP, provide a distinct advantage over solid-carrier-based systems. Recently, Iparraguirre et al. (2023) underscored the potential of algal extracts as a valuable bioresource for developing *Azospirillum*-based inoculants. The study examined the effects of *Macrocystis pyrifera* extracts at varying concentrations (3% and 30%) in formulations, focusing on their impact on the growth, viability, and survival of the *A. argentinense* inoculum strain. The results demonstrated that these algal extracts, which are rich in essential minerals and organic matter with a high carbon-to-nitrogen ratio, significantly enhanced bacterial growth and viability over a 28-day period. Notably, the pure algal extract exhibited a higher concentration of gibberellins. Moreover, the recovery of *A. argentinense* cells was markedly improved in both in vitro and in vivo assays using *Zea mays* seeds.

21.2.4.2.4 Inoculants Compatibility

The compatibility of *Azospirillum*-based inoculants with other microorganisms and chemical products used in agricultural production, such as fungicides and insecticides, is essential to ensure their effectiveness in promoting plant growth. A study on *A. argentinense* and *Pseudomonas* strains in lettuce revealed that strain compatibility significantly affects growth stimulation. While *A. argentinense* and *P. fluorescens* exhibited a synergistic effect, the pairing of *A. argentinense* with *P. protegens* inhibited growth, emphasizing the need for compatibility assessments (for further details, see Chap. 16). With these results the authors confirmed that co-inoculating bacterial species without evaluating their interactions may compromise their plant-stimulatory potential, leading to suboptimal inoculant performance (Díaz et al. 2023). A study on the compatibility of *A. brasilense* with *Bacillus* sp. in *Urochloa brizantha*, a perennial forage species, showed that co-inoculation was more effective than single inoculations. The combined application of both microorganisms significantly improved plant height, tiller number, leaf count, and biomass compared to individual treatments. These results highlight the synergistic potential of *A. brasilense* and *Bacillus* sp. in enhancing plant productivity (Terra et al. 2024). A study evaluating the compatibility of *A. brasilense* with the fungicide difenoconazole and the insecticide thiamethoxam in three wheat cultivars found contrasting effects. While thiamethoxam enhanced shoot and root growth and was compatible with *A. brasilense*, difenoconazole inhibited growth and showed antagonism toward the bacterium. Despite these interactions, seed germination and vigor remained unaffected by the treatments, indicating that wheat seed quality was preserved regardless of the use of fungicides, insecticides, or bacterial inoculation (Munareto et al. 2018). A study on the compatibility of *A. brasilense* with a common maize seed treatment pesticide in Brazil revealed its toxicity to the bacterium, increasing cell mortality within 24 h (Santos et al. 2020).

21.2.4.2.5 Additives and Inoculant Adaptation Strategies to Extreme Conditions

A microbial inoculant consists of one or more plant growth-promoting microorganisms (PGPMs) as the active ingredient, combined with a carrier material and various additives. The carrier must provide a suitable environment for the microorganisms and maintain microbial viability and fitness during storage, as well as on seed and in soil after inoculation (Mishra and Barolia 2020). The additives are protective molecules such as polymers that play a key role in this process by preserving water activity and serving as an additional source of carbon and energy for bacterial survival (Sivasakthivelan et al. 2023). Microbial inoculants can be formulated using either liquid or solid carriers. Liquid formulations are easy to handle and can be applied to seeds or used for foliar spraying (Santos et al. 2017). However, they often provide insufficient protection to bacterial cells, leading to early cell death (Bashan et al. 2014). Additives such as polymers (PVP, polyethylene glycol, sodium alginate, glycerol, Arabic gum, and trehalose), adjuvants (CMC, xanthan gum, and carrageenan), and surfactants are used to overcome this disadvantage (Brahmaprakash et al. 2020). Solid formulations used inorganic/organic or synthetic carriers such as peat, charcoal, clays, talc, wheat bran, vermiculite, perlite, and silicates (Diaz et al. 2023). These carriers are typically mixed with microorganisms and air-dried to maintain an optimal moisture level of 15–20%, which helps reduce microbial metabolic activity and prolong viability. Despite their benefits, solid carriers based on complex organic matter can exhibit high chemical variability, making it challenging to maintain batch-to-batch consistency (Diaz Rodriguez et al. 2025). Several strategies have been developed to optimize the adaptability of *Azospirillum*-based inoculants in extreme environments. One of them is to stimulate the production of polyhydroxybutyrate (PHB) by the bacteria within the inoculum. This reserve polymer allows bacteria to endure environmental stresses by serving as a source of carbon and energy (Tal and Okon 1985; Kadouri et al. 2003). To improve PHB synthesis, several carbon and nitrogen sources have been evaluated, and the highest quantities of PHB were produced when malic acid and ammonia chloride were used as carbon and nitrogen sources, respectively (Tal et al. 1990). The production of PHB in *A. brasilense* has been studied to understand its role in the adaptability of these bacteria. Martinez et al. (2014) analyzed PHB production in wild-type and mutant strains of *A. brasilense* under different NH_4Cl concentrations. The results showed variations in PHB accumulation, suggesting that this biopolymer may be linked to the bacteria's ability to adapt to stress conditions. The *phb*A and *phb*B genes are key components of the PHB biosynthesis pathway in bacteria. Martinez et al. (2023) investigated the effects of *phb*A and *phb*B gene overexpression in *A. brasilense* and found that it did not significantly increase PHB accumulation under normal growth conditions. However, under nutrient-limiting and stress conditions (e.g., osmotic or saline stress), the engineered strains exhibited a slight increase in PHB production compared to the wild-type strain. This increase correlated with improved bacterial viability and survival rates, suggesting that PHB plays a role in stress tolerance. These findings indicate that PHB production in

A. brasilense is regulated by additional mechanisms beyond phbA and phbB overexpression, highlighting the complexity of its biosynthesis and stress adaptation. Another study with the same strain compared the wild-type strain and a PHB-deficient mutant for metabolic versatility, stress resistance, survival in inoculants, and plant growth promotion. Both strains utilized carbon sources similarly, but the wild type had a shorter generation time. It also showed greater stress resistance and better survival in inoculants, though both strains colonized roots equally and had no differences in plant growth promotion. PHB production enhances survival in competitive environments but does not affect root colonization. (Kadouri et al. 2003). Another important survival mechanism is biofilm formation, which facilitates bacterial communication and enhances root colonization. Some studies suggest that bacteria can continuously produce biofilms under nutrient-rich conditions, while nutrient depletion may trigger a return to a motile state, allowing the bacteria to colonize new environments (O'Toole et al. 2000). Biofilm formation is particularly critical for *Azospirillum* spp., as it enhances root colonization. Bacteria migrate towards plant roots through chemotaxis, adhere to the root system, and proliferate in microcolonies stabilized by biofilms (Santi et al. 2013; Santos et al. 2017). Isolating and utilizing strains from extreme environments can enhance inoculant performance. For instance, certain extremophiles can tolerate various harsh conditions, making them potential candidates for inoculant development. (Kumar et al. 2020; Tiwari et al. 2024). Therefore, a broad screening of stress tolerance in *Azospirillum* strains is highly necessary, considering their potential for improving agricultural crops in semiarid regions (Sivasakthivelan et al. 2021).

21.3 Conclusion and Remarks

The use of *Azospirillum*-based inoculants in agriculture has gained increasing recognition due to their ability to enhance crop productivity while reducing reliance on chemical fertilizers. Their widespread adoption across South America, particularly in Brazil and Argentina, has demonstrated substantial yield improvements in key crops such as maize, wheat, and soybeans. Co-inoculation with rhizobia has further amplified these benefits, improving biological nitrogen fixation, stimulating root development, and enhancing plant resilience under stress conditions, particularly in drought-prone environments. Advances in formulation technologies, including liquid inoculants enriched with cell protectants such as trehalose and glycerol, have significantly improved bacterial viability, extending the shelf life and effectiveness of inoculants under field conditions. Additionally, alternative application methods, such as foliar and in-furrow inoculation, are being explored to overcome compatibility challenges with seed-applied pesticides and ensure more efficient microbial colonization. Emerging technologies, such as nanotechnology-based formulations, hold great potential for further improving the efficacy of *Azospirillum*-based products. The integration of nanoparticles, such as ZnO and TiO_2, with beneficial microbes has shown promise in enhancing plant growth, nutrient uptake, and stress tolerance. Similarly, precision agriculture techniques, including remote sensing and geospatial analysis, are being investigated to optimize inoculant application and assess their impact on crop yields with greater accuracy. Despite these advancements, challenges persist in ensuring consistent field performance across different environmental and soil conditions. Further research is needed to refine formulation techniques, improve bacterial stability during storage, and assess the compatibility of *Azospirillum* inoculants with commonly used agrochemicals. Looking ahead, the role of *Azospirillum* spp. in sustainable agriculture will continue to expand as efforts intensify to reduce the environmental footprint of conventional farming. Future studies should focus on optimizing the interactions between *Azospirillum* and other plant growth-promoting microbes to develop multi-strain inoculants with synergistic effects. Additionally, the application of artificial intelligence and big data analytics in agriculture could provide new insights into optimizing microbial inoculation strategies for site-specific conditions. As global agricultural systems shift toward regenerative practices, *Azospirillum*-based inoculants represent a promising biotechnological solution to improve crop productivity, promote soil health, and contribute to a more sustainable food production system.

In summary, there is no doubt that the use of *Azospirillum*-based inoculants will continue to grow in global agriculture. The changes introduced in the culture media, with nutrients more adapted to the strains used and the development of additives that extend the shelf life and the resistance to stress, contribute to better product performance, which encourages farmers to increasingly use this biological input, which performs several functions, in their crops.

References

Abbas SF, Bukhari MA, Raza MAS et al (2023) Seed nanopriming with zinc oxide improves wheat growth and photosynthetic performance in wheat under drought stress. Res Square. https://doi.org/10.21203/rs.3.rs-2353450/v1

Alves AK, Araújo MS, Chaves SF et al (2024) High throughput phenotyping in soybean breeding using RGB image vegetation indices based on drone. Sci Rep 14(1):32055

Ando Y, Zhao X (2006) Synthesis of carbon nanotubes by arc-discharge method. New Diamond Front Carbon Technol 16(3):123–138

Awasthi K (ed) (2021) Nanostructured zinc oxide: synthesis, properties and applications. Elsevier

Barbosa DD, Fernandes PD, Marcelino ADAL, Silva FDA, Dias MDS, Silva CRC & Santos RC (2021) Exogenous pyruvate mitigates the detrimental effects of water stress in contrasting peanut genotypes. Genet Mol Res 20(3):1–14

Bashan Y, de-Bashan LE (2015) Inoculant preparation and formulations for Azospirillum spp. In: Cassán FD, Okon Y, Creus CM (eds) Handbook for Azospirillum. Springer, New York, pp 469–486

Bashan Y, de-Bashan LE & Prabhu SR (2016) Superior polymeric formulations and emerging innovative products of bacterial inoculants for sustainable agriculture and the environment. In Agriculturally important microorganisms: Commercialization and regulatory requirements in Asia 15–46

Bashan Y, Hernandez JP, Leyva LA, Bacilio M (2002) Alginate microbeads as inoculant carrier for plant growth promoting bacteria. Biol Fertil Soils 35:359–368

Bashan Y, Holguin G, Lifshitz R (1993) Isolation and characterization of plant growth-promoting rhizobacteria. In: Glick BR, Thompson JE (eds) Methods in plant molecular biology and biotechnology. CRC Press, Boca Raton, pp 331–345

Bashan Y, de-Bashan LE, Prabhu SR et al (2014) Advances in plant growth-promoting bacterial inoculant technology: formulations and practical perspectives (1998–2013). Plant Soil 378:1–33

Benítez SV, Carrasco R, Giraldo JD et al (2024) Microbeads as carriers for Bacillus pumilus: a biofertilizer focus on auxin production. J Microencapsul 41:170–189. https://doi.org/10.1080/02652048.2024.2324812

Bhardwaj K (2023) Precision agriculture and data analytics. Modern Horizons Agric 89

Brahmaprakash GP, Sahu PK, Lavanya G (2020) Role of additives in improving efficiency of bioformulation for plant growth and development. In: Frontiers in soil and environmental microbiology. CRC Press, New York, pp 1–10

Brinker CJ, Scherer GW (2013) Sol-gel science: the physics and chemistry of sol-gel processing. Academic, London

Buzea C & Pacheco I (2017) Nanomaterial and nanoparticle: origin and activity. In Nanosci Plant Soil Syst 71–112

Carabineiro SAC (2014) Synthesis and applications of gold nanoparticles. Adv Nanotechnol 95

Cardozo P, Di Palma A, Martin S et al (2022) Improvement of maize yield by foliar application of Azospirillum brasilense Az39. J Plant Growth Regul:1–9

Cassán F, Coniglio A, López G et al (2020) Everything you must know about Azospirillum and its impact on agriculture and beyond. Biol Fertil Soils 56(7):897–915

Cerezini P, Kuwano BH, dos Santos MB, Terassi F, Hungria M & Nogueira MA (2016) Strategies to promote early nodulation in soybean under drought. Field Crop Res 196:160–167

Cesari AB, Paulucci NS, Yslas EI & Dardanelli MS (2020) Immobilization of Bradyrhizobium and Azospirillum in alginate matrix for long time of storage maintains cell viability and interaction with peanut. Appl Microbiol Biotechnol 104(23):10145–10164

Cejka J, Corma A, Zones S (eds) (2010) Zeolites and catalysis: synthesis, reactions and applications. Wiley, New York

Chen X, Mao SS (2007) Titanium dioxide nanomaterials: synthesis, properties, modifications, and applications. Chem Rev 107(7):2891–2959

Chakdar H & Pabbi S (2019) Microbial Bioinoculants for Sustainable Agriculture: Trends, Constraints, and Future Perspectives. In Microbes Sustain Dev Bioremediation 343–358

Chibeba AM, Kyei-Boahen S, de Fátima Guimarães M, Nogueira MA & Hungria M (2020) Towards sustainable yield improvement: field inoculation of soybean with Bradyrhizobium and co-inoculation with Azospirillum in Mozambique. Arch Microbiol 202(9):2579–2590

Choudhury A, Biswas A, Singh TP et al (eds) (2022) Smart agriculture automation using advanced technologies: data analytics and machine learning, cloud architecture, automation and IoT. Springer Nature

Cortés-Patiño S & Bonilla RR (2015) Polymers selection for a liquid inoculant of Azospirillum brasilense based on the Arrhenius thermodynamic model. Afr J Biotechnol 14(33): 2547–2553

Dayamani KJ (2010) Formulation and determination of effectiveness of liquid inoculants of plant growth promoting rhizobacteria. PhD thesis. University of Agricultural Sciences, Bangalore

de-Bashan L, Giraldo JD, Cruz-Barrera et al (2024) Enhancing the survival rate and effectiveness of plant growth-promoting bacteria through bio-encapsulation techniques. Biol Fertil Soils. https://doi.org/10.1007/s00374-024-01870-5

Demirci UB, Miele P, Yot PG (2016) Boron-based (nano-) materials: fundamentals and applications. Crystals 6(9):118

DeRosa MC (2010) Nanotechnology in fertilizers. Nat Nanotechnol 5:91

Díaz PR, Merlo F, Carrozzi L et al (2023) Lettuce growth improvement by Azospirillum argentinense and fluorescent Pseudomonas co-inoculation depends on strain compatibility. Appl Soil Ecol 189:104969

Díaz-Rodríguez AM, Parra Cota FI, Cira Chávez LA et al (2025) Microbial inoculants in sustainable agriculture: advancements, challenges, and future directions. Plants 14(2):191

Díaz-Zorita M & Fernández-Canigia MV (2009) Field performance of a liquid formulation of Azospirillum brasilense on dryland wheat productivity. Eur J Soil Biol 45(1):3–11

Díaz-Zorita M (2012) Soil productivity from the sandy pampas region under the present agricultural conditions. Agrociencia 16(3):84–88

Díaz-Zorita M, Canigia MVF, Bravo OÁ, Berger A & Satorre EH (2015) Field evaluation of extensive crops inoculated with Azospirillum sp. In Handbook for Azospirillum: technical issues and protocols 435–445

Döbereiner J, Day JM (1976) Associative symbiosis in tropical grasses: characterization of microorganisms and dinitrogen fixing sites. In: Newton WE, Nyman CJ (eds) Proceedings of the 1st international symposium on nitrogen fixation, vol 2. Washington State University Press, Pullman, pp 518–538

Döbereiner J, Baldani VLD, Baldani JI (1996) Como isolar e identificar bactérias diazotróficas de plantas não-leguminosas. Embrapa, Brasília

El-Beltagi HS, El-Waraky EA, Almutairi HH et al (2025) Morphophysiological and biochemical responses of cotton (Gossypium barbadense L.) to nano zinc (ZnO-NPs) and *Azospirillum* sp. under water deficit stress conditions. J Plant Nutr 48(2):326–344

Fukami J, Nogueira MA, Araujo RS et al (2016) Accessing inoculation methods of maize and wheat with *Azospirillum brasilense*. AMB Express 6:1–13

Fukami J, Ollero FJ, Megías M et al (2017) Phytohormones and induction of plant-stress tolerance and defense genes by seed and foliar inoculation with *Azospirillum brasilense* cells and metabolites promote maize growth. AMB Express 7:153. https://doi.org/10.1186/s13568-017-0453-7

Ghaffari-Moghaddam M, Hadi-Dabanlou R, Khajeh M et al (2014) Green synthesis of silver nanoparticles using plant extracts. Korean J Chem Eng 31:548–557

Giraldo JD, Garrido-Miranda KA, Schoebitz M (2023) Chitin and its derivatives: functional biopolymers for developing bioproducts for sustainable agriculture—a reality? Carbohydr Polym 299:120196. https://doi.org/10.1016/j.carbpol.2022.120196

Hungria M (2016) *Azospirillum*: Um velho novo aliado. Fertbio 2016: Rumo aos novos desafios

Hungria M, Nogueira M (2014) https://www.infoteca.cnptia.embrapa.br

Hungria M, Nogueira MA & Araujo RS (2013) Co-inoculation of soybeans and common beans with rhizobia and azospirilla: strategies to improve sustainability. Biol Fertil Soils 49(7): 791–801

Hungria M, Barbosa JZ, Rondina ABL et al (2022) Improving maize sustainability with partial replacement of N fertilizers by inoculation with *Azospirillum brasilense*. Agron J 114:2969–2980. https://doi.org/10.1002/agj2.21150

Iparraguirre J, Llanes A, Masciarelli O et al (2023) Formulation technology: Macrocystis pyrifera extract is a suitable support/medium for *Azospirillum brasilense*. Algal Res 69:102916

Kadouri D, Jurkevitch E, Okon Y (2003) Involvement of the reserve material poly-β-hydroxybutyrate in *Azospirillum brasilense* stress endurance and root colonization. Appl Environ Microbiol 69(6):3244–3250

Kah M, Kookana RS, Gogos A et al (2018) A critical evaluation of nanopesticides and nanofertilizers against their conventional analogues. Nat Nanotechnol 13:677–684

Karunathilake EMBM, Le AT, Heo S et al (2023) The path to smart farming: Innovations and opportunities in precision agriculture. Agriculture 13(8):1593

Kazi N, Deaker R, Wilson N, Muhammad K & Trethowan R (2016) The response of wheat genotypes to inoculation with Azospirillum brasilense in the field. Field Crop Res 196:368–378

Kumar V, Kumar S, Singh D (2020) Microbial polyhydroxyalkanoates from extreme niches: bioprospection status, opportunities and challenges. Int J Biol Macromol 147:1255–1267

Kumaresan G, Reetha D (2011) Survival of *Azospirillum brasilense* in liquid formulation amended with different chemical additives. J Phytol 3(10)

Larsen SC (2007) Nanocrystalline zeolites and zeolite structures: synthesis, characterization, and applications. J Phys Chem C 111(50):18464–18474

Maduraiveeran G, Jin W (2021) Carbon nanomaterials: synthesis, properties and applications in electrochemical sensors and energy conversion systems. Mater Sci Eng B 272:115341

Manivannan N, Aswathy S, Malaikozhundan B et al (2021) Nano-zinc oxide synthesized using diazotrophic *Azospirillum* improves the growth of mung bean, Vigna radiata. Int Nano Lett 11: 405–415

Martin-Díaz F, Baeza-Aranzaez J, Giraldo JD et al (2025) Innovative JetCutter technology to scale the production of a solid *Bacillus pumilus* biofertilizer to transit to sustainable agriculture. ACS Agric Sci Technol. https://doi.org/10.1021/acsagscitech.4c00605

Martínez Martínez MDLA, Martínez Cámara R, Soto Urzua L et al (2014) Stress resistance in *Azospirillum* strains overexpressing genes involved in poly-ß-hydroxybutyrate biosynthesis

Martínez MDLÁM, Urzúa LS, Carrillo YA et al (2023) Polyhydroxybutyrate metabolism in Azospirillum brasilense and its applications, a review. Polymers 15(14):3027

Martins GD, Xavier LCM, de Oliveira GP et al (2023) Using geospatial information to map yield gain from the use of *Azospirillum brasilense* in furrow. Agronomy 13(3):808

Mishra BK, Barolia SK (2020) Quality assessment of microbial inoculants as biofertilizer. Int J Curr Microbiol App Sci 9(10):3715–3729

Moreno-Galvan A, Rojas-Tapias DF, Bonilla R (2012) Development and evaluation of an alternative culture medium for mass cultivation of *Azospirillum brasilense* C16 using sequential statistical designs. Corpoica Ciencia y Tecnología Agropecuaria 13(2):201–220

Munareto JD, Martin TN, Müller TM et al (2018) Compatibility of Azospirillum brasilense with fungicide and insecticide and its effects on the physiological quality of wheat seeds. Semina: Ciências Agrárias 39(2):855–864

Nayana AR, Joseph BJ, Jose A et al (2020) Nanotechnological advances with PGPR applications. In: Hayat S, Pichtel J, Faizan M, Fariduddin Q (eds) Sustainable agriculture reviews 41. Springer, Cham. https://doi.org/10.1007/978-3-030-33996-8_9

Noman MT, Amor N, Petru M (2022) Synthesis and applications of ZnO nanostructures (ZONSs): a review. Crit Rev Solid State Mater Sci 47(2):99–141

O'Toole G, Kaplan HB, Kolter R (2000) Biofilm formation as microbial development. Ann Rev Microbiol 54(1):49–79

Puente ML, Zawoznik M, de Sabando ML et al (2019) Improvement of soybean grain nutritional quality under foliar inoculation with Azospirillum brasilense strain Az39. Symbiosis 77:41–47

Puente ML, Gualpa JL, Lopez GA, Molina RM, Carletti SM & Cassán FD (2018) The benefits of foliar inoculation with Azospirillum brasilense in soybean are explained by an auxin signaling model. Symbiosis 76(1):41–49

Puente ML, Zawoznik M, de Sabando ML, Perez G, Gualpa JL, Carletti SM & Cassán FD (2019) Improvement of soybean grain nutritional quality under foliar inoculation with Azospirillum brasilense strain Az39. Symbiosis 77(1):41–47

Prando AM, Oliveira AB, De Hungria M, Oliveira FT, De Harger N (2016) Transferência de tecnologia sobre inoculação em soja em parceria entre Embrapa e Emater. In: REUNIÃO LATINOAMERICANA DE RIZOBIOLOGIA - RELAR, 27., 2016, Londrina. Fortalecendo as parcerias Sul-Sul: anais. Curitiba: SBCS-NEPAR, 2016:308

Prando AM, Oliveira AB, De Lima D, Conte O, Harger N, Teixeira FT, Nogueira MA, Hungria M (2018) Ações de transferência de tecnologia sobre inoculação em soja, em parceria entre EMATER Paraná e Embrapa. In: CONGRESSO BRASILEIRO DE SOJA, 8., 2018, Goiânia. Inovação, tecnologias digitais e sustentabilidade da soja: anais. Brasília, DF: Embrapa, 2018: 77–79

Rahman IA, Padavettan V (2012) Synthesis of silica nanoparticles by sol-gel: size-dependent properties, surface modification, and applications in silica-polymer nanocomposites—a review. J Nanomater 2012(1):132424

Sadeghi B, Jamali M, Kia S et al (2010) Synthesis and characterization of silver nanoparticles for antibacterial activity

Shang Y, Hasan MK, Ahammed GJ, Li M, Yin H & Zhou J (2019) Applications of nanotechnology in plant growth and crop protection: a review. Molecules 24(14):2558

Sahu PK, Brahmaprakash GP (2016) Formulations of biofertilizers—approaches and advances. In: Singh DP, Singh HB, Prabha R (eds) Microbial inoculants in sustainable agricultural productivity, vol 2. Springer, New Delhi, pp 179–198
Santhosh GP (2015) Formulation and shelf life of liquid biofertilizer inoculants using cell protectants. Int J Res Biosci Agric Technol 7(2):243–247
Santi C, Bogusz D, Franche C (2013) Biological nitrogen fixation in non-legume plants. Ann Bot 111:743–767
Santini JM, Buzetti S, Teixeira Filho M et al (2018) Doses and forms of *Azospirillum brasilense* inoculation on maize crop. Revista Brasileira de Engenharia Agrícola e Ambiental 22:373–377
Santos MS, Hungria M, Nogueira MA (2017) Production of polyhydroxybutyrate (PHB) and biofilm by Azospirillum brasilense aiming at the development of liquid inoculants with high performance. Afr J Biotechnol 16(37):1855–1862
Santos MS, Rondina AB, Nogueira MA et al (2020) Compatibility of *Azospirillum brasilense* with pesticides used for treatment of maize seeds. Int J Microbiol 2020(1):8833879
Schoebitz M, López Belchí MD (2016) Encapsulation techniques for plant growth-promoting rhizobacteria. In: Bioformulations: for sustainable agriculture. Springer India, New Delhi, pp 251–265. https://doi.org/10.1007/978-81-322-2779-3_14
Schoebitz M, Simonin H, Poncelet D (2012) Starch filler and osmoprotectants improve the survival of rhizobacteria in dried alginate beads. J Microencapsul 29:532–538. https://doi.org/10.3109/02652048.2012.665090
Schoebitz M, López MD, Roldán A (2013) Bio-encapsulation of microbial inoculants for better soil-plant fertilization. A review. Agron Sustain Dev 33. https://doi.org/10.1007/s13593-013-0142-0
Sharma A, Karn RK, Pandiyan SK (2014) Synthesis of TiO_2 nanoparticles by sol-gel method and their characterization. J Basic Appl Eng Res 1(9):1–5
Sharma B, Tiwari S, Kumawat KC (2023) Nano-biofertilizers as bio-emerging strategies for sustainable agriculture development: potentiality and their limitations. Sci Total Environ 860: 160476
Sivasakthivelan P, Saranraj P (2013) *Azospirillum* and its formulations: a review. Int J Microbiol Res 4(3):275–287
Sivasakthivelan P, Saranraj P, Al-Tawaha ARM (2021) Adaptation of Azospirillum to stress conditions: a review. Adv Environ Biol 15(4):1–6
Sivasakthivelan P, Saranraj P, Sayyed RZ et al (2023) Inoculant production and formulation of Azospirillum species. In: Plant growth promoting microorganisms of arid region. Springer Nature, Singapore, pp 423–455
Tal S, Okon Y (1985) Production of the reserve material poly-β-hydroxybutyrate and its function in *Azospirillum brasilense* Cd. Can J Microbiol 31(7):608–613
Tal S, Smirnoff P, Okon Y (1990) The regulation of poly-β-hydroxybutyrate metabolism in *Azospirillum brasilense* during balanced growth and starvation. Microbiology 136:1191–1196
Tatiya S, Pandey M, Bhattacharya S (2020) Nanoparticles containing boron and its compounds—Synthesis and applications: a review. J Micromanufact 3(2):159–173
Terra LEDM, Santos MMD, Lopes MCS et al (2024) Co-Inoculation of *Azospirillum brasilense* and *Bacillus* sp. enhances biomass and photosynthetic efficiency in Urochloa brizantha. Agriculture 14(12)
Thai TH, Omari RA, Barkusky D et al (2020) Statistical analysis versus the M5P machine learning algorithm to analyze the yield of winter wheat in a long-term fertilizer experiment. Agronomy 10:1779
Tittabutr P, Payakapong W, Teaumroong N et al (2007) Growth, survival and field performance of bradyrhizobial liquid inoculant formulations with polymeric additives. ScienceAsia 33:69–77
Tiwari P, Bose SK, Park KI et al (2024) Plant-microbe interactions under the extreme habitats and their potential applications. Microorganisms 12(3):448
Torres D, Donadio F, López G et al (2022) Previous incubation of Bradyrhizobium japonicum E109 and Azospirillum argentinense Az39 (formerly A. brasilense Az39) improves the Bradyrhizobium-soybean symbiosis. J Soil Sci Plant Nutr 22(4):4669–4682

Usman M, Farooq M, Wakeel A et al (2020) Nanotechnology in agriculture: current status, challenges and future opportunities. Sci Total Environ 721:137778

Vendan RT, Thangaraju M (2007) Development and standardization of cyst based liquid formulation of *Azospirillum* bioinoculant. Acta Microbiol Immunol Hung 54(2):167–177

Verma HN, Singh P, Chavan RM (2014) Gold nanoparticle: synthesis and characterization. Vet World 7(2):72

Vithal N (2004) Development of liquid inoculant formulations for *Bradyrhizobium* sp. (Arachis), *Azospirillum lipoferum* and *Azotobacter chroococcum*. PhD thesis. University of Agricultural Sciences, Bangalore

Yadav A, Yadav K (2024) Exploring the effect of engineered nanomaterials on soil microbial diversity and functions: a review. J Environ Nanotechnol 13(1):48–62

Zayed MS (2016) Advances in formulation development technologies. In: Singh DP, Singh HB, Prabha R (eds) Microbial inoculants in sustainable agricultural productivity, vol 2. Springer, pp 219–237

Chapter 22
Innovative and Emerging Approaches in Quality Control of *Azospirillum* Inoculants

Mariana Puente, Julia Elena García, Lina Carmenza Valderrama Aguirre, Luciana Paula Di Salvo, Anahí Coniglio, María Eugenia Gallace, Dalila Giacobbe Boggio, Sofía Nievas, Santiago Otaiza-González, Enique Rodríguez Cáceres, Gisela Santella, Bertha Soriano-Bernilla, Gary Silva-Quipan, and Fabricio Dario Cassán

Abstract This chapter reviews recent advances and ongoing challenges in the quality control of *Azospirillum*-based inoculants, which are increasingly used in agriculture for their plant growth-promoting properties and contribution to sustainable production systems. Emphasis is placed on the quality control, outlining how both internal and independent laboratory processes are essential to ensure viability, purity, and efficacy of inoculants throughout manufacturing, storage, and application. The chapter highlights the development and implementation of standardized protocols in countries such as Argentina, Uruguay, Brazil, and Colombia, including the adoption of innovative methodologies like the drop plate technique and molecular assays for strain identification. Advanced microscopy-based techniques (optical,

M. Puente (✉) · J. E. García
Instituto Nacional de Tecnología Agropecuaria (INTA), Instituto de Microbiología y Zoología Agrícola, Buenos Aires, Argentina
e-mail: puente.mariana@inta.gob.ar

L. C. V. Aguirre
Universidad del Valle, Escuela de Ingeniería, Colombia

L. P. Di Salvo
Facultad de Agronomía, Departamento de Biología Aplicada y Alimentos, Cátedra de Microbiología Agrícola, Buenos Aires, Argentina, Universidad de Buenos Aires, Buenos Aires, Argentina

CONICET, Buenos Aires, Argentina

A. Coniglio
School of Biosciences, University of Nottingham, Sutton Bonington Campus, Nottingham, UK

M. E. Gallace
Facultad de Agronomía, Área de Recursos Naturales, Cátedra de Microbiología Agrícola, Universidad Nacional de La Pampa, La Pampa, Argentina

D. G. Boggio · G. Santella
Novozymes BioAg S.A., Pilar, Buenos Aires, Argentina

F. D. Cassán et al. (eds.), *The Definitive Handbook of Azospirillum*,
https://doi.org/10.1007/978-3-032-09222-9_22

SEM, and TEM) are discussed as powerful tools for the detailed characterization of *Azospirillum* morphology, viability, and contamination, enabling more accurate and reproducible quality assessments. The chapter also addresses agronomic efficiency evaluation, detailing both laboratory and field-based approaches and the key parameters used to measure inoculant performance, such as plant growth indices and nitrogen use efficiency. The impact of regulatory frameworks, country-specific requirements, and emerging trends in the development and deployment of *Azospirillum*-based inoculants are critically examined, with a focus on their implications for sustainability, productivity, and environmental stewardship in agriculture. This comprehensive overview underscores the need for ongoing research, harmonized standards, and interdisciplinary collaboration to advance the effective and safe use of microbial inoculants in global food systems.

Keywords Quality control · Cell viability · Standardization · Regulation · Purity

22.1 Introduction

22.1.1 Overview of the Importance of Quality Control of Azospirillum *Inoculants*

Inoculation with *Azospirillum* offers multiple benefits, and its adoption has grown significantly, positioning *Azospirillum* inoculants as one of the most widely used in agriculture. However, for these benefits to materialize, it is crucial to have a rigorous quality control system that ensures the inoculants maintain bacterial viability and activity during storage and application. In this context, it is important to distinguish between quality assurance and quality control. Quality assurance is carried out internally throughout the production process of the inoculants, ensuring that all quality criteria are met before advancing to the next stage. If these criteria are not met, production is canceled. On the other hand, quality control is performed by independent laboratories to verify whether the products intended for sale meet the quality standards (Stephens and Rask 2000).

S. Nievas · F. D. Cassán
Laboratorio de Fisiología Vegetal e Interacción Planta-Microorganismo, Instituto de Investigaciones Agrobiotecnológicas (INIAB-CONICET), Universidad Nacional de Río Cuarto, Río Cuarto, Córdoba, Argentina

S. Otaiza-González
Unidad de Estudios Agropecuarios, Instituto Nacional de Tecnología Agropecuaria (UDEA-INTA-CONICET), Córdoba, Argentina

E. R. Cáceres
Asesor Privado, Buenos Aires, Argentina

B. Soriano-Bernilla · G. Silva-Quipan
Laboratorio de Microbiología Ambiental, Departamento de Microbiología y Parasitología, Facultad de Ciencias Biológicas-Universidad Nacional de Trujillo, Trujillo, Perú

Several factors contribute to the quality of an inoculant, including the number of colony-forming units (CFU) at the time of manufacturing and expiration, the absence of contamination, an effective formulation, and an adequate shelf life, among others (Lupwayi et al. 2000). The literature shows that several authors have described methodologies to ensure the quality of inoculants; however, few countries have developed specific manuals focused on this topic.

In Argentina, REDCAI (Red de Control de Calidad de Inoculantes), a working group within the División de Microbiología Agropecuaria y Ambiental (DIMAyA) and part of the Asociación Argentina de Microbiología (AAM), has played a leading role in establishing standardized methodologies for the quality control of microbial inoculants.

REDCAI recently published the "Manual de Procedimientos Microbiológicos para la Evaluación de la Calidad de Inoculantes, Bioestimulantes y Biocontroladores" (Benintende et al. 2025), which includes a detailed section on quality control procedures for *Azospirillum* spp. inoculants. This section describes validated protocols for the quantification of *Azospirillum* spp. and the detection of contaminating microorganisms, both considered essential parameters for assessing inoculant quality and regulatory compliance. One of the key contributions of the manual is the standardization and validation of the drop plate method, proposed as a reliable and efficient alternative to the conventional spread plate technique for enumerating *Azospirillum* spp. (Fig. 22.1). This method improves reproducibility and facilitates routine microbiological procedures in quality control settings. The manual also includes a protocol for determining the viability of *Azospirillum* spp. on maize and wheat seeds (Fig. 22.2), a molecular assay for the specific identification of *A. argentinense* strain Az39, and an alternative method for the taxonomic identification of strains within the genus *Azospirillum* (Fig. 22.3) (Puente et al. 2025). For details see the Methodological Annex.

In Uruguay, the Instituto Nacional de Investigación Agropecuaria (INIA) has taken an active approach to inoculant quality control. In collaboration with the Ministerio de Ganadería, Agricultura y Pesca (MGAP), INIA implemented a control system that follows the guidelines established by REDCAI in Argentina (https://www.mgap.gub.uy). This system focuses on verifying the quality of the inoculants using protocols similar to those described in REDCAI, thereby standardizing and improving the effectiveness of the biological products used in agriculture. Furthermore, INIA conducts ongoing research to optimize control methodologies, ensuring that inoculants maintain their properties and benefits under the specific conditions of the country.

In Brazil, a "Manual de análises de bioinsumos para uso agrícola: Inoculantes" has been developed, outlining the necessary methodologies to ensure proper quality control of these biological inputs (Ferreira et al. 2024). In this context, Colombia has also made significant advancements in inoculant quality control. Research has been conducted to assess the quality of inoculants stored at different temperatures, using peat as a support. These studies are crucial for determining optimal storage conditions and ensuring the viability of bacteria in the inoculants. Additionally, factorial models for biofertilizer quality control have been proposed in the region, considering

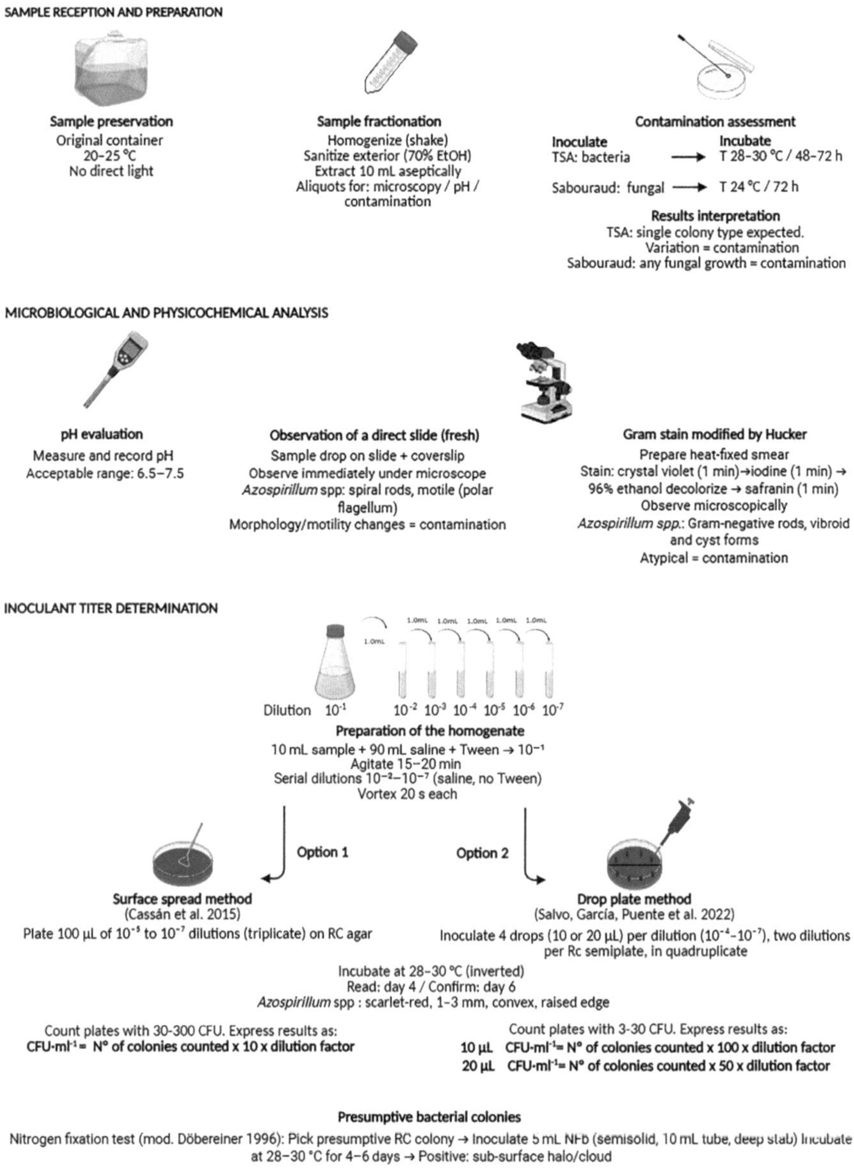

Fig. 22.1 Protocol for the evaluation of inoculants containing *Azospirillum* spp.

variables such as temperature and formulation type, to optimize both production and storage of these products (Aguilar 2020).

The Colombian Corporation of Agricultural Research (Agrosavia) has conducted research on native strains of *Azospirillum* in Guinea grass, demonstrating their

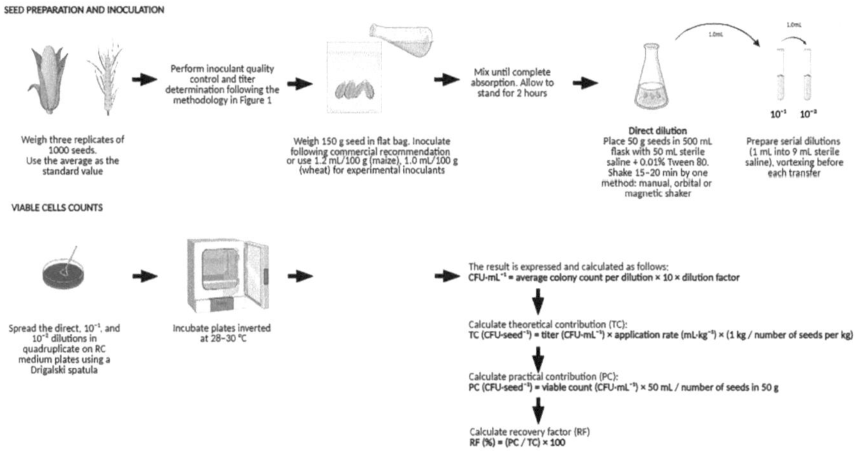

Fig. 22.2 Quantification of viable *Azospirillum* on maize and wheat seed

potential to reduce nitrogen fertilizer use and improve forage production (Cárdenas et al. 2010). Additionally, Colombian universities have studied the use of *Azospirillum brasilense* as a growth promoter in legume crops, contributing to agricultural sustainability (MADR 2023). The Ministry of Agriculture and Rural Development (MADR) has also promoted policies that encourage the efficient use of bio-inputs through agroecological programs (MADR 2024). However, the development of inoculants in the country still faces challenges, such as the need for greater investment in research and the uncertainty regarding the behavior of biological agents used (ANLA 2022). These initiatives underscore the importance of strengthening research and collaboration in the sector to promote more sustainable agriculture in Colombia.

Although the parameters used to determine the quality of inoculants are generally similar, each country must consider its own regulatory and normative requirements. These specificities can result in variations in quality standards and evaluation methods.

Therefore, it is essential that quality control protocols be adapted to local conditions and regulations, thereby ensuring the effectiveness and safety of inoculants in each specific context (Fig. 22.4).

This chapter addresses innovative and emerging approaches in the quality control of *Azospirillum*-based inoculants, focusing on advanced techniques and methodologies that enable a more precise and efficient evaluation of these products. The discussion emphasizes the application of cutting-edge microscopy-based techniques, which provide a deeper understanding of inoculant quality, effectiveness, and functionality. In particular, this technology contributes to more accurate and comprehensive assessments of *Azospirillum*-based inoculants, highlighting the importance of scientific advancements in improving quality control processes.

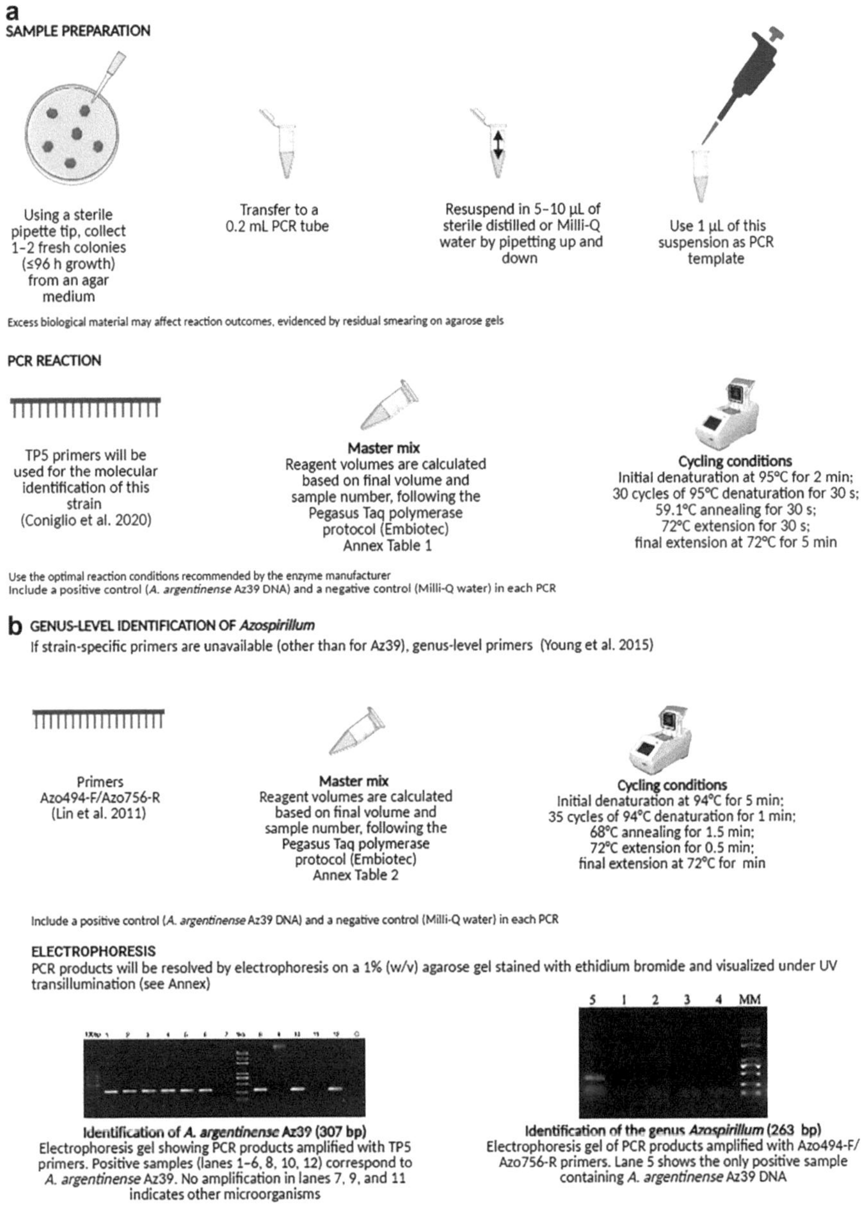

Fig. 22.3 Molecular identification and typing of *Azospirillum* spp.

Furthermore, the chapter delves into the agronomic evaluation of *Azospirillum*-based inoculants, exploring both the methods used to assess their agronomic efficiency and the factors that influence their performance under field conditions. A critical aspect discussed is the impact of country-specific regulations on the approval, distribution, and use of *Azospirillum*-based inoculants, shedding light on

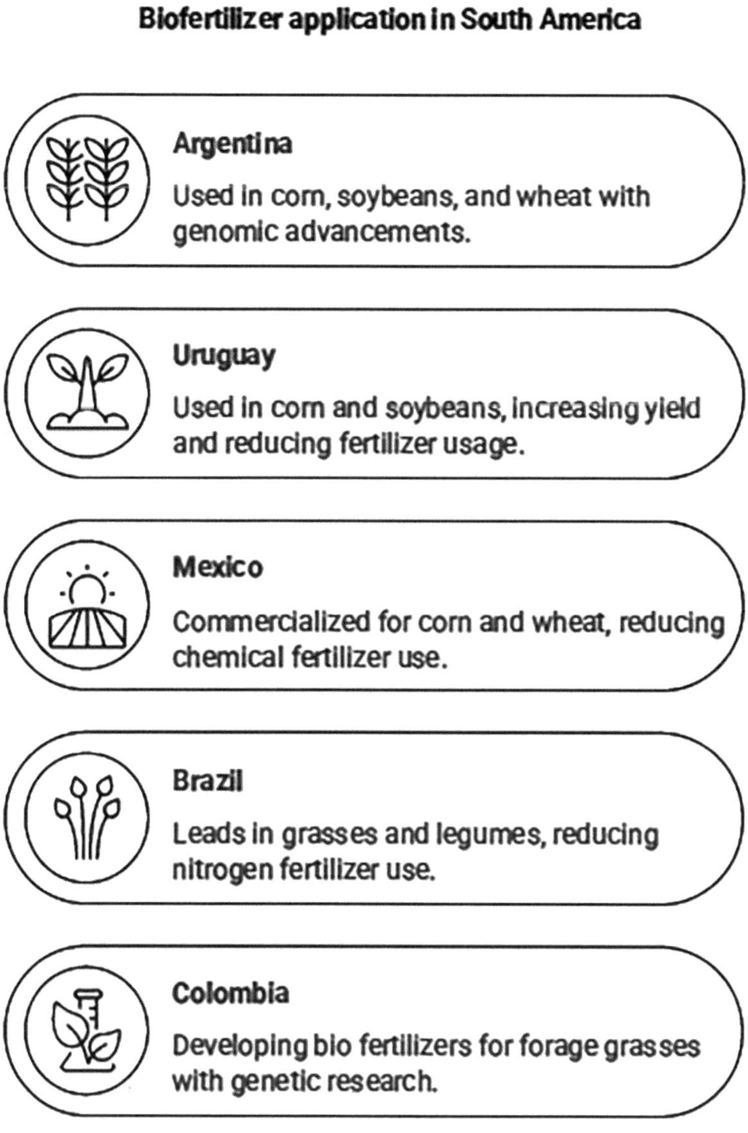

Fig. 22.4 Regional application of *Azospirillum*-based inoculants

the regulatory landscape across different regions and its implications for industry stakeholders.

Looking ahead, this chapter critically examines both current and emerging trends in the development of *Azospirillum*-based inoculants and explores their potential trajectories. It delves into how these trends are poised to shape the agricultural industry, with a particular emphasis on sustainability, productivity, and environmental impact. The chapter presents a forward-thinking perspective on the future of

Azospirillum-based inoculants, aiming to advance agricultural practices by addressing the dynamic challenges facing global food production systems.

22.2 New Frontiers in Quality Assessment

22.2.1 Microscopy-Based Techniques

When discussing microscopy as an essential tool for investigating the morphology of a sample, it is crucial to consider the resolution limit. Microscopy techniques involve the generation of amplified images using an excitation source, which significantly influences image quality, amplification capacity, cost, and sample specifications for analysis. The excitation sources used to stimulate samples can vary, including light, electrons, probes, sound waves, and others (Petry et al. 2019).

Microscopy-based assays are powerful tools for characterizing cellular phenotypes with both spatial and temporal resolution, and they can be applied to large-scale samples for screening genetic element libraries. However, strategies for conducting high-throughput microscopy experiments remain limited (Zhang et al. 2023).

The microscopic characterization, including Gram staining, cell size, shape, arrangement, and other morphological features, was performed on the bacterial isolates (Halimahtussadiyah et al. 2017). The genus *Azospirillum* is typically aerobic, Gram-negative, exhibits spiral motility, and measures 0.8–1.0 μm in diameter and 2–4 μm in length. Additionally, these bacteria contain intracellular polyhydroxybutyrate granules (Rasool et al. 2015).

A clear understanding of the strengths and limitations of the main microscopy techniques used in microbial analysis is essential when choosing the most appropriate protocol and aiming for methodological consistency. Among the most commonly applied methods are optical microscopy, scanning electron microscopy (SEM), and transmission electron microscopy (TEM). Optical microscopy remains a staple in microbiological studies due to its ease of use, affordability, and compatibility with a broad range of sample types. With magnification ranging from 1× to around 2000×, it allows for rapid visualization of general morphological features with minimal sample preparation. Its intuitive handling makes it especially useful for routine evaluations. Nonetheless, this technique is constrained by limited resolution and depth of field, which can be insufficient for observing fine structural details. SEM provides a substantial leap in magnification, reaching up to 100,000×. It excels at rendering high-resolution surface images and is particularly useful for observing external cell morphology and topography. In addition, SEM can perform elemental mapping, offering compositional insights. However, the technique requires specialized equipment, is relatively costly, and demands specific sample preparation, including dehydration, coating with conductive materials, and compatibility with vacuum conditions. TEM delivers the highest resolution among the three techniques, with magnifications up to 500,000×. It is particularly suited for examining internal ultrastructures at the nanometer scale, offering detailed insights into cellular components. The trade-offs include high operational and maintenance costs, elaborate sample processing, and the need for a tightly controlled analytical environment. In practice, the selection of a microscopy method should be aligned with the intended analytical goals, the nature of the sample, and the level of structural detail required. Each technique brings specific advantages but also practical constraints that must be weighed carefully to ensure accurate and meaningful observations.

22.3 Key Aspects in the Evaluation and Use of *Azospirillum*-Based Inoculants

22.3.1 Agronomic Efficiency Assessment

The agronomic efficiency of *Azospirillum*-based inoculants under different field conditions and their cost-effectiveness will determine which products are likely to be adopted by farmers and remain on the market (da Silva Medina et al. 2024). Therefore, evaluating the agronomic efficiency of these products is crucial. This section will review the methods used to assess agronomic efficiency, the indicators or parameters involved, and case studies where it has been evaluated. Additionally, it will examine key factors that influence agronomic efficiency, such as interactions with agrochemicals and the methods of inoculation, which must be considered during evaluation.

22.3.1.1 Methods for Assessing Agronomic Efficiency and Its Indicative Parameters

Early-stage assessments are suitable for screening to identify the most efficient strains in strain selection trials, as well as for determining optimal doses in crops that have not yet been evaluated. In germination assays, the parameters typically measured include: seed germination rate and seedling vigor parameters, such as total length of the seedlings (from the tip of the primary root to the top of the primary leaf), shoot length, and the length of the main root axis. These were evaluated using a ruler with millimeter graduations (Santos et al. 2020).

Since significant changes in plant root architecture are a main outcome of *Azospirillum* inoculation (Cassán et al. 2020), growth chamber or greenhouse assays provide accurate methods since the controlled environments, featuring sterile substrates, regulated temperature, and moisture levels, enhance repeatability and allow for precise measurement of inoculation effects on root systems (Vacheron et al. 2013).

However, to obtain product registration with the relevant authorities, it is necessary to demonstrate agronomic efficiency under field conditions (Neuhoff et al.

2024). When determining agronomic efficiency under field conditions, the parameters involved will depend on the crop. Generally, for extensive grain crops, the measured parameters include: dry biomass, stem diameter, plant height (or tassel insertion height for maize), SPAD index, thousand-grain weight, and grain yield (Hernandez et al. 2024; Maroniche et al. 2024). For example, Pereira et al. (2020) assessed agronomic efficiency of *Azospirillum* inoculation at the V4 maize stage (fourth expanded leaf), measuring parameters such as root volume, shoot dry biomass, and shoot nitrogen content (SNC) using the Kjeldahl method (ISO 2009). Regarding nitrogen use efficiency, Galindo et al. (2022) calculated the agronomic efficiency of applied nitrogen fertilizer (AE) using this method:

$$\text{AE}\left(\text{kg grain kg N applied}^{-1}\right) = (\text{grain yield at Nx} - \text{grain yield at N0})/\text{N level applied}$$

where Nx represents the nitrogen level applied, and N0 corresponds to the control treatment (without nitrogen applied in sidedressing). They demonstrated that inoculation with *A. brasilense* enhanced AE and nitrogen uptake in maize plants, leading to increased nitrogen accumulation in shoots, roots, and grains, resulting in a 7% increase in grain yield. The higher total nitrogen accumulation was mainly due to improved $N\text{-}NH_4^+$ accumulation in plant roots and shoots. They concluded that it is possible to reduce the nitrogen fertilization level by 50%.

22.3.1.2 Factors Affecting the Agronomic Efficiency of *Azospirillum*

The fungicides and insecticides used in seed treatments can negatively affect the survival of *Azospirillum* when applied to seeds and also reduce its efficiency in promoting plant growth. The degree of this impact depends on the toxicity of the product and duration of exposure and on the bacterial species (Santos et al. 2021). Maize is the second most treated crop with seed-applied pesticides, after soybean (Santos et al. 2020). This is why we reviewed the effects on *Azospirillum* survival of pesticides commonly used in maize crops.

Santos et al. (2020) evaluated the compatibility of Brazil's most common maize seed pesticide with *A. brasilense* strains Ab-V5 and Ab-V6 and its effects on early plant development. The pesticide increased *A. brasilense* cell mortality after 24 h of exposure. While seed germination and seedling growth were unaffected, the pesticide reduced root volume, dry weight, and root-hair incidence in greenhouse conditions. However, inoculation with *A. brasilense* mitigated these negative effects on root volume and root-hair incidence. Similar patterns of cell mortality were reported by Pereira et al. (2020), who observed survival rates of 65.87% on untreated seeds versus 13.56% on pesticide-treated seeds after 12 h of exposure.

The inoculation methods significantly impact the agronomic efficiency of *Azospirillum*-based inoculants. Seed treatment is the most common method for

Azospirillum inoculation in corn, but applying it to planting furrows or via foliar spray also shows positive effects (Santini et al. 2018). *Azospirillum* foliar sprays avoid competition with soil microbes, adverse soil conditions, and incompatibility with seed treatment agrochemicals (Cardozo et al. 2022). Bacteria likely enter passively through stomata, starting colonization in the substomatal chamber and spreading through intercellular spaces of the spongy chlorophyll tissue of the leaf mesophyll (Fukami et al. 2016). This method of application enables closer bacteria-plant contact, enhancing leaf colonization, nitrogen fixation, and phytohormone production (Preininger et al. 2018). Puente et al. (2019) demonstrated that the positive effects on plant growth through foliar application of the Az39 strain are due to both the presence of the Az39 cells as part of the colonization process and the *in situ* production of IAA. Inoculation with *A. argentinense* Az39 via corn seed was compared to foliar application at the V4 stage of corn. Az39 strain increased corn grain yield in all evaluated areas. Foliar application of this strain resulted in greater increases compared to seed treatment in Brazilian soils. While, in the Primavera do Leste-MT region, inoculation with Az39 in seeds and foliar application resulted in an 8% and 27% increase in corn yield, respectively, compared with the control treatment. In Londrina, the same treatments increased productivity by 15% and 27%, respectively, compared to the control (Hernandez et al. 2024). Similar results were obtained by García et al. (2024); although all inoculated treatments produced higher grain yields compared to the control, the greatest increase was also achieved with foliar application. The seed treatments increased by 3.5% the yield compared to the control, while foliar inoculation at V5-6 resulted in a 9.5% increase (average of two *Azospirillum* strains) in cornfield experiments in Barrow, Buenos Aires Province, Argentina.

22.3.2 Country-Specific Regulations

Available information on the regulatory requirements for the registration of products formulated with *Azospirillum* is scarce. Nevertheless, an overview of the countries where the relevant regulations are established and accessible is provided below.

In Argentina, the Servicio Nacional de Sanidad y Calidad Agroalimentaria (SENASA) is the authority responsible for ensuring compliance with technical regulations related to the development and formulation of bioinputs, overseeing their advancement and use within the agricultural sector. Annex II of Resolution N ° 2014/2025 outlines the "Requirements for the registration of bioinputs for plant nutrition, stimulation, amendments, substrates, and biological protectants". According to this annex, microorganism-based products that are imported or introduced into the country must undergo a prior risk assessment conducted by the competent authority to evaluate quarantine aspects and the potential impact on natural resources and public health. In the case of native microorganisms, the corresponding collection permit from the originating province must be provided. Furthermore, the regulation stipulates that microorganisms that are primary pathogens of animals, plants, or the environment cannot be registered. Additionally, microorganisms with taxonomic proximity to pathogens must demonstrate their safety through supporting scientific studies. Chapter 3 of Annex II specifies the requirements for registering microbial biological stimulants, which are defined as bioinputs containing at least one microorganism (fungus or bacterium) of the type PGPM (Plant Growth Promoting Microorganism), AMF (Arbuscular Mycorrhizal Fungi), or AMFB (Soil Bacteria Associated with Arbuscular Mycorrhizal Fungi).

The minimum concentration of *Azospirillum*-based products at the expiration date must be $1 \cdot 10^7$ CFU per mL or g of the formulation, with a shelf life ranging from six to twenty-four months. Additionally, the products must not exceed the established limits for pathogenic microorganisms: *Salmonella* spp. must be absent in 25 g or 25 mL, and *Escherichia coli* must be absent in 1 g or 1 mL of the product. A detailed description of the manufacturing process, the strain provision certificate, and the declaration of the composition of the substrate or carrier containing the microorganism must be provided. Furthermore, an efficacy testing report, in accordance with the agronomic efficacy protocol (if applicable), must be submitted along with supporting technical documentation regarding the product's specifications. This documentation should come from official institutions or peer-reviewed scientific publications. In the case of liquid formulations, the product must have an optimal pH for both the contained microorganisms and the plants.

The regulation of inoculants in Colombia, particularly those containing *Azospirillum*, is fundamental to promoting sustainable agricultural practices. The Instituto Colombiano Agropecuario (ICA), as the responsible entity, ensures the quality and safety of these products, framing their use within a context of sustainable agricultural development.

The Resolution 00375 of 2004 from the (ICA 2004) establishes the provisions for the registration and control of inoculants, defining them as products that enhance agricultural productivity through living microorganisms. This regulation mandates that products containing *Azospirillum* provide detailed information regarding their composition, efficacy, and safety before commercialization.

On the other hand, the Ministerio de Agricultura y Desarrollo Rural has implemented policies aimed at promoting the use of inoculants in the country. The Plan Nacional de Desarrollo (National Development Plan) recognizes the importance of these inputs in reducing reliance on chemical fertilizers and promoting more sustainable agricultural practices.

Despite the regulatory framework, significant challenges persist:

1. Lack of Awareness: Many farmers remain uninformed about the benefits and proper application of inoculants.
2. Limited Production: Small-scale production faces difficulties competing with conventional chemical fertilizers.
3. Research Needs: Further research on native microbiological reserves is required and their application in bio-inputs.

The regulation of inoculants with *Azospirillum* in Colombia represents a step towards more sustainable agriculture. However, strengthening training, local

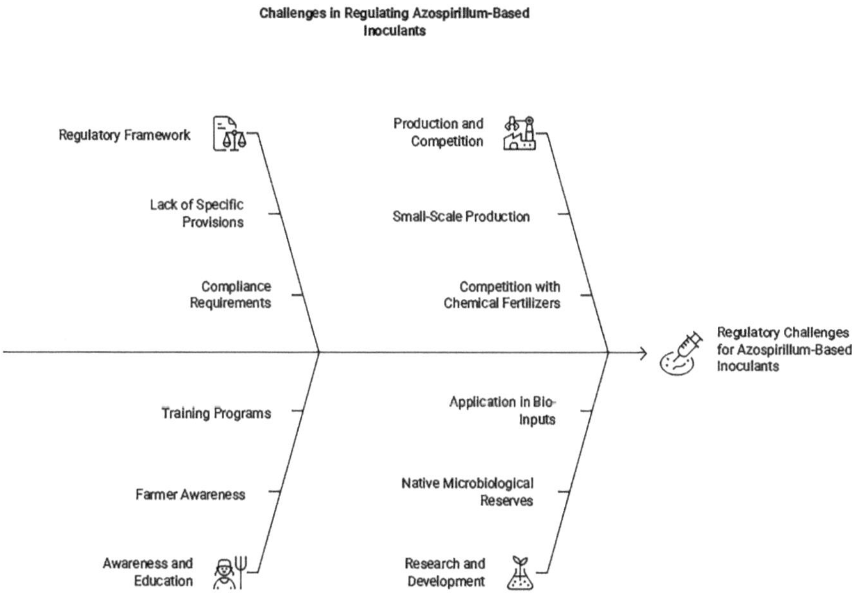

Fig. 22.5 Challenges in adopting bio-inoculants in agriculture

production, and research is crucial to maximize their potential within the agricultural sector (Fig. 22.5). Although several countries have established regulatory frameworks for microbial inputs, specific provisions addressing *Azospirillum* are limited or not clearly defined in most cases. A comparative overview of the current regulatory status of *Azospirillum*-based inoculants in countries of America is provided in Table 22.1.

22.4 Rethinking the Future

22.4.1 Trends and Future Directions

In Colombia, several studies have investigated the interaction of *Azospirillum* with herbicides and other inoculants, highlighting its potential in sustainable agriculture. A study published in AgriScientia examined the combination of *Azospirillum argentinense* Az39 with the herbicides acetochlor and dicamba, demonstrating that *Azospirillum* survived in the presence of these herbicides and utilized dicamba as a carbon source, although acetochlor reduced maize germination (Martin et al. 2023). Another article in Agronomía Colombiana analyzed how glyphosate and atrazine inhibit the growth of *Azospirillum brasilense*, finding that these substances negatively affect bacteria used in biological control and biological nitrogen fixation

Table 22.1 Current regulatory status of *Azospirillum*-based inoculants in American countries

Country	Bioinput designation	Regulatory entity	Current regulation	Specific regulation for *Azospirillum*
Argentina	Biological biostimulant	Servicio Nacional de Calidad y Seguridad Agroalimentaria https://www.argentina.gob.ar/senasa	Resolution 458/2025	Established
Bolivia	Biological	Servicio Nacional de Sanidad Agropecuaria e Inocuidad Alimentaria https://www.senasag.gob.bo	Law 830	Not established
Brazil	Inoculant	Ministerio de Agricultura, Ganadería y Abastecimiento https://www.gov.br/agricultura	Law 15070/2024	Established
Chile	Bioestimulant	Servicio Agrícola y Ganadero https://www.sag.gob.cl	Law 21349-2021	Not established
Colombia	Biological inoculant	Instituto Colombiano Agropecuario https://www.ica.gov.co	Resolution 00375/2004	Not established
Costa Rica	Microbial bioestimulant	Ministerio de Agricultura https://www.mag.go.cr	Decree 39733	Not established
Ecuador	Inoculant	Ministerio de Agricultura, Ganadería, Acuacultura y Pesca http://www.agrocalidad.gob.ec	Resolution 218	Not established
Guatemala	Fertilizer-related substances	Ministerio de Agricultura, Ganadería y Alimentación https://www.maga.gob.gt	Government Agreement 342-2010	Established
Honduras	Bioinput	Servicio Nacional de Sanidad e Inocuidad Agroalimentaria https://senasa.gob.hn	Agreement 450-2023	Not established
Mexico	Biofertilizer	Servicio Nacional de Sanidad, Inocuidad y Calidad Agroalimentaria https://www.gob.mx/senasica	NOM-021-FITO-1995	Not established
Nicaragua	Bioinput	Instituto Nicaragüense de Tecnología Agropecuaria https://inta.gob.ni	NTON 11 048-16	Not established
Panama	Bioinput	Instituto de Innovación Agropecuaria http://www.idiap.gob.pa	Under development	Not established
Paraguay	Inoculant	Servicio Nacional de Calidad y Sanidad Vegetal y de Semillas https://www.senave.gov.py	Resolution 564/10	Not established
Peru	Microbials for plant nutrition	Servicio Nacional de Sanidad Agraria https://www.senasa.gob.pe	Not established	Not established

(continued)

Table 22.1 (continued)

Country	Bioinput designation	Regulatory entity	Current regulation	Specific regulation for *Azospirillum*
United States	Bioestimulant	Environmental Protection Agency https://www.epa.gov	State-specific resolution	Not established
Uruguay	Biological inoculant	Ministerio de Ganadería, Agricultura y Pesca https://www.gub.uy	Resolution 4/013	Established

(Farias et al. 2021). Additionally, a study published in the Revista Sociedad Venezolana de Microbiología evaluated the effect of fertilizers and herbicides on the viability of *A. brasilense*, concluding that herbicides such as glyphosate and 2,4-D exhibited significant inhibitory effects (Romero-Perdomo et al. 2015). These findings underscore the necessity for further research on the interactions between bio-inputs and agrochemicals to promote more.

22.4.2 Long-Term Vision

The sequencing of the *Azospirillum* genome has allowed for a better understanding of its genetic and functional diversity, which has driven the development of more efficient strains. These advancements have facilitated the creation of improved formulations that enhance their survival and efficacy under several environmental conditions (Cassán et al. 2020; Dos Santos Ferreira et al. 2022) (Fig. 22.5).

References

Aguilar LA (2020) Producción masiva de *Azospirillum* spp., formulación, control de calidad y su uso en la agricultura: Revisión de Literatura

ANLA (2022) Protocolo técnico y normativo de bioinsumos, agroinsumos y controladores biológicos de origen natural (Technical and Regulatory Protocol of inoculants, Agro-inputs, and Biological Controllers of Natural Origin)

Benitende S, Perticari A, Rossi A, Toresani S (eds) (2025) Manual de procedimientos microbiológicos para la evaluación de calidad de inoculantes, bioestimulantes y biocontroladores, 1st edn. Asociación Argentina de Microbiología, Ciudad Autónoma de Buenos Aires. ISBN 978-987-48458-3-2

Cárdenas DM, Garrido MF, Bonilla RR et al (2010) Aislamiento e identificación de cepas de *Azospirillum* sp. en pasto guinea (*Panicum maximum* Jacq.) del Valle del Cesar. Pastos y Forrajes 33(3)

Cassán FD, Penna C, Creus CM et al (2015) Protocol for the quality control of Azospirillum spp. inoculants. In: Handbook for *Azospirillum*: technical issues and protocols, pp 487–499

Cassán F, Coniglio A, López G et al (2020) Everything you must know about *Azospirillum* and its impact on agriculture and beyond. Biol Fertil Soils 56:461–479

Coniglio NA, Lopez GA, Gualpa JL et al (2020) Development of sequence characterized amplified region markers for identification of *Azospirillum brasilense* Az39. RAM 52(1):50–60

da Silva Medina G, Rotondo R, Rodríguez GR (2024) Innovations in agricultural bio-inputs: commercial products developed in Argentina and Brazil. Sustainability 16(7):2763

Dos Santos Ferreira N, Coniglio A, Puente M et al (2022) Genome-based reclassification of *Azospirillum brasilense* Az39 as the type strain of *Azospirillum argentinense* sp. nov. Int J Syst Evol Microbiol 72(8):005475

Farias DIOA, Leite RDC, Ribeiro EA et al (2021) El glifosato y la atrazina inhiben el crecimiento de *Azospirillum brasilense*. Agron Colomb 39(1):113–120

Ferreira E, Nogueira M, Hungria M (2024) Manual de análises de bioinsumos para uso agrícola: inoculantes

Fukami J, Nogueira MA, Araujo RS et al (2016) Accessing inoculation methods of maize and wheat with *Azospirillum brasilense*. AMB Express 6:1–13

Galindo FS, Rodrigues WL, Fernandes GC et al (2022) Enhancing agronomic efficiency and maize grain yield with *Azospirillum brasilense* inoculation under Brazilian savannah conditions. Eur J Agron 134:126471

García JE, Zamora M, Giacobbe Boggio D et al (2024) The optimal growth stage of maize for *Azospirillum argentinense* foliar inoculation. Abstract book SAMIGE http://www.quimicaviva.qb.fcen.uba.ar/v23n3/samige2024/

Halimahtussadiyah R, Natsir M, Kurniawati D et al (2017) Isolation and identification of chitinolytic bacteria of pohara river of south east sulawesi and the optimization production of chitinase enzyme. AIP Conf Proc 020062-1

Hernández AG, da Silva EP, Ferreira PAA et al (2024) Inoculation methods and agronomic efficiency of *Azospirillum brasilense* Strain Az39 for corn crop under different brazilian edaphoclimatic conditions. Rev Bras Ciênc Agrár 19(1):e3609–e3609

ICA (2004) Resolución 00375 de 2004 (Resolution 00375 of 2004). Instituto Colombiano Agropecuario

Lin SY, Shen FT, Young CC (2011) Rapid detection and identification of the free-living nitrogen fixing genus *Azospirillum* by 16S rRNA-gene-targeted genus-specific primers. Antonie Van Leeuwenhoek 99:837–844

Lupwayi NZ, Olsen PE, Sande ES et al (2000) Inoculant quality and its evaluation. Field Crop Res 65(2-3):259–270

Maroniche G.A, Puente ML, García JE et al (2024) Phenogenetic profile and agronomic contribution of iAzospirillum argentinense/i Az39T, a reference strain for the South American inoculant industry. Microbiol Res 283:127650

Martin SM, Cardozo PG, Di Palma MA et al (2023) Impacto de herbicidas sobre el inoculante *Azospirillum argentinense* Az39 utilizado en cultivos de maíz. Agriscientia 40(1):2–2

Ministerio de Agricultura y Desarrollo Rural MADR (2023) Promoción del uso de bioinsumos Promotion of the use of bio-inputs. Ministerio de Agricultura y Desarrollo Rural

Ministerio de Agricultura y Desarrollo Rural MADR (2024) Plan Nacional de Desarrollo (National Development Plan). Ministerio de Agricultura y Desarrollo Rural

Neuhoff D, Neumann G, Weinmann M (2024) Testing plant growth promoting microorganisms in the field-a proposal for standards. Front Plant Sci 14:1324665

Pereira LC, de Carvalho C, Suzukawa AK et al (2020) Toxicity of seed-applied pesticides to *Azospirillum* spp.: an approach based on bacterial count in the maize rhizosphere. Seed Sci Technol 48(2):241–246

Petry R, Oliveira N, Alves A et al (2019) Chapter 2 - Nanomaterials properties of environmental interest and how to assess them. In: do Nascimento RF, Ferreira OP, De Paula AJ (eds)

Advanced nanomaterials, nanomaterials applications for environmental matrices. Elsevier, de Oliveira Sousa Neto V, pp 45–105

Preininger C, Sauer U, Bejarano A et al (2018) Concepts and applications of foliar spray for microbial inoculants. Appl Microbiol Biotechnol 102:7265–7282

Puente ML, Zawoznik M, de Sabando ML et al (2019) Improvement of soybean grain nutritional quality under foliar inoculation with *Azospirillum brasilense* strain Az39. Symbiosis 77:41–47

Maroniche G.A, Puente ML, García JE et al (2024) Phenogenetic profile and agronomic contribution of *Azospirillum argentinense* Az39T, a reference strain for the South American inoculant industry. Microbiol Res 283:127650

Puente ML, García JE, Di Salvo LP et al (2025) Sección *Azospirillum*. En: Benitende S, Perticari A, Rossi A, Toresani S (eds) Manual de procedimientos microbiológicos para la evaluación de calidad de inoculantes, bioestimulantes y biocontroladores. 1 Asociación Argentina de Microbiología, Ciudad Autónoma de Buenos Aires, p 138-159. ISBN 978-987-48458-3-2

Rasool L, Asghari M, Jamil A et al (2015) Identification of *Azospirillum* species from wheat rhizosphere. J Anim Plant Sci 25(4):1081–1086. ISSN: 1018-7081

Romero-Perdomo F, Criollo Campos PJ, Rodríguez AJ et al (2015) Efecto de fertilizantes y herbicidas sobre el crecimiento in vitro de *Azospirillum brasilense* C16. RSV 35(2):117–120

Salvo LPD, García JE, Puente ML et al (2022) The drop plate method as an alternative for *Azospirillum* spp viable cell enumeration within the consensus protocol of the REDCAI network. RAM 54(2):61–70

Santini JM, Buzetti S, Teixeira Filho M et al (2018) Doses and forms of *Azospirillum brasilense* inoculation on maize crop. RBEAA 22:373–377

Santos MS, Rondina AB, Nogueira MA et al (2020) Compatibility of *Azospirillum brasilense* with pesticides used for treatment of maize seeds. Int J Microbiol 2020(1):8833879

Santos MS, Rodrigues TF, Nogueira MA et al (2021) The challenge of combining high yields with environmentally friendly bioproducts: a review on the compatibility of pesticides with microbial inoculants. Agronomy 11(5):870

Stephens JHG, Rask HM (2000) Inoculant production and formulation. Field Crop Res 65:249–258

Vacheron J, Desbrosses G, Bouffaud ML et al (2013) Plant growth-promoting rhizobacteria and root system functioning. Front Plant Sci 4:356

Young CC, Lin SY, Shen FT et al (2015) Molecular tools for identification and characterization of plant growth promoting rhizobacteria with emphasis in *Azospirillum* spp. In: Handbook for *Azospirillum*: technical issues and protocols, pp 27–44

Zhang R, Huang Y, Li M et al (2023) High-throughput, microscopy-based screening and quantification of genetic elements. mLife 2(4):450–461

Part VI
Azospirillum spp. Beyond Agriculture

Chapter 23
Beyond Agriculture: Use of *Azospirillum* for Environmental Purposes

Blanca R. Lopez, Macario Bacilio, Maria Antonia Cruz Hernández, and Luz E. de-Bashan

Dedicated to the memory of Prof. Yoav Bashan, the founder of the Bashan Institute of Science and the leader of the Environmental Microbiology Group at CIBNOR for 28 years. He pioneered the study of Azospirillum brasilense *for environmental purposes.*

Abstract Research on *Azospirillum* for environmental restoration and remediation has not been extensively conducted. Four decades ago, pioneering studies revealed that these bacteria have a significant affinity for plants beyond grass. Subsequently, comprehensive studies using model strains demonstrated that *Azospirillum* retains its associative capacity, including survival in soil and colonization, as well as its growth-promoting effects such as nitrogen uptake, enhanced biomass and photosynthesis, and nutrient mobilization in interaction with desert plants, with which they have not co-evolved. These findings advanced the search for inoculants of *Azospirillum brasilense*, which, combined with native PGPB, in field testing in long-term studies, opened opportunities to develop methodological strategies for restoring degraded soils in arid regions. However, innovative techniques still need to be developed to evaluate specific indicators relevant to drylands. Studies on the

B. R. Lopez (✉)
The Bashan Institute of Science, Auburn, AL, USA

Centro de Investigaciones Biológicas del Noroeste (CIBNOR), La Paz, México
e-mail: bromero@cibnor.mx

M. Bacilio
Centro de Investigaciones Biológicas del Noroeste (CIBNOR), La Paz, México

M. A. C. Hernández
Centro de Biotecnología Genómica, Instituto Politécnico Nacional, Reynosa, Tamaulipas, México

L. E. de-Bashan
The Bashan Institute of Science, Auburn, AL, USA

Department of Entomology and Plant Pathology, Auburn University, Auburn, AL, USA

F. D. Cassán et al. (eds.), *The Definitive Handbook of Azospirillum*,
https://doi.org/10.1007/978-3-032-09222-9_23

bioremediation of contaminated sites featuring xenobiotics have identified strains capable of degrading hydrocarbons or heavy metals while retaining their plant growth-promoting traits.

Keywords Environment · Restoration · Bioremediation · Soil · Contamination

23.1 Introduction

In addition to its role in agriculture, *Azospirillum* has expanded its applications to include soil restoration and reclamation. This microorganism can thrive in harsh or nutrient-deficient environments and often establishes beneficial relationships with non-crop plants. Additionally, various strains of *Azospirillum* have been utilized for their ability to survive in contaminated environments. As a result, *Azospirillum* has shown promising outcomes in enhancing the fertility of degraded soils in arid regions and remediating soils contaminated with hydrocarbons and heavy metals. This chapter presents representative studies of *Azospirillum* outside of agricultural contexts, providing insights into its contributions to natural ecosystems and exploring its potential for environmental applications, particularly in drylands, where most research has been focused.

23.2 The Hidden Potential of Wild Plants as Sources of *Azospirillum*

From their initial isolation in 1925 until the mid-1970s, bacteria of this genus were routinely isolated from various soils, particularly from the roots of cultivated grain plants (Döbereiner et al. 1976). Later, plants different from cereal crops were explored in the search for *Azospirillum*. Several morphotypes of these bacteria were isolated by Döbereiner (1978) from the shameplant (*Mimosa pudica*), sisal (*Agave sisalana*), common bamboo (*Bambusa* sp.), and the bracken fern (*Pteridium aquilinum*). Then, *Azospirillum* strains were isolated from xerophytes such as *Stenocereus pruinosus* and *Stenocereus stellatus* (Mascarua-Esparza et al. 1988), as well as prickly pear (*Opuntia* sp.) (Rao and Venkateswarlu 1982; Mascarua-Esparza et al. 1988). Despite these efforts, few isolates have been found in wild plants, although some of them might be endangered or inhabit ecosystems needing protection or restoration.

Several decades ago, *Azospirillum* was thought to have a greater affinity for grass; however, increasing evidence has shown the opposite. Three decades ago, Bashan and Holguin (1995) demonstrated that *A. brasilense* could be established in the roots of non-gramineous weeds belonging to 18 botanical families. They found that adherence and active colonization are determined by bacterial cell mobility and the presence of plants, although this process is not plant specific. Pereg et al. (2016) found insufficient evidence on the affinity of *Azospirillum* strains for specific plants or cereal genotypes, urging more research on isolating species beyond cereals and cross-inoculation. Most growth promotion examples by *Azospirillum* in wild plants use strains isolated from crop plants. Early in the 1990s, Puente and Bashan (1993) provided the first evidence that *A. brasilense* strains can positively influence the growth of wild plants, specifically cacti from the Sonoran Desert of Northwest Mexico. This finding was later confirmed for various types of cacti, including tree-like forms, columnar and rock-dwelling species (Bashan et al. 1999; Puente et al. 2004b; Lopez et al. 2012).

23.3 Soil Development and Improvement of Soil Structure in Drylands

Erosion and weathering processes underlie soil formation and development (Frings and Buss 2019). In dry landscapes, water availability determines the pulses of biological activity; however, biological weathering contributes significantly to soil formation to a similar extent as physical and chemical weathering (Warke 2013). In the early stages of soil formation, rock weathering involves the disintegration of rock and the dissolution of minerals by the biological activity of lithobiontic microorganisms and other plant-associated microorganisms, which contribute to the formation of parent materials that support the development of incipient soils (Pointing and Belnap 2012; Lopez and Bacilio 2020). In arid regions of Mexico, the synergistic effect of cacti and their associated rhizobacteria on rock weathering was initially proposed by Puente et al. (2004a) as a fundamental phenomenon that accelerates soil formation in the subtropical Sonoran Desert. They observed abundant microorganisms such as bacteria, fungi, and actinobacteria residing in the rhizoplane of cacti such as the giant cardon cactus (*Pachycereus pringlei*), organ pipe cholla (*Stenocereus thurberi*), chain-link cholla (*Cylindropuntia cholla*), and the rock fig (*Ficus palmeri*) that live on rocks without apparent developed soil. Then, they inoculated the cardon cactus with rhizobacteria isolates, using *A. brasilense* as a positive control PGPB, and assessed plant growth and the removal of elements from the growing substrate, which consisted of pulverized rock. After one year of growth, the dry weight of the inoculated plants exceeded that of the non-inoculated ones, although the effect of *A. brasilense* was marginal. Native strains from the rhizoplane demonstrated more effective mobilization of P_2O_5 and Fe_2O_3 than *A. brasilense*; however, the latter showed a K_2O and MgO removal rate comparable to that of the bacteria isolated from the rhizoplane. Additionally, the pH of the substrates across treatments with rhizobacterial strains generally decreased, with differences observed among bacterial species. This finding partially corroborated the hypothesis that *A. brasilense* induces rhizosphere acidification through proton efflux from root cells, resulting from changes in membrane potential (Fig. 23.1a, Bashan 1990; Bashan and Levanony 1991). This process can enhance the acquisition of mineral nutrients, such as P (Kirk 1997). Later, Carrillo et al. (2002) observed that

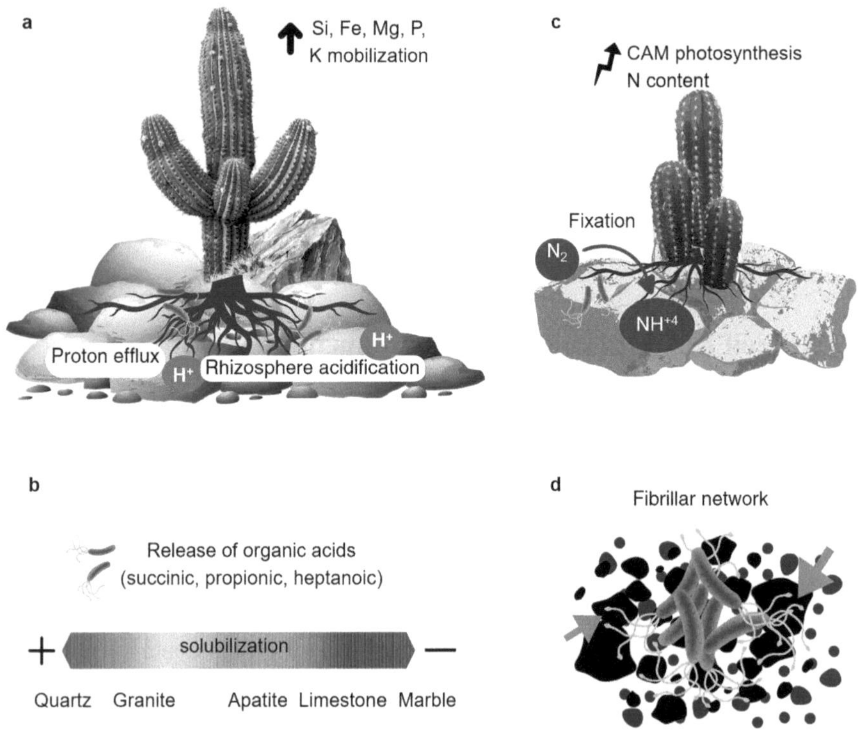

Fig. 23.1 Mechanisms through which *Azospirillum* contributes to soil development and enhances soil structure in drylands. (**a**) Rhizosphere acidification occurs through proton efflux from root cells due to changes in membrane potential and increased cation exchange and mobilization. (**b**) Mineral degradation and solubilization occur through the release of organic acids. (**c**) Nitrogen fixation provides the plant with assimilable nitrogen in an environment deficient in this nutrient. Enhanced nitrogen assimilation results in increased photosynthesis through the Crassulacean acid metabolism (CAM), characteristic of cacti. (**d**) Improvement of particle aggregation via a network of protein-based fibrillar connections

rhizosphere acidification increases when *P. pringlei* seedlings are inoculated with *A. brasilense*, and this effect can be enhanced by adding nitrogen, specifically ammonium.

During early soil development, the mobilization of nutrients to plants is a complex process that depends on factors such as the geological history of the rocks, their degree of weathering, mineral types, and chemical composition (Warke 2013). The rocky areas of the desert northwest of Mexico are home to a diversity of rock-dwelling cacti, such as the small rock-colonizing pioneer cactus *Mammillaria fraileana* (Lopez et al. 2009). This cactus hosts abundant populations of endophytic bacteria capable of dissolving rhyodacite and rhyolite rocks (Lopez et al. 2012) through dissolution mechanisms driven by the action of organic acids. In the case of *A. brasilense*, succinic, propionic, heptanoic, and gluconic acids could explain its ability to degrade in vitro quartz, granite, igneous rock, and, to a lesser degree, apatite, limestone, and marble (Fig. 23.1b, Puente et al. 2006). To investigate the role of cactus-endophyte interactions in the growth and mobilization of elements from a substrate supplemented with pulverized rock in the aerial plant compartment, Lopez et al. (2013) inoculated *M. fraileana* seedlings with some of their endophytes and used *A. brasilense* Cd as a control. At 6.5 months post-inoculation, they found endophytic colonization of all the inoculated bacteria. Moreover, *A. brasilense* induced the mobilization of Si, Fe, and Mg, and less K and Na, and stimulated the accumulation of nocturnal acidity in the tissues, which is indicative of CO_2 fixation by the CAM pathway, although at lower levels than endophytic bacteria isolated from the same cactus, such as *P. putida* M5TSA and *A. vinelandii* M2Per. Regarding biomass and total N, *A. brasilense* did not outperform any endophytes; however, it outdid non-inoculated plants. This emphasized the crucial role that nitrogen-fixing bacteria such as *Pseudomonas putida* M5TSA, *Azotobacter vinelandii* M2Per, and even *A. brasilense* could provide nitrogen to plants that inhabit massive rocks, where this nutrient is scarce (Fig. 23.1c). Furthermore, it highlighted the existence of a mechanism by which endophytic bacteria may stimulate the photosynthetic activity of cacti. More recently, Lopez and Bacilio (2020) suggested that, through the CAM pathway, cacti can also increase the exudation of organic acids, favoring the dissolution of minerals containing P, as well as the mobilization of elements that form metal complexes.

While arid zones exhibit significant soil diversity, most soils in these regions tend to be coarse-textured, loosely structured, and have poor horizon development (Warke 2013). In the 1980s and 1990s, systematic studies were conducted to evaluate the adhesion of *Azospirillum* to various soils and to assess its potential application to plants and soils in arid lands. The initial experiments with *A. brasilense* revealed low cell retention in sand, which was attributed to its limited adsorption surface area. However, a portion of the bacterial population remained adhered through a network of protein bridging (Bashan and Levanony 1989). Later, Bashan et al. (1991) noted that adding clay and organic matter to sandy soils enhanced the adsorption of bacterial cells of *A. brasilense*, which possess the ability to aggregate with sandy particles. In contrast to the mutant strain lacking aggregation ability, the aggregating strain exhibited adhesion through fibrillar connections between bacterial cells. Similarly, attachment between cells and sand or clay particles occurred through anchoring via single or multistranded fibrillar connections, forming a network of protein bridges (Fig. 23.1d). The role of proteins in forming organo-mineral associations was later explained by the high affinity of nitrogen compounds for mineral surfaces, as explained by Kleber et al. (2015). Thus, the strong adsorption between the fibrillar network and particles of sand, clay, and organic matter leads to the formation of stable aggregates. Therefore, Bashan and Holguin (1994) proposed that *Azospirillum* could be used in conjunction with indigenous plants from drylands to improve the soil structure of degraded soils. Subsequently, Bashan and Vazquez (2000) evaluated the survival of five *Azospirillum* species (*A. brasilense*, *A. lipoferum*, *A. halopraeferens*, *A. amazonense*, and *A. irakense*) in one semiarid soil and another artificial soil simulating the conditions of the former. They found that the survival of all strains was low and similar in both soil types. High levels of $CaCO_3$ (abundant in arid soils) and the abundance of fine and coarse sand were detrimental to the survival of *Azospirillum* species. However, adding organic matter improved their survival, confirming the feasibility of applying this strategy to soils from other arid regions.

23.4 Revegetation and Restoration of Fertility in Degraded Drylands

The environmental consequences of adopting inappropriate production models in inherently fragile ecosystems are reflected in land degradation and aridification, which are exacerbated by climate change (D'Odorico et al. 2013). One of the most effective strategies for restoring degraded soils in arid areas is revegetation with indigenous plants, as these are adapted to soils with low fertility and can survive with limited rainfall. Initial studies on the affinity of *Azospirillum* to colonize diverse plant species across a wide range of environmental conditions led to an exploration of its role in the recovery of degraded soils.

For revegetation to be viable, bacteria must survive long enough and have a measurable effect on plant growth. The initial studies on the use of *Azospirillum* in cacti employed the giant cardon, *P. pringlei*, as a model species since this tree-like cactus is a dominant species in the scrub vegetation of the southern Sonoran Desert in Mexico (Bashan et al. 2008). Puente and Bashan (1993) inoculated seeds of this cactus with two strains of *A. brasilense*, Cd and Sp-245, finding that the latter produced higher germination rates. Furthermore, both strains persisted in the rhizosphere of seedlings for up to 10 months post-inoculation, resulting in increased survival and growth, as indicated by height and volume. These results were foundational in suggesting that *Azospirillum* could be a good PGPB candidate for producing cardon plants in nurseries for reforestation purposes.

Other experiments also demonstrated that *A. brasilense* plays a crucial role in improving the establishment of several desert plants. Bashan et al. (1999) found that over 3.5 years, three species of columnar cacti—*P. pringlei*, *S. thurberi*, and *Lophocereus schottii*—showed higher survival rates and more rapid growth than the untreated controls. Notably, *A. brasilense* persisted in the rhizosphere of these cacti for a period of two years. The inoculation with *Azospirillum* to favor the establishment of long-living plants has practical implications because, as they grow slowly, it allows for obtaining useful plants for revegetation in less time than would occur in nature, as this process is limited by competition and predation in the desert (Bashan et al. 2008). Throughout their lifespan, cacti and other xerophytes develop extensive superficial root systems, making them ideal for preventing soil erosion and enhancing soil fertility (Bashan et al. 2008, 2009). However, the initial establishment and survival of these cacti in barren soils are limited not only by ecological and climatic factors but also by edaphic factors, such as the lower availability of water and nutrients. Therefore, facilitation processes through nurse plants are highly relevant in the desert.

Woody legumes are frequently nurse plants that form "resource islands" where the higher organic matter content, lower temperature, higher soil moisture, and the presence of mycorrhizal fungi (Bashan et al. 2000; Carrillo-Garcia et al. 2000) improve the availability of nutrients, primarily N and P (de-Bashan et al. 2022). Therefore, they are considered key landscape elements in nutrient recycling and the water cycle in arid zones (Torres-García et al. 2022). Resource islands formed by nurse legumes represent the dominant association in drylands; therefore, any tool that improves the establishment of nurse plants is desirable. Leyva and Bashan (2008) evaluated the affinity between *A. brasilense* and mesquite (*Prosopis articulata*), the primary resource island-forming species in the Baja California Peninsula, Mexico. They found that, naturally, the seedlings of this legume produce gluconic acid in their root exudates, one of the carbon sources preferred by *Azospirillum*, which induces gluconokinase activity. When inoculating the seedlings with *A. brasilense* Cd. and growing them in a hydroponic system, they detected significant activity of that enzyme in the roots, which showed abundant bacterial colonization by *Azospirillum*. Additionally, the inoculated seedlings exhibited greater size, a greener appearance, and higher weight compared to the non-inoculated seedlings.

In greenhouse experiments, *P. articulata* showed better establishment than other legumes frequently used in urban landscaping and reforestation in Northwestern Mexico and the Southwestern USA. Bashan et al. (2009) reported the effect of a mixture of PGPB (*A. brasilense* + *Bacillus pumilus*), the unidentified arbuscular mycorrhizal fungi (AMF) consisting mainly of *Glomus* sp., and a small amount of compost on the growth and physiological parameters of three legume trees. They found that, after 10 months of cultivation, in all the treatments, the three legumes mesquite (*P. articulata*), yellow palo verde (*Parkinsonia microphylla*), and blue palo verde (*Parkinsonia florida*) showed survival above 80%; however, mesquite was almost 100% and outperformed the other legumes in dry weight, stem diameter, number of branches per tree, and height.

In experiments involving the legume and cardon, growth promotion by *Azospirillum* was influenced by the soil environment of the resource island-forming nurse plant. In an 18-month experiment, Bacilio et al. (2006) found growth promotion in the inoculated plants; however, this effect was more pronounced in richer soils obtained from resource islands, where legume trees fostered other nurseries. However, adding compost to barren soil resulted in plant weight parameters similar to those of the islands, and the plants were larger and greener. These results indicated that adding compost to barren soils can create a soil similar to the scarce soils of resource islands, making the production of cardon seedlings viable for reforestation.

Longer-term field studies evaluated the growth of cardons growing beneath their nurse plants to simulate the resource island effect. The three legumes (*P. articulata*), yellow palo verde (*Parkinsonia microphylla*), and blue palo verde (*Parkinsonia florida*), were used. Soil treatments included inoculation with a mixture of PGPBP (*A. brasilense* + *Bacillus pumilus*), an AMF of *Glomus* sp., and a substrate supplemented with compost (Bashan et al. 2012). After 30 months of establishment, no significant changes in the survival of the different legumes were detected;

however, *P. articulata* maintained high survival rates thereafter. A decade after conducting the experiment, Moreno et al. (2017) reported that, the legume-cactus association is usually detrimental to the former, acting as a facilitator, but significantly increases the survival chances for the cardon cactus. These findings support the natural observation that cacti without a nurse are less likely to establish themselves successfully in the long term. In the same experimental area, Lopez-Lozano et al. (2016) conducted a comparative study of the restored areas, focusing on the association between cacti and nurse shrubs. They analyzed the nitrogen-fixing communities using PCR-DGGE of the *nif*H gene and assessed nitrogen fixation by acetylene reduction. The most significant finding was that, almost 10 years after restoration, the abundance of nitrogen-fixing bacteria was higher in the rhizosphere of cacti associated with their nurse shrubs, *P. articulata,* than in cacti growing separately. Furthermore, N_2 fixation was higher in restored sites than in undisturbed ones. Collectively, these studies demonstrate that the revegetation of degraded lands is achievable by mimicking the natural process of resource islands, particularly those consisting of mesquite (*P. articulata*) and the cardon cactus, which functions as an effective soil stabilizer and accelerates plant succession in barren areas where restoration would otherwise be nearly impossible (Bashan et al. 2008).

23.5 Desertified Regions and Other Degraded Lands

In arid and semiarid regions, desertification leads to a decline in land productivity across extensive areas (D'Odorico et al. 2013). In northeast China, degraded steppes have been overtaken by dunes, in which restoration efforts aim to recover their natural vegetation (Cao et al. 2021). Dune vegetation is strongly influenced by sand mobility and depends on the nutrients obtained through interactions with associated microorganisms, such as nitrogen fixers. Cao et al. (2021) studied whether changes in the structure of bacterial communities associated with the nitrogen cycle occurred during dune fixation with the legume *Caragana microphylla*. By analyzing the abundance of the nitrite (*nir*S and *nir*K) and nitrous oxide (*nos*Z) reductase genes using clone libraries and quantitative polymerase chain reaction, they found that the abundance of each gene increased with the age of the forests. Overall, the dominant clones were closely related to *Pseudomonas*, *Mesorhizobium*, *Rhizobium*, *Rhodopseudomonas*, *Azospirillum*, and *Cupriavidus*. *Pseudomonas* clones carrying the *nos*Z gene were more dominant in stabilized dunes than in non-vegetated shifting dunes. In contrast, *Mesorhizobium* clones carrying the *nir*K gene were more abundant in shifting dunes. Meanwhile, *Azospirillum* was the most dominant *nir*S bacterium in shifting sand-dune soils. This study provides one of the few indications that *Azospirillum* may play a significant role in soil denitrification of sandy soils; however, its occurrence in natural sandy soils remains unknown.

In degraded semiarid Mediterranean regions, the indigenous vegetation also serves as a viable alternative for restoring degraded lands. To facilitate revegetation, Schoebitz et al. (2014) utilized the shrub *Cistus albidus* plants and treated them with A. *brasilense* and *Pantoea dispersa* immobilized in clay and/or organic residues.

Their results showed the most significant enhancement of plant biomass in plants treated with the inoculant amended with organic residue. In the rhizosphere of plants from the same treatment, they found increased uptake of K and P, as well as enhanced enzymatic activities of dehydrogenase, urease, and protease, and higher total carbon, organic carbon, and microbial biomass carbon content, thereby improving soil quality.

In Brazil, around 60% of pasturelands show some level of degradation (Bolfe et al. 2024). The study by Hungria et al. (2016) suggested that revegetation with grasses inoculated with *A. brasilense* could aid in reclaiming these lands and utilizing the grass for forage production. In their two-year field experiment conducted at three sites, they assessed the effect of *A. brasilense* on the growth of *Brachiaria* spp. They found that in plants treated with *A. brasilense*, the nitrogen content in the shoots was comparable to that of a second nitrogen application. Therefore, using the inoculant represents a cost-effective option for obtaining *Brachiaria* pastures and contributing to the reclamation of these lands.

Mangroves in arid regions serve as ecosystems that act as oases, contributing to the biodiversity of their locations while also providing typical mangrove services, such as coastal protection, climate change mitigation, and enhanced water quality in their surroundings (Adame et al. 2020). Currently, the world's mangroves face risks from decreasing surface area and various environmental issues, including exposure to heavy metals and other pollutants (Adame et al. 2020; Hossain et al. 2022). There has been limited research on the potential of using plant growth-promoting bacteria (PGPB) to enhance the growth of tropical mangroves, and even less on *Azospirillum*, despite its proposed application in arid zone mangroves being suggested 25 years ago. Puente et al. (1999) analyzed the early colonization pattern in *Avicennia germinans* seedlings inoculated separately with *A. halopraeferens* and *A. brasilense*. It was found that both species survived in seawater and effectively colonized the rhizoplane of the seedlings, although with differences in aggregation patterns; while *A. brasilense* produced microaggregates, *A. halopraeferens* remained in individual cells covered by a thick film and still reached high populations in the rhizoplane. This study has opened up the possibility of enhanced production of mangrove plants in controlled conditions, supporting restoration programs; however, there are no recent studies in this regard.

23.6 Reclamation of Mine Tailings

Phytostabilization of metalliferous mine tailings offers an alternative solution to reduce health risks associated with atmospheric pollution from suspended particles and leaching into groundwater (Mendez and Maier 2008). In a greenhouse experiment, de-Bashan et al. (2010) examined whether *A. brasilense* Sp6 could survive in moderately acidic mine tailings collected in southern Arizona that were amended with compost while still promoting growth in quailbush. Using fluorescence in situ hybridization, the authors demonstrated successful root colonization and persistence of the bacteria in the rhizoplane for 60 days. Additionally, only the inoculated plants exhibited a significant increase in biomass. Moreover, PCR-DGGE revealed changes in the structure of rhizospheric bacterial communities between non-inoculated and inoculated plants.

The impact of arbuscular mycorrhizal fungi (AMF) in consortium with the nitrogen-fixing bacteria *Azospirillum* sp. and *Nostocedaphicum* on the growth of mouse-ear-hawkweed *Hieracium pilosella* and *Medicago sativa* in a Zn-Pb-rich site was evaluated by Ogar et al. (2015). The study results revealed that biomass, as well as dry shoot and root mass, significantly increased in trials where plants were inoculated with mycorrhizal fungi and nitrogen-fixing bacteria. The authors emphasized the potential to improve the revegetation of industrial sites contaminated with heavy metals by selecting symbionts that facilitate appropriate interactions.

Meyer et al. (2024) investigated the effect of an inoculant containing *A. brasilense* and the arbuscular mycorrhizal fungus (AMF) Rhizophagus intraradices on plant growth and soil biological activity in a coal mining area of southern Brazil that is undergoing reclamation with key pioneering herbaceous plants. The mixed plant community, comprising *Lablab purpureus*, *Paspalum notatum*, *Crotalaria juncea*, *Neonotonia wightii*, *Stylosanthes guianensis*, *Andropogon gayanus*, and *Trifolium repens*, was subjected to single or co-inoculation with both microorganisms. The authors found that gramineous plants dominated the initial stages of plant succession during reclamation. Inoculation or co-inoculation induced increased soil glomalin, acid phosphatase activity, and fluorescein diacetate activity. In addition, inoculated plants showed higher nitrogen (N) and phosphorus (P) content and yielded higher biomass, where *Crotalaria juncea* exhibited the highest production.

23.7 Methodological Strategies for the Recovery of Degraded Soils Assisted by *Azospirillum*

Only a few examples address the long-term potential of *Azospirillum* for solving environmental issues. The primary example is studies from Northwestern Mexico over the last three decades. For review, see Puente and Bashan (1993), Bashan et al. (2000), Carrillo-Garcia-Garcia et al. (2000), Bashan et al. (2002), Bashan et al. (2008), and Moreno et al. (2017). Key findings include a methodological framework for revegetating degraded soils with *Azospirillum*, summarized in Fig. 23.2. Strategies and expected outcomes are described below, including new tools to evaluate effects on plants and soil.

23.7.1 Selection of Plant Species for Revegetation

This strategy is based on the principles of ecological succession of desert plant communities dominated by shrub and succulent forms. During ecological facilitation of nurse plants to cacti, the nurse is represented by leguminous shrubs that accelerate

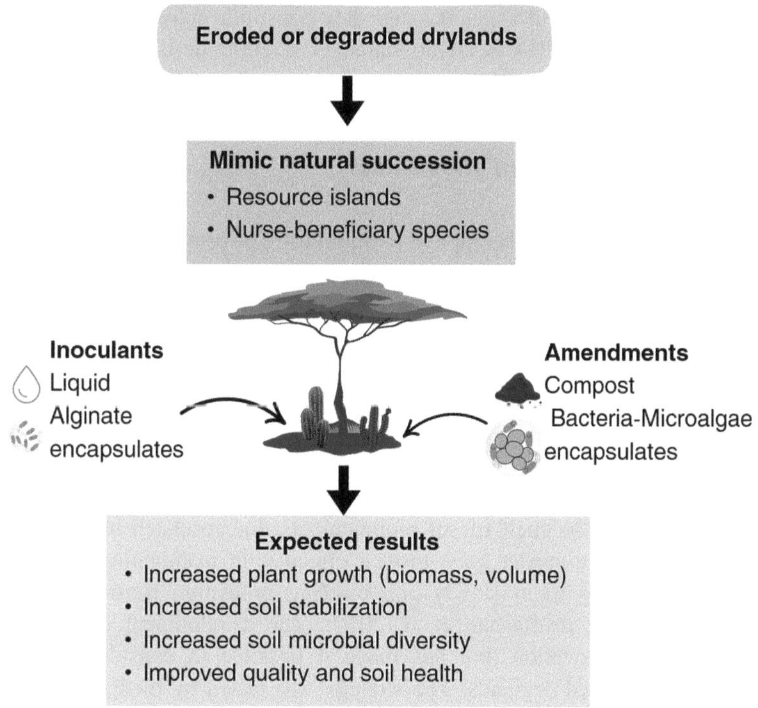

Fig. 23.2 Methodological framework and strategies for the revegetation of degraded soils assisted by *Azospirillum*

the succession process by growing faster than their beneficiary species (cactus) (Verdu et al. 2010). Under their canopy, legume trees usually develop resource islands with higher nutrient content, humidity, and microbiota than bare soil (de-Bashan et al. 2022). These conditions favor the growth of slow-growing species such as long-lived cactus (Carrillo-Garcia et al. 2000; Verdu et al. 2010).

23.7.2 Azospirillum *and Other Growth-Promoting Microorganisms*

A broad diversity of *Azospirillum* strains can be used due to their high affinity for associating with different botanical families. Although some isolates have been obtained from native desert plants, this process is laborious; thus, model strains have more experimental support. The inoculants can be supplemented with native PGPB strains tolerant to drought and high temperatures, such as *Bacillus pumilus* isolated from the roots of the giant cardon, *P. pringlei.* In addition, mycorrhizal fungi can stimulate a synergistic effect in the initial establishment of resource islands, which are the target in arid zones because they constitute active centers of nutrient recycling (Bashan et al. 2000; Carrillo-Garcia et al. 2000; de-Bashan et al. 2022).

23.7.3 Formulations of Inoculants

The microbial inoculant can be applied in liquid form or using a carrier at this stage. The vacuum infiltration technique has been effectively used for bacterial solutions. Using this procedure, *P. pringlei* seeds inoculated with *A. brasilense* sp-245 at 1.0×10^6 CFU mL^{-1} showed increased germination and colonization of seedlings ten months post-inoculation (Puente and Bashan 1993). In experiments with other columnar cacti, three annual inoculations were required to maintain a survival rate close to 75% after 3.5 years (Bashan et al. 1999), representing a significant advantage in the harsh conditions of the desert.

While liquid inoculants may have advantages for initial plant establishment, dry formulations extend the shelf life of bioinoculants. Encapsulated inoculants using alginate polymer as a carrier have proven effective in maintaining bacterial cell viability by releasing them slowly and offering protection from soil predators (Bashan et al. 2002; de-Bashan et al. 2010). The development of dry alginate inoculants is an innovation that has made it possible to consider the assisted restoration of degraded drylands. The strategy was tested by Bashan et al. (2002, 2009) on cacti and different legumes and consists of the application of dry alginate microencapsulates carrying *A. brasilense* Cd and *B. pumilus* at doses of 1×10^6 CFU g^{-1} soil and 1.2×10^6 CFU g^{-1} soil, respectively. To reinforce nutrient absorption, non-sterile native soil carrying AM fungi was added at the time of transplant. This inoculum is initially obtained from the soil of mature resource islands of *P. articulata* and subsequently propagated in sorghum as a trap plant to reach root colonization levels of 54.5%.

23.7.4 Soil Amendments

The experiments of Bashan and Vazquez (2000) demonstrated improved retention of *Azospirillum* in arid zone soils by adding small amounts of organic matter. This led to applying compost to desert soils to enhance their physicochemical properties, primarily water-holding capacity and nutrient availability (Bashan et al. 2004), to emulate conditions similar to resource islands and improve the establishment of cacti. Bacilio et al. (2006) analyzed the effect of dairy-wheat compost applied at dilutions of 1:2, 1:4, 1:8, 1:16, 1:32, 1:64, 1:128, and 1:256 v/v, with and without *A. brasilense* Cd. They observed that soil improved with compost at dilutions of 1:4 to 1:32, exhibiting superior effects on *P. pringlei* biomass production compared to rich resource island soils. However, it lost its effect at low dilutions (1:128 to 1:256). On the other hand, the inoculation effect was more evident at lower compost concentrations, and consistently, in soil directly from resource islands, the bacteria seemed to lose their effect.

Improvement of eroded desert soils has also been accomplished by adding alginate encapsulates containing *A. brasilense* and *Chlorella sorokiniana*, previously used to remove nutrients from wastewater. Lopez et al. (2013) report that the recovered co-immobilized material is dried at 40–43 °C for 24 h. They then took the resulting debris and applied it to the soil in which they aimed to grow a model plant, in this case, sorghum. Then they recovered the soil and recycled it to regrow plants, completing three cycles. When analyzing the roots by fluorescent in situ hybridization, they found that *A. brasilense* maintained its ability to colonize the tips of the roots after each of the three cultivation cycles. Then, denaturing gradient gel electrophoresis (DGGE) showed greater diversity and richness of bacterial communities at increasing cycles, which suggested positive changes in soil fertility.

23.7.5 Expected Results and Current Tools for Their Evaluation in the Field

Throughout the previous sections, it has been established that *Azospirillum* effectively assists in revegetation processes for the recovery of degraded soils. The main effects on plants include increased survival and establishment, greater biomass and height, and faster growth rates. Meanwhile, changes in the soil's physical, chemical, and biological properties indicate improvements in soil quality and health. However, under arid conditions, some effects on plants and soils often take a long time to become evident (de-Bashan et al. 2012). For some plant life forms, such as shrubs, the effect is relatively easy to measure; however, evaluating the effect on cacti remains a challenge. Conventionally, destructive techniques have been used to assess short-term effects (Lopez et al. 2012); however, for long-lived species, the only option is measuring biovolume (Moreno et al. 2017). New perspectives are emerging with the implementation of non-destructive technologies for in situ plant health assessment based on the Normalized Difference Vegetation Index (Zsebő et al. 2024). Regarding soil indicators, assessing nitrogen fixation through the abundance of the nifH gene has shown long-term changes in nitrogen fixation when using native trees and cacti (Lopez-Lozano et al. 2016); however, these need to be evaluated at shorter intervals. Currently, the analysis of microbial communities through high-throughput sequencing is helping to bridge the taxonomic and functional gaps. Nonetheless, validation with enzymatic activities (protease activity, microbial carbon biomass) remains challenging due to the low amount of organic matter in degraded soils.

23.8 Bioremediation

The toxicity caused by the excessive use of petroleum and the accumulation of heavy metals in the environment has raised global concern due to its detrimental effects (Wuana and Okieimen 2011; Kaushal and Wani 2015; Fan et al. 2017). Microbial bioremediation is a safe alternative that benefits from microorganisms' ability to adapt to environments contaminated by xenobiotics, using them as a source of carbon and energy (Rai et al. 2022). This section explores how *Azospirillum* spp's ability to adapt to challenging environments with limited nutrients (Pérez-Castañeda et al. 2011; Cui et al. 2022) makes it a promising candidate for research on remediation processes in environments contaminated with hydrocarbons and heavy metals.

23.8.1 Hydrocarbons Degradation

Azospirillum has mainly been studied as a plant growth promoter, yet growing evidence suggests its potential for xenobiotic degradation. In the 1990s, researchers tested strains isolated from crop rhizospheres for their tolerance to xenobiotics such as phenol and benzoate. Barkovsky et al. (1995) evaluated 31 strains of *Azospirillum* (21 *A. brasilense* and 10 *A. lipoferum*) for phenol and benzoate degradation. Nine strains degraded benzoate, including five *A. brasilense* and four *A. lipoferum*, while three *A. brasilense* strains degraded phenol. In 1993, López de Victoria and Lovell studied the chemotaxis of *A. lipoferum* 59b and two *A. brasilense* strains (Sp7 and CD) toward aromatic compounds such as benzoate and catechol, finding that *A. brasilense* CD was more sensitive to all tested compounds than *A. brasilense* Sp7, indicating strain specificity.

In 2010, Galazca et al. evaluated soils contaminated with anthracene, phenanthrene, and crude oil, as well as their effects on maize (*Zea mays*) plants inoculated with *Azospirillum* and *Pseudomonas stutzeri*. A reduction in hydrocarbon content was shown when these bacteria were introduced. Similarly, Cruz-Hernández et al. (2019) conducted a study on 19 strains of *A. brasilense* to assess their tolerance to phenanthrene, xylene, toluene, and naphthalene, as well as their biosurfactant production. The findings revealed that the strains were sensitive to the evaluated hydrocarbons despite producing biosurfactants.

Crude oil has been used as a carbon source for *Azospirillum*. Muratova et al. (2005) analyzed the tolerance and degradation of 33 *Azospirillum* strains using 1% light crude oil as the sole carbon source. After 14 days, *A. lipoferum* SR42 and *A. brasilense* SR80 degraded the oil by 57.5% and 56.5%, respectively. Additionally, *A. brasilense* SR80 showed positive associative capacity, being attracted to wheat exudates, colonizing roots, and maintaining its ability for IAA synthesis.

In a separate study, Miranda Martínez et al. (2007) evaluated a consortium of bacterial strains isolated from contaminated soils, including *A. brasilense* Cd and

A. halopraeferens. They analyzed population dynamics on sand inoculated with phenanthrene, both with and without German grass (*Echinochloa polystachya*). The study assessed the effects of bacterial consortium inoculation, non-inoculation, and the independent presence of each bacterium. After 60 days of inoculation, phenanthrene degradation was significantly higher in plants inoculated with *A. brasilense* Cd (57.02%) and in those inoculated with the bacterial consortium (59.01%), compared to the control group without plants (41.7%).

Al-Jawhary et al. (2022) evaluated *Azospirillum* spp., *Azotobacter* spp., and *Rhizobium* spp. for crude oil bioremediation in soil under laboratory conditions and found that *Azospirillum* spp. and *Rhizobium* spp. were more efficient in bioremediation compared to *Azotobacter* spp. The highest bacterial count was recorded in the soil treatment inoculated with *Rhizobium* spp. and crude oil during the first week, while the lowest count was observed in the control treatment. However, no statistically significant differences were found between treatments.

Possible synergistic effects of *Pleurotus ostreatus Florida* and *A. brasilense* were assessed by Pozdnyakova et al. (2022) during the degradation of polycyclic aromatic hydrocarbons (PAHs). Co-cultivation of these strains resulted in higher mycelial dry weights, increased bacterial colony-forming units, and hydrocarbon transformation. Therefore, they concluded that these strains exerted a mutually positive effect. Later, Nanekar and Kokitkar (2024) examined four bacterial strains, *Rhodococcus ruber* S9D2, *Azospirillum zeae* S13D1, *Pseudomonas chengduensis* S4P2, and *Pseudomonas donghuensis* S14P1, in relation to microbial enzyme activity associated with hydrocarbon degradation. It was demonstrated that *Pseudomonas* and *Azospirillum* exhibited enzymatic activity that could facilitate the degradation of complex petroleum hydrocarbons.

Most of the above-mentioned studies used strains isolated from plant-associated environments; however, around 2002, reports began to emerge on the isolation of *Azospirillum* strains from contaminated soils. For instance, Ecfkord et al. (2002) reported the isolation of five nitrogen-fixing strains, including one *A. brasilense*, from gasoline-contaminated soils in Antarctica. Alternatively, Nozawa et al. (2005) analyzed the bacterial community composition in microcosms with perchlorate and enriched with acetate or hydrogen to study perchlorate reduction rates and their effect on the microbial community. Partial sequencing of 16S rRNA genes recovered from the microcosms and phylogenetic analysis indicated the presence of *Azospirillum* spp. in both environments.

By 2008, Young et al. had reported the first isolation of an *Azospirillum* strain from oil-contaminated soil. Comparative analysis of the 16S rRNA gene sequences demonstrated that it belonged to a new *Azospirillum* species, which was named *Azospirillum rugosum*. Comparably, Lin et al. (2009) isolated a strain from tar residues, which, after characterization, was identified as a new species, *Azospirillum picis*. In the same way, Zhou et al. (2012) isolated a nitrogen-fixing strain, designated SgZ-5, from a microbial fuel cell (MFC), characterizing it as a novel species named *Azospirillum humicireducens*. Wu et al. (2020) reported two isolates, RWY-5-1-1T and ROY-1-1-2, obtained from a petroleum mixture in China. These strains were characterized phenotypically, genotypically, and chemotaxonomically, leading to their classification as a new species, *Azospirillum oleiclasticum*. Degradation assays and biosurfactant production analyses in NFB medium revealed that the petroleum degradation rate of the strain was 36.2% after 14 days of incubation. Additionally, stable emulsions were formed using diesel oil, kerosene, and soybean oil as carbon sources. This study demonstrated that these strains exhibited petroleum degradation capacity associated with plant growth promotion.

A plant microbiome-based bioremediation system for treating crude oil contamination was evaluated by Saeed et al. (2021). From ten PGPR strains isolated from contaminated soil, they selected *Pseudoarthrobacter phenanthrenivorans* (MS2) and *Azospirillum oryzae* (MS6) to be inoculated individually and in combination in the rhizospheric soil of maize along with hydrocarbons. The degradation efficiency of the consortium was higher than that of individual inoculations, in terms of both plant development parameters and hydrocarbon degradation. Nanekar and Kokitkar (2023) isolated bacterial strains from contaminated soils, selecting the two most efficient for growth evaluation in diesel. Through 16S rRNA gene sequencing and phylogenetic analysis, the strains were identified as *R. ruber* and *A. zeae*. *R. ruber* exhibited optimal growth at a diesel concentration of 2% at 37 °C and pH 7.0, whereas *A. zeae* showed optimal growth at a 1% diesel concentration, pH 6.0, and a temperature of 30 °C.

23.8.2 Heavy Metals

The inoculation of plants with PGPR has been proposed as an effective biotechnological tool to enhance crop yield and mitigate stress caused by heavy metals and metalloids. One of the widely studied species is *Azospirillum* spp. For instance, Langenbach et al. (1988) assessed the potential effects of Hg, Cd, Zn, and Pb on the growth and nitrogenase activity of *Azospirillum* spp. At 0.1 and 1 ppm concentrations, nitrogenase activity and bacterial growth were inhibited. Lead (1 ppm) inhibited 25% of acetylene reduction activity (ARA); however, bacterial growth was not affected by lead oxide or lead chloride. Zinc (1 ppm) inhibited ARA by 50%, while Hg and Cd inhibited the same 50% of ARA at lower concentrations (0.4 ppm). Similarly, Lyubum et al. (2006) investigated the effects of *A. baldaniorum* Sp245 on the morphological parameters of wheat (*Triticum aestivum* 'Saratovskaya 29') in a hydroponic system contaminated with arsenic (75, 750, and 7500 μg L^{-1}). Their findings showed that wheat plants inoculated with *Azospirillum* accumulated less arsenic than non-inoculated surface-sterilized plants.

Vezza et al. (2019) evaluated the viability of *A. brasilense* Cd in arsenic-contaminated agricultural soils. This strain tolerated arsenic concentrations commonly found in soil. Moreover, key bacterial traits such as root colonization, motility, exopolymer and biofilm formation, auxin production, siderophore synthesis, and nitrogen fixation were not significantly affected.

Ganesh et al. (2015) studied the toxic effects of chromium on the growth and yield of rice plants in chromium-contaminated soil inoculated with bacteria, including *Azospirillum*. The highest morphological and yield parameters were recorded in contaminated soil treated with *Azospirillum*, suggesting its potential for mitigating chromium toxicity. In addition, Vázquez et al. (2021) examined the effects of inoculating wheat plants containing Cd^{2+} with *Azospirillum argentinense* Az39. They observed that inoculated plants exhibited reduced cadmium and increased growth compared to non-inoculated plants. Also, in 2023, Cui et al. evaluated the combined application of *A. brasilense* and an exogenous H_2S signaling molecule on plant growth promotion and cadmium mitigation. Their findings demonstrated a reduction in Cd content, suggesting that this approach could be a viable strategy for lowering xenobiotic accumulation in crops.

In earlier studies, *Azospirillum* was used alone; however, there are studies on bacterial consortia containing *Azospirillum* strains. For instance, Belimov and Dietz (2000) evaluated the response of hydroponically grown barley seedlings inoculated with four rhizobacteria, including *Azospirillum lipoferum* 137, and treated with $CdCl_2$ (50 μM). Their results indicated that bacterial inoculation slightly stimulated root length and biomass in Cd-treated seedlings, partially alleviating Cd toxicity by enhancing nutrient uptake. Similarly, Belimov et al. (2004) reported resistance to lead (Pb) and cadmium (Cd) on the rhizobacteria *A. lipoferum* 137, *Arthrobacter mysorens* 7, and *Agrobacterium radiobacter* 10. In pot and field experiments, seed inoculation with bacteria improved barley growth and nutrient uptake from the contaminated soil. It also prevented Pb and Cd accumulation, mitigating their toxic effects.

Akond and Khan (2005) isolated five *Azospirillum* strains from rice root samples, identifying three as *A. brasilense* (N2, N18, and N21) and two as *Azospirillum amazonense* (N5 and N15). Nitrogen-fixing potential declined in all strains as heavy metal concentrations increased, with each strain showing decreased growth. Mercury and silver completely inhibited growth, while zinc inhibited strains N2 and N18. The study reported significant differences in heavy metal response patterns, even at the same concentration, impacting nitrogen-fixing potential and nitrogen-dependent growth in *Azospirillum* strains.

A. baldaniorum Sp245 and *A. brasilense* Sp7 were tested with heavy metals (Co^{2+}, Cu^{2+}, Zn^{2+}) by Kamnev et al. (2005). They found that in strain Sp7 heavy metals induced a higher accumulation of polyhydroxybutyrate (PHB) compounds. Conversely, strain Sp245 exhibited significantly lower metal uptake. The authors suggested that these behavioral differences could be attributed to their ability to adapt to stressful conditions due to their original ecological niches. It was also observed that adding Cu^{2+} or Cd^{2+} to the culture medium significantly reduced the levels of indole-3-acetic acid (IAA) produced by both bacterial strains.

Similarly, Arora et al. (2015) assessed the bioremediation potential of *Panicum virgatum* (switchgrass) in combination with fungi (*Glomus mossei*, *G. fasciculatum*, and *Gigaspora margarita*) and *Azospirillum* against lead and cadmium contamination. Growth parameters and bioremediation potential of fungi and *Azospirillum* were compared at different Pb and Cd concentrations. The results showed that fungi and *Azospirillum* enhanced root length, branching, surface area, and root and shoot biomass. The highest soil metal uptake was recorded after fungal and *Azospirillum* inoculation, while the lowest bioconcentration factor (BCF = 0.05) was observed at 10 ppm Cd. Consequently, switchgrass inoculated with fungi and *Azospirillum* could be an effective phytoremediator (phytoaccumulator/phytostabilizer). In a different study, Rojas et al. (2016) evaluated compost supplemented with heavy metals (As, Cd, Cr, Cu, Ni, Pb) and its effects on the growth of *A. brasilense* and *G. intraradices*. They observed that compost had no adverse effects on the growth of *A. brasilense* and *G. intraradices*, as both microorganisms were successfully isolated and identified in the plant rhizosphere. In the case of *A. brasilense*, the bacterium exhibited tolerance to high concentrations of arsenic (0.4375 mg kg^{-1}), lead (62.375 mg kg^{-1}), and copper (91.1875 mg kg^{-1}) but was susceptible to cadmium (0.5 mg kg^{-1}).

In research reported by Balakrishnan et al. (2017), different heavy metal concentrations (Cd, Cr, Cu, Ni, Pb, Fe, Mn, and Zn) were evaluated in soil and the rhizosphere of *Avicennia marina* in Indian mangroves, along with an analysis of metal-resistant bacteria. The results indicated that rhizosphere soil contained higher metal concentrations than non-rhizosphere soil. Compared to non-rhizosphere soil, Cd, Fe, Mn, and Zn levels were 6.0–16.7% higher, while Cr, Cu, Ni, and Pb were 1.7–2.8% higher. The analysis of heavy metal-resistant bacteria identified strains of *Pseudomonas*, *Azotobacter*, *Shewanella*, and *A. brasilense*. It was suggested that heavy metal-resistant bacteria could serve as indicators of heavy metal contamination and be utilized in the bioremediation of contaminated sites. Xu et al. (2018) evaluated *Bacillus subtilis* and *A. brasilense*, both abscisic acid (ABA)-producing bacteria, for their effect on cadmium accumulation in plants. Results showed significant ABA increases in wild-type *Arabidopsis* and reduced Cd levels in tissues, mitigating Cd toxicity and enhancing growth under exposure. This suggests these bacteria may lower Cd levels via ABA production.

Armendáriz et al. (2019) evaluated dual inoculation with *Bradyrhizobium japonicum* E109 and *A. argentinense* Az39 strains in arsenic-treated plants. An in vitro assay assessed bacterial population dynamics with 25 μM of AsV and AsIII. A synergistic effect occurred during co-cultivation, reducing mortality compared to each culture alone. In vivo assays showed enhanced plant growth under arsenic stress, with increased nodules and nitrogen content from dual inoculation. Additionally, arsenic translocation to aerial parts decreased, improving phytostabilization potential in soybean plants. These results suggest that dual inoculation with *B. japonicum* E109 and *A. argentinense* Az39 may effectively enhance the growth and yield of soybean plants exposed to arsenic.

Pan et al. (2019) evaluated the effect of the abscisic acid-producing bacteria *A. brasilense* and *Bacillus subtilis* on the growth of *Brassica chinensis* cultivated in cadmium-contaminated soil. They found that compared to inoculated control plants, biomass increased, and Cd concentrations decreased by 40–79% and 43–77%, respectively. Thus, applying ABA-producing bacteria could be an alternative strategy to enhance biomass production and improve the quality of plants grown in cadmium-contaminated soils.

In summary, xenobiotics can damage plants by inhibiting growth and reducing physiological and biochemical activities. The persistence and bioavailability of heavy metals and hydrocarbons depend on factors like environmental conditions,

Table 23.1. Examples of the role of *Azospirillum* alone or in interaction with plants in the degradation of xenobiotic compounds

Type of contaminant	Bacterial isolate (s)	Model plant	Trait/effect	References
Hydrocarbons				
Oil	*A. rugosum*	None	Natural adaptation to oil-contaminated soil	Young et al. (2008)
Tar residues	*A. picis*	None	Natural inhabitant of tar residues	Lin et al. (2009)
Petroleum mixture	*A. oleiclasticum* RWY-5-1-1T and ROY-1-1-2	None	Petroleum degradation and putative nitrogen fixation	Wu et al. (2020)
Phenol and benzoate degradation	21 strains of *A. brasilense*, 10 strains of *A. lipoferum*	None	Phenol and benzoate degradation	Barkovsky et al. (1995)
Benzoate and catechol	*A. lipoferum* 59b, *A. brasilense* strains Sp7 and CD	None	Strain-specific chemotaxis toward aromatic compounds	López de Victoria and Lovell (1993)
Anthracene, phenanthrene, and crude oil	*Azospirillum* and *P. stutzeri*	Maize *Zea mays*	Inoculation-induced hydrocarbon degradation	Galazca et al. (2010)
Phenanthrene, xylene, toluene, and naphthalene	19 strains of *A. brasilense*	None	Sensitivity to all compounds and surfactant production by the strains	Cruz-Hernández et al. (2019)
Crude oil	33 *Azospirillum strains*	Wheat *Triticum aestivum*	Oil degradation by *A. lipoferum* SR42 and *A. brasilense* SR80. The latter maintained PGPB traits (root colonization and IAA production)	Muratova et al. (2005)
Phenanthrene	Consortium of *A. brasilense* Cd and *A. halopraeferens*	German grass, *Echinochloa polystachya*	Phenanthrene degradation was stimulated by the presence of plants	Miranda Martínez et al. (2007)
Soil contaminated with crude oil	*Azospirillum* spp., *Azotobacter* spp., and *Rhizobium* spp.	None	*Azospirillum* spp. and *Rhizobium* spp. were more efficient in soil bioremediation	Al-Jawhary et al. (2022)
Polycyclic aromatic hydrocarbons	*P. ostreatus* Florida and *A. brasilense*	None	Co-cultivation induced degradation of aromatic compounds	Pozdnyakova et al. (2022)

(continued)

Table 23.1. (continued)

Type of contaminant	Bacterial isolate (s)	Model plant	Trait/effect	References
Hydrocarbons	*R. ruber* S9D2, *A. zeae* S13D1, *P. chengduensis* S4P2, and *P. donghuensis* S14P1	None	*A. zeae* and *P. chengduensis* showed enzyme activity associated with the degradation of complex petroleum hydrocarbons	Nanekar and Kokitkar (2024)
Gasoline-contaminated soils	*A. brasilense* strains	None	Inhabitants of soil contaminated with gasoline. Nitrogen fixers	Ecfkord et al. (2002)
Crude oil	*P. phenanthrenivorans MS2* and *A. oryzae MS6*	None	Degradation of crude oil was stimulated by the synergistic effect of both strains	Saeed et al. (2021)
Diesel	*R. ruber and A. zeae*	None	Enhanced growth of *Rhodococcus ruber* in 2% diesel	Nanekar and Kokitkar (2023)
Heavy metals				
Hg, Cd, Zn, and Pb	*Azospirillum* spp.	None	Nitrogenase activity was reduced by 25–50% at an average concentration of one ppm	Langenbach et al. (1988)
Ace	*A. baldaniorum* Sp245	Wheat *Triticum aestivum*	Plants inoculated with *Azospirillum* accumulated less arsenic than non-inoculated plants	Lyubum et al. (2006)
Ace	*A. brasilense* Cd	None	The strain tolerated moderate As concentrations in soils while PGPB traits (motility, IAA production, siderophore production, and biofilm formation) remained active	Vezza et al. (2019)
Cr	*Azospirillum* sp.	Rice *Oryza sativa*	Higher yield in plants inoculated with *Azospirillum*, mitigation of chromium toxicity	Ganesh et al. (2015)
CD	*A. argentinense* Az39	Wheat *Triticum aestivum*	Inoculated plants exhibited reduced cadmium and increased growth compared to non-inoculated plants	Vázquez et al. (2021)
CD	*A. brasilense* and a signal molecule	Pak choi *Brassica chinensis*	Inoculated plants exhibited reduced cadmium	Cui et al. (2022)
$CdCl_2$	Conscrtium of bacteria including *A. lipoferum* 137	Barley *Hordeum vulgare*		Belimov and Dietz (2000)

			Bacterial inoculation slightly stimulated root length, improved nutrient uptake, and biomass in Cd-treated seedlings	
Pb and Cd	*A. lipoferum* 137, *A. mysorens* 7, and *A. radiobacter* 10	Barley *Hordeum vulgare*	Resistance to Pb and Cd by the three strains. Inoculation of seeds induced growth and nutrient uptake and prevented heavy metal accumulation in the plants	Belimov et al. (2004)
Hg, Ag, and Zn	*A. brasilense* (N2, N18, and N21) and *A. amazonense* (N5 and N15)	None	Gradual decrease in nitrogenase-fixation potential as the metal concentration increased	Akond and Khan (2005)
Co, Cu, and Zn	*A. baldaniorum* Sp245 and *A. brasilense* Sp7	None	Increased metal content induced polyhydroxybutyrate accumulation in strain Sp7, while Sp245 showed lower metal uptake	Kamnev et al. (2005)
Pb and Cd	*Glomus mossei*, *G. fasciculatum*, *Gigaspora margarita*, and *Azospirillum* sp.	Switchgrass *Panicum virgatum*	Co-inoculation with fungi and *Azospirillum* increased root length, branching, surface area, root and shoot biomass, and induced metal uptake by plants	Arora et al. (2015)
Cd, Cr, Cu, Ni, Pb, Fe, Mn, and Zn	None	Black mangrove *Avicennia marina*	Rhizosphere soil contained higher metal concentrations than non-rhizosphere soil. *Pseudomonas, Azotobacter, Shewanella,* and *Azospirillum brasilense* were considered indicators of heavy metal contamination	Balakrishnan et al. (2017)
CD	*B. subtilis* and *A. brasilense*	*Arabidopsis*	Improved plant growth, mitigation of Cd toxicity mediated by Abscisic acid	Xu et al. (2018)
Ace	*B. japonicum* E109 *and A. argentinense* Az39	*Soybean Glycine max*	Dual inoculation in plants increased nodules and nitrogen content, reducing As translocation	Armendáriz et al. (2019)
CD	Abscisic acid producers: *A. brasilense* and *B. subtilis*	Pak choi *Brassica chinensis*	Plant biomass increased, and Cd content decreased in inoculated plants	Pan et al. (2019)

pH, and specific plant species. Microorganisms that degrade xenobiotics (Saeed et al. 2021) are scarce, making research on these strains crucial; *Azospirillum* strains show promise, as seen in Table 23.1. Their beneficial effects rely on viability and functionality in adverse conditions, especially in contaminated environments, which calls for further research in bioremediation applications.

23.9 Conclusions

In addition to its well-known benefits in agriculture, *Azospirillum* has potential applications for addressing environmental issues. It can help prevent soil erosion by promoting plant growth on barren and degraded lands that have lost their ability to regenerate. Moreover, *Azospirillum* can play a role in bioremediation efforts to decontaminate soils, leading to healthier environments. Various strains of *Azospirillum* have been researched and used as inoculants for different crops due to their growth-promoting properties and ability to thrive in degraded or contaminated environments. Their nitrogen-fixing capability, production of phytohormones, and use of contaminants as a source of carbon and energy contribute to significant ecological interactions between plants and microorganisms in hostile environments.

References

Adame MF, Reef R, Santini NS et al (2020) Mangroves in arid regions: ecology, threats, and opportunities. Estuar Coast Shelf Sci 248:106796

Al-Jawhari IF, Flayyih AI, AL-Mansor KJ et al (2022) Bioremediation of crude oil and its effect of residue in growth of wheat plants. GSCARR 13(1):188–199

Akond MA, Khan ZU (2005) Role of carbon sources and heavy metals on growth and nitrogen fixing potential of *Azospirillum* and the effect of *Azospirillum* strains on vegetative growth of rice. J Environ Sci Stud (Dhaka) 3:1–8

Armendariz AL, Talano MA, Nicotra MF et al (2019) Impact of double inoculation with *Bradyrhizobium japonicum* E109 and *A. brasilense* Az39 on soybean plants grown under arsenic stress. Plant Physiol Biochem 138:26–35

Arora K, Sharma S, Monti A (2015) Bio-remediation of Pb and Cd polluted soils by switchgrass: a case study in India. Int J Phytoremediation 18(7):704–709

Bacilio M, Hernandez J, Bashan Y (2006) Restoration of giant cardon cacti in barren desert soil amended with common compost and inoculated with *Azospirillum brasilense*. Biol Fertil Soils 43(1):112–119

Balakrishnan B, Sahu BK, Ranishree JK et al (2017) Assessment of heavy metal concentrations and associated resistant bacterial communities in bulk and rhizosphere soil of *Avicennia marina* of Pichavaram mangrove, India. Environ. Earth Sci 76(1)

Barkovskii AL, Korshunova VE, Pozdnyacova LI (1995) Catabolism of phenol and benzoate by *Azospirillum* strains. Appl Soil Ecol 2(1):17–24

Bashan Y (1990) Short exposure to *Azospirillum brasilense* Cd inoculation enhanced proton efflux of intact wheat roots. Can J Microbiol 36(6):419–425

Bashan Y, Holguin G (1994) Root-to-root travel of the beneficial bacterium *Azospirillum brasilense*. Appl Environ Microbiol 60(6):2120–2131

Bashan Y, Holguin G (1995) Inter-root movement of *Azospirillum brasilense* and subsequent root colonization of crop and weed seedlings growing in soil. Microb Ecol 29:269–281

Bashan Y, Levanony H (1989) Factors affecting adsorption of *Azospirillum brasilense* Cd to root hairs as compared with root surface of wheat. Can J Microbiol 35(10):936–944

Bashan Y, Levanony H (1991) Alterations in membrane potential and in proton efflux in plant roots induced by *Azospirillum brasilense*. Plant soil 137:99–103

Bashan Y, Vazquez P (2000) Effect of calcium carbonate, sand, and organic matter levels on mortality of five species of Azospirillum in natural and artificial bulk soils. Biol Fertil Soils 30(5–6):450–459

Bashan Y, Mitiku G, Whitmoyer RE et al (1991) Evidence that fibrillar anchoring is essential for *Azospirillum brasilense* Cd attachment to sand. Plant Soil 132:73–83

Bashan Y, Rojas A, Puente ME (1999) Improved establishment and development of three cactus species inoculated with *Azospirillum brasilense* transplanted into disturbed urban desert soil. Can J Microbiol 45(6):441–451

Bashan Y, Davis E, Carrillo-Garcia A et al (2000) Assessment of VA mycorrhizal inoculum potential in relation to the establishment of cactus seedlings under mesquite nurse-trees in the Sonoran Desert. Appl Soil Ecol 14(2):165–175

Bashan Y, Hernandez JP, Leyva LA et al (2002) Alginate microbeads as inoculant carriers for plant growth-promoting bacteria. Biol Fertil Soils 35:359–368

Bashan Y, Holguin G, De-Bashan LE (2004) Azospirillum-plant relationships: physiological, molecular, agricultural, and environmental advances (1997-2003). Can J Microbiol 50:521–577

Bashan Y, Puente ME, de-Bashan LE et al (2008) Environmental uses of plant growth-promoting bacteria. Mol Plant Microbe Interact 661(2):69–93

Bashan Y, Salazar B, Puente ME et al (2009) Enhanced establishment and growth of giant cardon cactus in an eroded field in the Sonoran Desert using native legume trees as nurse plants aided by plant growth-promoting microorganisms and compost. Biol Fertil Soils 45(6):585–594

Bashan Y, Salazar BG, Moreno M et al (2012) Restoration of eroded soil in the Sonoran Desert with native leguminous trees using plant growth-promoting microorganisms and limited amounts of compost and water. J Environ Manag 102:26–36

Belimov AA, Dietz K (2000) Effect of associative bacteria on element composition of barley seedlings grown in solution culture at toxic cadmium concentrations. Microbiol Res 155(2): 113–121

Belimov AA, Kunakova AM, Safronova VI et al (2004) Employment of rhizobacteria for the inoculation of barley plants cultivated in soil contaminated with lead and cadmium. Microbiology 73(1):99–106

Bolfe ÉL, Victoria DDC, Sano EE et al (2024) Potential for agricultural expansion in degraded pasture lands in Brazil based on geospatial databases. Land 13(2):200

Carrillo AE, Li CY, Bashan Y (2002) Increased acidification in the rhizosphere of cactus seedlings induced by *Azospirillum brasilense*. Naturwissenschaften 89:428–432

Carrillo-Garcia Á, Bashan Y, Rivera ED et al (2000) Effects of resource-island soils, competition, and inoculation with *Azospirillum* on survival and growth of *Pachycereus pringlei*, the giant cactus of the Sonoran Desert. Restor Ecol 8(1):65–73

Cao C, Zhang Y, Cui Z, Li H, Wang T, Ren Q (2021) Recovery of Soil-Denitrifying Community along a Chronosequence of Sand-Fixation Forest in a Semi-Arid Desertified Grassland. Forests 12:354

Cruz-Hernández MA, Jiménez-Andrade JM, Mendoza-Herrera A (2019) Characterization of the degradation potential of xenobiotic compounds by the rhizobacteria *Azospirillum brasilense*. Mex J Biotechnol 4(2):10–22

Cui Q, Liu D, Chen H et al (2022) Synergistic interplay between *Azospirillum brasilense* and exogenous signaling molecule H2S promotes Cd stress resistance and growth in pak choi (Brassica chinensis L.). J Hazard Mater 444:130425

D'Odorico P, Bhattachan A, Davis KF et al (2013) Global desertification: drivers and feedbacks. Adv Water Resour 51:326–344 de-Bashan LE, Hernández JP, Nelson KN (2010) Growth of quailbush in acidic, metalliferous desert mine tailings: effect of *Azospirillum brasilense* Sp6 on biomass production and rhizosphere community structure. Microbial Ecology 60(4):915–927

de-Bashan LE, Hernandez JP, Bashan Y (2012) The potential contribution of plant growth-promoting bacteria to reduce environmental degradation–A comprehensive evaluation. Appl Soil Ecol 61, 171–189

de-Bashan LE, Magallon-Servin P, Lopez BR et al (2022) Biological activities affect the dynamic of P in dryland soils. Biol Fertil Soils 58(2):105–119

Döbereiner J (1978) Influence of environmental factors on the occurrence of *Spirillum lipoferum* in soils and roots. Ecol Bull:343–352

Dobereiner J, Marriel IE, Nery M (1976) Ecological distribution of *Spirillum lipoferum* Beijerinck. Can J Microbiol 22(10):1464–1473

Eckford R, Cook FD, Saul D et al (2002) Free-living heterotrophic nitrogen-fixing bacteria isolated from fuel-contaminated Antarctic soils. Appl Environ Microbiol 68(10):5181–5185

Fan X, Zhang S, Mo X et al (2017) Effects of plant growth-promoting rhizobacteria and N source on plant growth and N and P uptake by tomato grown on calcareous soils. Pedosphere 27(6): 1027–1036

Frings PJ, Buss HL (2019) The central role of weathering in the geosciences. Elements 15(4): 229–234

Galazka A, Krol M, Perzynski A (2010) Bioremediation of crude oil derivatives in soils naturally and artificially polluted with the use of maize as the test plant part I. Pahs degradation. Acta Sci Pol Agric 9(3)

Ganesh KS, Sundaramoorthy P, Nagarajan M (2015) Organic soil amendments: potential source for heavy metal accumulation. WSN 16:28–39

Hossain MB, Masum Z, Rahman MS et al (2022) Heavy metal accumulation and phytoremediation potentiality of some selected mangrove species from the world's largest mangrove forest. Biology 11(8):1144

Hungria M, Nogueira MA, Araujo RS (2016) Inoculation of *Brachiaria* spp. with the plant growth-promoting bacterium *Azospirillum brasilense*: an environment-friendly component in the reclamation of degraded pastures in the tropics. Agric Ecosyst Environ 221:125–131

Kamnev AA, Tugarova AV, Antonyuk LP et al (2005) Effects of heavy metals on plant-associated rhizobacteria: comparison of endophytic and non-endophytic strains of *Azospirillum brasilense*. J Trace Elem Med Biol 19(1):91–95

Kaushal M, Wani SP (2015) Plant-growth-promoting rhizobacteria: drought stress alleviators to ameliorate crop production in drylands. Annals of Microbiology 66(1):35–42

Kirk GJD (1997) Changes in rice root architecture, porosity, and oxygen and proton release under phosphorus deficiency. New Phytol 135(2):191–200

Kleber M, Eusterhues K, Keiluweit M et al (2015) Mineral–organic associations: formation, properties, and relevance in soil environments. Adv Agron 130:1–140

Langenbach T, Nascimento A, Sarpa M (1988) Influence of heavy metals on nitrogen fixation and growth of *Azospirillum* strains. Rev Latinoam Microbiol 30:139–142

Leyva LA, Bashan Y (2008) Activity of two catabolic enzymes of the phosphogluconate pathway in mesquite roots inoculated with *Azospirillum brasilense* Cd. Plant Physiol Biochem 46(10): 898–904

Lin S, Young CC, Hupfer H, Siering C, Arun AB, Chen W, Lai W, Shen F, Rekha PD, Yassin AF (2009) *Azospirillum picis* sp. nov., isolated from discarded tar. Int J Syst Evol Microbiol 59(4): 761–765

Lopez BR, Bacilio M (2020) Weathering and soil formation in hot, dry environments mediated by plant–microbe interactions. Biol Fertil Soils 56(4):447–459

Lopez BR, Bashan Y, Bacilio M et al (2009) Rock-colonizing plants: abundance of the endemic cactus *Mammillaria fraileana* related to rock type in the southern Sonoran Desert. Plant Ecol 201(2):575–588

Lopez BR, Tinoco-Ojanguren C, Bacilio M et al (2012) Endophytic bacteria of the rock-dwelling cactus *Mammillaria fraileana* affect plant growth and mobilization of elements from rocks. Environ Exp Bot 81:26–36

Lopez BR, Bashan Y, Trejo A et al (2013) Amendment of degraded desert soil with wastewater debris containing immobilized *Chlorella sorokiniana* and *Azospirillum brasilense* significantly modifies soil bacterial community structure, diversity, and richness. Biol Fertil Soils 49(8): 1053–1063

Lopez-De-Victoria G, Lovell CR (1993) Chemotaxis of *Azospirillum* species to aromatic compounds. Appl Environ Microbiol 59(9):2951–2955

Lopez-Lozano NE, Carcaño-Montiel MG, Bashan Y (2016) Using native trees and cacti to improve soil potential nitrogen fixation during long-term restoration of arid lands. Plant Soil 403(1–2): 317–329

Lyubun YV, Fritzsche A, Chernyshova MP et al (2006) Arsenic transformation by *Azospirillum brasilense* Sp245 in association with wheat (*Triticum aestivum* L.) roots. Plant Soil 286:219–227

Mascarua-Esparza MA, Villa-Gonzalez R, Caballero-Mellado J (1988) Acetylene reduction and indoleacetic acid production by *Azospirillum* isolates from Cactaceous plants. Plant Soil 106: 91–95

Mendez MO, Maier RM (2008) Phytoremediation of mine tailings in temperate and arid environments. Rev Environ Sci Biotechnol 7:47–59

Meyer E, Stoffel SC, de Almeida AF et al (2024) *Rhizophagus intraradices* and *Azospirillum brasilense* improve growth of herbaceous plants and soil biological activity in revegetation of a recovering coal-mining area. Braz J Microbiol 55(3):2827–2837

Miranda-Martínez MR, Delgadillo-Martínez J, Alarcón A, Ferrera-Cerrato R (2007) Degradación de fenantreno por microorganismos en la rizósfera del pasto alemán. Terra Latinoam 25, 25–33

Moreno M, De-Bashan LE, Hernandez J et al (2017) Success of long-term restoration of degraded arid land using native trees planted 11 years earlier. Plant Soil 421(1–2):83–92

Muratova AY, Turkovskaya OV, Antonyuk LP et al (2005) Oil-oxidizing potential of associative rhizobacteria of the genus *Azospirillum*. Microbiology 74(2):210–215

Nanekar RD, Kokitkar SS (2023) Isolation, genetic identification and optimization of hydrocarbon degrading bacteria from petroleum contaminated soil in Raigad region. Ecol Environ Conserv 29, 1554–1561

Nanekar RD, Kokitkar SS (2024) Enzyme profiling of petroleum hydrocarbon-degrading bacteria isolated from oil contaminated soils. Ind J Sci Technol 17(23):2412–2420

Nozawa-Inoue M, Scow KM, Rolston DE (2005) Reduction of perchlorate and nitrate by microbial communities in vadose soil. Applied and Environmental Microbiology 71(7):3928–3934

Ogar A, Sobczyk Ł, Turnau K (2015) Effect of combined microbes on plant tolerance to Zn–Pb contaminations. Environ Sci Pollut Res Int 22:19142–19156

Pan W, Lu Q, Xu Q et al (2019) Abscisic acid-generating bacteria can reduce Cd concentration in pakchoi grown in Cd-contaminated soil. Ecotoxicol Environ Saf 177:100–107

Pereg L, de-Bashan LE, Bashan Y (2016) Assessment of affinity and specificity of *Azospirillum* for plants. Plant Soil 399:389–414

Pérez Castañeda LM, Hernández C, Mendoza Herrera A (2011) Variabilidad genética de aislamientos no-típicos de *Azospirillum brasilense* por análisis PCR-RFLP del ADN 16S ribosomal. Phyton (Buenos Aires) 80(1):27–34

Pointing SB, Belnap J (2012) Microbial colonization and controls in dryland systems. Nat Rev Microbiol 10(8):551–562

Pozdnyakova N, Muratova A, Turkovskaya O (2022) Degradation of polycyclic aromatic hydrocarbons by co-culture of *Pleurotus ostreatus* Florida and *Azospirillum brasilense*. Appl Microbiol 2(4):735–748

Puente ME, Bashan Y (1993) Effect of inoculation with *Azospirillum brasilense* strains on the germination and seedlings growth of the giant columnar Cardon cactus (*Pachycereus pringlei*). Symbiosis 15(1):49–60

Puente M, Holguin G, Glick BR et al (1999) Root-surface colonization of black mangrove seedlings by *Azospirillum halopraeferens* and *Azospirillum brasilense* in seawater. FEMS Microbiol Ecol 29(3):283–292
Puente ME, Li CY, Bashan Y (2004a) Microbial populations and activities in the rhizoplane of rock-weathering desert plants. II. Growth promotion of cactus seedlings. Plant Biol 6(05): 643–650
Puente ME, Bashan Y, Li CY et al (2004b) Microbial populations and activities in the rhizoplane of rock-weathering desert plants. I. Root colonization and weathering of igneous rocks. Plant Biol 6(05):629–642
Puente ME, Rodriguez-Jaramillo MC, Li CY et al (2006) Image analysis for quantification of bacterial rock weathering. J Microbiol Methods 64(2):275–286
Rai A, Belkacem M, Assadi I et al (2022) Bacteria in soil: promising bioremediation agents in arid and semi-arid environments for cereal growth enhancement. Appl Sci 12(22):11567
Rao AV, Venkateswarlu B (1982) Associative symbiosis of *Azospirillum lipoferum* with dicotyledonous succulent plants of the Indian desert. Can J Microbiol 28(7):778–782
Rojas Aparicio A, Vázquez Jacinto JM, Romero Gomezcaña N et al (2016) Evaluación de compost con presencia de metales pesados en el crecimiento de *Azospirillum brasilense* y *Glomus intraradices*. Rev Mex Cienc Agríc 7(8):2047–2054
Saeed M, Ilyas N, Arshad M, Sheeraz M, Ahmed I, Bhattacharya A (2021) Development of a plant microbiome bioremediation system for crude oil contamination. J Environ Chem Eng 9:105401
Schoebitz M, Mengual C, Roldán A (2014) Combined effects of clay immobilized *Azospirillum brasilense* and *Pantoea dispersa* and organic olive residue on plant performance and soil properties in the revegetation of a semiarid area. Sci Total Environ 466:67–73
Torres-García MT, Oyonarte C, Cabello J et al (2022) The potential of groundwater-dependent ecosystems to enhance soil biological activity and soil fertility in drylands. Sci Total Environ 826:154111
Vazquez A, Zawoznik M, Benavides MP et al (2021) *Azospirillum brasilense* Az39 restricts cadmium entrance into wheat plants and mitigates cadmium stress. Plant Sci 312:111056
Verdu M, Jordano P, Valiente-Banuet A (2010) The phylogenetic structure of plant facilitation networks changes with competition. J Ecol 98(6):1454–1461
Vezza ME, Nicotra MFO, Agostini E et al (2019) Biochemical and molecular characterization of arsenic response from *Azospirillum brasilense* Cd, a bacterial strain used as plant inoculant. Environ Sci Pollut Res Int 27(2):2287–2300
Warke PA (2013) Weathering in arid regions. In: Shroder J (ed) Treatise on geomorphology. Academic, San Diego, pp 197–227
Wu D, Zhang X, Liu H et al (2020) *Azospirillum oleiclasticum* sp. nov, a nitrogen-fixing and heavy oil degrading bacterium isolated from an oil production mixture of Yumen Oilfield. Syst Appl Microbiol 44(1):126171
Wuana RA, Okieimen FE (2011) Heavy metals in contaminated soils: a review of sources, chemistry, risks and best available strategies for remediation. ISRN Ecol 2011:1–20
Xu Q, Pan W, Zhang R et al (2018) Inoculation with *Bacillus subtilis* and *Azospirillum brasilense* produces abscisic acid that reduces Irt1-mediated cadmium uptake of roots. J Agric Food Chem 66(20):5229 5236
Young CC, Hupfer H, Siering C et al (2008) *Azospirillum rugosum* sp. nov., isolated from oil-contaminated soil. Int J Syst Evol Microbiol 58(4):959–963
Zhou S, Han L, Wang Y et al (2012) *Azospirillum humicireducens* sp. nov., a nitrogen-fixing bacterium isolated from a microbial fuel cell. Int J Syst Evol Microbiol 63(Pt_7):2618–2624
Zsebő S, Bede L, Kukorelli G et al (2024) Yield prediction using NDVI values from GreenSeeker and MicaSense Cameras at different stages of winter wheat phenology. Drones 8(3):88

Index

F. D. Cassan et al. (eds.), *The Definitive Handbook of Azospirillum*,
https://doi.org/10.1007/978-3-032-09222-9

MIX
Papier aus verantwortungsvollen Quellen
Paper from responsible sources
FSC® C105338

If you have any concerns about our products,
you can contact us on
ProductSafety@springernature.com

In case Publisher is established outside the EU,
the EU authorized representative is:
Springer Nature Customer Service Center GmbH
Europaplatz 3, 69115 Heidelberg, Germany

Printed by Libri Plureos GmbH
in Hamburg, Germany